T5-BCJ-217

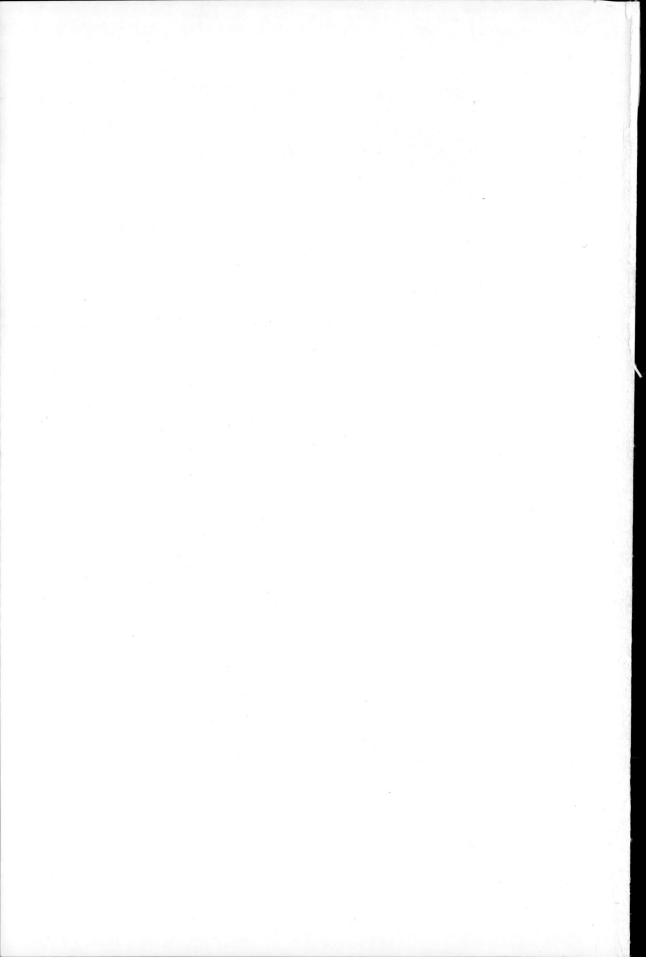

Heat Shock Response

Editor

Lutz Nover, Ph.D.
Department of Stress Research
Institute of Plant Biochemistry
Halle, Germany

CRC Press
Boca Raton Ann Arbor Boston London

Library of Congress Cataloging-in-Publication Data

Nover, Lutz.
Heat shock response/editor, Lutz Nover.
p. cm.
Includes bibliographical references.
ISBN 0-8493-4912-5
1. Heat shock proteins. 2. Heat—Physiological effect.
I. Title.
[DNLM: 1. Gene Expression Regulation. 2. Heat—adverse effects.
3. Heat-Shock Proteins. 4. Hyperthermia, Induced.
5. Transcription, Genetic. 6. Translation, Genetic. QU 55 N941h]
QP552.H43N68 1991
574.87′322—dc20
DNLM/DLC
for Library of Congress 89-22207
CIP

Direct all inquiries to CRC Press, Inc., 2000 Corporate Blvd., N.W., Boca Raton, Florida 33431.

International Standard Book Number 0-8493-4912-5

Library of Congress Card Number 89-22207
Printed in the United States

PREFACE

The effects of heat on living systems have been studied for well over a hundred years with historical roots in plant and animal physiology and organismal biology as well as microbiology. The efficacy of Coley's toxins in promoting tumor regression raised interest within the medical community at the turn of this century in the possible therapeutic benefits of heat, then associated with fever and inflammation. This interest is currently centered in the field of radiation oncology with efforts to achieve selective thermal killing of tumor cells and to control the complication of acquired thermotolerance. The current flurry of activity among molecular biologists is traceable to Ferruccio Ritossa's chance heating of *Drosophila* salivary tissue in the early 1960's and his subsequent demonstration that environmental stress in the readily manipulable forms of heat and chemical stressors can reproducibly alter gene expression in cells, the heat shock response. Since then, clues to the functions of heat shock proteins and the physiologic roles of the response have accumulated, slowly at first and more recently at a rapid pace. In 1975, Ashburner and colleagues provided a beacon for the field with the proposal that the response is homeostatic. This uncomplicated notion has survived the test of time and it is probably not premature to expand it now to indicate that the response is homeostatic and keyed to proteins. Evidence supporting roles for heat shock proteins in the folding or unfolding of other proteins, as solubilizers of damaged proteins, and as protein carriers is emerging. Research into the heat shock response is providing a focus for studies of gene regulation and for mechanistic studies of cellular responses to environmental stresss for investigators from broad fields of plant and animal biology including molecular and organismal biology, microbiology, physiology and neurobiology, and environmental and medical sciences.

The basis for this optimistic view of the heat shock field can be found in this volume. Dr. Nover and his colleagues have focused primarily on the past 15 years covering the cellular aspects of the heat shock response in prokaryotes and eukaryotes, but they also bring to this work an appreciation of the historical roots of the study of hyperthermia in organisms and a perspective on its future. Herein is the most thorough compilation of information on the cellular heat shock response available to date, and on this count alone, it is destined to become a landmark reference work. But this book contains much more. The reader will find critical analyses of data, unifying themes, discussions of discrepancies along with their possible causes and resolution, and valuable suggestions for future directions.

This timely volume will be very useful to investigators in a field that is expanding rapidly across many traditional disciplines. It will also serve as a useful entry vehicle for scientists who suddenly discover that environmental stress in the form of a physical agent such as heat, or an unsuspected chemical, or a biological agent such as a virus or microbial parasite has induced the heat shock response in their pet organisms.

Lawrence E. Hightower
Storrs, Connecticut

EDITOR

Lutz Nover, Ph.D., is head of the Department of Stress Research at the Institute of Plant Biochemistry in Halle, Germany.

Dr. Nover obtained his training in biology and biochemistry at the Martin Luther University in Halle, Germany, receiving his Ph.D. degree in 1970. His main objects of research were the partial synthesis and structure elucidation of cardiac glycosides, the biochemistry and regulation of secondary metabolism in fungi, and the molecular cell biology of heat shock and other stress response systems of plants.

Dr. Nover has published more than 50 research papers and authored or edited eight books.

CONTRIBUTORS

Dieter Neumann, Ph.D.
Department of Stress Research
Institute of Plant Biochemistry
Halle, Germany

Klaus-Dieter Scharf, Ph.D.
Department of Stress Research
Institute of Plant Biochemistry
Halle, Germany

ACKNOWLEDGMENTS

During the preparation of this manuscript I was kindly assisted by many co-workers from the heat shock group in Halle, especially G. Luckner, F. Müller-Uri, S. Rose, and W. Zott. The scientific legitimization derives from experimental work performed during many years of pleasant cooperation with Dieter Neumann, Dieter Scharf, and Uta zur Nieden. Many colleagues from the international heat shock family and good friends of our lab supplied the necessary literature, sent reprints, preprints, or unpublished material, and gave us the permission to reproduce material from their publications. These are V. Ya. Alexandrov (U.S.S.R.), B. Ames (U.S.), N. Amrhein (F.R.G.), A. P. Arrigo (Switzerland), B. G. Atkinson (Canada), C. L. Baszczynski (Canada), E. Baulieu (France), B. J. Benecke (F.R.G.), A. F. Bennet (U.S.), E. M. Berger (U.S.), Mariann Bienz (U.K.), H. Biessmann (U.S.), L. Björk (Sweden), J. J. Bonner (U.S.), Elida K. Boon-Niermeijer (Netherlands), G. Borbely (Hungary), B. P. Brandhorst (Canada), L. Browder (Canada), I. R. Brown (Canada), P. M. Candido (Canada), M. Chamberlin (U.S.), W. J. Gehring (Switzerland), E. W. Gerner (U.S.), Carol A. Gross (U.S.), K. Hahlbrock (F.R.G.), G. M. Hahn (U.S.), R. L. Hallberg (U.S.), M. Hecker (F.R.G.), J. J. Heikkila (Canada), L. Hightower (U.S.), T.-D. Ho (U.S.), T. Ishikawa (Japan), J. L. Key (U.S.), P. M. Kloetzel (F.R.G.), K. Kloppstech (F.R.G.), B. D. Korant (U.S.), K. W. Lanks (U.S.), A. Laszlo (U.S.), H. Laudien (F.R.G.), H. B. Le John (Canada), Susan Lindquist (U.S.), J. Lis (U.S.), H. F. Lodish (U.S.), M. Morange (France), R. I. Morimoto (U.S.), R. Nagao (U.S.), F. Neidhardt (U.S.), K. Ohtsuka (U.S.), Mary Lou Pardue (U.S.), C. S. Parker (U.S.), D. Pauli (Switzerland), H. R. B. Pelham (U.K.), Nancy S. Petersen (U.S.), Nora Plesofsky-Vig (U.S.), E. Raschke (F.R.G.), R. Rieger (F.R.G.), Claudina Rodrigues-Pousada (Spain), Marilyn M. Sanders (U.S.), C. Sato (Japan), M. Schlesinger (U.S.), F. Schoeffl (F.R.G.), Julie Silver (Canada), R. Sinibaldi (U.S.), Georghia Stephanou (Greece), J. R. Subjeck (U.S.), R. Tanguay (Canada), A. Tissieres (Switzerland), P. van Bergen en Henegouwen (Netherlands), Ruth A. van Bogelen (U.S.), G. van Dongen (Netherlands), L. van der Ploeg (U.S.), R. van Wijk (Netherlands), Elizabeth Vierling (U.S.), R. Voellmy (U.S.), W. J. Welch (U.S.), F. A. Wiegant (Netherlands), Jill Winter (U.S.), C. Wu (U.S.), R. Young (U.S.).

Many of the figures were reproduced with the kind permission of the authors and the copyright holders (see the credit lines with the figures).

Thanks to the kind cooperation of CRC Press and the help of Janice Morey, Jim Labeots, and Pat Roberson it was possible to update the text during proofreading.

Finally, I experienced the remarkable friendship of Larry Hightower (U.S.) who critiqued most parts of the crude manuscript and made essential suggestions for improvements and additions.

Lutz Nover
Halle

ABBREVIATIONS AND SPECIAL TERMS

Amino acid residues in Figures 2.8 through 2.15 are indicated by one-letter codes: A, alanine; C, cystein; D, aspartic acid; E, glutamic acid; F, phenylalanine; G, glycine; H, histidine, I, isoleucine; K, lysine; L, leucine; M, methionine; S, serine; T, threonine; V, valine; W, tryptophane; X, unknown residue or either one of the 20 residues; and Y, tyrosine.

Heat shock (hs)-induced genes and proteins are identified by hsp (for the genes) and HSP (for the proteins), with numbers indicating the relative molecular weights (M_r) of the proteins in kilodaltons (kDa). Thus, hsp70 refers to a gene coding for a HSP of the 70 kDa family. For a definition of the HSP families see Section 2.3. The terms hsc/HSC are applied to constitutively expressed members of the eukaryotic HSP families (see Section 2.2 on Nomenclature). Special sets of HSPs and their nomenclature are detailed in Table 2.4 for the yeast HSP70 family (SSA1—SSA4, SSB1/B2, SSC1, SSD1) and in Table 2.2 for the hs-induced genes of *E. coli* (dnaK/J, groEL/S, grpE, htpR = rpoH, lon, lysU, rpoD). Other stress proteins are designated by SP. P and pP, respectively, were used to indicate proteins and phosphoproteins in general. As with the hs genes and proteins, the genes and proteins induced by glucose deficiency (see Chapter 4) are abbreviated by grp and GRP, respectively.

Organisms are denoted by their Latin names with a two-letter code: At, *Arabidopsis thaliana;* BL, *Bremia lactucae* (fungus); Bm, *Bacillus megaterium* (bacilli); Bo, *Brassica oleracea* (cabbage); Ce, *Caenorhabditis elegans* (nematode); Chr, *Chenopodium rubrum* (goose foot); Cr, *Chlamydomonas reinhardi* (green alga); Cb, *Coxiella burnetii* (bacterium); Dd, *Dictyostelium discoideum* (slime mold); Dm, *Drosophila melanogaster* (fruit fly), Dh, *D. hydei*, and similarly for *D. pseudoobscura*, *D. simulans*, and *D. virilis*; Ec, *Escherichia coli* (coli bacterium); Gd, *Gallus domesticus* (chicken); Gm, *Glycine max* (soybean); Hs, *Homo sapiens* (man); Le, *Lycopersicon esculentum* (tomato); Lm, *Leishmania minor* (parasitic protozoon); Ml, *Mycobacterium leprae* (bacterium); Mm, *Mus musculus* (mouse); Nc, *Neurospora crassa* (fungus); Ph, *Petunia hybrida* (petunia); Ps, *Pisum sativum* (pea); Rn, *Rattus norvegicus* (rat); Sc, *Saccharomyces cerevisiae* (yeast); Ta, *Triticum aestivum* (wheat), Tb and Tc, *Trypanosoma brucei*, *T. cruzi* (trypanosomes); Xl, *Xenopus laevis* (claw-toed frog); Zm, *Zea mays* (corn).

Sequences and possible functions of promoter elements associated with hs genes are given in Sections 6.2 and 6.5. and Figure 6.6. These include the TATA-, Sp1-, and CAT-box. The element responsible for hs inducibility (HSE) is defined in Section 6.2. and Table 6.2. The HSE binding protein is referred to as hs transcription factor (HSF; see Section 7.5.). The *E. coli* transcription factor is a sigma subunit of RNA polymerase (σ^{32}) coded by the rpoH gene (former htpR gene; see Section 7.3.).

Reporter and marker genes are frequently used to analyze promoter activities and transformation. A collection of such genes used in connection with hs promoters is defined in Section 6.12: adh, cat, dhfr, ftz, gpt, gus, hgh, his3, lacZ, leu2, nptII, ry, tk, tnp, ura3, and w. They are mainly used in connection with Table 6.1. Gene fusion constructs, e.g., Dm-hsp70 (P/L) x cat gene, are composed of the organism (*Drosophila melanogaster*), the piece of the hs gene (hsp70 promoter-leader regions) and the reporter gene used (the bacterial gene for chloramphenicol acetyltransferase, cat). Details are given in the legend to Table 6.1 and in Section 6.1.

Further abbreviations include CH, cycloheximide; G_0, G_1, G_2, M, S, phases of the cell cycle (see Section 15.1); pre-mRNP and pre-rRNP, precursor of messenger ribonucleoprotein and ribosomal ribonucleoprotein particles; rRNA, rRNP, rprotein, precursors or parts of ribosome biosynthesis (see Chapter 10); and RNAPI, II, III, eukaryotic RNA polymerases transcribing ribosomal genes (I), protein coding genes (II), and genes coding for tRNA, 5SRNA, respectively (III).

TABLE OF CONTENTS

Introduction

INTRODUCTION

L. Nover

The flexibility of biological structures and their building blocks are the basis for their function and efficiency in adaptation to highly divergent metabolic and environmental conditions. Temperature is one of the most important factors deeply influencing the cellular networks of catalytic and regulatory interactions.

Strategies of long-term adaptation of endothermic and ectothermic organisms are not in the scope of this book. They are thoroughly dealt with in various reviews and books, e.g., by Alexandrov,[2,3] Brock,[9,10] Hochachka and Somero,[20] Kogut,[23] Langridge,[25] Laudien,[26] Ljungdahl,[28] Precht et al.,[30] and Prosser.[31] Because of multifold connections and partial overlaps of the responses, other stress factors are frequently included in the considerations.[1,8,22,27,28a,29,34]

With rare exceptions, where an excursus of the organismic level seems appropriate, the heat shock response (HSR) as described in the frame of this book comprises the cellular aspects only. The complex, but transient reprogramming of all cellular activities serves the protection of sensitive structures from heat damage and, consequently, the rapid and complete recovery after the stress period. Although the onset of molecular biology can be clearly traced back to the short but seminative report of Ritossa[32] in 1962 on the selective induction of new sites of gene activity after heat treatment of *Drosophila* larvae, the scientific roots are much older.

In 1864, the famous German plant physiologist Julius Sachs[33] published a concise paper on the upper temperature limit of the vegetation. By heating whole plants or part of them to 45 to 51°C, he observed a transient block of the protoplasma streaming and changes of membrane permeability. He documented differences between the high intrinsic heat sensitivity of young, newly unfolded leaves compared to old leaves or leaves in the bud stage, which are much less sensitive. He also investigated the reversibility of these cellular heat shock (hs) effects. Summarizing, he interpreted his results as evidence that the increased temperature overcomes molecular forces which maintain the internal organization of cells. The intracellular structures, which are so important for life, collapse without significant changes of the external form.[33] The intensive interest of plant physiologists in the field continued for many years providing evidence for seasonal and circadian variation of heat sensitivity, for induced thermotolerance, and for the role of phytohormones in this phenomenon.[2-4,14,24,35,39,40]

Considerations of the peculiarities of thermophilic microorganisms with respect to membrane structures and intracellular organization led Brock[9,10] to the definition of an upper temperature limit for growth of different classes of ectothermic organisms. The limit of 50 to 60°C found for insects, crustacean, vascular plants, fungi, and algae may be connected with the thermoinstability of their organellar membranes and cytoskeletal systems.

Almost simultaneously with Julius Sachs' publication, another important impulse resulted from the first report in Germany of Busch[11] (1866) on the complete regression of a sarcoma of the face after repeated attack of erysipelas, i.e., a local inflammation caused by infection of skin or mucous membranes by *Streptococcus erysipelatis*. Thirty years later, Coley[13] summarized successful therapeutic experience with 47 tumor patients after accidental or deliberate infections with *S. erysipelatis*. He also used an extract of the bacterium, later known as Coley's toxin, for direct injection into tumor tissue. In continuation of the original observation by Busch,[11] local heating was applied by Westermark[37] to patients with inoperable cervix carcinomas, and in 1903 Jensen[21] reported on the inhibitory effect of hyperthermia (5 min 47°C) on transplantation efficiency of mouse tumor explants.

The rapid progress of experimental hyperthermia using animal tumor models was reviewed in 1927 by Westermark.[38] Although today hyperthermia must be considered a valuable tool for therapy of inoperable tumors, routine application, mostly in combination with radiotherapy and/or chemotherapy,[6,7,12,15,17,18,36] still needs much technical and logistic improvements. However, it is noted that considerable parts of our knowledge on different aspects of the HSR, as are described in the following chapters, result from a great number of studies on hyperthermic killing of cancer cells and the inherent problem of induced thermotolerance.

A third area of early interest in cellular hs effects was concerned with alterations of gene expression long before heat shock proteins (HSPs) were known. In 1919 Alsop[5] reported on abnormal development of the nervous system of chick embryos after hs. But it was the famous German geneticist Goldschmidt,[19] who laid the foundation of the modern developmental genetics by studies on the temperature- and stage-dependent induction of phenocopies in *Drosophila* larvae. In his experimental data published in 1935, about 500,000 individuals with hs-induced developmental abnormalties, which phenotypically correspond to genetically defined mutants, confirmed his earlier concept on the interaction of genes in the developmental program of organisms. The extension of these studies to vertebrates, including man, led German[16] 50 years later to a very similar "embryonic stress hypothesis" of teratogenesis: "...induction of the heat shock response in the mammalian embryo during the critical period of organogenesis can alter the established program of activation and inactivation of genetic loci essential for normal intrauterine development, the result being anatomic malformation."

Tracing the historical roots of a scientific field helps in the understanding of its origins and remarkable stages of development. However, only as a result of our present day knowledge are these origins and their founders adequately recognized. Their experimental data can be reconsidered, extended, and, possibly, put on a new basis. Hopefully, the following coordinate series of reviews on the essential aspects of the cellular HSR of prokaryotic and eukaryotic organisms not only illustrate the enormous wealth of information accumulated in the preceding 15 years, but they also increase our appreciation of the roots and help to point out many unsolved problems.

REFERENCES

1. **Adams, C. and Rinne, R. W.,** Stress protein formation: Gene expression and environmental interaction with evolutionary significance, *Int. Rev. Cytol.*, 79, 305—315, 1982.
2. **Alexandrov, V. Ya.,** *Cell, Molecules and Temperature*, Springer-Verlag, Berlin, 1977.
3. **Alexandrov, V. Ya.,** Cell reparation of non-DNA injury, *Int. Rev. Cytol.*, 60, 223—269, 1979.
4. **Alexandrov, V. Ya., Lomagin, A. G., and Feldman, N. L.,** The responsive increase in thermo-stability of plant cells, *Protoplasma*, 69, 417—458, 1970.
5. **Alsop, F. M.,** The effect of abnormal temperatures upon the developing nervous system in the chick embryo, *Anat. Rec.*, 15, 307—332, 1919.
6. **Anghileri, L. J. and Robert, J., Eds.,** *Hyperthermia in Cancer Treatment*, Vol. 1—3, CRC Press, Boca Raton, FL, 1986.
7. **Arcangeli, G., Benassi, M., Cividalli, A., Lovisolo, G. A., and Mauro, F.,** Radiotherapy and hyperthermia: analysis of clinical results and identification of prognostic variables, *Cancer*, 60, 950—956, 1987.
8. **Atkinson, B. G. and Walden, D. B., Eds.,** *Changes in Eukaryotic Gene Expression in Response to Environmental Stress*, Academic Press, Orlando, FL, 1985.
9. **Brock, Th. D.,** *Thermophilic Microorganisms and Life at High Temperatures*, Springer-Verlag, New York, 1978.
10. **Brock, Th. D.,** Life at high temperatures, *Science*, 230, 132—138, 1985.
11. **Busch, W.,** Über den Einfluss welchen heftigere Erysipeln zuweilen auf organisierte Neubildungen ausüben, *Verh. Naturforsch. Preuss. Rhein. Westphal.*, 23, 28—30, 1866.

12. **Cavaliere, R., Ciocatto, E. C., Giovanelia, B. C., Heidelberger, C., Johnson, R. O., Margottini, M., Mondovi, B., Moricca, G., and Rossi-Fanelli, A.**, Selective heat sensitivity of cancer cells, *Cancer,* 20, 1351—1381, 1967.
13. **Coley, W. B.**, The treatment of malignant tumors by repeated inoculations of erysipels. With a report of ten original cases, *Am. J. Med. Sci.,* 105, 487—511, 1893.
14. **Engelbrecht, L. and Mothes, K.**, Weitere Untersuchungen zur experimentellen Beeinflussung der Hitzewirkung bei Blättern von *Nicotiana rustica, Flora,* 154, 279—298, 1964.
15. **Field, S. B. and Bleehen, N. M.**, Hyperthermia in the treatment of cancer, *Cancer Treat. Rev.,* 6, 63—94, 1979.
16. **German, J.**, Embryonic stress hypothesis of teratogenesis, *Am. J. Med.,* 76, 293—301, 1984.
17. **Gerner, E. W.**, Thermal dose and time-temperature factors for biological responses to heat shock, *Int. J. Hyperthermia,* 4, 319—328, 1987.
18. **Gerweck, L. E.**, Hyperthermia in cancer therapy: The biological basis and unresolved questions, *Cancer Res.,* 45, 3408—3414, 1985.
19. **Goldschmidt, R.**, Gen und Außeneigenschaft (Untersuchungen an Drosophila) I. u. II. *Z. Indukt. Abstamm. Vererbungsl.,* 69, 38—69, 70—131, 1935.
20. **Hochachka, P. W. and Somero, G. N.**, *Biochemical Adaptation,* Princeton University Press, Princeton, NJ, 1984.
21. **Jensen, C. O.**, Experimentelle Untersuchungen über Krebs bei Mäusen. *Zentralbl. Bakteriol.,* 34, 28—122, 1903.
22. **Key, J. L. and Kosuge, T., Eds.**, *Cellular and Molecular Biology of Plant Stress,* Alan R. Liss, New York, 1985.
23. **Kogut, M.**, Are there strategies of microbial adaptation to extreme environments?, *Trends Biochem. Sci.,* 1, 15—18 and 2, 1—4, 1980.
24. **Lange, O. L. and Schwemmle, B.**, Untersuchungen zur Hitzeresistenz vegetativer und blühender Pflanzen von Kalanchoe blossfeldiana, *Planta,* 55, 208—225, 1960.
25. **Langridge, J.**, Biochemical aspects of temperature response, *Annu. Rev. Plant Physiol.,* 14, 441—462, 1963.
26. **Laudien, H., Ed.**, Temperature Relations in Animals and Man, Biona Report No. 4, G. Fischer, Stuttgart, 1986.
27. **Levitt, J.**, Frost, drought, and heat resistance, *Protoplasmatologia,* Bd. VIII/6. Springer, Wien, 1958.
28. **Ljungdahl, L. G.**, Physiology of thermophilic bacteria, *Adv. Microb. Physiol.,* 19, 150—243, 1979.

28a. **Nover, L., Neumann, D., and Scharf, K.-D., Eds.**, *Heat Shock and Other Stress Response Systems of Plants,* Springer Verlag, Berlin, 1990.

29. **Ort, D. R. and Boyer, J. S.**, Plant productivity, photosynthesis, and environmental stress, in *Changes in Eukaryotic Gene Expression in Response to Environmental Stress,* Atkinson, B. G. and Walden, D. B., Eds., Academic Press, Orlando, FL, 1985, 279—313.
30. **Precht, H., Christophersen, J., Hensel, H., and Larcher, W.**, *Temperature and Life,* Springer-Verlag, Berlin, 1973.
31. **Prosser, C. L., Ed.**, *Molecular Mechanisms of Temperature Adaptation,* American Association of Adv. Science Publishers, 1984.
32. **Ritossa, F.**, A new puffing pattern induced by heat shock and DNP in Drosophila, *Experientia,* 18, 571—573, 1962.
33. **Sachs, J.**, Über die obere Temperatur-Gränze der Vegetation, *Flora,* 47, 5—12, 24—29, 33—39, 64—75, 1864.
34. **Sachs, M. M. and Ho, T.-H. D.**, Alteration of gene expression during environmental stress in plants, *Annu. Rev. Plant Physiol.,* 37, 363—376, 1986.
35. **Sapper, I.**, Versuche zur Hitzeresistenz der Pflanzen, *Planta,* 23, 518—556, 1935.
36. **Suit, H. D.**, Hyperthermic effects on animal tissues, *Radiology,* 123, 483—487, 1977.
37. **Westermark, F.**, Über die Behandlung des ulcerirenden Cervix carcinoms mittels konstanter Wärme, *Zbl. Gynäkol.,* 1335—1339, 1898.
38. **Westermark, N.**, The effect of heat upon rat-tumors, *Skand. Arch. Physiologie,* 52, 257—322, 1927.
39. **Yarwood, C. E.**, Acquired tolerance of leaves to heat, *Science,* 134, 941—942, 1961.
40. **Yarwood, C. E.**, Adaptation and sensitization of bean leaves to heat, *Phytopathology,* 54, 936—940, 1964.

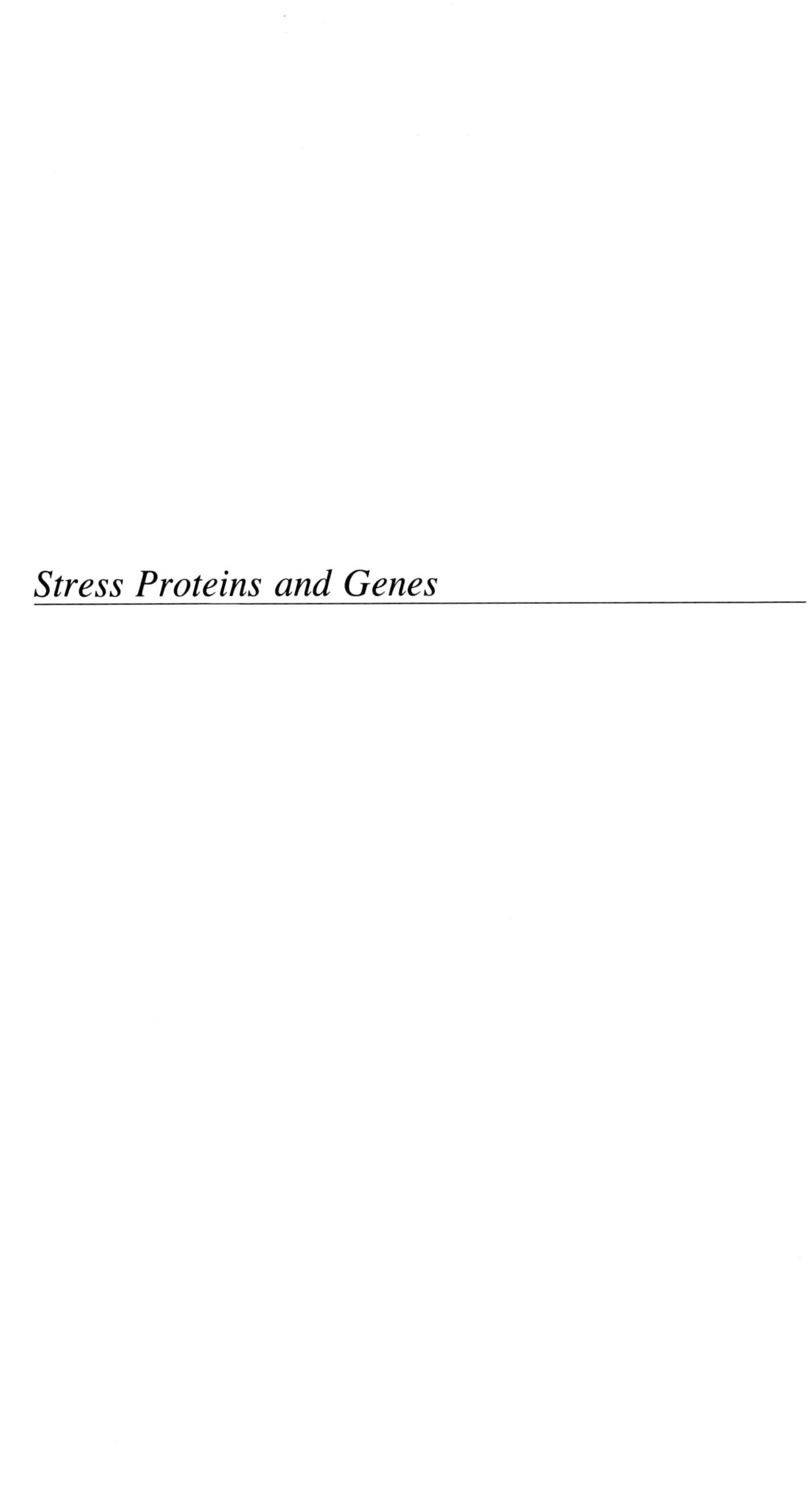

Stress Proteins and Genes

Chapter 1

INDUCERS OF HSP SYNTHESIS: HEAT SHOCK AND CHEMICAL STRESSORS

L. Nover

TABLE OF CONTENTS

1.1. THE PHYSIOLOGICAL HEAT SHOCK REGIME

The heat shock (hs) response is a general property of all living organisms. The hyperthermic threshold is cell- and organism specific and depends on the normothermic state. A convenient method to elucidate the optimum hs temperature range is the analysis of heat shock protein (HSP) synthesis. Relevant data for many organisms are compiled in Table 2.1.

Transitions from normothermic to hyperthermic conditions are common experiences especially for microorganisms, plants, and poikilothermic animals. For experimental purposes three types of hs regimes are broadly used:

1. The immediate, permanent hs requires mild to moderate temperature elevation and is especially suited to study changes at the transition between the normothermic and hyperthermic states.
2. A short, severe hs pulse followed by several hours of recovery is sufficient to trigger HSP synthesis, which proceeds equally well under normothermic conditions. This procedure results in a system with minimum disturbances of other cellular activities but with elevated levels of HSPs (preinduced cells). This type of conditioning is the main basis for studying the phenomena of acquired thermotolerance (see Chapter 17).
3. More representative of natural conditions is a gradual increase of the temperature. Different aspects of the complex hs response are not expressed simultaneously but rather in a temperature-dependent sequence. A survey of changes in *Drosophila* cells (Figure 1.1) shows a continuum of effects on different parts of gene expression, but also on cell ultrastructure, cell survival, and developmental processes. Threshold and optimum temperatures for distinct parts of the hs response differ by more than 10°C. This is also true for the pattern of HSP synthesis. HSP83 is a low temperature HSP (26 to 33°C), whereas synthesis of small HSPs and HSP70 is optimum at 35°C and 37 to 38°C, respectively.

Although sufficient experimental data are lacking for other systems, the situation is evidently basically similar. Failure to detect essential aspects of the hs responses in systems other than *Drosophila* may simply reflect ignorance about the optimum temperature conditions. At the upper limit of the hyperthermic gradient (Figure 1.1) is the transition to irreversible cell damage, usually avoided by the experimentalist. However, this is the temperature range used for selective cell killing during hyperthermic treatment of cancer (see Chapter 20).

In the presentation of data used in Figure 1.1, a decisive factor, time, was omitted. It is an experimental advantage of heat as stressor that many processes can be triggered almost without delay provided the optimum conditions are applied. However, the length of the stress period thoroughly influences the outcome. On the one hand, the extent of hs-induced changes is frequently characterized by a time-dose relationship. For evident reasons, this matter is of particular importance for hyperthermic cell killing and is briefly discussed in Chapter 20. On the other hand, the term "hs response" is used to describe a highly active process of adaptation with continuous changes in the cellular state. Early or immediate parts (activation of hs genes, polysome decay, inhibition of pre-rRNP processing, collapse of cytoskeleton, cell cycle block) can be distinguished from more delayed or late parts (decline of pre-rRNA transcription, mRNA degradation, accumulation of HSP and autorepression of HSP synthesis, recovery of control protein synthesis, induced thermotolerance). It is almost trivial to note that the absolute time scale for "early" and "late" is again organism specific. Recovery of control protein synthesis during a moderate hs takes few minutes in *Escherichia coli* but several hours in plant and animal cells.

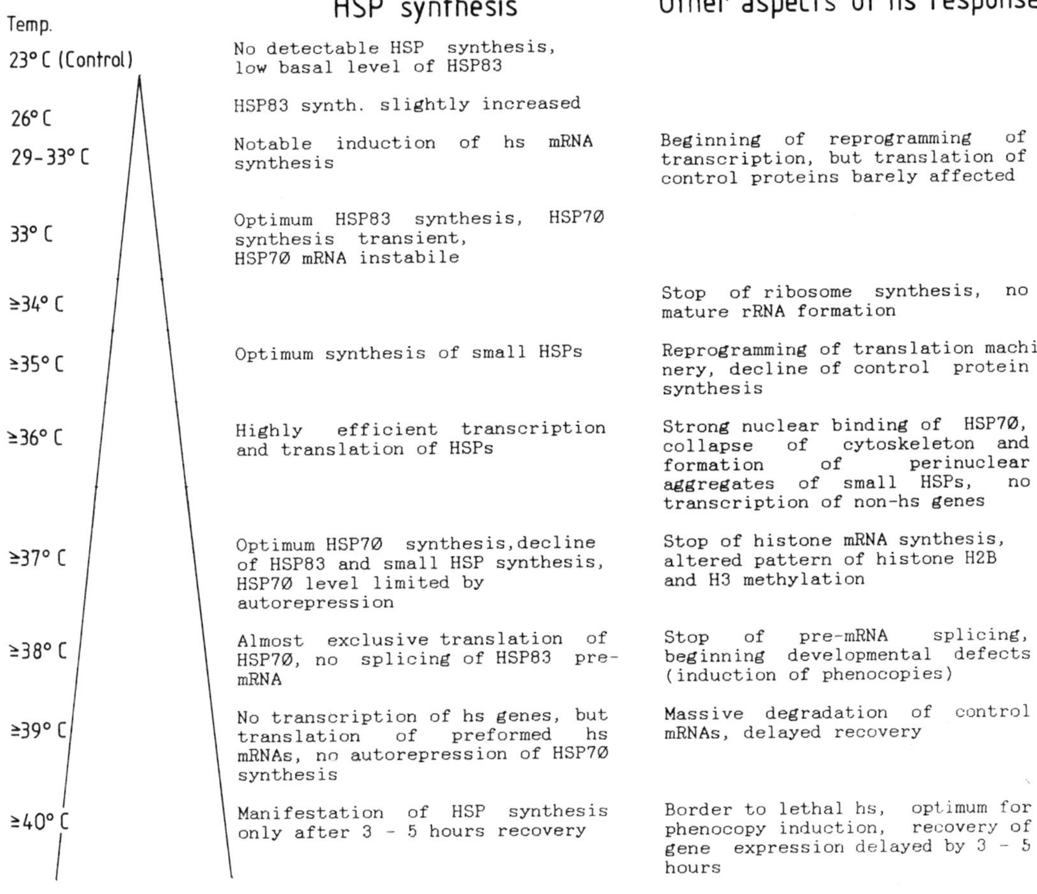

FIGURE 1.1. Temperature-dependent changes of gene expression in *Drosophila*. The figure demonstrates that the heat shock (hs) response in *Drosophila* comprises a continuum of temperature dependent changes at all levels of gene expression. Data, compiled from Ballinger and Pardue,[10a] Leicht et al.,[156a] Brandt and Milcarek,[23a] Chomyn et al.,[43] Desrosier and Tanguay,[57a] DiDomenico et al.,[58] Lindquist,[164a] Spradling et al.,[226a] Velazquez and Lindquist,[243] Yost and Lindquist,[261a] include results from cell cultures as well as from embryonic and adult tissues, i.e., the absolute temperature threshold given on the left may in fact vary with the cell type or culture condition.[164a] Furthermore as a result of induced thermotolerance, this threshold usually increases by 1 to 2°C in conditioned cells (see Chapter 17).

So far almost neglected by the molecular biologist are truly late changes after long-term continuous or interrupted hs periods as they are encountered by plants under natural conditions or experimentally used for hyperthermic virus elimination (see Chapter 19). Adaptation of bacterial cells to continuous life at supranormal temperature conditions is well known.[26,165,197] Effective use of hs-regulatory elements in biotechnology will require corresponding investigations also for animal and plant cells (see Section 6.7).

For practical reasons, hs experiments with vertebrates and plants are usually performed with isolated cell systems or organ cultures. It was important to demonstrate that increase of the ambient temperature of cells in their natural, i.e., organismic, surrounding has similar effects. For plants this was done by analyzing changes of leaf temperature[74] or essential aspects of the hs response under field conditions.[31,127,132a,138] In vertebrates mild whole body hyperthermia [26b,27,54,55,59,69,70,232,246] or experimentally induced fever after application of pyrogens, e.g., of bacterial lipopolysaccharides,[98a] of lysergic acid diethylamide (LSD) or amphetamine[26b,34,51,69] are effective for induction of HSP synthesis and other parts of the stress response. LSD and amphetamine are "organismic" inducers. Because their action is mediated by the fever response, they are ineffective *in vitro* with isolated tissues.[44] Local hyperthermia as the result of diseased states can also explain the increased synthesis of HSP70 after microsurgical cuts in the brain,[27a] in chondrocytes isolated from patients with osteoarthritis[144,167] or in Alzheimer-afflicted brain tissue.[86] Use of whole body stress treatments with pregnant rats, mice, and guinea pigs[64,74a,75] as well as with rat embryos *in vitro*[175a,246a] demonstrated the particularly high sensitivity of embryonal stages even to short periods of hyperthermia or to low doses of toxic compounds (see Section 18.1).

1.2. CHEMICAL STRESSORS INDUCING HSP SYNTHESIS

1.2.1. SURVEY OF SYSTEMS AND STRESSORS

The discovery of hs-induced changes of gene activity in salivary glands of *Drosophila buschkii* by Ritossa[202] already revealed the equivalence of chemical and physical stressors. Formation of hs puffs on polytene chromosomes was also observed after application of 1 m*M* DNP or 10 m*M* salicylate. In the following years, a bewildering multiplicity of chemical stressors were found to exert similar effects, initially with *Drosophila* salivary glands and cell cultures as experimental objects, but soon also with vertebrate, yeast, plant, bacteria, and other systems. The compilation in Table 1.1 is based on the early summary by Ashburner and Bonner[7] for *Drosophila* and the extended version in Nover.[187] It includes results with bacteria (B), (*Escherichia coli, Salmonella typhimurium, Bacillus subtilis*), with the unicellular ciliate *Tetrahymena* (T), with fungi (F) (*Saccharomyces cerevisiae, Neurospora crassa*, and *Dictyostelium discoideum*), with insects (I) (mainly *Drosophila* species, but also *Chironomus thummi* and *Sarcophaga bullata*) and with vertebrates (avian and mammalian cells).

Other systems successfully induced with a limited number of chemical stressors are

1. Plants: 50 μM arsenite and 1m*M* Cd^{2+} in soybean seedlings and tobacco cell cultures,[63a,87,134,163] each 5 m*M* of Zn^{2+}, Cu^{2+}, Cd^{2+}, or arsenite in Petunia leaves,[255d] each 1 m*M* of Cd^{2+}, *p*-fluoro-phenylalanine or azetidine carboxylic acid, 7×10^{-5} *M* of Hg^{2+} or 5×10^{-4} *M* of arsenite in tomato cell cultures,[188a] water limitation, wounding or treatment with abscisic acid of maize mesocotyls.[101]
2. *Xenopus*: injection of abnormal proteins into oocytes,[3] cell isolation and establishment of cultures (culture shock),[256] 6% ethanol or 10 to 50 μM arsenite.[102]

Investigations of chemical stressors frequently encounter experimental difficulties not observed with hs. The rapidity, selectivity, and extent of induction are usually much lower. Finding of the appropriate, narrow concentration range effective as stressor but not yet causing irreversible damage is essential. Moreover, contrasting to hs, it usually takes time after addition of the toxic compound to establish the metabolic stress situation[37,52,114,256] and, later, to eliminate the stressor from the cell. To avoid some of the inherent problems, also connected with a strongly diminished precursor uptake in the presence of many chemical stressors, the experimental protocol is usually adapted to the hs regime 2; i.e., after treatment of cells with a given stressor, altered protein synthesis is investigated in the following recovery period. Our experience with plant cell cultures demonstrate that, with few exceptions, reliable characterization of chemical stressors may require two-dimensional protein electrophoresis, Northern analysis of mRNAs, and, if possible, also the use of HSP antibodies. It is worth noticing that the overwhelming parts of chemical stressors active in *Drosophila* species[7] were described as inducers of hs puffs in isolated salivary glands, i.e., in a well defined experimental system using microscopic inspection of polytene chromosomes. However, in most cases, HSP synthesis itself was not investigated.

A potential pitfall of inducer studies is the heavy metal contamination of biological

TABLE 1.1
Inducers of Heat Shock Protein Synthesis

Inducer	Organism				
	B	**F**	**T**	**I**	**V**
Oxidizing agents and drugs affecting respiration and energy metabolism					
Amytal (0.3 m*M*)	—	—	—	155	—
Anoxia and recovery	177, 224	24, 126	88	6, 28, 141, 203	162, 214
Antimycin A (sat.)	—	—	—	155	148
Arsenate (10 m*M*)	—	—	—	28	137
Arsenite (10—100 μ*M*)	—	126, 251a	1	141, 155, 245	9, 15, 29, 75, 97, 100, 106, 142, 147, 148, 154a, 158—160, 160a, 162, 254
Azide (3 m*M*)	—		—	28, 155, 203, 230	—
Carbonyl cyanide m-chlorophenylhydrazone (0.1 m*M*)	—	—	—	—	148, 160a
2,4-Dinitrophenol (0.1—1 m*M*)	—	251, 251a	—	28, 65, 141, 155, 198, 202, 220	—
H_2O_2 (0.05—1 m*M*)	177, 200, 224, 235	—	—	48, 166, 230, 251a	—
Hydroxylamine (10 m*M*)	—	—	—	210	—
Menadione (1 m*M*)	—	—	—	155	—
Methylene blue (10 m*M*)	—	—	—	155	—
Oligomycin + KCN	—	—	—	155	148
Rotenone (sat.)	—	—	—	155	—
Anion transporters					
4,4′-Diisothiocyanostilbene 2,2′ disulfonate (100 μ*M*)	—	—	—	216	—
4-Acetamido 4′-isothiocyanostilbene 2,2′-disulfonate (100 μ*M*)	—	—	—	216	—
Flufenamic acid (100 μ*M*)	—	—	—	216	—
Mefenamic acid (100 μ*M*)	—	—	—	216	—
Niflumic acid (100 μ*M*)	—	—	—	216	—

TABLE 1.1 (continued)
Inducers of Heat Shock Protein Synthesis

Inducer	Organism				
	B	F	T	I	V
K^+-ionophores					
Dinactin (0.1—1 μ*M*)	—	—	—	198	—
Trinactin	—	—	—	13	—
Valinomycin (10 μ*M*)	—	—	—	198	—
Transition series metals					
Cd^{2+} (10—100 μ*M*)	242	—	—	52	37, 41, 99, 106, 158, 159, 162, 175b
Cu^{2+} (0.1—1 m*M*)	—	—	—	—	9, 37, 106, 157—159
Hg^{2+} (10 μ*M*)	—	—	—	—	9, 158, 159
Zn^{2+} (0.1 m*M*)	—	—	—	—	9, 37, 41, 99, 106, 159, 175b
Chelating drugs					
Disulfiram (0.2 μ*M*)	—	—	—	—	37, 158, 159
8-Hydroxyquinoline	—	—	—	—	29, 159
Ketoxal bisthiosemicarbazone (0.3 μ*M*)	—	—	—	—	158, 159
o-Phenanthroline	—	—	—	—	159
Salicylate (10 m*M*)	—	—	—	202, 203	—
Sulfhydryl reagents					
p-Chloromercuribenzoate (5 μ*M*)	—	—	—	—	37, 158
Diamide (0.3 m*M*)	—	—	—	—	69a, 106, 154a
Iodoacetamide (10 μ*M*)	—	—	—	—	4, 37, 158
Amino acid analogs					
Azetidine-2-carboxylic acid (5 m*M*)	—	—	—	58	161, 237, 250
Canavanine (0.1—1 m*M*)	79	92	123, 255	58	4, 37, 41, 97, 106, 133, 161
Ethionine	—	—	255	—	—
p-Fluorophenylalanine (0.2 m*M*)	—	92	255	—	106
Hydroxynorvaline (0.1—1 m*M*)	—	—		—	133

β-Hydroxyleucine	—	—	—	58	—
O-Methyl threonine (0.1—1 m*M*)	—	—	—	—	133
Norvaline (3 mg/ml)	98, 200	—	—	—	—
L-Threo-α-amino-β-chlorobutyric acid	—	—	—	—	4
Serine hydroxamate (3 mg/ml)	98	—	—	—	—
Histidinol	—	—	206	—	—
Inhibitors of gene expression					
Chloramphenicol (1 m*M*)	—	—	—	12	—
1-Chloro-β,D-ribofuranosyl benzimidazol	—	—	—	11	—
Gentianamycin (2 mg/ml)	—	—	—	—	61
Paromomycin (1 mg/ml)	—	79a	255	—	—
Puromycin	242	—	255	—	97, 106, 154
Streptomycin	79	—	—	—	—
Tetracycline	—	—	255	—	—
Cycloheximide (10 ng/ml)	—	79a	—	—	—
Inhibitors of DNA topoisomerase II					
Coumeromycin	143, 239	—	—	—	—
Nalidixic acid	239, 242	—	—	—	—
Novobiocin	—	—	—	209	—
Teniposide	—	—	—	209	—
Drugs affecting membrane structure					
Ethanol (2—6%)	5, 177, 224, 242	169, 195	—	235	160, 160a, 91
Propanol	—	—	—	—	91
Butanol	—	—	—	—	91
Pentanol	—	—	—	—	91
Octanol	—	—	—	—	91
Lidocaine	—	—	—	—	91, 160a
Procaine	—	—	—	—	91
Digitonin, saponins, Sarkosyl, Triton X-100, Nonidet P-40	—	—	—	182	—
4-Hydroxynonenal (150 μM)	—	—	—	—	34a
Steroid hormones					
Dexamethasone (0.1 m*M*, under conditions of glucose deprivation)	—	—	—	23, 32	130, 131
Diethylstilbestrol (10 μ*M*)	—	—	—	23, 32	—
Ecdysterone (1 μ*M*)	—	—	—	23, 42, 118, 119	—
Hydrocortisone	—	—	—	33	—
Methyltestosterone (10 μ*M*)	—	—	—	23, 32	—
Estradiol	—	—	—	—	70a, 176, 197

TABLE 1.1 (continued)
Inducers of Heat Shock Protein Synthesis

Inducer	Organism				
	B	F	T	I	V
Infection with DNA viruses					
λ-phage	10, 60, 140	—	—	—	—
Adenovirus	—	—	—	—	124, 183, 218, 255b
Herpes simplex virus	—	—	—	—	185
Polyoma virus	—	—	—	—	135
SV 40	—	—	—	—	135, 190, 207, 235a, 240
Cytomegalovirus	—	—	—	—	45a
Antineoplastic drugs					
1,3-bis-(2-chloroethyl)-1-nitrosourea (BCNU, 200 μ*M*)	—	—	—	—	211
1-(2-chloroethyl)-3-cylohexyl-1-nitrosourea (CCNU, 200 μ*M*)	—	—	—	—	211
Teratogens, carcinogens, mutagens					
Coumarin (0.1—1 m*M*)	—	—	—	23, 32	—
Diphenylhydantoin (0.1—1 m*M*)	—	—	—	23, 32	—
Pentobarbital (0.1—1 m*M*)	—	—	—	23, 32	—
Tolbutamide (0.1—1 m*M*)	—	—	—	23, 32	—
5-Azacytidine (30 m*M*)	—	—	—	23, 32	—
Thalidomide (1 m*M*)	—	—	—	23, 32	—
Diethylnitrosamine	—	—	—	—	40
2-Acetylaminofluorene	—	—	—	—	40
Methylmethane sulfonate (200 μg/ml)	—	—	—	—	68a
Varia					
Abnormal proteins	79, 121, 191a	—	—	108, 109, 129	63b, 133, 154a, 238
Ammonium chloride (10 m*M*)	—	—	—	210	—
Amphetamine-ind. hyperthermia	—	—	—	—	34, 59
Ether (60 min)	—	—	—	23	—

Lysergic acid diethylamide-ind. hyperthermia	—	—	—	—	44, 51, 69
Dicoumarol (1 m*M*)	—	—	—	203	—
Hemin (erythroid cells)	—	—	—	—	221
High pH (pH 8.7)	243b	—	—	—	252
2-Heptyl-4-hydroxyquinoline -*N*-oxide (0.25 mg/ml)	—	—	—	155	—
Calcitonin	—	—	—	—	4a
Prostaglandin J_2 (4 μg/ml)	—	—	—	—	189a, 209a
Uridine (10 m*M*)	—	—	—	210	150
Vitamin B-6	—	—	—	141	—
Methylcholanthrene	—	—	—	—	240
Wounding	—	—	—	—	4, 53, 100, 106, 253
Isolation and cultivation of rat embryonic hearts	—	—	—	—	107b
Myocardial necrosis	—	—	—	—	253b
Serum stimulation, mitogens, lymphokines	—	—	—	—	66a, 260, 261
Recovery from long-term glucose deprivation	—	—	—	—	212, 252
Ischemia and recovery (blood deprivation)	—	—	—	—	34, 59, 94, 122, 171a
Cell cycle arrest, high cell density, nutrient starvation	83, 242	24, 116, 126, 145, 263	88, 93	—	189a
N-Tosyl-L-phenylalanine chloromethylketone (TPCK)	—	—	255	—	—
Cigarette smoke or 10 m*M* nicotine or catechol	—	—	—	—	114
UV irradiation	143	—	—	—	27b, 255c

Note: Inducers other than hs were compiled and discussed by Ashburner and Bonner,[7] Burdon,[30] Lanks,[149] Li,[160a] and Nover.[187] Groups of organisms are abbreviated as follows: B (bacteria, mainly *Escherichia coli* and *Salmonella typhimurium*), F (fungi, yeast, and *Neurospora*), T (*Tetrahymena*), I (insects, mainly *Drosophila*), and V (vertebrates). Hyphens indicate that no data are available. Numbers refer to the list of references at the end of the chapter. Arrangements of inducers in groups are only meant for easier orientation (see text). Their classification into general or hs-like inducers and specific inducers is discussed in the text.

fluids (serum, growth factor preparations). Whelan and Hightower[252a] described the inducing effect of a commercial medium used for chicken embryo fibroblast culture. They detected trace amounts of Zn^{2+}, which are sufficient to cause induction of HSP synthesis, if the medium is deprived of Zn-chelating amino acids, especially of cysteine and histidine. It is not unlikely that the earlier report of Levinson et al.[159a] on induced HSP synthesis after amino acid deprivation of chicken embryo fibroblasts is based on the same effect. The results also emphasize another important problem with chemical stressors: the outcome is thoroughly influenced by the availability, uptake, and intracellular state of the inducer.

The classification of chemical stressors in Table 1.1 is useful for rapid orientation, but cannot serve as a clue for their mode of action. It may be even misleading, because the concentrations used for HSP induction, e.g., by ethanol or drugs affecting respiration and energy metabolism, are by one to three orders of magnitude higher than those required for inhibition of respiration.[148,149] Principally the same argument holds true for 1 m*M* chloramphenicol (CAP) inducing HSP synthesis in *Drosophila*.[12] At this concentration, CAP causes depolarization of plasma membranes.[14] At any rate, protein synthesis or respiratory function of mitochondria are apparently not the primary target of hs or chemical stressors as suggested earlier (see Section 1.5).

Although intrinsic technical difficulties limit the value of negative results, examples are given to indicate possible organism or group specificity. In the early characterization of Cd^{2+}, Cu^{2+}, Zn^{2+}, and Hg^{2+} as HSP-inducing stressors in chicken[158,159] other metal ions were found inactive, e.g., Co^{2+}, Ni^{2+}, Fe^{2+}, Fe^{3+}, Mn^{2+}, Pt^{2+}, and Pb^{2+}. But in *Neurospora crassa* even Cd^{2+}, Zn^{2+}, and Cu^{2+} are non-inducing.[125] The list of inducers may be particularly restricted for the cellular slime mold *Dictyostelium*. Arsenite, canavanine, ethanol, cycloheximide, antimycin, and 2,4-dinitrophenol stopped growth, but did not affect HSP synthesis.[208] Burdon[30] reports unpublished results on the failure to induce HSP synthesis in mammalian cells using cycloheximide, H_2O_2, azide, or dinitrophenol. These findings correlate with a note of Lanks[149] that a wide spectrum of chemical stressors affecting mitochondrial metabolism and found positive in *Drosophila* (Table 1.1) were not active in vertebrates. Finally, a striking example of tissue specificity concerns chicken embryonic lens cells, in which even universal inducers, like Cd^{2+} and arsenite, failed to increase HSP synthesis.[46]

1.2.2. HEAT SHOCK-LIKE INDUCERS

Contrasting to hs, few of the chemical stressors were found positive in a broader range of organisms. It is very likely that most or all representatives of this group are true hs-like inducers, i.e., their activity depends on the hs-signal transformation chain and the presence of the HSE promoter element (see Section 1.4). Such inducers are

1. Arsenite in yeast, *Neurospora*, *Drosophila*, *Xenopus*, plants and vertebrates
2. Abnormal proteins in *Escherichia coli*, *Drosophila*, and *Xenopus*
3. Amino acid analogs, e.g., canavanine, azetidine carboxylic acid, *p*-fluoro-phenylalanine, in yeast, plants, *Tetrahymena*, *Drosophila*, and vertebrates
4. Cd^{2+} and similar metal ions in *E. coli*, *Drosophila*, plants and vertebrates
5. Ethanol in bacteria, yeast, *Xenopus*, *Drosophila*, and vertebrates
6. Anoxia and recovery in bacteria, *Tetrahymena*, yeast, *Drosophila*, and vertebrates
7. Puromycin in bacteria, *Tetrahymena* and vertebrates

Direct evidence for the hs-like mode of action of chemical stressors comes from different types of experiments:

1. Similar to hs, the whole set of HSPs, characteristic of the particular organism, is

induced. This is clearly evident from two-dimensional analyses of the HSP patterns in tomato cell cultures after treatment with heavy metals, e.g., Cd^{2+}, Zn^{2+}, and Hg^{2+}, with arsenite or amino acid analogs, e.g., azetidine carboxylic acid or *p*-fluorophenylalanine.[211a] Using seven class-specific cDNA probes, Edelmann et al.[63a] reported similar results by analyzing HSP mRNAs levels after hs, arsenite, or $CdCl_2$ induction of soybean seedlings.

Due to the peculiarities of the object, the corresponding observations on the activity of chemical stressors in *Drosophila* were mostly done with salivary gland polytene chromosomes. Though the actual analyses of the HSP patterns are frequently lacking, all hs puffs are induced by arsenite, azide, dicoumarol, dinactin, dinitrophenol, H_2O_2, menadione, methylene blue, rotenone, recovery from anoxia, salicylate, valinomycin, and different detergents (digitonin, saponin, Sarkosyl, TritonX100).[7,182]

2. An appropriate set of hs test promoter/reporter gene constructs can be used for transformation of expression systems (see Chapter 6). Comparative testing of induction by hs and chemical stressors should give essentially the same results, and deletion of the hs activator element (HSE) from the test promoter should abolish inducibility. Basically this type of experiment was performed with hs and arsenite as stressors by Kay et al.[132] (mouse cells), Mirault et al.[175] (COS cells), and by Ananthan et al.[3] using *Xenopus* oocytes injected with abnormal proteins. Using *E. coli,* Goff and Goldberg[79] reported on the induction of the test gene (hs [lon] promoter × lac Z gene) by canavanine, puromycin, streptomycin, or accumulation of the human plasminogen activator protein. The effect was abolished in $htpR^-$ strains, i.e., in bacteria lacking a functional hs activator protein (σ-factor).
3. New assay methods (see Section 7.4) allow the direct analysis of the key step in the activation of hs genes, i.e., the binding of the activated transcription factor (HSF) to the promoter region.[257,258] Footprint analyses and gel retardation assays were used by Mosser et al.[180a] to detect active HSF in HeLa cells stressed by hs, Cd^{2+}, or azetidine carboxylic acid. In the same way, Cajone et al.[34a] documented HSF activation in HeLa cells after treatment with hydroxynonenal, a major product of lipid peroxidation and an inducer of HSP synthesis. Finally, Zimarino and Wu[262] characterized the hsp83 gene activation in *Drosophila* S2 cells by hs, dinitrophenol, or salicylate. The three stressors gave similar results, although hs activation is probably faster than activation by the two chemical stressors. This delay is even more pronounced for hs-like stressors not acting by denaturation of preexisting proteins but by synthesis of abnormal proteins. These include RNA viral infection of vertebrate cells, incorporation of amino acid analogs, premature chain termination by puromycin as well as ethanol or streptomycin which affect the translation fidelity. Typically, hs gene induction by this subgroup of hs-like stressors is abolished in the presence of cycloheximide,[58,133,238] which is not true for hs and other chemical stressors.[6,262]

1.2.3. NON-HEAT SHOCK-LIKE INDUCERS

This group is composed of specific inducers whose effects on gene activation is not mediated by the hs activator element (HSE), described in Chapters 6 and 7. Characteristically, expression of only few or even a single HSP is affected. Examples are

1. Induction of small HSPs in *Drosophila* by developmental factors, e.g., ecdysterone.[42,118,119,236] The promoter element (ecdysterone box), interacting with the putative hormone receptor complex, is separate from the HSE.[45,78,113,139,152,173,178,201,201a] Further details on the developmental control of HSP synthesis, also in other organisms, are discussed in Section 8.4.
2. Presumably closely related to this developmental activation is the induction of small

HSPs, mainly of HSPs 22 and 23, by teratogens when applied to *Drosophila* embryonic cell cultures.[23,32] Also steroid compounds, like dexamethasone, diethylstilbestrol, and methyltestosterone, proved to be teratogenic in this *Drosophila* assay.

3. Depending on the target tissue (mammary carcinoma cells or uterine or brain tissue), induction of HSP90, HSP70, or HSP25-27 by steroid hormones (mainly estrogen) was reported by Fuqua et al.,[70a] Ramachandran et al.,[197] and Mobbs et al.[176] The selective increase of HSP90 levels after induction with steroid hormones may be connected with its function in complex with the inactive steroid receptor protein (see Section 16.8).
4. Coexistence of hs inducibility and control by mitogens in a complex promoter region is typical for the human hsp70 gene but probably also for related genes of other mammalian species (see Figure 6.6). Separate promoter elements control hs-enhanced transcription and stimulation by serum factors, the adenovirus E1A protein or by inherent cell cycle-dependent activator proteins.[66a,117,174,218,254a,260,261] Induction of HSP70 synthesis in vertebrates and *Xenopus* by wounding or disruption of cell-cell contacts may in fact reflect this mode of cell cycle control.[4,53,100,106,253,256]

For the majority of inducers contained in Table 1.1, a classification as hs-like or non-hs-like stressor is not yet possible. The complexity of the problem is best illustrated by the increase of HSPs 89 and 70 synthesis in vertebrate cells after long-term glucose deprivation followed by addition of glucose, dexamethasone, or uridine.[130,131,150,212,252] Elements of the cell cycle control or, alternatively, accumulation and increased degradation of aberrant, underglycosylated proteins, i.e., the hs type of control, may be responsible for the induction during recovery from glucose deprivation.

Basically similar arguments hold true for the HSP70 induction by wounding in vertebrates[4,53,100,106] and even in plants.[101] Using hsp70 antisense RNA, Brown et al.[27a] visualized the selective and highly localized activity increase of this hs gene after micro-surgical cuts into the rat cerebral cortex.

Two groups of specialized inducers were not included in Table 1.1. In vertebrate cells, glucose deprivation and several compounds interfering with protein glycosylation induce synthesis of glucose-regulated proteins (GRPs see Table 4.1). This includes also a number of RNA viruses frequently cited to induce stress proteins in vertebrate cells.[47,73,120,194,196] On the other hand, in *Drosophila* a prominent hs puff, not coding for a new protein but for a special class of hs-RNA, can be induced selectively by vitamin B6, colchicine, or benzamide application (see Chapter 5).

1.2.4. ADDITIONAL STRESS PROTEINS

It is not surprising that induction of stress-related proteins (SP) by chemical compounds is frequently not restricted to HSP. A prominent example is a SP32 (M_r 32 to 40 kDa) found in vertebrate cells after induction with hs, arsenite but also with ethanol, transition series metals, iodoacetamide, diethylmaleate, hydrogen peroxide, UVA irradiation, or disulfiram.[37,56,134a,148,215b,217a] However, a similar arsenite-induced stress protein was also reported for *Neurospora*[126] and *Tetrahymena*.[1]

Induction of SP32 by a considerable number of HSP-inducing chemical stressors is of particular interest because the same protein was recognized as an indicator for tumor-promoting activity of many structurally unrelated compounds:[110-112] these include indole alkaloids or polyacetates, the phorbol ester TPA, alkylating agents like methyl methane sulfonate, and *N*-methyl-*N'*-nitro-*N*-nitrosoguanidine, sodium deoxycholate, the protease inhibitors TPCK and TLCK, and the carcinogen diaminobenzidine. In fact, SP32 synthesis in mouse 3T3 cells is also mildly stimulated by hs (10 min 45°C, Hiwasa and Sakiyama[110]). On the other hand, the four transition series metal ions, active as HSP inducers in vertebrate cells, are also well known tumor promoters in Syrian hamster embryo.[204] In all likelihood, this SP32 represents the heme oxygenase identified in human and rat cells.[134a,217a] The rat, but not the human, ho gene contains a functional heat shock element in its promoter region (see Table 6.2).[216a]

In plants, the whole spectrum of stress response proteins is much more diversified than in animals. A considerable number of general stress proteins and metabolites belong to different but partially overlapping stress domains.[188a] One striking example is the soybean HSP26, which is unrelated to any other plant HSP so far known. It is coded by an intron-containing gene and induced by hs, Cd^{2+}, Cu^{2+}, arsenite, the phytohormones abscisic acid, and 2,4-dichlorophenoxyacetic acid or by osmotic stress.[54a]

In addition to induced HSP synthesis, stress by intoxication with heavy metals is connected with increased formation of metallothioneins in animals and yeast.[63,99,168] Interestingly, in plants and many fungi it is not metallothionein, but rather the synthesis of specialized glutathione derivatives with extremely high metal binding affinity called phytochelatins, that accompanies heavy metal stress.[80,81,81a] The key enzyme, phytochelatin synthase, is not noticeably induced but rather is activated in the presence of heavy metals, e.g., Cd^{2+}, Ag^{+}, Bi^{3+}, Pb^{2+}, Zn^{2+}, and Hg^{2+}.[81b]

A special group of chemical stressors concerns the induction of two prominent constitutive proteins in the endoplasmic reticulum (ER) of vertebrate cells (GRP94 and GRP78). They are discussed separately in Chapter 4 (Table 4.1). However, some of the hs-like inducers also increase GRP synthesis, an effect presumably connected with perturbations of glycoprotein synthesis (see Figure 4.1).

1.2.5. OTHER ASPECTS OF THE HEAT SHOCK RESPONSE

Although increased HSP synthesis is a very convenient test for a chemical stressor to be included in Table 1.1, it should be kept in mind that hs triggers a complex reprogramming of many cellular activities. Characterization of chemical compounds in this respect is very limited.

Among the most stress-sensitive parts of the cell is the nucleolus (see Section 10.5). During hs there is an immediate breakdown of ribosome biosynthesis connected with disaggregation of the nucleolar structure and accumulation of HSP70 in this compartment. Consequently, nuclear/nucleolar binding of HSP70 and ultrastructural changes of the nucleolus were used as additional criteria for characterization of chemical stressors. Although the results are not unequivocal, it is evident that anoxia, arsenite and amino acid analogs, e.g., azetidine carboxylic acid, exert hs-like effects on the nuclear compartment of *Drosophila* and mammalian cells.[46,241,243,249] The outcome depends on the severity of stress,[245,247] i.e., on the concentration of chemical stressor and duration of treatment. In the case of analogs recompartmentation of newly formed HSP70 is only observed in the recovery period because stress proteins synthesized in the presence of the drug are nonfunctional.[249]

Similarities and differences between hs and chemical stress response systems are also apparent when comparing effects of hs, arsenite, and amino acid analogs in rat embryo fibroblasts[247,249] and in chicken embryo fibroblasts.[46] HSP synthesis, translation reprogramming and recovery, changes of the cytoskeleton, and intracellular localization of HSPs were studied. Thomas and Mathews[238] reported very similar effects of hs and azetidine carboxylic acid on HSP synthesis, polysome decay, translational shut off of control protein synthesis, and RNA synthesis in HeLa cells. The same is true for a group of anion transporters tested in *Drosophila*.[216] Increased HSP synthesis is connected with other prominent parts of the hs response, e.g., cytoskeletal rearrangement, repression of DNA and control protein synthesis, dephosphorylation of ribosomal protein S6.

1.3. INTERACTION OF STRESSORS AND MODULATORS OF HSP SYNTHESIS

Considering the network of changes observed under stress conditions, it is not surprising that the extent of HSP synthesis is notably repressed or enhanced by the metabolic state of the affected cells. Lanks[149] summarized our knowledge about such "modulators" of HSP synthesis. Most prominent with respect to their inhibitory effect in vertebrate cells are conditions connected with enhanced synthesis of glucose-regulated proteins (Table 4.1), i.e., glucose deprivation, addition of insulin, tunicamycin, 2-deoxyglucose, mercaptho-ethanol or Ca^{2+}-ionophore A23 187, long-term anoxia and low extracellular pH.[149,252] Restoration of full level hs inducibility after glucose deprivation can be achieved by addition of pyruvate, uridine, or dexamethasone.[130,131,150] The induction of HSP synthesis in chicken cells by high extracellular pH[252] (see Table 1.1) may be related to early observations of Ashburner and Bonner[7] that hs-induced puffing of polytene chromosomes is inhibited if *Drosophila melanogaster* salivary glands are immersed in nutrient medium at pH 4.4 instead of pH 6.8. However, the intracellular pH value in salivary gland cells decreases during hs at 35°C from pH 7.38 to 6.91.[62] A similar result was reported for yeast. HSP induction by hs, arsenite, or 2,4-DNP coincides with a rapid decrease of intracellular pH, and, at least for dinitrophenol, much higher doses of the stressor are required for equal inducing effects if the extracellular pH is increased from 5.8 to 7.3.[251,251a]

An interesting class of inhibitors of HSP synthesis in chicken embryo fibroblasts was described by Hightower et al.[107a] and Edington et al.[63b] 1 *M* Glycerol, erythritol or 85 to 99% D_2O severely diminished induction of HSP synthesis by hs, arsenite, or Cu^{2+}. Maximum induction could only be achieved by drastically increasing the level of stressors. The inhibitory effect was confirmed by White and White[253b] when studying the influence of glycerol on HSP70 synthesis in rat brain slices. Polyhydroxy alcohols and D_2O are known as thermoprotectants. They presumably act by minimizing the damaging effects of heat or chemical stressors on proteins or other cellular components, i.e., they diminish the level of the putative intracellular stress signal.

A couple of positive modulators of HSP synthesis were described. Most of them are active only in distinct cells or under restricted metabolic conditions. Thus, 10 ng/ml vitamin D protects human peripheral blood monocytes but not fibroblasts from heat damage, an effect which is connected with increased HSP synthesis.[44a,195a] On the other hand, the promoting activity of dexamethasone, pyruvate, or uridine on HSP synthesis in vertebrates is restricted to glucose-starved cells. Related to this are results with more general modulators of HSP synthesis. Bleomycin, an antitumor glycopeptide antibiotic, was shown to stimulate hs-induced HSP synthesis in *Aspergillus* and *Drosophila*.[227] In vertebrate cells cytotoxicity of the antibiotic, broadly used in cancer therapy, is increased by hyperthermia.[90,164,244] A remarkable sensitization of mouse cells to mild hs was achieved by 4 to 7 h pretreatment with interferon α or β.[176b] The stimulation of HSP70 synthesis at 42°C (Figure 1.2B) is achieved by interferon doses which are effective also in inhibition of EMC virus replication (Figure 1.2A). It coincides with repression of actin translation (Figure 1.2B). This type of stress sensitization is even observed when HSP70 synthesis is induced by arsenite. The IFN effect is due to an increased transcription of hs promoters and to a stabilization of HSP mRNAs.[62b]

Mutually synergistic effects of a similar type were also reported for the interaction of mild stress conditions, which separately have very low inducing activity. Examples are hs and DNP in *Drosophila* salivary glands,[65] hs and ethanol in mouse lymphocytes,[205] or hs and arsenite in *Xenopus* kidney epithelial cells.[102] Figure 1.3 shows that 10 μ*M* arsenite or hs at 30°C are poor inducers, but taken together, they evoke maximum induction of HSP70 and HSP30 synthesis and the corresponding mRNA levels. As already mentioned for bleomycin, the principal of mutual stress potentiation applies also to the application of cytotoxic drugs together with local hyperthermia in cancer therapy (Section 20.3). A number of the applied antitumor drugs may be effective as hs sensitizers as well. This should be investigated in more detail.

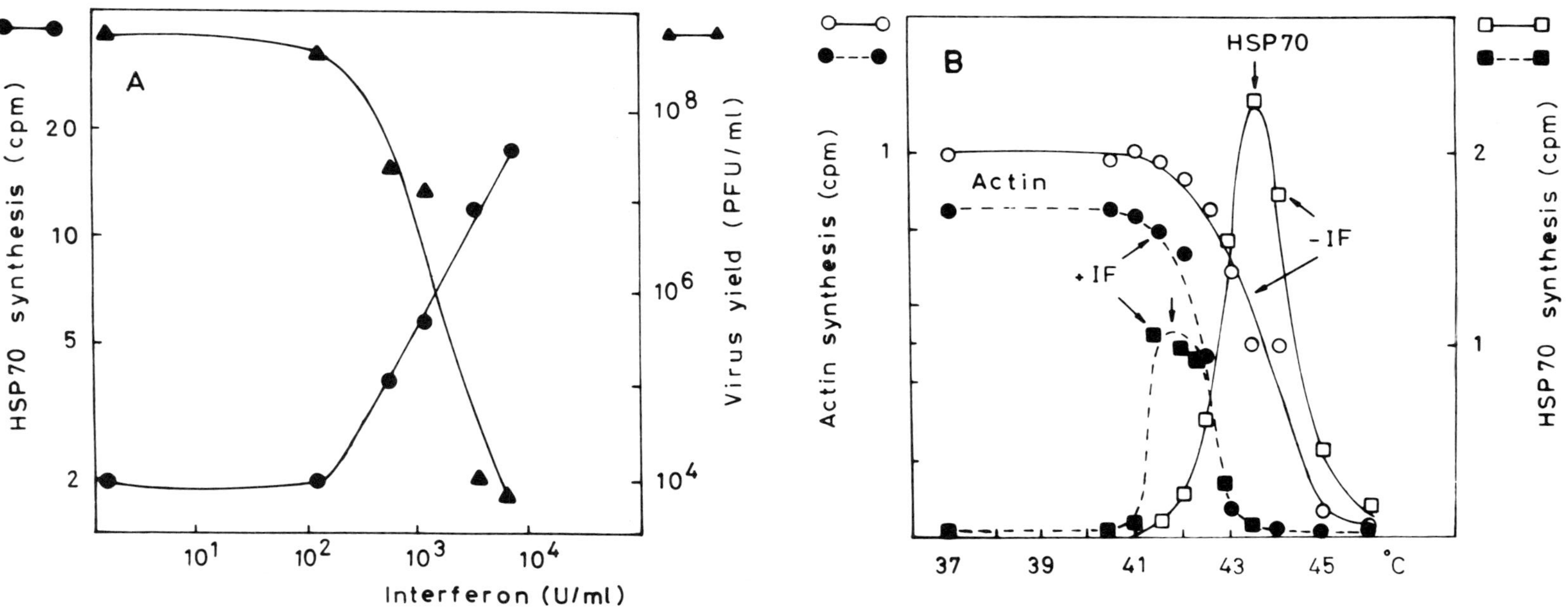

FIGURE 1.2. Influence of interferon on the heat shock (hs)-induced HSP70 synthesis and the decay of actin synthesis in mouse cells. (A) The increasing potential for HSP70 synthesis after 45 min induction at 42°C. Subsequent labeling for 3 h in the recovery period correlates with the decrease of the encephalomyocarditis virus yield at interferon doses 10^2 U/ml. The addition of interferon α or β was 20 h before the experiment. (B) The interferon effect due to a temperature sensitization. In the presence of 3×10^3 U/ml the optimum of HSP70 inducibility and, correspondingly, the decay of actin synthesis are shifted by almost 2°C to lower temperatures. (Modified from Morange, M., *J. Cell. Physiol.*, 127, 417—422, 1986.)

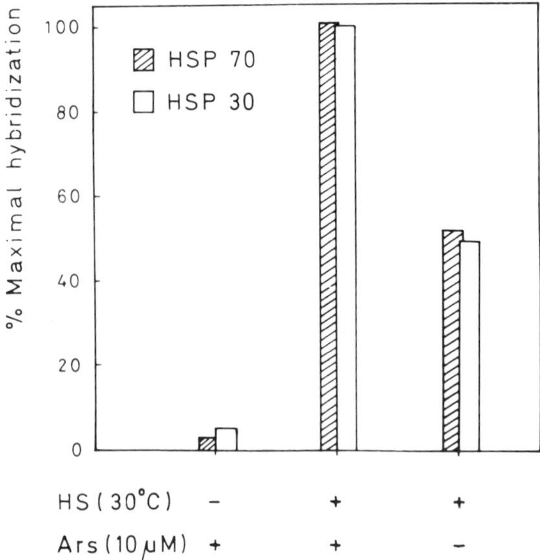

FIGURE 1.3. Synergistic effects of stressors for the induction of HSP mRNAs in *Xenopus* kidney epithelial cells. *X. laevis* A6 cells were exposed to a 2-h treatment with a mild hs (30°C) and/or with low doses of arsenite. Using Northern blots, mRNA levels for HSP70 and HSP30 were analyzed with appropriate clones.[16b] Inductive effect of the two mild stress regimes in combination equals or even surpasses the effect of each stressor at optimum activity, i.e., hs at 35°C or arsenite at 75 μ*M* concentration. (Modified from Heikkila, J. J., Darash, S. P., Mosser, D. D., and Bols, N. C., *Biochem. Cell Biol.*, 65, 310—316, 1987.)

Stress response modulators in plants are phytohormones and related compounds.[191b] Thus, abscisic acid, triadimefon, cytokinins, and brassinosteroids were reported to improve survival or protein synthesis after an hs period. So far there is no evidence that increased HSP synthesis is responsible for these effects.

1.4. SEARCH FOR THE HEAT SHOCK SIGNAL SYSTEM

1.4.1. ELEMENTS OF THE SIGNAL SYSTEM AND THE ROLE OF ABNORMAL PROTEINS

Control of gene expression by extracellular physical (e.g., heat and light) or chemical signals (e.g., nutrients, hormones, and stressors) are usually mediated by more or less elaborate signal systems. These include receptor proteins, proteins involved in signal transformation and eventually in the generation of intracellular "second messengers", and, finally, the regulatory protein which triggers the response. Specificity, velocity and extent of the response are thoroughly influenced by the signal system.[16a,37a,186,234]

Investigations on the signal system involved in HSP synthesis induced by hs or chemical stressors were stimulated by the early suggestion of Kelley and Schlesinger,[133] Hightower,[105] Collins and Hightower[47] that accumulation of abnormal proteins may be a decisive factor. Although we are still far from a detailed understanding, essential elements of the signal system were defined and a number of experimental findings support the central role of abnormal proteins:

1. Promoter sequencing and functional analysis led Pelham[192] to the definition of a 14

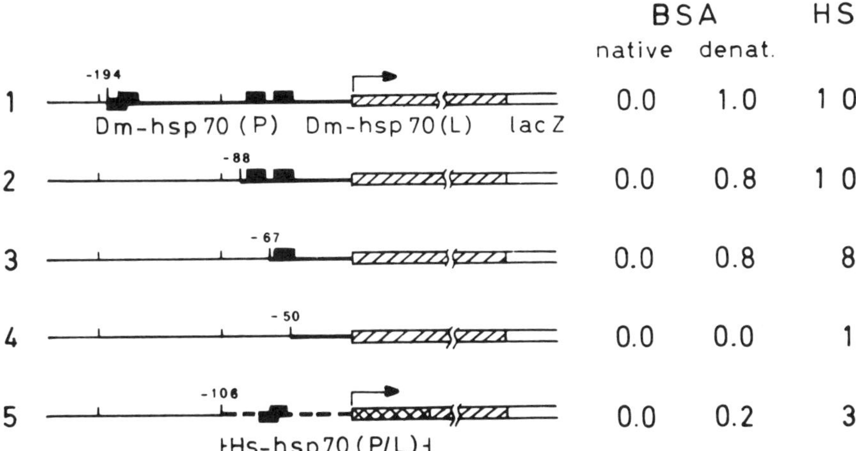

FIGURE 1.4. Injection of denatured proteins into *Xenopus* oocytes induce transcription from heat shock (hs) promoters. *Xenopus* oocytes were injected with hsp70 (P/L) × lac Z fusion genes together with native or denatured bovine serum albumin (BSA). β-Galactosidase was induced by hs treatment (HS) or by injection of native or denatured BSA. Constructs 1 to 4 contain 194, 88, 67, and 50 bp of 5′ noncoding sequences of the Dm-hsp70 promoter, whereas construct 5 represents a hybrid of the human hsp70 P/L region fused at bp +105 to bp of the Dm-hsp70 leader region. HSE elements are indicated by boxes, the transcription start by arrows. (Modified from Ananthan, J., Goldberg, A. L., and Voellmy, R., *Science,* 232, 522—524, 1986.)

nucleotide hs promoter element (HSE) responsible for hs inducibility.[17,192,193] Later, it was shown to be present in the promoter region of all hs-inducible genes of eukaryotic cells (see Chapter 6).[18,188]

2. The second essential element is the corresponding HSE-specific DNA binding protein, the HSF. Details of HSF detection, purification and functional analysis are described in Chapter 7. HSF is a constitutive protein, which undergoes a hs-induced transition in its activity as already shown by Craine and Goldberg[52b] in 1981. The bacterial analog of HSF is the product of gene htpR, coding for a hs-specific σ subunit of RNA polymerase (see Section 7.1).
3. It is worth noticing that the threshold temperature for HSP induction greatly varies with the organism investigated (see Table 2.1). The difference between the normothermic and hyperthermic state is largely determined by the cell-specific elements of the signal transformation chain, e.g., by the HSF activity. Using the *Drosophila* hsp70 promoter as the universal test system,[188] hs induction was achieved at 25°C in sea urchin eggs,[171] at 34 to 36°C in *Xenopus* oocytes,[17,151,172] at 36 to 37°C in *Drosophila* (see Figure 1.1), at 40°C in tobacco cells,[225,226] at 41°C in mosquito cells,[16] and at 42 to 43°C in mammalian cells.[2,49,50,179,192] The hs threshold temperature is genetically fixed and cannot be changed by long-term preadaptation at low temperatures.[180]
4. Injection of abnormal proteins into *Xenopus* oocytes[3] is sufficient to activate transcription from a *Drosophila* or human hsp70 promoter (Figure 1.4.). But abnormal proteins can also be generated *in situ,* e.g., by incorporation of amino acid analogs[58,133,238] by antibiotics causing mistranslation or premature termination,[79a,97,106,154] or by protein denaturation. Thus, HSP induction in Chinese hamster ovary cells by diamide, arsenite, or the cross-linker dithiobis(succinimidyl propionate) coincides with formation of high-molecular-weight protein complexes.[154a] On the other hand, addition of deuterium oxide or 1*M* glycerol strongly inhibits activation of hs genes (hsp70 and hsp90) in chicken embryo cells. This protection from stress-dependent protein damage at 44°C can be overcome by raising the temperature to 49°C.[63b]

Induction of hs genes by the accumulation of abnormal proteins was also observed in *E. coli* transformed with a malE-lacZ fusion gene,[121] with a human plasminogen activator gene,[79] or with a truncated gene coding for the N-terminal domain of the λ-repressor.[191a] Evidently, it is the level of aberrant protein and not its degradation that is responsible for the inducing effect. Similar to the role of the λ-phage cIII-protein,[10] stabilization of the short-lived hs-specific sigma factor by interaction with unfolded proteins may be the underlying mechanism.[191a,255a]

Recently, Seufert and Jentsch[215a] described a yeast mutant defective in the genes for the two hs-inducible ubiquitin-conjugating enzymes UBC4 and UBC5. Due to the inefficient degradation of short-lived and abnormal proteins in the double mutant (ubc4/ubc5), a constitutive, high level of HSP synthesis and thermotolerance are observed.

Particularly striking is the cell-specific induction of HSPs in the indirect flight muscle of *Drosophila* transformed in the germ line with mutant actin genes.[108,109,128] The specificity results from the particular developmental control of actin gene expression. In this respect, it is an interesting speculation that differences in the generation of abnormal proteins in the cytoplasm and in the ER/Golgi system, respectively, may be the basis for the classification of chemical stressors into HSP and GRP inducers, respectively,[105a] (see Figure 4.1). This concept was recently supported by experiments with mutant virus strains causing accumulation of malfolded haemaglutinin in the ER of infected simian cells and, consequently, induction of GRPs 78 and 94.[142a]

5. The close connection between perturbations of protein degradation and the hs-induction signal was impressively demonstrated by Finley et al.[67] using a mouse cell line with a temperature sensitive ubiquitin transferase system. Adenosine triphosphate (ATP)-dependent, covalent tagging of proteins with the 76 amino acid residue ubiquitin is the start reaction for protein degradation in all types of eukaryotic organisms. At the nonpermissive temperature, which by itself is not inducing, massive stimulation of HSP synthesis is observed in this mutant. However, the simple explanation that ubiquitin conjugation is the primary target of hs is not valid. Microinjection of ubiquitin into HeLa cells[38,39] confirmed the results about drastic alterations of the ubiquitin pathway after hs and about reduced protein breakdown,[181] but the ubiquitin conjugation system itself was found intact (see also Chapter 12). These results agree with an earlier report[104] of a tenfold increased level of ubiquitin-conjugated protein in mammalian cells treated with amino acid analogs.
6. The apparent urgency of removing the damaged proteins from hs cells may explain why certain proteins directly related to the proteolytic system are hs-induced. This is true for a protease of bacteria (see Table 2.2) and for hs-induced polyubiquitin genes of eukaryotes (Sections 2.3.5 and 12.3).[22,40a,68,68b,154b,180b,189,190a,234a,243a] The hs-specific polyubiquitin gene of yeast, induced by starvation, hs and amino acid analogs, is essential for survival under stress conditions.[68]

1.4.2. STABILITY VS. MODIFICATION OF THE HEAT SHOCK TRANSCRIPTION FACTOR

A model for the hs signal transformation chain[18,181a] involves the existence of a constitutively expressed, short-lived HSF, whose degradation is diminished or stopped if the proteolytic system is overloaded with aberrant proteins. These may be generated by heat denaturation, by chemical stressors interfering with SH-group metabolism (heavy metals, arsenite, iodoacetamide, *p*-chloromercuribenzoate), or by treatments generating oxygen free radicals (recovery from anoxia, addition of hydrogen peroxide) as well as by translation of aberrant proteins, e.g., after RNA viral infection, addition of ethanol, puromycin, or amino acid analogs.

Although numerous experimental data support the intricate role of the proteolytic system in the hs-related signal transformation, there is no evidence for a short-lived transcription factor in eukaryotes. Functional characterization and purification of HSF from *Drosophila* and HeLa and yeast cells (see Section 7.6) revealed constitutive proteins, whose activity state rather than amount varies with the stress conditions. In yeast, HSF activation is connected with changes of the phosphorylation state[223] and in *Drosophila* and plant cells repeated cycles of activation/deactivation of HSF by hs and recovery can be achieved in the presence of cycloheximide.[211a,262]

However, in contrast to eukaryotic cells the results with *E. coli* are in good agreement with the Pelham model. HSF of *E. coli* is a special σ-subunit of RNA polymerase with M_r 32 kDa.[82] σ^{32} is short-lived with a half-life of approximately 4 min.[84] Under hs or ethanol stress a 17-fold rise of the σ^{32} level results from increased synthesis and decreased degradation.[66,231] Changing levels of σ^{32} correlate with the increase of transcription from *E. coli* hs promoters (see Sections 7.1 to 7.3). Stabilization of σ^{32} against degradation is also achieved by complex formation with the λ-phage cIII protein, which evidently acts as a protein-specific antiprotease.[10] This mechanism explains the induction of hs genes after lysogenic infection of *E. coli* cells by phage λ.

Unlike the situation in bacteria, the actual link between stress-induced changes of the proteolytic system and the corresponding activation of HSF remains to be found in eukaryotes. The original model of Munro and Pelham[181a] is evidently too simple. This was impressively demonstrated by Parker-Thornburg and Bonner,[191] who selected 21 *Drosophila* mutants with constitutive expression of HSP coding genes. Only 1 mutant in their collection had a general expression, whereas the other 20 had a tissue-specific expression of HSPs. The high frequency of these gain-of-mutations points to the multiplicity of genes affected. However, the simple explanation of a tissue-specific formation of aberrant proteins seems not valid. P-element transformation of *Drosophila* with a great number of genes coding for mutant proteins expressed in different tissues did not induce HSP synthesis. In fact, this is also true for the mutant actin genes (act 88F) expressed in the indirect flight muscle cells.[108,109,128] Only very few of the strongly antimorphic mutants causing flightless individuals are also strong inducers of HSPs.[128] Thus, we are presumably confronted with a multiplicity of subtle perturbations in metabolic homeostasis resulting in stress activation of hs genes, possibly complicated by species-specific differences in response to a given inducer. The search for the missing link in the signal transformation chain goes on.

1.5. A NETWORK OF SIGNAL SYSTEMS

With the few exceptions noted above, inducers compiled in Table 1.1 were characterized solely by their capability to evoke puffing of hs loci in *Drosophila* and/or to stimulate HSP synthesis. The complex pattern of cellular reprogramming during the hs response almost certainly requires other signal systems in addition to that involved in hs gene activation. Unfortunately, investigations in this direction are practically lacking. Thus, we can only summarize the relevant points to stimulate corresponding activities. The "need" for additional hs-induced signal transformation chains is evident when considering the rare cases with translational induction of HSP synthesis (Section 8.6), the rapid rearrangements of cytoskeletal systems (Section 14.3), polysomal breakdown, dephosphorylation of ribosomal small subunit and selectivity of the translation apparatus (Chapter 11), the immediate stop of nucleolar preribosome processing (Section 10.2), the block of DNA synthesis and cell cycle activities (Chapter 15), and early, HSP-independent stages of induced thermotolerance (Section 17.5).

A remarkable number of metabolic systems, frequently characterized as parts of signal transformation chains[16a,37a] are more or less affected by hs. However, it is important to notice that some of the results, that are described in the following, stem from studies on hyperthermic treatment of cancer cells, i.e., they were obtained under conditions of severe hs used to kill cells.

1.5.1. MITOCHONDRIAL RESPIRATION

The observation that many chemical stressors, inducing hs puffs in *Drosophila*, are respiratory inhibitors led to the early suggestion that mitochondria are primary targets of stress attack generating the signal for hs induction.[13,156,199,219] A couple of respiratory enzymes were claimed to be hs induced (see data compiled in Table 2.6). However, though members of the HSP60 and HSP70 families are localized in mitochondria (Section 16.5), no major HSP is a respiratory enzyme. As discussed above, concentrations of respiratory drugs used for hs inductions are orders of magnitude higher than those required for inhibition of mitochondrial activities. Finally, respiration-deficient mutant cell lines of yeast[170,251] and vertebrates[148] or rice embryos under anoxic conditions[176a] exhibit normal induction of HSP synthesis. Thus, mitochondrial respiration is evidently not involved in the signal transformation. In support of these findings Drummond and Steinhardt[62a] made a careful comparison of HSP synthesis and SH-group content in *Drosophila* Kc cells after induction by hs or reoxygenation following anoxia. No changes of the GSH/GSSG ratio were connected with high levels of HSP synthesis after hs induction. In contrast, severe oxidative stress with reduced glutathione (GSH) and fourfold increased GSSG but only moderate induction of HSP synthesis were found during reoxygenation.

1.5.2. BREAKDOWN OF CYTOSKELETON

The almost immediate collapse of cytoskeletal systems in the perinuclear region (Section 14.3), is evidently not required or involved in induction of hs genes, at least not in mammalian cells.[248] However, a role for changes at the translation level (Section 11.5.1) or for the formation of cytoplasmic HSP aggregates (Section 16.7) is very likely.

1.5.3. INTRACELLULAR Ca^{2+} LEVELS

Results of the relation between hs phenomena and Ca^{2+} levels are equivocal. HS treatment of *Drosophila* larval salivary glands coincides with a tenfold increase of intracellular Ca^{2+} levels,[62] but a direct role for HSP induction can be excluded. In vertebrate cells, a transient increase of Ca^{2+} influx and subsequently of the intracellular Ca^{2+} levels[228,229] is connected with an improved heat resistance. In contrast, a continuous long-term overload of the cytosol with the cation led to toxic effects and enhanced heat sensitivity. This may explain the findings of Lamarche et al.[146] that long-term Ca^{2+}-deprivation of rat hepatoma cells strongly increases thermoresistance, but the cells become refractory to HSP induction. In contrast to this Kim and Lee[136] found the hs inducibility unaffected in hamster fibroblasts deprived from Ca^{2+} by EGTA treatment, i.e., under conditions inducing GRP synthesis (see Chapter 4).

Landry et al.[148a] summarized the seemingly contradictory results. Not the extracellular Ca^{2+} but rather its redistribution between intracellular pools may be decisive for different aspects of the hs response. On the one hand, there is no doubt that HSP90 and HSP70, bind calmodulin, which controls its interaction with cytoskeletal proteins in a Ca^{2+}-dependent manner (see Sections 2.3.1 and 2.3.2). On the other hand, rapid changes in phosphorylation level of cytoplasmic proteins including HSP27 may depend on a Ca^{2+}-activated protein kinase. But in any case, the triggering of HSP synthesis is not affected by Ca^{2+}.

1.5.4. PHOSPHOINOSITIDE METABOLISM

Closely related to and presumably preceding the Ca^{2+} influx[16a,37a] are hs induced changes of phosphoinositide metabolism in CHO and mouse 3T3 cells.[36,228] Already, after less than a minute, release of inositol triphosphate was observed, followed by an increase of phosphorylation of polyphosphoinositides. These results emphasize the important role of membranes as primary targets of the heat stress possibly with pleiotropic effects on other parts of the hs response.

1.5.5. POLYAMINE LEVELS

Early observations on the depletion of heat-shocked CHO cells of spermine and spermidine were correlated with the block of DNA synthesis.[76] Investigation of the biochemical mechanism[95,96] revealed a hs-induced spermidine N^1-acetyltransferase and subsequent oxidative degradation of acetylspermidine. Inhibition of endogenous polyamine oxidase confers partial resistance against heat damage. The generation of hydrogen peroxide and aldehydes by polyamine oxidation may relate hyperthermia to oxidative stress. Glutathione depletion potentiates the toxic effects of polyamine oxidation,[95,96,175b] whereas elevated levels of glutathione are observed in thermotolerant cells.[176,184] The enhancement of hyperthermic cell killing by exogenous addition of 10^{-5} *M* spermine or spermidine[77] is presumably caused by oxidative degradation of these compounds, i.e., it is due to the potentiation of hyperthermic cell damage by oxidative stress.

1.5.6. CYCLIC AMP

A transient, marked increase of cyclic adenosine monophosphate (cAMP) levels in CHO cells was reported by Calderwood et al.,[35] but this "cAMP response" is evidently not related to HSP synthesis. Addition of the stable analog dibutyryl-cAMP enhanced the thermoresistance of CHO and HeLa cells without increasing HSP synthesis.[35,97a] Hs-stimulated adenylate cyclase activity was also observed in the slime mold *Dictyostelium discoideum.*[89]

The situation in yeast is evidently opposite to that in animal cells. Essential aspects of the hs response (synthesis of three major HSPs, acquisition of thermotolerance and transient cell cycle arrest) depend on decreased levels of cAMP-dependent protein phosphorylation.[217] They are inhibited in cells treated with cAMP or in a mutant cell line with a cAMP-independent protein kinase. Moreover, a second mutant with defective adenylate cyclase shows constitutive thermotolerance and synthesis of HSPs 72A, 72B, and 41 (see Section 8.4). The restriction of the effect to a small group of HSPs, including also polyubiquitin,[234c] demonstrates that decrease of the cAMP level is evidently not part of the general, hs-dependent induction system of HSP synthesis in yeast. However, it is intriguing that the HSF is a phosphorylated protein, whose activation is correlated with an increased phosphorylation level (see Figure 7.5).

1.5.7. HEAT SHOCK AND OXIDATION STRESS: ROLE OF DIADENOSINE TETRAPHOSPHATE (Ap_4A)

Ames and co-workers[20,152a] defined a new bacterial "alarmone", diadenosine 5′, 5‴-P_1, P_4-tetraphosphate (Ap_4A), proposed to mediate oxidative and heat stress induced changes of cell metabolism in bacteria. However, this concept may not be valid in this general form. On the one hand, induction of HSP synthesis in *E. coli* proceeds also without increase of Ap_4A levels and, vice versa, elevation of Ap_4A is not always connected with HSP synthesis.[19,242] Investigations of Bochner et al.[21] provided evidence for a possible role of Ap_4A in autorepression of HSP synthesis in *E. coli* by the DnaK protein (see Section 8.7).

The stimulating concept of Ap_4A as the stress alarmone of course led to the detection of similar effects also in eukaryotic cells, e.g., in the slime mold *Physarum polycephalum*,[72] in yeast,[57] and, after severe hs, also in *Xenopus* oocytes.[85] Interestingly, injection of Ap_4A into oocytes was able to potentiate the inducing effect of a preceding mild hs.[85] However, inducers of HSP synthesis in *Drosophila* (hs, $CdCl_2$) had no or very moderate and delayed effects on the cellular levels of Ap_4A and related dinucleotides.[25] The same negative outcome is also true for ethanol and hs treatments of mouse 3T3 and Chinese hamster lung cells[215] and for hs induction of *Artemia salina.*[174a] Summarizing, Ap_4A may have stress-related effects in bacteria, but it is certainly not a general stress "alarmone" of eukaryotic cells.

The evidence against a role of Ap_4A in the heat shock signal transformation system does not argue against an interrelation of hs and oxidation stress with respect to distinct parts of the response. Oxidation stress in plants is secondary to other stressful situations, e.g., hs, intoxication with heavy metals, cold, or osmotic stress. Lipid peroxidation and oxidation of SH-group-containing proteins may be the consequence. Induction of scavenging enzymes (superoxide dismutase, catalase) is usually part of the protection mechanism (for a summary, see Nover et al.[188a]). Evidence for this derives from the following observations:

1. A number of chemical stressors causing oxidation stress were characterized as inducers of HSP synthesis in different organisms, e.g., hydrogen peroxide, menadione, heavy metals, and UV irradiation (see Table 1.1).
2. In yeast and *Neurospora crassa,* catalase activity is markedly increased after hs or treatment with hydrogen peroxide, menadione, or the herbicide paraquat.[18a,41a]
3. Hs, paraquat treatment, or pathogen infection cause the same modification of the translation apparatus in maize, i.e., binding of a 57-kDa protein to polysomes. Isolated p57 inhibits *in vitro* translation with maize polysomes.[259a]
4. Vitamin D_3 treatment triggers erythrophagocytosis by human monocyte macrophages. Evidently, the generation of active oxygen connected with phagocytosis leads to the induction of HSP synthesis, including synthesis of heme oxygenase (SP32). The effect is inhibited by *N*-(2′-mercaptoethyl)-1,3-propanediamine, indicating that glutathione depletion is involved.[44a]

1.5.8. GUANOSINE 3′-DIPHOSPHATE 5′-DIPHOSPHATE (ppGpp)

In bacteria, many stress conditions connected with reduced growth rates coincide with increased levels of ppGpp or related compounds. During the stringent response,[71] many "housekeeping functions" of the bacterial cell are reduced (e.g., ribosome biosynthesis, tRNA synthesis), while expression of other genes is induced. Among them are also HSP coding genes in *E. coli*,[83] in *Bacillus subtilis*,[200] and cyanobacteria.[233] The hs-induced increase of ppGpp[200,233] connects the two bacterial stress response systems. In *B. subtilis* induction of HSP synthesis by amino acid analogs or oxygen deprivation is mediated by the ppGpp system, i.e., it is only observed in $relA^+$ strains, whereas hs induction proceeds equally well in $relA^+$ and relA strains.[98] The results were confirmed by van Bogelen et al.[242] with *E. coli*. Although HSP induction by some stressors is connected with increased levels of ppGpp (ethanol, $CdCl_2$, H_2O_2), this is not true for hs induction.

REFERENCES

1. **Amaral, M. D., Galego, L., and Rodrigues-Pousada, C.,** Stress response of *Tetrahymena pyriformis* to arsenite and heat shock: Differences and similarities, *Eur. J. Biochem.*, 171, 463—470, 1988.
2. **Amin, J., Mestril, R., Lawson, R., Klapper, H., and Voellmy, R.,** The heat shock consensus sequence is not sufficient for hsp70 gene expression in *Drosophila melanogaster, Mol. Cell. Biol.*, 5, 197—203, 1985.
3. **Ananthan, J., Goldberg, A. L., and Voellmy, R.,** Abnormal proteins serve as eukaryotic stress signals and trigger the activation of heat shock genes, *Science,* 232, 522—524, 1986.
4. **Anderson, N. L., Giometti, C. S., Gemmell, M. A., Nance, S. L., and Anderson, N. G.,** A two-dimensional electrophoretic analysis of the heat-shock-induced proteins of human cells, *Clin. Chem.*, 28, 1084—1092, 1982.

4a. **Andrus, L., Altus, M. S., Pearson, D., Grattan, M., and Nagamine, Y.,** The hsp 70 mRNA accumulates in LLC-PK1 pig kidney cells treated with calcitonin but not with 8-bromo-cAMP, *J. Biol. Chem.*, 263, 6183—6187, 1988.

5. **Arnosti, D. N., Singer, V. L., and Chamberlin, M. J.,** Characterization of heat shock in *Bacillus subtilis, J. Bacteriol.*, 168, 1243—1249, 1986.

6. **Ashburner, M.,** Patterns of puffing activity in salivary gland chromosomes of *Drosophila.* V. Responses to environmental treatments, *Chromosoma,* 31, 356—376, 1970.

7. **Ashburner, M. and Bonner, J. J.,** The induction of gene activity in *Drosophila* by heat shock, *Cell,* 17, 241—254, 1979.

8. **Atkinson, B. G. and Walden, D. B., Eds.,** *Changes in Eukaryotic Gene Expression in Response to Environmental Stress,* Academic Press, Orlando, FL, 1985.

9. **Atkinson, B. G., Cunningham, T., Dean, R. L., and Somerville, M.,** Comparison of the effects of heat shock and metal ion stress on gene expression in cells undergoing myogenesis, *Can. J. Biochem. Cell Biol.*, 61, 404—413, 1983.

10. **Bahl, H., Echols, H., Straus, D. B., Court, D., Crowl, R., and Georgopoulos, C. P.,** Induction of the heat shock response of *E. coli* through stabilization of sigma-32 by the phage lambda cIII protein, *Genes Dev.*, 1, 57—64, 1987.

10a. **Ballinger, D. G. and Pardue, M. L.,** The control of protein synthesis during heat shock in *Drosophila* cells involves altered polypeptide elongation rates, *Cell,* 33, 103—114, 1983.

11. **Barettino, D., Morcillo, G., and Diez, J. L.,** Induction of heat shock Balbiani rings after RNA synthesis inhibition in polytene chromosomes of *Chironomus thummi, Chromosoma,* 87, 507—518, 1982.

12. **Behnel, H. J.,** Comparative study of protein synthesis and heat shock puffing activity in *Drosophila* salivary glands treated with chloramphenicol, *Exp. Cell Res.*, 142, 223—228, 1982.

13. **Behnel, H. J. and Seydewitz, H. H.,** Changes of the membrane potential during formation of heat shock puffs induced by ion carriers in *Drosophila* salivary glands, *Exp. Cell Res.*, 127, 133—141, 1980.

14. **Behnel, H. J. and Wekbart, G.,** Induced stabilization of the transmembrane potential of *Drosophila* cells by heat shock and periodic applications of chloramphenicol, *J. Cell Sci.*, 87, 197—201, 1987.

15. **Bensaude, O., Babinet, C., Morange, M., and Jacob, F.,** Heat shock proteins, first major products of cygotic gene activity in mouse embryo, *Nature,* 305, 331—333, 1983.

16. **Berger, E. M., Marino, G., and Torrey, D.,** Expression of *Drosophila* hsp70-cat hybrid gene in *Aedes* cells induced by heat shock, *Somat. Cell Mol. Genet.*, 11, 371—378, 1985.

16a. **Berridge, Y. J.,** Inositol trisphosphate and diacylglycerol two interacting second messengers, *Annu. Rev. Biochem.*, 56, 159—193, 1987.

16b. **Bienz, M.,** Developmental control of the heat shock response in *Xenopus, Proc. Natl. Acad. Sci. U.S.A.*, 81, 3138—3142, 1984.

17. **Bienz, M. and Pelham, H. R. B.,** Expression of a *Drosophila* heat-shock protein in *Xenopus* oocytes: Conserved and divergent regulatory signals, *EMBO J.*, 1, 1583—1588, 1982.

18. **Bienz, M. and Pelham, H. R. B.,** Mechanisms of heat-shock gene activation in higher eucaryotes, *Adv. Genet.*, 24, 31—72, 1987.

18a. **Bilinski, T., Krawiec, Z., Litwinska, J., and Blaszcynski, M.,** Mechanisms of oxygen toxicity as revealed by studies of yeast mutants with changed response to oxidative stress, in *Oxy-Radicals in Molecular Biology and Pathology,* Cerutti, P. A., Fridovich, I., and McCord, J. M., Eds., Alan R. Liss, New York, 1988, 109—123.

19. **Bloom, M., Skelly, S., Van Bogelen, R., Neidhardt, F., Brot, N., and Weissbach, H.,** In vitro effect of the *Escherichia coli* heat-shock regulatory protein on expression of heat shock genes, *J. Bacteriol.*, 166, 380—384, 1986.

20. **Bochner, B. R., Lee, P. C., Wilson, S. W., Cutler, Ch. W., and Ames, B. N.,** AppppA and related adenylylated nucleotides are synthesized as a consequence of oxidation stress, *Cell,* 37, 225—232, 1984.

21. **Bochner, B. R., Zylicz, M., and Georgopoulos, C.,** *Escherichia coli* Dnak protein possesses a 5′-nucleotidase activity that is inhibited by AppppA, *J. Bacteriol.*, 168, 931—935, 1986.

22. **Bond, U. and Schlesinger, M. J.,** Ubiquitin is a heat shock protein in chicken embryo fibroblasts, *Mol. Cell. Biol.*, 5, 949—956, 1985.

23. **Bournias-Vardiabasis, N. and Buzin, C. H.,** Developmental effects of chemicals and the heat chock response in *Drosophila* cells, *Teratog. Carcinog. Mutag.*, 6, 523—537, 1986.

23a. **Brandt, C. and Milcarek, C.,** Heat shock induced alterations in polyadenylate metabolism in *Drosophila melanogster, Biochemistry,* 19, 6152—6158, 1980.

24. **Brazzell, C. and Ingolia, T. D.,** Stimuli that induce a yeast heat shock gene fused to β-galactosidase, *Mol. Cell. Biol.*, 4, 2573—2579, 1984.

25. **Brevet, A., Plateau, P., Best-Belpomme, M., and Blanquet, S.,** Variation of Ap4A and other dinucleotide polyphosphates in stressed *Drosophila* cells, *J. Biol. Chem.*, 260, 15566—15570, 1985.

25a. **Brock, T. D.,** Life at high temperatures, *Science,* 230, 132—138, 1985.

26. **Brock, T. D.,** *Thermophilic Microorganismus and Life at High Temperatures,* Springer-Verlag, New York, 1978.

26a. **Brown, I. R.,** Modification of gene expression in the mammalian brain after hyperthermia, in *Gene Expression in Brain,* Zomzely-Neurath, C. and Walker, W. A., Eds., John Wiley & Sons, New York, 1985, 157—171.

26b. **Brown, I. R. and Rush, S. J.,** Expression of heat shock genes (HSP70) in the mammalian brain — distinguishing constitutively expressed and hyperthermia-inducible messenger RNA species, *J. Neurosci. Res.,* 25, 14—19, 1990.

27. **Brown, I. R., Lowe, D. G., and Moran, L. A.,** Expression of heat shock genes in fetal and maternal rabbit brain, *Neurochem. Res.,* 10, 1277—1284, 1985.

27a. **Brown, M. F., Rush, S., and Ivy, G. O.,** Induction of a heat shock gene at the site of tissue injury in the rat brain, *Neuron,* 2, 1559—1564, 1989.

27b. **Brunet, S. and Giacomoni, P. U.,** Heat shock messenger RNA in mouse epidermis after UV irradiation, *Mutation Res.,* 219, 217—224, 1989.

28. **Bultmann, H.,** Induction of heat shock puff by hypoxia in polytene foot pad chromosomes of *Sarcophaga bullata, Chromosoma,* 93, 358—366, 1986.

29. **Burdon, R. H., Slater, A., McMahon, M., and Cato, A. C. B.,** Hyperthermia and the heat-shock proteins of HeLa cells, *Br. J. Cancer,* 45, 953—963, 1982.

30. **Burdon, R. H.,** Heat shock and the heat shock proteins, *Biochem. J.,* 240, 313—324, 1986.

31. **Burke, J. J., Hatfield, J. L., Klein, R. P., and Mullet, J. E.,** Accumulation of heat shock proteins in field-grown cotton, *Plant Physiol.,* 78, 394—398, 1985.

32. **Buzin, C. H. and Bournias-Vardiabasis, N.,** Teratogens induce a subset of small heat shock proteins in *Drosophila* primary embryonic cell cultures, *Proc. Natl. Acad. Sci. U.S.A.,* 81, 4075—4079, 1984.

33. **Caggese, C., Bozzetti, M., Palumbo, G., and Barsanti, P.,** Induction by hydrocortisone-21-sodium succinate of the 70K heat-shock polypeptide in isolated salivary glands of *Drosophila melanogaster* larvae, *Experientia,* 39, 1143—1144, 1983.

34. **Cairo, G., Bardella, L., Schiaffonati, L., and Bernelli-Zazzera, A.,** Synthesis of heat shock proteins in rat liver after ischemia and hyperthermia, *Hepatology,* 5, 357—361, 1985.

34a. **Cajone, F., Salina, M., and Benelli-Zazzera, A.,** 4-Hydroxynonenal induces a DNA-binding protein similar to the heat-shock factor, *Biochem. J.,* 262, 977—979, 1989.

35. **Calderwood, S. K., Stevenson, M. A., and Hahn, G. M.,** Cyclic AMP and the heat shock response in Chinese hamster ovary cells, *Biochem. Biophys. Res. Commun.,* 126, 911—916, 1985.

36. **Calderwood, S. K., Stevenson, M. A., and Hahn, G. M.,** Heat stress stimulates inositol trisphosphate release and phosphorylation of phosphoinositides in CHO and BALB/C 3T3 cells, *J. Cell. Physiol.* 130, 369—376, 1987.

37. **Caltabiano, M. M., Koestler, T. P., Poste, G., and Greig, R. G.,** Induction of 32- and 34-kDa stress proteins by sodium arsenite, heavy metals, and thiol-reactive agents, *J. Biol. Chem.,* 261, 13381—13387, 1986.

37a. **Carafoli, E.,** Intracellular calcium homeostasis, *Annu. Rev. Biochem.,* 56, 395—433, 1987.

38. **Carlson, N. and Rechsteiner, M.,** Microinjection of ubiquitin: Intracellular distribution and metabolism in HeLa cells maintained under normal physiological conditions, *J. Cell Biol.,* 104, 537—546, 1987.

39. **Carlson, N., Rogers, S., and Rechsteiner, M.,** Microinjection of ubiquitin: Changes in protein degradation in HeLa cells subjected to heat-shock, *J. Cell Biol.,* 104, 547—555, 1987.

40. **Carr, B. I., Huang, T. H., Buzin, C. H., and Itakura, K.,** Induction of heat shock gene expression without heat shock by hepatocarcinogens and during hepatic regeneration in rat liver, *Cancer Res.,* 46, 5106—5111, 1986.

40a. **Christensen, A. H. and Quail, P. H.,** Sequence analysis and transcriptional regulation by heat shock of polyubiquitin transcripts from maize, *Plant Mol. Biol.,* 12, 619—632, 1989.

41. **Cervera, J.,** Induction of self-tolerance and enhanced stress protein synthesis in L-132 cells by cadmium chloride and by hyperthermia, *Cell Biol. Int. Rep.,* 9, 131—142, 1985.

41a. **Chary, P., and Natvig, D. O.,** Evidence for three differentially regulated catalase genes in *Neurospora crassa*: Effects of oxidative stress, heat shock, and development, *J. Bacteriol.,* 171, 2646—2652, 1989.

42. **Cheney, C. M. and Shearn, A.,** Developmental regulation of *Drosophila* imaginal disk proteins: Synthesis of a heat-shock protein under nonheat-shock conditions, *Dev. Biol.,* 95, 325—330, 1983.

43. **Chomyn, A., Moller, G., and Mitchell, H. K.,** Patterns of protein synthesis following heat shock in pupae of *Drosophila melanogaster, Dev. Genet.,* 1, 77—95, 1979.

44. **Clark, B. D. and Brown, I. R.,** A retinal heat shock protein is associated with elements of the cytoskeleton and binds to calmodulin, *Biochem. Biophys. Res. Commun.,* 139, 974—981, 1986.

44a. **Clerget, M. and Polla, B. S.,** Erythrophagocytosis induces heat shock protein synthesis by human monocytes macrophages, *Proc. Natl. Acad. Sci. U.S.A.,* 87, 1081—1085, 1990.

45. **Cohen, R. S. and Meselson, M.,** Separate regulatory elements for the heat-inducible and ovarian expression of the *Drosophila* hsp26 gene, *Cell,* 43, 737—746, 1985.

45a. **Colberg-Poley, A. M. and Santomenna, L. D.,** Selective induction of chromosomal gene expression by human cytomegalovirus, *Virology,* 166, 217—228, 1988.

46. **Collier, N. C. and Schlesinger, M. J.,** Induction of heat-shock proteins in the embryonic chicken lens, *Exp. Eye Res.,* 43, 103—117, 1986.

47. **Collins, P. L. and Hightower, L. E.,** Newcastle disease virus stimulates the cellular accumulation of stress (heat shock) mRNAs and proteins, *J. Virol.,* 44, 703—707, 1982.

48. **Compton, J. L. and McCarthy, J. B.,** Induction of the *Drosophila* heat shock response in isolated polytene nuclei, *Cell,* 14, 191—201, 1978.

49. **Corces, V. and Pellicer, A.,** Identification of sequences involved in the transcriptional control of a *Drosophila* heat-shock gene, *J. Biol. Chem.,* 259, 14812—14817, 1984.

50. **Corces, V., Pellicer, A., Axel, R., and Meselson, M.,** Integration, transcription, and control of a *Drosophila* heat shock gene in mouse cells, *Proc. Natl. Acad. Sci. U.S.A.,* 78, 7038—7042, 1981.

51. **Cosgrove, J. W. and Brown, I. R.,** Heat shock protein in mammalian brain and other organs after a physiologically relevant increase in both temperature induced by D-lysergic acid diethylamide, *Proc. Natl. Acad. Sci. U.S.A.,* 80, 569—573, 1983.

52. **Courgeon, A. M., Maisonhaute, C., and Best-Belpomme, M.,** Heat-shock proteins are induced by cadmium in *Drosophila* cells, *Exp. Cell Res.,* 153, 515—521, 1984.

52a. **Courgeon, A. M., Rollet, E., Becher, J., Maisonhaute, C., and Best-Belpomme, M.,** Hydrogen peroxide (H_2O_2) induces actin and some heat-shock proteins in *Drosophila* cells, *Eur. J. Biochem.,* 171, 163—170, 1988.

52b. **Craine, B. L. and Kornberg, T.,** Activation of the major *Drosophila* heat-shock genes in vitro, *Cell,* 25, 671—681, 1981.

53. **Currie, R. W. and White, F. P.,** Trauma-induced protein in rat tissues: A physiological role for a ''heat shock'' protein?, *Science,* 214, 72—73, 1981.

54. **Currie, R. W. and White, F. P.,** Characterization of the synthesis and accumulation of a 71-kilodalton protein induced in rat tissues after hyperthermia, *Can. J. Biochem. Cell Biol.,* 61, 438—446, 1983.

54a. **Czarnecka, E., Nagao, R. T., Key, J. L., and Gurley, W. B.,** Characterization of Gm-hsp26-A, a stress gene encoding a divergent heat shock protein of soybean: Heavy-metal-induced inhibition of intron processing, *Mol. Cell. Biol.,* 8, 1113—1122, 1988.

55. **Dean, R. L. and Atkinson, B. G.,** Synthesis of heat shock proteins in quail red blood cells following brief, physiologically relevant increases in whole body temperature, *Comp. Biochem. Physiol.,* 81B, 185—191, 1985.

56. **De Jong, W. W., Hoekman, W. A., Mulders, J. W. M., and Bloemendal, H.,** Heat shock response of the rat lens, *J. Cell Biol.,* 102, 104—111, 1986.

57. **Denisenko, O. N.,** Synthesis of diadenosine-5′,5‴-p1,p3-triphosphate in yeast at heat shock, *FEBS Lett.,* 178, 149—152, 1984.

57a. **Desrosier, R. and Tanguay, R. M.,** The modification in the methylation patterns of H2B and H3 after heat shock can be correlated with the inactivation of normal gene expression, *Biochem. Biophys. Res. Commun.,* 133, 823—829, 1985.

58. **Didomenico, B. J., Bugaisky, G. E., and Lindquist, S.,** The heat shock response is self-regulated at both the transcriptional and posttranscriptional levels, *Cell,* 31, 593—603, 1982.

59. **Dienel, G. A., Kiessling, M., Jacewicz, M., and Pulsinelli, W. A.,** Synthesis of heat shock proteins in rat brain cortex after transient ischemia, *J. Cerebr. Blood Flow Metab.,* 6, 505—510, 1986.

60. **Drahos, D. J. and Hendrix, R. W.,** Effect of bacteriophage lambda on the synthesis of groE protein and other *E. coli* proteins, *J. Bacteriol.,* 149, 1050—1063, 1982.

61. **Dreano, M., Fouillet, X., Brochot, J., Vallet, J.-M., Michel, M.-L., Rungger, D., and Bromley, P.,** Heat-regulated expression of the hepatitis B virus surface antigen in the human Wish cell line, *Virus Res.,* 8, 43—59, 1987.

62. **Drummond, I. A. S., McClure, S. A., Poenie, M., Tsien, R. Y., and Steinhardt, R. A.** Large changes in intracellular pH and calcium observed during heat shock are not responsible for the induction of heat shock proteins in *Drosophila melanogaster, Mol. Cell. Biol.,* 6, 1767—1775, 1986.

62a. **Drummond, I. A. S. and Steinhardt, R. A.,** The role of oxidative stress in the induction of *Drosophila* heat-shock proteins, *Exp. Cell Res.,* 173, 439—449, 1987.

62b. **Dubois, M. F., Mezger, V., Morange, M., Ferrieux, C., Lebon, P., and Bensaude, O.,** Regulation of the heat-shock response by interferon in mouse L cells, *J. Cell. Physiol.,* 137, 102—109, 1988.

63. **Durnam, D. M. and Palmiter, R. D.,** Induction of metallothionein-I mRNA in cultured cells by heavy metals and iodoacetate: Evidence for gratuitous inducers, *Mol. Cell. Biol.,* 4, 484—491, 1984.

63a. **Edelman, L., Czarnecka, E., and Key, J. L.,** Induction and accumulation of heat shock-specific poly-(A+) RNAs and proteins in soybean seedlings during arsenite and cadmium treatments, *Plant Physiol.,* 86, 1048—1056, 1988.

63b. **Edington, B. V., Whelan, S. A., and Hightower, L. E.,** Inhibition of heat shock (stress) protein induction by deuterium oxide and glycerol: Additional support for the abnormal protein hypothesis of induction, *J. Cell. Physiol.,* 139, 219—228, 1989.

64. **Edwards, M. J.,** The experimental production of arthrogryposis multiplex congenita in guinea pigs by maternal hyperthermia during gestation, *J. Pathol.*, 104, 221—229, 1971.
65. **Ellgaard, E. G.,** Similarities in chromosomal puffing induced by temperature shocks and dinitrophenol in *Drosophila, Chromosoma,* 37, 417—422, 1972.
66. **Erickson, J. W., Vaughn, V., Walter, W. A., Neidhaardt, F. C., and Gross, C. A.,** Regulation of the promoters and transcripts of rpoH, the *Escherichia coli* heat shock regulatory gene, *Genes Devel.*, 1, 419—432, 1987.
66a. **Ferris, D. K., Harel-Bellan, A., Morimoto, R. I., Welch, W. J., and Farrar, W. L.,** Mitogen and lymphokine stimulation of heat shock proteins in T lymphocytes, *Proc. Natl. Acad. Sci. U.S.A.*, 85, 3850—3854, 1988.
67. **Finley, D., Ciechanover, A., and Varshavsky, A.,** Thermolability of ubiquitin-activating enzyme from the mammalian cell cycle mutant ts85, *Cell,* 37, 43—55, 1984.
68. **Finley, D., Oezkaynak, E., and Varshavsky, A.,** The yeast polyubiquitin gene is essential for resistance to high temperatures, starvation and other stresses, *Cell,* 48, 1035—1046, 1987.
68a. **Fornace, A. J., Alamo, I., Hollander, M. C., and Lamoreaux, E.,** Induction of heat shock protein transcripts and B2 transcripts by various stresses in Chinese hamster cells, *Exp. Cell Res.*, 182, 61—74, 1989.
68b. **Fornace, A. J., Alamo, I., Hollander, M. C., and Lamoreaux, E.,** Ubiquitin mRNA is a major stress-induced transcript in mammalian cells, *Nucl. Acids Res.*, 17, 1215—1230, 1989.
69. **Freedman, M. S., Clark, B. D., Cruz, J. F., Gurd, J. W., and Brown, I. R.,** Selective effects of LSD and hyperthermia on the synthesis of synaptic proteins and glycoproteins, *Brain Res.*, 207, 129—145, 1981.
69a. **Freeman, M. L., Scidmore, N. C., Malcolm, A. W., and Meredith, M. J.,** Diamide exposure, thermal resistance, and synthesis of stress (heat shock) proteins, *Biochem. Pharmacol.*, 36, 21—29, 1987.
70. **Fujio, N., Hatayama, T., Kinoshita, H., and Yukioka, M.,** Induction of four heat shock proteins and their mRNAs in rat after wholebody hyperthermia, *J. Biochem.*, 101, 181—187, 1987.
70a. **Fuqua, S. A. W., Blumsalingaros, M., and McGuire, W. L.,** Induction of the estrogen-regulated 24K protein by heat shock, *Cancer Res.*, 49, 4126—4129, 1989.
71. **Gallant, Y. A.,** Stringent control in *E. coli, Annu. Rev. Genet.*, 13, 393—415, 1979.
72. **Garrison, P. N., Mathis, S. A., and Barnes, L. D.,** In vivo levels of diadenosine tetraphosphate and adenosine tetraphosphoguanosine in *Physarum polycephalum* during the cell cycle and oxidative stress, *Mol. Cell. Biol.*, 6, 1179—1186, 1986.
73. **Garry, R. F., Ulug, E. T., and Bose, H. R.,** Induction of stress proteins in Sindbis virus- and vesicular stomatitis virus-infected cells, *Virology,* 129, 319—332, 1983.
74. **Gates, D. M., Alderfer, R., and Taylor, E.,** Leaf temperature of desert plates, *Science,* 159, 994—995, 1968.
74a. **German, J.,** Embryogenic stress hypothesis of teratogenesis, *Am. J. Med.*, 76, 293—301, 1984.
75. **German, J., Louie, E.,and Banerjee, D.,** The heat-shock response in vivo: Experimental induction during mammalian organogenesis, *Teratog. Cancerog. Mutag.*, 6, 555—562, 1986.
76. **Gerner, E. W., and Russel, D. H.,** The relationship between polyamine accumulation and DNA replication in synchronized Chinese hamster ovary cells after heat shock, *Cancer Res.*, 37, 482—489, 1977.
77. **Gerner, E. W., Holmes, D. K., Stickney, D. G., Noterman, J. A., and Fuller, D. J. M.,** Enhancement of hyperthermia-induced cytotoxicity by polyamines, *Cancer Res.*, 40, 432—438, 1980.
78. **Glaser, R. L., Wolfner, M. F., and Lis, J. T.,** Spatial and temporal pattern of hsp 26 expression during normal development, *EMBO J.*, 5, 747—754, 1986.
79. **Goff, S. A. and Goldberg, A. L.,** Production of abnormal proteins in *E. coli* stimulates transcription of lon and other heat shock genes, *Cell,* 41, 587—595, 1985.
79a. **Grant, C. M., Firoozan, M., and Tuite, M. F.,** Mistranslation induces the heat-shock response in the yeast *Saccharomyces cerevisiae, Mol. Microbiol.*, 3, 215—220, 1989.
80. **Grill, E., Winnacker, E.-L., and Zenk, M. H.,** Phytochelatins: The principal heavy-metal complexing peptides of higher plants, *Science,* 230, 674—676, 1985.
81. **Grill, E., Winnacker, E.-L., and Zenk, M. H.,** Phytochelatins, a class of heavy-metal-binding peptides from plants, are functionally analogous to metallothioneins, *Proc. Natl. Acad. Sci. U.S.A.*, 84, 439—443, 1987.
81a. **Grill, E.,** Phytochelatins in plants, in *Metal Ion Homeostasis,* Hamer, D. H. and Winge, D. R., Eds., Alan R. Liss, New York, 1989, 283—300.
81b. **Grill, E., Loeffler, S., Winnacker, E.-L., and Zenk, M. H.,** Phytochelatins, the heavy-metal-binding peptides of plants, are synthesized from glutathione by a specific gammaglutamylcysteine dipeptidyl transpeptidase (phytochelatine synthase), *Proc. Natl. Acad. Sci. U.S.A.*, 86, 6838—6842, 1989.
82. **Grossman, A. D., Erickson, J. W., and Gross, C. A.,** The htpR gene product of *E. coli* is a sigma factor for heat-shock promoters, *Cell,* 38, 383—390, 1984.
83. **Grossman, A. D., Taylor, W. E., Burton, Z. F., Burgess, R. R., and Gross, C. A.,** Stringent response in *Escherichia coli* induces expression of heat shock proteins, *J. Mol. Biol.*, 186, 357—365, 1985.

84. **Grossman, A. D., Straus, D. B., Walter, W. A., and Gross, C. A.,** Sigma 32 synthesis ion regulate the synthesis of heat shock proteins in *Escherichia coli, Genes Devel.*, 1, 179—184, 1987.
85. **Guedon, G., Sovia, D., Ebel, J. P., Befort, N., and Remy, P.,** Effect of diadenosine tetraphosphate microinjection on heat shock protein synthesis in *Xenopus laevis* oocytes, *EMBO J.*, 4, 3743—3749, 1985.
86. **Guillemette, J. G., Wong, L., McLachlan, D. R. C., and Lewis, P. N.,** Characterization of messenger RNA from the cerebral cortex of control and Alzheimer-afflicted brain, *J. Neurochem.*, 47, 987—997, 1986.
87. **Gurley, W. B., Czarnecka, E., Nagao, R. T., and Key, J. L.,** Upstream sequences required for efficient expression of a soybean heat shock gene, *Mol. Cell. Biol.*, 6, 559—565, 1986.
88. **Guttman, S. D., Glover, C. V. C., Allis, C. D., and Gorovsky, M. A.,** Heat shock, deciliation and release from anoxia induce the synthesis of the same set of polypeptides in starved *T. pyriformis, Cell*, 22, 299—307, 1980.
89. **Hagmann, J.,** Caffeine and heat shock induce adenylate in *Dictyostelium discoideum, EMBO J.*, 5, 3437—3441, 1986.
90. **Hahn, G. M., Braun, J., and Har-Kedar, I.,** Thermochemotherapy: Synergism between huperthermia (42-43 grad) and adriamycin (or bleomycin) in mammalian cell inactivation, *Proc. Natl. Acad. Sci. U.S.A.*, 72, 937—940, 1975.
91. **Hahn, G. M., Shiu, E. C., West, B., Goldstein, L., and Li, G. C.,** Mechanistic implications of the induction of thermotolerance in Chinese hamster cells by organic solvents, *Cancer Res.*, 45, 4138—4143, 1985.
92. **Hall, B. G.,** Yeast thermotolerance does not require protein synthesis, *J. Bacteriol.*, 156, 1363—1365, 1983.
93. **Hallberg, R. L., Kraus, K. W., and Findly, R. C.,** Starved *Tetrahymena thermophila* cells that are unable to mount an effective heat shock response selectively degrade their rRNA, *Mol. Cell. Biol.*, 4, 2170—2179, 1984.
94. **Hammond, G. L., Lai, Y.-K., and Markert, C. L.,** Diverse forms of stress lead to new patterns of gene expression through a common and essential metabolic pathway, *Proc. Natl. Acad. Sci. U.S.A.*, 79, 3485—3488, 1982.
95. **Harari, P. M., Fuller, D. J. M., and Gerner, E. W.,** Heat shock stimulates polyamine oxidation by two distinct mechanisms in mammalian cell cultures, *Int. J. Radiat. Oncol.*, 16, 457—487, 1989.
96. **Harari, P. M., Tome, M. E., and Gerner, E. W.,** Heat shock-induced polyamine oxidation in mammalian cells, *J. Cell Biol.*, 103, 175a, 1986.
97. **Hatayama, T., Honda, K., and Yukioka, M.,** HeLa cells synthesize a specific heat shock protein upon exposure to heat shock at 42°C but not at 45°C, *Biochem. Biophys. Res. Commun.*, 137, 957—963, 1986.
97a. **Hatayama, T., Honda, K., and Yukioka, M.,** Effects of sodium butyrate and dibutyryl cyclic AMP on thermosensitivity of HeLa cells and their production of heat shock proteins, *Biochem. Int.*, 13, 793—798, 1986.
98. **Hecker, M., Richter, A., Schroeter, A., Wolfel, L., and Mach, F.,** Synthese von Hitzeschockproteinen nach einer Aminosaeure- und Sauerstofflimitation in *Bacillus subtilis* recA+- und recA-Staemmen, *Z. Naturforsch.*, C42, 941—947, 1987.
98a. **Heikkila, J. J. and Brown, I. R.,** Hyperthermia and disaggregation of brain polysomes induced by bacterial pyrogen, *Life Sci.*, 2, 347—352, 1979.
99. **Heikkila, J. J., Schultz, G. A., Iatrou, K., and Gedamu, L.,** Expression of a set of fish genes following heat or metal ion exposure, *J. Biol. Chem.*, 257, 12000—12005, 1982.
100. **Heikkila, J. J. and Schultz, G. A.,** Different environmental stresses can activate the expression of a heat shock gene in rabbit blastocysts, *Gamete Res.*, 10, 45—56, 1984.
101. **Heikkila, J. J., Papp, J. E. T., Schultz, G. A., and Bewley, D.,** Induction of heat shock protein messenger RNA in maize mesocotyls by water stress, abscisic acid and wounding, *Plant Physiol.*, 76, 270—274, 1984.
102. **Heikkila, J. J., Darasch, S. P., Mosser, D. D., and Bols, N. C.,** Heat and sodium arsenite act synergistically on the induction of heat shock gene expression in *Xenopus laevis* A6 cells, *Biochem. Cell Biol.*, 65, 310—316, 1987.
103. **Heikkila, J. J., Ovsenek, N., and Krone, P.,** Examination of heat shock protein mRNA accumulation in early *Xenopus leavis* embryos, *Biochem. Cell Biol.*, 65, 87—94, 1987.
104. **Hershko, A., Eytan, E., Ciechanover, A., and Haas, A. L.,** Immunochemical analysis of the turnover of ubiquitin-protein conjugates in intact cells. Relationship to break-down of abnormal proteins, *J. Biol. Chem.*, 257, 13964—13970, 1982.
105. **Hightower, L. E.,** Cultured animal cells exposed to amino acid analogues or puromycin rapidly synthesize several polypeptides, *J. Cell. Physiol.*, 102, 407—427, 1980.
105a. **Hightower, L. E.,** personal communication.

106. **Hightower, L. E. and White, F. P.,** Cellular response to stress: Comparison of a family of 71—73 kilodalton proteins rapidly synthesized in rat tissue slices and canavanine-treated cells in culture, *J. Cell. Physiol.,* 108, 261—275, 1981.
107. **Hightower, L. E. and White, F. P.,** Preferential synthesis of rat heat-shock and glucose-regulated proteins in stressed cardiovascular cells, in *Heat Shock from Bacteria to Man,* Schlesinger, M. J., Ashburner, M., and Tissieres, A., Eds., Cold Spring Harbor Laboratory, Cold Spring Harbor, NY, 1982, 369—377.
107a. **Hightower, L. E., Guidon, P. T., Whelan, S. A., and White, C. N.,** Stress responses in avian and mammalian cells, in *Changes in Eukaryotic Gene Expression in Response to Environmental Stress,* Atkinson, B. G. and Walden, D. B., Eds., Academic Press, Orlando, FL, 1985, 369—377.
107b. **Higo, H., Higo, K., Satow, Y., and Lee, J.-Y.,** Induction of an hsp70-like protein during organ culture of rat embryonic heat, *Biochem. Int.,* 15, 727—734, 1987.
108. **Hiromi, Y. and Hotta, Y.,** Actin gene mutations in *Drosophila* heat shock activation in the direct flight muscles, *EMBO J.,* 4, 1681—1687, 1985.
109. **Hiromi, Y., Okamoto, H., Gehring, W. J., and Hotta, Y.,** Germline transformation with *Drosophila* mutant actin genes induces constitutive expression of heat shock genes, *Cell,* 44, 293—301, 1986.
110. **Hiwasa, T. and Sakiyama, S.,** Increase in the synthesis of a 32,000-molecular-weight protein in BALB/C 3T3 cells after treatment with tumor promoters, chemical carcinogens, metal salts and heat shock, *Cancer Res.,* 46, 2474—2481, 1986.
111. **Hiwasa, T., Fujimura, S., and Sakiyama, S.,** Tumor promoters increase the synthesis of a 32,000-dalton protein in BALB/c 3T3 cells, *Proc. Natl. Acad. Sci. U.S.A.,* 79, 1800—1804, 1982.
112. **Hiwasa, T., Fujiki, H., Sugimura, T., and Sakiyama, S.,** Increase in the synthesis of a Mr 32,000 protein in BALB/c 3T3 cells treated with tumor-promoting indole alkaloids or polyacetates, *Cancer Res.,* 43, 5951—5955, 1983.
113. **Hoffman, E. and Corces, V.,** Sequences involved in temperature and ecdysterone-induced transcription are located in separate regions of a *Drosophila melanogaster* heat shock gene, *Mol. Cell. Biol.,* 6, 663—673, 1986.
114. **Hunt, L. A.,** Sidestream cigaret smoke exposure of mouse cells induces cell stress/heat shock-like proteins, *Toxicology,* 39, 259—273, 1986.
115. **Iida, H. and Yahara, I.,** Yeast heat-shock protein of Mr 48,000 is an isoprotein of enolase, *Nature,* 315, 688—690, 1985.
116. **Iida, H. and Yahara, I.,** Durable synthesis of high molecular weight heat shock proteins in GO cells of the yeast and other eukaryotes, *J. Cell Biol.,* 99, 199—207, 1984.
117. **Imperiale, M. J., Kao, H.-T., Feldman, L. T., Nevins, J. R., and Strickland, S.,** Common control of the heat shock gene and early adenovirus genes: Evidence for a cellular E1A-like activity, *Mol. Cell. Biol.,* 4, 867—874, 1984.
118. **Ireland, R. C. and Berger, E. M.,** Synthesis of low molecular weight heat shock peptides stimulated by ecdysterone in a cultured *Drosophila* cell line, *Proc. Natl. Acad. Sci. U.S.A.,* 79, 855—859, 1982.
119. **Ireland, R. C., Berger, E., Sirotkin, K., Yund, M. A., Osterbur, D., and Fristrom, J.,** Ecdysterone induces the transcription of four heat-shock genes in *Drosophila* S3 cells and imaginal disks, *Dev. Biol.,* 93, 498—507, 1982.
120. **Isaka, T., Yoshida, M., Owada, M., and Toyoshima, K.,** Alterations in membrane polypeptides of chick embryo fibroblasts induced by transformation with avian sarcoma viruses, *Virology,* 65, 226—237, 1975.
121. **Ito, K., Akiyama, Y., Yura, T., and Shiba, K.,** Diverse effects of the MalE-Lacz hybrid protein on *Escherichia coli* cell physiology, *J. Bacteriol.,* 167, 201—204, 1986.
122. **Jacewicz, M., Kiessling, M., and Pulsinelli, W. A.,** Selective gene expression in focal cerebral ischemia, *J. Cereb. Blood Flow Metab.,* 6, 263—272, 1986.
123. **Jones, K. A. and Findly, R. C.,** Induction of heat shock proteins by canavanine in *Tetrahymena.* No change in ATP levels measured in vivo by NMR, *J. Biol. Chem.,* 261, 8703—8707, 1986.
124. **Kao, H.-T. and Nevins, J. R.,** Transcriptional activation and subsequent control of the human heat shock gene during adenovirus infection, *Mol. Cell. Biol.,* 3, 2058—2065, 1983.
125. **Kapoor, M.,** A study of the effect of heat shock and metal ions on protein synthesis in *Neurospora crassa* cells, *Int. J. Biochem.,* 18, 15—30, 1986.
126. **Kapoor, M. and Lewis, J.,** Alteration of the protein synthesis pattern in *Neurospora crassa* cells by hyperthermal and oxidative stress, *Can. J. Microbiol.,* 33, 162—168, 1987.
127. **Kappen, L. and Lösch, R.,** Diurnal patterns of heat tolerance in relation to CAM, *Z. Pflanzenphysiol.,* 114, 87—96, 1984.
128. **Karlik, C. C., Coutu, M. D., and Fyrberg, E. A.,** A nonsense mutation within the act 88F actin gene disrupts myofibril formation in *Drosophila* indirect flight muscles, *Cell,* 38, 711—719, 1984.
129. **Karlik, C. C., Saville, D. L., and Fyrberg, E. A.,** Two missense alleles of the *Drosophila melanogaster* act88F actin gene are strongly antimorphic but only weakly induce synthesis of heat shock proteins, *Mol. Cell. Biol.,* 7, 3084—3091, 1987.

130. **Kasambalides, E. J. and Lanks, K. W.,** Dexamethasone can modulate glucose-regulated and heat shock protein synthesis, *J. Cell. Physiol.*, 114, 93—98, 1983.

131. **Kasambalides, E. J. and Lanks, K. W.,** Antagonistic effects of insulin and dexamethasone on glucose-regulated and heat shock protein synthesis, *J. Cell. Physiol.*, 123, 283—287, 1985.

132. **Kay, R. J., Boissy, R. J., Russnak, R. H., and Candido, E. P. M.,** Efficient transcription of a *Caenorhabditis elegans* heat shock gene pair in mouse fibroblasts is dependent on multiple promoter elements which can function bidirectionally, *Mol. Cell. Biol.*, 6, 3134—3143, 1986.

132a. **Kee, S. C. and Nobel, P. S.,** Concomitant changes in high temperature and heat-shock proteins in desert succulents, *Plant Physiol.*, 80, 596—598, 1986.

133. **Kelley, P. M. and Schlesinger, M. J.,** The effect of amino acid analogues and heat shock on gene expression in chicken embryo fibroblasts, *Cell,* 15, 1277—1286, 1978.

134. **Key, J. L., Kimpel, J., Vierling, E., Lin, C.-Y., Nagao, R. T., Czarnecka, E., and Schoeffl, F.,** Physiological and molecular analyses of the heat shock response in plants, in *Changes in Eukaryotic Gene Expression in Response to Environmental Stress,* Atkinson, B. G. and Walden, D. B., Eds., Academic Press, Orlando, FL, 1985, 327—348.

134a. **Keyse, St. M. and Tyrrell, R. M.,** Heme oxygenase is the major 32-kDa stress protein induced in human skin fibroblasts by UVA radiation, hydrogen peroxide, and sodium arsenite, *Proc. Natl. Acad. Sci. U.S.A.*, 86, 99—103, 1989.

135. **Khandjian, E. W. and Tuerler, H.,** Simian virus 40 and polyoma virus induce synthesis of heat shock proteins in permissive cells, *Mol. Cell. Biol.*, 3, 1—8, 1983.

136. **Kim, K. S. and Lee, A. S.,** The effect of extracellular Ca^{2+} and temperature on the induction of the heat-shock and glucose-regulated proteins in hamster fibroblasts, *Biochem. Biophys. Res. Commun.*, 140, 881—888, 1986.

137. **Kim, Y.-J., Shuman, J., Sette, M., and Przybyla, A.,** Arsenate induces stress proteins in cultured rat myoblasts, *J. Cell Biol.*, 96, 393—400, 1983.

138. **Kimpel, J. A. and Key, J. L.,** Presence of heat shock messenger RNA species in field grown soybeans *Glycine max, Plant Physiol.*, 79, 672—678, 1985.

139. **Klemenz, R. and Gehring, W. J.,** Sequence requirement for expression of the *Drosophila melanogaster* heat shock protein hsp22 gene during heat shock and normal development, *Mol. Cell. Biol.*, 6, 2011—2019, 1986.

140. **Kochan, J. and Murialdo, H.,** Stimulation of grpE synthesis in *E. coli* by bacteriophage lambda infection, *J. Bacteriol.*, 149, 1166—1170, 1982.

141. **Koninkx, J. F. J. G.,** Protein synthesis in salivary glands of *Drosophila hydei* after experimental gene induction, *Biochem. J.*, 158, 623—628, 1976.

142. **Kothary, R. K. and Candido, E. P. M.,** Induction of a novel set of polypeptides by heat shock or sodium arsenite in cultured cells of rainbow trout, *Salmo gairdnerii, Can. J.. Biochem.*, 60, 347—355, 1982.

142a. **Kozutsumi, J., Segal, M., Normington, K., Gething, M. J., and Sambrock, J.,** The presence of malfolded proteins in the endoplasmatic reticulum signals the induction of glucose-regulated proteins, *Nature,* 332, 462—464, 1988.

143. **Krueger, J. H. and Walker, G. C.,** GroEL and dnak genes of *Escherichia coli* are induced by UV irradiation and nalidixic acid in an $htpR^+$-dependent fashion, *Proc. Natl. Acad. Sci. U.S.A.*, *81,* 1499—1503, 1984.

144. **Kubo, T., Towle, C. A., Mankin, H. J., and Treadwell, B. V.,** Stress-induced proteins in chondrocytes from patients with osteoarthritis, *Arthritis Rheum.*, 28, 1140—1145, 1985.

145. **Kurtz, S. and Lindquist, S.,** Changing patterns of gene expression during sporulation in yeast, *Proc. Natl. Acad. Sci. U.S.A.*, 81, 7323—7327, 1984.

146. **Lamarche, S., Chretien, P., and Landry, J.,** Inhibition of the heat shock response and synthesis of glucose-regulated proteins in calcium-deprived rat hepatoma cells, *Biochem. Biophys. Res. Commun.*, 131, 868—876, 1985.

147. **Landry, J. and Chretien, P.,** Relationship between hyperthermia-induced heat shock proteins and thermotolerance in Morris hepatoma cells, *Can. J. Biochem. Cell Biol.*, 61, 428—437, 1983.

148. **Landry, J., Chretien, P., De Muys, J. M., and Morais, R.,** Induction of thermotolerance and heat shock protein synthesis in normal and respiration-deficient chick embryo fibroblasts, *Cancer Res.*, 45, 2240—2247, 1985.

148a. **Landry, J., Crete, P., Lamarche, S., and Chretien, P.,** Activation of calcium-dependent processes during heat shock role in cell thermoresistance, *Radiat. Res.*, 113, 426—436, 1988.

149. **Lanks, K. W.,** Modulators of the eukaryotic heat shock responses, *Exp. Cell Res.*, 165, 1—10, 1986.

150. **Lanks, K. W., Gao, J.-P., and Kasambalides, E. J.,** Nucleosides restore heat resistance and suppress glucose-regulated protein synthesis by glucose-deprived L929 cells, *Cancer Res.*, 48, 1442—1446, 1987.

151. **Lawson, R., Mestril, R., Schiller, P., and Voellmy, R.,** Expression of heat shock-beta-galactosidase hybrid genes in cultured *Drosophila* cells, *Mol. Gen. Genet.*, 198, 116—124, 1984.

152. **Lawson, R., Mestril, R., Luo, Y., and Voellmy, R.,** Ecdysterone selectively stimulates the expression of a 23,000-Da heat shock protein-beta-galactosidase hybrid gene in cultured *Drosophila* cells, *Dev. Biol.,* 110, 321—330, 1985.

152a. **Lee, P. C., Bochner, B. R., and Ames, B. N.,** AppppA, heat-shock stress, and cell oxidation, *Proc. Natl. Acad. Sci. U.S.A.,* 80, 7496—7500, 1983.

153. **Lee, A. S., Delegeane, A. M., Baker, V., and Chow, P. C.,** Transcriptional regulation of two genes specifically induced by glucose starvation in a hamster mutant fibroblast cell line, *J. Biol. Chem.,* 258, 597—603, 1983.

154. **Lee, Y. J. and Dewey, W. C.,** Induction of heat shock proteins in Chinese hamster ovary cells and development of thermotolerance by intermediate concentrations of puromycin, *J. Cell. Physiol.,* 132, 1—11, 1987.

154a. **Lee, K.-J. and Hahn, G. M.,** Abnormal protein as the trigger for the induction of stress responses: Heat, diamide and sodium arsenite, *J. Cell. Physiol.,* 136, 411—420, 1988.

154b. **Lee, H., Simon, J. A., and Lis, J. T.,** Structure and expression of ubiquitin genes of *Drosophila melanogaster, Mol. Cell. Biol.,* 8, 4727—4735, 1988.

155. **Leenders, H. J. and Berendes, H. D.,** The effect of changes in the respiratory metabolism upon genome activity in *Drosophila.* I. The induction of gene activity, *Chromosome,* 37, 433—444, 1972.

156. **Leenders, H. J., Derkson, J., Maas, P. M. J. M., and Berendes, H. D.,** Selective induction of a giant puff in *Drosophila hydei* by vitamin B6 and derivatives, *Chromosoma,* 41, 447—460, 1973.

156a. **Leicht, B. G., Biessmann, H., Palter, K. P., and Bonner, J. J.,** Small heat shock proteins of *Drosophila* associate with the cytoskeleton, *Proc. Natl. Acad. Sci. U.S.A.,* 83, 90—94, 1986.

157. **Leone, A., Pavlakis, G. N., and Hamer, D. H.,** Menkes' disease: Abnormal metallothionein gene regulation in response to copper, *Cell,* 40, 301—309, 1985.

158. **Levinson, W., Idriss, J., and Jackson, J.,** Metal-binding drugs induce synthesis of four proteins in normal cells, *Biol. Trace Element Res.,* 1, 15—23, 1979.

159. **Levinson, W., Oppermann, H., and Jackson, J.,** Transition series metals and sulfhydryl reagents induce the synthesis of four proteins in eukaryotic cells, *Biochim. Biophys. Acta,* 606, 170—180, 1980.

159a. **Levinson, W., Kravitz, S., and Jackson, J.,** Amino acid deprivation induces synthesis of four proteins in chick embryo cells, *Exp. Cell Res.,* 130, 459—463, 1980.

160. **Li, G. C.,** Induction of thermotolerance and enhanced heat shock protein synthesis in Chinese hamster fibroblasts by sodium arsenite and by ethanol, *J. Cell. Physiol.,* 115, 116—122, 1983.

160a. **Li, G. C. and Laszlo, A.,** Thermotolerance in mammalian cells: A possible role for heat shock proteins, in *Changes in Eukaryotic Gene Expression in Response to Environmental Stress,* Atkinson, B. G. and Walden, D. B., Eds., Academic Press, Orlando, FL, 1985, 227—254.

161. **Li, G. C. and Laszlo, A.,** Amino acid analogs while inducing heat shock proteins sensitive CHO cells to thermal damage, *J. Cell. Physiol.,* 122, 91—97, 1985.

162. **Li, G. C., Shrieve, D. C., and Werb, Z.,** Correlations between synthesis of heat shock proteins and development of tolerance to heat and to adriamycin in Chinese hamster fibroblasts: Heat shock and other inducers, in *Heat Shock from Bacteria to Man,* Schlesinger, M. J., Ashburner, M., and Tissieres, A., Eds., Cold Spring Harbor Laboratory, Cold Spring Harbor, NY, 1982, 395—404.

163. **Lin, C.-Y., Roberts, J. K., and Key, J. L.,** Acquisition of thermotolerance in soybean seedlings, *Plant Physiol.,* 74, 152—160, 1984.

164. **Lin, P. S., Hefter, K., and Jones, M.,** Hyperthermia and bleomycin schedules on V79 Chinese hamster cell cytotoxicity in vitro, *Cancer Res.,* 43, 4557—4561, 1983.

164a. **Lindquist, S.,** Varying patterns of proteins synthesis in *Drosophila* during heat shock: Implication for regulation, *Dev. Biol.,* 77, 463—479, 1980.

165. **Ljungdahl, L. G.,** Physiology of thermophilic bacteria, *Adv. Microb. Physiol.,* 19, 150—243, 1979.

166. **Love, J. D., Vivino, A. A., and Minton, K. W.,** Hydrogen peroxide toxicity may be enhanced by heat shock gene induction in *Drosophila, J. Cell. Physiol.,* 126, 60—68, 1986.

167. **Madreperla, S. A., Louwerenburg, B., Mann, R. W., Towle, C. A., Mankin, H. J., and Treadwell, B. V.,** Induction of heat-shock protein synthesis in chondrocytes at physiological temperatures, *J. Orthopaed. Res.,* 3, 30—35, 1985.

168. **Maytin, E. V. and Young, D. A.,** Separate glucocorticoid, heavy metal, and heat shock domains in thymic lymphocytes, *J. Biol. Chem.,* 258, 12718—12722, 1983.

169. **McAlister, L. and Finkelstein, D. B.,** Heat shock proteins and thermal resistance in yeast, *Biochem. Biophys. Res. Commun.,* 93, 819—824, 1980.

170. **McAlister, L., Strausberg, S., Kulaga, A., and Finkelstein, D. B.,** Altered patterns of protein synthesis induced by heat shock of yeast, *Curr. Genet.,* 1, 63—74, 1979.

171. **McMahon, A. P., Novak, T. J., Britten, R. J., and Davidson, E. H.,** Inducible expression of a cloned heat shock fusion gene in sea urchin embryos, *Proc. Natl. Acad. Sci. U.S.A.,* 81, 7490—7494, 1984.

171a. **Mehta, H. B., Popovich, B. K., and Dillmann, W. H.,** Ischemia induces changes in the level of mRNAs coding for stress protein 71 and creatine kinase M, *Circ. Res.,* 63, 512—517, 1988.

172. **Mestril, R., Rungger, D., Schiller, P., and Voellmy, R.,** Identification of a sequence element in the promoter of the *Drosophila melanogaster* hsp23 gene that is required for its heat activation, *EMBO J.*, 4, 2971—2976, 1985.
173. **Mestril, R., Schiller, P., Amin, J., Klapper, H., Ananthan, J., and Voellmy, R.,** Heat shock and ecdysterone activation of the *Drosophila melanogaster* hsp-23 gene; a sequence element implied in developmental regulation, *EMBO J.*, 5, 1667—1673, 1986.
174. **Milarski, K. L. and Morimoto, R. I.,** Expression of human HSP70 during the synthetic phase of the cell cycle, *Proc. Natl. Acad. Sci. U.S.A.*, 83, 9517—9522, 1986.
174a. **Miller, D. and McLenhan, A. G.,** Changes in intracellular levels of Ap3A in cysts and larvae of *Artemia* do not correlate with changes in protein synthesis after heat-shock, *Nucl. Acids Res.*, 14, 6031—6040, 1986.
175. **Mirault, M. E., Southgate, R., and Delwart, E.,** Regulation of heat-shock genes: A DNA sequence upstream of *Drosophila* hsp70 genes is essential for their induction in monkey cells, *EMBO J.*, 1, 1279—1285, 1982.
175a. **Mirkes, P. E.,** Hyperthermia induced heat shock response and thermotolerance in postimplantation rat embryos, *Dev. Biol.*, 119, 115—123, 1987.
175b. **Misra, S., Zafarullah, M., Price-Haughey, J., and Gedamu, L.,** Analysis of stress-induced gene expression in fish cell lines exposed to heavy metals and heat shock, *Biochim. Biophys. Acta*, 1007, 325—333, 1989.
175c. **Mitchell, J. B., Russo, A., Kinsella, T. J., and Glatstein, E.,** Glutathione elevation during thermotolerance induction and thermosensitization by glutathione depletion, *Cancer Res.*, 43, 987—991, 1983.
176. **Mobbs, C. V., Romano, G. J., Schwartz-Giblin, S., and Pfaff, D. W.,** Biochemistry of a steroid-regulated mammalian mating behavior; heat shock proteins and secretion, enkephalin and GABA, in *Neural Control of Reproductive Function*, Lakowski, J. M., Perez-Polo, J. R., and Rassin, D. K., Eds., Alan R. Liss, New York, 1989, 95—116.
176a. **Mocquot, B., Ricard, B., and Pradet, A.,** Rice embryos can express heat-shock genes under anoxia, *Biochimie*, 69, 677—681, 1987.
176b. **Morange, M., Dubois, M. F., Bensaude, O., and Lebon, P.,** Interferon pretreatment lowers the threshold for maximal heat-shock response in mouse cells, *J. Cell. Physiol.*, 127, 417—422, 1986.
177. **Morgan, R. W., Christman, M. F., Jacobson, F. S., Storz, G., and Ames, B. N.,** Hydrogen peroxide-inducible proteins in *Salmonella typhimurium* overlap with heat shock and other stress proteins, *Proc. Natl. Acad. Sci. U.S.A.*, 83, 8059—8063, 1986.
178. **Morganelli, C. M., Berger, E. M., and Pelham, H. R. B.,** Transcription of *Drosophila* small hsp-tk hybrid genes is induced by heat shock and by ecdysterone in transfected *Drosophila* cells, *Proc. Natl. Acad. Sci. U.S.A.*, 82, 5865—5869, 1985.
179. **Morris, T., Marashi, F., Weber, L., Hickey, E., Greespan, D., Bonner, J., Stein, J., and Stein, G.,** Involvement of the 5′-leader sequence in coupling the stability of a human H3 histone mRNA with DNA replication, *Proc. Natl. Acad. Sci. U.S.A.*, 83, 981—985, 1986.
180. **Mosser, D. D., Heikkila, J. J., and Bols, N. C.,** Temperature ranges over which rainbow trout fibroblasts survive and synthesize heat-shock proteins, *J. Cell. Physiol.*, 128, 432—440, 1986.
180a. **Mosser, D. D., Theodorakis, N. G., and Morimoto, R. I.,** Coordinate changes in heat shock element-binding activity and hsp70 gene transcription rates in human cells, *Mol. Cell. Biol.*, 8, 4736—4744, 1988.
180b. **Mueller-Taubenberger, A., Hagmann, J., Noegel, A., and Gerisch, G.,** Ubiquitin gene expression in *Dictyostelium* is induced by heat and cold shock, cadmium, and inhibitors of protein synthesis, *J. Cell Sci.*, 90, 51—58, 1988.
181. **Munro, S. and Pelham, H. R. B.,** Use of peptide tagging to detect proteins expressed from cloned genes: Deletion mapping functional domains of *Drosophila* hsp70, *EMBO J.*, 3, 3087—3093, 1984.
181a. **Munro, S. and Pelham, H.,** What turns on heat shock genes?, *Nature*, 317, 477—478, 1985.
182. **Myohara, M. and Okada, M.,** Digitonin treatment activates specific genes including the heat-shock genes in salivary glands of *Drosophila melanogaster, Dev. Biol.*, 130, 348—355, 1988.
182a. **Nelson, D. R. and Killeen, K. P.,** Heat shock proteins of vegetative and fruiting *Myxococcus xanthus* cells, *J. Bacteriol.*, 168, 1100—1106, 1986.
183. **Nevins, J. R.,** Induction of the synthesis of a 70,000 dalton mammalian heat shock protein by the adenovirus ElA gene product, *Cell*, 29, 913—919, 1982.
184. **Nieto-Sotelo, J. and Ho, T. H. D.,** Effect of heat shock on the metabolism of glutathione in maize roots, *Plant Physiol.*, 82, 1031—1035, 1986.
185. **Notarianni, E. L. and Preston, C. M.,** Activation of cellular stress protein genes by herpes simplex virus temperature-sensitive mutants which overproduce immediate early polypeptides, *Virology*, 123, 113—122, 1982.
186. **Nover, L.,** Molecular basis of cell differentiation, in *Cell Differentiation*, Nover, L., Luckner, M., and Parthier, B., Eds., Springer-Verlag, Berlin, 1982, 99—254.
187. **Nover, L., Ed.,** *Heat Shock Response of Eukaryotic Cells*, Springer-Verlag, Berlin, 1984.

188. **Nover, L.,** Expression of heat shock genes in homologous and heterologous systems, *Enzyme Microb. Technol.*, 9, 130—144, 1987.

188a. **Nover, L., Neumann, D., and Scharf, K. D., Eds.,** *Heat Shock and Other Stress Response Systems of Plants,* Springer-Verlag, Berlin, 1990.

189. **Özkaynak, E., Finley, D., Solomon, M. J., and Varshavsky, A.,** The yeast ubiquitin genes: A family of natural gene fusions, *EMBO J.*, 6, 1429—1439, 1987.

189a. **Ohno, K., Fukushima, M., Fujiwara, M., and Narumiya, S.,** Induction of 68,000-dalton heat shock proteins by cyclopentenone prostaglandins. Its association with prostaglandin-induced G-1 block in cell cycle progression, *J. Biol. Chem.*, 263, 19764—19770, 1988.

190. **Omar, R. A. and Lanks, K. W.,** Heat shock protein synthesis and cell survival in clones of normal and Simian virus 40-transformed mouse embryo cells, *Cancer Res.*, 44, 3976—3982, 1984.

190a. **Ovsenek, N. and Heikkila, J. J.,** Heat shock-induced accumulation of ubiquitin messenger RNA in *Xenopus laevis* embryos is developmentally regulated, *Dev. Biol.*, 129, 582—585, 1988.

191. **Parker-Thornburg, J. and Bonner, J. J.,** Mutations that induce the heat shock response of *Drosophila, Cell,* 51, 763—772, 1987.

191a. **Parsell, D. A. and Sauer, R. T.,** Induction of a heat shock-like response by unfolded protein in *Escherichia coli:* Dependence on protein level not protein degradation, *Genes Dev.*, 3, 1226—1232, 1989.

191b. **Parthier, B.,** Phytohormones and other stress modulators, in *Heat Shock and Other Stress Response Systems of Plants,* Nover, L., Neumann, D., and Scharf, K.-D., Eds., Springer-Verlag, Berlin, 1990, 82—86.

192. **Pelham, H. R. B.,** A regulatory upstream promoter element in the *Drosophila* hsp70 heat-shock gene, *Cell,* 517—528, 1982.

193. **Pelham, H. R. B. and Bienz, M.,** A synthetic heat-shock promoter element confers heat-inducibility on the herpes simplex virus thymidine kinase gene, *EMBO J.*, 1, 1473—1477, 1982.

194. **Peluso, R. W., Lamb, R. A., and Choppin, P. W.,** Infection with paramyxoviruses stimulates synthesis of cellular polypeptides that are also stimulated in cells transformed by Rous sarcoma virus or deprived of glucose, *Proc. Natl. Acad. Sci. U.S.A.*, 75, 6120—6124, 1978.

195. **Plesset, J., Palm, C., and McLaughlin, C. S.,** Induction of heat shock proteins and thermotolerance by ethanol in *Saccharomyces cerevisiae, Biochem. Biophys. Res. Commun.*, 108, 1340—1345, 1982.

195a. **Polla, B. S., Healy, A. M., Wojno, W. C., and Krane, S. M.,** Hormone 1α, 25-dihydroxy vitamin D3 modulate heat shock response in monocytes, *Am. J. Physiol.*, 252, 640—649, 1987.

196. **Pouyssegur, J., Shiu, R. P. C., and Pastan, I.,** Induction of two transformation-sensitive membrane polypeptides in normal fibroblasts by a block in glycoprotein synthesis or glucose deprivation, *Cell,* 11, 941—947, 1977.

196a. **Quinn, P. J.,** The fluidity of cell membranes and its regulation, *Progr. Biophys. Mol. Biol.*, 38, 1—104, 1981.

197. **Ramachandran, C., Catelli, M. G., Schneider, W., and Shyamala, G.,** Estrogenic regulation of uterine 90-kilodalton heat shock protein, *Endocrinology,* 123, 956—961, 1988.

197a. **Rechsteiner, M.,** Ubiquitin mediated pathways for intracellular proteolysis, *Annu. Rev. Cell. Biol.*, 3, 1—30, 1987.

198. **Rensing, L.,** Effects of 2,4-dinitrophenol and dinactin on heat-sensitive and ecdysone-specific puffs of *Drosophila* salivary gland chromosomes in vitro, *Cell Diff.*, 2, 221—228, 1973.

199. **Rensing, L., Olomski, R., and Drescher, K.,** Kinetics and models of the *Drosophila* heat shock system, *Biosystems,* 15, 341—356, 1982.

200. **Richter, A. and Hecker, M.,** Heat-shock proteins in *Bacillus subtilis.* A two-dimensional gel electrophoresis study, *FEMS Microbiol. Lett.*, 36, 69—71, 1986.

201. **Riddihough, G. and Pelham, H. R. B.,** Activation of the *Drosophila* hsp 27 promoter by heat shock and ecdysone involves independent and remote regulatory sequences, *EMBO J.*, 5, 1653—1658, 1986.

201a. **Riddihough, G. and Pelham, H. R. B.,** An ecdysone response element in the *Drosophila* hsp 27 promoter, *EMBO J.*, 6, 3729—3734, 1987.

202. **Ritossa, F.** A new puffing pattern induced by heat shock and DNP in *Drosophila, Experientia,* 18, 571—573, 1962.

203. **Ritossa, F.,** Experimental activation of specific loci in polytene chromosomes of *Drosophila, Exp. Cell Res.*, 35, 601—607, 1964.

204. **Rivedal, E. and Sanner, T.,** Metal salts as promoters on in vitro morphological transformation of hamster embryo cells initiated by benzo(a)pyrene, *Cancer Res.*, 41, 2950—2953, 1981.

205. **Rodenhiser, D., Jung, J. H., and Atkinson, B. G.,** The synergistic effect of hyperthermia and ethanol on changing gene expression of mouse lymphocytes, *Can. J. Genet. Cytol.*, 28, 1115, 1987.

206. **Ron, A. and Wheatley, D. N.,** Stress proteins are induced in *Tetrahymena pyriformis* by histidinol but not in mammalian (L-929) cells, *Exp. Cell Res.*, 153, 158—166, 1984.

207. **Rose, T. M. and Khandjian, E. W.,** A 105 000-dalton antigen of transformed mouse cells is a stress protein, *Can. J. Biochem. Cell Biol.*, 63, 1258—1264, 1985.

208. **Rosen, E., Sivertsen, A., Firtel, R. A., Wheeler, S., and Loomis, W. F.,** Heat shock genes of *Dictyostelium,* in *Changes in Eukaryotic Gene Expression in Response to Environmental Stress,* Atkinson, B. G. and Walden, D. B., Eds., Academic Press, Orlando, FL, 1985, 257—278.

209. **Rowe, T. C., Wang, J. C., and Liu, L. F.,** In vivo localization of DNA topoisomerase cleavage sites on *Drosophila* heat shock chromatin, *Mol. Cell. Biol.,* 6, 985—992, 1986.

209a. **Santoro, M. G., Garaci, E., and Amici, C.,** Prostaglandins with antiproliferative activity induce the synthesis of a heat shock protein in human cells, *Proc. Natl. Acad. Sci. U.S.A.,* 86, 8407—8411, 1989.

210. **Scalenghe, F. and Ritossa, F.,** The puff inducible in region 93D is responsible for the synthesis of the major "heat shock" polypeptide in *Drosophila melanogaster, Chromosoma,* 63, 317—327, 1977.

211. **Schaefer, E. L., Morimoto, R. I., Theodorakis, N. G., and Seidenfeld, J.,** Chemical specificity for induction of stress-response genes by DNA-damaging drugs in human adenocarcinoma cells, *Carcinogenesis,* 9, 1733—1738, 1988.

211a. **Scharf et al.,** unpublished.

212. **Sciandra, J. J. and Subjeck, J. R.,** Effects of glucose on protein synthesis and thermosensitivity in Chinese hamster ovary cells, *J. Biol. Chem.,* 258, 12091—12093, 1983.

213. **Sciandra, J. J. and Subjeck, J. R.,** Heat shock proteins and protection of proliferation and translation in mammalian cells, *Cancer Res.,* 44, 5188—5194, 1984.

214. **Sciandra, J. J., Subjeck, J. R., and Hughes, C. S.,** Induction of glucose-regulated proteins during anaerobic exposure and of heat-shock proteins after reoxygenation, *Proc. Natl. Acad. Sci. U.S.A.,* 81, 4843—4847, 1984.

214a. **Schlesinger, M. J., Ashburner, M., and Tissieres, A., Eds.,** *Heat Shock from Bacteria to Man,* Cold Spring Harbor Laboratory, Cold Spring Harbor, NY, 1982.

215. **Segal, E. and Le Pecq, J.-B.,** Relationship between cellular diadenosine 5′,5‴-P1,P4-tetraphosphate level, cell density, cell growth stimulation and toxic stresses, *Exp. Cell Res.,* 167, 119—126, 1986.

215a. **Seufert, W. and Jentsch, S.,** Ubiquitin-conjugating enzymes UBC4 and UBC5 mediate selective degradation of short-lived and abnormal proteins, *EMBO J.,* 9, 543—550, 1990.

215b. **Shelton, K. R., Egle, P. M., and Todd, J. M.,** Evidence that glutathione participates in the induction of stress protein, *J. Biol. Chem.,* 261, 1935—1940, 1986.

216. **Sherwood, A. C., John-Alder, K., and Sanders, M. M.,** Anion transport is linked to heat shock induction, in *Stress-Induced Proteins,* Pardue, M. L., Feramisco, J. R., and Lindquist, S., Eds., Alan R. Liss, New York, 1989, 117—128.

216a. **Shibahara, S., Mueller, R. M., and Taguchi, H.,** Transcriptional control of rat heme oxygenase by heat shock, *J. Biol. Chem.,* 262, 12889—12892, 1987.

217. **Shin, D.-Y., Matsumoto, K., Iida, H., Uno, I., and Ishikawa, T.,** Heat shock response of *Saccharomyces cerevisiae* mutants altered in cyclic AMP-dependent protein phosphorylation, *Mol. Cell. Biol.,* 7, 244—250, 1987.

217a. **Shuman, J. and Przybyla, A.,** Expression of the 31-kDa stress protein in rat myoblasts and hepatocytes, *DNA,* 7, 475—482, 1988.

218. **Simon, M. C., Kitchener, K., Kao, H.-T., Hickey, E., Weber, L., Voellmy, R., Heintz, N., and Nevins, J. R.,** Selective induction of human heat shock gene transcription by the adenovirus E1A gene products, including the 12S E1A product, *Mol. Cell. Biol.,* 7, 2884—2890, 1987.

219. **Sin, Y. T.,** Induction of puffs in *Drosophila* salivary gland by mitochondrial factor(s), *Nature,* 228, 159—160, 1975.

220. **Singh, O. P. and Gupta, J. P.,** Differential induction of chromosome puffs in 2 cell types of *Melanagromyza obtusa, Chromosoma,* 91, 359—362, 1985.

221. **Singh, M. K. and Yu, J.,** Accumulation of heat shock-like protein during differentiation of human erythroid cell line K562, *Nature,* 309, 631—633, 1984.

222. **Sorger, P. K. and Pelham, H. R. B.,** The glucose-regulated protein grp94 is related to heat shock protein hsp90, *J. Mol. Biol.,* 194, 341—344, 1987.

223. **Sorger, P. K. and Pelham, H. R. B.,** Purification and characterization of a heat shock element binding protein from yeast, *EMBO J.,* 6, 3035—3042, 1987.

224. **Spector, M. P., Aliabadi, Z., Gonzales, T., and Forster, J. W.,** Global control in *Salmonella typhimurium:* Two-dimensional electrophoretic analysis of starvation-, anaerobiosis-, and heat shock-inducible proteins, *J. Bacteriol.,* 168, 420—424, 1986.

225. **Spena, A. and Schell, J.,** The expression of a heat-inducible chimeric gene in transgenic tobacco plants, *Mol. Gen. Genet.,* 206, 436—440, 1987.

226. **Spena, A., Hain, R., Ziervogel, U., Saedler, H., and Schell, J.,** Construction of a heat-inducible gene for plants. Demonstration of heat-inducible activity of the *Drosophila* hsp70 promoter in plants, *EMBO J.,* 4, 2739—2743, 1985.

226a. **Spradling, A., Pardue, M. L., and Penman, S.,** Messenger RNA in heat-shocked *Drosophila* cells, *J. Mol. Biol.,* 109, 559—587, 1977.

227. **Stephanou, G. and Demopoulos, N. A.,** Heat shock phenomena and *Aspergillus nidulans*. II. Combined effects of heat and bleomycin to heat shock protein synthesis, survival rate and induction of mutations, *Curr. Genet.*, 12, 443—448, 1987.

228. **Stevenson, M. A., Calderwood, S. K., and Hahn, G. M.,** Rapid increases in inositol trisphosphate and intracellular calcium after heat shock, *Biochem. Biophys. Res. Commun.*, 137, 826—833, 1986.

229. **Stevenson, M. A., Calderwood, S. K., and Hahn, G. M.,** Effect of hyperthermia (45°C) on calcium flux in Chinese hamster ovary HA-1 fibroblasts and its potential role in cytotoxicity and heat resistance, *Cancer Res.*, 47, 3712—3717, 1987.

230. **Strand, D. J. and McDonald, J. F.,** Copia is transcriptionally responsive to environmental stress, *Nucl. Acids Res.*, 13, 4401—4410, 1985.

231. **Straus, D. B., Walter, W. A., and Gross, C. A.,** The heat shock response of *Escherichia coli* is regulated by changes in the concentration of sigma 32, *Nature*, 329, 348—351, 1987.

232. **Subjeck, J. R., Sciandra, J. J., and Shyy, T. T.,** Analysis of the expression of the two major proteins of the 70 kilodalton mammalian heat shock family, *Int. J. Radiat. Biol.*, 47, 275—284, 1985.

233. **Suranyi, G., Korcz, A., Palfi, Z., and Borbely, G.,** Effects of light deprivation on RNA synthesis, accumulation of guanosine 3′(2′)-diphosphate 5′-diphosphate, and protein synthesis in heat-shocked *Synechococcus* sp. strain PCC 6301, a cyanobacterium, *J. Bacteriol.*, 169, 632—639, 1987.

234. **Sutherland, E. W.,** Studies on the mechanism of hormone action, *Science*, 177, 401—408, 1972.

234a. **Swindle, J., Ajioka, J., Eisen, H., Sanwal, B., Jacquemot, C., Browder, Z., and Buck, G.,** The genomic organization and transcription of the ubiquitin genes of *Trypanosoma cruzi*, *EMBO J.*, 7, 1121—1127, 1988.

234b. **Taglicht, D., Padan, E., Oppenhein, A. B., and Schuldiner, S.,** An alkaline shift induces the heat shock response in *Escherichia coli*, *J. Bacteriol.*, 169,885—887, 1987.

234c. **Tanaka, K., Matsumoto, K., and Tohe, A.,** Dual regulation of the expression of the polyubiquitin gene by cyclic AMP and heat shock in yeast, *EMBO J.*, 7, 495—502, 1988.

235. **Tanguay, R. M.,** Genetic regulation during heat shock and function of heat-shock proteins: A review, *Can. J. Biochem. Cell Biol.*, 61, 387—394, 1983.

235a. **Taylor, I. C. A., Solomon, W., Weiner, B. M., Paucha, E., Bradley, M., and Kingston, R. E.,** Stimulation of the human heat shock protein 70 promotor in vitro by Simian virus 40 large T-antigen, *J. Biol. Chem.*, 264, 16160—16164, 1989.

236. **Thomas, S. R. and Lengyel, J. A.,** Ecdysteroid-regulated heat-shock gene expression during *Drosophila melanogaster* development, *Dev. Biol.*, 115, 434—438, 1986.

237. **Thomas, G. P. and Mathews, M. B.,** Control of polypeptide chain elongation in the stress response: A novel translation control, in *Heat Shock from Bacteria to Man*, Schlesinger, M. J., Ashburner, M., and Tissieres, A., Eds., Cold Spring Harbor Laboratory, Cold Spring Harbor, NY, 1982, 207—213.

238. **Thomas, G. P. and Mathews, M. B.,** Alterations of transcription and translation in HeLa cells exposed to amino acid analogs, *Mol. Cell. Biol.*, 4, 1063—1072, 1984.

239. **Travers, A. A. and Mace, H. A. F.,** The heat shock phenomenon in bacteria — a protection against DNA relaxation, in *Heat Shock from Bacteria to Man*, Schlesinger, M. J., Ashburner, M., and Tissieres, A., Eds., Cold Spring Harbor Laboratory, Cold Spring Harbor, NY, 1982, 127—130.

240. **Ullrich, S. J., Robinson, E. A., Law, L. W., Willingham, M., and Appella, E.,** A mouse tumor-specific transplantation antigen is a heat shock-related protein, *Proc. Natl. Acad. Sci. U.S.A.*, 83, 3121—3135, 1986.

241. **Van Bergen en Henegouwen, P. M. P., Berbers, G., Linnemans, W. A. M., and Van Wijk, R.,** Cytoplasmic and nuclear localization of the 84,000 dalton heat shock protein in mouse neuroblastoma cells, *Eur. J. Cell. Biol.*, 43, 469—478, 1987.

242. **Van Bogelen, R. A., Kelley, P. M., and Neidhardt, F. C.,** Differential induction of heat shock, SOS, and oxidation stress regulons and accumulation of nucleotides in *Escherichia coli*, *J. Bacteriol.*, 169, 26—32, 1987.

243. **Velazquez, J. M. and Lindquist, S.,** HSP70: Nuclear concentration during environmental stress and cytoplasmic storage during recovery, *Cell*, 36, 655—662, 1984.

243a. **Vierstra, R. D., Burke, T. J., Callis, J., Hatfield, P. M., Jabben, M., Shanklin, J., and Sullivan, M. L.,** Characterization of the ubiquitin-dependent proteolytic pathway in higher plants, in *The Ubiquitin System*, Schlesinger, M. and Hershko, A., Eds., Cold Spring Harbor Press, Cold Spring Harbor, NY, 1988, 119—125.

244. **Vig, B. K.,** Hyperthermic enhancement of chromosome damage and lack of effect on sister-chromatid exchange induced by bleomycin in Chinese hamster cells in vitro, *Mutat. Res.*, 61, 309—317, 1979.

245. **Vincent, M. and Tanguay, R. M.,** Different intracellular distributions of heat shock and arsenite-induced proteins in *Drosophila* Kc cells. Possible relation with the phosphorylation and translocation of a major cytoskeletal protein, *J. Mol. Biol.*, 162, 365—378, 1982.

246. **Voellmy, R. and Bromley, P. A.,** Massive heat-shock polypeptide synthesis in late chicken embryos: Convenient system for study of protein synthesis in highly differentiated organisms, *Mol. Cell. Biol.*, 2, 479—483, 1982.

246a. **Walsh, D. A., Klein, N. W., Hightower, L. E., and Edwards, N. J.,** Heat shock and thermotolerance during early rat embryo development, *Teratology*, 36, 181—191, 1987.

247. **Welch, W. J. and Feramisco, J. R.,** Nuclear and nucleolar localization of the 72,000 dalton heat shock protein in heat-shocked mammalian cells, *J. Biol. Chem.*, 259, 4501—4513, 1984.

248. **Welch, W. J. and Feramisco, J. R.,** Disruption of the three cytoskeletal networks in mammalian cells does not affect transcription, translation, or protein translocation changes induced by heat shock, *Mol. Cell. Biol.*, 5, 1571—1581, 1985.

249. **Welch, W. J. and Suhan, J. P.,** Cellular and biochemical events in mammalian cells during and after recovery from physiological stress, *J. Cell Biol.*, 103, 2035—2052, 1986.

250. **Welch, W. J., Garrels, J. I., Thomas, G. P., Lin, J. J. C., and Feramisco, J. R.,** Biochemical characterization of the mammalian stress proteins and identification of two stress proteins as glucose- and calcium ionophore-regulated proteins, *J. Biol. Chem.*, 258, 7102—7111, 1983.

251. **Weitzel, G., Pilatus, U., and Rensing, L.,** Similar dose response of heat-shock protein synthesis and intracellular pH change in yeast, *Exp. Cell Res.*, 159, 252—256, 1985.

251a. **Weitzel, G., Pilatus, U., and Rensing, L.,** The cytoplasmic pH, ATP content and total protein synthesis rate during heat-shock protein inducing treatments in yeast, *Exp. Cell Res.*, 170, 64—79, 1987.

252. **Whelan, S. A. and Hightower, L. E.,** Differential induction of glucose-regulated and heat shock proteins: Effects of pH and sulfhydryl-reducing agents on chicken embryo cells, *J. Cell. Physiol.*, 125, 251—258, 1985.

252a. **Whelan, S. A. and Hightower, L. E.,** Induction of stress proteins in chicken embryo cells by low-level zinc contamination in amino acid-free media, *J. Cell. Physiol.*, 122, 205—209, 1985.

253. **White, F. P. and Currie, R. W.,** A mammalian response to trauma: The synthesis of a 71-kD protein, in *Heat Shock from Bacteria to Man*, Schlesinger, M. J., Ashburner, M., and Tissieres, A., Eds., Cold Spring Harbor Laboratory, Cold Spring Harbor, NY, 1982, 379—386.

253a. **White, F. P. and White, S. R.,** Isoproterenol induced myocardial necrosis is associated with stress protein synthesis in rat heart and thoracic aorta, *Cardiovasc. Res.*, 20, 512—515, 1986.

253b. **White, F. P. and White, S. R.,** Effects of polyhydroxy alcohols on protein synthesis in brain slices, *J. Neurochem.*, 48, 1560—1565, 1987.

254. **Wiegant, F. A. C., Van Bergen en Henegouwen, P. M. P., Van Dongen, A. A. M. S., and Linnemans, W. A. M.,** Stress-induced thermotolerance of the cytoskeleton, *Cancer Res.*, 47, 1674—1680, 1987.

255. **Wilhelm, J. M., Spear, P., and Sax, C.,** Heat-shock proteins in the protozoan *Tetrahymena:* Induction by protein synthesis inhibition and possible role in carbohydrate metabolism, in *Heat Shock from Bacteria to Man*, Schlesinger, M. J., Ashburner, M., and Tissieres, A., Eds., Cold Spring Harbor Laboratory, Cold Spring Harbor, NY, 1982, 309—314.

255a. **Wilkison, W. O. and Bell, R. M.,** sn-Glycerol-3-phosphate acyltransferase tubule formation is dependent upon heat shock proteins (htpR), *J. Biol. Chem.*, 263, 14505—14510, 1988.

255b. **Williams, G. T., McClanahan, T. K., and Morimoto, R. I.,** E1a transactivation of the human HSP70 promoter is mediated through the basal transcription complex, *Mol. Cell. Biol.*, 9, 2574—2587, 1989.

255c. **Williams, K. J., Landgraf, B. E., Whiting, N. L., and Zurlo, J.,** Correlation between the induction of heat shock protein70 and enhanced viral reactivation in mammalian cells treated with ultraviolet light and heat shock, *Cancer Res.*, 49, 2735—2742, 1989.

255d. **Winter, J., Wright, R., Duck, N., Gasser, C., Fraley, R., and Shah, D.,** The inhibition of Petunia hsp70 mRNA processing during $CdCl_2$ stress, *Mol. Gen. Genetics*, 211, 315—319, 1988.

256. **Wolffe, A.-P., Glover, J. F., and Tata, J. R.,** Culture shock. Synthesis of heat-shock-like proteins in fresh primary cell cultures, *Exp. Cell Res.*, 154, 581—590, 1984.

257. **Wu, C.,** Activating protein factor binds in vitro to upstream control sequences in heat shock gene chromatin, *Nature*, 311, 81—84, 1984.

258. **Wu, C.,** Two protein-binding sites in chromatin implicated in the activation of heat-shock genes, *Nature*, 309, 229—234, 1984.

259. **Wu, F. S., Park, Y.-C., Roufa, D., and Martonosi, A.,** Selective stimulation of the synthesis of an 80 000-dalton protein by calcium ionophores, *J. Biol. Chem.*, 256, 5309—5312, 1981.

259a. **Wu, C. H., Warren, H. L., Sitaraman, K., and Tsai, C. Y.,** Translational alterations in maize leaves responding to pathogen infection, paraquat treatment, or heat shock, *Plant Physiol.*, 86, 1323—1329, 1988.

260. **Wu, B. J. and Morimoto, R. I.,** Transcription of the human hsp70 gene is induced by serum stimulation, *Proc. Natl. Acad. Sci. U.S.A.*, 82, 6070—6074, 1985.

261. **Wu, B. J., Kingston, R., and Morimoto, R. I.,** Human HSP70 promoter contains at least two distinct regulatory domains, *Proc. Natl. Acad. Sci. U.S.A.*, 83, 629—633, 1986.

261a. **Yost, H. J. and Lindquist, S.,** RNA splicing is interrupted by heat shock and is rescued by heat shock protein synthesis, *Cell*, 45, 185—193, 1986.

262. **Zimarino, V. and Wu, C.,** Induction of sequence-specific binding of *Drosophila* heat shock activator protein without protein synthesis, *Nature,* 327, 727—730, 1987.
263. **Zuker, C., Cappello, J., Chisholm, R. L., and Lodish, H. F.,** A repetitive *Dictyostelium* gene family that is induced during differentiation and by heat shock, *Cell,* 34, 997—1005, 1983.

Chapter 2

HEAT SHOCK PROTEINS

L. Nover and K.-D. Scharf

TABLE OF CONTENTS

2.1. SURVEY OF EXPERIMENTAL SYSTEMS WITH INDUCED HSP SYNTHESIS

2.1.1. ANALYSIS OF HSP SYNTHESIS

The discovery of heat shock proteins (HSPs) in *Drosophila*[385] and the following demonstration of the newly formed mRNAs specifically hybridize to the corresponding puffs observed by Ritossa[330] laid the foundation for similar investigations in many other species.[248,249,260,368,369] Three factors decisively facilitated the development:

1. The remarkable selectivity of the heat shock (hs) response may result in protein patterns with about 90% of the radioactively labeled precursor amino acid being incorporated into HSPs. This is illustrated in Figure 2.1 for *Drosophila* (A) and tomato cell cultures (B), respectively. Especially the more physiological procedure of a gradual heating, as contrasted to an immediate hs, provides optimum conditions for massive HSP synthesis at temperatures where practically no other proteins are labeled. As is detailed in the following section, in most cells a group of "large" HSPs in the M_r range 70 to 90 kDa can be distinguished from a group of "small" HSPs with M_r of 17 to 30 kDa.
2. Simultaneously, there were two important developments in the electrophoretic analysis of proteins. Laemmli[202b] introduced electrophoresis in SDS-containing polyacrylamide gels, and O'Farrell[300] combined this technique with isoelectric focusing to give a two-dimensional separation with an unprecedented resolution of many proteins in a single run. The potential of the method for analysis of HSP patterns is shown in Figure 2.2. After heating tomato cell cultures for 3 h to 39°C, a labeled protein sample was separated by a two-dimensional electrophoresis using the O'Farrell/Laemmli method.[296] The group of small HSPs (open arrow heads in the autoradiograph B with numbers 16 to 29) can be separated into a remarkable multiplicity of major and minor species.
3. Rapid and preferential synthesis of HSPs is combined with a relatively high stability. Hence, they may accumulate in fairly large amounts, which facilitate isolation and characterization. In tomato cell cultures, this is particularly striking because of a massive formation of cytoplasmic particles (hs granules, see Section 16.7) containing the mass of small HSPs and part of the HSP70. They can be easily isolated and purified. After Coomassie staining (Figure 2.2C) the protein composition shows that they are mainly composed of HSPs. Structural binding of HSPs is a general stress-dependent phenomenon probably of all eukaryotic cells (see Chapter 16) and may decisively influence the analytical result.

Experimental systems with hs induced synthesis of HSPs are surveyed in Table 2.1. With more than 220 entries the table comprises data of about 140 different species starting with bacteria and extending to eukaryotic microorganisms, to different animals, to green algae, and, finally, to plants. It is reasonable to state that all types of organisms respond to hyperthermic conditions by synthesis of hs proteins. Moreover, with few exceptions (see Section 8.3) all cells of a multicellular organism respond in a basically similar manner. However, the hyperthermic threshold and temperature range greatly vary with the type of organism and its normothermic state. Illustrative examples are *Escherichia coli* (37° → 42 to 46°C), psychrophilic bacteria from arctic regions (0° → 10 to 30°C), thermophilic bacteria (37° → 60°C), yeast (22° → 36°C), *Drosophila* (25° → 33 to 37°C), sea urchin (20° → 25 to 30°C), *Xenopus* (20° → 30 to 35°C), chicken (40.5° → 44 to 45°C), human (37° → 42 to 44°C), and plants (25° → 35 to 42°C). Details of this physiological hs response range, which appear to be genetically fixed for each organism, are discussed in Section 1.1.

Despite the availability of excellent analytical techniques, a considerable number of

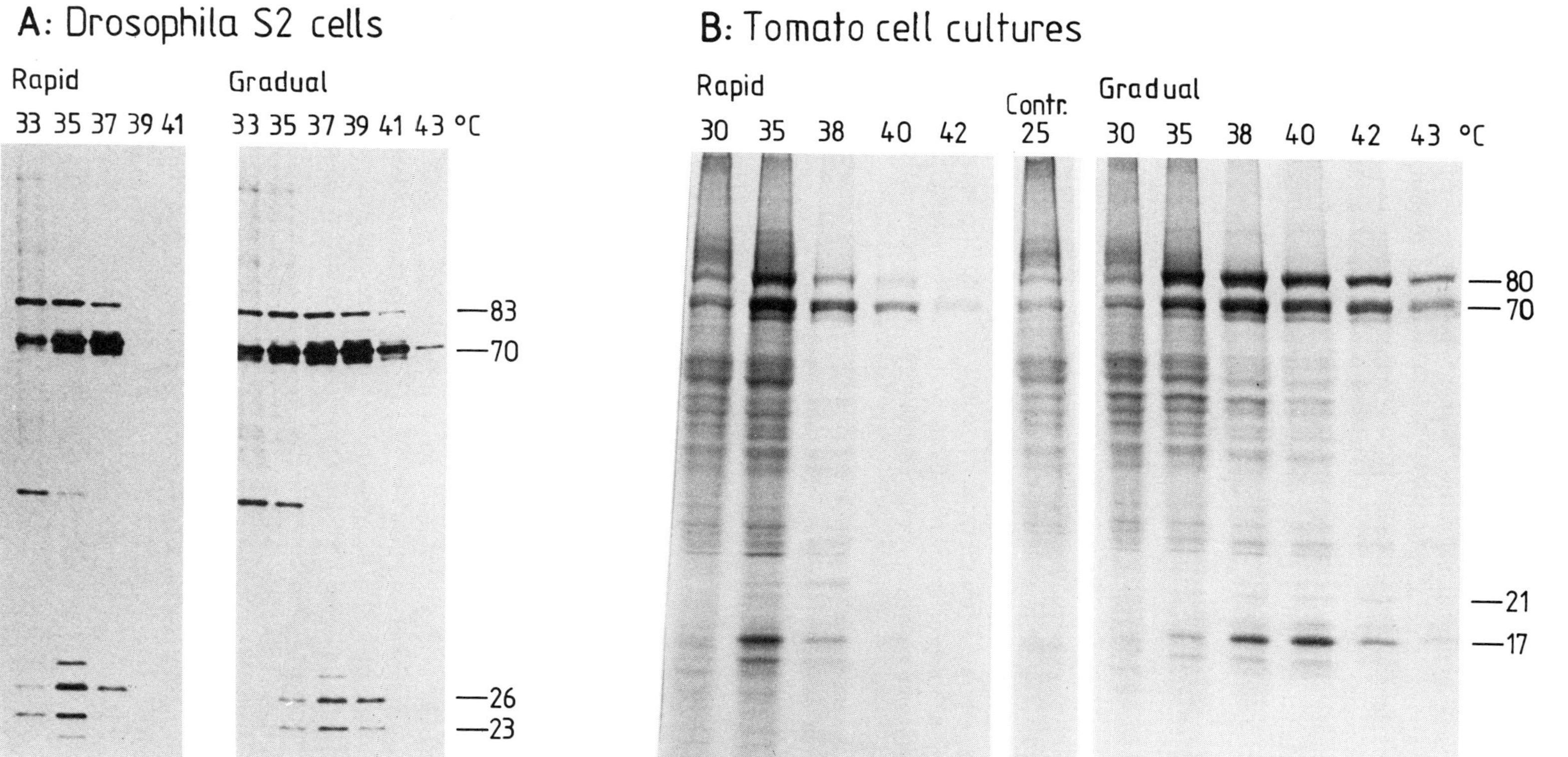

FIGURE 2.1. HSP synthesis in cell cultures of *Drosophila* (A) and tomato (B). Proteins were labeled for 30 min with ^{3}H-leucine (A) or for 1 h with ^{35}S-methionine (B) and separated by SDS polyacrylamide gel electrophoresis. Cells were either immersed directly in a water bath of the indicated temperatures (rapid heating) or warmed gradually (2°C/15 min for *Drosophila* S2 cells, 2°C/h for tomato suspension cultures). Synthesis of the large and small HSPs (M_r indicated at the margins) is markedly improved at higher hs temperatures, if cells were allowed to adapt gradually. (Reprinted from Lindquist, S., *Dev. Biol.*, 77, 463, 1980; and Scharf, K.-D. and Nover, L., *Cell*, 30, 427, 1982. With permission.)

recent papers, are very poor in the documentation of induced HSPs. Above all, two-dimensional separations are frequently lacking, elaboration of the optimum labeling conditions is neglected, and the M_r given for a particular organism vary greatly, even by the same group of research workers. To avoid confusion, the HSPs are included in Table 2.1 with the M_r values reported in the original paper. Whenever possible, a standardized set of major HSPs found in a given type of organism is indicated in the subheading, e.g., for *Drosophila,* bird, rodent, human, and plant cells. In addition, a collection of two-dimensional separations documents typical patterns of induced HSPs in bacteria (Figure 2.3A to D), eukaryotic microorganisms (Figure 2.4A to D), animals (Figure 2.6A to D), and plants (Figure 2.7A to D). Additional information and references to data in Table 2.1 are summarized in the legends of Figures 2.3 to 2.7.

2.1.2. HSPs OF *E. COLI* AND OTHER BACTERIA (FIGURE 2.3)

The survey of HSPs in bacteria comprises the two closely related enteric bacteria *Escherichia coli* (A) and *Salmonella typhimurium* (C) as well as a the cyanobacterium *Synechococcus* sp. (B) and the sporulating *Bacillus subtillis* (D). The extensive characterization of the hs response in *E. coli*[284,444] led to the identification of several HSPs with respect to their cellular functions (Table 2.2), which, however, are not necessarily hs-related functions. Among them are a specialized lys-tRNA synthetase (LysU), an ATP-dependent protease (Lon), and two RNA polymerase sigma subunits, i.e., RpoD (σ^{70}) and RpoH (σ^{32}). Sequential increase, at first of σ^{32} and then of σ^{70} synthesis, as well as the accumulation of the Lon and DnaK proteins are essential elements of the control network of HSP synthesis in *E. coli* (see Section 7.3).[138b]

Remarkably complex are the cellular functions influenced by the major HSPs of *E. coli,* i.e., by the GrpE protein and by the two pairs of subunits, each containing a large and small subunit, coded by the DnaK/J and the GroEL/S operons. They were originally defined as essential host cell contributions to phage replication and morphogenesis.[133,162a,284,306,413] In fact, they are required for optimum growth under all conditions, irrespective of stress either by phage infection, hs, or others. In most cases, cooperation between different HSPs is necessary, and the activity of DnaK and GroEL needs ATP.

1. Replication of the λ phage DNA involves transient formation of a primosome at the origin. Ordered assembly and disassembly of this DNP complex needs interaction with the host cell DnaK, DnaJ, and GrpE proteins. But their catalytic function in the replication initiation extends also to low-copy-number plasmids and probably to the bacterial chromosome itself.[1d,58a,111,177c,183b,384a,433a] Furthermore, a comparable role of DnaK may even be envisaged for initiation of transcription. At the least, immune precipitates of sigma subunits of *E. coli* (σ^{32} and σ^{70}) always contain DnaK protein.[359b] It is tempting to speculate that the rapid release of sigma subunits from the RNA polymerase core after initiation-complex formation and/or their reincorporation into free RNA polymerase is facilitated by DnaK.
2. Cell lysis by action of the phage ΦX174 E protein on the bacterial membrane is affected differentially by hs. Initially, hs-induced membrane instability is connected with increased lysis. Following this, synthesis of DnaK/J, GroEL/S, and GrpE proteins contributes to membrane stabilization. Mutants of all five hs genes are particularly sensitive to the lysis effect by the ΦX174 E protein.[443a]
3. Genomic stress, e.g., by UV irradiation of *E. coli,* leads to the induction of the error-prone repair system (SOS-repair). Interestingly, overexpression of the umuDC operon, as part of this repair system, results in a transient cold sensitivity of the bacteria, i.e., inability to grow at or below 30°C. Depending on the particular regulatory situation that causes umuDC overexpression, cold sensitivity can be overcome by mutation

of the dnaK, lon, groE, or rpoH genes. Though the mechanism is unknown, the hs genes evidently participate in the regulation of SOS-repair and the formation of the UmuD/C-DNA complexes.[111a]

4. Mutants in dnaK, dnaJ, grpE, or groEL are defective in the energy-dependent proteolysis as measured by removal of puromycin-peptides or β-galactosidase fragments.[374a] How these proteins act together to maintain proteolytic homeostasis is unknown.
5. The particular role of the heptameric 20S complex of the GroEL/S proteins in phage head assembly[162a] is only part of a more general function in the assembly of multimeric protein complexes and protein topogenesis (see Section 12.2). On the one hand, the 20S GroE complex was found associated with pre-β-lactamase as a prerequisite for an ATP-dependent lactamase secretion. Remarkably, export of other proteins is not affected by groE mutants, e.g., of the maltose-binding protein, alkaline phosphatase, lipoprotein, or the ompA/F products.[46a,202a] On the other hand, an essential role for effective protein assembly was demonstrated by Goloubinoff et al.[137a] using *E. coli* transformed with the *Anacystis nidulans* genes for ribulosebisphosphate carboxylase. Formation of the active enzyme of eight large and eight small subunits depends on increased amounts of GroE complex, i.e., it can be stimulated by hs. This ATP-dependent help of GroE or other HSPs in protein assembly may also explain why the incorporation of a plasmid-encoded RNA polymerase β-subunit into the holoenzyme is defective in rpoH mutants.[183a]

Immunological techniques were used to identify the Lon-protease (HSP94), the DnaK protein (HSP69), and the GroEL protein (HSP62) as major hs-induced proteins in other bacteria as well, i.e., in *Salmonella typhimurium* (C69 and C56; see Figure 2.3C),[267] in *Caulobacter crescentus* (HSPs92, 70 and 62),[138,327] in *Pseudomonas aeruginosa* (HSPs76 and 61),[1b] in *Bacillus subtilis* (HSPs94, 76, and 66),[6] and the HSP62 as GroEL in *Mycobacterium smegmatis.*[354a] As outlined in the following section, some bacterial HSPs show a remarkable degree of sequence homology with their eukaryotic counterparts. These are HtpG as a member of the HSP90 family (see Figure 2.8),[25] DnaK of the HSP70 family (see Figure 2.10),[24] and GroEL of the HSP60 family (see Figure 2.11).[251]

Characterization of different stress factors in *E. coli*[399] and *S. typhimurium*[267,367] showed only partial overlaps of the corresponding protein patterns. In *E. coli* inducible proteins of three independently regulated regulons were defined: (1) the hs regulon (rpoH-controlled), (2) the oxidative stress regulon (oxyR-controlled), and (3) the SOS regulon (lexA-controlled). Most similar to the hs-induced pattern are those observed after addition of ethanol or during recovery from puromycin. Other stressors, e.g., nalidixic acid, $CdCl_2$, H_2O_2, amino acid starvation, inducc various parts of the three regulons. They are evidently part of a stressor-specific regulatory entity tentatively called a stimulon.[399]

2.1.3. HSPs OF EUKARYOTIC MICROORGANISMS AND YEAST (FIGURE 2.4)

This group is represented by *Aspergillus nidulans* (Figure 2.4A), the ergot fungus *Claviceps purpurea* (Figure 2.4B), the oomycete *Achlya klebsiana* (Figure 2.4C), and the protozoan ciliate *Tetrahymena pyriformis* (Figure 2.4D). In all cases the prominent HSPs are found in the 70- to 90-kDa region, whereas the group of small HSPs is highly variable. There are essentially no small HSPs in *Aspergillus* (Figure 2.4A), one dominant protein (HSP23) in ergot (Figure 2.4B), two minor HSPs of 27 kDa in *Achlya* (Figure 2.4C), and a complex group of eight to ten major HSPs between 25 and 32 kDa in *Tetrahymena* (Figure 2.4D).

It should be mentioned that the set of HSPs represented in Figure 2.4 evidently fails to reveal the organellar HSPs. The detection and isolation of a mitochondrial HSP60 from

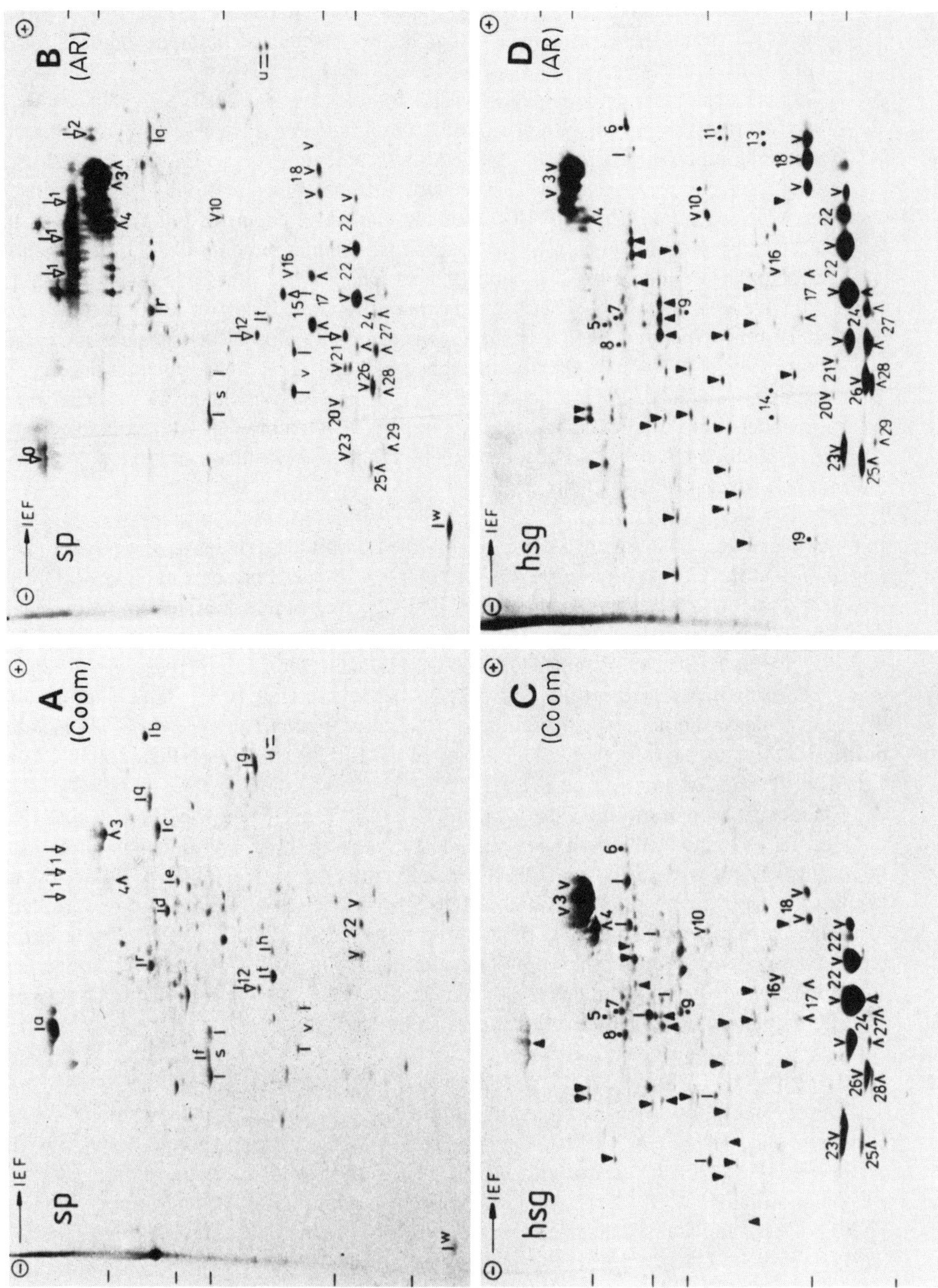

A
(Coom)
sp
B
(AR)
sp
C
(Coom)
hsg
D
(AR)
hsg
IEF

FIGURE 2.2. Two-dimensional separation of proteins from heat-shocked tomato cell cultures. Cultures were heat-shocked for 3 h at 39°C and labeled in the last 90 min with [$^{-35}$S]-methionine. Proteins of the 165,000 × g supernatant (soluble proteins, A, B) and of the heat shock granules (hsg, C, D) sedimented by 1 h centrifugation at 60,000 × g[297] (see Section 16.7) were separated by 2D-electrophoresis with isoelectric focusing in the first (IEF) and SDS-polyacrylamide gradient gel electrophoresis in the second dimension. A, C represent the Coomassie-stained gels; B, D represent autoradiographs. Heat shock proteins, (HSPs) are numbered (open arrows, open arrow heads). Control proteins with discontinued (a to g) and continued synthesis (q to w), respectively, are marked in A, B with vertical bars. hsg-specific proteins not detected in A, B were labeled in C, D (closed arrow heads for control proteins; dots for additional minor HSPs). (From Nover, L. and Scharf, K.-D., *Eur. J. Biochem.*, 139, 303, 1984. With permission.)

No.	HSP	M_r (kDa)	pI	hsg	No.	HSP	M_r (kDa)	pI	hsg
0	HSP95	95	6.5	−	17		21.5	5.8, 6.0	+
1	HSP80	80	5.4—6.2	−	18	HSP21	21.2	5.1, 5.3, 5.4	+ +
2		74	5.0	−	20		18.5	6.3	(+)
3	HSP70	70	5.25—5.5	+	21		18.0	6.0, 6.1	(+)
4	HSP68	68	5.5—5.7	+	22	HSP17	16.8	5.4, 5.6, 5.9	+ + +
10	HSP36	35.5	5.3—5.5	+	23		16.8	6.6	+ + +
12		28.7	6.1	−	24	HSP16	16.5	5.9, 6.1	+ + +
15		25	5.9	−	25		15.0	6.7	+ + + +

Note: Main representatives of HSPs are identified with their numbers, M_r, isoelectric points, and relative extent of structural binding to the hsg fraction (−, not bound; +, 10%; + +, 20 to 30%; + + +, 50%; + + + +, 80% of total HSP).

TABLE 2.1
Survey on Systems Characterized for Induced Heat Shock Protein Synthesis*

Organism/cell	Hs conditions	HSPs induced (M_r in kDa)	Other inducers	Ref.
Bacteria				
Escherichia coli	5—10 min, 37° → 42—46°C; 1D, 2D	HSPs 94, 84, 66, 62.5, 56.5, 33, 25, 21.5, 21, 15.4, 14.7, 13.5, 10 (see Figure 2.3A and Table 2.2)	Eth, Pur, NA, H_2O_2, Cd	281, 282, 391, 399, 442a, 444
Salmonella typhimurium	10 min, 37° → 42°C; 2D	HSPs 89, 79, 69, 64, 56, 52, 46, 31, 24, 23 (see Figure 2.3C)	H_2O_2, Eth	267, 367
Bacillus subtilis	10 min, 37° → 50—52°C; 2D	HSPs see Figure 2.3D	Norvaline, H_2O_2	329
	10—20 min, 37° → 46—50°C; 1D	HSPs 87, 82, 76, 66, 31, 29, 22, 21, 19, 17, 16, 13	Eth	6
Growing and sporulating stages	15 min, 30° → 43°C; 1D	HSPs 84, 69, 48, 32, 22	—	387
Bacillus psychrophilus and other arctic psychrophilic bacilli	1—4 h, 0° → 10—32°C; 1D	HSPs 89, 82, 79, 77, 72, 70, 66, 45, 44, 40, 39, 38, 37, 36, 35, 34, 33, 30, 29, 28, 27, 24, 23, 22	—	247
Bacillus subtilis, B. megaterium, B. cereus	10 min, 37° → 48°C; 1D	Major HSP with M_r 66 kDa, minor HSP between 14 to 97 kDa	—	374b
Bacteroides fragilis	2—10 min, 37° → 48°C; 1D	HSPs 125, 80, 74, 65, 56, 52, 20	*Not* induced by Eth, H_2O_2, UV	138a
Caulobacter crescentus	5—10 min, 30° → 42°C; 1D, 2D	HSPs 92, 70, 68, 62, 54, 48, 46, 41, 34, 28, 24, 23, 20, 13	H_2O_2	138, 327

* The hs conditions describe the duration and extent of temperature increase as well as the analysis of labeled protein mixtures by one-dimensional and two-dimensional (1D, 2D) electrophoresis. HSPs are listed in the usual way with their apparent molecular weights (Mr) derived by the particular author(s) from SDS gel electrophoresis. Because of considerable discrepancies between data reported by different groups, the actual set of HSPs found in a given type of organism may be indicated for better orientation (see e.g., human, birds, plants). If tested together with hs, HSP induction by chemical stressors is included (see Table 1.1): arsenite (Ars), carbonyl cyanide m-chlorophenylhydrazone (CCCP), 2, 4-dinitrophenol (DNP), Cd^{2+} (Cd), Cu^{2+} (Cu), Hg^{2+} (Hg), Zn^{2+} (Zn), ethanol (Eth), canavanine (Can), azetidin carboxylic acid (AzC), p-fluoro-phenylalanine (pFP), puromycin (Pur), anoxia and recovery (N_2), p-chloromercuribenzoate (pCMB)). Further abbreviations are for low molecular weight (lmw) (HSPs) and for whole body hyperthermia, (WBH).

Methylophilus methylotrophus	30° → 40°C; 1D	HSPs 94, 83, 78, 63, 60, 36, 29, 20, 16, 14, 13	5% methanol or ethanol	423a
Mycobacterium smegmatis	60 min, 30° → 46°C; 1D	HSPs 95, 72, 62, 40, 32, 28, 23, 16	—	354a
Neisseria gonorrhoeae	10 min, 37° → 43°C; 1D, 2D	Maximum 37 HSP identified	4% Eth	192c
Pseudomonas aeruginosa	10 min, 30° → 45°C; 1D	HSPs 103, 95, 90, 86, 81, 76, 74, 61, 50, 43, 40, 35, 27.5, 26, 21, 17.8, 15.7	Eth	1b
Rhodomicrobium vannielii	2 h, 40°C; 1D, 2D	HSPs 67, 34, 26, 23, 20, 18	—	350a
Clostridium acetobutylicum	30 min, 28° → 45°C; 1D, 2D	HSPs 83, 74, 68, 62, 49, 36, 22, 18	—	382a
Sulfolobus acidocaldarius, *S.* spec.	10 min, 70° → 85°C; 1D, 2D	HSPs 92, 74, 66, 33, 22, 16, 14.5	—	177a, 390a
Thiobacillus ferrooxidans	10 min, 30° → 41°C; 1D	HSPs 92, 74, 66, 33, 22, 16, 14.5	Eth	177a
Anabaena spec. strain 2—31	45°C; 1D, 2D	HSPs 92, 82, 75, 65, 32, 23, 19	Salt, osmotic stress	21a
Legionella pneumophila	60 min, 30° → 42°C; 1D	HSPs 85, 78, 70, 60, 17	Novobiocin, patulin, puromycin	220a
Acholeplasma laidlawii, *Mycoplasma capricolum*	20 min, 32° → 42°C; 1D	HSPs 92, 90, 68, 56, 49, 39, 35, 29, 27, 20		100a
Anacystis nidulans	60 min, 30°→45—50°C; 1D	HSPs 77, 71, 69, 60, 59, 16, 13.5, 12.5	—	238
Synechococcus sp. (cyanobacterium)	30—120 min, 39° → 47°C; 1D, 2D	HSPs 91, 79, 78, 74, 65, 64, 61, 49, 45, 24, 22, 18, 16, 14, 12, 11 (see Figure 2.3B)	—	52
Phormidium laminarium (thermoph. cyanobact.)	1—3 h, 45° → 55°C; 1D	HSPs 89, 86, 72, 66, 63, 47, 43, 38, 33	—	290
Myxococcus xanthus (vegetative cells, starvation-induced fruiting cells, sporulating cells)	15—90 min, 28° → 36—42°C; 1D	HSPs 92, 85, 75, 69, 65, 63, 60, 55, 52, 50, 47, 45, 42, 39, 36, 25, 24.5, 23, 22, 20, 18.5, 16, 14.5	—	285
Streptomyces lividans	15 min, 30° → 43°C; 1D	HSPs 87, 76, 53, 46	—	387
Zymomonas mobilis	10 min, 30° → 45°C; 2D	HSPs 96, 72, 58, 48, 42, 38, 35, 33, 21, 16	Eth	253, 254
Halobacterium volcani	60 min, 37° → 60°C; 1D, 2D	HSPs 98, 91, 85, 79, 21	—	99
Halobacterium trapanicum	60 min, 37° → 60°C; 1D	HSPs 105, 97, 87, 74, 44, 30	—	99

TABLE 2.1 (continued)
Survey on Systems Characterized for Induced Heat Shock Protein Synthesis*

Organism/cell	Hs conditions	HSPs induced (M_r in kDa)	Other inducers	Ref.
Halobacterium marismortui	60 min, 37° → 60°C; 1D	HSPs 100, 96, 86, 76, 45, 28	—	99
Halobacterium halobium	60 min, 37° → 60°C; 1D	HSPs 98, 92, 25	—	99
Halobacterium salinarium	60 min, 37° → 60°C; 1D	HSPs 98, 92, 90, 26	—	99
Eukaryotic microorganisms				
Protozoa				
Tetrahymena pyriformis	15—30 min, 25° → 34°C; 1D, 2D	HSPs 92, 70—75, 58, 46, 35, 25—29 (see Figure 2.4D)	Arsenite	3, 125, 131
	1 h, 33.8°C; 1D, 2D	HSPs 91, 75, 73 plus 10 HSPs 23—30	Deciliation, recovery from anoxia	148, 149
	40 min, 28° → 34—36°C; 1D	HSPs 120, 90, 80, 72, 46, 35, 31, 29	—	126
Tetrahymena thermophila	30° → 40°C; 1D	HSPs 91, 80, 73, 60, 45, 28	—	125
	0.5 h, 30° → 40°C; 1D, 2D	HSPs 80, 73, 58 plus 1mw HSPs 25—34 kDa	—	154, 250
	20 min, 30° → 39°C; 1D, 2D	HSPs 85, 84, 74 plus 5 lmw HSPs of 30—34 kDa	see Table 1.1	435a
Giardia lamblia	10—20 min, 37° → 43°C; 1D	HSPs 100, 83, 70, 30	Eth	229a
Trypanosoma cruzi	27° → 37—41°C; 1D, 2D	HSPs 100, 83, 71, 65, 62, 45, 44, 43, 40, 35, 34	—	1a
Leishmania mexicana	10 h, 24° → 34—37°C; 1D	HSPs 83, 70, 68, 27, 26, 23, 22	—	170, 350d
Naegleria gruberi (amebo-flagellate)	25° → 38°C; 1D	HSPs 96, 70, 68	—	417
Mycophyta				
Physarum polycephalum	2 h, 22° → 32°C; 1D	HSPs 105, 82, 74, 69	—	439
Dictyostelium discoideum (growing and developing cells)	1 h, 22° → 30—34°C; 1D, 2D	HSPs 82, 70, 60, 43	—	233, 234, 234a, 333
Polysphondylium pallidum (cells at different developmental stages)	1—3 h, 23° → 31°C; 1D	HSPs 105, 103, 87, 74, 70, 33, 32	—	128

Blastocladiella emersonii (germinating, growing and sporulating cells)	30 min, 27° → 36—38°C; 1D, 2D	HSPs 84, 82, 76, 70, 60, 50, 45, 39, 30, 25, 24, 17	Sporulation	48
Achlya ambisexualis (water mold)	10—30 min, 30° → 35°C; 1D	HSPs 78, 70, 46, 44, 36, 34, plus 3 lmw HSPs	—	150
	60 min, 28° → 37°C; 1D, 2D	HSPs 96, 85, 74, 70, 43 plus 23—28 kDa HSPs	—	356
Achlya klebsiana	60 min, 28° → 37°C; 1D, 2D	HSPs 96, 85, 72, 70, 69, 68, 60, 52, 26 (see Figure 2.4C)	—	219
Melampsora lini (flax rust, uredosporelings)	2 h, 17° → 30°C; 1D	HSPs 84, 71, 43.5, 30.5, 19.5, 18, 17	—	352
Neurospora crassa	1 h, 28° → 48°C; 1D, 2D	HSPs 95, 81, 68, 41, 25, 16	Arsenite	179, 180, 181, 182
	1 h, 28° → 45°C; 1D	HSPs 98, 83, 67, 38, 34, 30	—	318a, 319
	25° → 42°C; 1D	HSPs 99, 81, 69	—	86
Aspergillus nidulans	60 min, 37° → 43°C; 1D	HSPs 110, 98, 95, 85, 40, 29, 23 (see Figure 2.4A)	—	289, 371
Claviceps purpurea (ergot fungus)	2 h, 25° → 40°C; 2D	HSPs 105, 87, 78, 71, 58, 24.3 (see Figure 2.4B)	—	289
Yeasts				
Saccharomyces cerevisiae (bakers yeast)	15—60 min, 23° → 36°C; 1D, 2D	HSPs 100, 90, 79, 69, 56, and 38 (see Figure 2.5)	—	245, 246, 319a
	20—40 min, 22° → 37°C; 2D	20 major plus 20 minor HSPs identified	—	256, 257
	30 min, 25° → 38°—42°C; 1D	HSPs 96, 84, 70, 48, 46, 35, 26	—	316
	30 min, 23° → 36°C; 2D	HSPs 89, 86, 77, 75, 71, 49, 48, 46, 27, 25	—	172, 353a
	90 min, 23° → 40°C; 1D	HSPs 98, 85, 70, 65, 56, 44, 33	DNP	425
	30 min, 23° → 39°C; 2D	Discrimination of 6 members of HSP70 family and two HSP90 (see Figure 2.5)	—	431
Histoplasma capsulatum	60 min, 25° → 37—40°C; 1D, 2D	HSPs 92, 83, 78, 70, 31, 24	—	352a
Candida albicans	20 min, 23° → 37°C or 37° → 41—46°C; 1D	HSPs 98, 85, 81, 76, 54, 34, 26, 18	—	98, 446b
Fonsecaea pedrosi	5 min, 23° → 37—40°C; 1D	HSPs 84, 70, 63, 52.5, 50, 48.5, 42	—	171

TABLE 2.1 (continued)
Survey on Systems Characterized for Induced Heat Shock Protein Synthesis*

Organism/cell	Hs conditions	HSPs induced (M_r in kDa)	Other inducers	Ref.
Psychrotrophic yeasts				
Trichosporon pullulans	20—50 min, 5° → 26—29°C; 1D	HSPs 94, 87, 82, 80, 78, 74, 66, 62, 42, 36, 33	Recovery from anoxia	34
Sporobolomyces salmonicolor	1—3 h, 10° → 31—34°C; 1D	HSPs 112, 100, 85, 80, 74, 41	Recovery from anoxia	34
Animals				
Coelenteratae				
Hydra attenuata	90 min, 17° → 30—33°C; 1D	HSPs 90, 80, 60, 28	Cd, Azide	53
Molluscs				
Aplysia californica (nervous tissue)	30 min, 34—37°C; 1D, 2D	HSPs 110, 90, 87, 70	—	140
Lymnea stagnalis (water snail, larvae)	1 h, 25° → 35—38°C; 1D	HSPs 100, 87, 70, 68, 65, 37.5, 16, 13.5	Eth	51
Aurelia aurita (jelly fish, larvae, and adults)	1—2 h, at 34°C; 1D	HSPs 93, 83, 70/68, 45, 39	—	44
Squid (nervous tissue)	2—4 h, 20° → 30°C; 1D	HSPs 95, 68 (traversin)	—	394
Nematodes				
Caenorhabditis elegans	4 h, 29—35°C; 1D	HSPs 81, 70, 41, 38, 29, 19, 18, 16	—	362
Crustaceae				
Artemia salina (brine shrimp, cysts and nauplius larvae)	5 min, 40°C; 1D, 2D	HSPs 89, 70, 68, 31	—	257a
Insects				
Drosophila melanogaster (fruit fly)	Basic patterns of HSPs 83, 70, 68, 27, 26, 23, 22 (see Figure 2.6A)			
Larval salivary glands, brain, Malpighian tubes	20 min, 25° → 37.5°C; 1D	HSPs 83, 70, 68 plus small HSPs	—	385
Larval salivary glands	40 min, 25° → 37°C; 2D	HSPs 83, 40, HSP isoforms in 68—78 kDa region, HSP 34 and small HSPs	—	64
	20 min, 25° → 37°C; 1D	HSPs 83, 70/68, 43, 27, 26, 23, 22	N_2	224

Larval salivary glands, Kc cells, testis and ovaries of adult, embryos	20—30 min, 25° → 36°C; 1D	HSPs 84, 68—75 complex, 34, 27, 26, 23, 22	—	415
Ovaries embryos, and primary embryonic cells	1 h, 25° → 37°C; 2D	HSPs 83 and 4 isoforms of the 68—72 kDa complex	—	308
Early embryonic stages (preblastoderm, blastula, gastrula stages)	15—40 min, 22° → 37°C; 1D, 2D	HSPs 81, 70/68, 28, 26, 23, 21	—	38, 116
Pupae (thoracic epithelium, brain)	40 min, 40.2°C then recovery; 1D	HSPs 84, 70, 68, 34, 27, 26, 23, 22	—	76
Cell cultures and imaginal discs	1 h, 25° → 37°C; 1D	HSPs 83, 70, 68, 27, 26, 23, 22	Ecdysterone	35, 73b, 176a, 447
Schneider cells, Kc cells	1 h, 25° → 38°C; 2D	HSP 83, 40 HSP isoforms in 68—70 kDa region, HSP 34 and small HSPs	—	64
	1 h, 23° → 33—38°C; 1D	HSPs 82, 73, 72, 70, 68, 28, 26, 23, 22	—	230, 249, 373c
	1—2 h, 25° → 37°C; 1D	HSPs 83, 70, 27, 26, 23, 22	Ars	381
	2 h, 25° → 37°C; 1D, 2D	HSPs 80, 70, 68, 36, 26, 25, 23, 22	Cd	88
Drosophila hydei				
Larval salivary glands	2 h, 25° → 37°C; 1D	HSPs 70, 67, 40, 35, 26, 25, 20	DNP, Ars, N_2	54, 195
	20 min, 25° → 37°C; 1D	HSPs 70, 67, 38, 26, 25, 20	N_2	224
Larval salivary glands, tissue culture, embryos	1 h, 25° → 37°C; 1D, 2D	HSPs 70, 67, 38, 33, 26, 25, 20	—	363
Drosophilia virilis, D. funebris, D. buschkii, D. ananassae, D. simulans (larval salivary glands)	20 min, 25° → 37°C;	6—7 HSPs, major HSP always HSP70	—	224
Calpodes ethlius (butterfly, larval silk gland, fat body, wing disc, and central nervous tissue)	1—2 h, 22° → 37°C, hs *in situ* and with isol. tissues; 1D, 2D	HSPs 95, 81, 74, 26, 22 (2—5 isoforms)	—	104
Sarcophaga bullata (fly, pupal foot pad cells)	1—2 h, 25° → 37°—44°C; 1D	HSPs 115, 82, 65	N_2, Ars, DNP, Azide	59, 60
Chironomus thummi	1 h, 18° → 30—36°C; 1D	HSPs 86, 76, 70, 28, 26	—	69

TABLE 2.1 (continued)
Survey on Systems Characterized for Induced Heat Shock Protein Synthesis*

Organism/cell	Hs conditions	HSPs induced (M_r in kDa)	Other inducers	Ref.
Locusta migratoria, larval epidermis	30° → 45—47°C; 1D, 2D	HSPs 85, 70 (complex) plus 4 HSPs 18—22	—	22a
Ceratitis capitata (Mediterranean fruit fly)	30—90 min, 25° → 34—39°C; 1D	HSPs 87, 69, 34, 20, 16, 14, 13, 12	—	1, 372
Aedes albopictus (mosquito, cell culture)	1—2 h, 28° → 37°C; 1D	HSPs 90, 82, 76, 66	—	70
	30 min, 25° → 41°C; 1D	HSPs 83, 70, 38(?), 28, 27, 25, 23	—	36
Plodia interpunctella (lepidopteran, cell culture)	30 min, 25° → 37°C; 1D	HSPs 83, 70, 68, 28, 24, 23 and 18(?)	—	36
Echinoderms				
Paracentrotus lividus	1 h, 20° → 31°C; 1D	HSPs 79, 72/70, 50, 47, 44, 42	—	330b, 330c
Arbacia lixula (embryo)	1 h, 20° → 31°C; 1D	HSPs 70/72	—	330c
Sphaerechinus granularis (embryo)	1 h, 20° → 31°C; 1D	HSPs 70/72	—	330c
Strongylocentrotus purpuratus	2 h, 15° → 25°C; 2D	HSPs 90, 70, 50, 49, 40, 38 (see Figure 2.6B)	—	32
Arbacia punctulata (blastula, gastrula stage embryos)	1 h, 20° → 31°C; 1D	HSPs 80, 70	—	160
Fishes				
Pimephales promelas, FHM cells	10 min, 18°—35°C; 1D	HSPs 95, 82, 70, 40, 32	—	135b, 213b
Tilapia mossambica, *T. nilotica* (ovary cells)	15 min, 31° → 40°C plus recov.; 1D, 2D	HSPs 87, 70, 60, 44, 27	—	73a
Salmo gairdneri (rainbow trout)				
Embryo fibroblast culture	2 h, 22° → 28—30°C; 1D	HSPs 100, 87, 70/68, 39, 27, 19	Ars	196, 269
Chinook salmon embryo cells	2 h, 20° → 24—28°C; 1D	HSPs 95, 84, 70, 65, 51, 46, 28	Zn, Cd	132a, 158

Amphibia				
Eurycea bislineata, Desmognatus ochrophaeus (salamander, brain, skin, stomach, liver)	1 h, heating from 15°C to 2°C below critical thermal maximum; 1D	HSPs 70 and 30	—	118
Rana catesbiana (bullfrog, lung epithelial and epidermal cells)	2 h, 20° → 32—36°C; 1D, 2D	HSPs 65, 25	—	186
Xenopus laevis (clawtoed frog)	(see Table 8.1 and Figure 8.3)			
Oocyte	20 min, 20° → 32—34°C; 1D, 2D	Only HSP 70	—	40
	30—60 min, 19° → 35°C; 1D, 2D	HSPs 160, 70—65, ~48, ~45, 30 plus lmw HSPs	Injection of Ap_4A	146
Early embryonic and developmental stages	20—80 min, 22° → 35—36°C; 1D	HSPs 87, 76, 70—68, 57, 42, 35	—	56, 159, 291
Kidney epithelial cells	2 h, 32—34°C; 2D	HSPs 75, 68, 59, 43, 38	—	186
	2 h, 22° → 35°C; 1D, 2D	HSPs 100, 87, 73, 70, 59, 54, 51, plus 16 lmw HSPs of 28—32 kDa	Ars	100, 160a
Erythroid cells	2 h, 33°C; 1D	HSPs 87, 70, 65, 62, 30—32	—	436
Hepatocytes, testis cells	12 h, 34°C; 1D, 2D	HSPs 85, 70, 30	Tissue isolation	438
Birds	Basic pattern of HSPs: HSPs 89, 70, 35, 25 (see Figure 2.6D)			
Chicken embryo				
Blastula	1 h, 44°C; 2D			
Epiblast region	1 h, 44°C; 2D	HSPs 89, 70, 24, 18	—	446
Hypoblast region	1 h, 44°C; 2D	Traces of HSP 70	—	446
Fibroblast culture	4 h, 45°C; 1D	HSPs 100, 70, 35, 25	Cu, Cd, Zn, Hg, Can, chelating drugs	221, 222
	2 h, 42.5—44°C; 1D	HSPs 85, 70, 25	—	409
	1 h, 45°C; 2D	HSPs 83, 68, 25	Ars	421
	1 h, 45°C; 1D	HSPs 95, 76, 22	Amino acid analogs	184
	30 min, 45—47°C; 1D	HSPs 98, 70, 35, 29	Ars, oligomycin, antimycin A, CCCP	210
Embryo cells	1 h, 44°C; 1D	HSPs 88, 71/72, 23	High extracellular pH, recovery from glucose deprivation	433
Myotube culture	2 h, 45°C; 1D	HSPs 81, 65, 25	—	20, 21
Lens culture	1.5 h, 45°C; 1D	HSPs 89, 70, 24	—	82
Neural retina culture	5 h, 43°C; 1D	HSPs 85, 70	Ars	68a

TABLE 2.1 (continued)
Survey on Systems Characterized for Induced Heat Shock Protein Synthesis*

Organism/cell	Hs conditions	HSPs induced (M_r in kDa)	Other inducers	Ref.
Chicken embryos (heart, liver, lung, kidney, gut, brain)	WBH, 2 h, 44°C; 1D	HSPs 85, 70, 25	—	409
Chicken adult				
Reticulocyte culture	150 min, 45°C; 1D, 2D	HSPs 90, 70, 24	—	16
	30 min, 44—45°C; 1D, 2D	HSP70	—	267a
Lymphoblastoid cells	1—5 h, 45°C; 1D, 2D	HSPs 89, 70, 23, 22	—	23
Quail *(Coturnix coturnix japonica)*				
Embryos	*In ovo* hyperthermia, 1—2 h, 44—45°C; 1D, 2D	HSPs 88, 82, 64, 25	—	16
Myotube culture	2 h, 44°C; 1D, 2D	HSPs 88, 82, 64, 55	Ars, Cu, Zn	15
Myoblast culture	1 h, 45°C; 1D, 2D	HSPs 94, 88, 82, 64, 25	—	13
Adults, red blood cells	WBH, 1—3 h, 43°C; 1D, 2D	HSPs 90, 70, 26	—	105
Adults, red blood cell cultures	1 h, 43°C; 1D, 2D	HSPs 90, 70, 26	—	105
Rodents	Basic pattern of HSPs: HSPs 110, 90, 73, 72, 32, 28, plus GRPs 94, 78			
Rat				
Embryos	30 min, 37° → 42—43°C; 2D	HSPs 82, 78, 69, 39, 34, 33.5, 31.5, 28	—	259
Brain cortex	30 min, WBH at 42°C; 2D	HSPs 110, 93, 70/68 complex	Recov. from ischemia	110
Liver, kidney, spleen	15 min, WBH at 42°C; 1D, 2D	HSPs 100, 85, 71, 70	Ischemia of liver, hepatectomy	130, 130a
Liver	WHB by administration of amphetamine (41.3°C)	HSPs 89, 70	Recovery from ischemia	65
Brain, heart, liver, lung, spleen, bladder, adrenals, kidney, thymus, pancreas, testis, muscle, aorta, fat, salivary gland, tongue, bone, esophagus, skin	WBH 15 min, 42.5°C (rectal) plus 2—3 h recovery; 2D	HSP71	Wounding	96
Liver cell culture	1 h, 37° → 42°C; 1D	HSPs 110, 89, 77, 68	Cd	71

Cerebellar neurons	20 min, 37° → 45°C plus recovery; 2D	HSPs 65(=70!), 47, 18	—	432
Hepatocytes, epithelial cells, Morris hepatoma cells	10—60 min, 37° → 43°C; 1D	HSPs 107, 89, 70, 68, 27	Ars, Eth	207, 208, 209
Thymocytes	1—4 h, 37° → 42°C; 2D	Main HSPs 115, 89, 73, 70, 68, totally 68 HSPs	—	242, 243
Lens culture (also lung, muscle, brain, liver)	30 min, 37° → 43—45°C plus recovery; 1D, 2D	HSPs 85, 73, 71, 32, 16	Ars	106
Embryo fibroblasts (Rat-1 cells)	1.5 h, 37° →42°C or 30 min, → 45°C plus recovery; 1D, 2D	HSPs 110, 90, 73, 72, 32, 28, plus GRPs 100, 80	Ars, AzC	263, 429
Cerebellum cells	10 min, 37° → 45°C plus recovery; 2D	HSP71	Cd	311
Myoblast	20 min, 37° → 45°C plus recovery; 1D, 2D	HSPs 130, 74, 72, 25, plus GRP 100	Ars	189
Heart tissue	*In situ* hs 41 min, 37° → 42°C; *in vitro* translation; 2D	HSP71, two isoforms	Ischemia	155
L 132 cells	1 h, 37° → 42°C	HSPs 89, 68	Cd, Zn, Can	71
Mammary carcinoma cells	4—5 h, gradual heating 37° → 42°C or 20 min, → 45°C plus recovery; 1D	HSPs 112, 90, 70, 22	—	389
Morris hepatoma cells	1—3 h, 37° → 43°C; 1D	HSPs 107, 89, 70, 68, 27	—	205
Reuber hepatoma cells	30 min, 37° → 43°C plus recovery; 1D, 2D	HSPs 100, 84, 70, 68, 65, 28	—	396b, 401, 434
Gerbil				
Fibroma cells	5 h, 37° → 42°C; 1D	HSPs 110, 90, 73, 72, plus GRPs 100, 80	AzC	429
Mouse				
Granulosa cells from Grafian follicles	20—40 min, 37° → 43°C; 2D	HSPs 110, 89, 68, 22	—	94b
Embryo fibroblasts, teratocarcinoma cells	10 min, 37° → 44°C plus recovery; 1D	HSPs 105, 89, 70, 68	—	266
Leg muscle (anterior tibialis)	Microwave heating in *situ* to 43°C then 4 h, 37°C; 2D	HSPs 89, 68, 66	—	378
Embryo blastocysts	10 min. 37° → 44°C; 2D	HSPs 89, 68, 66(?)	—	153, 266
Embryo fibroblasts and SV40 transformed MEF	1—4 h, 37° → 43°—45°C; 2D	HSPs 85, 70	—	303, 304

TABLE 2.1 (continued)
Survey on Systems Characterized for Induced Heat Shock Protein Synthesis*

Organism/cell	Hs conditions	HSPs induced (M_r in kDa)	Other inducers	Ref.
Neuroblastoma cells	30—60 min, 37° → 43°C plus recovery; 1D	HSPs 100, 84, 70, 68	—	396b,c 434, 435
Lymphoid cells (10 cell lines)	10 min, 37° → 42—44°C plus recovery; 1D	HSPs 89, 70, 58	Ars	18
Squamous carcinoma	Hs *in situ* 15 min, → 43°C, labeling at 37°C in *vitro* 1D, 2D	HSPs 110, 87, 70, 68	—	226
B16 melanoma cells	30 min, 37° → 43°C; 1D	HSPs 110, 89, 70, 68	—	5
Leukemia cells	2 h, 37° → 42°C; 1D	HSPs 110, 90, 70	—	63
Mammary carcinoma cells	12 min, 37° → 45°C plus recovery; 1D, 2D	HSPs 110, 89, 68, 66, GRP 78	—	378
Mastocytoma cells	4 h, 37° → 42°C; 1D, 2D	HSPs 100, 85, 69, 68, 23	—	301
L cells	5 min, 37° → 45°C plus recovery; 1D	HSPs 120, 95, 93, 76, 75	Can	184
	10 min, 37° → 44°C plus recovery; 1D, 2D	HSPs 89, 70, 68	—	235
	2 h, 37° → 41.5°C; 1D	HSPs 85, 70	—	212
Rabbit				
Blastocysts	2 h, 37° → 43°C; 1D	HSPs 95, 70, 28	Ars, injury	160
Brain	WBH by treatment with LSD amphetamine or bacteriae pyrogens (42°C); 1D	HSPs 95, 74	—	57, 80, 87, 129
Lymphocytes	1 h, 37° → 41—43°C; 1D, 2D	HSPs 110, 100, 90, 70, 65, 26	—	331a
Retina culture	30 min, 37° → 43°C; 2D	HSP74 and hmw HSPs	—	81
Syrian hamster fibroblasts	1 h, 36.5° → 44°C; 1D, 2D	HSPs 64, 62, 25	—	14
Chinese hamster				
Ovary fibroblasts (CHO cells)	12 min, 37° → 45°C plus recovery; 1D, 2D	HSPs 110, 89, 68, 66	—	376
	5 h, 37° → 40°C or 20 min, 45°C plus recovery; 1D	HSPs 107, 89, 70, 68	—	211, 321
	5 h, 37° → 42°C; 1D	HSPs 110, 90, 73, 72, plus GRPs 94 and 78	AzC	429
	1—4 h, 41°C or 6 min, 37° → 46°C plus recovery; 1D	HSPs 97, 87, 70, 31, 26	Can, AzC, N_2, Eth	225, 227

Baby hamster, Kidney cells	5 min, 37° → 45°C plus recovery; 1D	HSPs 120, 95, 93, 76	Can	184
	30 min, 37° → 45°C plus recovery; 1D	HSPs 110, 90, 73, 72	—	263
	5 h, 37° → 42°C; 1D	HSPs 110, 90, 73, 72, plus GRPs 100 and 80	AzC	429
Calf				
Chondrocyte culture	40 min, 37° → 45°C	HSPs 110, 70	—	237
Dog				
Heart, perfused	1 h, 37° → 42°C; 2D	HSP71 (5 isoforms)	—	204
Monkey				
African green monkey kidney cell culture	90 min, 43.5°C; 1D	HSPs 92, 70/72	—	241
Human	Basic pattern of HSPs: HSPs 110, 90, 73, 72 plus GRPs 94, 78 (see Figure 2.6C)			
Polymorphonuclear leukocyte culture	1 h, 37° → 42°C; 1D	HSPs 85, 70	—	119
Peripheral blood monocytes	20 min, 37° → 43—45°C plus 2 h recovery; 1D, 2D	HSPs 83, 70, 48—50, 27, plus basic lmw HSPs	Synthesis enhanced by vitamin D_3	319b
Erythroid cells K562	37° → 43°C; 2D	HSP70 isoforms	Haemin-induced differentiation	358
Lymphocytes	20 min, 37° → 45°C plus recovery; 2D	HSPs 69, 66, 57, 52, 49, 47, 45, 24, 20, 17	—	432, 331a
Lymphoblastoid cell culture	10—30 min, 37° → 45°C; 2D	HSPs 110, 90, 70 (3 isoforms)	Iodoacetamide, valine analog	5
Lymphoid cells	10 min, 37° → 42—44°C plus recovery; 1D	HSPs 89, 70, 68	—	18
Melanoma cells	3—21 h, 37° → 40—42°C; 1D, 2D	HSPs 86, 70—72, 26	—	107, 124
	30—60 min, 37° → 43°C; 1D	HSPs 100, 90, 72	Zn, Cu, Cd, Ars, pCMB, Can, AzC, disulfiram	66
293 cells	40 min, 37° → 45°C; 1D	HSPs 110, 90, 73, 72	—	263
Brain tumor cells, skin	16 h, 37° → 42°C; 2D	HSP70 isoforms	—	328
Fibroblast culture	10 min, 37° → 46°C plus recovery; 1D	HSPs 116, 85, 70	—	392
Chondrocyte culture	Osteoarthritis patients	HSP70	—	200
Epidermoid carcinoma cells	1 h, 36.5° → 44°C; 1D, 2D	HSPs 64, 62, 25	—	14

TABLE 2.1 (continued)
Survey on Systems Characterized for Induced Heat Shock Protein Synthesis*

Organism/cell	Hs conditions	HSPs induced (M_r in kDa)	Other inducers	Ref.
HeLa cells	15 min, 45°C or 2 h, 42°C; 2D	HSPs 105, 85, 73, 70, plus GRPs 92, 78	Eth, Ars, Pur, Can, Cu	157
	10 min, 45°C; 1D, 2D	HSPs 100, 72—74, 37	Ars	61, 361
	3 h, 42°C; 1D, 2D	HSPs 80, 70(5 isoforms), 27	—	164
	2 h, 42°C; 1D, 2D	HSPs 110, 90, 72/73, plus GRPs 100, 80	AzC	429
	1.5 h, 43°C; 2D	HSPs 110, 90, 70, 28, plus GRPs 100, 80	—	263, 428
	20 min, 45°C; 2D	HSPs 110, 90, 72, 73, 26, 24, plus GRPs 100, 75	—	115
Normal and malignant lung cells	1 h, 37° → 43°C and 1 h, 37° → 41°C, respectively; 1D	HSPs 100, 90, 70	—	393
Endometrium, decidua	2 h, 37° → 41°C; 1D, 2D	HSPs 94, 88, 70, 27	—	332
Phycophyta (algae)				
Euglena gracilis	1 h, 25° → 37°C; 1D, 2D	HSPs 94, 80, 70, 67, 65, 15	—	3a, 289
Nitzschia alba (diatom)	1 h, 26° → 35°C; 2D	HSPs 95, 88, 85, 81, 70, 61, 38, 30, 24	—	204a
Acetabularia mediterranea	2 h, 34°C; 1D	HSP70 (coded in chloroplasts?)	—	194
Chlamydomonas reinhardi	90 min, 40—42°C; 1D	HSPs 70, 36, 22	Light	193, 290a, 410b
Chlorella pyrenoidosa	2 h, 25° → 37°C; 1D, 2D	HSPs 100, 93, 83, 71, 22, 16.6	—	289
Chlorella protothecoides	1 h, 25° → 35—37°C; 1D	HSPs 96, 91, 80, 74, 70, 68, 66, 44, 19	—	396a
Volvox carteri	1 h, 32° → 42.5°C; 1D	HSPs 97, 88, 77, 73, 67, 60	—	192
Plants	Basic patterns of HSPs: HSPs 95, 80, 70—75 plus many lmw HSPs (Figure 2.7)			
Monocotyledonous plants				
Lilium longiflorum germinating pollen, leaf disks	10 min, 25° → 45°C plus recovery or 30 min, 39°C; 1D, 2D	HSPs 94, 80, 75, 65, 41, 28	—	167, 401a
Sorghum bicolor (seedling)	2 h, 35° → 45°C; 1D	HSPs 94, 80, 70, 18	—	167a, 198b, 305
Gladiolus × *gandavensis*, cormels	25° → 40°C; 1D, 2D	HSPs 102, 100, 88, 80, 75, 72, 68, plus several HSPs 14—22	—	133b
Tradescantia paludosa, shoot apex	2 h, 25° → 40°C; 2D	HSPs 94, 82, 72, 21 plus 10 lmw HSPs	—	441

Saccharum officinarum (sugarcane, cultured cells)	2 h, 25° → 36—38°C; 1D, 2D	HSPs 97, 90, 80, 70, 18.6, 18.4	—	263b
Pennisetum americanum (Pearl millet, seedling)	3 h, 28° → 45—47°C; 1D, 2D	HSPs 92, 80, 70, plus 17 lmw HSPs	—	167a, 187, 239
Panicum miliaceum	3 h, 25° → 45°C; 2D	HSPs 70 plus 23 lmw HSPs	—	239
Hordeum vulgare (barley)				
Aleurone layer	3 h, 25° → 40°C; 1D	HSPs 105, 101, 87, 84, 71—76, 34, plus 17—19 kDa complex	—	33
Root	1 h, 25° → 40°C; 1D	HSPs 103, 99, 83, 80, 75, 70, 62, 59, 54, 48, 40, 32, plus HSPs 16—17	—	280
Coleoptile	1 h, 25° → 40°C; 1D	HSPs 103, 99, 83, 80, 75, 70, plus HSPs 16—17	—	280
Seedling, root, coleoptile	4 h, 25° → 34—37°C; 1D, 2D	HSPs 100, 94, 85, 70, 47, 22	—	240
Triticum aestivum (wheat)				
Seedling	3 h, 25° → 40°C; 2D	HSP70 plus 12 lmw HSPs	—	239
	5 h, 25° → 41°C; 2D	HSPs 110, 95, 80, 75, plus 25 lmw HSPs	—	448
Root, coleoptile	1 h, 25° → 40°C; 1D	HSPs 103, 99, 83, 80, 75, 70, 59, 55, 48, 40, 32, plus 1mw HSPs 16—17	—	280
Leaves	1 h, 22° → 34—37°C; 1D, 2D	HSPs 94, 92, 85, 83, 70, 62, 54, 42, 33, 26, 22, 17, 16	—	198c
Oryza sativa (rice, seedling)	3 h, 25° → 40°C; 2D	HSP70 plus 15 lmw HSPs	—	239, 263a
Zea mays (maize)				
Seedling	3 h, 28° → 40—45°C; 1D, 2D	HSPs 92, 80, 70, plus 24 lmw HSPs	—	187, 239
	2 h, 25° → 41—45°C; 1D, 2D	HSPs 108, 89, 85, 81, 78, 74, 72, 60, 18	—	325, 359
	1—2 h, 25° → 40°C; 1D	HSPs 94, 83, 80, 72, 25, 21, plus 16—19 kDa complex	—	2
Seedling (plumule, radicle, mesocotyle, leaf)	1 h, 25° → 42°C; 1D, 2D	HSPs 108, 89, 84, 76, 73, 23, and 18	—	29, 30, 85

TABLE 2.1 (continued)
Survey on Systems Characterized for Induced Heat Shock Protein Synthesis*

Organism/cell	Hs conditions	HSPs induced (M_r in kDa)	Other inducers	Ref.
Seedlings of 10 cultivars	2 h, 25° → 42°C; 1D	HSPs 93, 89, 84, 76, 73, 22, and 18 kDa complex	—	442
Coleoptile	Slow heating to 41°C; 2D	HSPs 82, 76, 23.4, plus 15 HSPs 15—21 (see Figure 2.7A)	—	289
Immature, developing pollen	2.5 h, 25° → 38°C; 1D	HSPs, 102, 84, 82, 74, 56, 46, 18; no HSPs in mature pollen	—	129a
Dicotyledonous plants				
Glycine max (soybean)				
Seedling	2 h, 25° → 40—43°C; 1D	HSPs 94, 80, 75, 72, 71, 19, plus 16 kDa complex	—	2
	3 h, 25° → 40°C; 1D, 2D	HSPs 95, 80, 70, plus 27 lmw HSPs	Ars, Cd	188, 239
Suspension culture	1 h, 25° → 40°C; 1D	HSPs 94, 80, 75, 37, 24, plus HSPs 16—19	—	2, 28
Green cell culture	1—2 h, 25° → 37—40°C; 1D, 2D	HSPs 92, 84, 70, 27 plus 25 lmw HSPs	—	405
Phaseolus vulgaris (bean, devel. seeds)	6 h, 43°C; 1D	HSPs 95, 70/68, plus lmw HSPs	—	79
Pisum sativum (pea, seedling)	3 h, 25° → 35—40°C; 1D, 2D	HSPs 93, 80, 70, plus 22 lmw HSPs	—	187, 239
Cajanus cajan (pigeon pea, roots)	3 h, 25° → 40—45°C; 1D	HSPs 81, 70, 60, 58, 55, 53, 48, 37 plus lmw HSPs	—	192a
Vigna unguiculata (cowpea, leaves)	4 h, 25° → 40°C; 1D	HSPs 95, 80, 70—75, 18	—	102
Vigna sinensis	3 h, 25° → 35—40°C; 1D	HSPs 96, 90, 75 plus lmw HSPs	—	198b
Arabidopsis thaliana	4 h, 25° → 40°C; 1D, 2D	HSPs 100, 90, 83, 73, 23 plus HSPs 16—18	—	201a 440a

Brassica napus, B. campestris, B. juncea (hypocotyls, leaves, flower buds)	75 min, 23° → 40°C; 1D, 2D	HSPs 100, 78, 70 plus 10—15 HSPs 18—27	—	29a
Brassica oleraceae (broccoli)	1—2 h, 20° → 37—40°C; 1D, 2D	HSPs 90, 88, 86, 74, 69, 66, 47, 43, 42, 29, 27, 23, 21, 19, 18, 12	—	122
Peganum harmala, cell cultures	Slow heating to 40.5°C; 1D, 2D	HSPs 126, 95, 81, 72, 23, 19.5, plus 24 HSPs 15—18 (Figure 2.7b)	—	289
Lycopersicon peruvianum (Peruvian tomato)				
Suspension culture	1—3 h, 25° → 37—39°C; 1D, 2D	HSPs 95, 80, 75, 70, 36.5, plus 20 HSPs of 14—30 kDa (Figure 2.2)	Ars, Cd, Hg,A_zC, pFP	289, 296, 338
Protoplasts	1 h, 25° → 37° plus 2 h, at 39°C; 2D	Same as in suspension culture		288
Nicotiana tabacum (tobacco)				
Germinating pollen, leaves	30 min, 27° → 39°C plus recovery; 1D, 2D	HSPs 95, 85, 65 plus lmw HSPs	—	401a
Mesophyll protoplasts and 2 d old protoplast cultures	2 h, 25° → 40°C; 2D	HSPs 120, 100, 75, 26, 25, plus 10 lmw HSPs in the 18—20 kDa region	—	252
Suspension culture	2 h, 25° → 39—43°C; 1D	HSPs 110, 100, 90, 85.5, 82, 76, 63, 53, 37, plus lmw HSPs	—	28
Leaves	4 h, 25° → 40°C; 1D	HSPs 95, 80, 70—75, 18	—	103
Nicotiana plumbaginifolia (tobacco)				
Protoplasts	3 h, 25° → 40°C; 1D	HSPs 100, 96, 86, 72/74, 67, 50, 30, plus 18—26 complex	—	326
Gossypium hirsutum (cotton)				
Leaves	3 h, 25° → 40°C or growth at 40°C; 1D	HSPs 100, 94, 89, 75, 60, 58, 37, 21	—	62
Seedlings	3 h, 28° → 43°C; 1D, 2D	HSPs 92, 80, 70, plus 15 lmw HSPs	—	187

TABLE 2.1 (continued)
Survey on Systems Characterized for Induced Heat Shock Protein Synthesis*

Organism/cell	Hs conditions	HSPs induced (M_r in kDa)	Other inducers	Ref.
Daucus carota (carrot, callus, protoplasts, embryos, plantlets)	30—180 min, 25° → 38°C; 1D	HSPs 95, 88, 74 (complex), 45, 37 and multiple bands in the lmw region	—	317
Spinacia oleracea (Spinach, leaves)	2 h, 26° → 40°C; 1D	HSPs 107, 85, 74, 31, 27, 21, 19	—	149a
Helianthus annuus (sunflower, seedling)	3 h, 25° → 40°C; 2D	HSP70 plus 23 lmw HSPs	—	239
Petroselinum crispum (parsley, cell culture)	Slow heating to 40.5°C; 2D	HSPs 81, 71, 34, 22, 19, plus 15 HSPs 14-17	—	289, 417a
Digitalis lanata, cell culture	2 h, 25° → 40°C; 2D	HSPs 89, 78, 71, 40, 25, 21, plus 25 HSPs 15—19	—	289

TABLE 2.2
Survey of *Escherichia coli* Heat Shock and Related Proteins

No.	Position in 2-D gel (see Figure 2.3A)	MW[a]	Gene (map position)	Specific inducers[b]	Function, properties (see also text)	Ref.
1	H 94.0	87,000	lon(10)	NA, ST	ATP-dependent protease, DNA-stimulated ATPase homotetramer, no homology to any known HSP family	74a, 132, 136, 314
2	F 84.1	—	htpM	NA, Cd	—	133, 284
3	B 83.0	70,263	rpoD(67)		Control σ-factor of RNA polymerase	46, 144, 382
4	B 66.0	69,121	dnaK(0.3)	NA, Cd, HP, UV	Belongs to HSP70 family (see Figure 2.10) ATPase, 5′ nucleotidase, role for autorepression of hs response, involved in λ DNA replication and NA synthesis, required for growth and cell division under all temperature conditions	24, 47, 58b, 306, 337, 383, 451, 452
5	C 62.5	71,492	htpG(11.1)	NA, Cd	Homology to euk. HSP90 family (see Figure 2.8)	25
6	D 60.5	—	lysU(92.0)		lys-tRNA synthetase form II	398
7	B 56.5	62,883	groEL(93.5)	NA, UV	Homology to euk. HSP60 of mitchondria and chloroplasts (Figure 2.11), ATPase, associates with GroES to heptameric complex, role in phage assembly and NA synthesis	162, 162a, 251, 413
8	H 26.5	40,973	dnaJ(0.3)		λ DNA replication, NA synthesis	26, 213a, 301a
9	B 25.3	21,668	grpE	NA, Cd	DNA replication	5a, 230c
10	F 21.5	—	htpL		—	133, 284
11	G 21.0	—	htpO	NA	—	133, 284
12	C 15.4	10,670	groES(93.5)	NA, HP	Associates with GroEL to heptameric complex, ATPase, role in NA synthesis, and phage assembly	21b, 384, 413
13	C 14.7	—	htpE	NA, Cd, HP	—	133, 284
14	G 13.5	—	htpN	AC	—	133, 284
15	F 10.1	—	htpK		—	133, 284
16	F 33.4	32,381	rpoH (=htpR, ~75)		hs-Specific σ-factor of RNA polymerase, promoter consensus sequence (see Figure 7.2)	46, 145, 206, 386, 414, 445

[a] The relative mobilities in 2D-electrophoretic analyses define the Mr of the HSP as indicated by the α-numeric code in the second column (see Figure 2.3A). If complete sequence data are available the true molecular weights are given in the third column.

[b] Most of the *E. coli* HSPs are also induced by λ-phage infection (see Section 7.3), 4 to 10% ethanol and recovery from 200 μg/ml puromycin. More specific inducers are NA, nalidixic acid (172 μ*M*), Cd, cadmium chloride (600 μ*M*), HP, hydrogen peroxide (150 μ*M*), ST, isoleucin restriction connected with ppGpp synthesis, AC, 6-amino-7-chloro-5,8-dioxoquinoline (5 μg/ml), UV, irradiation with UV light. For reference see Krueger and Walker,[199] van Bogelen et al.[399]

Data summarized in References 133, 281—284 and 444.

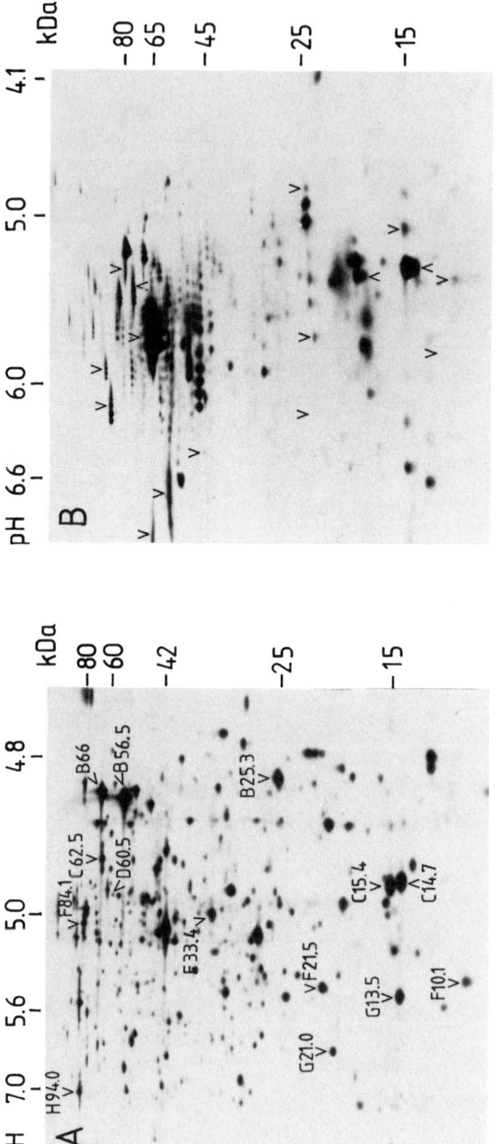

FIGURE 2.3. (top)

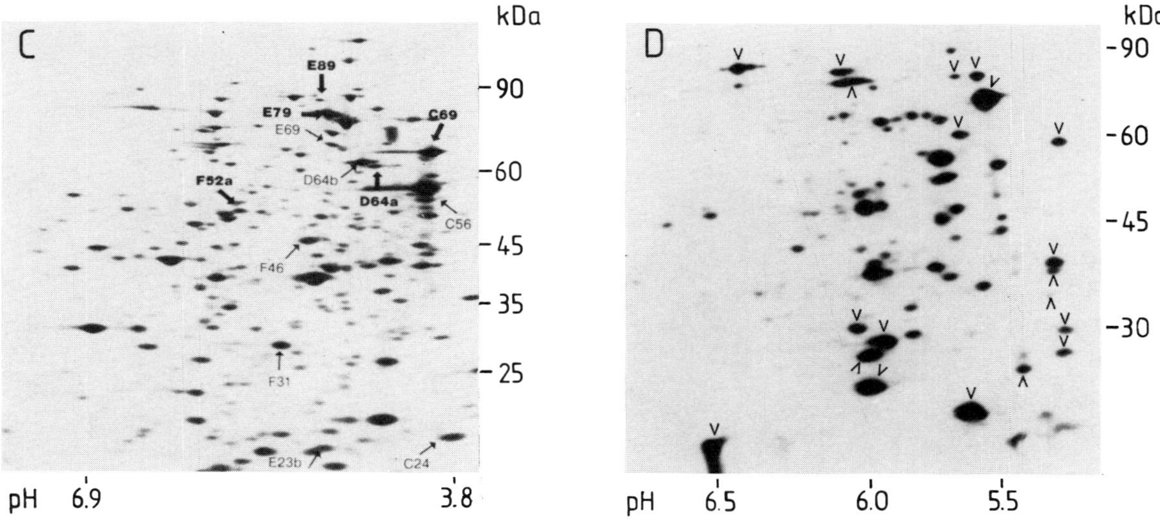

FIGURE 2.3. Heat shock-induced proteins in bacteria: *Escherichia coli* (A), *Synechococcus* sp. (B), *Salmonella typhimurium* (C), and *Bacillus subtilis* (D). Fluorographs and autoradiographs of proteins labeled under hs conditions (see Table 2.1) are shown. Positions of molecular weight standards are shown on the right margins. The pH gradients of the first-dimension gels are indicated at the upper and lower margins, respectively. HSPs are marked by open arrow heads (A, B, D) or arrows (C), and eventually identified by a binary code number. For *E. coli* (A) these refer to the code used in Table 2.2. (Reprinted from Tsuchido et al.[391] (A), Borbely et al.[53] (B), Morgan et al.[267] (C), and Richter and Hecker [329] (D). With permission.)

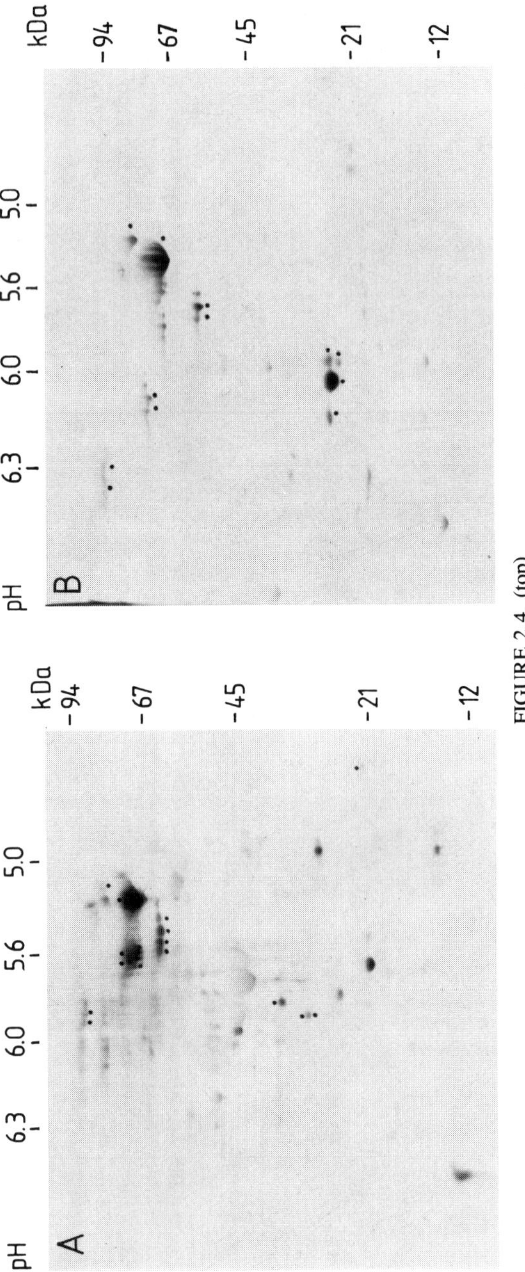

FIGURE 2.4. (top)

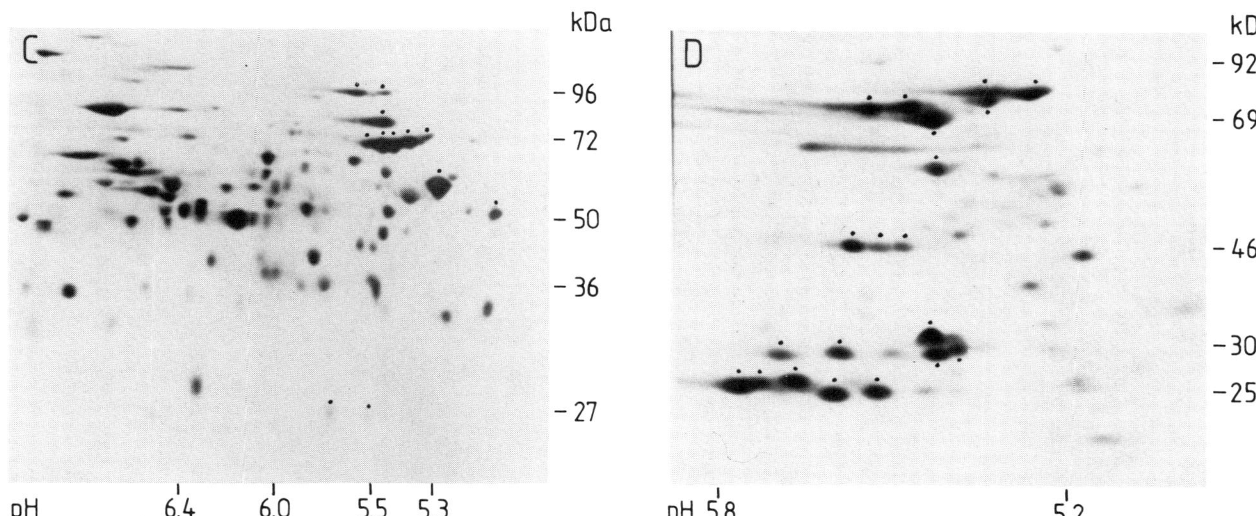

FIGURE 2.4. Heat shock-induced proteins in eukaryotic microorganisms: *Aspergillus nidulans* (A), *Claviceps purpurea* (B), *Achlya klebsiana* (C), and *Tetrahymena pyriformis* (D). For explanations see legend to Figure 2.3. HSPs are marked by dots. [Reprinted from Neumann et al.[289] (A, B), LeJohn and Braitwaite[219] (C), and Amaral et al.[3](D). With permission.]

Tetrahymena mitochondria[250,251] opened a series of investigations on similar proteins in chloroplasts and mitochondria of other organisms as well (see Section 2.3.3). In addition to the HSP60,[170a] *Neurospora crassa* mitochondria contain an HSP32 (Figure 2.15) that belongs to the HSP20 family.[289] It can be anticipated that similar proteins will also be found in mitochondria of other eukaryotes.

The HSP pattern in yeast includes two HSP90,[52a,123] a complex group of nine HSP70-type proteins (see Table 2.4, and Figures 2.5 and 2.10), a mitochondrial representative of the HSP60 family (Figure 2.11),[323a] a single small HSP of 26 kDa[53a,202,313,379a] (Figure 2.12), plus a group of intermediate-sized HSPs in the 35- to 60-kDa region. Among them are proteins expressed also in resting cells[172] and two glycolytic enzymes (see Table 2.6). HSP48 has enolase[173] and HSP46 has phosphoglycerate kinase activity.[316] Finally, there are a 118-kDa glycoprotein induced by hs, sulfur starvation, or cAMP deficiency[395b,396] and a 66.2-kDa STI1 gene induced by hs and canavanine. The latter is essential for growth at high temperatures but unrelated to any member of the HSP70 family. It functions presumably as a mediator of the hs response.[292a]

2.1.4 HSPs OF ANIMALS (FIGURE 2.6)

Examples of induced HSP synthesis in animals include *Drosophila* (Figure 2.6A), sea urchin (Figure 2.6B), human (HeLa) cells (Figure 2.6C), and chicken lymphocytes (Figure 2.6D). In *Drosophila*, the largest HSP with 83 kDa is expressed at 25°C as well as under hs conditions.[75] The HSP70 complex[64,92,415] contains the dominant, constitutively expressed HSCs 70 and 72, whose synthesis is reduced by the hs treatment and the newly formed HSP70 and 68. Of the group of four small HSPs only the two acidic ones are represented in Figure 2.6A (HSPs 23 and 22). Due to their basicity, HSPs 27 and 26 moved out of the gel. Three further small HSP-coding genes were sequenced, but their protein products were not yet identified. These are HSP26.5[19] (gene 1), HSP12.6 coded by the intron-containing gene 2,[310] and HSP19[309] (gene 3). Together with the other four small HSP-coding genes they are part of the locus 67B of chromosome 3 of *Drosophila melanogaster* (Figure 3.3). Another minor HSP of *Drosophila* with 34 kDa, frequently observed in one-dimensional gels (see references given in Table 1.1), is also highly basic.[64] Besides other minor HSPs of intermediate size (44 to 66 kDa), it can be detected by two-dimensional electrophoresis among the *in vitro* translation products using a rabbit reticulocyte lysate and poly(A)$^+$ RNA from hs salivary glands.[64] The gene coding for HSP34 was not yet identified.

In embryos of sea urchin *(Strongylocentrotus purpuratus)*, small HSPs are essentially lacking. The picture is dominated by HSPs 90, 70, 50, 40, and 38 (Figure 2.6B). The hs-induced patterns of protein synthesis in different tissues of the claw-toed frog *Xenopus laevis* are documented in Table 8.1. Dominant in A6 kidney epithelial cells (Figure 8.3) are HSPs 100, 87, 62, five representatives of the HSP68 — 73 complex, and 15 proteins in the 30-kDa region.[100] Minor HSPs with 28, 37, and 40 to 43 kDa are also observed. Many further references as well as details on the tissue specificity are given in Section 8.3.

The HSP pattern of vertebrate cells is very similar (Figures 2.6C and D). Synthesis of the two constitutively expressed HSP90 is enhanced by hs, whereas in the HSP70 region generally two types of proteins can be discriminated: a slightly larger, constitutive form (72 to 73 kDa) with increased synthesis during hs and one or two newly formed HSP70 and 71. The small HSP of 25 to 27 kDa in mammals and 24 kDa in chicken is presumably coded by a single gene, i.e., multiple forms arise by posttranslational phosphorylation (Table 2.5). Because of its extremely low or lacking methionine content, the small HSP of vertebrates was frequently not detected when methionine was used for labeling. The vertebrate pattern of stess proteins is completed by the minor nucleolar HSP110 and the two glucose-regulated proteins GRPs95 and 78 (see also Figure 4.2). However, similar to other organisms, many more minor HSPs were detected, e.g., by using extensive two-dimensional analysis of proteins from rat thymocytes.[242,243]

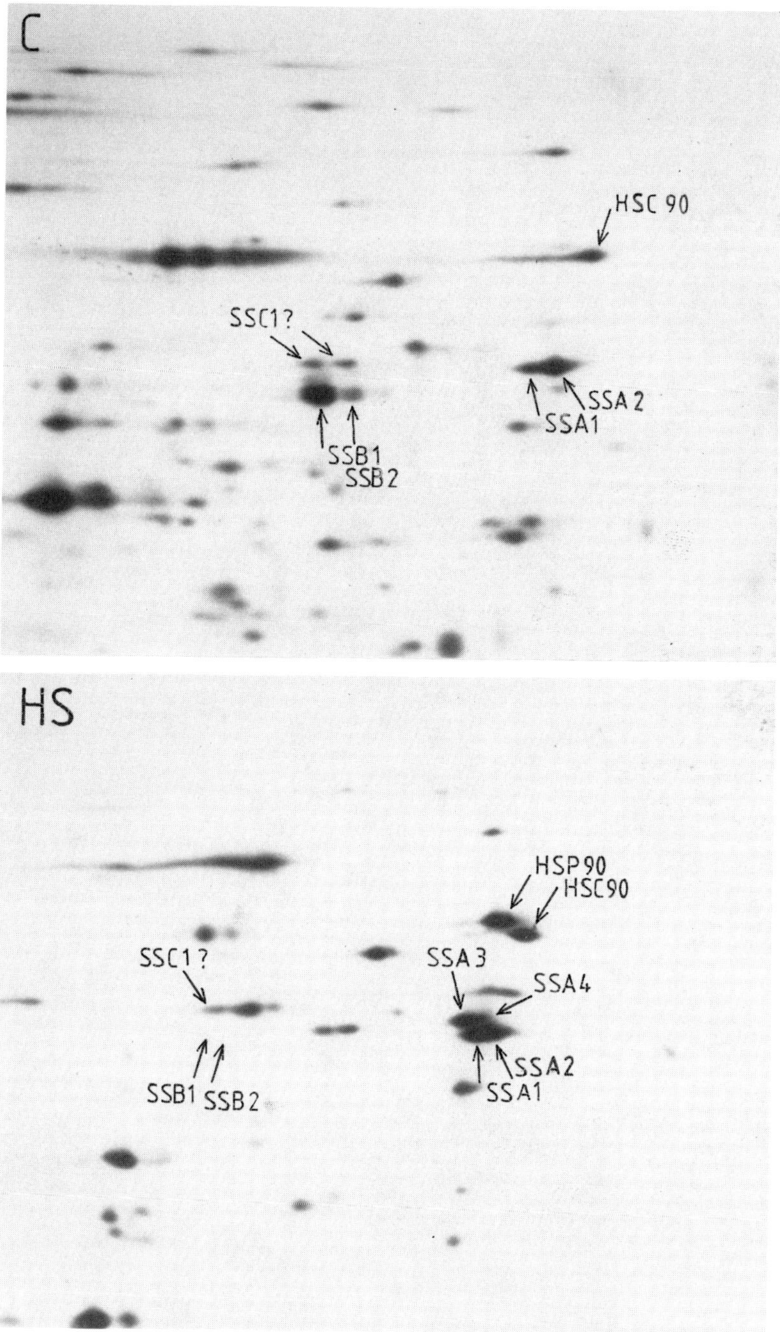

FIGURE 2.5. Two-dimensional separation of yeast proteins labeled at 23°C (C) and after 15 min heat shock at 39°C (HS) respectively. Only parts of the autoradiographs with the HSPs70 and 90 are shown. The members of the HSP70 family are marked in correspondence to Table 2.4. It is evident that, similar to other organisms, HSP90 is accompanied by a HSC90. (From Werner-Washburne, M., Stone, D. E., and Craig, E. A., *Mol. Cell Biol.*, 7, 2568, 1987. With permission.)

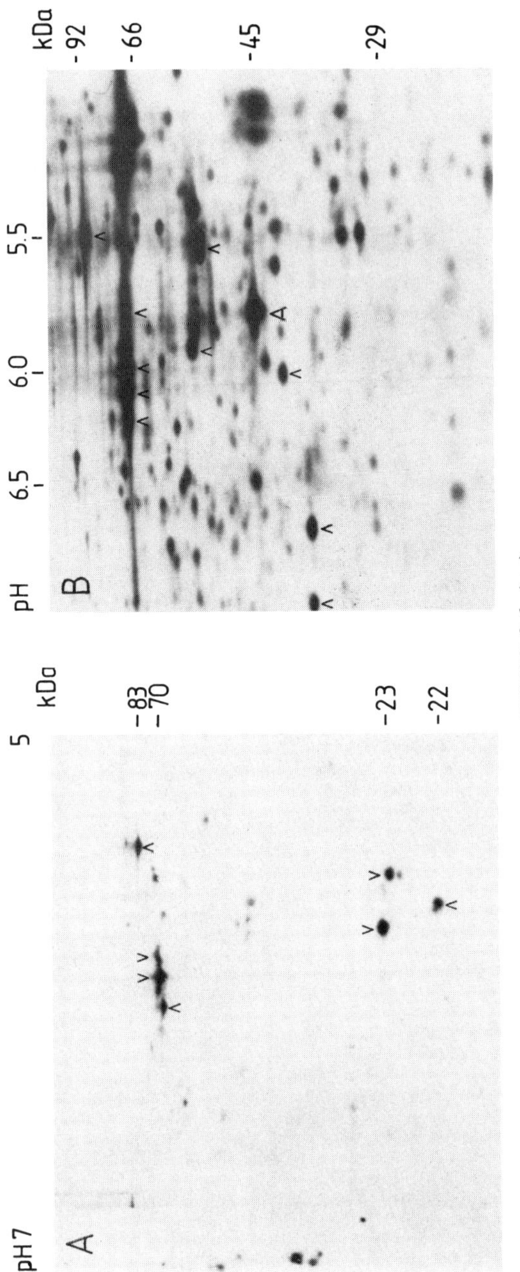

FIGURE 2.6. (top)

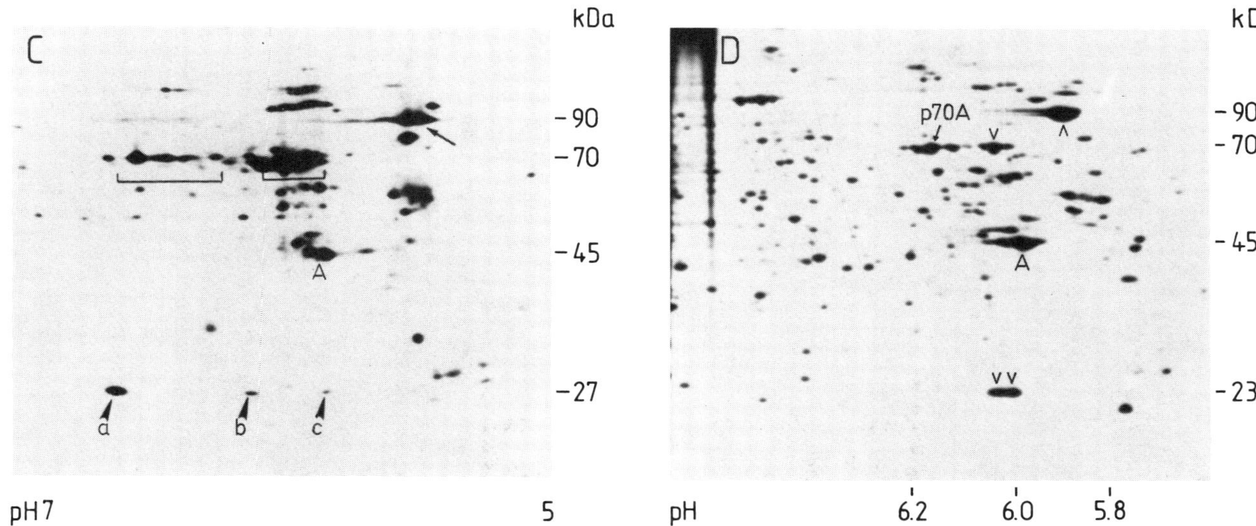

FIGURE 2.6. Heat shock (hs)-induced proteins in animal cells: *Drosophila melanogaster* (A), *Strongylocentrotus purpuratus* (B), HeLa cells (C) and chicken lymphocytes (D). For explanations, see legend to Figure 2.3. Heat shock proteins are marked by open arrowheads (A, B, D). In part C, the arrow points to HSP90 and the brackets mark the HSP70 complex, whereas closed arrowheads indicate positions of HSP27 (a) and its phosphorylated derivatives (b, c). A in parts B to D marks position of actin. (From Palter et al.[307] (A), Bedard and Brandhorst[32] (B), Arrigo and Welch [10] (C) and Morimoto and Fodor[267a] (D). With permission.)

The HSP25-27 of mammalian cells deserves special note as an indicator of malignant transformation:

1. Under normal temperature conditions, it was found to accumulate up to 1% of the total protein content in various mouse tumor cell cultures, when growth is retarded.[33a,130b]
2. Depending on the cell type, its phosphorylation level increases in response to growth factors, tumor promoters, and tumor necrosis factor alpha,[8a,10,330a,426] as well as under conditions of heat shock or arsenite stress.[78a,211a,330a]
3. In human breast cancer, an estrogen-induced protein with M_r 24 kDa evidently represents the HSP27. It may also be identical with the p29 found as a component of the estrogen receptor complex.[129b,247b,389a]
4. Similar to mouse tumor lines, HSP27 and its phosphorylated derivatives are prominent polypeptide markers of human acute lymphoblastic leukemia cells.[374]

2.1.5. HSPs OF PLANTS (FIGURE 2.7)

Besides the large HSPs with M_r 70 and 80 kDa plus eventually a short-lived HSP95, the protein pattern synthesized in hs-induced plant cells is dominated by the highly complex group of small HSPs as shown for maize coleoptiles (*Zea mays,* Figure 2.7A), soybean hypocotyl (*Glycine max,* Figure 2.7B), tomato ovaries (*Lycopersicon esculentum,* Figure 2.7C), and cell cutures of *Peganum harmala* (Figure 2.7D). In most cases 20 to 30 newly formed proteins are detected in this region. The major representatives are closely related proteins in the 17-kDa region.

Most of the small plant HSPs belong to a HSP20 subfamily forming cytoplasmic multimeric complexes with changing aggregation tendency during hs and recovery (see Sections 2.3.4 and 16.7). But few of them are constituents of chloroplasts, e.g., HSP21 and HSP22 in tomato (nos. 17 and 18 in Figure 2.2; for details see Section 16.5). In addition, plants contain two types of HSP60 in chloroplasts and mitochondria, respectively,[161a,319c] a DnaK-like HSP70 in mitochondria,[289,289a] and probably three different constitutive members of the HSP70 family in chloroplasts.[240a] Reports on whole sets of organelle-encoded HSPs in tropical species,[191a,198b] e.g., in pigeon pea (*Cajanus cajan*) and cowpea (*Vigna sinensis*), remain to be confirmed, especially by cloning and analysis of the corresponding hs genes in chloroplast or mitochondrial DNA. In all cases where the respective information is available (see Section 16.5), the organellar stress proteins are encoded in the nucleus and translated in the cytoplasm as precursor proteins.

2.2. NOMENCLATURE OF HEAT SHOCK PROTEINS

Initially their definition appeared very simple because it was mainly based on the results with a single organism, *Drosophila melanogaster.*[295] HSPs are newly formed proteins observed after hs induction or treatment with a number of chemical stressors. Their expression depends on the transcription of hs genes.

With the rapidly increasing data from other systems, including bacteria, the situation became much more complex. It turned out that frequently HSPs are coded by oligo- or multigene families combining various members with divergent regulatory behavior. The ''necessity'' for cells to express HSPs or their related proteins also under non-hs conditions (see details in Chapter 8) is brought about by two means: (1) complex promoters connected with HSP coding genes may harbor regulatory elements for hs induction, developmental, or cell cycle control as well as for constitutive expression (see examples given in Figure 6.6 and discussion in Sections 8.1 to 8.5). (2) Different members of a gene family are responsive to different regulatory demands. Derived from the particular situation in *Dro-*

sophila the term heat shock cognate (hsc) genes was introduced[92] to characterize genes with constitutive expression of proteins closely related to HSP70. These are HSC70 and HSC72[307] (Table 2.3) whose synthesis is depressed under hs, but whose abundance always surpasses that of HSP70.

Although similar examples were reported in other organisms as well, e.g., for the HSP70 families in yeast (Table 2.4) and vertebrates cloning and sequencing of a number of hs genes with high constitutive and hs-maintained or enhanced expression showed, in several cases, that they contain complex promoters with hs elements incorporated. Hence, they are true HSP-coding genes (see promoters 2a, 5, and 27 in Figure 6.6). From the biological point of view, it can be anticipated that this mode of maintenance transcription during hs or other stress periods may also apply to a number of "control" genes whose continued expression is essential for survival. For the sake of convenience, we maintained the term hsc (HSC) to indicate *constitutively expressed* hs or related genes/proteins belonging to one of the HSP families (see below). On the other hand, due to a posttranscriptional block of intron splicing (see Section 9.5) even the expression of "classical" hs genes, e.g., of the hsp83 gene of *Drosophila*, can be depressed at temperatures $\geq$ 37°C, when HSP70 synthesis is just optimum. This situation probably also applies to other intron-containing hs genes (see Figure 3.1 and Section 9.5).

A reasonable escape from the difficulties with an appropriate definition follows the nomenclature originally derived for *E. coli*:[284] in this organism all HSPs are constitutive proteins, whose synthesis is transiently enhanced under hs conditions, because their genes belong to a common regulatory unit (regulon). The coordinating regulatory gene (rpoH) codes for a hs-specific σ-subunit of RNA polymerase (see Section 7.1). Following this, members of the eukaryotic "hs regulon" are genes whose transcription is triggered, enhanced, or maintained due to the presence of a hs promoter element recognized by the hs transcription factor (see Chapters 6 and 7).

2.3 THE HSP FAMILIES

Following a suggestion of the participants of a Cold Spring Harbor Meeting,[308a] it appears reasonable to order hs and related proteins into distinct classes charactarized by sequence homology and possibly functional relatedness. Starting with the seminal review by Ashburner and Bonner[12] in 1979, many reviews and books document the rapid progress in this field, e.g., those by Schlesinger et al. (1982),[342] Neidhardt et al. (1984),[284] Nover (1984),[295] Atkinson and Walden (1985),[16a] Burdon (1986),[60a] Lindquist (1986),[230a] Subjeck and Shyy (1986),[378a] Pelham (1986, 1988, and 1989),[312-312b] Bienz and Pelham (1987),[41] Lindquist and Craig (1988),[230b] Schöffl et al. (1988),[346a] Nagao and Key (1989),[275a] Neumann et al. (1989),[289] Pardue et al. (1989),[308b] Tomasovic (1989),[389a] and Georgopoulos et al. (1990).[132b]

2.3.1. THE HSP90 FAMILY

Representatives with known sequence are compiled in Figure 2.8. This family includes abundant cytoplasmic proteins with increased synthesis under hs conditions, e.g., the HSP90 of human,[165a,324,443a] mouse,[264] chicken,[42] and yeast,[123] the *Drosophila* HSP83[45,152] as well as the plant HSP80.[359a] In *E. coli,* the homologous protein is the htpG gene product.[25] Other members of the HSP90 family are the larger of the two glucose-regulated proteins (GRPs) (see Chapter 4), i.e., GRP94 of chicken and mouse,[201,244,364] so well as an abundant, constitutively expressed surface glycoprotein of the protozoan parasite *Trypanosoma cruzi.*[113] Sequence homology to representatives of the HSP90 family were also reported for two tumor-specific transplantation antigens of mouse cells.[395] There are at least two independently regulated forms of HSP90 in vertebrates[27,165a,357] and yeast.[52a,123]

Originally, the conservation of structure between members of the HSP90 family was

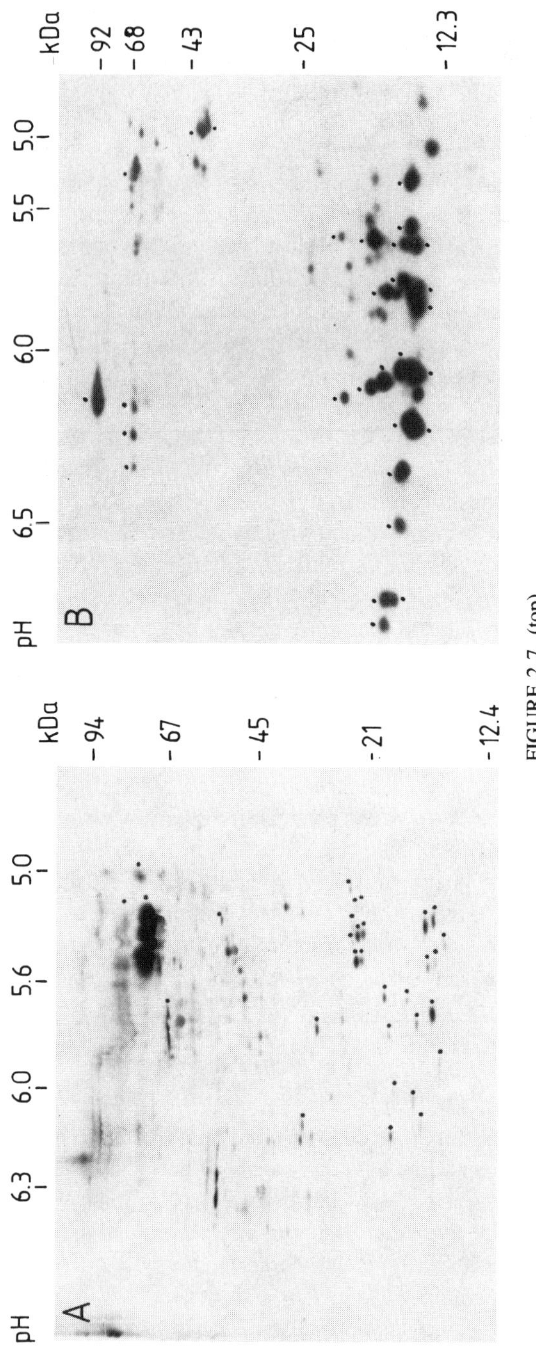

FIGURE 2.7. (top)

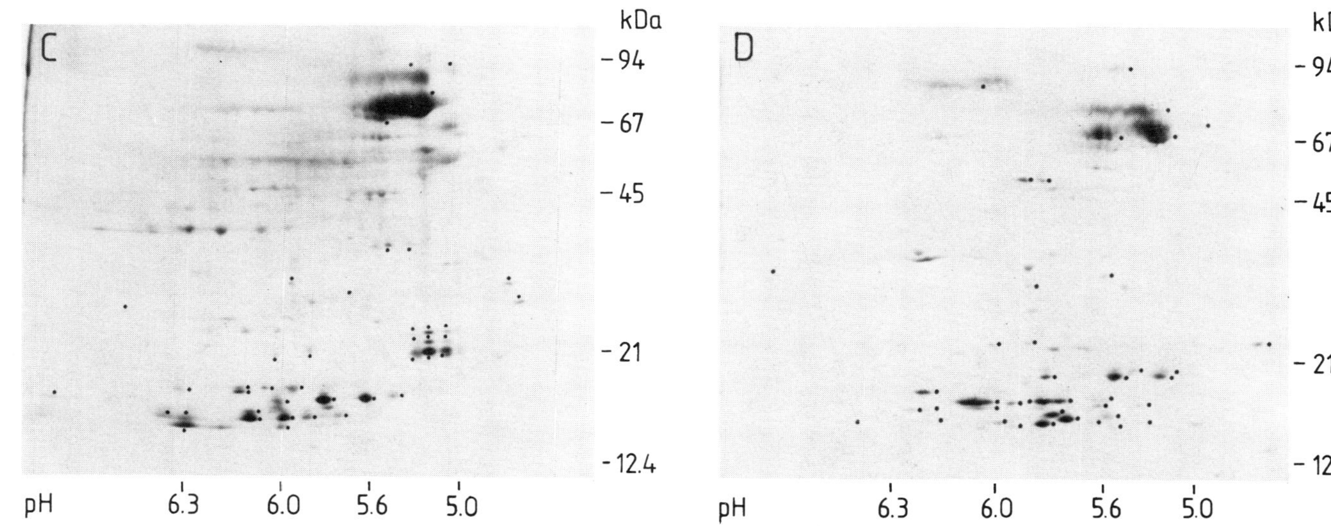

FIGURE 2.7. Heat shock-induced proteins in plants: *Zea mays* (corn) coleoptiles (A), *Glycine max* (soybean) hypocotyl (B), *Lycopersicon esculentum* (tomato) ovary (C), and *Peganum harmala* cell cultures (D). For explanations see legend to Figure 2.3. HSPs are marked by dots. [From Neumann et al.[289] (A,C,D) and Mansfield and Key.[239] (B). With permission.]

TABLE 2.3
Members of the *Drosophila* HSP70 Family

Protein	Gene(s)	Locus (3rd chromosome)	Homology to hsp70 (%)[a]	Relative abundance of transcripts[b] 25°C	37°C	Synthesis[b] 25°C	37°C	Protein abundance[b] (2 h 37°C)
HSP70[c]	hsp70	87A7, 87C1	100	1	≥300	—	+ + + +	+ +
HSP68	hsp68	95D	85			—	+ +	+ +
HSC72	hsc72	?	?			+ +	(+)	+ +
HSC70[c]	hsc4[a]	88E	82	100	200	+ + +	+	+ + + +
—	hscl[a]	70C	75	3	3	?	?	—
—	hsc2[a]	87D	77	1	1	?	?	—

[a] Only the 5′ parts of genes were sequenced. The homology estimate to the hsp70 genes is based on this information. Genes hscl and hsc2 contain introns of 1.7 and 0.65 kb, respectively, inserted in codon 65, and between codons 55 and 56, respectively.
[b] These data were analyzed with adult flies of *Drosophila melanogaster* Oregon strain.
[c] For sequences see Figure 2.10

Data compiled from References 91, 92, 166, 307, 404, 415.

detected by immunological techniques. Antibodies to chicken HSP90[185,342a] cross-reacted with similar proteins in *Drosophila,* frog, rodent, and human cells, but did not detect the HSP90 proteins in plant tissue. On the other hand, antibodies to the *Drosophila* HSP83 did not cross-react with members of the HSP90 family from avian and mammalian cells.[67] These results can be explained by inspection of the general structure of the HSP90 family. Figure 2.9 summarizes the extent of sequence homology with reference to the theoretical consensus HSP90 sequence shown in Figure 2.8. The amino acid residues are grouped in a mosaic with regions of very high homology to regions without homology at the C- and N-termini. The two large GRPs are characterized by two extensions at the N-terminus and one at the C-terminus. Their localization in the ER/Golgi system is reflected by a hydrophobic leader sequence of the precursor proteins and the C-terminal −KDEL required for retention of the proteins in the lumen of the ER.[274,364] In contrast to this, all cytoplasmic HSP90 proteins end up with a conserved −MEEVD pentapeptide, which corresponds to the −I(V)EEVD at the C-terminus of all eukaryotic HSP70 proteins. The only exception is the constitutively expressed HSC70 of tomato (no. 15 in Figure 2.10). It is a remarkable observation that transfer of the −EEVD codons to the 3′ end of β-galactosidase mRNA allows selective translation in a heat-shocked rabbit reticulocyte lysate under physiological ion conditions.[107a] If this result can be confirmed *in vivo* and with other translation systems, it provides an interesting new insight into translational control in general (see Section 11.5).

Two charged regions of the HSP90 proteins are marked with K/E in Figure 2.9. The larger one centered around amino acid residue 250 is flanked by two proline residues and contains on the average 55 polar amino acid residues out of a total of 70. Computer modeling of the secondary structure in this region, based on the chicken HSP90 sequence, revealed the potential to form a negatively charged α-helix.[42] This "DNA-like" structure may be essential for shielding the DNA binding site of the inactive steroid hormone receptors found associated with HSP90 (see Section 16.8). In support of this hypothesis, it is worth noting that this K/E region is partially deleted and distorted in those members of the HSP90 family that are not complexed with hormone receptors, i.e., GRPs94 (nos. 5 and 6 in Figure 2.8), the plant HSP80 (nos. 8 and 9), the *Trypanosoma* surface protein (no. 11), and the *E. coli* HtpG (no. 12).

Besides the association with hormone receptors, several others types of soluble protein complexes have been detected. But presently, we have no idea about the function of HSP90 in these complexes or about their relevance for stress-related phenomena. Thus, in mammalian cells, HSP90 as well as GRP94 bind to actin in a calmodulin-dependent manner under polymerizing conditions. They may help to cross-link actin filaments.[197a,294] Furthermore, HSP90 evidently participates in the formation of transient cytoplasmic complexes with various retroviral oncogene products required for maturation of the latter (see Table 16.1 and Figure 16.8). Finally, in different mammalian cells, HSP90 is coprecipitated with antibodies specific for α- and β-tubulin;[337] in addition, by use of antibodies to HSP90 and tubulin, a strict colocalization of both proteins in interphase and mitotic cells was demonstrated.[324b]

Cytoplasmic HSP90 proteins evidently contain highly hydrophobic regions at their surface. HSP90 from calf uterus and rat liver were purified by hydrophobic affinity chromatography through phenyl-sepharose CL-4B columns. Part of the hydrophobic region was shielded by interaction with steroid receptor but probably also with calmodulin or $pp60^{src}$, i.e., hydrophobic interactions may be essential for complexation of HSP90 with different cytoplasmic proteins.[177]

Except for the localization of GRP94 in the ER/Golgi system, the members of the HSP90 family are abundant cytoplasmic proteins without spectacular redistribution during the stress period.[67,75,185,203,397] But careful investigations of Lanks[211b] revealed an hs-dependent oligomerization *in vivo* as well as *in vitro,* if purified mouse HSP90 was incubated for one hour at 43°C in the presence of detergent. Similar to the interaction with heterologous proteins, this aggregation tendency may depend on the hydrophobic domains and thus may interfere with the normal functions of HSP90 in the heterologous complexes. This interesting observation needs further investigation.

As detailed in Table 2.5, HSPs90 are phosphoproteins, whereas GRPs94 are glycoproteins. Dougherty et al.[112] and Lees-Miller and Anderson[217b] identified a type II casein kinase of rat liver cells phosphorylating HSP90 at conserved serine residues. Another protein kinase (PK) probably acting on HSP90 is the double-stranded DNA-activated PK, typically observed after virus infection. The enzyme from Hela cells phosphorylates two N-terminal threonine residues. In both cases, a possible relation to an endogenous PK acting on HSP90 remains to be demonstrated. But it is worth noticing that a heat-shock-induced PK activity phosphorylating HSP90 was detected in HeLa cell lysates.[217c] Identification with any one of the two PKs mentioned above is lacking. Matts and Hurst[241a] reported evidence that HSP90 might act as a "regulatory subunit" of the hemin-regulated PK in rabbit reticulocytes. HSP90 is released and phosphorylated(?) upon activation of the eIF-2α kinase by heme-deficiency, hs, or oxidants (see Section 11.5.3).

2.3.2. THE HSP70 FAMILY

2.3.2.1. Protein Structure

Figure 2.10 presents a compilation of 21 proteins belonging to this family (for references, see legend to Figure 2.10). The collection includes heat-shock-induced and constitutive members as well as a group of GRP78 proteins from rat, chicken, and yeast (nos. 6 to 8). The two bacterial DnaK proteins (nos. 20 and 21) are markedly similar to the yeast mitochondrial SSC1 protein (no. 19). Meanwhile, similar DnaK-like proteins were characterized from mitochondria of parasitic protozoa,[120a,350b] the nematode *Caenorhabditis elegans,*[163a] and human cells.[220b] Interestingly, each one DnaK-like protein was identified in the chloroplasts and mitochondria of plants[240a,289a] and in the green alga *Euglena gracilis.*[3a] Evidence for this particular subfamily of HSP70 proteins was derived from sequence comparison[120a,163a,289a,380b] and immunological data,[3a,220b,240a] respectively.

Similar to the preceding presentation of the HSP90 proteins, the homology considerations

```
 0 Consensus aa  MW      MXXXXXXXXX XXXXXXXXXX XXXXXXXXXX XXXXXXXXXX XXXXXXXXXX XXXXXXXXXX XXXXXXXXXX XXXETFAFQA EIXQLMSLII NTFYSNKEIF
 1 Hs-HSP90a 732 84,622 .--------- ---------- ---------- ---------- ---------- -------PEE TQTQDQPMEE EEV....... ..A....... ..........
 2 Hs-HSP90b 724 83,303 .--------- ---------- ---------- ---------- ---------- -------PEE -----VHHGE EEV....... ..A....... ..........
 3 Mm-HSP90  724 84,000 .--------- ---------- ---------- ---------- ---------- ---------- --PEEVHHGE EEV....... ..A....... ..........
 4 Gd-HSP90  728 84,112 .--------- ---------- ---------- ---------- ---------- --------PE AVQTQDQPME EEV....... ..A....... ..........
 5 Mm-GRP94  802 92,475 .RVLWVLGLC CVLLTFGFVR ADDEVDVDGT VEEDLGKSRE GSRTDDEVVQ REEEAIQLDG LNASQIRELR EKS.K..... .vNRm.K... .sL.K.....
 6 Gd-GRP94  795 91,555 .KSAWALALA CTLLLAASVT AEE-VDVDAT VEEDLGKSRE GSRTDDEVVQ REEEAIQLDG LNASQIKEIR EKS.K..... .vNRm.K... .sL.K.....
 7 Dm-HSP83  717 81,858 .--------- ---------- ---------- ---------- ---------- ---------- ---------P EEA....... ..A....... ..........
 8 Zm-HSP82  715 81,929 .--------- ---------- ---------- ---------- ---------- ---------- ASADVHMAGG AET....... ..N..l.... ..........
 9 Bo-HSP80  699 80,056 .--------- ---------- ---------- ---------- ---------- ---------- ---------- ADA....... ..N..l.... ..........
10 Sc-HSP90  709 81,419 .--------- ---------- ---------- ---------- ---------- ---------- ---------- -AS...E... ..T....... ..V.......
11 Tc-p85    704 80,500 .--------- ---------- ---------- ---------- ---------- ---------- ---------- --T....... ..N....... ..........
12 Ec-HtpG   624 71,492 .--------- ---------- ---------- ---------- ---------- ---------- ---------- KGQ..RG..S .vK..lH.m. HsL.......

 0 LRELISNASD ALDKIRYESL TDPSKLDXXX ELXIRIIPDK XXRTLTIVDT GIGMTKADLI NNLGTIAKSG TKAFMXXXXE ALXAGAXXDI SMIGQFGVGF YSAYLVADKV XVISKHNDDE
 1 ......S... .......... .......SGK ..H.Nl..N. QD.A...... .......... .......... ....----. ..Q...--.. .......... ........e. T..t......
 2 .......... .......... .......SGK ..K.D...NP QE....l... .......... .......... ....----. ..Q...--.. .......... ........e. V..R......
 3 .......... .......... .......SGK ..K.D...NP QE....l... .......... .......... ....----. ..Q...--.. .......... ........e. V..t......
 4 ......S... .......... .......SGK d.K.Nl..N. HD........ .........v .......... ....----. ..Q...--.. .......... ........e. T..t......
 5 .......... ......LI.. ..ENA.AGNE ..Tvk.KC.. EKNL.HvT.. .v...rEe.v K......... .SE.lNKMT. .QED.QS-TS El........ ...f...... I.T....N.T
 6 .......... ......LI.. ..ENA.AGNE ..Tvk.KC.. EKNM.HvT.. ......Ee.. K......... .SE.lNKMT. mQDDSQS-TS El........ ...f....r. I.T....N.T
 7 .......... .......... .......SGK ..Y.kl..N. TAG....i.. ......S..v .......... ....----. ..Q...--.. .......... .......... T.T..N....
 8 .......... ......f... ..K....AQP ..F..lv... ASk..s.i.s .v....S..v .......r.. ..E..----. ..A...T-.v .......... .........r. M.Tt......
 9 ......S... ......f... ..K....GQP ..F.H..... ANN....i.s .........v .......r.. ..E..----. ..A...--.v .......... .......... I.Tt......
10 .......... ........K.. s..KQ.etEP d.F...T.K. EQkV.E.R.s .......e.. .......... ....----. ..S...--.v .......... ..lf....r. Q....S....
11 ......S... .C.....Q.. .NQAV.GDES H.R..vv... ANk...vE.. ........e.v .......r.. ....----. ..E..G--.m .......... .........r. T.v..N....
12 .......... .a..l.fRA. sN.DLYeGDG ..Rv.vSF.. DK.T...S.N .v...rDev. Dh........ ..S.l----. S.GSDQAK.S Ql........ ...ii..... T.RtrAAGeK

 0 QYIXXXWESX AGGSFTVXAD XXGEXLGRGT KIXLHLKEDQ TEYLEERRVK ELVKKHSQFI GYPIXLWVEK EREKEVSDDE EEEEKXEEKE EEXXEXEEKE XPKIEDVGSD EEDDKEXXEK
 1 ..a---...S .......RT. T-..Pm.... .vI....... .........i. .i........ ....T.f... ..d....... A..KEDK.E. K.KE.K.SEd K.E....... ..ee.-KDGD
 2 ..a---...S .......R.. H-..Pi.M.. .vI....... .......... .v........ ....T.yl.. ......i... A....G-... ..DKdD...- -......... .....-SGKd
 3 ..a---...S .......RX. H-..Pi.... .vI....... .......... .v.....X.. ....T.yl.. ?.....i... A....G-... ..?DK.?...- -......... .....-SGKd.
 4 ..a---...S .......Rl. N-..P..... .vI....... .........i. .i........ ....R.f... ..d......d .A...E.... .KEEKT.d.- -.E....... ..ee.KDGd.
 5 .H.---...D -SNE.s.I.. PR.NT..... T.T.V...eA sd...LDTi. N..r.Y.... Nf..Yv.SS. TETV.EPLe. d.AA.-.... .----Sdd-. -AAv.e---- -----.EE..
 6 .H.---...D -SNE.s.ID. PR.NT..... T.T.V...eA sd...LDT.. N....Y.... Nf..Yv.SS. TETV.EPVe. ..A-...... .----TddN. -AAv..---- -----.EE..
 7 ..v---...S .......R.. NS-.P..... ..V.Yi.... .d....Ski. .i.N...... ....K.L... .......... Add..K-.Gd .KKEMETd-. -......... .dA..KDKd-
 8 ..v---...Q .......TH. TT..Q..... ..T.F..d.. L.......l. d......E.. S...Y..T.. TT...i.... ...dNK--.. .--------- -GDv.e.--- dde..dTKd.
 9 ..v---...Q .......TR. TS..S..... .mV....... L.......l. d......E.. S...S..i.. TI...i.... d....---.d .--------- -G.v.e.--- d.e-..KE..
10 ...---...N .......Tl. EVN.Ri.... IlR.F..d.. L.....k.i. .vi.r..E.v a...Q.V.T. .V....PIP. ..KKDE.K.d ..KKdEdd.K -..l.e.--- ---.-.EE..
11 A.T---...S ...t...TPT PDCdLK-... r.V....... Q.......l. d.i....E.. ..D.E.M... AT....t.ed .d.-----AA ATKN.EG.-. -..v.e.K-- --..A.EG..
12 PENGVF...A gE.Ey..-.. ITK.D--... E.T...r.GE D.f.ddW..r SiiS.Y.DH. aL.v------ ---------- -.I..R...d ---------- ---G.T.I.- ----------

 0 KKKTKKVKEX XXDXEELNKT KPIWTRNPDD ITQEEYGEFY KSLSNDWEDH LAVKHFSVEG QLEFRALLFV PRRAPFDLFE NXXXKKKNNI KLYVRRVFIM DXCEDLIPEY LNFVKGVVDS
 1 ...K..i..K YI.Q...... .......... ..n....... ...t...... .......... .......... .......... .R--...... .......... .N..e..... ...ir.....
 2 ......i..K YI.Q...... .......... .......... ...t...... .......... .........i .......... .K--...... .......... .S.de..... ...ir.....
 3 ......i..K YI.Q...... .......... .......... ...t...... .......... .........i .......... .K--...... .......... .S.de..... ...ir.....
 4 ...K.-i..K YI.E...... .......... ..n....... ...t...... .......... .......... .......... .R--...... .......... .N..e..... ...mr.....
 5 .P.....EKT VW.W.Lm.DI ....Q.PSKe VEEd..KA.. ..F.KeSd.P m.YI..ta.. EvT.kSi... .TS..RG..d EYGS..SDY. .........T .DFH.mm.K. ..........
 6 .P.....EKT VW.W.Lm.DI ....Q.PSKe VEEd..KA.. .tF.KeHd.P m.YI..ta.. EvT.kSi... .NS..RG..d EYGS..SDF. .........T .DFH.mm.K. ..........
 7 A..K.Ti..K YTeD...... .......... .s........ ...t...... .......... .........i ...T...... .Q--..r... .......... .N........ ...m......
 8 S..K.....V SHeWVQi..Q ....L.K.ee ..Rd..aS.. ...t...... .......... ....k.i... .........d T--R..L... .......... .N..e....w .G........
 9 ...KI.E--V SHeWdLv..Q ....M.K.ee .NK...aA.. ........e. .......... ....k.v... .k.......d T--K..P... .......... .N...i.... .G....i...
10 .P.......E VQeI...... ..l.....S. ......NA.. ..i......P .Y........ .......i.i .k........ SK--...... .........T .EA......w .S........
11 .........V TQeFVVQ..H ..l...D.K. v.K...aA.. .Ai.....eP .ST....... .......i... .k.....m.. PS--..r... .......... .N....C..w .A..r.....
12 ---------- ---W.Ki..A QAl....KSe ..D...K... .HiAH.fN.P .TWS.NR... KQ.yTS..yi .SQ..W.mW- .R--DH.HGl ....Q..... .DA.QFm.N. .R..r.li..
```

```
 0  EDLPLNISRE MLQQXKILKV IRKNLVKKXL ELFXEIAEDX KENYKKFYEX FSKNLKLGIH EDSXNRXRLA ELLRYXXXHS SXSGDEMTSL XDYVSRMKEX QKXIYYITGE SKKXVENSPF
 1  .......... ....S..... ........C. ...T.l...- .........Q ....i..... ...Q..Kk.S .....---yt .A.....V.. K..Ct....N ..H....... t.DQ.A..A.
 2  .......... ....S..... .....i..C. ...S.l...- .........A .......... ...T..R..S .....---.t .Q........ Se.......T ..S....... ..EQ.A..A.
 3  .......... ....S..... .....i..C. ...S.l...- .........A .......... ...T..R..S .....---.t .Q........ Se.......T ..S....... ..EQ.A..A.
 4  .......... ....S..... ........C. ...T.l...- .........Q ....i..... ...Q..Kk.S .....---Yt .A.....V.. K..Ct....N ..Hv...... t.DQ.A..A.
 5  d....v.... T..H.l.... ...K..r.T. dmIKK..de- .Y.-DT.wKE .GT.i...vI ..HS..T... K...f---Q. .HHSTdi... DQ..E....K .DK..fmA.S .r.Ea.S...
 6  d....v.... T..H.l.... ...K..r.T. dmIKK...e- .Y.-DT.wKE .GT.v...vI ..HS..T... K...f---Q. .HHESNl... DQ..E....K .DK..fmA.A .r.Ea.S...
 7  .......... ....N.v... ........Tm ..IE.lT..- ........dQ ........v. ...N..Ak.. dF..f---.t .A...dFC.. A.......dN ..Hv.f.... ..DQ.S..A.
 8  d......... T..N...... ........Ci .m.F....N- .dD.A...dA ....i..... ...Q..Ak.. d....---.. tK....T... K...t....G ..D....... .r.A......
 9  .......... T..N...... ........Cm ...F....N- ..D.N....A .......... ...Q..Tki. .....---.. tK....l... K...t....G .ED.f..... ...A......
10  .....l.... ....N..m.. .....i..Li .a.N.....- S.qfE...SA ....i...v. ..tQ..AA.. K....---N. tK.V..l... T...t..P.H ..N....... .L.A..K...
11  .......... N..N...... .....i..A. ...E....N- ..D......Q .G..v..... ...A..Kk.m K...f---.. .E..ed..t. K...t....G ..C...v..d ...Kl.T...
12  S....v.... i..DSTvTrN l.NA.T.rV. QmLEKl.K.D A.K.QT.wQQ .GLV..E.PA ..FA.QEAi. K...fAST.t DS.AQT-V.. E........G .EK......d .YAAaKS..H

 0  VERLRKKGFE VLYMTEPIDE YCIQQLKEFD GKXLVSVTKE GLELXXXEDE EEKKKREEXK AKFEXLCKLM KDILXDKKVE KVVVSNRLVD SPCCIVTSQY GWSANMERIM KAQALXXXRD
 1  .....H.L.. .i..I..... ..v......e ..T....... ....--P... .....Q..K. T..N...i.. ....-e.... .........T ........T. ..t....... .....---..
 2  ...v..r... .v........ ..v....... ..S....... ....--P... .....M..S. ....N..... .e..-..... ..Ti.....S ........T. ..t....... .....---..
 3  ...v..r... .v........ ..v....... ..S....... ....--P... .....M..S. ....N..... .e..-..... ..Ti.....S ........T. ..t....... .....---..
 4  .....H.L.. .i..I..... ..v......e ..T....... ....--P... .....Q..K. ....N...i. ....-e.... .........T ........T. ..t....... .....---..
 5  ....L...y. .i.l...v.. ....A.P... ..RFQN.A.. .vKF--D.S. KT.ES..ATE KE..P.LNW. ..KaLK..i. .a...q..Te ...Al.A... ...G...... ....YQTGk.
 6  ....L...y. .i.l...v.. ....A.P... ..RFQN.A.. .vKF--E.S. KS.ES..ALE KE..P.LNW. ..KaLK..i. .a.l.q..TQ ...Al.A... ...G...... ....YQTGk.
 7  ...vkAr... .v........ .V..H...yK ..Q....... ....--P... S......D.. ....S..... .S..-.N... .......... .........f .......... .....---..
 8  l...k...y. ..f.VdA... .AvG....y. ..K...a... ..K.DDEd.. .A..r...R. Kr..E...vi ....G.r-.. .....D.i.. ....l..GE. ..t....... .....---..
 9  l.k.k...y. ....VdA... .A.G.....e ..K...a... ..K.EES... -K...-..L. E...G...vi ..v.G..-.. ..i..D.v.. ....l..GE. ..t....... .....---..
10  ldA..A.N.. ..fl.d.... .AFT.....e ..T..Di..d -F..--E.Td ...AE..KEI KEy.P.T.al .e..G.Q-.. .....Yk.l. A.A-.R.G.f .......... .....---..
11  i.Qa.rr... ..f....... .Vm..v.d.e D.KFaCl... .vHF--E.T. ....Q...E. TAy.R...a. ..v.G..-.. .....E..aT ...Il...Ef ....H..Q.. rN...---..
12  l.L.....I. ..LlsdR... WMmnY.T... ..PFQ..s.V DESGEKLA.. VdESAk.AE. .LTPFiDr-v .Al.Ger-.K D.RltH..T. t.AIvS.DAD EM.TQ.AKLF A.AgQ---kV

 0  XSTMXGYMXX KKTLEINPDH PIVEELRKRA EADKNXDKSV XKDLVVLLFE TALLSSGFSL EDPQTFXXRI XRMIKLGLXI DEDEXXXEEX EXXXXDXXXX LEGDEDXSXM EEVD*
 1  N...-...AA ..H....... S.i.T..Qk. .....-.... -....i..y. .......... ......HAN. Y.......G. ...dPTAddT SAAVTeEMPP L...d.T.R. ....*
 2  N...-...MA ..H....... ....T..Qk. .....-..A. -......... .......... ......HSN. Y.......G. ....VAA..P NAAVP.EIPP ......A.R. ....*
 3  N...-...MA ..H....... ....T..Qk. .....-..A. -......... .......... ......HSN. Y.......G. ....VTA..P SAAVP.EIPP ......A.R. ....*
 4  N...-...AA ..H....... S.i.T..Qk. .....-.... -....i..y. .......... ......HAN. Y.......G. ...dTAA..A SPAVTeEMPP ....d.T.R. ....*
 5  I..NY-.aSQ ...F....R. .liRdmLr.i KE.ED-..t. -M..a.v... ..T.R..yL. P.TKAyGD.. E..lr.S.N. .PeAQVE..P .EEPE.TSED a.DS.QDEGE .MDAGTEEEE
 6  I..NY-.aSQ ...F....R. .liKdmLr.v KENED-..t. -S..a.v... ..T.R..yM. P.TKEyGD.. E..lr.S.N. .L.AKVE..P .E-PE.AAEE a.Q..eEVDA dAE.SE----
 7  TA..-...AG ..Q....... ....T..Qk. d....-..A. -....i.... .S........ dS..VHAS.. Y.......G. ....PMTTDD AQSAG.APSL v.DT..A.H. ....*
 8  -.s.Sa..SS ...m.....N G.m....... ...r.-.... -....l.... .....t.... d..nT.AA.. H..l....N. ...AAAD.dA dMPAL.EGAA E.-----.K. ....*
 9  -.s.G...SS ...m....eN A.md...... .....-.... -....l.... .....t.... dE.nT.GS.. H..l....S. .d.dAVEAdA .MPPLeDDAD a..----.K. ....*
10  -.s.SS..SS ...F..S.KS ..iK..k..v dEGGAQ..t. -...TK..y. .....t.... de.Ts.AS.. N.L.S...N. ....ETETAP .ASTAAPVEE vPA.---tE. ....*
11  -.s.Sa..MS ...m....A. ...K..kr.v ...E.-..A. -....Y...d ....t...t. d..TsyAE.. H.......Sl .dedNGN..A .PAAAVPAEP vA.T---.S. .N..*
12  PEV------- .YIF.l.... Vl.----... -..TE-.EAK FSEW.E..Ld Q...AERGt. ...nL.---. R..NQ.LvS*

 5  EETEKESTE- KDEL*
 6  --TQKESTDV KDEL*
```

FIGURE 2.8. The HSP90 family: sequence comparison and basic structure of proteins. Sequences were derived from the following sources: (1), human HSP90α;[165a] (2) human HSP90β;[324] (3), mouse HSP90;[264] (4), chicken HSP90;[43] (5), mouse GRP94;[244] (6), chicken GRP94;[201] (7), *Drosophila* HSP83;[45] (8), maize HSP80;[359a] (9), cabbage HSP80;[359a] (10), yeast HSP90;[123] (11), *Trypanosoma cruzi* surface glycoprotein p85;[113] (12), *Escherichia coli* HtpG = C62.5.[25] The consensus sequence derived from all entries is given above. The C-termini of the cytoplasmic proteins (1 to 4, 7 to 11) are formed by the conserved MEEVD sequence, whereas the two GRPs end with KDEL, characteristic of ER proteins (see Chapter 4). The amino acid residues are indicated by the one-letter code (see Abbreviations and Special Terms at the front of the book). For comparison of sequences 1 to 12 with the hypothetical consensus HSP90 identical residues are indicated by dots, lacking residues by hyphens, and homologous substitutions within groups of similar amino acid residues by lower case letters (with A = I = L = M = V, F = Y = W, S = T, R = K, D = E, and N = Q). (From Neumann, D. et al., *Biol. Zentralbl.*, 108, 1—156, 1989. With permission.)

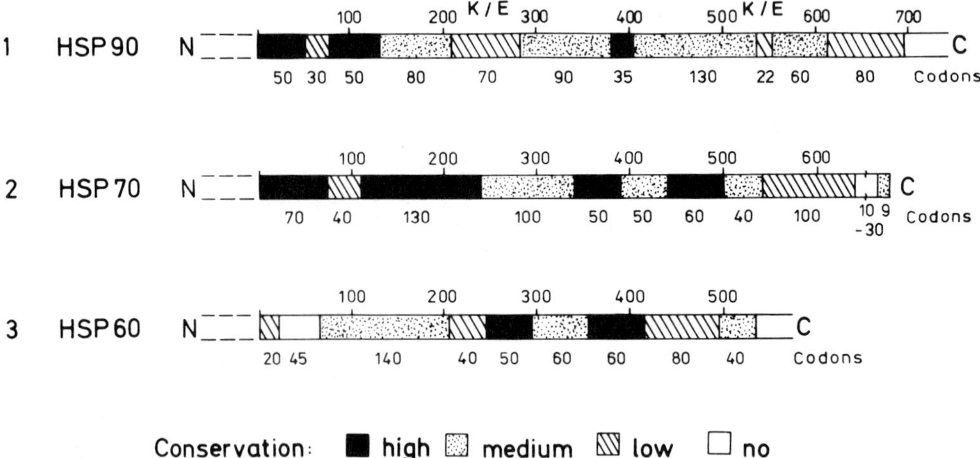

FIGURE 2.9. Extent of sequence conservation between members of the HSP90, HSP70, and HSP60 families. Proteins are symbolized by block diagrams derived from data contained in Figures 2.8, 2.10, and 2.11. The large HSPs represent a mosaic of different regions with high, medium, low, or lacking sequence conservation between individual members of the same family. K/E in the HSP90 structure denotes highly hydrophilic regions with 55 polar amino acid residues out of a total of 70 for the larger and 18 polar residues out of 20 for the smaller K/E region. (From Neumann, D., et al., *Biol. Zentralbl.*, 108, 1—156, 1989. With permission.)

are based on a consensus HSP70,whose sequence was derived by comparison of the other representatives of the family. For all cytoplasmic HSP70 proteins of eukaryotes, conservation is very high except for the 150 amino acid residues at the C-terminus (Figure 2.9), but even for the bacterial and mitochondrial DnaK-like proteins the overall homology to the eukaryotic members of the HSP70 family is in the range of 50%. Close inspection of the length- and sequence-variable C-terminal part of the eukaryotic HSP70 (Numbers 1 to 5 and 9 to 18 of Figure 2.10) reveals three interesting features. The highly hydrophobic sequence contains up to 50% glycine residues and ends in a conserved nonapeptide—SGPTIEEVD. The last five residues correspond to the −MEEVD terminus of the HSP90 family (Section 2.3.1). In contrast to this, the GRP78 end with −KDEL or −HDEL (yeast KAR2 protein), the hydrophobic glycine-rich sequence is lacking, and an extension at the N-terminus evidently functions as leader sequence mediating the cotranslational transfer of the protein into the ER.[273,312b]

The remarkable conservation of HSP70 proteins and their genes (Figures 2.9 and 2.10) considerably facilitated the analysis of their hs-induced expression and/or the isolation of corresponding genes from different kinds of organisms. In most cases, probes of the *Drosophila* hsp70 genes were used and isolated about ten years ago by Craig et al.,[90] Livak et al.,[232] Mirault et al.,[261] and Schedl et al.[340] Thus, cross hybridization was found with hsp70-specific DNA or RNA from mammals[236,265] plants,[289,331] nematodes,[362] fishes,[158,197] snail,[140] *Tetrahymena,*[3] and the mold *Blastocladiella emersonii.*[48] Bardwell and Craig[24] used HSP70 probes from *Drosophila,* yeast, and *E. coli* to always identify the same genomic DNA fragments of the archaebacterium *Methanosarcina barkeri.* By means of heteroduplex analysis, Moran et al.[265] provided early evidence for sequence homology between mouse, yeast, and *Drosophila* hsp70 genes. Finally, Lowe and Moran[236] and Lowe et al.[236] used a *Drosophila* DNA probe to hybrid-select mRNAs of the hsp70 gene family from mouse cells.

At the protein level, the high conservation within the HSP70 family was demonstrated by Voellmy et al.[410] comparing peptide maps of HSP70 of chicken, human, and *Drosophila* cells. Generation of antibodies to the major chicken HSPs (HSPs90, 70, and 24)[185,342a] allowed analyzing of cross-reactivity with other species. The anti-HSP70 detected similar proteins in different kinds of organisms, including yeast, *Dictyostelium,* maize, nematode, *Drosophila, Xenopus,* and mammalian species. Similar to this result, a monoclonal antibody to the *Drosophila* HSP70[404] cross-reacts with constitutive and/or hs-induced members of the HSP70 family in sea urchin, nematode, chicken, human cells, and plants.[230a,440a] On the other hand, antibodies raised against the *E. coli* DnaK protein detect the mitochondrial HSP70 in human and calf cells,[220b] the HSP78 in the chloroplast stroma of pea, spinach, and maize,[3a,240a] and the DnaK-like proteins in chloroplasts and mitochondria of *Euglena gracilis.*[3a]

Probably all members of the HSP70 family are ATP-binding proteins. The bacterial DnaK protein exhibits ATPase and autophosphorylating activities.[222a,451] Based on the ATP-binding, Welch and Feramisco[427] developed a convenient method for purification of the mammalian HSP70 proteins by affinity chromatography through ATP-agarose columns. This method was later extended to all types of HSP70 proteins, e.g., to the bovine and rat brain clathrin uncoating ATPase[139a,343] (HSC72), to its microsomal relative GRP78, to the mouse HSC70 and a constitutively expressed p70 of spermatogenic cells[1c] as well as to cytosolic members of the yeast HSP70 family[72a,94] (see Table 2.4). Moreover, all members of the *Drosophila* HSP70 family (HSPs 68, 70, HSCs 70, and 72)[31] and the DnaK-like proteins of chloroplasts and mitochondria[3a,220a] are accessible to purification by ATP-affinity chromatography.

2.3.2.2. Intracellular Localization and Function

Intracellular localization of HSP70-type proteins point to four main sites of action (see summary and references in Table 16.1):

1. GRP78 of vertebrates reside in the lumen of the ER (Sections 4.1 and 4.3). They evidently have essential functions for the stabilization of polypeptides subunits on their way to maturation and assembly into multimeric protein complexes. In addition, GRP78 was found associated with aberrant proteins in the ER, e.g., with a mutant SV40 large T-antigen introduced into the ER by fusing the C-terminal 694 amino acid residues to the 15 amino acid residue leader sequence of the influenza virus hemaglutinin.[351] It helps to retain aberrant proteins in the ER.[111b]
2. HSP70 and related proteins are tightly bound components of the microtubule and the intermediate size filament (IF) cytoskeletal system (see Table 16.1). Before their identification as members of the HSP70 family, they were called microtubule associated protein (MAP) and β-intermexin, respectively.[81a,139a,168,228,277,277a,301,420,424,432] It is tempting to speculate that the rapid and reversible reorganization of cytoskeletal systems (see Section 14.3) and/or their mutual interconnection, including the plasma membrane, may depend on organizing or interfacing proteins of the HSP70-type.[168,277,277a,308]
3. The function of the HSC71-73 (HSP70) in the reorganization of the cytoskeletal systems is very reminiscent of its role as clathrin-uncoating ATPase.[55,72a,101,102,343,344,396] Only the association with the particular clathrin arrangement in coated vesicles or the free clathrin triskelions trigger the ATPase activity of HSC71 necessary for uncoating.[163b,334] Freeze-etch techniques allow visualization of the complex of clathrin triskelia with three molecules of HSC72, which is stabile in the presence of EDTA or ATP analogs but dissociates in the presence of Mg-ATP.[163b]
4. During hs, members of the HSP70 family are bound to the nucleus (for references see Table 16.1), in vertebrate and plant cells mainly to the nucleolus (Figures 16.1 and 16.3). As detailed in Chapter 10, the nucleolar locale of HSP70 is tightly correlated with the protection of the machinery for ribosome biosynthesis and its recovery after the stress period. It is remarkable that also the constitutive members of the HSP70 family are bound to nuclear structures in heat-shocked *Drosophila*[307] and vertebrate cells.[255,396,428] On the one hand, synthesis of a distinct HSC70 species in synchronized

```
 6 RN-GRP78   NTE:  KFTVVAAALL LLCAVRAEEE DKKEDVGT
 7 Gd-GRP78   NTE:  RHLLLALLLL GGARADDEEK KEDVGT
 8 Sc-GRP78   NTE:  FFNRLSAGKL LVPLSVVLYA LFVVILPLQN SFHSSNVLVR GADDVENYGT
19 Sc-SSC1    NTE:  LAAKNILNRS SLSSSFRIAT RLQSTKVQGS

 0 CONSENSUS                 MXXXXKGPAV GIDLGTTYSC VGVFQHGKVE IIANDQGNRT TPSYVAFTXD TERLIGDAAK NQVAMNPTNT
 1 Hs-HSP70   640 70,008     .A---KAA.. .......... .......... .......... ........-. .......... ....l..Q..
 2 Hs-HSC71   646 70,899     .S---KGP.. .......... .......... .......... ........-. .......... ..........
 3 Rn-HSC73   646 70,865     .S---KGP.. .......... .......... .......... ........-. .......... ..........
 4 Mm-HSC70   633 69,734     .S--ARGP.i .......... .......... .......... ........-. .......... ..........
 5 Mm-HSC72   646 70,830     .S---KGP.. .......... .......... .......... ........-. .......... ..........
 6 Rn-GRP78   654 72,341     .-(NTE)-v. .......... ....KN.r.. .........I ........Pe G......... ..lTS..E..
 7 Gd-GRP78   652 72,020     .-(NTE)-v. .......... ....KN.r.. .........I ........Pe G......... ..lTS..E..
 8 Sc-GRP78   682 74,462     .-(NTE)-vi .......... .a.MKN..T. .l..e....I ........-. D......... ....a..Q..
 9 Gd-HSP70   634 70,000     .S--GKGP.i .......... .......... .......... ........-. .......... ..........
10 Xl-HSP70   647 70,914     .A--TKGV.. .......... .......... .......... ........-. .......... .......Q..
11 Ce-HSP70A  640 69,096     .S---KHN.. .......... ....M..... .......... ........-. .......... .......H..
12 Dm-HSP70   641 70,258     .------P.i .......... ...y...... .N.Y...... ........-. s...N.eP.. .......R..
13 Dm-HSC70   651 70,780     .S---KAP.. .......... .......... .......... ........-. .......... ........q.
14 Zm-HSP70   645 70,361     .A-KSEGP.i .......... ...w..Dr.. .......... .....G..-. .G........ ..........
15 Le-HSC70   666 73,495     .AGKGEGP.i .......... ...w..Dr.. .......... .....G..-. .......... ....l..I..
16 Ph-HSP70   651 71,238     .AGKGEGP.i .......... ...w..Dr.. .......... .....G..-. .......... .......I..
17 Tb-HSP70   661 71,370     .---TYEG.i .......... ...w.NEr.. .......... ........-. s......... ..........
18 Sc-HSP70   643 70,065     .S-----K.. .......... .aH.ANDr.d .......... ...f....-. .......... ..a....s..
19 Sc-SSC1    654 70,622     .-(NTE)-vi .......N.a .aiMEGKVPK ..E.AE.S.. ...V....Ke G...v.IP.. R.avv..E..
20 Ec-DnaK    638 69,121     .-----GKii .......N.. .aiMDGTTPR vlE.AE.D.. ...Ii.y.Q. G.T.v.QP.. R.avT..Q..
21 Bm-DnaK    605 64,931     .S-----Kii .......N.. .a.LEG.EPK v.P.PE.... ...V...K-N G..Qv.ev.. R.aiT..-..
 0 CONSENSUS                 MXXXXKGPAV GIDLGTTYSC VGVFQHGKVE IIANDQGNRT TPSYVAFTXD TERLIGDAAK NQVAMNPTNT

 0 VFDAKRLIGR KFDDPVVQSD MKHWPFKVVX DGGXKPKIQV EYKGXETKXF XPEEISSMVL TKMKEIAEAY LGXXVTXAVV TVPAYFNDSQ
 1 .......... ..G....... ......Q.iN ..D-...v.. S...-...A. Y......... .......... ..YP..N..i ..........
 2 .......... r...A..... ......M..N .a.-r..v.. ....-...S. Y...v..... .......... ..KT..N... ..........
 3 .......... r...A..... ......M..N .a.-r..v.. ....-...S. Y...v..... .......... ..KT..N... ..........
 4 i..R...... ..e.AT.... .......r..S e..-...v.. ....-.M.T. F......... .......... ..GK.QS..i ..........
 5 .......... r...A..... ......M..N Na.-r..v.. ....-...S. Y...v..... .......... ..KT..N... ..........
 6 .......... TwN..S..Q. i.FL.....E KKT-..Y... dIG.GQ..T. A.....A... .....T.... ..KK..H... ........A.
 7 .......... TWN..S..Q. i.YL.....E KKa-..H... dVG.GQ..T. A.....A... .....T.... ..KK..H... ........A.
 8 i..i.....L .yN.RS..K. i..L..N..N KD.-..AvE. SV..-.K.V. T.....G.i. G...Q...D. ..TK..H... ........A.
 9 i......... .y...T.... .......r..N e..-...v.. ....-.M.T. F......... .......... ..KK.ET..i ..........
10 .......... ..N.....C. l.....Q..S .E.-...vK. ....-.E.S. F......... .....T.... ..HP..N..i ..........
11 .......... .....a.... ........iS AE.A...v.. ....-.N.I. T......... L...KT...f .EPT.KD... ...T......
12 .......... .y...KiAE. .........S ...-....G. ....-.s.R. A......... .....T.... ..ESi.D..i ..........
13 i......... ....Aa.... ......E..S AD.A....E. T.KDEKKTF. -......... .....T.... ..KT..N..i ..........
14 .......... r.SS.a...S ..L..SrHl- GL.D..M.VF N...-.E.Q. AAGG...... I......... ..STiKN... ..........
15 .......... r.S.AS..E. ..L.....iP GP.D..M.V. T...-.E.E. AA........ .........f ..ST.KN... ..........
16 .......... r.S..S.... i.L.....iP GP.D..M.V. T...-.E.Q. AA........ .......... ..TTiKN... ..........
17 .......... ..S.S..... .........T K.DD..V... Qfr.-...T. N......... L....v..S. ..KQ.AK... ..........
18 .......... N.N..E..A. ...f...liD VD.-..Q... .f..-...N. T..Q..P.F. G....T..S. ..AK.ND... ..........
19 l.AT...... r.e.AE..R. i.QV.y.i.K HSNGDAW--. .Ar.-Q.--y S.AQ.GGF.. N....T.... ..KP.KN... ..........
20 l.Ai...... r.Q.EE..R. vSIM...iiA ADNGDAW--. .V..-Q-.MA P.-Q..AE.. K...KT..D. ..EP..E..i ........A.
21 iICv..Nm.T    .NK.CAe [illegible] --.KQYTP.- .M-------- -----.Aii. QNl.GY..E. ..EP..K..i ........AE
 0 VFDAKRLIGR KFDDPVVQSD MKHWPFKVVX DGGXKPKIQV EYKGXETKXF XPEEISSMVL TKMKEIAEAY LGXXVTXAVV TVPAYFNDSQ

 0 RQATKDAGTI AGLNVLRIIN EPTAAAIAYG LDKKXXXXGX ERNVLIFDLG GGTFDVSILT IEDGXIXXXX FEVKATAGDT HLGGEDFDNR
 1 ........V. .......... .......... ..rTGK--.- .......... .......... .d..-.---- .......... ..........
 2 .......... .......... .......... ....VG--a- .......... .......... ....-.---- ....S..... ..........
 3 .......... .......... .......... ....VG--a- .......... .......... ....-.---- ....S..... ..........
 4 .......... T......... .......... ....GCAG.- .k........ .......... ....-.---- ....S..... ..........
 5 .......... .......... .......... ....V---.A .......... .......... ....-.---- ....S..... ..........
 6 .......... .....m.... .......... ...rE---.- .k.i.v.... .......l.. .dN.-v---- ...V..N... ........q.
 7 .......... .....m.... .......... ...rE---.- .k.i.v.... .......l.. .dN.-v---- ...V..N... ........Q.
 8 .......... ........v. .......... ...SDK---- .Hqiivy... .......l.s ..N.-v---- ...Q..S... ........Yk
 9 .......... T....m.... .......... ....GTRA.- .k........ .......... ....-.---- ....S..... ..........
10 ........Vl ....i..... .......... ...GAR--.- .Q........ .......... .d..-.---- .......... ..........
11 ........A. .......... .......... ....GH--.- .......... .......... ....-.---- ....S..... ..........
12 ........H. .......... ......l... ...NLK--.- .......... .......... .de.Sl---- ...rS..... ..........
13 .......... ..P..P.... .......... ....AV--.- .......... .........s .d..-.---- ....S..... ..........
14 ........V. .....m.... .......... ....ATSS.- .k........ .......l.. ..e.-.---- .......... ..........
15 ........V. S....m.... .......... ....ATSA.- .k........ .......l.. ..e.-.---- .......... ..........
16 ........V. .....m.... .......... ....ASSA.- .kM....... .......l.. ..e.-.---- .......... ..........
17 .......... ...E.....A .......... ...ADE--.K .......... ......tl.. .dG.-.---- ......N... ..........
18 .......... .......... .......... ....----.K .EH....... .......l.F ....-.---- .......... ..........
19 ........Q. v......vv. ......l... .e.SDS---- -kV.av.... .....i...D .dN.-v---- ....S.N... ........IY
20 ........R. ...E.K.... ......l... ...GT---.N -.Tiavy... .....i..iE .deV-DGEKT ...L..N... ........S.
21 ........K. ...E.E.... ......l... .e.TDE---- dQT..vy... .........E lG..-v---- ...r.....N R...d...qV
 0 RQATKDAGTI AGLNVLRIIN EPTAAAIAYG LDKKXXXXGX ERNVLIFDLG GGTFDVSILT IEDGXIXXXX FEVKATAGDT HLGGEDFDNR

 0 MVNHFVXEFK RKHKXKDISX NKRALRRLRT ACERAKRTLS SSTQASIEXX XIDSLFEGXI DFYTSITRAR FEELNADLFR GTLEPVEKAL
 1 l.....E... ....-....Q ....v..... .......... ...G..l.-- -.......-. .......... ....CS.... S.........
 2 .....iA... ....-....E ....v..... .......... ........-- -....y..-. .......... .......... ...d......
 3 .....iA... ....-....E ....v..... .......... ........-- -....y..-. .......... .......... ...d......
 4 ..S.LaE... ....-...GP ....v..... .......... ........-- -....y..-v .......... .......... ..........
 5 .....iA... ....-....E ....v..... .......... ........-- -....y..-. .......... .......... ...d......
 6 vmE..iKLy. k.TG-..vRK DN..vQk..R EV.k...A.. .QH..R..-- -.e.F...-E ..SEtl...k .....m.... s.mK..Q.v.
 7 vmE..iKLy. k.TG-..vRK DN..vQk..R EV.k...A.. .QH..R..-- -.e.F...-E ..SEtl...k .....m.... S.mK..Q.v.
 8 i.RQLiKA.. k..G-I.v.D .Nk..Ak.kR EA.k...A.. .QMSTR..-- -...FVd.-. .LSEtl...k .....l...k K..K....v.
 9 ...R..E... B...-rKNAB ....v..... .....r.... ........-- -.......-. .......... .......... ..........
10 ......E... ....-...GQ .......... ..d....... ..s.....-- -.......-. ....A..... ....CS.... ..........
11 .....CA... ....-..lAS .P........ .....NE... ..C.....-- -.......-. ....N..... ....C..... S.md....S.
12 l.T.LaE... ..Y.-..lRS .P........ .A........ ...E.t..-- -..A....-Q ....Kvs... ....C.N... N..Q......
13 l.T...Q... ....-..ltT .......... .......... ........-- -.......-T .......... .......... S.md......
14 ......Q... ..N.-....G .P........ .......... .tA.Tt..-- -.......-. ..TPRSs... .....m.... KCm.....C.
15 ......H... ....-...tG .P........ .......... .tA.Tt..-- -....y..-v ...st..... .....m.... KCm.....C.
16 ......Q... ..N.-....G .P........ .......... .tA.Tt..-- -....y..-. ...st..... .....m.... KCm.....C.
17 l.A..TE... ..N.G..l.S .L........ .......... .AA..t..-- -..A...N-. ..QAt..... ....CG.... ...Q...rv.
18 l....iQ... ..N.-..l.T .Q........ .......... ..A.T.v.-- -.......-. .......... ....C..... S..d....v.
19 llREI.SR.. TETG-I.lEN DrM.iQ.i.E .A.k..IE.. .tVSTE.NLP F.TADAS.PK HINMKFs..Q ..T.T.P.Vk R.vd..K...
20 li.YL.E... kDQG-I.lRN DPL.mQ..kE .A.k..IE.. .AQ.TDvNLP Y.TADAT.PK HMNIKv...k L.S.VE..VN Rsi..lKV..
21 iiDYL.A... kENG-V.l.K D.M..Q..kD .A.k..kD.. GV.STQ.SLP F.TAgEA.Pl NLEV.ls..k .d..S.G.VE R.mA..RQ..
 0 MVNHFVXEFK RKHKXKDISX NKRALRRLRT ACERAKRTLS SSTQASIEXX XIDSLFEGXI DFYTSITRAR FEELNADLFR GTLEPVEKAL
```

FIGURE 2.10. The HSP70 family: sequence comparison and basic structure of proteins. Sequences are derived from the following sources: (1), human HSP70;[169] (2), human HSC71;[117] (3), rat HSC73;[302,365] (4), mouse testis HSC70;[446a] (5), mouse HSC72;[133a] (6), rat GRP78;[72,273] (7), chicken GRP78;[373b] (8), yeast GRP78 = KAR2;[294a,332a] (9), chicken HSP70;[268] (10), *Xenopus* HSP70;[39] (11), *Caenorhabditis elegans* (nematode) HSP70;[362a] (12), *Drosophila* HSP70;[175,183,388] (13), *Drosophila* HSC70;[360a] (14), maize HSP70;[331] (15), *Lycopersicon esculentum* (tomato) HSC70;[436a] (16), *Petunia hybrida* HSP70;[437] (17), *Trypanosoma brucei* HSP70;[135] (18), yeast HSP70 = SSA1;[360] (19), yeast mitochondrial SSC1;[94] (20), *Escherichia coli* HSP70 = Dnak;[24] (21), *Bacillus megaterium* HSP70 = Dnak.[380a] For further explanations see the caption to Figure 2.8. The sequences of the three GRP78 and of the yeast SSC1 GRP78 contain N-terminal

```
 0 RDAKLDKSQI HDIVLVGGST RIPKVQKLLX QDFFNGKELN KSINPDEAVA YGAAVQAA   SGDKSENVQD LLLLDVTPLS LGIETAGGVM
 1 .......A.. ..l....... .........- ......rd.. .........g ........   M......... ......A... ..l.......
 2 .......... .......... ....i....- .......... .......... ........   .......... .......... ..........
 3 .......... .......... ....i....- .......... .......... ........   .......... .......... ..........
 4 .......G.. Qe........ ....i....- .......... .......... ........   I......... .......... ..........
 5 .......... .......... ....i....- .......... .......... ........   .......... .......... ..........
 6 E.SD.K..D. De........ ....i.Q.v- Ke......PS rG........ .......g   ...QDTG--. .v....C..t .....v....
 7 E.SD.K..D. De........ ....i.Q.v- Ke......PS rG........ .......g   ...QDTG--. .v....C..t .....v....
 8 Q.SG.e.KDv D......... ......Q..E S-y.D..KaS .G........ Y......g   ..e--.G.E. .v....NA.t .....T....
 9 .......G.. Qe........ ....i....- .......... .......... ........   M......... .......... ..........
10 .......... .e........ .........- ......r... .......... ........   M......... ......A... ..l.......
11 ....m....v .......... .........- S.L.S..... ........l. ........   ......A... ......A... ..........
12 N...m..G.. .......... ......S..- .e..H..N.. L......... ........   ...Q.GKi.. v..v..A... ..........
13 ........V. .......... ......r..- ..L....... .......... .......... H....QE... .......... ..........
14 ....m...sv ..v....... ......Q.-- .........C .......... .......... ..eGN.RS-. .......... ..l.......
15 ....m...Tv ..v....... ......QvaM TN.......C .......... .......... ..eGN.K... .......... ..l.......
16 ....m...Sv ..v....... ......Q..- .........C .......... .......... ..eGN.K... .......... ..l.....G.
17 Q...m..RAv ..v....... .....MQ.v- S...G..... .........? .......F.. t.G..KQTEG ......A..t ..........
18 .........v De........ ........v- T.y.....P. r......... .......... t..E.SKT.. ......A... ..........
19 k..G.ST.D. Sevl....Ms .m...VETvK SL.--..dPS .Av....... I.....g.v. ..e----.T. v......... .....l...F
20 Q..G.SV.D. D.vi....Q. .m.M...Kv- Ae..-...PR .Dv....... I.....ggv. t..----.K. v......... .....m....
21 k..G.SA.El DKvi...... ...A..Dai- KKET-.QdP. .Gv....v.. L...i.ggv. t..----.K. vv........ .....m...F
 0 RDAKLDKSQI HDIVLVGGST RIPKVQKLLX QDFFNGKELN KSINPDEAVA YGAAVQAAIL SGDKSENVQD LLLLDVTPLS LGIETAGGVM

 0 TVLIKRNTTI PTKQTQIFST YSDNQPGVLI QVYEGERAMR TKDNNLLGKF ELSGIPPAPR GVPQIEVTFD IDANGILNVS AXDKSTGKXN
 1 .a.....s.. ........t. .......... .........- ........r. ...C.....- .......... .........t .TKD....A.
 2 .......... ......T.t. .......... .........- .......... ..t....... .......... .......... .V......E.
 3 .......... ......T.t. .......... .........- .......... ..t....... .......... .......... .V......E.
 4 .P........ ......T.t. .....SS..v .........- .......... d.t....... .......... .........t .A......E.
 5 .......... ......TLt. .......... .........- .......... ..t....... .......... .......... .V......E.
 6 .K..P...Vv ...Ks..... A.....T.T. K......Pl- ....H...T. d.t....... .........e ..v....R.t .E..G..NK.
 7 .K..P...Vv ...Ks..... A.....T.T. K......Pl- T...H...T. d.t....... .........e ..v....R.t .E..G..NK.
 8 .P......A. ...Ks..... AV....T.m. K........- s......... ..t....... .........A l......K.. .T..G...SE
 9 .a........ ......T.t. .....SS..v .........- .......... d.t....... .......... .......... .V......E.
10 .......... ......S.t. .......... ..f......- .......... .......... .......... .......... .Ve..s..Q.
11 .a........ ...TA.T.t. .......... .........- .......... .......... .......... .......... .T......AK
12 .K..E..CR. .C...KT... ........S. .........- .....a..T. d......... .......... l......... .KeM....AK
13 s......... ......T.I. .......... .........- .......... .......... .......... .........t .Ler..N.E.
14 ....P..... ...KE.v... .......... ........-. .......... .......... .....T.... ..v.N..... .E..t..QK.
15 ....P..... ...KE.v... .......... ..f....Ra. .r........ ...v...... v....T.C.. .......... .E..t..QK.
16 ....P..... ...KE.v... .......... ........-. .......... .......... .....T.C.. .......... .E..t..QK.
17 .a........ ...Ks..... ........H. ..f....T.- ...CH...T. d......... .......... l......S.. .Ee.G...R.
18 .K..P..s.. S..KFE.... .A........ ..f.....-k ......W... .......... .........Y v.S....I.. .Ve.G...S.
19 .R..P..... ...Ks..... AAAG.TS.E. R.fQ...El- Vr..K.i.N. T.A......k .......... ...D..i... .R..A.N.DS
20 .T..Ak.... ...Hs.v... AE...SA.T. H.LQ...K-. AA..KS..Q. N.D..N.... .m........ ...D..H... .K..Ns..Eq
21 .K..E..... ..SKs.v... AA.S.TA.D. n.LQ...P.- sA..KT..r. Q.tD...... .......s.. ..K...v..R .K.LG.N.Eq
 0 TVLIKRNTTI PTKQTQIFST YSDNQPGVLI QVYEGERAMR TKDNNLLGKF ELSGIPPAPR GVPQIEVTFD IDANGILNVS AXDKSTGKXN

 0 KITITNDKGR LSKEEIERMV QEAEKYKAED EKXRERIXAK NXLESYAFNM KXTVEDXEKL XGKLXDEDKX KILDKCXEII SWLDXNQXAX
 1 .......... .......... .......... .VQ...vS.. .A........ .SA...-.G. K..iSeA..K .v....Q.v. ....A.TL.-
 2 .......... ....d..... .......... ..Q.dkvSS. .S........ .A....-... Q..iN....Q ......N... N...K..T.-
 3 .......... ....d..... .......... ..Q.dkvSS. .S........ .A....-... Q..iN....Q ......N... ....K..T.-
 4 .......... ...dd.d... ....r..S.. .AN.d.vA.. .Av...Ty.i .Q....-... R..iSeQ..N ......Q.v. N...R..M.-
 5 .......... ....d..... .......... ..Q.dkvSS. .S........ .A....-... Q..iN....Q ......N... ....K..T.-
 6 .......QN. .tP....... nd...fAE.. K.Lk..iDTr .E.....ySl .NQiG.K... G...SP...E TmEKAVE.K. E..eSH.D.-
 7 .......QN. .tP....... nd...fAE.. K.Lk...D.r .E.....ySl .NQiG.K... G...SS...E T.EKAVE.K. E..eSH.D.-
 8 S......... .tQ...d... E....fAS.. ASIkAkvESr .K..N..hSl .NQ.NG-D-. GE..Ee...E Tl..AANdvl E...D.FETA
 9 .......... ...dd.d... .......... .AN.d.vG.. .S....Ty.. .Q....-... K..iS.Q..Q .v....Q.v. ....R..M.-
10 .......... ....d..k.. .........d. dAQ...vD.. .A.......l .SM...-.Nv K..iS....R T.Se..TQv. ...eN..L.-
11 Q.......D. F..dd..... n........d. .AQkd..G.. .G.......l .Q.i..-... KD.iSP...K ..E...D..l K...S..T.-
12 N...K..... ..QA..d... n......AD.. ..H.Q..TSr .A....v..v .Qs..--QAP A...DeA..N Sv....N.T. R...S.TT.-
13 .......... ....d..... n......rN.. ..Qk.T.A.. .G....C... .A.lde-dN. KT.iS.S.rT T.....N..- K...A..L.-
14 .......... .......k.. .......... .EVkKkvD.. .A..N..y.. rN.iK.-d.i AS..PA...K ..E.AVDGa. ....S..L.-
15 .......... .......k.. .......... .ELkKkvE.. .S..N..y.. rN..K.EKI- GS..SSd..K ..E.AVDQa. ...eS..L.-
16 .......... .......... .......S.. .ELkKkvE.. .A..N..y.. rN.iK.DKI- NSQ.SAA..K r.E.AID.a. K...N..L.-
17 Q.V....... ...Ad..... Sd.A..E... KAHV??.D.. .G..N...S. .N.iN.-PNv A...D.A..N AvTTAVE.al R..ND..E.-
18 .......... ....d..k.. A....f.E.. ..ESQ..AS. .Q...I.ySl .N.iSe---a GD..EQA..D TvTK-AE.T. ....S.TT.-
19 S..vAGSS.- ..EN...Q.. nd...f.SQ. .ARkQA.ETA .KadQL.NDT ENslKe---F E..vDKAeAQ .vR.QITSlK ELvARV.Gg-
20 ....KASS.- .NEd..Qk.. Rd..ANAEA. R.FE.LvQTr .QGdHLlHST rKQ..e---a GD..PAd..T A.ESALTAlE tA.KGEDK.-
21 A...KSST.- ..Dd..d... K...-EN.dA d.Q.KEEVE- --.RNE.Dql VF.T.KTL.D LEGKVe.AE- -vTKAKEAKD ALKAAIEKN-
 0 KITITNDKGR LSKEEIERMV QEAEKYKAED EKXRERIXAK NXLESYAFNM KXTVEDXEKL XGKLXDEDKX KILDKCXEII SWLDXNQXAX

 0 EKEEFEHKQK ELEXVCNPII TKLYQXXXXX XXXXXXXXXX XXXXXXXXXX XXXXXXXXXX XXSGPTIEEV D*
 1 ..d.....R. ...Q...... sG...GAGGP GPGGFGAQGP KGGSG----- ---------- --........ .*
 2 .......Q.. ...K...... .....SAGGM PGGMPGGFPG GGAPPSGGAS ---------- --........ .*
 3 .......Q.. ...K...... .....SAGGM PGGMPGGFPG GGAPPSGGAS ---------- --........ .*
 4 ..d.y..... ...R...... s....GGPGG GGSS------ ---------- ---------- --G....... .*
 5 .......Q.. ...K...... .....SAGGM PGGMPGGFPG GGAPPSGGAS ---------- --........ .*
 6 dI.d.KA.K. ...EiVQ... s...GSGGPP PTGE------ ---------- ---------- --EDTSEKDE L*
 7 dI.d.KS.K. ...E.VQ..v s...GSAGPP PTGE------ ---------- ---------- --EEAAEKDE L*
 8 IA.d.dE.FE S.SK.AY..T s...GGADGS GAADYDDEDE DDD------- ---------- --GDYFEHDE L*
 9 ....y..... ...Kl....v .....GAGGA GAGGS----- ---------- ---------- --G....... .*
10 ....yAFQ.. d..K..Q... .....--GGV PGGVPGGMPG SSCGAQARQG GN-------- --........ .*
11 ......SQ.. d..GlAK.Dl s....SAGGA PPGAAPGGAA GGAG------ ---------- ---....... .*
12 .....d..ME ..TRH.S..m ..mH.QGAGA AGG-PGANCG QQAGGFGGY- ---------- --....v... .*
13 ArRSTSTArR NGR-..ATRS LPSLYQGAGF -PPGGMPGGG GGMPGAAGAA GAAGAGGA-- ---....... .*
14 .V....D.M. ...Gi..... A.m.-?GEGA GMGAAAGMDE DAPSGGSG-- ---------- --A..K.... .*
15 .Vd...D.M. ...Gi..... A.m..GAGGD AGVPMDDDAP PSGVAVQDLR LRRLIKRLIK ILV.FIFISI V*
16 .Ad...D.M. ...Si..... A.m..GGAGG ATMDEDGPSV GGSAGSQTG- ---------- --A..K.... .*
17 SL..yN.r.. ...G..A..l s.m..GMGGG DGPGGMPEGM PGGMPGGMPG GMGGGMGGAA AS...Kv... .*
18 S....dD.L. ..QDiA...m s....AGGAP GGAAGGASGG FGGGAPPAPA PEA------- --E...v... .*
19 .EVNA.ELKT KT.ElQTSSm KLFE.LYKND SNNNNNNNGN ---------- ---------- --NAESGETK Q*
20 AI.AKMQELA QvSQKLME.a QQQHAQQQTA GADASANNAK DDDVVDA--- ---------- --EFEEVKDK K*
21 dL..IKA.KD ..QEiVqAlT V...EQAQQA QQAGEQGAQN DDVVDAE--- ---------- --FEEVNDDK K*
 0 EKEEFEHKQK ELEXVCNPII TKLYQXXXXX XXXXXXXXXX XXXXXXXXXX XXXXXXXXXX XXSGPTIEEV D*
```

Figure 2.10 (continued).

extensions (NTE) removed during entry into the ER (GRP) or mitochondria (SSC1). At the C-terminus the GRP end with the KDEL or HDEL tetrapeptide sequence typical of luminal ER proteins. The variable C-terminal parts of the other eukaryotic members of the HSP70 family are frequently highly enriched in $(X)_{1-3}$ GG$(X)_{1-3}$GG motives and end in the conserved SGPTIEEVD nonapeptide. (Modified from Neumann, D., et al., *Biol. Zentralbl.*, 108, 1—156, 1989. With permission.)

TABLE 2.4
Multigene Family Coding for Yeast HSP70 Proteins

Group	Genes[a] (previous names)	Expression[b] 23°C (constit.)	37°C (hs-induced)	Properties and functions
A	SSA1 (YG 100) SSA2 (YG 102)	+ + + +	+ + + + +	Relative amount of HSP70 A1/A2 about 1:4 at 23°C; A1/A2-defective strains grow slowly at 30°C and stop growing at 37°C. They exhibit high intrinsic resistance to cell killing at 52°C, probably due to increased constit. expression of A3/A4. Transient induction of A1/A2 at 23°C when cultures approach stationary phase. A1/A2 proteins, but probably also the other members of the HSP70 family, bind ATP. Protein transport into ER and mitochondria is disrupted in A1/A2/A4-defective strains (sequence contained in Figure 2.10).
	SSA3 (YG 106) SSA4 (YG 107)	(+) (+)	+ + + + + +	No defects observed with A3/A4-defective strains (growth, sporulation, induced thermotolerance at 23, 30, or 37°C)→both proteins are dispensible except in A1/A2-defective strains. A3 expressed constit. in late stationary phase cultures.
B	SSB1 SSB2	+ + + + + +	(+) (+)	Proteins coded by B1/B2 can not substitute for A1/A2. B1/B2 double mutants are cold-sensitive (growth) (for sequence data see Reference 294a, 360b).
C	SSC1	+	+ + +	C1 protein is essential for vegetative growth; transient, tenfold hs-induction; produced as preprotein transported into mitochondria; gene localized on chromosome number 10, adjacent to cyc 1 gene (for sequence see Figure 2.10).
D	SSD1	+	+ + +	So far no effect of elimination of D1 gene observed.
E	KAR2	+ +	+ + +	The GRP78 homolog of yeast; localized in the perinuclear ring of ER membranes; induced also by tunicamycin, A23187 and 2-deoxyglucose; for sequence, see Figure 2.10.

[a] Nucleotide sequence homology between individual genes is high within one group, i.e., 96% (A1 vs. A2), 92% (B1 vs. B2) and 84% for A1/A2 vs. A2/A4, but only 63% group A vs. group B and 50% for group A vs. groups C.and D.[94]

[b] Expression analyzed by promotor fusion experiments (see Table 6.1, Nos. A40 to A41a), Northern analysis and 2D-electrophoretic separation of proteins.[94, 431] For details of the developmental control see Section 8.4.

Data were compiled from References 93—94a, 120, 230b, 431, 431a.

HeLa cells is increased transiently during S-phase of the cell cycle. It migrates to the nucleus and returns to the cytoplasm in the G2-phase.[255] On the other hand, Welch and Mizzen[428] isolated rat brain HSC73 (clathrin-uncoating ATPase), conjugated it with a fluorescing rhodamine chromophore and microinjected the labeled protein into rat embryo fibroblasts. During hs a selective binding to the nuclear compartment was observed.

Two nuclear binding states can be defined in vertebrate cells:[82a,223] part or all of the nucleoplasmic HSP70 can be washed out by nonionic detergent, whereas the tightly bound HSP70 is mainly in the nucleolus and released only in an ATP-dependent manner.[82a,223] Analysis of deletion proteins of the *Drosophila* HSP70 protein expressed in transformed monkey COS cells showed that the highly conserved N-terminal part harbors the ATP-recognition sites, whereas the C-terminal part is required for nucleolar binding.[223,272] These two basic domains were also discriminated by chymotryptic digestion of the clathrin uncoating HSC71: a 44-kDa N-terminal part contains the ATPase activity, which acts clathrin independent if the C-terminal substrate-binding domain is lacking.[72a]

Translocation of HSP70 and other large proteins to the nucleus proceeds in two steps: first, the binding by recognition of the nuclear targeting signal by the pore complex, and second, ATP-dependent translocation.[289b] Four different types of targeting signals were described. Common to three of them is a cluster of four to six basic amino acid residues.[289b,443] The only sequence of this kind conserved in many HSP70 proteins is marked with a thick bar in Figure 2.10. It starts at about residue 250, and, most remarkably, the following 25 amino acid residues exhibit homology with a group of DNA-binding proteins, including transcription factors of mammals and nematodes with the POU homeobox,[131a,163] the yeast GCN4, and the avian c-jun proteins.[410a]

```
HSP70 cons.:  -KRKHKK-DISXN-KRALRRLRTACERAKRT-
human Oct1 :  -RRRKKRTSIETNIRVALEKSF--LENQKPT-(405)
yeast GCN4 :  -KRARNTEAARRSRARKLQRMKQLEDKVEEL-(260)
avian c-jun : -KRMRNRIAASKSRKRKLERIARLEEKVKTL-(252)
```

Identical or homologous amino acid residues with respect to the HSP70 consensus sequence are underlined. This part of the POU homeobox may in fact represent the nuclear targeting signal. At least, transfer of the corresponding sequence motif of the human HSP70 to the N-terminus of the usually cytoplasmic pyruvate kinase of chicken directs this enzyme to the nucleus of monkey COS cells transformed with the corresponding hybrid gene construct.[98a] Finally, it is worth noting that marked deviations from the consensus sequence in this region are observed for all HSP70 members that are not translocated to the nucleus, i.e., the GRPs78 found in the ER (nos. 6 to 8 of Figure 2.10) and the prokaryotic or organellar HSPs70 (nos. 19 to 21).

Stevenson and Calderwood[373a] revealed another intriguing property of this HSP70 sequence element. The right half

```
-KRALRRLRTACERAKRT(SSS)-
```

exhibits homology to well-known calmodulin-binding domains, e.g., of muscle myosin light chain kinase. In fact, the purified mouse HSC70 interacts with calmodulin in a Ca^{2+}-dependent manner, and a synthetic oligopeptide of the indicated structure inhibits calmodulin binding and hence activity of cyclic nucleotide phosphodiesterase.

It is essential to point to a discrepancy of results on possible functional domains of the HSP70 obtained with two different methods. Deletion analyses with *Drosophila* and human hsp70 genes transferred to HeLa and monkey cells were reported.[223,255a,272] In agreement with the sequence data (see below), the N-terminal part, or more precisely the region of amino acid residues 122 to 264,[255a] harbors the ATP-binding site. But, in contrast to the putative nuclear targeting signal between amino acid residues 250 and 275, the C-terminal half of the molecule (amino acid residues 351 to 641) was found to be necessary and sufficient for stress-dependent transport to the nucleus/nucleolus. A decision on the contributions of both parts of the molecule can only be derived from mutants generated by site-directed mutagenesis.

A general concept for the mechanism of action of HSP70-related proteins under normothermic and hyperthermic conditions was put forward by Pelham[312] (see also review by Bienz and Pelham[41]). Starting with the observation by Lewis and Pelham[223] on an ATP-dependent release of HSP70 from isolated nuclei of COS cells, Pelham[312,312b] suggests that HSP70 in the cytoplasmic and nuclear compartment or GRP78 in the ER may help to solubilize high molecular weight aggregates of proteins exposing hydrophobic regions.[273] Examples are preribosomes, precursors of multimeric proteins, e.g., IgH chains, aberrant proteins or partially denatured proteins formed under hs conditions. The ATP-dependent disaggregation of these complexes may help in protein assembly, but also in the reactivation or degradation of their constituent proteins.

The ATP-dependent release of HSP70 from its tight nuclear binding sites suggests a type of shuttle mechanism for removal of unwanted proteins in the recovery period. In fact, an unusual accumulation of nuclear matrix proteins is observed in heat-shocked vertebrate cells (see Section 9.3). This also includes a considerable number of oncogene products.[121,231,370] Interestingly, some of them were found associated with proteins of the HSP70 family, e.g. p53, SV40 large T-antigen, mutant polyomavirus medium T-antigen.[81b,126a,165b,315,375,418] However, the role of this interaction with oncogene products may, by far, exceed the stress and recovery period. The immortalizing nuclear oncogene p53 was always found associated with constitutively expressed members of the HSP70 family. Mutant proteins of p53 with higher affinity for HSC70 exhibit greatly extended half-lives and increased transforming capacity.[81b,126a,165a,315] Even if the p53 gene was transferred to *E. coli* the oncogene product was detected as complex with the resident bacterial HSP70 (DnaK), and this complex could be dissociated *in vitro* by addition of ATP.[81b]

The evident role of HSP70-type proteins in stabilizing other proteins is reminiscent of the early suggestion of a function for HSPs (Minton et al.[258]): "We suggest that one of the functions of heat shock proteins, may be to stabilize other proteins kinetically in a . . . nonspecific fashion." Recent evidence in support of this comes from experiments with the HSP70-defective yeast mutants generated by Craig and co-workers (see Table 2.4). Representatives of the SSA family play an essential role for import of immature precursor proteins into the ER and mitochondria,[74b,108] and purified SSA1/SSA2 proteins function as the cytosolic factor required for preprotein import by yeast mitochondria.[274a] But analogous results with HSC72 and protein import into dog pancrease microsomes were also reported by Zimmermann et al.[447a] HSP70-type proteins may help to keep the proteins in an unfolded state, necessary for their efficient transport through membranes, or alternatively, they may act as ATP-dependent unfoldase.[312a,312b] At any rate, members of the HSP70 family exhibit high salt-resistant (1*M* NaCl) affinity for short unfolded peptides attached to Affigel columns. Km values may be as low as 12 μM, and the release from this affinity column is brought about by ATP.[127a] Steric accessibility of these peptides in partially unfolded proteins may be decisive for their interaction with HSP70.

When inspecting the collection of HSP70 proteins in Figure 2.10, a potential site for interaction with ATP is marked (brackets) by homology to the well-characterized ATP binding sites of protein kinases (PK):[155a,416a]

$$\begin{array}{ll} \text{PK} & :-\genfrac{}{}{0pt}{}{L}{I}GXGX\genfrac{}{}{0pt}{}{F}{Y}GXV\genfrac{}{}{0pt}{}{Y}{W}\text{-7-15aa-}VA\genfrac{}{}{0pt}{}{V}{I}KX\genfrac{}{}{0pt}{}{LK}{TR}- \\ \text{HSP70:} & -\underline{LG}G\underline{G}T\underline{F}DVSI\text{-7aa----}FE\underline{VK}ATA- \end{array}$$

The envisaged functions of HSP70 are evidently very close to the role of clathrin-uncoating ATPase or the catalytic activities of members of the HSP70 family in the reorganization of cytoskeletal systems. It can be anticipated that in most cases a substrate-dependent ATPase activity is involved. However, this concept may even extend to the other two large HSP families. The GroEL-type protein of *E. coli* as well as its homologs of chloroplasts and mitochondria have ATPase activity and help to assemble high molecular weight protein complexes (see Section 2.3.3), whereas the members of the HSP90 family are bound to transient protein complexes in the cytoplasm or function as GRP94 in the ER (see Section 2.3.1). But the latter have no ATPase activity.

Members of the HSP70 family can be modified by methylation and/or phosphorylation. This is summarized in Table 2.5. GRP78 is a major substrate for ADP-ribosylation in

TABLE 2.5
Modification of Heat Shock and Related Stress Proteins

Protein	Modification/system	Ref.
HSP90 family		
GRP94	Glycoprotein of the ER/Golgi system in vertebrates, may be phosphorylated	84, 229, 354b, 429
HSP90	Phosphoprotein in vertebrates modified at Thr and Ser residues	58, 185, 217a, b, 416
HSP80	Phosphoprotein in tomato cell cultures	296
HSP70 family		
GRP78	Phosphorylated/ADP-ribosylated protein in ER of vertebrates	68, 214, 429
	Cotranslational methylation of Lys residues (= SP83) in chicken	421, 422
HSP70	Phosphorylation of Thr and Ser residues in *Dictyostelium*	234a
	Phosphorylated and methylated in tomato cell cultures	296
	Methylated in *Drosophila* cells during recovery	7
	Methylated at Lys and Arg residues, chicken	421, 422, 419
	Autophosphorylation of DnaK in *E.coli*	451
	Ca^{2+}-dependent *in vitro* phosphorylation of human mitochondrial HSP70	222b
	Autophosphorylation of DnaK in *E. coli*	451
HSP20 family		
HSP25—27	Two Ser-phosphorylated isoforms in mammalian cells, phosphorylation increased by hs or treatment with arsenite, tumor promoters, or tumor necrosis factor-α	8a, 10, 33a, 130b, 189, 190, 330a, 426
HSP26/27	10—20% of total amount phosphorylated in *Drosophila*	331b
HSP17	Methylated in tomato cell cultures	296
HSP47	A phosphorylated membrane sialoglycoprotein of vertebrates, preferentially induced by mild hs, represents the major collagen-binding protein (see Section 2.3.6)	276, 276a

vertebrate cells.[68] A peculiarity detected in rat liver and brain is the close association, of HSP70 with palmitic or stearic acid.[147,147a] An autoproteolytic activity of HSP70 was reported by Mitchell et al.[262] for *Drosophila,* CHO, and mouse cells. However, it remains to be shown that this is a property of the native, highly purified protein itself, which is now available by ATP-affinity chromatography. In this respect, it is intriguing that also the clathrin uncoating ATPase (HSC71-73) shows self-degradation.[396]

The identification of the 73-kDa poly(A)-binding protein of HeLa cells as a member of the HSP70 family[347] is evidently wrong. At least in yeast, where both coding genes are well defined, there is no relation between the hsp70 gene family and the gene coding for the poly(A)-binding protein.[336] Moreover, purified rat HSP70 has no poly(A)-binding activity.[41]

2.3.2.3. Multiplicity of HSP70 Proteins

In most cases, the multiplicity of the HSP70 family in different organisms is not yet fully analyzed. Mammalian cells contain at least eight different proteins (see Figure 6C and D), i.e., the GRP78, the dominant constitutively expressed and moderately hs-induced clathrin-uncoating ATPase (HSC71 to 73), a mitochondrial DnaK-like protein,[220b] three types of HSC70 specifically expressed during sperm cell development (Section 8.4)[1c,198a] and two hs-induced HSP70, one of which is also under cell cycle control.[191,235,255,273,357,423] The estimate of even 10 to 20 genes from Southern analyses indicates the existence of a considerable number of pseudogenes.[271,302]

In the nematode *Caenorhabditis elegans,* one abundantly transcribed and hs-induced HSP70A is accompanied by a constitutively expressed GRP78 (HSP70C) and the mitochondrial DnaK-like HSP70F. In addition, there is one pseudogene derived from the hsp70A gene.[163a,362,362a]

In *Drosophila,* two types of hs-induced proteins of 70 kDa (HSP70 and 68) are opposed to two related dominant constitutive proteins HSCs70 and 72 plus two minor constitutively expressed genes, whose protein products were not yet identified (Table 2.3). The four proteins are detected by a monoclonal antibody generated by Velazquez et al.,[404] which evidently recognizes a common domain also in related proteins of other species. Despite an increase in the level of hsc4 transcripts during hs, translation is reduced, but the abundance of this protein always significantly surpasses that of HSP70.[307] Both 70-kDa proteins (HSP70 and HSC70) may exert similar functions. They bind to cytoskeletal elements and migrate to the nucleus after stress treatment.[307,402,403]

The properties of nine members of the yeast HSP70 family are summarized in Table 2.4. Their protein products are identified in Figure 2.5. The original nomenclature of the genes[93,176] (YG100 to 107) was revised.[431] The yeast HSP70 family can be classified into group A (four members with constitutive and/or hs-induced synthesis), group B (two members with hs-repressed synthesis), group C (a transiently induced HSP with mitochondrial localization), the yeast GRP78, and the not yet identified SSD1. The differential effects of control vs. hs conditions on gene activity were analyzed by means of promoter fusions to the *E. coli* lacZ gene and testing the β-galactosidase levels in transformed yeast cells.[120,431a] To define the possible functions of the individual members of this protein family, the corresponding genes were made defective by site-directed insertion mutagenesis.[89,89a,93-94a,431] The results are summarized in the last column of Table 2.4. As already mentioned, recent analysis of SSA mutants demonstrated that these members of the HSP70 family are essential for intracellular protein topogenesis.[74b,108,274a] All members of the yeast HSP70 family show a complex pattern of developmental control of expression (see Section 8.4).[431a]

The complexity of the plant HSP70 family is far from clear. Two intron-containing genes from maize and *Petunia,* respectively, were sequenced.[331,437] But it is evident that separate genes code for the constitutive and hs-induced proteins of the HSP70 complex.[114a] Careful inspection of the corresponding regions of two-dimensional gels of the tomato proteins (Figure 2.2) shows that, in addition to the dominant protein(s) with M_r 70 kDa, there are the slightly more basic HSP68 and one or two species of an HSP75. Our investigations with corresponding antibodies showed that there is no cross-reaction between the three protein types of the plant HSP70 complex. The HSP68 is a mitochondrial protein of the DnaK type.[289,289a]

Five to eight members of the HSP70 family were detected by two-dimensional analyses of *in vivo* and *in vitro* labeled products from *Arabidopsis.*[440a] Cloning and partial sequencing of two adjacent genes revealed introns inserted in codon 72. Gene 1 is constitutively expressed and moderately hs-induced. Gene 2 presumably represents a pseudogene. The multiplicity of the plant HSP70 family evidently also involves three different constitutive members in chloroplasts, one in the outer envelope membrane and two in the stroma. By immunological methods, the HSC78 detected in pea chloroplasts was shown to be a DnaK-like protein.[3a,240a]

2.3.3. THE HSP60 FAMILY

In contrast to the other three classes, the HSP60 family is a new class of mitochondrial HSPs with homology to the dominant hs-induced bacterial groEL product. Figure 2.11 summarizes sequence information for the mitochondrial HSP60 of mammals[177b,314a] and yeast,[323a] for the corresponding protein of wheat chloroplasts,[161a] and for four bacterial GroEL proteins.[161a,354,408a] Though the overall homology is not as extended as for the preceding HSP90 and HSP70 families, the typical mosaic of highly conserved and less conserved parts is observed as well (Figure 2.9).

The new class of HSP60 proteins was originally detected by McMullin and Hallberg.[250,251] In *Tetrahymena thermophila,* a constitutive mitochondrial protein (HSP58) is synthesized in the cytosol, transported to the mitochondria, and exists as a homooligomeric complex with hollow core morphology and 12.5 nm diameter. In the early stages of the hs responsive, its synthesis transiently increases. Accumulation of HSP58 correlates with the severity of the stress treatment. Antibodies to HSP58 revealed similar proteins in mitochondria of yeast (p64), *Xenopus* (p60), maize (p62), and human cells (p59). Moreover, there is cross-reaction with the bacterial GroEL protein. Similar to the 12.5 nm particles of *Tetrahymena,* homooligomeric 20S complexes were also isolated from mitochondria of yeast,[74,251] *Neurospora*[170a] and corn.[319c] Heat shock induction was demonstrated for the HSP60 of yeast, corn and human cells.[251,319c,415a]

The 20S aggregates of HSP60 are very much reminiscent of the peculiar structure of the GroEL complex in bacteria: each of the seven subunits of GroES and GroEL form a heptameric complex with a morphology similar to the mitochondrial HSP60 particles.[162,162a] This complex is bound to the bacterial membrane and functions as the morphogenetic core for phage head assembly. It is remarkable to notice that the well-known plant chloroplastic protein associated with unassembled large subunit complexes of ribulose-bisphosphate carboxylase[274b] evidently represents the chloroplast member of the HSP60 family[161a] (see Figure 2.11, no. 2). Thus, plants have two immunologically differentiated members of the HSP60 family, one in chloroplasts and one in mitochondria.[319c]

It was already mentioned that the functional identity of the *E. coli* GroEL with the chloroplast HSP60 was established by Goloubinoff et al.[137a] (see review by Roy[334a]). The assembly of ribulose-bisphosphate carboxylase in *E. coli,* transformed with the *Anacystis nidulans* rubisco genes, requires GroEL. A similar function of HSP60 as "molecular chaperonin" in protein assembly was also demonstrated for yeast mitochondria.[74,304a] The catalytic activity of HSP60 in mitochondria needs ATP and the typical 20S particles. In a ts mutant (Mif4), formation of high-molecular-weight HSP60 aggregates at the nonpermissive temperature is connected with defects in the assembly of three different protein complexes (ornithine transcarbamylase, F_1-ATPase, and cytochrome b_2).[74]

The proteins described by McMullin and Hallberg[250,251] may be identical with the HSP52-62 detected in plant mitochondria[279,293,359] and chloroplasts.[198b] However, HSPs in the 60-kDa region were also found in *Xenopus,*[100,291] rainbow trout fibroblasts,[269] in human embryo fibroblasts, and BHK cells.[213] Their possible relation to the GroEL family of proteins remains to be investigated. In two cases, there is evidence that HSP60 may be coded for and synthesized by plant mitochondria[359] and chloroplasts, respectively.[198b] But these results remain to be confirmed by identification of the corresponding genes.

2.3.4. THE HSP20 FAMILY

Unlike the three "large HSP" families with proteins of 90, 70, and 60 kDa, the multiplicity of the low molecular weight (lmw) HSPs is much more variable between different organisms. Despite this heterogeneity also in size, they are collected under the family term HSP20. There is evidently a single lmw HSP in yeast (HSP26),[53a,313,379a] chicken (HSP24),[82,82a] and mammalian cells (HSP27).[10,165] However, 4 major small HSPs in *Drosophila*[19,174,366] (HSP27, 26, 23, 22); 8 to 10 small HSPs in *Dictyostelium,*[233,234] or *Xenopus*[100] (Figure 8.3), and generally, more than 20 lmw HSPs are detected in plants[239,289,296] (Figures 2.2 and 2.7). The amino acid sequences of 23 small HSPs from yeast, *Drosophila,* the nematode *Caenorhabditis elegans,* human, and plants are compiled in Figure 2.12. In the C-terminal part of all genes a 36 amino acid residue sequence element is remarkably conserved, especially if the patterns of hydrophilic and hydrophobic residues (hydropathy index) are considered. This part of the HSP20 proteins is detailed in Figures 2.13 and 2.14. Interestingly, the vertebrate α-crystallin contains a structurally related domain,[174] which corresponds to the third plus part of the fourth exon of the coding gene.[400] Elements of this sequence are even found in a 40-kDa major egg antigen of a parasitic worm *Schistosoma mansoni* living in liver and intestine of mammals[286] and in a mycobacterial HSP of 16.6

```
0 CONSENSUS AA  MW      MXXXXXXXXX XXXXXXXXXX XXXXXXAKXI KFDXDARXXL XRGVNXLADA VKVTLGPKGR NVVLEKSWGA
1 Hs-HSP60  573 61,049  .LRLPTVFRQ MRPVSRVLAP HLTRAY..Dv ..GA...ALm LQ..DL.... .A..m..... T.ii.Q...S
2 Ta-HSP60  543 57,516  G--------- ---------- ----AD..E. A..QKS.AA. QA..EK..N. .G.....r.. ....dE-y.N
3 Sc-HSP60  572 60,830  .LRSSVVRSR ATLRPLLRRA YSS---H.El ..GVeG.AS. Lk..ET..e. .Aa....... ..li.QPf.P
4 Mt-GroEL  540 56,751  .--------- ---------- ------..T. Ay.Ee..RG. E..l.A.... .......... ......K...
5 Ml-GroEL  540 56,695  .--------- ---------- ------..T. Ay.Ee..RG. E..l.S.... .......... ......K...
6 Ec-GroEL  547 57,140  .--------- ---------- -----A..Dv ..GN...VKm L....V.... .......... ....d..f..
7 Cb-htpB   552 58,249  .--------- ---------- -----A..Vl ..SHevLHAm S...EV..N. .......... ....d..f..

0 PTITNDGVSI AKEIELEDKY ENIGAXLVKE VAXKTNDXAG DGTTTATVLA QAIIKEGLRX VAAGANPVXL KRGIEKAVXA VIEELXKXAK
1 .Kv.K...tv ..S.d.K... K....K..Qd ..NN..eE.. .......... RS.A...FEK iSK.....Ei r..vML..D. ..A..K.QS.
2 .KvV....t. .rA...ANPM ..a..A.ir. ..S....S.. ......C... RE...L.iLS .TS.....S. .k..d.T.QG l....ErK.r
3 .K..K...TV ..S.V.K..f ..m..K.lQ. ..S...eA.. ....s....G R..FT.SvkN ....c..mD. r..SQV..EK ...F.SANK.
4 .......... ........P. .K...E.... ..K..D.V.. .......... ..lR.....N .......lG. ........EK .T.T.L.G..
5 .......... ........P. .K...E.... ..K..D.V.. .......... ..lv.....N .......lG. ........DK .T.T.L.D..
6 ....K....v .r.......f ..m..Qm... ..S.A..A.. ..NN...... ....T...KA ....m..mD. ....d...T. av...KALSV
7 ....K....v .........f ..m..Qm... ..Sr.S.D.. .......... ...lVd.ikA .i..m..mD. ....d...T. avA..K.IS.

0 PVXTXEXIAQ VATISAXGDX SIGXLIAEAM DKVGXEGVIT VEDGXTFXXX LELXEGMXFD RGYISPYFVT XXEXXXAELE XPYILLXDKK
1 ..T.P.E... ......N..K E..Ni.Sd.. K...RK.... .K..K.LNDE ..iI...K.. .........iN TSKGQKC.FQ DA.v..Se..
2 ..KGSGD.KA ..s...GN.E L..Am..d.i ....Pd..ls i.SSSs.ETT vdvE...EI. ......Q... NL.KSIv.F. NARvliT.Q.
3 EiT.S.E... ......N..S Hv.K.l.S.. e...K..... iRe.R.LEDE ..vT...R.. ..f.....i. DPKSSKv.F. K.Ll..Se..
4 E.E.K.Q..A T.A...-..Q ...D...... ....N..... ..eSN..GLQ ...T...R.. k....G.... DP.RQE.V.. D.....VSS.
5 E.E.K.Q..A T.A...-..Q ...D...... ....N..... ..eSN..GLQ ...T...R.. k....G.... DA.RQE.V.. E.....VSS.
6 .CSDSKA... .G....TS.E tv.K...... ....K..... ....TGLQDE .dvV...Q.. ...l....iN KP.TGAv... S.f...A...
7 .CKDQKA... .G....NS.K ...Di..... e...K..... ....SGLENA ..vV...Q.. ...l....iN NQQNMS.... N.f...V...

0 ISSIKDILPL LEXVXXAGKP LLIIAEDVEG EALATLVVNK IRGXVKVVAV KAPGFGDRRK AMLQDMAILT GGXVISEEXL GLXLENAXLX
1 ....QS.v.a ..IaNAHR.. .v......d. ...S...l.r lkVGlQ.... .......N.. NQ.K....a. ..A.FG..G. T.N..DvQPH
2 .t...e.i.. ..QTTQLRC. .F.v...iT. .......... l..IiN.a.i ...S..e... .v...i..v. .AEYlAKd-. ..Lv...TvD
3 ....Q....a ..ISNQSRr. ........d. ....ACil.. l..Q...c.. .......N.. NTiG.i.v.. ..T.Ft..-. D.KP.qCTiE
4 v.tv..l... ..K.IG.... .......... ...S...... ...TF.S... .......... .......... ..Q.....-v ..T....D.S
5 v.tv..l... ..K.IQ...S .......... ...S...... ...TF.S... .......... .......... .AQ.....-v ..T...TD.S
6 ..N.rem..v ..A.AK.... .......... ......a...T ...I...a.. .......... .....i.T.. ..T.....-i .mE..K.T.E
7 ..N.reli.. ..N.AKS.r. ..v....i.. .........N ...V...a.. .......... .....i.v.. ..K.....-v ..S..A.S.D

0 XLGKARKVVV TKDDTTIVEG AGDXDAIXGR VAQIRXEIEX SDSDXYDREK LQERLAKLAG GVAVIKVGAA TEVELKERKX RIEDALXATR
1 D...vGE.i. ....AMllK. K..KAQ.EK. iQE.IEQldV tT.e-.ek.. .N......SD ....l...GT sd..vN.k.D .vT....N...
2 Q..T...iTi HQTT..liAD .ASK.E.QA. ...lkK.lSE t..l-..S.. .A..i...S. .........T ..T..Ed.QL .....KN..F
3 N..SCDSiT. ..e..v.lN. S.PKe..QE. iE..kGS.dI tTtNS.ck.. ........S. .....r..G. s...vG.k.D .Yd...N...
4 L......... ...e...... ...T...A.. .....Q...N ....-..... .......... ......a... .........H .....vRNAk
5 L........m ...e...... ...T...A.. .....T...N ....-..... .......... ......a... .........H .....vRNAk
6 D..Q.kr..i N..T...id. v.eEA..Q.. .....QQ..E AT..-..... ....v..... .......... ....m..k.A .v....H...
7 D..S.kr... .......id. S..AGD.KN. .E...K..EN .S..-..k.. .......... .......... ....m..k.A .v....H...

0 AAVEEGIVAG GGVALLXXXX XAPALDXLXX XNXDXXVGAX IIKRALEAPL XQIAXNXGVE PXVVAEKVXX XXXXXXXXGY NAXTGEYEDL
1 ........l. ..C...R--- Ci....S.TP A.E.QKi.iE ....T.Ki.A MT..K.A... GSliv..iMQ S---SSEV.. D.MA.dfVNm
2 ..i.....P. ..a.YVHLST Yv..iKE-TI EDH.ERl..D ..Qk..Q..A SL..N.A... GE..i..iKE S---EWEM.. ..M.DK..N.
3 .......lP. ..T..vK--- ASRv..EvVV D.F.QKl.vD ..r..iTR.a K..iE.A.E. GS.iiG.lID EYGDDFAK.. D.SKS..T.m
4 .......... ...T..Q--- A..T..E.K- LEG.EAT..N .v.V...... K...F.S.l. .G......RN L---PAGH.L ..Q..V....
5 .......... ...T..Q--- A.....K.KL TGDeAT-..N .v.V...... K...F.S.m. .G......RN L---SVGH.L ..A.......
6 ......v... ......iR--- V.SK.AD.RG Q.E.QN..iK vaL..m.... R..vL.C.E. .S...NT.KG G---DGNY.. ..A.E..GNm
7 ......v.P. .....iRVLK S---..SvEV E.E.QR..vE .ar..mAYP. S..vK.T..Q AA...d..LN H--KDVNY.. ..A....G.m

0 LXXGVIDPXK VTRSALQNAA SVAGLMLTTE AVVXDKPKXX XAXAXGAXXM GGMGGGMGGM GGMX*
1 vEK.i...T. .V.t..LD.. G..S.lT.A. V..TeI..EE KDPGM..--. .......GG- -..F*
2 iES.....A. ...C...... ..S.mv...Q .i.Ve...PK PKV.EP.EGQ l--------- --SV*
3 .AT.i...F. .V..G.VD.S G..S.la... vaiV.A.EPP A.A--..--- ...P...P.. P..M*
4 .AA..a..V. .......... .i...F.... ...A...EKE K.SvP.GGD. ...------- --DF*
5 .KA..a..V. .......... .i...FT.-. ...A...EKT A.P.SDPT-- .....-.--- --DF*
6 iDM.il..T. .......Y.. ......i... Cm.T.L..ND -.ADL..AG- -.....-... ...M*
7 iEM.il..T. ...t...... .i....i... Cm.TeA..KK EESmP.GGD. .....-.... ...M*
```

FIGURE 2.11. Sequence comparison of seven members of the HSP60 family. For explanations and details of presentation see legend to Fig. 2.8. Sequences are from: (1) human mitochondrial HSP60;[177b] (2) wheat chloroplast HSP60 = binding protein of ribulose-biophosphate carboxylase large subunit;[161a] (3) yeast mitochondrial HSP60;[323a] (4—7) bacterial GroEL proteins from *Myco-bacterium tuberculosis* (4), *M. leprae* (5),[354] *E. Coli* (6),[161a] and *Coxiella burnetii* (7).[408a] (From Neumann, D. et al., *Biol. Zentralbl.*, 108, 1—156, 1989. With permission.)

kDa.[205a,286a] Augusteyn et al.[17] prepared monoclonal antibodies against this conserved domain of bovine α-crystallin and found cross-reactivity with *Drosophila* small HSPs.

Careful sequence analyses of many members of the HSP20 family confirmed a single 35 to 37 amino acid residue conserved domain for all proteins (see legend to Figure 2.12).[289] It is followed by a short variable stretch of amino acid residues forming the C-terminus. For the yeast HSP26, this is the only homology, whereas other members of this family are characterized by further, group-specific domains extending towards the N-terminus. This is true for a 40 to 42 amino acid residue region specific for all animal HSP20 proteins (nematodes, *Drosophila,* and vertebrates) and for a corresponding region of 48 to 52 residues specific for members of the HSP20 family in plants. Even the bacterial p16.6 belongs to the latter group.[205a,286a] Finally, there are two further homology regions at the N-terminus, different for plant cytoplasmic HSP15 to 18 and the organellar HSPs (Figure 2.15), respectively. Thus the mosaic structure of HSP20 proteins, though much more complicated, is basically similar to the structure of the large HSPs (Figure 2.9).

Homology to α-crystallin and also the similarity of isoelectric points for the *Drosophila* small HSPs with two basic (HSP26, 27, pI 7.5) and two acidic proteins (HSP23, 22, pI 5.8 to 5.9) led Ingolia and Craig[174] to the proposal that the tendency for aggregate formation[8,9,11,11a,114,151,218,348] is bound to the α-crystallin-like features. However, similar high molecular weight complexes of small HSPs are also observed in vertebrates[10,11a,82a] yeast,[333a] and plants.[287,288,289,297,298] These particles of an estimated molecular weight of 500,000 Da sediment in the range of 12- to 16S. Under hs they form even larger perinuclear aggregates in connection with cytoskeletal elements[83,298] which are presumably involved in transient segregation of untranslated mRNAs.[298] Further details are discussed in Sections 11.4 and 16.7.

Small HSPs may be modified by phosphorylation or methylation (Table 2.5). The phosphorylation state of the mammalian HSP27 is highly variable depending on the addition of serum factors, tumor promoters, or on the stress conditions.[10,33a,426]

The great multiplicity of plant small HSPs can be grouped into three subfamilies. This was documented by Schöffl and Key[345,345a] by cross-hybridization data using different cDNA clones of soybean. Meanwhile nine members of subfamily I genes were sequenced (Figure 2.12). They code for the dominant small plant HSPs with M_r 17 kDa.[97,276,323,346] In addition, the structure of one unrelated subfamily VI protein (HSP17.9 D, Figures 2.12 and 2.13) was recently analyzed.

Members of a third subfamily with M_r 21 to 24 kDa are localized in chloroplasts.[134,193,350,379,406,407] They are prominent, newly formed HSPs, synthesized in the cytoplasm and transported to the chloroplast. In pea and soybean, corresponding precursor proteins with a 5- to 6-kDa transit peptide at the N-terminus were characterized and shown to be taken up *in vitro* by isolated chloroplasts.[193,406,407] The coding parts of two genes from *Chlamydomonas*[141] and pea,[407] respectively, are shown in Figure 2.15. In contrast to the pea gene, no transit peptide sequence could be identified in the *Chlamydomonas* gene, and, correspondingly, no precursor protein was found in this case.[193,350] The tight membrane binding of HSP21 in heat-shocked pea and *Chlamydomonas*[134,193,350] and evidence for a protective effect on photosynthetic membranes suggest a direct role of HSP21 in this respect. The photosystem II complex may be particularly affected.[134,350] It is interesting to notice that in the chloroplast-free tomato cell cultures, the corresponding HSP21 is contained in proplastids, but so is a prominent cytoplasmic HSP which becomes part of the hs granules (see Figure 2.2 and table given in legend). It will be interesting to learn, whether this is a peculiarity of these cell cultures, or whether HSP21 plays a dual role. Recently, a gene coding for a HSP32 of *Neurospora crassa* mitochondria was sequenced[318] (see Figure 2.15). Similar to the chloroplastic counterparts the protein is synthesized in the cytoplasm, but there is no evidence for a preprotein with an N-terminal transit peptide.

Two lmw plant HSPs were reported that, because they lack sequence homology, cannot be assigned to the HSP20 family. One is a general stress-inducible HSP26 of soybean (see Section 1.2.4).[96a,97a,152a] The other is an HSP25 detected in heat-shocked photoautotrophic *Chenopodium rubrum* cell cultures. The sequence of 204 amino acid residues contains a potential transmembrane domain at the N-terminus, whereas the C-terminus is very distantly related to other plant HSP20 proteins.[194a]

2.3.5. THE UBIQUITIN (HSP8.5) FAMILY

Because of its low molecular weight, hs induction of ubiquitin in eukaryotic cells remained undetected for many years. A remarkable indicator of profound disturbances in proteolytic homeostasis (see Sections 1.4 and 12.2) is the fact that, in bacteria as well as in eukaryotes, key components of the proteolytic systems are members of the hs-regulated

```
1: Hs-HSP27     MW=22,327   199 AA
     MTERRVPFSL LRGPSWDPFR DWYPHSRLFD QAFGLPRLPE EWSQWLGGSS WPGYVRPLPP
     AAIESPAVAA PAYSRALSRQ LSSGVSEIRH TADRWRVSLD VNHFAPDELT VKTKDGVVEI
     TGKHEERQDE HGYISRCFTR KYTLPPGVDP TQVSSSLSPE GTLTVEAPMP KLATQSNEIT
     IPVTFESRAQ LGGRSCKIR

2: Sc-HSP26      MW=23,877   214 AA
     MSFNSPFFDF FDNINNEVDA FNRLLGEGGL RGYAPRRQLA NTPAKDSTGK EVARPNNYAG
     ALYDPRDETL DDWFDNDLSL FPSGFGFPRS VAVPVDILDH DNNYELKVVV PGVKSKKDID
     IEYHQNKNQI LVSGEIPSTL NEESKDKVKV KESSSGKFKR VITLPDYPGV DADNIKADYA
     NGVLTLTVPK LKPQKDGKNH VKKIEVSSQE SWGN

3: Ce-HSP16-1    MW=16,301   145 AA
     MSLYHYFRPA QRSVFGDLMR DMAQMERQFT PVCRGSPSES SEIVNNDQKF AINLNVSQFK
     PEDLKINLDG HTLSIQGEQE LKTEHGYSKK SFSRVILLPE DVDVGAVASN LSEDGKLSIE
     APKKEAIQGR SIPIQQAPVE QKTSE

4: Ce-HSP16-2    MW=16,220   145 AA
     MSLYHYFRPA QRSVFGDLMR DMALMERQFA PVCRISPSES SEIVNNDQKF AINLNVSQFK
     PEDLKINLDG RTLSIQGEQE LKTDHGYSKK SFSRVILLPE DVDVGAVASN LSEDGKLSIE
     APKKEAVQGR SIPIQQAIVE EKSAE

5: Ce-HSP16-48   MW=16,299   143 AA
     MLMLRSPFSD SNVLDHFLDE ITGSVQFPYW RNADHNSFNF SDNIGEIVND ESKFSVQLDV
     SHFKPEDLKI ELDGRELKIE GIQEKKSEHG YSKRSFSKMI LLPEDVDLTS VKSAISNEGK
     LQIEAPKKTN SSRSIPINFV AKH

6: Ce-HSP16-41   MW=16,018   143 AA
     MLMLRSPYSD SNALDHFLDE LTGSVQFPYW RNADHNSFNF SDNIGEIVND ESKFSVQLDV
     SHFKPENLKI KLDGRELKIE GIQETKSEHG YLKRSFSKMI LLPEDADLPS VKSAISNEGK
     LQIEAPKKTN SSRSIPINFV AKH

7: Dm-HSP27      MW=23,458   212 AA
     MSIIPLLHLA RELDHDYRTD WGHLLEDDFG FGVHAHDLFH PRRLLLPNTL GLGRRRYSPY
     ERSHGHHNQM SRASGGPNAL LPAVGKDGFQ VCMDVSQFKP NELTVKVVDN TVVVEGKHEE
     REDGHGMIQR HFVRKYTLPK GFDPNEVVST VSSDGVLTLK APPPPSKEQA KSERIVQIQQ
     TGPAHLSVKA PAPEAGDGKA ENGSGEKMET SK

8: Dm-HSP26      MW=22,994   208 AA
     MSLSTLLSLV DELQEPRSPI YELGLGLHPH SRYVLPLGTQ QRRSINGCPC ASPICPSSPA
     GQVLALRREM ANRNDIHWPA TAHVGKDGFQ VCMDVAQFLP SELNVKVVDD SILVEGKHEE
     RQDDHGHIMR HFVRRYKVPD GYKAEQVVSQ LSSDGVLTVS IPKPQAVEDK SKERIIQIQQ
     VGPAHLNVKA NESEVKGKEN GAPNGKDK

9: Dm-HSP23      MW=20,399   186 AA
     MANIPLLLSL ADDLGRMSMV PFYEPYYCQR QRNPYLALVG PMEQQLRQLE KQVGASSGSS
     GAVSKIGKDG FQVCMDVSHF KPSELVVKVQ DNSVLVEGNH EEREDDHGFI TRHFVRRYAL
     PPGYEADKVA STLSSDGVLT IKVPKPPAIE DKGNERIVQI QQVGPAHLNV KENPKEAVEQ
     DNGNDK

10:Dm-HSP22      MW=19,733   174 AA
     MRSLPMFWRM AEEMARMPRL SSPFHAFFHE PPVWSVALPR NWQHIARWQE QELAPPATVN
     KDGYKLTLDV KDYSELKVKV LDESVVLVEA KSEQQEAEQG GYSSRHFLGR YVLPDGYEAD
     KVSSSLSDDG VLTISVPNPP GVQETLKERE VTIEQTGEPA KKSAEEPKDK TASQ

11: Dm-HS-G1     MW=26,560   238 AA
     MSLIPFILDL AEELHDFNRS LAMDIDDSAG FGLYPLEATS QLPQLSRGVG AWECNDVGAH
     QGSVGGHRSI AIIRTIVWPE PRLLAAISRW WSWKRNWAIR ARPGQAARPV ANGASKSAYS
     VVNRNGFQVS MNVKQFAANE LTVKTIDNCI VVEGQHDEKE DGHGVISRHF IRKYILPKGY
     DPNEVHSTLS SDGILTVKAP QPLPVVKGSL ERQERIVDIQ QISQQQKDKD AHRQSRQR
```

FIGURE 2.12. Sequences of members of the HSP20 family. Sequences are derived from the following sources: (1), human HSP 27;[165] (2), yeast HSP 26;[379a] (3-6), nematode HSP 16;[178,335] (7-10), *Drosophila* HSPs 27, 26, 23 and 22;[366] (11), *Drosophila* gene 1 protein;[19] (12), *Drosophila* gene 2 protein;[310] (13), *Drosophila* gene 3 protein;[309] (14-17), soybean HSP 17.5;[97,276,346] (18,19), soybean HSPs 17.9 and 18.5;[323] (20,21), *Arabidopsis* HSPs 17.4 and 18.2;[161,380b] (22), pea HSP18;[407b] and (23) wheat HSP17.[247a] The C-terminal homology region, characteristic of all

12: Dm-HS-G2 MW=12,600 111 AA
MATYEQVKDV PNHPDVYLID VRRKEELQQT GFIPASINIP LDELDKALNL DGSAFKNKYG
RSKPEKQSPI IFTCRSGNRV LEAEKIAKSQ GYSNVVIYKG SWNEWAQKEG L

13: Dm-HS-G3 MW=18,792 169 AA
MPDIPFVLNL DSPDSMYYGH DMFPNRMYRR LHSRQHHDLD LHTLGLIARM GAHAHHLVAN
KRNGELAALS RGGASNLQGN FEVHLDVGLF QPGELTVKLV NECIVVEGKH EEREDDHGHV
SRHFVPAVSA AQGVRFGCHC FHFVGGWSSQ YHGSTISFQG GAQGAHHTH

14: Gm-HSP17.5E MW=17,535 154 AA
MSLIPGFFGG RRSNVFDPFS LDMWDPFKDF HVPTSSVSAE NSAFVSTRVD WKETPEAHVF
KADIPGLKKE EVKVQIEDDR VLQISGERNV EKEDKNDTWH RVERSSGKFT RRFRLPENAK
VNEVKASMEN GVLTVTVPKE EVKKPDVKAI EISG

15: Gm-HSP17.5M MW=17,546 153 AA
MSLIPSIFGG RRSNVFDPFS LDVWDPFKDF HFPTSLSAEN SAFVNTRVDW KETPEAHVFE
ADIPGLKKEE VKVQIEDDRV LQISGERNLE KEDKNDTWHR VERSSGNFMR RFRLPENAKV
EQVKASMENG VLTVTVPKEE VKKPDVKAIE ISG

16: Gm-HSP17.6L MW=17,572 154 AA
MSLIPSIFGG PRSNVFDPFS LDMWDPFKDF HVPTSSVSAE NSAFVNTRVD WKETQEAHVL
KADIPGLKKE EVKVQIEDDR VLQISGERNV EKEDKNDTWH RVDRSSGKFM RRFRLPENAK
VEQVKACMEN GVLTVTIPKE EVKKSDVKPI EISG

17: Gm-HSP17.3B MW=17,347 153 AA
MSLIPSFFGG RRSSVFDPFS LDVWDPFKDF PFPSSLSAEN SAFVSTRVDW KETPEAHVFK
ADIPGLKKEE VKLEIQDGRV LQISGERNVE KEDKNDTWHR VERSSGKLVR RFRLPENAKV
DQVKASMENG VLTVTVPKEE IKKPDVKAID ISG

18: Gm-HSP17.9D MW=17,862 159 AA
MDFRVMGLES PLFHTLQHMM DMSEDGAGDN KTHNAPTWSY VRDAKAMAAT PADVKEYPNS
YVFEIDMPGL KSGDIKVQVE DDNLLLICGE RLRDEEKEGA KYLRMERRVG KLMRKFVLPE
NANTDAISAV CQDGVLSVTV QKLPPPEPKK PRTIQVKVA

19: Gm-HSP18.5C MW=18,486 161 AA
MSLIPNFFGG RRNNVFDPFS LDVWDPFLDF PFPNTLSSAS FPEFSRENSA FVSTRVDWKE
TPEAHVFKAD IPGLKKEEVK VQIEDDKVLQ ISGERNVEKE DKNDTWHRVE RSSGKFMRRF
RLPENAKVEQ VKASMENGVL TVTVPKEEVK KPDVKAIEIS G

20: At-HSP17.4 MW=17,602 157 AA
MSLIPSIFGG RRTNVFDPFS LDVFDPFEGF LTPSGLANAP AMDVAAFTNA KVDWRETPEA
HVFKADLPGL RKEEVKVEVE DGNILQISGE RSNENEEKND KWHRVERSSG KFTRRFRLPE
NAKMEEIKAS MENGVLSVTV PKVPEKKPEV KSIDISG

21: At-HSP18.2 MW=18,133 161 AA
MSLIPSIFGG RRSNVFDPFS QDLWDPFEGF FTPSSALANA STARDVAAFT NARVDWKETP
EAHVFKADLP GLKKEEVKVE VEDKNVLQIS GERSKENEEK NDKWHRVERA SGKFMRRFRL
PENAKMEEVK ATMENGVLTV VVPKAPEKKP QVKSIDISGA N

22: Ps-HSP18 MW=18,099 158 AA
MSLIPSFFSG RRSNVFDPFS LDVWDPLKDF PFSNSSLSAS FPRENPAFVS TRVDWKETPE
AHVFKADLPG LKKEEVKVEV EDDRVLQISG ERSVEKEDKN DEWHRVERSS GKFLRRFRLP
ENAKMDKVKA SMENGVLTVT VPKEEIKKAE VKSIEISG

23: Ta-HSP17 MW=16.876 151 AA
MSIVRRSNVF DPFADLWADP FDTFRSIVPA ISGGSSETAA FANARVDWKE TPEAHVFKVD
LPGVKKEEVK VEVEDGNVLV VSGERSREKE DKNDKWHRVE RSSGKFVRRF RLPEDAKVEE
VKAGLENGVL TVTVPKAEVK KPEVKAIEIS G

Figure 2.12 (continued).

members of the HSP20 family, is underlined (see also Figure 2.15).[289] The consensus sequence of this region with 34 to 37 amino acid residues can be given as follows:

$$-{}^{S}_{T}XXX{}^{F}_{Y}XRXXXLP{}^{D}_{E}X--{}^{LD}_{VE}X{}^{D}_{E}X{}^{V}_{I}X{}^{A}_{S}X{}^{L}_{V}X-{}^{D}_{E}G{}^{V}_{I}L{}^{TVTV}_{SLSI}PK-$$

Drosophila genes 2 and 3 evidently do not belong to the HSP20 family.

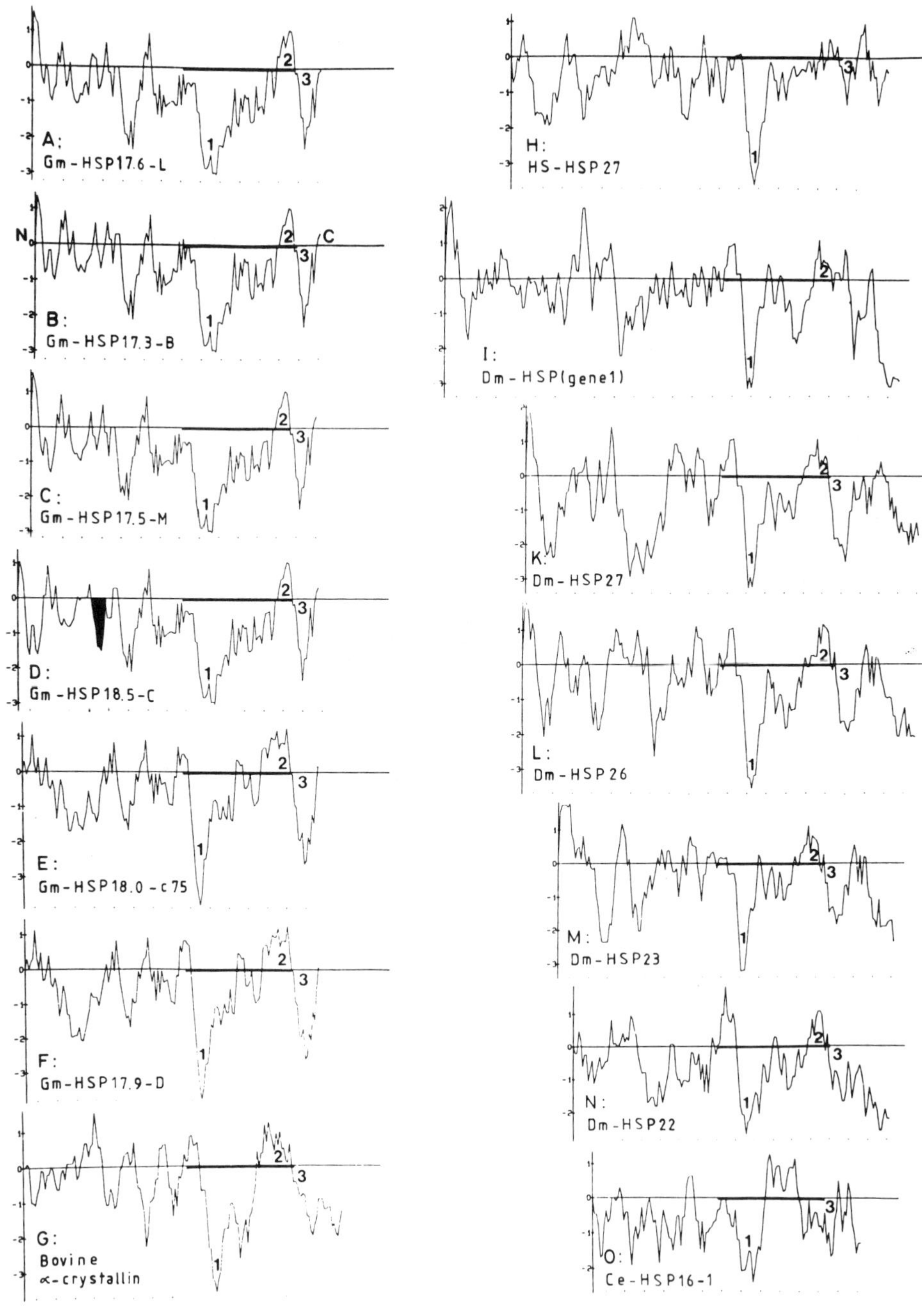

FIGURE 2.13. Hydropathy indices of small HSPs from soybean (A to F), human (H), *Drosophila* (I to N) and nematodes (O). Sequences of the genes and the corresponding references are given in Figure 2.12. The regions with a limited homology between all genes of the HSP20 family are aligned (thick bar). The characteristic pattern of a hydrophilic domain (1) followed by a small hydrophobic (2) and another hydrophilic domain at the C-terminus is illustrated in an enlarged version in Figure 2.14. The same type of basic structure is observed in bovine α-crystallin (G). (Compiled by Raschke, E., Molekulare Analyse verschiedener Gene für kleine Hitzeschockproteine der Sojabohne [*Glycine max* (L.) Merrill], Thesis, University of Bielefeld (FRG), 1987.)

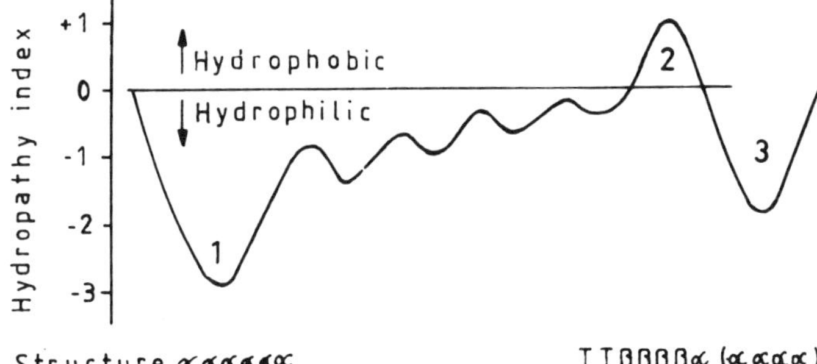

FIGURE 2.14. Structural features of the C-terminal part of small heat shock proteins. A consensus sequence was derived for the homologous C-terminal part of small HSPs. The hydropathy index shows the typical sequence of a hydrophilic, a hydrophobic, and another hydrophilic domain. Structural elements derived for these regions are α-helix (α), random coil (T), and β-sheet (β). (From Raschke, E., Molekulare Analyse verschiedener Gene für kleine Hitzeschockproteine der Sojabohne [*Glycine max* (L.) Merrill], Thesis, University of Bielefeld (FRG), 1987.)

```
A:Cr-HSP20        MW=16,800   157 AA
     MALSNYVFGN SAADPFFTEM DRAVNRMINN ALGVAPTSAG KAGHTHAPMD IIESPTAFEL
     HADAPGMGPD DVKVELQEGV LMVTGERKLS HTTKEAGGKV WRSERTAYSF SRAFSLPENA
     NPDGITAAMD KGVLVVTVPK REPPAKPEPK RIAVTGA

B:Ps-HSP21        MW=26,276   332 AA
     MAQSVSLSTI ASPILSQKPG SSVKSTPPCM ASFPLRRQLP RLGLRNVRAQ AGGDGDNKDN
     SVEVHRVNKD DQGTAVERKP RRSSIDISPF GLLDPWSPMR SMRQMLDTMD RIFEDAITIP
     GRNIGGGEIR VPWEIKDEEH EIRMRFDMPG VSKEDVKVSV EDDVLVIKSD HREENGGEDC
     WSRKSYSCYD TRLKLPDNCE KEKVKAELKD GVLYITIPKT KIERTVIDVQ IQ

C:Nc-HSP30        MW=25,262   228 AA
     MALFPRGFYG SYGSDPSFTN LFRLLDDFDT YTREVQGSAP ETGSRRHTQP TRTFSPKFDV
     RETEQTYELH GELPGIDRDN VQIEFTDPQT IVIRGRVERN YTAGTPPAQV AGVLTEKGEP
     HSPAAHHATV EDDVDEDNRS VATTATGANN QNNQQVAQRA SAPTTEEKPK APAEKYWVSE
     RSIGEFSRTF NFPGRVDQNA VSASLNNGIL TITVPKAKKH ETIRIAIN
```

FIGURE 2.15. Amino acid sequences of three organellar proteins of the HSP20 family. A: *Chlamydomonas reinhardi* HSP20;[141] B: *Pisum sativum* HSP21;[407] and C: *Neurospora crassa* HSP30.[318] All three proteins are encoded in the nucleus, synthesized in the cytoplasm, and transported into chloroplasts (A, B) and mitochondria, respectively (C). The sequence of the pea protein (B) contains a 46 to 47 amino acid residue basic transit peptide, which is cleaved off during uptake into the chloroplast. Precursor forms of the other two proteins are not known. Homology with each other and with other members of the HSP20 family is found in the C-terminal HSP20 domain (underlined as in Figure 2.12) and other parts in the N-terminal half.[289]

gene families. Covalent modification with branched polyubiquitin chains is a major tag for proteins to be degraded.[324a,341a]

In most organisms studied in detail, ubiquitin is coded by small multigene families with constitutively expressed and hs-induced members. Frequently the hs-induced transcription unit is a polyubiquitin gene coding for a covalently linked polyubiquitin that is processed posttranslationally to the corresponding monomers.[341,341a] Ubiquitin was characterized as a heat-shock-induced protein in human cells,[22,127b,341] in chicken,[49,50] *Xenopus laevis* embryos,[305a] *Drosophila melanogaster,*[214a] plants,[79a,407c] yeast,[299,127,380c] *Dictyostelium,*[270a] and trypanosomes.[308a] An exception to this rule is the polyubiquitin genes of the nematode *Caenorhabditis elegans,* which is not induced by hs.[135a]

2.3.6. SPECIAL HSPs

Besides the five major families of HSPs and their related proteins there are several other HSPs:

1. In vertebrates a dominant nucleolar protein HSP110 is also induced by hs.[355,377,430]
2. A minor HSP with 32 kDa of vertebrates is strongly induced in the presence of arsenite, heavy metals, sulfhydryl reagents, 10 μ*M* heme, and others (see Section 1.2). It was identified as heme oxygenase (see Table 2.6). The rat ho gene contains a single functional heat shock element in the promoter region.[270,353]
3. The human complement protein factor B (90-kDa glycoprotein) coded by the class III gene region of the MHC complex on chromosome 6 is increased in the course of an inflammation or tissue injury as part of the acute phase response.[312c] The corresponding promoter region contains three isolated hs elements[440] (see Table 6.2). It is worth noting that, in addition to many other genes involved in human diseases, this class III region also contains the hsp70A gene (see Figure 6.6, no. 3).[338a]
4. Because of their role in cell cycle control and malignant transformation, hs- or arsenite-induced expression of cellular oncogenes (c-myc and c-fos) in different mammalian cells is remarkable.[65a,113a,145a] Usually the induction of these genes is transient and precedes expression of the normal hs genes. A time sequence of induction (c-fos mRNA → c-myc mRNA → hsp70 mRNA) was also reported during regression of the rat ventral prostate gland after withdrawal of androgens.[63a] Because the v-myb and c-myc proteins, by interaction with other cellular proteins, increase hsp70 expression in different vertebrate cells,[178a,190a,192b] it can be speculated that an hs-independent, delayed trigger mechanism may exist to maintain elevated levels of HSP70 synthesis under certain stress conditions.
5. In vertebrate cells, a constitutive 47-kDa phospho-sialoglycoprotein of the ER membrane is induced by hs.[240b,276,276a] Synthesis of this HSP47 also increases during differentiation of mouse teratocarcinoma cells after addition of retinoic acid or dibutyryl cAMP,[201b] and it decreases after transformation of vertebrate cells with Rous sarcoma virus or Simian virus 40.[276c] In all cases, corresponding changes at the mRNA level were observed. The intriguing property of this unusual, transformation-sensitive stress protein is its capability to bind to extracellular matrix proteins, e.g., to gelatin or to native collagen types I and IV, and to the major serum protein of the developing fetus α-fetoprotein.[276-276b] HSP47 is synthesized as a preproprotein of 42 kDa with an N-terminal leader peptide. Cleavage of the leader peptide gives a 41-kDa intermediate, which is glycosylated to the mature 47-kDa protein. A function in the intracellular transporting of collagen or fetuin precursors[276b] but also in substrate adhesion can be envisaged, if a localization in the outer membrane is also assumed. In support of the latter hypothesis, Martin and Regan[240b] reported on the induction of HSP47 in rat glial cells by the teratogenic anticonvulsant drug valproate, which inhibits mitosis and increases substrate adhesion. Valproate induction is specific for glial cells, whereas induction by mild hs (42°C) was found in all types of rat cells tested.
6. The intimate connection between stress treatments and the induction of developmental processes in the cellular slime mold *Dictyostelium* was documented by Rosen et al.[332a] and Zuker et al.[449,450] Synthesis of transcripts from a 4.9-kb transposon, existing in about 40 copies per cell, is usually programmed during development of the fruiting body induced by nutrient starvation or other stresses, but it can also be triggered in vegetative cells by hs, e.g., by 30 min at 30°C.

   An almost perfect match of the consensus hs box (see Table 6.2) was mapped upstream of a putative transcription start site. Corresponding transcripts were found as differently sized poly(A)-containing-RNAs in the cytoplasm. Their protein-coding capacity is unknown (see also Sections 5.1 and 5.2 for hs-induced RNAs in *Drosophila*). Interestingly, one out of 17 actin genes of *Dictyostelium* also contains a hs promoter element.[449] Recently, another example was added to this list.[237a] A 1-kb mRNA coding for an extremely hydrophobic membrane protein and usually synthesized under developmental control, is also formed after induction by hs (15 min 30°C), 100 μ*M* $CdCl_2$, or cold shock (5 h 4°C).

2.4. HEAT-SHOCK-INDUCED ENZYMES

The search for cellular functions of HSPs also included the question on hs-inducible enzymes. In a series of papers Leenders and co-workers reported on increased activities of mitochondrial enzymes in *Drosophila hydei*, e.g., of NADH dehydrogenase, α-glycerophosphate dehydrogenase, or tyrosine aminotransferase.[195,217] The activity increase was dependent on RNA and protein synthesis and correlated with induction of distinct hs puffs in *D. hydei*. However, when critically summarizing these results, Ashburner and Bonner[12] pointed out that there is no conclusive evidence for an enzymatic function of anyone of the major *Drosophila* HSPs. In fact, careful analysis of the activity changes of α-glycerophosphate dehydrogenase confirmed the requirement of RNA and protein synthesis, but showed that the enzyme itself is not hs induced. Rather the decrease of its Km value results from altered modification, presumably by deacetylation.[412] These results stress the well known fact that increasing enzyme activities, even if dependent on concomitant RNA and protein synthesis, can not be taken as proof for hs-induced *de novo* synthesis of these enzymes. It remains an intriguing question, however, what heat inducible protein factor brings about this altered enzyme activity. The same arguments hold true for reports on hs-induced synthesis of acid RNase in *Tetrahymena*[381a] and of a ribosome-associated RNase as well as of a microsomal protease in yeast.[142,349]

In contrast to the former, induction of enzymes as the response to hs is well documented for *E. coli, Neurospora crassa,* yeast (enolase, phosphoglycerate kinase), and mammalian species (Table 2.6). In several cases compiled in Table 2.6, the effects are apparent only after several hours of recovery, e.g. 2′,5′ oligoadenylate synthase in mouse DBK cells,[77,78] ornithine decarboxylase in rat tissues,[109] peroxidase,[373] and *N*-methyltransferase in plants.[220] However, disregarding the ATPase activity of HSP70- and probably also of HSP60-type proteins (see Section 2.3.2), it is still correct to state that the major HSPs of eukaryotic cells are structural or regulatory proteins rather than enzymes. This does not exclude an essential role of the hs-induced enzymes in the network of protective mechanisms during the heat shock response. The matter needs much more attention.

TABLE 2.6
Heat Shock Induced Enzymes

Organism	Enzyme	Remarks	Ref.
Bacteria			
Escherichia coli	Lys-tRNA synthetase	Product of lysU gene (see Table 2.2)	398
	Mn-dependent superoxide dismutase	Slow induction by 1 h at 48°C, only in presence of O_2	320
	Lon-protease	See Table 2.2, product of Lon gene	136, 137, 314
	Serine protease (Htr A)	Indispensible at >42°C	230d
Bacillus subtilis	Lon-protease	Cross-reactivity with antibodies to *E. coli* lon protease	6
Caulobacter crescentus	Lon-protease	Cross-reactivity with antibodies to *E. coli* lon protease	327
Eukaryotic microorganisms			
Saccharomyces cerevisiae (yeast)	Proteinase (B?)		142, 143
	Catalase T	Induced by hs at 37°C	42
	Ribosomal RNase	Also induced by glucose deprivation, increase inhibited by CH	349
	Enolase	Corresponds to HSP48, promoter reveals no heat shock element	173, 394a
	Phosphoglycerate kinase	Corresponds to HSP46, only induced in glucose-rich medium, promoter contains functional HSE (see Table 6.2)	316—316b
	Ubiquitin-conjugating enzymes UBC4, UBC5	Tagging of short-lived or abnormal proteins with polyubiquitin	350c
Neurospora crassa	Peroxidase	1 h at 48°C confers resistance to 2 m*M* H_2O_2, increased peroxidase and thermotolerance levels correlated	182—182b
	Catalase	Isozymes 1 and 2 induced by hs (48°C), menadione, paraquat, or H_2O_2	73
Tetrahymena spec.	Acid RNase	Increase inhibited by CH	381a
Animals			
Drosophila spec.	Malate dehydrogenase, acetylcholinesterase	Induced in ovaries of different species by 30 min 33°C + 30 min 40°C + 30 min 25°C	138c, 139

TABLE 2.6 (continued)
Heat Shock Induced Enzymes

Organism	Enzyme	Remarks	Ref.
Drosophila hydei	NADH dehydrogenase	Coded by puff 4-81B, increase requires transcription and translation	215, 217
	Tyrosine amino transferase	Coded by puff 2-48BC, immunoprecipitation with antibodies reveal increased amounts	54, 216, 217
	α-Glycerophosphate dehydrogenase	Change of Km, no new synthesis	411, 412
Xenopus laevis embryo	Glyceraldepyde-3-phosphate dehydrogenase	HSP35, minor isozyme	292
Vertebrates	Heme oxygenase (SP32)	Induced 10 to 20-fold at 42°C, dependent on HSE in promoter, inhibited by actinomycin D (see Section 2.3.6)	188a, 270, 353, 354c
Mouse mammary carcinoma cells	Nuclease	Increase inhibited by CH	322
Mammals	Clathrin uncoating ATPase	HSC71-73 (see HSP70 family, Section 2.3.2)	72a, 396
CHO cells	Spermidine acetyltransferase	Increase inhibited by CH but not by actinomycin D	156
Mouse DBK cells	2′5′oligoadenylate synthase	1 h 44.5°C plus 6 h recovery, inhibited by actinomycin D	77, 78
Mouse embryo cells	Superoxide dismutase, glutathione peroxidase	Increased activities after 15 min 45°C plus 6 h recovery, correlates with level of induced thermotolerance	303
Rat brain, liver, kidney	Ornithine decarboxylase	Increase during recovery from hyperthermia and chemical stressors, inhibited by actinomycin D	109
Rabbit synovial fibroblasts	Metallo-proteinases (collagenase, stromelysin)	3 h at 42°C, excreted enzymes, may contribute to destruction of connective tissue during rheumatoid arthritis	399a
Plants			
Cucumber seedling	Peroxidase	Slow increase, detected after 12 h recovery from 40 s at 50°C	373
Barley leaves	*N*-methyl transferase	Induced by long-term, moderate hs, involved in gramine synthesis	220

REFERENCES

1. **Alahiotis, S. N.,** Heat shock proteins. A new view on the temperature compensation, *Comp. Biochem. Physiol.*, 75B, 379—387, 1983.

1a. **Alcina, A., Ursainqui, A., and Carasco, L.,** The heat shock response in *Trypanosoma cruzi, Eur. J. Biochem.*, 172, 121—127, 1988

1b. **Allan, B., Linseman, M., Macdonald, L. A., Lam, J. S., and Kropinski, A. M.,** Heat shock response of *Pseudomonas aeruginosa, J. Bacteriol.*, 170, 3668—3674, 1988.

1c. **Allen, R. L., O'Brien, D. A., and Eddy, E. M.,** A novel hsp70-like protein (p70) is present in mouse spermatogenic cells, *Mol. Cell. Biol.*, 8, 828—832, 1988.

1d. **Alfano, C. and McMacken, R.,** Heat-shock protein mediated disassembly of nucleoprotein structures is required for the initiation of bacteriophage-lambda DNA replication, *J. Biol. Chem.*, 264, 10709—10718, 1989.

2. **Altschuler, M. and Mascarenhas, J. P.,** Heat shock proteins and effects of heat shock in plants, *Plant Mol. Biol.*, 1, 103—115, 1982.

3. **Amaral, M. D., Galego, L., and Rodrigues-Pousada, C.,** Stress response of *Tetrahymenia pyriformis* to arsenite and heat shock: Differences and similarities, *Eur. J. Biochem.*, 171, 463—470, 1988.

3a. **Amir-Shapira, D., Leustek, T., Dalie, B., Weissbach, H., and Brot, N.,** Hsp70 proteins, similar to *Escherichia coli* DnaK, in chloroplasts and mitochondria of *Euglena gracilis, Proc. Natl. Acad. Sci. U.S.A.*, 87, 1749—1752, 1990.

4. **Anderson, N. L., Giometti, C. S., Gemmell, M. A., Nance, S. L., and Anderson, N. G.,** A two-dimensional electrophoretic analysis of the heat-shock-induced proteins of human cells, *Clin. Chem.*, 28, 1084—1092, 1982.

5. **Anderson, R. L., Tao, T. W., Betten, D. A., and Hahn, G. M.,** Heat shock protein levels are not elevated in heat-resistant B16 melanoma cells, *Radiat. Res.*, 105, 240—246, 1986.

5a. **Ang, D., Chandrasekhar, G. N., Zylicz, M., and Georgopoulos, C.,** *Escherichia coli* grpE gene codes for heat shock protein B25.3, essential for both lambda DNA replication at all temperatures and host growth at high temperature, *J. Bacteriol.*, 167, 25—29, 1986.

6. **Arnosti, D. N., Singer, V. L., and Chamberlin, M. J.,** Characterization of heat shock in *Bacillus subtilis, J. Bacteriol.*, 168, 1243—1249, 1986.

7. **Arrigo, A.-P.,** Acetylation and methylation patterns of core histones are modified after heat or arsenite treatment of *Drosophila* tissue culture cells, *Nucl. Acids Res.*, 11, 1389—1404, 1983.

8. **Arrigo, A.-P.,** Cellular localization of HSP23 during *Drosophila* development and following subsequent heat shock, *Dev. Biol.*, 122, 39—48, 1987.

8a. **Arrigo, A.-P.,** Tumor necrosis factor induces the rapid phosphorylation of the mammalian heat shock protein hsp28, *Mol. Cell. Biol.*, 10, 1276—1280, 1990.

9. **Arrigo, A.-P. and Ahmad-Zadeh, C.,** Immunofluorescence localization of a small heat shock protein (hsp23) in salivary gland cells of *Drosophila melanogaster, Mol. Gen. Genet.*, 184, 73—79, 1981.

10. **Arrigo, A.-P. and Welch, W. J.,** Characterization and purification of the small 28,000 dalton mammalian heat shock protein, *J. Biol. Chem.*, 262, 15359—15369, 1987.

11. **Arrigo, A.-P., Fakan, S., and Tissieres, A.,** Localization of the heat shock-induced proteins in *Drosophila melanogaster* tissue culture cells, *Dev. Biol.*, 78, 86—103, 1980.

11a. **Arrigo, A.-P., Suhan, J. P., and Welch, W. J.,** Dynamic changes in the structure and intracellular locale of the mammalian low-molecular-weight heat shock protein, *Mol. Cell. Biol.*, 8, 5059—5071, 1988.

12. **Ashburner, M. and Bonner, J. J.,** The induction of gene activity in *Drosophila* by heat shock, *Cell*, 17, 241—254, 1979.

13. **Atkinson, B. G.,** Synthesis of heat-shock proteins by cells undergoing myogenesis, *J. Cell Biol.*, 89, 666—671, 1981.

14. **Atkinson, B. G. and Pollock, M.,** Effect of heat shock on gene expression in human epidermoid carcinoma cells (strain KB) and in primary cultures of mammalian and avian cells, *Can. J. Biochem.*, 60, 316—327, 1982.

15. **Atkinson, B. G., Cunningham, T., Dean, R. L., and Somerville, M.,** Comparison of the effects of heat shock and metal-ion stress on gene expression in cells undergoing myogenesis, *Can. J. Biochem. Cell Biol.*, 61, 404—413, 1983.

16. **Atkinson, B. G., Dean, R. L., and Blaker, T. W.,** Heat shock induced changes in the gene expression of terminally differentiating avian red blood cells, *Can. J. Genet. Cytol.*, 28, 1053—1063, 1987.

16a. **Atkinson, B. G. and Walden, D. B., Eds.,** *Changes in Eukaryotic Gene Expression in Response to Environmental Stress*, Academic Press, Orlando, FL, 1985.

17. **Augusteyn, R. C., Boyd, A., and Kelly, L.,** Immunological comparison of heat-shock proteins and α-crystallin, *Curr. Eye. Res.*, 5, 759—762, 1986.

18. **Aujame, L.,** Murine plasmacytoma constitute a class of natural heat-shock variants in which the major inducible hsp-68 gene is not expressed, *Can. J. Genet., Cytol.,* 28, 1064—1075, 1987.
19. **Ayme, A. and Tissieres, A.,** Locus 67B of *Drosophila melanogaster* contains seven, not four, closely related heat shock genes, *EMBO J.,* 4, 2949—2954, 1985.
20. **Bag, J.,** Regulation of heat-shock protein synthesis in chicken muscle culture during recovery from heat shock, *Eur. J. Biochem.,* 135, 373—378, 1983.
21. **Bag, J.,** Recovery of normal protein synthesis in heat-shocked chicken myotubes by liposome-mediated transfer of mRNAs, *Can. J. Biochem. Cell Biol.,* 63, 231—235, 1985.
21a. **Bhagwat, A. A. and Apte, S. K.,** Comparative analysis of proteins induced by heat shock, salinity and osmotic stress in the nitrogen-fixing cyanobacterium *Anabaena* sp. strain L-31, *J. Bacteriol.,* 171, 5187—5189, 1989.
21b. **Baird, P. N., Hall, L. M., and Coates, A. R. M.,** A major antigen from *Mycobacterium tuberculosis* which is homologous to the heat shock proteins groES from *E. coli* and the htpA gene product of *Coxiella burneti, Nucl. Acids Res.,* 16, 9047, 1988.
22. **Baker, Z. T. and Board, P. G.,** The human ubiquitin gene family structure of a gene and pseudogenes from the UbB subfamily, *Nucl. Acids Res.,* 15, 443—463, 1987.
22a. **Baldaia, L., Maisonhaute, C., Porcheron, P., and Best-Belpomme, M.,** Effect of heat shock on protein synthesis in *Locusta migratoria* epidermis, *Arch. Insect Biochem. Physiol.,* 4, 225—231, 1987.
23. **Banerji, S. S., Theodorakis, N. G., and Morimoto, R. J.,** Heat shock-induced translational control of HSP70 and globin synthesis in chicken reticulocytes, *Mol. Cell. Biol.,* 4, 2437—2448, 1984.
24. **Bardwell, J. C. A. and Craig, E. A.,** Major heat shock gene of *Drosophila* and *Escherichia coli* heat-inducible dnaK gene are homologous, *Proc. Natl. Acad. Sci. U.S.A.,* 81, 848—852, 1984.
25. **Bardwell, J. C. A. and Craig, E. A.,** Eukaryotic Mr 83,000 heat shock protein has a homologue in *Escherichia coli, Proc. Natl. Acad. Sci. U.S.A.,* 84, 5177—5181, 1987.
26. **Bardwell, J. C. A., Tilly, K., Craig, E., King, J., Zylicz, M., and Georgopoulos, C.,** The nucleotide sequences of *Escherichia coli* K12 dnaJ + gene. A gene that encodes a heat-shock protein, *J. Biol. Chem.,* 261, 1782—1785, 1986.
27. **Barner, J. V., Bensaude, O., Morange, M., and Babinet, C.,** Mouse 89 kD heat shock protein. Two peptides with distinct developmental regulation, *Exp. Cell Res.,* 170, 186—194, 1987.
28. **Barnett, T., Altschuler, M., McDaniels, C. N., and Mascarenhas, J. P.,** Heat shock induced proteins in plant cells, *Dev. Genet.,* 1, 331—340, 1980.
29. **Baszczynski, C. L.,** Immunochemical analysis of heat-shock protein synthesis in maize (*Zea mays* L.), *Can. J. Genet. Cytol.,* 28, 1076—1087, 1986.
29a. **Baszczynski, C. L.,** Gene expression in *Brassica* tissues and species following heat shock, *Biochem. Cell Biol.,* 66, 1303—1311, 1988.
30. **Baszczynski, C. L., Walden, D. B., and Atkinson, B. G.,** Regulation of gene expression in corn *(Zea mays* L.) by heat shock, *Can. J. Biochem.,* 60, 569—579, 1982.
31. **Baulieu, J. F. and Tanguay, R. M.,** Members of the *Drosophila* HSP70 family share ATP-binding properties, *Eur. J. Biochem.,* 172, 341—347, 1988.
32. **Bedard, P. A. and Brandhorst, B. P.,** Translational activation of maternal mRNA encoding the heat-shock protein hsp90 during sea urchin embryogenesis, *Dev. Biol.,* 117, 286—293, 1986.
33. **Belanger, F. C., Brodl, M. R., and Ho, T.-H. D.,** Heat shock causes destabilization of specific mRNAs and destruction of endoplasmic reticulum in barley aleurone cells, *Proc. Natl. Acad. Sci. U.S.A.,* 83, 1354—1358, 1986.
33a. **Benndorf, R., Kraft, R., Otto, A., Stahl, J., Boehm, H., and Bielka, H.,** Purification of the growth-related protein p25 of the Ehrlich ascites tumor and analysis of its isoforms, *Biochem. Int.,* 17, 225—234, 1988.
34. **Berg, G. R., Inniss, W. E., and Heikkila, J. J.,** Stress proteins and thermotolerance in psychotrophic yeasts from artic environments, *Can. J.Microbiol.,* 33, 383—389, 1987.
35. **Berger, E. M.,** The regulation and function of small heat-shock protein synthesis, *Dev. Genet.,* 4, 255—265, 1984.
36. **Berger, E. M., Marino, G., and Torrey, D.,** Expression of *Drosophila* hsp70-cat hybrid gene in *Aedes* cells induced by heat shock, *Somat. Cell Mol. Genet.,* 11, 371—378, 1985.
37. **Berger, E. M., Vitek, M. P., and Morganelli, C. M.,** Transcript length heterogeneity at the small heat-shock protein genes of *Drosophila, J. Mol. Biol.,* 186, 137—148, 1985.
38. **Bergh, S. and Arking, R.,** Developmental profile of the heat shock response in early embryos of *Drosophila, J. Exp. Zool.,* 231, 379—391, 1984.
39. **Bienz, M.,** *Xenopus* hsp70 genes are constitutively expressed in injected oocytes, *EMBO J.,* 3, 2477—2483, 1984.
40. **Bienz, M. and Gurdon, J. B.,** The heat-shock response in *Xenopus* oocytes is controlled at the translational level, *Cell,* 29, 811—819, 1982.

41. **Bienz, M. and Pelham, H. R. B.,** Mechanisms of heat-shock gene activation in higher eucaryotes, *Adv. Genet.*, 24, 37—72, 1987.
42. **Bilinski, T., Krawiec, Z., Litwinska, J., and Blaszczynski, M.,** Mechanisms of oxygen toxicity as revealed by studies of yeast mutants with changed response to oxidative stress, in *Oxy-radicals in Molecular Biology and Pathology*, Cerutti, P. A., Fridovich, I., and McCord, J. M., Eds., Alan R. Liss, New York, 1988, 109—123.
43. **Binart, N., Chambraud, B., Dumas, B., Rowlands, D. A., Bigogne, C., Levin, J. M., Garnier, J., Baulieu, E.-E., and Catelli, M.-G.,** The cDNA-derived amino acid sequence of chick heat shock protein Mr 90,000 (HSP90) reveals a "DNA like" structure: Potential site of interaction with steroid receptors, *Biochem. Biophys. Res. Commun.*, 159, 140—147, 1989.
44. **Black, R. E. and Bloom, L.,** Heat shock proteins in *Aurelia (Cnidaria, Scyphozoa), J. Exp. Zool.*, 230, 303—308, 1984.
45. **Blackman, R. K. and Meselson, M.,** Interspecific nucleotide sequence comparisons used to identify regulatory and structural features of the *Drosophila* hsp82 gene, *J. Mol. Biol.*, 188, 499—516, 1986.
46. **Bloom, M., Skelly, S., Van Bogelen, R., Neidhardt, F., Brot, N., and Weissbach, H.,** In vitro effect of the *Escherichia coli* heat-shock regulatory protein on expression of heat shock genes, *J. Bacteriol.*, 166, 380—384, 1986.
46a. **Bochkareva, E. S., Lissin, N. M., and Girshovich, A. S.,** Transient association of newly synthesized unfolded proteins with the heat-shock groEL protein, *Nature*, 336, 254—257, 1988.
47. **Bochner, B. R., Zylicz, M., and Georgopoulos, C.,** *Escherichia coli* Dnak protein possesses a 5′-nucleotidase activity that is inhibited by AppppA, *J. Bacteriol.*, 168, 931—935, 1986.
48. **Bonato, M. C. M., Silva, A. M., Gomes, S. L., Maia, J. C. C., and Juliani, M. H.,** Differential expression of heat-shock proteins and spontaneous synthesis of HSP70 during the life cycle of *Blastocladiella emersonii, Eur. J. Biochem.*, 163, 211—220, 1987.
49. **Bond, U. and Schlesinger, M. J.,** Ubiquitin is a heat shock protein in chicken embryo fibroblasts, *Mol. Cell. Biol.*, 5, 949—956, 1985.
50. **Bond, U. and Schlesinger, M. J.,** The chicken ubiquitin gene contains a heat shock promoter and express an unstable mRNA in heat-shocked cells, *Mol. Cell. Biol.*, 6, 4602—4610, 1986.
51. **Boon-Niermeijer, E. K., Tuyl, M., and Van De Scheur, H.,** Evidence for two states of thermotolerance, *Int. J. Hyperthermia*, 2, 93—105, 1986.
52. **Borbely, G., Suranyi, G., Korcz, A., and Palfi, Z.,** Effect of heat shock on protein synthesis in the cyanobacterium *Synechococcus* sp. strain PCC 6301, *J. Bacteriol.*, 161, 1125—1130, 1985.
52a. **Borkovich, K. A., Farrelly, F. W., Finkelstein, D. B., Taulieu, J., and Lindquist, S.,** Hsp82 is an essential protein that is required in higher concentrations for growth of cells at higher temperatures, *Mol. Cell. Biol.*, 9, 3913—3930, 1989.
53. **Bosch, T. C. G., Krylow, S. M., Bode, H. R., and Steele, R. E.,** Thermo-tolerance and synthesis of heat shock proteins: These responses are present in *Hydra attenuata* but absent in *Hydra oligactis, Proc. Natl. Acad. Sci. U.S.A.*, 85, 7927—7931, 1988.
53a. **Bossier, P., Fitch, I. T., Boucherie, H., and Tuite, M. F.,** Structure and expression of a yeast gene encoding the small heat-shock protein hsp26, *Gene*, 78, 323—330, 1989.
54. **Brady, T. and Belew, K.,** Pyridoxine induced puffing (II-48C) and synthesis of a 40 kD protein in *Drosophila hydei* salivary glands, *Chromosoma*, 82, 89—98, 1981.
55. **Braell, W. A., Schlossman, D. M., Schmid, S. L., and Rothman, J. E.,** Dissociation of clathrin coats coupled to the hydrolysis of ATP: role of an uncoating ATPase, *J. Cell Biol.* 99, 734—741, 1984.
56. **Browder, L. W., Pollock, M., Heikkila, J. J., Wilkes, J., Wang, T., Krone, P., Ovsene, N., and Kloc, M.,** Decay of the oocyte-type heat shock response of *Xenopus laevis, Dev. Biol.*, 124, 191—199, 1987.
57. **Brown, I. R.,** Hyperthermia induces the synthesis of a heat shock protein by polysomes isolated from the fetal and neonatal mammalian brain, *J. Neurochem.*, 40, 1490—1493, 1983.
58. **Brugge, J. S., Erikson, E., and Erikson, R. L.,** The specific interaction of the Rous sarcoma virus transforming protein, pp60src, with two cellular proteins, *Cell*, 25, 363—372, 1981.
58a. **Bukan, B. and Walker, G. C.,** dnaK52 mutants of *Escherichia coli* have defects in chromosome segregation and plasmid maintenance at normal growth temperature, *J. Bacteriol.*, 171, 6030—6038, 1989.
58b. **Bukan, B. and Walker, G. C.,** Cellular defects caused by deletion of the *Escherichia coli* dnaK gene indicate roles for heat shock protein in normal metabolism, *J. Bacteriol.*, 171, 2337—2346, 1989.
59. **Bultmann, H.,** Heat shock responses in polytene foot pad cells of *Sarcophaga bullata, Chromosoma*, 93, 347—357, 1986.
60. **Bultmann, H.,** Induction of a heat shock puff by hypoxia in polytene foot pad chromosomes of *Sarcophaga bullata, Chromosoma*, 93, 358—366, 1986.
60a. **Burdon, R. H.,** Heat shock and the heat shock proteins, *Biochem. J.*, 240, 313—324, 1986.

61. **Burdon, R. H., Slater, A., McMahon, M., and Cato, A. C. B.,** Hyperthermia and the heat-shock proteins of HeLa cells, *Br. J. Cancer,* 45, 953—963, 1982.

62. **Burke, J. J., Hatfield, J. L., Klein, R. P., and Mullet, J. E.,** Accumulation of heat shock proteins in field-grown cotton, *Plant Physiol.,* 78, 394—398, 1985.

63. **Burns, C. P., Lambert, B. J., Haugstad, B. N., and Guffy, M. M.,** Influence of rate of heating on thermosensitivity of L1210 leukemia cells: membrane lipids and Mr 70,000 heat shock protein, *Cancer Res.,* 46, 1882—1887, 1986.

63a. **Buttyan, R., Zakeri, Z., Lockshin, R., and Wolgemuth, D.,** Cascade induction of c-fos, c-myc, and heat shock 70K transcripts during regression of the rat ventral prostate gland, *Mol. Endocrinol.,* 2, 650—657, 1988.

64. **Buzin, C. H. and Petersen, N. S.,** A comparison of the multiple *Drosophila* heat shock proteins in cell lines and larval salivary glands by two-dimensional gel electrophoresis, *J. Mol. Biol.,* 158, 181—201, 1982.

65. **Cairo, G., Bardella, L., Schiaffonati, L., and Bernelli-Zazzera, A.,** Synthesis of heat shock proteins in rat liver after ischemia and hyperthermia, *Hepathology,* 5, 357—361, 1985.

65a. **Cajone, F., Salina, M., and Bernelli-Zazzera, A.,** C-myc gene expression in heat-adapted and heat-shocked cells, *Cell Biol. Int. Rep.,* 12, 549—554, 1988.

66. **Caltabiano, M. M., Koestler, T. P., Poste, G., and Greig, R. G.,** Induction of 32- and 34-kDA stress proteins by sodium arsenite, heavy metals, and thiol-reactive agents, *J. Biol. Chem.,* 261, 13381—13387, 1986.

67. **Carbajal, M. E., Duband, J. L., Lettre, F., Valet, J. P., and Tanguay, R. M.,** Cellular localization of *Drosophila* 83-kilodalton heat shock protein in normal, heat-shocked, and recovering cultured cells with a specific antibody, *Biochem. Cell Biol.,* 64, 816—825, 1986.

68. **Carlsson, L. and Lazarides, E.,** ADP-ribosylation of the Mr 83,000 stress-inducible and glucose-regulated protein in avian and mammalian cells: Modulation by heat shock and glucose starvation, *Proc. Natl. Acad. Sci. U.S.A.,* 80, 4664—4668, 1983.

68a. **Carr, A. and De Pomerai, D. I.,** Stress protein and crystallin synthesis during heat shock and transdifferentiation of embryonic chick neural retina cells, *Dev. Growth Differ.,* 27, 435—445, 1985.

69. **Carretero, M. T., Carmona, M. J., Morcillo, G., Barettino, D., and Diez, J. L.,** Asynochronous expression of heat-shock genes in *Chironomus thummi, Biol. Cell,* 56, 17—21, 1986.

70. **Carvalho, M. G. C. and Rebello, M. A.,** Induction of heat shock proteins during the growth of *Aedes albopictus* cells, *Insect Biochem.,* 17, 199—206, 1986.

71. **Cervera, J.,** Induction of self-tolerance and enhanced stress protein synthesis in L-132 cells by cadmium chloride and by hyperthermia, *Cell Biol. Int. Rep.,* 9, 131—142, 1985.

72. **Chang, S. C., Wooden, S. K., Nakaki, T., Kim, Y. K., Lin, A. Y., Kung, L., Attenello, J. W., and Lee, W. S.,** Rat gene encoding the 78-kDa glucose-regulated protein GRP78: Its regulatory sequences and the effect of protein glycosylation on its expression, *Proc. Natl. Acad. Sci. U.S.A.,* 84, 680—684, 1987.

72a. **Chappell, T. G., Welch, W. J., Schlossman, D. M., Palter, K.B., Schlesinger, M. J., and Rothman, J. E.,** Uncoating ATPase is a member of the 70 kilodalton family of stress proteins, *Cell,* 45, 3—13, 1986.

73. **Chary, P. and Natvig, D. O.,** Evidence for three differentially regulated catalase genes in *Neurospora crassa*: Effects of oxidative stress, heat shock, and development, *J. Bacteriol.,* 171, 2646—2652, 1989.

73a. **Chen, R., Lanzon, L. M., DeRocher, A. E., and Vierling, E.,** Accumulation, stability and localization of a major chloroplast heat shock protein, *J. Cell Biol.,* submitted.

73b. **Cheney, C. M. and Shearn, A.,** Developmental regulation of *Drosophila* imaginal disk proteins: Synthesis of a heat-shock protein under nonheat-shock conditions, *Dev. Biol.,* 95, 325—330, 1983.

74. **Cheng, M. Y., Hartl, F.-U., Martin, J., Pollock, R. A., Kalonsek, F., Neupert, W., Hallberg, E. M., Hallberg, R. L., and Horwich, A. L.,** Mitochondrial heat-shock protein hsp60 is essential for assembly of proteins imported into yeast mitochondria, *Nature,* 337, 620—625, 1989.

74a. **Chin, D. T., Goff, S. A., Webster, T., Smith, T., and Goldberg, A. L.,** Sequence of the lon gene in *Escherichia coli*: A heat-shock gene which encodes the ATP-dependent protease La, *J. Biol. Chem.,* 263, 11718—11728, 1988.

74b. **Chirico, W. J., Waters, M. G., and Blobel, G.,** 70K heat shock related proteins stimulate protein translocation into microsomes, *Nature,* 332, 805—810, 1988.

75. **Chomyn, A. and Mitchell, H. K.,** Synthesis of the 84,000 dalton protein in normal and heat shocked *Drosophila melanogaster* cells as detected by specific antibody, *Insect Biochem.,* 12, 105—114, 1982.

76. **Chomyn, A., Moller, G., and Mitchell, H. K.,** Patterns of protein synthesis following heat shock in pupae of *Drosophila melanogaster, Dev. Genet.,* 1, 77—95, 1979.

77. **Chousterman, S., Chelbi-Alix, M. K., and Thang, M. N.,** Heat-shock induced regulation of 2′,5′-oligoadenylate synthetase, *Prog. Clin. Biol. Res.,* 202, 67—74, 1985.

78. **Chousterman, S., Chelbi-Alix, M. K., and Thang, M. N.,** 2′,5′-Oligoadenylate synthetase expression is induced in response to heat shock, *J. Biol. Chem.,* 262, 4806—4811, 1987.

78a. **Chretien, P. and Landry, J.,** Enhanced constitutive expression of the 27-kDa heat shock proteins in the heat-resistant variants from Chinese hamster cells, *J. Cell. Physiol.,* 137, 157—166, 1988.
79. **Chrispeels, M. and Greenwood, J. S.,** Heat stress enhances phytohemaglutinin synthesis but inhibits its transport out of the endoplasmic reticulum, *Plant Physiol.,* 83, 778—784, 1987.
79a. **Christensen, A. H. and Quail, P. H.,** Sequence analysis and transcriptional regulation by heat shock of polyubiquitin transcripts from maize, *Plant Mol. Biol.,* 12, 619—632, 1989.
80. **Clark, B. D. and Brown, I. R.,** Protein synthesis in mammalian retina following the intravenous administration of LSD, *Brain Res.,* 247, 97—104, 1982.
81. **Clark, B. D. and Brown, I. R.,** Induction of a heat shock protein in the isolated mammalian retina, *Neurochem. Res.,* 11, 269—279, 1986.
81a. **Clark, B. D. and Brown, I. R.,** A retinal heat shock protein is associated with elements of the cytoskeleton and binds to calmodulin, *Biochem. Biophys. Res. Commun.,* 139, 974—981, 1986.
81b. **Clark, C. F., Cheng, K., Frey, A. B., Stein, R., Hinds, P. W., and Levien, A. J.,** Purification of complexes of nuclear oncogene p53 with rat and *Escherichia coli* heat shock proteins. In vitro dissociation of hsc70 and DnaK from murine p53 by ATP, *Mol. Cell. Biol.,* 8, 1206—1215, 1988.
82. **Collier, N. C. and Schlesinger, M. J.,** Induction of heat-shock proteins in the embryonic chicken lens, *Exp. Eye Res.,* 43, 103—117, 1986.
82a. **Collier, N. C. and Schlesinger, M. J.,** The dynamic state of heat shock proteins in chicken embryo fibroblasts, *J. Cell Biol.,* 103, 1495—1507, 1986.
83. **Collier, N. C., Heuser, J., Levy, M. A., and Schlesinger, M. J.,** Ultrastructural and biochemical analysis of the stress granule in chicken embryo fibroblasts, *J. Cell Biol.,* 106, 1131—1139, 1988.
84. **Collins, P. L. and Hightower, L. E.,** Newcastle disease virus stimulates the cellular accumulation of stress (heat shock) mRNAs and proteins, *J. Virol.,* 44, 703—707, 1982.
85. **Cooper, P., Ho, T.-H. D., and Hauptmann, R. M.,** Tissue specificity of the heat-shock response in maize, *Plant Physiol.,* 75, 431—441, 1984.
86. **Cornelius, G. and Rensing, L.,** Circadian rhythm of heat shock protein synthesis of *Neurospora crassa, Eur. J. Cell Biol.,* 40, 130—132, 1986.
87. **Cosgrove, J. W. and Brown, I. R.,** Heat shock protein in mammalian brain and other organs after a physiologically relevant increase in body temperature induced by D-lysergic acid diethylamide, *Proc. Natl. Acad. Sci. U.S.A.,* 80, 569—573, 1983.
88. **Courgeon, A. M., Maisonhaute, C., and Best-Belpomme, M.,** Heat-shock proteins are induced by cadmium in *Drosophila* cells, *Exp. Cell Res.,* 153, 515—521, 1984.
88a. **Craig, E. A.,** Personal communication.
89. **Craig, E. A. and Jacobsen, K.,** Mutations of the heat inducible 70 kilodalton genes of yeast confer temperature sensitive growth, *Cell,* 38, 841—849, 1984.
89a. **Craig, E. A. and Jacobsen, K.,** Mutations in cognate genes of *Saccharomyces cerevisiae* hsp70 result in reduced growth rates at low temperatures, *Mol. Cell. Biol.,* 5, 3517—3524, 1985.
90. **Craig, E. A., McCarthy, B. J., and Wadsworth, S. C.,** Sequence organization of two recombinant plasmids containing genes for the major heat shock-induced protein of *D. melanogaster, Cell,* 16, 575—588, 1979.
91. **Craig, E. A., Ingolia, T., Slater, M., and Manseau, L.,** *Drosophila* and yeast multigene families related to the *Drosophila* heat shock genes, in *Heat Shock from Bacteria to Man,* Schlesigner, M. J., Ashburner, M. and Tissieres, A., Eds., Cold Spring Harbor Laboratory, Cold Spring Harbor, N.Y., 1982, 11—18.
92. **Craig, E. A., Ingolia, T. D., and Manseau, L. J.,** Expression of *Drosophila* heat-shock cognate genes during heat-shock and development, *Dev. Biol.,* 99, 418—426, 1983.
93. **Craig, E. A., Slater, M. R., Boorstein, W. R., and Palter, K.,** Expression of the *S. cerevisiae* hsp70 multigene family, *UCLA Symp. Mol. Cell. Biol.,* 30, 659—667, 1985.
93a. **Craig, E. A., Kramer, T., and Kosic-Smithers, J.,** SSC1, a member of the 70-kDa heat shock protein multigene family of *Saccharomyces cerevisiae,* is essential for growth, *Proc. Natl. Acad. Sci. U.S.A., 84,* 4156—4160, 1987.
94. **Craig, E. A., Kramer, J., Shilling, J., Werner-Washburne, M., Holmes, S., Kosic-Smithers, J., and Nicolet, C. M.,** SSC1, an essential member of the yeast HSP70 multigene family, encodes a mitochondrial protein, *Mol. Cell. Biol.,* 9, 3000—3008, 1989.
94a. **Craig, E., Boorstein, W., Park, H.-O., Stone, D., and Nicolet, C.,** Complex regulation of three heat inducible HSP70 related genes in *Saccharomyces cerevisiae,* in *Stress-Induced Proteins,* Pardue, M. L., Feramisco, J. R., and Lindquist, S., Eds., Alan R. Liss, New York, 1989, 51—61.
94b. **Curci, A., Bevilacqua A., and Mangia, F.,** Lack of heat-shock response in preovulatory mouse oocytes, *Dev. Biol.,* 123, 154—160, 1987.
95. **Currie, R. W. and White, F. P.,** Trauma-induced protein in rat tissues; A physiological role for a ''heat shock'' protein?, *Science,* 214, 72—73, 1981.
96. **Currie, R. W. and White, F. P.,** Characterization of the synthesis and accumulation of a 71-kilodalton protein induced in rat tissues after hyperthermia, *Can. J. Biochem. Cell Biol.,* 61, 438—446, 1983.

96a. **Czarnecka, E., Edelman, L., Schoeffl, F., and Key, J. L.,** Comparative analysis of physical stress responses in soybean seedlings using cloned heat shock cDNAs, *Plant Mol. Biol.,* 3, 45—58, 1984.

97. **Czarnecka, E., Gurley, W. B., Nagao, R. T., Mosquera, L. A., and Key, J. L.,** DNA sequence and transcript mapping of a soybean gene encoding a small heat shock protein, *Proc. Natl. Acad. Sci. U.S.A.,* 82, 3726—3730, 1985.

97a. **Czarnecka, E., Nagao, R. T., Key, J. L., and Gurley, W. B.,** Characterization of Gm-hsp26-A, a stress gene encoding a divergent heat shock protein of soybean: Heavy-metal-induced inhibition of intron processing, *Mol. Cell. Biol.,* 8, 1113—1122, 1988.

98. **Dabrowa, N. and Howard, D. H.,** Heat shock and heat stroke proteins observed during germination of the blastoconidia of *Candida albicans, Infect. Immunol.,* 44, 537—539, 1984.

98a. **Dang, C. V. and Lee, W. F.,** Nuclear and nucleolar targeting sequences of c-erb-A, c-myb, N-myc, p53, HSP70, and HIV tat proteins, *J. Biol. Chem.,* 264, 18019—18023, 1989.

99. **Daniels, C. J., McKee, A. H. Z., and Doolittle, W. F.,** Archaebacterial heat-shock proteins, *EMBO J.,* 3, 745—749, 1984.

100. **Darasch, S., Mosser, D. D., Bols, N. C., and Heikkila, J. J.,** Heat shock gene expression in *Xenopus laevis* A6 cells in response to heat shock and sodium arsenite, *Biochem. Cell Biol.,* 66, 862—870, 1988.

100a. **Dascher, C. C., Poddar, S. K., and Maniloff, J.,** Heat shock response in mycoplasmas, genome-limited organisms, *J. Bacteriol.,* 172, 1823—1827, 1990.

101. **Davis, J. Q. and Bennett, V.,** Human erythrocyte clathrin and clathrin-uncoating protein, *J. Biol. Chem.,* 260, 14850—14856, 1985.

102. **Davis, J. Q., Dansereau, D., Johnstone, R. M., and Bennett, V.,** Selective externalization of an ATP-binding protein structurally related to the clathrin-uncoating ATPase/heat shock protein in vesicles containing terminal transferrin receptors during reticulocyte maturation, *J. Biol. Chem.,* 261, 15368—15371, 1986.

103. **Dawson, W. O. and Grantham, G. L.,** Inhibition of stable RNA synthesis and production of a noval RNA in heat-stressed plants, *Biochem. Biophys. Res. Commun.,* 100, 23—30, 1981.

104. **Dean, R. L. and Atkinson, B. G.,** The acquisition of thermal tolerance in larvae of *Calpodes ethlius (Lepidoptera)* and the *in situ* and *in vitro* synthesis of heat-shock proteins, *Can. J. Biochem. Cell Biol.,* 61, 472—479, 1983.

105. **Dean, R. L. and Atkinson, B. G.,** Synthesis of heat shock proteins in quail red blood cells following brief, physiologically relevant increases in whole body temperature, *Comp. Biochem. Physiol.,* 81B, 185—191, 1985.

106. **De Jong, W. W., Hoekman, W. A., Mulders, J. W. M., and Bloemendal, H.,** Heat shock response of the rat lens, *J. Cell Biol.,* 102, 104—111, 1986.

107. **Delpino, A., Mileo, A. M., Mattei, E., and Ferrini, U.,** Characterization of the heat shock response in M-14 human melanoma cells continuously exposed to supranormal temperatures, *Exp. Mol. Pathol.,* 45, 128—141, 1986.

107a. **Denisenko, O. N. and Yarchuk, O. B.,** Regulation of LacZ mRNA translatability in a cell free system at heat shock by the last four sense codons, *FEBS Lett.,* 247, 251—254, 1989.

108. **Deshaies, R. J., Koch, B. D., Werner-Washburne, M., Craig, E. A., and Schekman, R.,** A subfamily of stress proteins facilitates translocation of secretory and mitochondrial precursor polypeptides, *Nature,* 332, 800—805, 1988.

109. **Dienel, G. A. and Cruz, N. F.,** Induction of brain ornithine decarboxylase during recovery from metabolic, mechanical, thermal, or chemical injury, *J. Neurochem.,* 42, 1053—1061, 1984.

110. **Dienel, G. A., Kiessling, M., Jacewicz, M., and Pulsinelli, W. A.,** Synthesis of heat shock proteins in rat brain cortex after transient ischemia, *J. Cereb. Blood Flow Metab.,* 6, 505—510, 1986.

111. **Dodson, M., McMacken, R., and Ehols, M.,** Specialized nucleoprotein structures at the origin of replication of bacteriophage lambda, *J. Biol. Chem.,* 264, 10719—10725, 1989.

111a. **Donnelly, C. E. and Walker, G. C.,** groE mutants of *Escherichia coli* are defective in umuDC-dependent UV mutagenesis, *J. Bacteriol.,* 171, 6117—6125, 1989.

111b. **Dorner, A. J., Krane, M. G., and Kaufman, R. J.,** Reduction of endogenous GRP78 levels improves secretion of a heterologous protein in CHO cells, *Mol. Cell. Biol.,* 8, 4063—4070, 1988.

112. **Dougherty, J. J., Rabideau, D. A., Iannotti, A. M., Dullivan, W. P., and Toft, D. O.,** Identification of the 90 kDa substrate of rat liver type II casein kinase with the heat shock protein which binds steroid receptors, *Biochim. Biophys. Acta,* 927, 74—80, 1987.

113. **Dragon, E. A., Sias, S. R., Kato, E. A., and Gabe, J. D.,** The genome of *Trypanosoma cruzi* contains a consitutively expressed, shock protein, *Mol. Cell. Biol.,* 7, 1271—1275, 1987.

113a. **Dragunow, M., Currie, R. W., Robertson, H. A., and Faull, R. L. M.,** Heat shock induces c-fos protein-like immunoreactivity in glial cells in adult rat brain, *Exp. Neurol.,* 106, 105—109, 1989.

114. **Duband, J. L., Lettre, F., Arrigo, A. P., and Tanguay, R. M.,** Expression and localization of hsp-23 in unstressed and heat-shocked *Drosophila* cultured cells, *Can. J. Genet. Cytol.,* 28, 1088—1092, 1987.

114a. **Duck, N., McCormick, S. H., and Winter, J.,** Heat shock protein hsp70 cognate gene expression in vegetative and reproductive organs of *Lycopersicon esculentum, Proc. Natl. Acad. Sci. U.S.A.,* 86, 3674—3678, 1989.

115. **Duncan, R. and Hershey, J. W. B.,** Heat shock-induced translational alterations in HeLa cells, *J. Biol. Chem.,* 259, 11882—11889, 1984.

116. **Dura, J. M.,** Stage dependent synthesis of heat shock induced proteins in early embryos of *Drosophila melanogaster, Mol. Gen. Genet.,* 184, 381—385, 1981.

117. **Dworniczak, B. and Mirault, M.-E.,** Structure and expression of a human gene coding for a 71 kd heat shock 'cognate' protein, *Nucl. Acids Res.,* 15, 5181—5198, 1987.

118. **Easton, D. P., Rutledge, P. S., and Spotila, J. R.,** Heat shock protein induction and induced thermal tolerance are independent in adult salamanders, *J. Exp. Zool.,* 241, 263—267, 1987.

119. **Eid, N. S., Kravath, R. E., and Lanks, K. W.,** Heat shock protein synthesis by human polymorphonuclear cells, *J. Exp. Med.,* 165, 1448—1452, 1987.

120. **Ellwood, M. S. and Craig, E. A.,** Differential regulation of the 70K heat shock gene and related genes in *Saccharomyces cerevisiae, Mol. Cell. Biol.,* 4, 1454—1459, 1984.

120a. **Engman, D. M., Kirchhoff, L. V., and Donelson, J. E.,** Molecular cloning of mtp70, a mitochondrial member of the hsp70 family, *Mol. Cell. Biol.,* 9, 5163—5168, 1989.

121. **Evan, G. I. and Hancock, D. C.,** Studies on the interaction of the human c-myc protein with cell nuclei: p62c-myc as a member of a discrete subset of nuclear proteins, *Cell,* 43, 253—261, 1985.

122. **Fabijanski, S., Altosaar, I., and Arnison, P. G.,** Heat shock response of *Brassica oleracea* L. (Broccoli), *J. Plant Physiol.,* 128, 29—38, 1987.

123. **Farrelly, F. W. and Finkelstein, D. B.,** Complete sequence of the heat shock-inducible HSP90 gene of *Saccharomyces cerevisiae, J. Biol. Chem.,* 259, 5745—5751, 1984.

124. **Ferrini, U., Falcioni, R., Delpino, A., Cavaliere, R., Zupi, G., and Natali, P. G.,** Heat-shock proteins produced by two human melanoma cell lines: Absence of correlation with thermosensitivity, *Int. J. Cancer,* 34, 651—655, 1984.

125. **Findly, R. C., Gillies, R. J., and Shulman, R. G.,** *In vivo* phosphorous-31 nuclear magnetic resonance reveals lowered ATP during heat shock of *Tetrahymena, Science,* 219, 1223—1225, 1983.

126. **Fink, K. and Zeuthen, E.,** Heat shock proteins in *Tetrahymena* studied under growth conditions, *Exp. Cell Res.,* 128, 23—30, 1980.

126a. **Finlay, C. A., Hinds, P. W., Tan, T.-H., Eliyahn, D., Oren, M., and Levine, A. J.,** Activating mutations for transformation by p53 produce a gene product that forms an hsc70-p53 complex with an altered half-life, *Mol. Cell. Biol.,* 8, 531—539, 1988.

127. **Finley, D., Oezkaynak, E., and Varshavsky, A.,** The yeast polyubiquitin gene is essential for resistance to high temperatures, starvation and other stresses, *Cell,* 48, 1035—1046, 1987.

127a. **Flynn, G. C., Chappell, T. G., and Rothman, J. E.,** Peptide binding and release by proteins implicated as catalysts of protein assembly, *Science,* 245, 385—390, 1989.

127b. **Fornace, A. J., Alamo, I., Hollander, M. C., and Lamoreaux, E.,** Ubiquitin mRNA is a major stress-induced transcript in mammalian cells, *Nucl. Acids Res.,* 17, 1215—1230, 1989.

128. **Francis, D. and Lin, L.,** Heat shock response in a cellular slime mold, *Polysphondylium pallidum, Dev. Biol.,* 79, 238—242, 1980.

129. **Freedman, M. S., Clark, B. D., Cruz, T. F., Gurd, J. W., and Brown, I. R.,** Selective effects of LSD and hyperthermia on the synthesis of synaptic proteins and glycoproteins, *Brain Res.,* 207, 129—145, 1981.

129a. **Frova, C., Taramino, G., and Binelli, G.,** Heat-shock proteins during pollen development in maize, *Dev. Genet.,* 10, 324—332, 1989.

129b. **Fuqua, S. A. W., Blumsalingaros, M., and McGuire, W. L.,** Induction of the estrogen-regulated 24K protein by heat shock, *Cancer Res.,* 49, 4126—4129, 1989.

130. **Fujio, N., Hatayama, T., Kinoshita, H., and Yukioka, M.,** Induction of four heat shock proteins and their mRNAs in rat after whole-body hyperthermia, *J. Biochem.,* 101, 181—187, 1987.

130a. **Fujio, N., Hatayama, T., Kinoshita, H., and Yukioka, M.,** Induction of mRNA for heat shock proteins in livers of rats after ischemia and partial hepatectomy, *Mol. Cell. Biochem.,* 77, 173—178, 1987.

130b. **Gaestel, M., Benndorf, R., Strauss, M., Schunk, W.-H., Kraft, R., Otto, A., Boehm, H., Stahl, J., Drabsch, H., and Bielka, H.,** Molecular cloning, sequencing and expression in *Escherichia coli* of the 25-kDa growth-related protein of Ehrlich ascites tumor and its homology to mammalian stress proteins, *Eur. J. Biochem.,* 179, 209—213, 1989.

131. **Galego, L. and Rodrigues-Pousada, C.,** Regulation of gene expression in *Tetrahymena pyriformis* under heat-shock and during recovery, *Eur. J. Biochem.,* 149, 571—588, 1985.

131a. **Garcia-Blanco, M. A., Clerc, R. G., and Sharp, P. A.,** The DNA-binding homeo domain of the Oct-2 protein, *Genes Dev.,* 3, 739—745, 1989.

132. **Gayda, R. C., Stephens, P. E., Hewick, R., Schoemaker, J. M., Dreyer, W. J., and Markovitz, A.,** Regulatory region of the heat shock-inducible capR(lon) gene: DNA and protein sequences, *J. Bacteriol.*, 162, 271—275, 1985.

132a. **Gedamu, L., Culham, B., and Heikkila, J. J.,** Analysis of the temperature-dependent temporal pattern of heat-shock-protein synthesis in fish cells, *Biosci. Rep.*, 3, 647—648, 1983.

132b. **Georgopoulos, C., Tissieres, A., and Morimoto, R., Eds.,** *Stress Proteins in Biology and Medicine,* Cold Spring Harbor Press, Cold Spring Harbor, NY, 1990.

133. **Georgopoulos, C., Tilly, V., Ang, D., Chandrasekhar, G. N., Fayet, O., Spence, J., Ziegelhoffer, T., Liberek, K., and Zylicz, M.,** The role of the *Escherichia coli* heat shock proteins in bacteriophage lambda growth, in *Stress-Induced Proteins,* Pardue, M. L., Feramisco, J. R., and Lindquist, S., Eds., Alan R. Liss, New York, 1989, 37—47.

133a. **Giebel, L. B., Dworniczak, B. P., and Bautz, E. K. F.,** Developmental regulation of a constitutively expressed mouse mRNA encoding a 72-kDa heat shock-like protein, *Dev. Biol.*, 125, 200—207, 1988.

133b. **Ginzburg, C. and Salomon, R.,** The effect of dormancy on the heat shock response in *Gladiolus gandavensis* cormels, *Plant Physiol.*, 81, 259—267, 1986.

134. **Glasczinski, H., Ohad, I., and Kloppstech, K.,** Temperature-dependent binding to the thylakoid membranes of nuclear-coded chloroplast heat-shock proteins, *Eur. J. Biochem.*, 73, 579—583, 1988.

135. **Glass, D. J., Polveres, R. J., and Van Der Ploeg, L. H. T.,** Conserved sequences and transcription of the hsp70 gene family in *Trypanosoma brucei, Mol. Cell. Biol.*, 6, 4657—4666, 1986.

135a. **Graham, R. W., Jones, D., and Candido, E. P. M.,** UbiA, the major polyubiquitin locus in *Caenorhabditis elegans,* has unusual structural features and is constitutively expressed, *Mol. Cell. Biol.*, 9, 268—277, 1989.

135b. **Goeroek, T.,** Detection of heat shock proteins in a fish cell-line, *Biona Report,* 4, 155—162, 1986.

136. **Goff, S. A., Casson, L. P., and Goldberg, A. L.,** Heat shock regulatory gene htpR influences rates of protein degradation and expression of the lon gene in *Escherichia coli, Proc. Natl. Acad. Sci. U.S.A.*, 81, 6647—6651, 1984.

137. **Goff, S. A. and Goldberg, A. L.,** Production of abnormal proteins in *E. coli* stimulates transcription of lon and other heat shock genes, *Cell,* 41, 587—595, 1985.

137a. **Goloubinoff, P., Gatenby, A. A., Lorimer, G. H.,** GroE heat shock proteins promote assembly of foreign prokaryotic ribulose bisphosphate carboxylase oligomers in *Escherichia coli, Nature,* 337, 44—47, 1989.

138. **Gomes, S. L., Juliani, M. H., Maia, J. C. C., and Silva, A. M.,** Heat shock protein synthesis during development in *Caulobacter crescentus, J. Bacteriol.*, 168, 923—930, 1986.

138a. **Goodman, H. J. K., Strydom, E., and Woods, D. R.,** Heat shock stress in *Bacteroides fragilis, Arch. Microbiol.*, 142, 362—364, 1985.

138b. **Gottesman, S.,** Genetics of proteolysis in *Escherichia coli, Annu. Rev. Genet.*, 23, 163—198, 1989.

138c. **Goulielmos, G. N. and Alahiotis, S. N.,** Induction of malate dehydrogenase and acetylcholinesterase by 20-hydroxyecdysone and heat shock in *Drosophila* ovaries, *Insect Biochem.*, 19, 393—399, 1989.

139. **Goulielmos, G., Kilias, G., and Alahiotis, S. N.,** Adaptation of *Drosophila* enzymes to temperature. V. Heat shock effect on the malate dehydrogenase of *Drosophila melanogaster, Comp. Biochem. Physiol.*, 85B, 229—234, 1986.

139a. **Green, L. A. D. and Liem, R. K. H.,** Beta-internexin is a microtubule-associated protein identical to the 70-kDa heat shock cognate protein and the clathrin uncoating ATPase, *J. Biol. Chem.*, 264, 15210—15215, 1989.

140. **Greenberg, S. G. and Lasek, R. L.,** Comparison of labelled heat-shock proteins in neuronal and nonneuronal cells of *Aplysia californica, J. Neurosci.*, 5, 1239—1245, 1985.

141. **Grimm, B., Ish-Shalom, D., Even, D., Glaczinski, H., Ottersbach, P., Ohad, I., and Kloppstech, K.,** The nuclear-coded chloroplast 22-kDa heat-shock protein of *Chlamydomonas.* Evidence for translocation into the organelle without a processing step, *Eur. J. Biochem.*, 182, 539—546, 1989.

142. **Gross, T. and Schulz-Harder, B.,** Heat-shock induction of ribonuclease and protease in yeast, *H.-S. Z. Physiol. Chemie,* 364, 1132—1133, 1983.

143. **Gross, T. and Schulz-Harder, B.,** Induction of a proteinase by heat-shock in yeast, *FEMS Microbiol. Lett.*, 33, 199—203, 1986.

144. **Gross, C. A., Grossman, A. D., Liebke, H., Walther, W., and Burgess, R. R.,** Effects of the mutant sigma allele rpoD800 on the synthesis of specific macromolecular components of the *E. coli* K12 cell, *J. Mol. Biol.*, 172, 283—300, 1984.

145. **Grossman, A. D., Erickson, J. W., and Gross, C. A.,** The htpR gene product of *E. coli* is a sigma factor of heat-shock promoters, *Cell,* 38, 383—390, 1984.

145a. **Gubits, R. M. and Fairhurst, J. L.,** C-fos mRNA levels are increased by the cellular stressors, heat shock and sodium arsenite, *Oncogene,* 3, 163—168, 1988.

146. **Guedon, G., Sovia, D., Ebel, J. P., Befort, N., and Remy, P.,** Effect of diadenosine tetraphosphate microinjection on heat shock protein synthesis in *Xenopus laevis* oocytes, *EMBO J.*, 4, 3743—3749, 1985.

147. **Guidon, P. T. and Hightower, L. E.,** Purification and initial characterization of the 71-kilodalton rat heat-shock protein and its cognate as fatty acid binding proteins, *Biochemistry,* 25, 3231—3239, 1986.

147a. **Guidon, P. T. and Hightower, L. E.,** The 73 kilodalton heat shock cognate protein purified from rat brain contains nonesterified palmitic and stearic acids, *J. Cell. Physiol.,* 128, 239—245, 1986.

148. **Guttman, S. D. and Gorovsky, M. A.,** Cilia regeneration in starved *Tetrahymena:* An inducible system for studying gene expression and organelle biosynthesis, *Cell,* 17, 307—317, 1979.

149. **Guttman, S. D., Glover, C. V. C., Allis, C. D., and Gorovsky, M. A.,** Heat shock, deciliation and release from anoxia induce the synthesis of the same set of polypeptides in starved *T. pyriformis, Cell,* 22, 299—307, 1980.

149a. **Guy, C. L., Niemi, K. J., and Brambl, R.,** Altered gene expression during cold acclimation of spinach, *Proc. Natl. Acad. Sci. U.S.A.,* 82, 3673—3677, 1985.

150. **Gwynne, D. I. and Brandhorst, B. P.,** Alterations in gene expression during heat shock of *Achlya ambisexualis, J. Bacteriol.,* 149, 488—493, 1982.

151. **Haass, C., Falkenburg, P. E., and Kloetzel, P.-M.,** The molecular organization of the small heat shock proteins in *Drosophila,* in *Stress-Induced Proteins,* Pardue, M. L., Feramisco, J. R., and Lindquist, S., Eds., Alan R. Liss, New York, 1989, 175—185.

152. **Hacket, R. W. and Lis, J. T.,** Localization of the hsp83 transcript within a 3292 nucleotide-sequence from the 63B heat shock locus of *Drosophila melanogaster, Nucl. Acids Res.,* 11, 7011—7030, 1983.

152a. **Hagen, G., Uhrhammer, N., and Guilfoyle, T. J.,** Regulation of expression of an auxin-induced soybean sequence by cadmium, *J. Biol. Chem.,* 263, 6442—6446, 1988.

153. **Hahnel, A. C., Gifford, D. J., Heikkila, J. J., and Schultz, G. A.,** Expression of the major heat shock protein (hsp70) family during early mouse development, *Teratolog. Cancer. Mutag.,* 6, 493—510, 1986.

154. **Hallberg, R. L., Kraus, K. W., and Hallberg, E. M.,** Induction of acquired thermotolerance in *Tetrahymena thermophila:* Effects of protein synthesis inhibitors, *Mol. Cell. Biol.,* 5, 2061—2069, 1985.

155. **Hammond, G. L., Lai, Y.-K., and Markert, C. L.,** Diverse forms of stress lead to new patterns of gene expression through a common and essential metabolic pathway, *Proc. Natl. Acad. Sci. U.S.A.,* 79, 3485—3488, 1982.

155a. **Hannink, M. and Donoghue, D. J.,** Lysine residue 121 in the proposed ATP-binding site of the v-mos protein is required for transformation, *Proc. Natl. Acad. Sci. U.S.A.,* 82, 7894—7898, 1985.

156. **Harari, P. M., Tome, M. E., and Gerner, E. W.,** Heat shock-induced polyamine oxidation in mammalian cells, *J. Cell Biol.,* 103, 175a, 1986.

157. **Hatayama, T., Honda, K., and Yukioka, M.,** HeLa cells synthesize a specific heat shock protein upon exposure to heat shock at 42°C but not at 45°C, *Biochem. Biophys. Res. Commun.,* 137, 957—963, 1986.

158. **Heikkila, J. J., Schultz, G. A., Iatrou, K., and Gedamu, L.,** Expression of a set of fish genes following heat or metal ion exposure, *J. Biol. Chem.,* 257, 12000—12005, 1982.

159. **Heikkila, J. J., Kloc, M., Bury, J., Schultz, G. A., and Browder, L. W.,** Acquisition of the heat shock response and thermotolerance during early development of *Xenopus laevis, Dev. Biol.,* 107, 483—489, 1985.

160. **Heikkila, J. J., Browder, L. W., Gedamu, L., Nickells, R. W., and Schultz, G. A.,** Heat-shock gene expression in animal embryonic systems, *Can. J. Genet. Cytol.,* 28, 1093—1105, 1986.

160a. **Heikkila, J. J., Darasch, S. P., Mosser, D. D., and Bols, N. C.,** Heat and sodium arsenite act synergistically on the induction of heat shock gene expression in *Xenopus laevis* A6 cells, *Biochem. Cell Biol.,* 65, 310—316, 1987.

161. **Helm, K. W. and Vierling, E.,** An Arabidopsis thaliana cDNA clone encoding a low molecular weight heat shock protein, *Nucl. Acids Res.,* 17, 7995, 1989.

161a. **Hemmingsen, S. M., Woolford, C., van der Vies, S. M., Tilly K., Dennis, D. T., Georgopoulos, C. P., Hendrix, R. W., and Ellis, R. J.,** Homologous plant and bacterial proteins chaperone oligomeric protein assembly, *Nature,* 333, 330—335, 1988.

162. **Hendrix, R. W.,** Purification and properties of groE, a host protein involved in bacteriophage assembly, *J. Mol. Biol.,* 129, 375—392, 1979.

162a. **Hendrix, R. W. and Tsui, L.,** Role of the host in virus assembly: cloning of the *Escherichia coli* groE gene and identification of its protein product, *Proc. Natl. Acad. Sci. U.S.A.,* 75, 136—139, 1978.

163. **Herr, W., Sturm, R. A., Clerc, R. G., Corcoran, L. M., Baltimore, D., Sharp, P. A., Ingraham, H. A., and Horvitz, M. G.,** The POU domain: a large conserved region in the mammalian pit-1, oct-1, oct-2, and *Caenorhabditis elegans* unc-86 gene products, *Genes Devel.,* 2, 1513—1516, 1988.

163a. **Heschl, M. F. P. and Baillie, D. L.,** Characterization of the hsp70 multigene family of *Caenorhabditis elegans, DNA,* 8, 233—243, 1989.

163b. **Heuser, J. and Steer, C. J.,** Trimeric binding of the 70-kDa uncoating ATPase to the vertices of clathrin triskelia: A candidate intermediate in the vesicle uncoating reaction, *J. Cell Biol.,* 109, 1457—1466, 1989.

164. **Hickey, E. D. and Weber, L. A.,** Modulation of heat shock polypeptide synthesis in HeLa cells during hyperthermia and recovery, *Biochemistry,* 21, 1513—1521, 1982.

165. **Hickey, E., Brandon, S. E., Sadis, S., Smale, G., and Weber, L. A.,** Molecular cloning of sequences encoding the human heat-shock proteins and their expression during hyperthermia, *Gene,* 43, 147—154, 1986.

165a. **Hickey, E., Brandon, S. E., Smale, G., Lloyd, D., and Weber, L. A.,** Sequence and regulation of a gene encoding a human 89-kilodalton heat shock protein, *Mol. Cell. Biol.,* 9, 2615—2626, 1989.

165b. **Hinds, P. W., Finlay, C. A., Frey, A. B., and Levine, A. J.,** Immunological evidence for the association of p53 with a heat shock protein, hsc70, in p53-plus-ras-transformed cell lines, *Mol. Cell. Biol.,* 7, 2863—2869, 1987.

166. **Holmgren, R., Livak, K., Morimoto, R., Freund, R., and Meselson, M.,** Studies of cloned sequences from four *Drosophila* heat shock loci, *Cell,* 18, 1359—1370, 1979.

167. **Hong-Qi, Z., Croes, A. F., and Linskens, H. F.,** Qualitative changes in protein synthesis in germinating pollen of *Lilium longiflorum* after heat shock, *Plant Cell Environ.,* 7, 689—691, 1984.

167a. **Howarth, C.,** Heat shock proteins in Sorghum bicolor and Pennisetum americanum I. Genotypic and developmental variation during seed germination, *Plant Cell Environ.,* 12, 471—478, 1989.

168. **Hughes, E. N. and August, J. T.,** Coprecipitation of heat shock proteins with a cell surface glycoprotein, *Proc. Natl. Acad. Sci. U.S.A.,* 79, 2305—2309, 1982.

169. **Hunt, C. and Morimoto, R. J.,** Conserved features of eukaryotic hsp70 genes revealed by comparison with the nucleotide sequence of human hsp70, *Proc. Natl. Acad. Sci. U.S.A.,* 82, 6455—6459, 1985.

170. **Hunter, K. W., Cook, C. L., and Hayunga, E. G.,** Leishmanial differentiation *in vitro:* Induction of heat shock proteins, *Biochem. Biophys. Res. Commun.,* 125, 755—760, 1984.

170a. **Hutchinson, E. G., Tichelaar, W., Hofhaus, G., Weiss, H., and Leonard, K. R.,** Identification and electron microscopic analysis of a chaperonin oligomer from *Neurospora crassa* mitochondria, *EMBO J.,* 8, 1485—1490, 1989.

171. **Ibrahim-Granet, O. and De Bievre, C.,** Etude des proteines heat shock et heat-stroke chez *Fonsecaea pedrosoi*: Agent de chromomycose, *Bull. Soc. Fr. Mycol. Med.,* 15, 213—220, 1986.

172. **Iida, H. and Yahara, I.,** Durable synthesis of high molecular weight heat shock proteins in GO cells of the yeast and other eukaryotes, *J. Cell Biol.,* 99, 199—207, 1984.

173. **Iida, H. and Yahara, I.,** Yeast heat-shock protein of Mr 48,000 is an isoprotein of enolase, *Nature,* 315, 688—690, 1985.

174. **Ingolia, T. D. and Craig, E. A.,** Four small *Drosophila* heat shock proteins are related to each other and to mammalian crystallin, *Proc. Natl. Acad. Sci. U.S.A.,* 79, 2360—2364, 1982.

175. **Ingolia, T. D., Craig, E. A., and McCarthy, B. J.,** Sequence of three copies of the gene for the major *Drosophila* heat shock induced protein and their flanking regions, *Cell,* 21, 669—679, 1980.

176. **Ingolia, T. D., Slater, M. R., and Craig, E. A.,** *Saccharomyces cerevisiae* contains a complex multigene family related to the major heat shock-inducible gene of *Drosophila, Mol. Cell. Biol.,* 2, 1388—1398, 1982.

176a. **Ireland, R. C. and Berger, E. M.,** Synthesis of low molecular weight heat shock peptides stimulated by ecdysterone in a cultured *Drosophila* cell line, *Proc. Natl. Acad. Sci. U.S.A.,* 79, 855—859, 1982.

177. **Iwasaki, M., Saito, H., Yamamoto, M., Korach, K. S., Hirogome, T., and Sugano, H.,** Purification of heat shock protein90 from calf uterus and rat liver and characterization of the highly hydrophobic region, *Biochim. Biophys. Acta,* 992, 1—8, 1989.

177a. **Jerez, C. A.,** The heat shock response in meso- and thermo-acidophilic chemolitotrophic bacteria, *FEMS Microbiol. Lett.,* 56, 289—294, 1988.

177b. **Jindal, S., Dudani, A. K., Singh, B., Harley, C. B., and Gupta, R. S.,** Primary structure of a human mitochondrial protein homologous to the bacterial and plant chaperonins and to the 65-kd mycobacterial antigen, *Mol. Cell. Biol.,* 9, 2279—2283, 1989.

177c. **Johnson, C., Chandrasekhar, G. N., and Georgopoulos, C.,** *Escherichia coli* DnaK and GrpE heat shock proteins interact both *in vivo* and *in vitro, J. Bacteriol.,* 171, 1590—1596, 1989.

178. **Jones, D., Russnak, R. H., Kay, R. J., and Candido, E. P. M.,** Structure expression and evolution of a heat shock gene locus in *Caenorhabditis elegans* that is flanked by repetitive elements, *J. Biol. Chem.,* 261, 12006—12015, 1986.

178a. **Kaddurah-Daouk, R., Greene, J. M., Baldwin, A. S., and Kingston, R. E.,** Activation and repression of mammalian gene expression by the c-myc protein, *Genes Dev.,* 1, 347—357, 1987.

179. **Kapoor, M.,** A study of the heat-shock response in *Neurospora crassa, Int. J. Biochem.,* 15, 636—649, 1983.

180. **Kapoor, M.,** A study of the effect of heat shock and metal ions on protein synthesis in *Neurospora crassa* cells, *Int. J. Biochem.,* 18, 15—30, 1986.

181. **Kapoor, M. and Chow, A. W. L.,** A two-dimensional immunoelectrophoretic analysis of the heat-shock response exhibited by *Neurospora crassa* cells, *Can. J. Biochem. Cell. Biol.,* 62, 691-698, 1984.

182. **Kapoor, M. and Lewis, J.,** Alteration of the protein synthesis pattern in *Neurospora crassa* cells by hyperthermal and oxidative stress, *Can. J. Microbiol.,* 33, 162—168, 1987.

182a. **Kapoor, M. and Sreenivasan, G. M.,** The heat shock response of *Neurospora crassa*: Stress-induced thermotolerance in relation to peroxidase and superoxide dismutase levels, *Biochem. Biophys. Res. Commun.*, 156, 1097—1102, 1988.

182b. **Kapoor, M., Sreenivasan, G. M., Goel, N., and Lewis, J.,** Development of thermotolerance in *Neurospora crassa* by heat shock and other stresses eliciting peroxidase induction, *J. Bacteriol.*, 172, 2798—2801, 1990.

183. **Karch, F., Toeroek, I., and Tissieres, A.,** Extensive regions of homology in front of the two hsp 70 heat shock variant genes in *Drosophila melanogaster, J. Mol. Biol.*, 148, 219—230, 1981.

183a. **Kashlev, M. V., Gragerov, A. I., and Nikiforov, V. G.,** Heat shock response in *Escherichia coli* promotes assembly of plasmid encoded RNA polymerase beta-subunit into RNA polymerase, *Mol. Gen. Genet.*, 216, 469—474, 1989.

183b. **Kawasaki, Y., Wada, C., and Yura, T.,** Roles of *Escherichia coli* heat shock proteins DnaK, DnaJ and GrpE in mini-F plasmid replication, *Mol. Gen. Genet.*, 220, 277—282, 1990.

184. **Kelley, P. M. and Schlesinger, M. J.,** The effect of amino acid analogues and heat shock on gene expression in chicken embryo fibroblasts, *Cell*, 15, 1277—1286, 1978.

185. **Kelley, P. M. and Schlesinger, M. J.,** Antibodies to two major chicken heat shock proteins cross-react with similar proteins in widely divergent species, *Mol. Cell. Biol.*, 2, 267—274, 1982.

186. **Ketola-Pirie, C. A. and Atkinson, B. G.,** Cold- and heat-shock induction of new gene expression in cultured amphibian cells, *Can. J. Biochem. Cell Biol.*, 61, 462—471, 1983.

187. **Key, J. L., Czarnecka, E., Lin, C.-Y., Kimpel, J., Mothershed, C., and Schoeffl, F.,** A comparative analysis of the heat shock response in crop plants, in *Current Topics in Plant Biochemistry and Physiology*, Randall, D. D., Blevins, D. G., Larsons, R. L., and Rapp, B. J., Eds., University of Missouri Press, Columbia, 1983, 107—118.

188. **Key, J. L., Lin, C. Y., and Chen, Y. M.,** Heat shock proteins of higher plants, *Proc. Natl. Acad. Sci., U.S.A.*, 78, 3526—3530, 1981.

188a. **Keyse, St. M. and Tyrrell, R. M.,** Heme oxygenase is the major 32-kDa stress protein induced in human skin fibroblasts by UVA radiation, hydrogen peroxide, and sodium arsenite, *Proc. Natl. Acad. Sci. U.S.A.*, 86, 99—103, 1989.

189. **Kim, Y.-J., Shuman, J., Sette, M., and Przybyla, A.,** Arsenate induces stress proteins in cultured rat myoblasts, *J. Cell Biol.*, 96, 393—400, 1983.

190. **Kim, Y.-J., Shuman, J., Sette, M., and Przybyla, A.,** Nuclear localization and phosphorylation of three 25-kilodalton rat stress proteins, *Mol. Cell. Biol.*, 4, 468—474, 1984.

190a. **Kingston, R. E., Baldwin, A. S., and Sharp, P. A.,** Regulation of heat shock protein 70 gene expression by c-myc, *Nature*, 312, 280—282, 1984.

191. **Kioussis, J., Cato, A. C. B., Slater, A., and Burdon, R. H.,** Polypeptides encoded by polyadenylated and non-polyadenylated messenger RNAs from normal and heat shocked HeLa cells, *Nucl. Acids Res.*, 9, 5203—5214, 1981.

192. **Kirk, M. M. and Kirk, D. L.,** Translational regulation of protein synthesis, in response to light, at a critical stage of *Volvox* development, *Cell*, 41, 419—428, 1985.

192a. **Kishore, R. and Upadhyaya, K. C.,** Heat shock proteins of pigeon pea *(Cajanus cajan), Plant Cell Physiol.*, 29, 517—521, 1988.

192b. **Klempnauer, K.-H., Arnold, H., and Biedenkapp, H.,** Activation of transcription by v-myb: evidence for two different mechanisms, *Genes Dev.*, 3, 1582—1589, 1989.

192c. **Klimpel, K. W. and Clark, V. L.,** The heat shock response of type 1 and type 4 *Gonococci, Sex. Transm. Dis.*, 16, 141—147, 1989.

193. **Kloppstech, K., Meyer, G., Schuster, G., and Ohad, I.,** Synthesis, transport and localization of a nuclear coded 22-kd heat-shock protein in the chloroplast membranes of peas and *Chlamydomonas reinhardi, EMBO J.*, 4, 1901—1909, 1985.

194. **Kloppstech, K., Ohad, I., and Schweiger, A.-G.,** Evidence for an extranuclear coding site for a heat-shock protein in *Acetabularia, Eur. J. Cell Biol.*, 42, 239—245, 1986.

194a. **Knack, G. and Kloppstech, K.,** cDNA sequence of a heat-inducible protein of *Chenopodium* sharing little homology with other heat shock proteins, *Nucl. Acids Res.*, 17, 5380, 1989.

195. **Koninkx, J. F. J. G.,** Protein synthesis in salivary glands of *Drosophila hydei* after experimental gene induction, *Biochem. J.*, 158, 623—628, 1976.

196. **Kothary, R. K. and Candido, E. P. M.,** Induction of a novel set of polypetides by heat shock or sodium arsenite in cultured cells of rainbow trout, *Salmo gairdnerii, Can. J. Biochem*, 60, 347—355, 1982.

197. **Kothary, R. K., Jones, D., and Candido, E. P. M.,** 70-Kilodalton heat shock polypeptides from rainbow trout: Characterization of cDNA sequences, *Mol. Cell. Biol.*, 4, 1785—1791, 1984.

197a. **Koyasu, S., Nishida, E., Miyata, Y., Sakai, H., and Yahara, I.,** Hsp-100, a 100-kDa heat shock protein, is a Ca^{2+}-calmodulin-regulated actin-binding protein, *J. Biol. Chem.*, 264, 15083—15087, 1989.

198. **Koyasu, S., Nishida, E., Kadowaki, T., Matsuzaki, F., Iida, K., Harada, F., Kasuga, M., Sakai, H., and Yahara, I.,** Two mammalian heat shock proteins, HSP90 and HSP100, are actin-binding proteins, *Proc. Natl. Acad. Sci. U.S.A.*, 83, 8054—8058, 1986.

198a. **Krawczyk, Z., Szymik, N., and Wisniewski, J.,** Expression of hsp70-regulated gene in developing and degenerating rat testis, *Mol. Biol. Rep.*, 12, 35—41, 1987.

198b. **Krishnasamy, S., Mannar-Mannan, R., Krishnan, M., and Gnanam, A.,** Heat shock response of the chloroplast genome in *Vigna sinensis, J. Biol. Chem.*, 263, 5104—5109, 1988.

198c. **Krishnan, M., Nguyen, H. T., and Burke, J. J.,** Heat shock protein synthesis and thermal tolerance in wheat, *Plant Physiol.*, 90, 140—145, 1989.

199. **Krueger, J. H. and Walker, G. C.,** groEL and dnaK genes of *Escherichia coli* are induced by UV irradiation and nalidixic acid in an htpR$^+$-dependent fashion, *Proc. Natl. Acad. Sci. U.S.A.*, 81, 1499—1503, 1984.

200. **Kubo, T., Towle, C. A., Mankin, H. J., and Treadwell, B. V.,** Stress-induced proteins in chondrocytes from patients with osteoarthritis, *Arthritis Rheum.*, 28, 1140—1145, 1985.

201. **Kulomaa, M. S., Weigel, N. L., Kleinsek, O. A., Beattie, W. G., Connely, O. M., March, C., Zarucki-Schulz, T., Schrader, W. T., and O'Malley, B. W.,** Amino acid sequence of a chicken heat shock protein derived from the complementary DNA nucleotide sequence, *Biochemistry*, 25, 6244—6252, 1986.

201a. **Kurkela, S., Franck, M., Heino, P., Lang, V., and Palva, E. T.,** Cold induced gene expression in *Arabidopsis thaliana* L., *Plant Cell Rep.*, 7, 495—498, 1988.

201b. **Kurkinen, M., Taylor, A., Garrels, J. I., and Hogan, B. L. M.,** Cell surface-associated proteins which bind native type IV collagen or gelatin, *J. Biol. Chem.*, 259, 5915—5922, 1984.

202. **Kurtz, S., Rossi, J., Petko, L., and Lindquist, S.,** An ancient developmental induction: Heat-shock proteins induced in sporulation and oogenesis, *Science*, 231, 1154—1157, 1986.

202a. **Kusukawa, N., Yura, T., Ueguchi, C., Akiyama, Y., and Ito, K.,** Effects of mutations in heat shock genes groES and groEL on protein export in *Escherichia coli, EMBO J.*, 8, 3517—3521, 1989.

202b. **Laemmli, U.K.,** Cleavage of structural proteins during the assembly of the head of bacteriophage T4, *Nature*, 227, 680—685, 1970.

203. **Lai, B.-T., Chin, N. W., Stanek, A. E., Keh, W., and Lanks, K. W.,** Quantitation and intracellular localization of the 85K heat shock protein by using monoclonal and polyclonal antibodies, *Mol. Cell. Biol.*, 4, 2802—2810, 1984.

204. **Lai, Y. K., Havre, P. A., and Hammond, G. L.,** Heat shock stress initiates simultaneous transcriptional and translational changes in the dog heart, *Biochem. Biophys. Res. Commun.*, 134, 166—171, 1986.

204a. **Lai, Y.-K., Li, C. W., Hu, C. H., and Lee, M. L.,** Quantitative and qualitative analyses of protein synthesis during heat shock in the marine diatom *Nitzschia alba (Bacillariophyceae), J. Phycol.*, 24, 509—514, 1988.

205. **Lamarche, S., Chretien, P., and Landry, J.,** Inhibition of the heat shock response and synthesis of glucose-regulated proteins in calcium-deprived rat hepatoma cells, *Biochem. Biophys. Res. Commun.*, 131, 868—876, 1985.

205a. **Lamb, F. I., Singh, N. B., and Colston, M. J.,** Thc spccific 18-kilodalton antigcn of *Mycobacterium leprae* is present in *Mycobacterium habana* and functions as a heat shock protein, *J. Immunol.*, 144, 1922—1925, 1990.

206. **Landick, R., Vaughn, V., Lau, E. T., Van Bogelen, R. A., Erickson, J. W., and Neidhardt, F. C.,** Nucleotide sequence of the heat shock regulatory gene of *E. coli* suggests its protein product may be a transcription factor, *Cell*, 38, 175—182, 1984.

207. **Landry, J. and Chretien, P.,** Relationship between hyperthermia-induced heat shock proteins and thermotolerance in Morris hepatoma cells, *Can. J. Biochem. Cell Biol.*, 61, 428—437, 1983.

208. **Landry, J., Bernier, D., Chretien, P., Nicole, L. M., Tanguay, R. M., and Marceau, N.,** Synthesis and degradation of heat shock proteins during development and decay of thermotolerance, *Cancer Res.*, 42, 2457—2461, 1982.

209. **Landry, J., Chretien, P., Bernier, D., Nicole, L. M., Marceau, N., and Tanguay, R. M.,** Thermotolerance and heat shock proteins induced by hyperthermia in rat liver cells, *Int. J. Radiat. Oncol.*, 8, 59—62, 1982.

210. **Landry, J., Chretien, P., De Muys, J. M., and Morais, R.,** Induction of thermotolerance and heat shock protein synthesis in normal and respiration-deficient chick embryo fibroblasts, *Cancer Res.*, 45, 2240—2247, 1985.

211. **Landry, J., Samson, S., and Chretien, P.,** Hyperthermia-induced cell death, thermotolerance, and heat shock proteins in normal, respiration-deficient, and glycolysis-deficient Chinese hamster cells, *Cancer Res.*, 46, 324—327, 1986.

211a. **Landry, J., Crete, P., Lamarche, S., and Chretien, P.,** Activation of calcium-dependent processes during heat shock. Role in cell thermoresistance, *Radiat. Res.*, 113, 426—436, 1988.

211b. **Lanks, K. W.,** Temperature-dependent oligomerization of HSP85 *in vitro, J. Cell Physiol.,* 140, 601—607, 1989.

212. **Lanks, K. W., Hitti, I. F., and Chin, N. W.,** Substrate utilization for lactate and energy production by heat-shocked L929 cells, *J. Cell. Physiol.,* 127, 451—456, 1986.

213. **LaThangue, N. E., Shriver, K., Dawson, C., and Chan, W. L.,** Herpes simplex virus infection causes the accumulation of a heat-shock protein, *EMBO J.,* 3, 267—277, 1984.

213a. **Lathigra, P. R., Young, D. B., Sweetser, D., and Young, R. A.,** A gene from *Mycobacterium tuberculosis* which is homologous to the DnaJ heat shock protein of *Escherichia coli, Nucl. Acids Res.,* 16, 1636, 1988.

213b. **Laudien, H., Dietl, J. U., and Schmidt, J.,** Heat-hardening and protein biosynthesis in tissue cultures from *Pimephales promelas (Pisces, Cuprinidae), J. Therm. Biol.,* 8, 431—432, 1983.

214. **Lee, A. S., Bell, J., and Ting, J.,** Biochemical characterization of the 94- and 78-kilodalton glucose-regulated proteins in hamster fibroblasts, *J. Biol. Chem.,* 259, 4616—4621, 1984.

214a. **Lee, H., Simon, J. A., and Lis, J. T.,** Structure and expression of ubiquitin genes of *Drosophila melanogaster, Mol. Cell. Biol.,* 8, 4727—4735, 1988.

215. **Leenders, H. J. and Berendes, H. D.,** The effect of changes in the respiratory metabolism upon genome activity in *Drosophila* I. The induction of gene activity, *Chromosoma,* 37, 433—444, 1972.

216. **Leenders, H. J., Derkson, J., Maas, P. M. J. M., and Berendes, H. D.,** Selective induction of a giant puff in *Drosophila hydei* by vitamin B6 and derivatives, *Chromosoma,* 41, 447—460, 1973.

217. **Leenders, H. J., Berendes, H. D., Helmsing, P. J., Derkson, J., and Koninkx, Y. F. G.,** Nuclear-mitochondrial interactions in the control of mitochondrial respiratory metabolism, *Sub-Cell. Biochem.,* 3, 119—147, 1974.

217a. **Lees-Miller, S. P. and Anderson, C. W.,** The human double-stranded DNA-activated protein kinase phosphorylates the 90-kDa heat-shock protein, Hsp90-alpha at two NH2-terminal threonine residues, *J. Biol. Chem.,* 264, 17275—17280, 1989.

217b. **Lees-Miller, S. P. and Anderson, C. W.,** Two human 90-kDa heat shock proteins (hsp90) are phosphorylated *in vivo* at conserved serines that are phosphorylated *in vitro* by casein kinase II, *J. Biol. Chem.,* 264, 2431—2437, 1989.

217c. **Legagneux, V., Dubois, M. F., Morange, M., and Bensaude, O.,** Phosphorylation of the 90 kDa heat shock protein in heat shocked HeLa cell lysates, *FEBS Lett.,* 231, 417—420, 1988.

218. **Leicht, B. G., Biessmann, H., Palter, K. B., and Bonner, J. J.,** Small heat shock proteins of *Drosophila* associate with the cytoskeleton, *Proc. Natl. Acad. Sci. U.S.A.,* 83, 90—94, 1986.

219. **LeJohn, H. B. and Braithwaite, E. E.,** Heat and nutritional shock-induced proteins of the fungus *Achlya* are different and under independent transcriptional control, *Can. J. Biochem. Cell Biol.,* 62, 837—846, 1984.

220. **LeLand, T. J. and Hanson, A. D.,** Induction of a specific N-methyl-transferase enzyme by long-term heat stress during barley leaf growth, *Plant Physiol.,* 79, 451—457, 1985.

220a. **Lema, M. W., Brown, A., Butler, C. A., and Hoffman, P. S.,** Heat-shock response in *Legionella pneumophila, Can. J. Microbiol.,* 34, 1148—1153, 1988.

220b. **Leustek, T., Dalie, B., Amir-Shapira, D., Brot, N., and Weissbach, H.,** A member of the hsp70 family is localized in mitochondria and resembles *Escherichia coli* DnaK, *Proc. Natl. Acad. Sci. U.S.A.,* 86, 7805—7808, 1989.

221. **Levinson, W., Oppermann, H., and Jackson, J.,** Transition series metals and sulfhydryl reagents induce the synthesis of four proteins in eukaryotic cells, *Biochim. Biophys. Acta,* 606, 170—180, 1980.

222. **Levinson, W., Idriss, J., and Jackson, J.,** Metal-binding drugs induce synthesis of four proteins in normal cells, *Biol. Trace Element Res.,* 1, 15—23, 1979.

223. **Lewis, M. and Pelham, H. R. B.,** Involvement of ATP in the nuclear and nucleolar functions of the 70 kD heat shock protein, *EMBO J.,* 4, 3137—3143, 1985.

224. **Lewis, M., Helmsing, P. J., and Ashburner, M.,** Parallel changes in puffing activity and patterns of protein synthesis in salivary glands of *Drosophila, Proc. Natl. Acad. Sci. U.S.A.,* 72, 3604—3608, 1975.

225. **Li, G. C. and Laszlo, A.,** Amino acid analogs while inducing heat shock proteins sensitize CHO cells to thermal damage, *J. Cell. Physiol.,* 122, 91—97, 1985.

226. **Li, G. C. and Mak, Y.,** Induction of heat shock protein synthesis in murine tumors during the development of thermotolerance, *Cancer Res.,* 45, 3816—3824, 1985.

227. **Li, G. C. and Werb, Z.,** Correlation between synthesis of heat shock proteins and development of thermotolerance in Chinese hamster fibroblasts, *Proc. Natl. Acad. Sci. U.S.A.,* 79, 3218—3222, 1982.

228. **Lim, L., Hall, C., Leung, T., and Whatley, S.,** The relationship of the rat brain 68kDa microtubule-associated protein with synaptosomal plasma membranes and with the *Drosophila* 70kDa heat-shock protein, *Biochem. J.,* 224, 677—680, 1984.

229. **Lin, J. J.-C., Welch, W. J., Garrels, J. I., and Feramisco, J. R.,** The association of the 100-kD heat-shock protein with the Golgi apparatus, in *Heat Shock from Bacteria to Man,* Schlesinger, M. J., Ashburner, M., and Tissieres, A., Eds., Cold Spring, Harbor Laboratory, Cold Spring Harbor, NY, 1982, 267—273.

229a. **Lindley, T. A. Chakraborty, P. R., and Edlind, T. D.,** Heat shock and stress response in *Giardia lamblia, Mol. Biochem. Parasitol.*, 28, 135—143, 1988.

230. **Lindquist, S.,** Varying patterns of protein synthesis in *Drosophila* during heat shock: implications for regulation, *Dev. Biol.*, 77, 463—479, 1980.

230a. **Lindquist, S.,** The heat-shock response (review), *Annu. Rev. Biochem.*, 55, 1151—1191, 1986.

230b. **Lindquist, S. and Craig, E. A.,** The heat-shock proteins, *Annu. Rev. Genet.*, 22, 631—677, 1988.

230c. **Lipinska, B., King, J., Ang, D., and Georgopoulos, C.,** Sequence analysis and transcriptional regulation of the *Escherichia coli* grpE gene encoding a heat shock protein, *Nucl. Acids Res.*, 16, 7545—7562, 1988.

230d. **Lipinska, B., Zylicz, M., and Georgopoulos, C.,** The HtrA (DegP) protein, essential for *Escherichia coli* survival at high temperatures, is an endopeptidase, *J. Bacteriol.*, 172, 1791—1797, 1990.

231. **Littlewood, T. D., Hancock, D. C., and Evan, G. I.,** Characterization of a heat shock-induced insoluble complex in the nuclei of cells, *J. Cell Sci.*, 88, 65—72, 1987.

232. **Livak, K. J., Freund, R., Schweber, M., Wensing, P. C., and Meselson, M.,** Sequence organization and transcription at two heat shock loci in *Drosophila, Proc. Natl. Acad. Sci. U.S.A.*, 75, 5613—5617, 1978.

233. **Loomis, W. F. and Wheeler, S.,** Heat shock response of *Dictyostelium, Dev. Biol.*, 79, 399—408, 1980.

234. **Loomis, W. F. and Wheeler, S.,** Chromatin-associated heat shock proteins in *Dictyostelium, Dev. Biol.*, 90, 412—418, 1982.

234a. **Loomis, W. F., Wheeler, S., and Schmidt, J. A.,** Phosphorylation of the major heat shock protein of *Dictyostelium discoideum, Mol. Cell. Biol.*, 2, 484—489, 1982.

235. **Lowe, D. G. and Moran, L. A.,** Proteins related to the mouse L-cell major heat shock protein are synthesized in the absence of heat shock gene expression, *Proc. Natl. Acad. Sci. U.S.A.*, 81, 2317—2321, 1984.

236. **Lowe, D. G., Fulford, W. D., and Moran, L. A.,** Mouse and *Drosophila* genes encoding the major heat shock protein (hsp70) are highly conserved, *Mol. Cell. Biol.*, 3, 1540—1543, 1983.

237. **Madreperla, S. A., Louwerenburg, B., Mann, R. W., Towle, C. A., Mankin, H. J., and Treadwell, B. V.,** Induction of heat-shock protein synthesis in chondrocytes at physiological temperatures, *J. Orthopaed. Res.*, 3, 30—35, 1985.

237a. **Maniak, M. and Nellen, W.,** A developmentally regulated membrane protein gene in *Dictyostelium discoideum* is also induced by heat shock and cold shock, *Mol. Cell. Biol.*, 8, 153—159, 1988.

238. **Mannan, R. M., Krishnan, M., and Gnanam, A.,** Heat-shock proteins of cyanobacterium *Anacystis nidulans, Plant Cell Physiol.*, 27, 377—381, 1986.

239. **Mansfield, M. A. and Key, J. L.,** Synthesis of the low molecular weight heat shock proteins in plants, *Plant Physiol.*, 84, 1007—1017, 1987.

240. **Marmiroli, N., Restivo, F. M., Odoardi Stanca, M., Terzi, V., Giovanelli, B., Tassi, F., and Lorenzoni, C.,** Induction of heat shock proteins and acquisition of thermotolerance in barley seedlings (*Hordeum vulgare* L.), *Genet. Agrar.* 40, 9—25, 1986.

240a. **Marshall, J. S., Derocher, A. E., Keegstra, K., and Vierling, E.,** Identification of hsp70 homologues in chloroplasts, *Proc. Natl. Acad. Sci. U.S.A.*, 87, 374—378, 1990.

240b. **Martin, M. L. and Regan, C. M.,** The anticonvulsant sodium valproate specifically induces the expression of a rat glial heat shock protein which is identified as the collagen type IV receptor, *Brain Res.*, 459, 131—137, 1988.

241. **Matthopoulos, D. P.,** Heat shock induces variably the major heat shock proteins of CV1 clones, *FEBS Lett.*, 195, 169—173, 1986.

241a. **Matts, R. L. and Hurst, R.,** Evidence for the association of the heme-regulated eIF-kinase with the 90-kDa heat shock protein in rabbit reticulocyte lysate *in situ, J. Biol. Chem.*, 264, 15542—15547, 1989.

242. **Maytin, E. V. and Young, D. A.,** Separate glucocorticoid, heavy metal and heat shock domains in thymic lymphocytes, *J. Biol. Chem.*, 258, 12718—12722, 1983.

243. **Maytin, E. V., Colbert, R. A., and Young, D. A.,** Early heat shock proteins in primary thymocytes. Evidence for transcriptional and translational regulation, *J. Biol. Chem.*, 260, 2384—2392, 1985.

244. **Mazzarella, R. A. and Green, M.,** ERp99, an abundant, conserved glycoprotein of the endoplasmic reticulum, is homologous to the 90-kDa heat shock protein (hsp90) and the 94-kDa glucose regulated protein (GRP94), *J. Biol. Chem.*, 262, 8875—8883, 1987.

245. **McAlister, L. and Finkelstein, D. B.,** Heat shock proteins and thermal resistance in yeast, *Biochem. Biophys. Res. Commun.*, 93, 819—824, 1980.

246. **McAlister, L., Strausberg, S., Kulaga, A., and Finkelstein, D. B.,** Altered patterns of protein synthesis induced by heat shock of yeast, *Curr. Genet,* 1, 63—74, 1979.

247. **McCallum, K. L., Heikkila, J. J., and Inniss, W. E.,** Temperature-dependent pattern of heat shock protein synthesis in psychrophilic and psychrotrophic microorgansims, *Can. J. Microbiol.*, 32, 516—521, 1986.

247a. **McElwain, E. F. and Spiker, S.,** A wheat cDNA clone which is homologous to the 17kd heat-shock protein gene family of soybean, *Nucl. Acids Res.*, 17, 1764, 1989.

247b. **McGuire, S. E., Fuqua, S. A. W., Naylor, S. L., Helin-Davis, D. A., and McGuire, W. L.,** Chromosomal assignments of human 27-kDa heat shock protein gene family, *Somat. Cell Mol. Genet.,* 15, 167—171, 1989.

248. **McKenzie, S. L. and Meselson, M.,** Translation *in vitro* of *Drosophila* heat-shock messages, *J. Mol. Biol.,* 117, 279—283, 1977.

249. **McKenzie, S. L., Henikoff, S., and Meselson, M.,** Localization of RNA from heat-induced polysomes at puff sites in *Drosophila melanogaster, Proc. Natl. Acad. Sci. U.S.A.,* 72, 1117—1121, 1975.

250. **McMullin, T. W. and Hallberg, R. L.,** A normal mitochondrial protein is selectively synthesized and accumulated during heat shock in *Tetrahymena thermophila, Mol. Cell. Biol.,* 7, 4414—4423, 1987.

251. **McMullin, T. W. and Hallberg, R. L.,** A highly evolutionarily conserved mitochondrial protein is substructurally related to the protein encoded by the *E. coli* groEL gene, *Mol. Cell. Biol.,* 8, 371—380, 1988.

252. **Meyer, J. and Chartier, Y.,** Long-lived and short-lived heat shock proteins in tobacco mesophyll protoplasts, *Plant Physiol.,* 72, 26—32, 1983.

253. **Michel, G. P. F., Azoulay, T., and Starka, J.,** Ethanol effect on the membrane protein pattern of *Zymomonas mobilis, Ann. Inst. Pasteur Microbiol.,* 136A, 174—180, 1985.

254. **Michel, G. P. F. and Starka, J.,** Effect of ethanol and heat stresses on the protein pattern of *Zymomonas mobilis, J. Bacteriol.,* 165, 1040—1042, 1986.

255. **Milarski, K. L. and Morimoto, R. I.,** Expression of human HSP70 during the synthetic phase of the cell cycle, *Proc. Natl. Acad. Sci. U.S.A* 83, 9517—9522, 1986.

255a. **Milarski, K. L. and Morimoto, R. I.,** Mutational analysis of the human hsp70 protein: Distinct domains for nucleolar localization and adenosine triphosphate binding, *J. Cell Biol.,* 109, 1947—1962, 1989.

256. **Miller, M. J., Xuong, N. H., and Geiduschek, E. P.,** A response of protein synthesis to temperature shift in the yeast *Saccharomyces cerevisiae, Proc. Natl. Acad. Sci. U.S.A.,* 76, 5222—5225, 1979.

257. **Miller, M. J., Xuong, N. H., and Geiduschek, E. P.,** Quantitative analysis of the heat shock response of *Saccharomyces cerevisiae, J. Bacteriol.,* 151, 311—327, 1982.

257a. **Miller, D. and McLennan, A. G.,** The heat shock response of the cryptobiotic brine shrimp *Artemia.* II. Heat shock proteins, *J. Therm. Biol.,* 13, 125—134, 1988.

258. **Minton, K. W., Karmin, P., Hahn, G. M., and Minton, A. P.,** Nonspecific stabilization of stress-susceptible proteins by stress-resistant proteins: A model for the biological role of heat shock proteins, *Proc. Natl. Acad. Sci. U.S.A.,* 79, 7107—7111, 1982.

259. **Mirkes, P. E.,** Hyperthermia induced heat shock response and thermotolerance in postimplantation rat embryos, *Dev. Biol.,* 119, 115—123, 1987.

260. **Mirault, M.-E., Goldschmidt-Clermont, M., Moran, L., Arrigo, A. P. and Tissieres, A.,** The effect of heat shock on gene expression in *Drosophila melanogaster, Cold Spring Harb. Symp. Quant. Biol.,* 42, 819—827, 1978.

261. **Mirault, M. E., Goldschmidt-Clermont, M., Artavanis-Tsakonas, S., and Schedl, P.,** Organization of the multiple genes for the 70,000-dalton heat shock protein in *Drosophila melanogaster, Proc. Natl. Acad. Sci. U.S.A.,* 76, 5254—5258, 1979.

262. **Mitchell, H. K., Petersen, N. S., and Buzin, C. H.,** Self-degradation of heat shock proteins, *Proc. Natl. Acad. Sci. U.S.A.,* 82, 4969—4973, 1985.

263. **Mizzen, L. A. and Welch, W. J.,** Characterization of the thermotolerant cell I. Effects on protein synthesis activity and the regulation of HSP70 expression, *J. Cell Biol.,* 106, 1105—1116, 1988.

263a. **Mocquot, B., Ricard, B., and Pradet, A.,** Rice embryos can express heat-shock genes under anoxia, *Biochimie,* 69, 677—681, 1987.

263b. **Moisyadi, S. and Harrington, H. M.,** Characterization of the heat shock response in cultured sugarcane cells. 1. Physiology of the heat shock response and heat shock protein synthesis, *Plant Physiol.,* 90, 1156—1162, 1989.

264. **Moore, S. K., Kozak, C., Robinson, E. A., Ullrich, S. J., and Appella, E.,** Cloning and nucleotide sequence of the murine hsp 84 cDNA and chromosome assignment of related sequences, *Gene,* 56, 29—40, 1987.

265. **Moran, A., Chauvin, M., Kennedy, M. E., Korri, M., Lowe, D. G., Nicholson, R. C., and Perry, M. D.,** The major heat-shock protein (hsp70) gene family: Related sequences in mouse, *Drosophila,* and yeast, *Can. J. Biochem. Cell Biol.,* 61, 488—499, 1983.

266. **Morange, M., Diu, A., Bensaude, O., and Babinet, C.,** Altered expression of heat shock proteins in embryonal carcinoma and mouse early embryonic cells, *Mol. Cell. Biol.,* 4, 730—735, 1984.

267. **Morgan, R. W., Christman, M. F., Jacobson, F. S., Storz, G., and Ames, B. N.,** Hydrogen peroxide-inducible proteins in *Salmonella typhimurium* overlap with heat shock and other stress proteins, *Proc. Natl. Acad. Sci. U.S.A.,* 83, 8059—8063, 1986.

267a. **Morimoto, R. and Fodor, E.,** Cell-specific expression of heat shock proteins in chicken reticulocytes and lymphocytes, *J. Cell Biol.*, 99, 1316—1323, 1984.

268. **Morimoto, R., Hunt, C., Huang, S.-Y., Berg, K. L., and Banerji, S. S.,** Organization, nucleotide sequence and transcription of the chicken HSP70 gene, *J. Biol. Chem.*, 261, 12692—12699, 1986.

269. **Mosser, D. D., Heikkila, J. J., and Bols, N. C.,** Temperature ranges over which rainbow trout fibroblasts survive and synthesize heat-shock proteins, *J. Cell. Physiol.*, 128, 432—440, 1986.

270. **Müller, R. M., Taguchi, H., and Shibahara, S.,** Nucleotide sequence and organization of the rat heme oxygenase gene, *J. Biol. Chem.*, 262, 6795—6801, 1987.

270a. **Mueller-Taubenberger, A., Hagmann, J., Noegel, A., and Gerisch, G.,** Ubiquitin gene expression in *Dictyostelium* is induced by heat and cold shock, cadmium, and inhibitors of protein synthesis, *J. Cell Sci.*, 90, 51—58, 1988.

271. **Mues, G. I., Munn, T. Z., and Raese, J. D.,** A human gene family with sequence homology to *Drosophila melanogaster* Hsp70 heat shock genes, *J. Biol. Chem.*, 261, 874—877, 1986.

272. **Munro, S. and Pelham, H. R. B.,** Use of peptide tagging to detect proteins expressed from cloned genes: Deletion mapping functional domains of *Drosophila* hsp70, *EMBO J.*, 3, 3087—3093, 1984.

273. **Munro, S. and Pelham, H. R. B.,** An hsp70-like protein in the ER: Identity with the 78 kd glucose-regulated protein and immunoglobulin heavy chain binding protein, *Cell,* 46, 291—300, 1986.

274. **Munro, S. and Pelham, H. R. B.,** A C-terminal signal prevents secretion of luminal ER proteins, *Cell,* 48, 899—907, 1987.

274a. **Murakami, H., Pain, D., and Blobel, G.,** 70-kD heat shock-related protein is one of at least two distinct cytosolic factors stimulating protein import into mitochondria, *J. Cell Biol.*, 107, 2051—2057, 1988.

274b. **Musgrove, J. E. and Ellis, R. J.,** The Rubisco large subunit binding protein, *Philos. Trans. R. Soc. London Ser. B,* 313, 419—428, 1986.

275. **Nagao, R. T., Czarnecka, E., Gurley, W. B., Schoeffl, F., and Key, J. L.,** Genes for low-molecular-weight heat shock proteins of soybeans: Sequence analysis of a multigene family, *Mol. Cell. Biol.*, 101, 1323—1331, 1985.

275a. **Nagao, R. T. and Key, J. L.,** Heat shock protein genes of plants, in *Cell Cultural and Somatic Cell Genetics of Plants,* Vol. 6, Schell, J. and Vasil, T. K., Eds., Academic Press, Orlando, FL, 1989, 297—328.

276. **Nagata, K., Saga, S., and Yamada, K. M.,** A major collagen-binding protein of chick embryo fibroblasts is a novel heat shock protein, *J. Cell Biol.*, 103, 223—230, 1986.

276a. **Nagata, K., Hirayoshi, K., Obara, M., Saga, S., and Yamada, K. M.,** Biosynthesis of a novel transformation-sensitive heat-shock protein that binds to collagen. Regulation by messenger RNA levels and *in-vitro* synthesis of a functional precursor, *J. Biol. Chem.*, 263, 8344—8349, 1988.

276b. **Nakai, A., Hirayoshi, K., Saga, S., Yamada, K. M., and Nagata, K.,** The transformation-sensitive heat shock protein (hsp47) binds specifically to fetuin, *Biochem. Biophys. Res. Commun.*, 164, 259—264, 1989.

276c. **Nakai, A., Hirayoshi, K., and Nagata, K.,** Transformation of BALB/3T3 cells by Simian virus-40 causes a decreased synthesis of a collagen-binding heat-shock protein (Hsp47), *J. Biol. Chem.*, 265, 992—999, 1990.

277. **Napolitano, E. W., Pachter, J. S., and Liem, R. K. H.,** Intracellular distribution of mammalian stress proteins. Effects of cytotoxicity of 42°C and 45°C hyperthermia in cultured Chinese hamster cells, *Int. J. Radiat. Biol.*, 44, 475—481, 1983.

277a. **Napolitano, E. W., Pachter, J. S., Chin, S. S. M., and Liem, R. K. H.,** Beta-Internexin, a ubiquitous, intermediate filament-associated protein, *J. Cell Biol.*, 101, 1323—1331, 1985.

278. **Napolitano, W. E., Pachter, J. S., and Liem, R. K. H.,** Intracellular distribution of mammalian stress proteins. Effects of cytoskeletal-specific agents, *J. Biol. Chem.*, 262, 1493—1504, 1987.

279. **Nebiolo, C. M. and White, E. M.,** Corn mitochondrial protein synthesis in response to heat shock, *Plant Physiol.*, 79, 1129—1132, 1985.

280. **Necchi, A., Pogna, N. E., and Mapelli, S.,** Early and late heat shock proteins in wheats and other cereal species, *Plant Physiol.*, 84, 1378—1384, 1987.

281. **Neidhardt, F.C. and Van Bogelen, R. A.,** Positive regulatory gene for temperature-controlled proteins in *Escherichia coli, Biochem. Biophys. Res. Commun.*, 100, 894—900, 1981.

282. **Neidhardt, F. C., Van Bogelen, R. A., and Lau, E. T.,** The high-temperature regulon of *Escherichia coli*, in *Heat Shock from Bacteria to Man,* Schlesinger, M. J., Ashburner, M. and Tissieres, A., Eds., Cold Spring Harbor Laboratory, Cold Spring Harbor, NY, 1982, 139—145.

283. **Neidhardt, F. C., Van Bogelen, R. A., and Lau, E. T.,** Molecular cloning and expression of a gene that controls the high-temperature regulon of *Escherichia coli, J. Bacteriol.*, 153, 597—603, 1983.

284. **Neidhardt, F. C., Van Bogelen, R. A., and Vaughn, V.,** The genetics and regulation of heat-shock proteins, *Annu. Rev. Genet.*, 18, 295—329, 1984.

285. **Nelson, D. R. and Killeen, K. P.,** Heat shock proteins of vegetative and fruiting *Myxococcus xanthus* cells, *J. Bacteriol.*, 168, 1100—1106, 1986.

286. **Nene, V., Dunne, D. W., Johnson, K. S., Taylor, D. W., and Cordingley, J. S.,** Sequence and expression of a major egg antigen from *Schistosoma mansoni*. Homologies to heat shock proteins and alpha-crystallins, *Mol. Biochem. Parasitol.*, 21, 179—188, 1986.

286a. **Nerland, A. H., Mustafa, A. S., Sweetser, D., Godal, T., and Young, R. A.,** A protein antigen of *Mycobacterium leprae* is related to a family of small heat shock proteins, *J. Bacteriol.*, 170, 5919—5921, 1988.

287. **Neumann, D., Scharf, K.-D., and Nover, L.,** Heat shock induced changes of plant cell ultrastructure and autoradiographic localization of heat shock proteins, *Eur. J. Cell Biol.*, 34, 254—264, 1984.

288. **Neumann, D., Zur Nieden, U., Manteuffel, R., Walter, G., Scharf, K.-D., and Nover, L.,** Intracellular localization of heat shock proteins in tomato cells cultures, *Eur. J. Cell Biol.*, 43, 71—81, 1987.

289. **Neumann, D., Nover, L., Parthier, B., Rieger, R., Scharf, K.-D., Wollgiehn, R., and zur Nieden, U.,** Heat shock and other stress response systems of plants, *Biol. Zentralbl.*, 108, 1—156, 1989.

289a. **Neumann, D. et al.,** in preparation.

289b. **Newmeyer, D. D. and Forbes, D. J.,** Nuclear import can be separated into distinct steps *in vitro:* Nuclear pore binding and translocation, *Cell*, 52, 641—653, 1988.

290. **Nicholson, P., Osborn, R. W., and Howe, C. J.,** Induction of protein synthesis in response to ultraviolet light, nalidixic acid and heat shock in the cyanobacterium *Phormidium laminosum, FEBS Lett.*, 221, 110—113, 1987.

290a. **Nicholson, P. and Howe, C. J.,** Stress-induced protein synthesis in *Chlamydomonas reinhardtii, FEMS Microbiol. Lett.*, 60, 283—287, 1989.

291. **Nickells, R. W. and Browder, L. W.,** Region-specific heat-shock protein synthesis correlates with a biphasic acquisition of thermotolerance in *Xenopus laevis* embryos, *Dev. Biol.*, 112, 391—395, 1985.

292. **Nickells, R. W. and Browder, L. W.,** A role for glyceraldehyde-3-phosphate dehydrogenase in the development of thermotolerance in *Xenopus laevis* embryos, *J. Cell Biol.*, 107, 1901—1910, 1988.

292a. **Nicolet, C. M. and Craig, E. A.,** Isolation and characterization of STI1, a stress-inducible gene from *Saccharomyces cerevisiae, Mol. Cell. Biol.*, 9, 3638—3646, 1989.

293. **Nieto-Sotelo, J. and Ho, T.-H. D.,** Absence of heat shock protein synthesis in isolated mitochondria and plastids from maize, *J. Biol. Chem.*, 262, 12288—12292, 1987.

294. **Nishida, E., Koyasu, S., Sakai, H., and Yahara, I.,** Calmodulin-regulated binding of the 90 kDa heat shock protein to actin filaments, *J. Biol. Chem.*, 261, 16033—16037, 1986.

294a. **Normington, K., Kohno, K., Kozutsumi, Y., Gething, M.-J., and Sambrook, J.,** *S. cerevisiae* encodes an essential protein homologous to mammalian BiP, *Cell*, 57, 1223—1236, 1989.

295. **Nover, L., Ed.** *Heat Shock Response of Eukaryotic Cells*, Springer-Verlag, Berlin, 1984.

296. **Nover, L. and Scharf, K.-D.,** Synthesis, modification and structural binding of heat shock proteins in tomato cell cultures, *Eur. J. Biochem.*, 139, 303—313, 1984.

297. **Nover, L., Scharf, K.-D., and Neumann, D.,** Formation of cytoplasmic heat shock granules in tomato cell cultures and leaves, *Mol. Cell. Biol.*, 3, 1648—1655, 1983.

298. **Nover, L., Scharf, K.-D., and Neumann, D.,** Cytoplasmic heat shock granules are formed from precursor particles and contain a specific set of mRNAs, *Mol. Cell. Biol.*, 9, 1298—1308, 1989.

298a. **Nover, L., Neumann, D., and Scharf, K.-D., Eds.,** *Heat Shock and Other Stress Response Systems of Plants*, Springer-Verlag, Berlin, 1990.

299. **Özkaynak, E., Finley, D., Solomon, M. J., and Varshavsky, A.,** The yeast ubiquitin genes: A family of natural gene fusions, *EMBO J.*, 6, 1429—1439, 1987.

300. **O'Farrell, P. H.,** High resolution two-dimensional electrophoresis of protein, *J. Biol. Chem.*, 250, 4007—4021, 1975.

301. **Ohtsuka, K., Tanabe, K., Nakamura, H., and Sato, C.,** Possible cytoskeletal association of 69,000 and 68,000-Dalton-heat shock proteins and structural relations among heat shocks proteins in murine mastocytoma cells, *Radiation Res.*, 108, 34—42, 1986.

301a. **Okhi, M., Tamara, F., Nishimura, S., and Uchida, H.,** Nucleotide sequence of the *Escherichia coli* dnaJ gene and purification of the gene product, *J. Biol. Chem.*, 261, 1778—1781, 1986.

302. **O'Malley, K., Mauron, A., Barchas, J. D., and Kedes, L.,** Constitutively expressed rat mRNA encoding a 70-kilodalton heat-shock-like protein, *Mol. Cell. Biol.*, 5, 3476—3483, 1985.

303. **Omar, R. A. and Lanks, K. W.,** Heat shock protein synthesis and cell survival in clones of normal and simian virus 40-transformed mouse embryo cells, *Cancer Res.*, 44, 3976—3982, 1984.

304. **Omar, R. A., Yano, S., and Kikkawa, Y.,** Antioxidant enzymes and survival of normal and Simian virus 40-transformed mouse embryo cells after hyperthermia, *Cancer Res.*, 47, 3473—3476, 1987.

304a. **Ostermann, J., Horwich, A. L., Neupert, W., and Hertl, F.-U.,** Protein folding in mitochondria requires complex formation with hsp60 and ATP hydrolysis, *Nature*, 341, 125—130, 1989.

305. **Ougham, H. J. and Stoddart, J. L.,** Synthesis of heat-shock protein and acquisition of thermotolerance in high-termperature tolerant and high-temperature susceptible lines of Sorghum, *Plant Sci.*, 44, 163—168, 1986.

305a. **Ovsenek, N. and Heikkila, J. J.**, Heat shock-induced accumulation of ubiquitin messenger RNA in *Xenopus laevis* embryos is developmentally regulated, *Dev. Biol.*, 129, 582—585, 1988.

306. **Paek, K.-H. and Walker, G. C.**, *Escherichia coli* dnaK null mutants are inviable at high temperature, *J. Bacteriol.*, 169, 283—290, 1987.

307. **Palter, K. B., Watanabe, M., Stinson, L., Mahowald, A. P., and Craig, E. A.**, Expression and localization of *Drosophila melanogaster* hsp70 cognate proteins, *Mol. Cell. Biol.*, 6, 1187—1203, 1986.

308. **Pan, B. T. and Johnston, R. M.**, Selective externalization of the transferrin receptor by sheep reticulocytes *in vitro*, *J. Biol. Chem.*, 259, 9776—9782, 1984.

308a. **Pardue, M. L.**, The heat shock response in biology and human disease: a meeting review, *Genes Dev.*, 2, 783—785, 1988.

308b. **Pardue, M. L., Feramisco, J. R., and Lindquist, S., Eds.**, *Stress-Induced Proteins*, Alan R. Liss, New York, 1989.

309. **Pauli, D. and Tonka, C.-H.**, A new *Drosophila* heat shock gene from locus 67B is expressed during embryogenesis and pupation, *J. Mol. Biol.*, 198, 233—240, 1987.

310. **Pauli, D., Tonka, C.-H., and Ayme-Southgate, A.**, An unusual split *Drosophila* heat shock gene expressed during embryogenesis, pupation and in testis, *J. Mol. Biol.*, 200, 47—53, 1988.

311. **Pearce, B. R., Dutton, G. R., and White, F. P.**, Induction of a stress protein in developing cell cultures of the rat cerebellum, *J. Neurochem.*, 41, 291—294, 1983.

312. **Pelham, H. R. B.**, Speculations on the functions of the major heat shock and glucose-regulated proteins, *Cell*, 46, 959—961, 1986.

312a. **Pelham, H.**, Heat shock proteins. Coming in from the cold, *Nature*, 332, 776—777, 1988.

312b. **Pelham, H. R. B.**, Heat shock and the sorting of luminal ER proteins, *EMBO J.*, 8, 3171—3176, 1989.

312c. **Perlmutter, D. H.**, Distinct mediators and mechanisms regulate human acute phase gene expression, in *Stress-Induced Proteins*, Pardue, M. L., Feramisco, J. R., and Lindquist, S., Eds., Alan R. Liss, New York, 1989, 257—274.

313. **Petko, L. and Lindquist, S.**, Hsp 26 is not required for growth at high temperatures, nor for thermotolerance, spore development, or germination, *Cell*, 45, 885—894, 1986.

314. **Phillips, T. A., Van Bogelen, R. A., and Neidhardt, F. C.**, Lon gene product of *Escherichia coli* is a heat-shock protein, *J. Bacteriol.*, 159, 283—287, 1984.

314a. **Picketts, D. J., Mayanil, C. S. K., and Gupta, R. S.**, Molecular cloning of a Chinese hamster mitochondrial protein related to the 'chaperonin' family of bacterial and plant proteins, *J. Biol. Chem.*, 264, 12001—12008, 1989.

315. **Pinhasi-Kimhi, O., Michalovitz, D., Ben-Zeev, A., and Oren, M.**, Specific interaction between the p53 cellular tumor antigen and major heat shock proteins, *Nature*, 320 182—185, 1986.

316. **Piper, P. W., Curran, B., Davies, M. W., Lockheart, A., and Reid, G.**, Transcription of the phosphoglycerate kinase gene of *Saccharomyces cerevisiae* increases when fermentative cultures are stressed by heat-shock, *Eur. J. Biochem.*, 161, 525—531, 1986.

316a. **Piper, P. W., Curran, B., Davies, M. W., Hirst, K., Lockheart, A., and Seward, K.**, Catabolite control of the elevation of pgk messenger RNA levels by heat shock in *Saccharomyces cerevisiae*, *Mol. Microbiol.*, 2, 353—362, 1988.

316b. **Piper, P. W., Curran, B., Davies, M. W., Hirst, K., Lockheart, A., Ogden, J. E., Stanway, C. A., Kingsman, A. J., and Kingsman, S. M.**, A heat shock element in the phosphoglycerate kinase gene promoter of yeast, *Nucl. Acids Res.*, 16, 1333—1348, 1989.

317. **Pitto, L., Loschiavo, F., Gioliano, G., and Terzi, T.**, Analysis of heat-shock protein pattern during somatic embryogenesis of carrot, *Plant Mol. Biol.*, 2, 231—237, 1983.

318. **Plesofsky-Vig, N. and Brambl, R.**, personal communication.

318a. **Plesofsky-Vig, N. and Brambl, R.**, Heat shock response of *Neurospora crassa:* Protein synthesis and induced thermotolerance, *J. Bacteriol.*, 162, 1083—1091, 1985.

319. **Plesofsky-Vig, N., and Brambl, R.**, Two developmental stages of *Neurospora crassa* utilize similar mechanisms for recovery, *Mol. Cell. Biol.*, 7, 3041—3048, 1987.

319a. **Plesset, J., Palm, C., and McLaughlin, C. S.**, Induction of heat shock proteins and thermotolerance by ethanol in *Saccharomyces cerevisiae*, *Biochem. Biophys. Res. Commun.*, 108, 1340—1345, 1982.

319b. **Polla, B. S., Healy, A. M., Wojno, W. C., and Krane, S. M.**, Hormone 1α, 25-dihydroxy vitamin D3 modulates heat shock response in monocytes, *Am. J. Physiol.*, 252, 640—649, 1987.

319c. **Prasad, T. K. and Hallberg, R. L.**, Identification and metabolic characterization of the *Zea mays* mitochondrial homology of the *Escherichia coli* groEL protein, *Plant Mol. Biol.*, 12, 609—618, 1989.

320. **Privalle, C. T. and Fridovich, I.**, Induction of superoxide dismutase in *Escherichia coli* by heat shock, *Proc. Natl. Acad. Sci. U.S.A.*, 84, 2723—2726, 1987.

321. **Przybytkowski, E., Bates, J. H. T., Bates, D. A., and Mackillop, W. J.**, Thermal adaptation in CHO cells at 40°C: The influence of growth conditions and the role of heat shock proteins, *Radiat. Res.*, 107, 317—331, 1986.

322. **Ralhan, R. and Johnson, G. S.,** Destabilization of cytoplasmic mouse mammary tumor RNA by heat shock: Prevention by cycloheximide pretreatment, *Biochem. Biophys. Res. Commun.*, 137, 1028—1033, 1986.

322a. **Raschke, E.,** Molekulare Analyse verschiedener Gene füer kleine Hitzeschockproteine der Sojabohne *(Glycine max* (L.) Merrill), Thesis, University of Bielefeld (FRG), 1987.

323. **Raschke, E., Baumann, G., and Schoeffl, F.,** Nucleotide sequence analysis of soybean small heat shock protein genes belonging to two different multigene families, *J. Mol. Biol.*, 199, 549—557, 1988.

323a. **Reading, D. S., Hallberg, R., and Myers, A. M.,** Characterization of the yeast HSP60 gene coding for a mitochondrial assembly factor, *Nature,* 337, 655—659, 1989.

324. **Rebbe, N. F., Ware, J., Betina, R. M., Modrich, P., and Stafford, D. W.,** Nucleotide sequence of a cDNA for a member of the human 90kDa heat-shock protein family, *Gene,* 53, 235—245, 1987.

324a. **Rechsteiner, M.,** Ubiquitin-mediated pathways for intracellular proteolysis, *Annu. Rev. Cell Biol.*, 3, 1—30, 1987.

324b. **Redmond, T., Sanchez, E. R., Bresnick, E. H., Schlesinger, M. J., Toft, D. O., Pratt, W. B., and Welsh, M. J.,** Immunofluorescence colocalization of the 90-kDa heat-shock protein and microtubules in interphase and mitotic mammalian cells, *Eur. J. Cell Biol.*, 50, 66—75, 1989.

325. **Rees, C. A. B., Hogan, N. C., Walden, D. B., and Atkinson, B. G.,** Identification of mRNAs encoding low molecular mass heat shock protein in maize *(Zea mays* L.*), Can. J. Genet. Cytol.*, 28, 1106—1114, 1987.

326. **Restivo, F. M., Tassi, F., Maestri, E., Lorenzoni, C., Puglisi, P. P., and Marmiroli, N.,** Identification of chloroplast associated heat-shock proteins in *Nicotiana plumbaginifolia* protoplasts, *Curr. Genet.*, 11, 145—151, 1986.

327. **Reuter, S. H. and Shapiro, L.,** Assymetric segregation of heat shock proteins upon cell division in *Caulobacter crescentus, J. Mol. Biol.*, 194, 653—662, 1987.

328. **Richter, W. W. and Issinger, O.-G.,** Differential heat shock response of primary human cell cultures and established cell lines, *Biochem. Biophys. Res. Commun.*, 141, 46—52, 1986.

329. **Richter, A. and Hecker, M.,** Heat-shock proteins in *Bacillus subtilis.* A two-dimensional gel electrophoresis study, *FEMS Microbiol. Lett.*, 36, 69—71, 1986.

330. **Ritossa, F.,** A new puffing pattern induced by heat shock and DNP in *Drosophila, Experientia,* 18, 571—573, 1962.

330a. **Robaye, B., Hepburn, A., Lecocq, R., Fiers, W., Boeynaems, J. M., and Dumont, J. E.,** Tumor necrosis factor alpha induces the phosphorylation of 28kDa stress proteins in endothelial cells—Possible role in protection against cytotoxicity, *Biochem. Biophys. Res. Commun.*, 163, 301—308, 1989.

330b. **Roccheri, M. C., Sconzo, G., Di Bernhardo, M. G., Albanese, I., Di Carlo, M., and Giudice, G.,** Heat shock proteins in sea urchin embryos. Territorial and intracellular location, *Acta Embryol. Morphol. Exp.*, 2, 91—99, 1981.

330c. **Roccheri, M. C., Sconzo, G., La Rosa, M., Oliva, D., Abrignani, A., and Giudice, G.,** Response to heat shock of different sea urchin species, *Cell Differ.*, 18, 131—135, 1986.

331. **Rochester, D. E., Winter, J. A., and Shah, D. M.,** The structure and expression of maize genes encoding the major heat shock protein, hsp70, *EMBO J.*, 5, 451—458, 1986.

331a. **Rodenhiser, D., Jung, J. H., and Atkinson, B. G.,** Mammalian lymphocytes: Stress-induced synthesis of heat-shock proteins in vitro and in vivo, *Can. J. Biochem. Cell Biol.*, 63, 711—722, 1985.

331b. **Rollet, E. and Best-Belpomme, M.,** HSP 26 and 27 are phosphorylated in response to heat shock and ecdysterone in *Drosophila melanogaster* cells, *Biochem. Biophys. Res. Commun.*, 141, 426—434, 1986.

332. **Ron, A. and Birkenfeld, A.,** Stress proteins in the human endometrium and decidua, *Hum. Reprod.*, 2, 277—280, 1987.

332a. **Rose, M. D., Misra, L. M., and Vogel, J. P.,** KAR2, a karyogamy gene, is the yeast homolog of the mammalian BiP/GRP78 genes, *Cell,* 57, 1211—1221, 1989.

332b. **Rosen, E., Sivertsen, A., and Firtel, R. A.,** An unusal transposon encoding heat shock inducible and developmentally regulated transcripts in *Dictyostelium, Cell,* 35, 243—251, 1983.

333. **Rosen, E., Sivertsen, A., Firtel, R. A., Wheeler, S., and Loomis, W. F.,** Heat-shock genes of *Dictyostelium,* in *Changes in Eukaryotic Gene Expression in Response to Environmental Stress,* Atkinson, B. G. and Walden, D. B., Eds., Academic Press, Orlando, FL, 1985, 257—278.

333a. **Rossi, J. M. and Lindquist, S.,** The intracellular location of yeast HSP26 varies with metabolism, *J. Cell Biol.*, 108, 425—439, 1989.

334. **Rothman, J. E. and Schmid, S. L.,** Enzymatic recycling of clathrin from coated vesicles, *Cell,* 46, 5—9, 1986.

334a. **Roy, H.,** Rubisco assembly: A model system for studying the mechanism of chaperonin action, *Plant Cell,* 1, 1035—1042, 1989.

335. **Russnak, R. H. and Candido, E. P. M.,** Locus encoding a family of small heat shock genes in *Caenorhabditis elegans:* Two genes duplicated to form a 3.8-kilobase inverted repeat, *Mol. Cell. Biol.*, 5, 1268—1278, 1985.

336. **Sachs, A. B., Bond, M. W., and Kornberg, R. D.,** A single gene from yeast for both nuclear and cytoplasmic polyadenylate-binding proteins: Domaine structure and expression, *Cell,* 45, 827—835, 1986.

337. **Saito, H. and Uchida, H.,** Organization and expression of the dnaJ and dnaK genes of *Escherichia coli* K-12, *Mol. Gen. Genet.,* 164, 1—8, 1978.

338. **Sanchez, E. R., Redmond, T., Scherrer, L. C., Bresnick, E. H., Welsh, M. J., and Pratt, W. B.,** Evidence that the 90-kDa heat shock protein is associated with tubulin-containing complexes in L cell cytosol and in intact PtK cells, *Mol. Endocrinol.,* 2, 756—760, 1988.

338a. **Sargent, C. A., Dunham, I., Trowsdale, J., and Campbell, R. D.,** Human major histocompatibility complex contains genes for the major heat shock protein HSP70, *Proc. Natl. Acad. Sci. U.S.A.,* 86, 1968—1972, 1989.

339. **Scharf, K.-D. and Nover, L.,** Heat shock induced alterations of ribosomal protein phosphorylation in plant cell cultures, *Cell,* 30, 427—437, 1982.

340. **Schedl, P., Artavanis-Tsakonas, S., Steward, R., Gehring, W. J., Miraŭlt, M.-E., Goldschmidt-Clermont, M., Moran, L., and Tissieres, A.,** Two hybrid plasmids with *D. melanogaster* DNA sequences complementary to mRNA coding for the major heat shock protein, *Cell,* 14, 921—929, 1978.

341. **Schlesinger, M. J. and Bond, U.,** Ubiquitin genes, in *Oxford Survey of Eukaryotic Genes,* 4, 77, 1987.

341a. **Schlesinger, M. and Hershko, A., Eds.,** *The Ubiquitin System, Curr. Commun. Mol. Biol.,* Cold Spring Harbor Press, Cold Spring Harbor, NY, 1988.

342. **Schlesinger, M. J., Ashburner, M., and Tissieres, A., Eds.,** *Heat Shock from Bacteria to Man,* Cold Spring Harbor Laboratory, Cold Spring Harbor, NY, 1982.

342a. **Schlesinger, M. J., Kelley, P. M., Aliperti, G., and Malfer, C.,** Properties of three major chicken heat-shock proteins and their antibodies, in *Heat Shock from Bacteria to Man,* Schlesinger, M. J., Ashburner, M., and Tissieres, A., Eds., Cold Spring Harbor Laboratory, Cold Spring Harbor, NY, 1982, 243—250.

343. **Schlossmann, D. M., Schmid, S. L., Braell, W. A., and Rothman, J. E.,** An enzyme that removes clathrin coats: Purification of an uncoating ATPase, *J. Cell Biol.,* 99, 723—733, 1984.

344. **Schmid, S. L., Braell, W. A., and Rothman, J. E.,** ATP catalyzes the sequestration of clathrin during enzymatic uncoating, *J. Biol. Chem.* 260, 10057—10062, 1985.

345. **Schöffl, F. and Key, J. L.,** An analysis of mRNAs for a group of heat shock proteins of soybean using-cloned cDNAs, *J. Mol. Appl. Genet.,* 1, 301—314, 1982.

345a. **Schöffl, F. and Key, J. L.,** Identification of a multigene family for small heat shock proteins in soybean and physical characterization of one individual gene coding region, *Plant Mol. Biol.,* 2, 269—278, 1983.

346. **Schöffl, F., Raschke, E., and Nagao, R. T.,** The DNA sequence analysis of soybean heat shock genes and identification of possible regulatory promoter elements, *EMBO J.,* 3, 2491—2497, 1984.

346a. **Schöffl, F., Baumann, G., and Raschke, E.,** The expression of heat shock genes—A model for environmental stress response, in *Plant Gene Research, Vol. V,* Verma, D. P. S. and Goldberg, B., Eds., Springer-Verlag, Berlin, 1988, 253—273.

347. **Schönfelder, M., Horsch, A., and Schmid, H.-P.,** Heat shock increases the synthesis of the poly(A)-binding protein in HeLa cells, *Proc. Natl. Acad. Sci. U.S.A.,* 82, 6884—6888, 1985.

348. **Schuldt, C. and Kloetzel, P. M.,** Analysis of cytoplasmic 19S ring-type particles in *Drosophila* which contain hsp23 at normal growth temperature, *Dev. Biol.,* 110, 65—74, 1985.

349. **Schulz-Harder, B.,** Heat shock induction of a ribonuclease in *Saccharomyces cerevisiae, FEMS Microbiol. Lett.,* 17, 23—26, 1983.

350. **Schuster, G., Even, D., Kloppstech, K., and Ohad, I.,** Evidence for protection by heat-shock proteins against photoinhibition during heat-shock, *EMBO J.,* 7, 1—6, 1988.

350a. **Scott, N. W. and Dow, C. S.,** The influence of temperature stress on protein synthesis in the *Rhodomicrobium vannielii* RM5 and in a rifampicin-resistant mutant R82, *FEMS Microbiol. Lett.,* 48, 147—152, 1987.

350b. **Searle, S., Campos, A. J. R., Coulson, R. M. R., Spithill, T. W., and Smith, D. F.,** A family of heat shock protein 70-related genes are expressed in the promastigotes of *Leishmania major, Nucl. Acids Res.,* 17, 5081—5095, 1989.

350c. **Seufert, W. and Jentsch, S.,** Ubiquitin-conjugating enzymes UBC4 and UBC5 mediate selective degradation of short-lived and abnormal proteins, *EMBO J.,* 9, 543—550, 1990.

350d. **Shapira, M., McEwen, J. G., and Jaffe, C. L.,** Temperature effects on molecular processes which lead to stage differentiation in *Leishmania, EMBO J,* 7, 2895—2901, 1988.

351. **Sharma, S., Rodgers, L., Brandsma, J., Gething, M.-J., and Sambrook, J.,** SV40 T antigen and the exocytotic pathway, *EMBO J.,* 4, 1479—1489, 1985.

352. **Shaw, M., Boasson, R., and Scrubb, L.,** Effect of heat shock on protein synthesis in flax rust uredosporelings, *Can. J. Bot.,* 63, 2069—2076, 1985.

352a. **Shearer, G., Birge, C. H., Yuckenberg, P. D., Kobayashi, G. S., and Medoff, G.,** Heat-shock proteins induced during the mycelial-to-yeast transitions of strains of *Histoplasma capsulatum, J. Gen. Microbiol.,* 133, 3375—3382, 1987.

353. **Shibahara, S., Mueller, R. M., and Taguchi, H.**, Transcriptional control of rat heme oxygenase by heat shock, *J. Biol. Chem.*, 262, 12889—12892, 1987.

353a. **Shin, D.-Y., Matsumoto, K., Iida, H., Uno, I., and Ishikawa, T.**, Heat shock response of *Saccharomyces cerevisiae* mutants altered in cyclic AMP-dependent protein phosphorylation, *Mol. Cell. Biol.*, 7, 244—250, 1987.

354. **Shinnick, T. M., Sweetser, D., Thole, J. E. R., Van Embden, J. D. A., and Young, R. A.**, The etiologic agents of leprosy and tuberculosis share immunoreactive protein antigen with the vaccine strain *Mycobacterium bovis* BCG, *Infect. Immun.*, 55, 1932—1935, 1987.

354a. **Shinnick, T. M., Vodkin, M. H., and Williams, J. C.**, The *Mycobacterium tuberculosis* 65-kilodalton antigen is a heat shock protein which corresponds to common antigen and to the *Escherichia coli* GroEL protein, *Infect. Immunol.*, 56, 446—451, 1988.

354b. **Shiu, R. P. C., Pouyssegur, J., and Pastan, I.**, Glucose depletion accounts for the induction of two transformation-sensitive membrane proteins in *Rous sarcoma* virus-transformed chick embryo fibroblasts, *Proc. Natl. Acad. Sci. U.S.A.*, 74, 3840—3844, 1977.

354c. **Shuman, J. and Przybyla, A.**, Expression of the 31-kDa stress protein in rat myoblasts and hepatocytes, *DNA*, 7, 475—482, 1988.

355. **Shyy, T. T., Subjeck, J. R., Heinaman, R., and Anderson, G.**, Effect of growth state and heat shock on nucleolar localization of the 110,000-Da heat shock protein in mouse embryo fibroblasts, *Cancer Res.*, 46, 4738—4745, 1986.

356. **Silver, J. C., Andrews, D. R., and Pekkala, D.**, Effect of heat shock on synthesis and phosphorylation of nuclear and cytoplasmic proteins in the fungus *Achlya, Can. J. Biochem. Cell Biol.*, 61, 447—455, 1983.

357. **Simon, M. C., Kitchener, K., Kao, H.-T., Hickey, E., Weber, L., Voellmy, R., Heintz, N., and Nevins, J. R.**, Selective induction of human heat shock gene transcription by the adenovirus E1A gene products, including the 12S E1A product, *Mol. Cell. Biol.*, 7, 2884—2890, 1987.

358. **Singh, M. K. and Yu, J.**, Accumulation of heat shock-like protein during differentiation of human erythroid cell line K562, *Nature*, 309, 631—633, 1984.

359. **Sinibaldi, R. M. and Turpen, T.**, Heat shock protein is encoded within mitochondria of higher plants, *J. Biol. Chem.*, 260, 15382—15385, 1985.

359a. **Sinibaldi, R. and Brunke, P.** Personal communication.

359b. **Skelly, S., Fu, C.-F., Dalie, B., Redfield, B., Coleman, T., Brot, N., and Weissbach, H.**, Antibody to sigma 32 cross-reacts with DnaK: Association of DnaK protein with *Escherichia coli* RNA polymerase, *Proc. Natl. Acad. Sci. U.S.A.*, 85, 5497—5501, 1988.

360. **Slater, A. and Craig, E. A.**, submitted.

360a. **Slater, M. J. and Craig, E. A.**, The SSA1 and SSA2 genes of yeast *Saccharomyces cerevisiae, Nucl. Acids Res.*, 17, 805—806, 1989.

360b. **Slater, M. R. and Craig, E. A.**, The SSB1 heat shock cognate gene of the yeast *Saccharomyces cerevisiae, Nucl. Acids Res.*, 17, 4891, 1989.

361. **Slater, A., Cato, A. C. B., Sillar, G. M., Kioussis, J., and Burdon, R. H.**, The pattern of protein synthesis induced by heat shock of HeLa cells, *Eur. J. Biochem.*, 17, 341—346, 1981.

362. **Snutch, T. P. and Baillie, D. L.**, Alterations in the pattern of gene expression following heat shock in the nematode *Caenorhabditis elegans, Can. J. Biochem. Cell Biol.*, 61, 480—487, 1983.

362a. **Snutch, T. P., Heschl, M. F. P., and Baillie, D. L.**, The *Caenorhabditis elegans* hsp70 gene family: A molecular genetic characterization, *Gene*, 64, 241—255, 1988.

363. **Sondermeijer, P. J. A. and Lubsen, N. H.**, Heat-shock peptides in *Drosophila hydei* and their synthesis *in vitro, Eur. J. Biochem.*, 88, 331—339, 1978.

364. **Sorger, P. K. and Pelham, H. R. B.**, The glucose-regulated protein grp94 is related to heat shock protein hsp90, *J. Mol. Biol.*, 194, 341—344, 1987.

365. **Sorger, P. K. and Pelham, H. R. B.**, Cloning and expression of a gene becoding hsc73, the major hsp70-like protein in unstressed rat cells, *EMBO J.*, 6, 993—998, 1987.

366. **Southgate, R., Ayme, A., and Voellmy, R.**, Nucleotide sequence analysis of the *Drosophila* small heat shock gene cluster at locus 67B, *J. Mol. Biol.*, 165, 35—57, 1983.

367. **Spector, M. P., Aliabadi, Z., Gonzales, T., and Forster, J. W.**, Global control in *Salmonella typhimurium:* Two-dimensional electrophoretic analysis of starvation, anaerobiosis-, and heat shock-inducible proteins, *J. Bacteriol.*, 168, 420—424, 1986.

368. **Spradling, A., Penman, S., and Pardue, M. L.**, Analysis of *Drosophila* mRNA by in situ hybridization: Sequences transcribed in normal and heat shocked cultured cells, *Cell*, 4, 395—404, 1975.

369. **Spradling, A., Pardue, M. L., and Penman, S.**, Messenger RNA in heat-shocked *Drosophila* cells, *J. Mol. Biol.*, 109, 559—587, 1977.

370. **Staufenbiel, M. and Deppert, W.**, Different structural systems of the nucleus are targets for SV40 large T antigen, *Cell*, 33, 173—181, 1983.

371. **Stephanou, G. and Demopoulos, N. A.**, Heat shock phenomena in *Aspergillus nidulans*. II. Combined effects of heat and bleomycin to heat shock protein synthesis, survival rate and induction of mutations, *Curr. Genet.*, 12, 443—448, 1987.

372. **Stephanou, G., Alahiotis, S. N., Mormaras, V. J., and Christodoulou, C.**, Heat shock response in *Ceratitis capitata, Comp. Biochem. Physiol. B,* 74B, 425—432, 1983.

373. **Stermer, B. A. and Hammerschmidt, R.**, Heat shock induces resistance to *Cladosporium cucumerinum* and enhances peroxydase activity in cucumbers, *Physiol. Plant Pathol.*, 25, 239—249, 1984.

373a. **Stevenson, M. A. and Calderwood, S. K.**, Members of the 70-kilodalton heat shock protein family contain a highly conserved calmodulin-binding domain, *Mol. Cell. Biol.*, 10, 1234—1238, 1990.

373b. **Stoeckle, M. Y., Sugano, S., Hampe, A., Vashistha, A., Pellman, D., and Hanafusa, H.**, 78-Kilodalton glucose-regulated protein is induced in Rous sarcoma virus-transformed cells independently of glucose deprivation, *Mol. Cell. Biol.*, 8, 2675—2680, 1988.

373c. **Storti, R. V., Scott, M. P., Rich, A., and Pardue, M. L.**, Translational control of protein synthesis in response to heat shock in *D. melanogaster* cells, *Cell,* 22, 825—834, 1980.

374. **Strahler, J. R., Kuick, R., Eckerskorn, C., Lottspeich, F., Richardson, B. C., Fox, D. A., Stoolman, L. M., Hanson, C. A., Nichols, D., Tueche, H. J., and Hanash, S. M.**, Identification of two related markers for common acute lymphoblastic leukemia as heat shock proteins, *J. Clin. Invest.*, 85, 200—207, 1990.

374a. **Straus, D. B., Walter, W. A., and Gross, C. A.**, *Escherichia coli* heat shock gene mutants are defective in proteolysis, *Genes Devel.*, 2, 1851—1858, 1988.

374b. **Streips, U. N. and Polio, F. W.**, Heat shock proteins in *Bacilli, J. Bacteriol.*, 162, 434—437, 1985.

375. **Stuerzbecher, H.-W., Chumakov, P., Welch, W. J., and Jenkins, J. R.**, Mutant p53 proteins bind hsp72/73 cellular heat shock-related proteins in SV40-transformed monkey cells, *Oncogene,* 1, 201—211, 1987.

376. **Subjeck, J. R., Sciandra, J. J., and Johnson, R. J.**, Heat shock proteins and thermotolerance: a comparison of induction kinetics, *Br. J. Radiol.*, 55, 579—584, 1982.

377. **Subjeck, J. R., Shyy, T., Shen, J., and Johnson, R. J.**, Association between the mammalian 110,000-dalton heat-shock protein and nucleoli, *J. Cell Biol.*, 97, 1389—1395, 1983.

378. **Subjeck, J. R., Sciandra, J. J., and Shyy, T. T.**, Analysis of the expression of the two major proteins of the 70 kilodalton mammalian heat shock family, *Int. J. Radiat. Biol.*, 17, 275—284, 1985.

378a. **Subjeck, J. R. and Shyy, T. T.**, Stress protein systems of mammalian cells, *Am. J. Physiol.*, 250, C1—C17, 1986.

379. **Süß, K. H. and Jordanov, I. T.**, Biosynthetic cause of *in vivo* acquired thermotolerance of photosynthetic light reactions and metabolic responses of chloroplasts to heat stress, *Plant Physiol.*, 81, 192—199, 1986.

379a. **Susek, R. E. and Lindquist, S. L.**, Hsp26 of *Saccharomyces cerevisiae* is homologous to the superfamily of small heat shock proteins, but is without a demonstrable function, *Mol. Cell. Biol.*, 9, 5265—5271, 1989.

380. **Sussman, M. D. and Setlow, P.**, Nucleotide sequence of a *Bacillus megaterium* gene homologous to the dnak gene of *Escherichia coli, Nucl. Acids Res.*, 15, 3923, 1987.

380a. **Swindle, J., Ajioka, J., Eisen, H., Sanwal, B., Jacquemot, C., Browder, Z., and Buck, G.**, The genomic organizaiton and transcription of the ubiquitin genes of *Trypanosoma cruzi, EMBO J.*, 7, 1121—1127, 1988.

380b. **Takahashi, T. and Komeda, Y.**, Characterization of two genes encoding small heat-shock proteins in *Arabidopsis thaliana, Mol. Gen. Genet.*, 219, 365—372, 1989.

380c. **Tanaka, K., Matsumoto, K., and Toh-E, A.**, Dual regulation of the expression of the polyubiquitin gene by cyclic AMP and heat shock in yeast, *EMBO J,* 7, 495—502, 1988.

381. **Tanguay, R. M., Camato, R., Lettre, F., and Vincent, M.**, Expression of histone genes during heat shock and in arsenite-treated *Drosophila* Kc cells, *Can. J. Biochem. Cell Biol.*, 61, 414—420, 1983.

381a. **Tarnowka, M. A. and Yuyama, S.**, Heat shock dependent fluctuations of RNase activity during the cell cycle of synchronized *Tetrahymena, J. Cell. Physiol.*, 95, 85—93, 1978.

382. **Taylor, W. E., Straus, D. B., Grossman, A. D., Burton, Z. F., Gross, C. A., and Burgess, R. R.**, Transcription from heat-inducible promoter causes heat shock regulation of the sigma subunit of *E. coli* RNA polymerase, *Cell,* 38, 371—381, 1984.

382a. **Terracciano, J. S., Rapaport, E., and Kashket, E. R.**, Stress- and growth phase-associated proteins of *Clostridium acetobutylicum, Appl. Environ. Microbiol.*, 54, 1989—1995, 1988.

383. **Tilly, K., McKittrick, N., Zylicz, M., and Georgopoulos, C.**, The dnaK protein modulates the heat-shock response of *Escherichia coli, Cell,* 34, 641—646, 1983.

384. **Tilly, K., Van Bugelen, R. A., Georgopoulos, C., and Neidhardt, F. C.**, Identification of the heat-inducible protein C 15.4 as the groES gene product in *Escherichia coli, J. Bacteriol.*, 154, 1505—1507, 1983.

384a. **Tilly, K. and Yarmolinsky, M.**, Participation of *Escherichia coli* heat shock proteins DnaJ, DnaK and GrpE in P1 plasmid replication, *J. Bacteriol.*, 171, 6025—6029, 1989.

385. **Tissieres, A., Mitchell, H. K., and Tracy, U. M.,** Protein synthesis in salivary glands of *D. melanogaster*. Relation to chromosome puffs, *J. Mol. Biol.*, 84, 389—398, 1974.

386. **Tobe, T., Ito, K., and Yura, T.,** Isolation and physical mapping of temperature-sensitive mutants defective in heat-shock induction of proteins in *Escherichia coli, Mol. Gen. Genet.*, 195, 10—16, 1984.

387. **Todd, J. A., Hubbard, T. J. P., Travers, A. A., and Ellar, D. J.,** Heat-shock proteins during growth and sporulation of *Bacillus subtilis, FEBS Lett.*, 188, 209—214, 1985.

388. **Török, I. and Karch, R.,** Nucleotide sequences of heat shock activated genes in *Drosophila melanogaster*. I. Sequences in the regions of the 5′ and 3′ ends of the hsp70 gene in the hybrid plasmid 56H8, *Nucl. Acids Res.*, 8, 3105—3123, 1980.

389. **Tomasovic, S. P., Steck, P. A., and Heitzman, D.,** Heat-stress proteins and thermal resistance in rat mammary tumor cells, *Radiat. Res.* 95, 399—413, 1985.

389a. **Tomasovic, S. P.,** Functional aspects of the mammalian heat-stress protein response, *Life Chem. Rep.*, 7, 33—63, 1989.

390. **Topol, J., Ruden, D. M., and Parker, C. S.,** Sequences required for *in vitro* transcriptional activation of a *Drosophila* hsp 70 gene, *Cell*, 42, 527—537, 1985.

390a. **Trent, J. D., Osipink, J., and Pinkan, T.,** Acquired thermotolerance and heat shock in the extremely thermophilic archaebacterium *Sulfolobus* sp. strain B12, *J. Bacteriol.*, 172, 1478—1484, 1990.

391. **Tsuchida, T., Van Bogelen, R. A., and Neidhardt, F. C.,** Heat shock response in *Escherichia coli* influences cell division, *Proc. Natl. Acad. Sci. U.S.A.*, 83, 6959—6963, 1986.

392. **Tsuji, Y., Ishibashi, S., and Ide, T.,** Induction of heat shock proteins in young and senescent human diploid fibroblasts, *Mech. Ageing Dev.*, 36, 155—160, 1986.

393. **Tsukeda, H., Maekawa, H., Izumi, S., and Nitta, K.,** Effect of heat shock on protein synthesis by normal and malignant human lung cells in tissue culture, *Cancer Res.*, 11, 5188—5192, 1981.

394. **Tytell, M., Greenberg, S. S., and Lasek, R. J.,** Heat shock-like protein is transferred from glia to axon, *Brain Res.*, 363, 161—164, 1986.

394a. **Uemura, J., Shiba, T., Paterson, M., Jigami, Y., and Tanaka, H.,** Identification of a sequence containing the positive regulatory region of *Saccharomyces cerevisiae* gene EN01, *Gene*, 45, 67—75, 1986.

395. **Ullrich, S. J., Robinson, E. A., Law, L. W., Willingham, M., and Appella, E.,** A mouse tumor-specific transplantation antigen is a heat shock-related protein, *Proc. Natl. Acad. Sci. U.S.A.*, 83, 3121—3125, 1986.

396. **Ungewickell, E.,** The 70-kd mammalian heat shock proteins are structurally and functionally related to the uncoating protein that releases clathrin triskelia from coated vesicles, *EMBO J.*, 4, 3385—3391, 1985.

396a. **Valliammai, T., Gnanam, A., and Dharmalingam, K.,** Heat shock response in *Chlorella protothecoides*, *Plant Cell Physiol.*, 28, 975—986, 1987.

396b. **Van Bergen en Henegouwen, P. M. P., Jordi, W. J. R. M., Van Dongen, G., Ramaekers, F. C. S., Amesz, H., and Linnemans, W. A. M.,** Studies on a possible relationship between alterations in the cytoskeleton and induction of heat shock protein synthesis in mammalian cells, *Int. J. Hyperthermia*, 1, 69—83, 1985.

396c. **Van Bergen en Henegouwen, P. M. P. and Linnemans, W. A. M.,** Heat shock gene expression and cytoskeleton alterations in mouse neuroblastoma cells, *Exp. Cell Res.*, 171, 367—375, 1987.

397. **Van Bergen en Henegouwen, P. M. P., Berbers, G., Linnemans, W. A. M., and Van Wijk, R.,** Cytoplasmic and nuclear localization of the 84,000 dalton heat shock protein in mouse neuroblastoma cells, *Eur. J. Cell Biol.*, 43, 469—478, 1987.

398. **Van Bogelen, R. A., Vaugh, V., and Neidhardt, F. C.,** Gene for heat-inducible Lysyl tRNA synthetase (lysU) maps near cadA in *Escherichia coli, J. Bacteriol.*, 153, 1066—1068, 1983.

399. **Van Bogelen, R. A., Kelley, P. M., and Neidhardt, F. C.,** Differential induction of heat shock, SOS, and oxidation stress regulons and accumulation of nucleotides in *Escherichia coli, J. Bacteriol.*, 169, 26—32, 1987.

399a. **Vance, B. A., Kowalski, C. G., and Brinckerhoff, C. E.,** Heat shock of rabbit synovial fibroblasts increases expression of mRNAs for two metalloproteinases, collagenase and stromelysin, *J. Cell Biol.*, 108, 2037—2043, 1989.

400. **Van Den Heuvel, R., Hendriks, W., Quax, W., and Bloemendal, H.,** Complete structure of the hamster alpha A crystallin gene. Reflection of an evolutionary history by means of exon shuffling, *J. Mol. Biol.*, 185, 273—284, 1985.

401. **Van Dongen, G., Geilenkirchen, W., Van Rijn, J., and Van Wijk, R.,** Increase of thermoresistance after growth stimulation of resting Reuber H35 hepatoma cells, *Exp. Cell Res.*, 166, 427—441, 1986.

401a. **van Herpen, M. M. A., Reijnen, W. H., Schrauwen, J. A. M., Degroot, P. F. M., Jager, J. W. H., and Wullems, G. J.,** Heat shock proteins and survival of germinating pollen of *Lilium longiflorum* and *Nicotiana tabacum, Plant Physiol.*, 134, 345—351, 1989.

402. **Velazquez, J. M. and Lindquist, S.,** hsp70: Nuclear concentration during environmental stress and cytoplasmic storage during recovery, *Cell*, 36, 655—662, 1984.

403. **Velazquez, J. M., DiDomenico, B. J., and Lindquist, S.,** Intracellular localization of heat shock proteins in *Drosophila, Cell,* 20, 679—689, 1980.

404. **Velazquez, J., Sonoda, S., Bugaisky, G. E., and Lindquist, S.,** Is the major *Drosophila* heat shock protein present in cells that have not been heat shocked?, *J. Cell Biol.,* 96, 286—290, 1982.

404a. **Verma, R., Iida, H., and Pardee, A. B.,** Modulation of expression of the stress-inducible P118 of *Saccharomyces cerevisiae* by the cyclic AMP II. A study of P118 expression in mutants of the cyclic AMP cascade, *J. Biol. Chem.,* 263, 8576—8582, 1988.

404b. **Verma, R., Iida, H., and Pardee, A. B.,** Identification of a novel stress-inducible glycoprotein in *Saccharomyces cerevisiae, J. Biol. Chem.,* 263, 8569—8575, 1988.

405. **Vierling, E. and Key, J. L.,** Ribulose 1,5-bisphosphate carboxylase synthesis during heat shock, *Plant Physiol.* 78, 155—162, 1985.

406. **Vierling, E., Mishkind, M. L., Schmidt, G. W., and Key, J. L.,** Specific heat shock proteins are transported into chloroplasts, *Proc. Natl. Acad. Sci. U.S.A.,* 83, 361—365, 1986.

407. **Vierling, E., Nagoa, R. T., De Rocher, A. E., and Harris, L. M.,** Heat shock protein localized to chloroplasts is a member of a eukaryotic superfamily of heat shock proteins, *EMBO J.,* 7, 575—582, 1988.

407a. **Vierling, E., Harris, L. M., and Chen, Q.,** The major low molecular weight heat shock protein in chloroplasts shows antigenic conservation among diverse higher plant species, *Mol. Cell. Biol.,* 9, 461—468, 1989.

407b. **Vierling, E.,** personal communication.

407c. **Vierstra, R. D., Burke, T. J., Callis, J., Hatfield, P. M., Jabben, M., Shanklin, J., and Sullivan, M. L.,** Characterization of the ubiquitin-dependent proteolytic pathway in higher plants, in *The Ubiquitin System,* Schlesinger, M. and Hershko, A., Eds., Cold Spring Harbor Press, Cold Spring Harbor, NY, 1988, 119—125.

408. **Vincent, M. and Tanguay, R. M.,** Heat-shock induced proteins present in the cell nucleus of *Chironomus tentans* salivary gland, *Nature,* 281, 501—503, 1979.

408a. **Vodkin, M. H. and Williams, J. C.,** A heat shock operon in *Coxiella burnetii* produces a major antigen homologous to a protein in both mycobacteria and *Escherichia coli, J. Bacteriol.,* 170, 1227—1234, 1988.

409. **Voellmy, R. and Bromley, P. A.,** Massive heat-shock polypeptide synthesis in late chicken embryos: convenient system for study of protein synthesis in highly differentiated organisms, *Mol. Cell. Biol.,* 2, 479—483, 1982.

410. **Voellmy, R., Bromley, P., and Kocher, H. P.,** Stuctural similarities between corresponding heat shock proteins from different eukaryotic cells, *J. Biol. Chem.,* 258, 3516—3522, 1983.

410a. **Vogt, P. K., Bos, T. J., and Doolittle, R. F.,** Homology between the DNA-binding domain of the GCN4 regulatory protein of yeast and the carboxyl-terminal region of a protein coded for by the oncogene jun, *Proc. Natl. Acad. Sci. U.S.A.,* 84, 3316—3319, 1987.

410b. **von Gromoff, E. D., Treier, U., and Beck, C. F.,** Three light-inducible heat shock genes of *Chlamydomonas reinhardtii, Mol. Cell. Biol.,* 9, 3911—3918, 1989.

411. **Vossen, J. G. H. M., Leenders, H. J., Derksen, J., and Jeucken, G.,** Chromosomal puff induction in salivary glands from *Drosophila* hydei by arsenite, *Exp. Cell Res.,* 109, 277—283, 1977.

412. **Vossen, J. G. H. M., Leenders, H. J., and Knoppien, W. G.,** A change in the affinity of larval α-glycerophosphate dehydrogenase in relation to the activity of the heat shock genes of *Drosophila hydei, Insect Biochem.,* 13, 349—359, 1983.

413. **Wada, M. and Itikawa, H.,** Participation of *Escherichia coli* K-12 groE gene products in the synthesis of cellular DNA and RNA, *J. Bacteriol.,* 157, 694—696, 1984.

414. **Wada, C., Akiyama, Y., Ito, K., and Yura, T.,** Inhibition of F plasmid replication in htpR mutants of *Escherichia coli* deficient in sigma 32 protein, *Mol. Gen. Genet.* 203, 208—213, 1986.

415. **Wadsworth, S. C.,** A family of related proteins is encoded by the major *Drosophila* heat shock gene family, *Mol. Cell. Biol.,* 2, 286—292, 1982.

415a. **Waldinger, D., Subramanian, A. R., and Cleve, H.,** The polymorphic human chaperonine protein HuCha60 is a mitochondrial protein sensitive to heat shock and cell transformation, *Eur. J. Cell. Biol.,* 50, 435—441, 1989.

416. **Walker, A. I., Hunt, T., Jackson, R. J., and Anderson, C. W.,** Double-stranded DNA induces the phosphorylation of several proteins including the 90,000 mol. wt. heat-shock protein in animal cell extracts, *EMBO J.,* 4, 139—145, 1985.

416a. **Walker, J. E., Saraste, M., Runswick, M. J., and Gay, N. J.,** Distantly regulated sequences in the alpha and beta-subunits of ATP synthase, myosin, kinase and other ATP-requiring enzymes and a common nucleotide binding fold, *EMBO J,* 1, 945—951, 1982.

417. **Walsh, C.,** Appearance of heat shock proteins during the induction of multiple flagella in *Naegleria gruberi, J. Biol. Chem.,* 255, 2629—2632, 1980.

417a. **Walter, M.,** The induction of phenylpropanoid biosynthetic enzymes by ultraviolet light or fungal elicitor in cultured parsley cells is overridden by a heat shock treatment, *Planta Med.,* 177, 1—8, 1989.

418. **Walter, G., Carbone, A., and Welch, W. J.,** Medium tumor antigen of polyomavirus transformation-defective mutant NG59 is associated with 73-kilodalton heat shock protein, *J. Virol.*, 61, 405—410, 1987.
419. **Wang, C. and Lazarides, E.,** Arsenite-induced changes in methylation of the 70,000 dalton heat shock proteins in chicken embryo fibroblasts, *Biochem. Biophys. Res. Commun.*, 119, 735—743, 1984.
420. **Wang, C., Asai, D. J., and Lazarides, E.,** The 68,000-dalton neuro-filament-associated polypeptide is a component of nonneuronal cells and of skeletal myofibrils, *Proc. Natl. Acad. Sci. U.S.A.*, 77, 1541—1545, 1980.
421. **Wang, C., Gomer, R. H., and Lazarides, E.,** Heat shock proteins are methylated in avian and mammalian cells, *Proc. Natl. Acad. Sci. U.S.A.*, 78, 3531—3535, 1981.
422. **Wang, C., Lazarides, E., O'Connor, C. M., and Clarke, S.,** Methylation of chicken fibroblast heat shock proteins at lysyl and arginyl residues, *J. Biol. Chem.*, 257, 8356—8362, 1982.
423. **Watowich, S. S. and Morimoto, R. I.,** Complex regulation of heat shock and glucose responsive genes in human cells, *Mol. Cell. Biol.*, 8, 393—405, 1988.
423a. **Watt, P. W. and North, M. J.,** The stress-shock response of the bacterium *Methylophilus methylotrophus*, *FEBS Lett.*, 215, 295—299, 1987.
424. **Weatherbee, J. A., Luftig, R. B., and Weiling, R. R.,** Purification and reconstitution of HeLa cell microtubules, *Biochemistry*, 19, 4116—4123, 1980.
425. **Weitzel, G., Pilatus, U., and Rensing, L.,** The cytoplasmic pH, ATP content and total protein synthesis rate during heat-shock protein inducing treatments in yeast, *Exp. Cell Res.*, 170, 64—79, 1987.
426. **Welch, W. J.,** Phorbol ester, calcium ionophore, or serum added to quiescent rat embryo fibroblast cells all result in the elevated phosphorylation of two 28,000-Dalton mammalian stress proteins, *J. Biol. Chem.*, 260, 3058—3062, 1985.
427. **Welch, W. J. and Feramisco, J. R.,** Rapid purification of mammalian 70,000-dalton stress proteins: Affinity of the proteins for nucleotides, *Mol. Cell. Biol.*, 5, 1229—1237, 1985.
428. **Welch, W. J. and Mizzen, L. A.,** Characterization of the thermotolerant cell: II. Effects on the intracellular distribution of HSP70, intermediate filaments, and snRNP's, *J. Cell Biol.*, 106, 1117—1130, 1988.
429. **Welch, W. J., Garrels, J. I., Thomas, G. P., Lin, J. J. C., and Feramisco, J. R.,** Biochemical characterization of the mammalian stress proteins and identification of two stress proteins as glucose- and calcium ionophore-regulated proteins, *J. Biol. Chem.*, 258, 7102—7111, 1983.
430. **Welch, W. J., Garrels, J. I., and Feramisco, J. R.,** The mammalian stress proteins, in *Heat Shock from Bacteria to Man*, Schlesinger, M. J., Ashburner, M., and Tissieres, A., Eds., Cold Spring Harbor Laboratory, Cold Spring Harbor, NY, 1982, 257—266.
431. **Werner-Washburne, M., Stone, D. E., and Craig, E. A.,** Complex interactions among members of an essential subfamily of hsp70 genes in *Saccharomyces cerevisiae*, *Mol. Cell. Biol.*, 7, 2568—2577, 1987.
431a. **Werner-Washburne, M., Becker, J., Kosic-Smithers, J., and Craig, E. A.,** Yeast Hsp70 RNA levels vary in response to the physiological status of the cell, *J. Bacteriol.*, 171, 2680—2688, 1989.
432. **Whatley, S. A., Leung, T., Hall, C., and Lim, L.,** The brain 68-kilodalton microtubule-associated protein is a cognate form of the 70-kilodalton mammalian heat-shock protein and is present as a specific isoform in synaptosomal membranes, *J. Neurochem.*, 47 1576—1583, 1986.
433. **Whelan, S. A. and Hightower, L. E.,** Differential induction of glucose-regulated and heat shock proteins: Effects of pH and sulfhydryl-reducing agents on chicken embryo cells, *J. Cell. Physiol.*, 125, 251—258, 1985.
433a. **Wickner, S. H.,** Three *Escherichia coli* heat shock proteins are required for P1 plasmid DNA replication: Formation of an active complex between *Escherichia coli* DnaJ protein and the P1 initiator protein, *Proc. Natl. Acad. Sci. U.S.A.*, 87, 2690—2694, 1990.
434. **Wiegant, F. A. C., Van Bergen en Henegouven, P. M. P., Van Dongen, A. A. M. S., and Linnemans, W. A. M.,** Stress-induced thermotolerance of the cytoskeleton, *Cancer Res.*, 47, 1674—1680, 1987.
435. **Wiegant, F. A. C., Van Bergen en Henegouven, P. M. P., and Linnemans, W. A. M.,** Studies on the mechanism of heat shock induced cytoskeletal reorganization in mouse neuroblastoma N2A cells, Proefschrift, University of Utrecht, 1987, 141—158.
435a. **Wilhelm, J. M., Spear, P., and Sax, C.,** Heat-shock proteins in the protozoan *Tetrahymena*: Induction by protein synthesis inhibition and possible role in carbohydrate metabolism, in *Heat Shock from Bacteria to Man*, Schlesinger, M. J., Ashburner, M., and Tissieres, A., Eds., Cold Spring Harbor Laboratory, Cold Spring Harbor, NY, 1982, 309—314.
435b. **Wilkison, W. O. and Bell, R. M.,** sn-Glycerol-3-phosphate acyltransferase tubule formation is dependent upon heat shock proteins (htpR), *J. Biol. Chem.*, 263, 14505—14510, 1988.
436. **Winning, R. S. and Browder, L. W.,** Changes inheat shock protein synthesis and HSP70 gene transcription during erythropoiesis of *Xenopus laevis*, *Dev. Biol.*, 128, 111—120, 1988.
436a. **Winter, J. and Duck, N.,** personal communication.
437. **Winter, J., Wright, R., Duck, N., Gasser, C., Fraley, R., and Shah, D.,** The inhibition of Petunia hsp 70 mRNA processing during $CdCl_2$ stress, *Mol. Gen. Genet.*, 211, 315—319, 1988.

438. **Wolffe, A.-P., Glover, J. F., and Tata, J. R.,** Culture shock. Synthesis of heat-shock-like proteins in fresh primary cell cultures, *Exp. Cell. Res.,* 154, 581—590, 1984.

439. **Wright, M. and Tollon, Y.,** Induction of heat-shock proteins at permissive growth temperatures in the plasmodium of the myxomycete *Physarum polycephalum, Eur. J. Biochem.,* 127, 49—56, 1982.

440. **Wu, L., Morley, B. J., and Campbell, R. D.,** Cell-specific expression of the human complement protein factor B gene: Evidence for the role of two distinct 5′-flanking elements, *Cell,* 48, 331—342, 1987.

440a. **Wu, C. H., Caspar, T., Browse, J., Lindquist, S., and Somerville, C.,** Characterization of an HSP70 cognate gene family in *Arabidopsis, Plant Physiol.,* 88, 731—740, 1988.

441. **Xiao, C. M. and Mascarenhas, J. P.,** High temperature-induced thermotolerance in pollen tubes of *Tradescantia* and heat-shock proteins, *Plant Physiol.,* 78, 887—890, 1985.

442. **Yacoob, R. K. and Filion, W. G.,** Temperature-stress response in maize: A comparison of several cultivars, *Can. J. Genet. Cytol.,* 28, 1125—1131, 1986.

442a. **Yamamori, T. and Yura, T.,** Genetic control of heat-shock protein synthesis and its bearing on growth and thermal resistance in *Escherichia coli* K-12, *Proc. Natl. Acad. Sci. U.S.A.,* 79, 860—864, 1982.

443. **Yamasaki, L., Kanda, P., and Lanford, R. E.,** Identification of four nuclear transport signal-binding proteins that interact with diverse transport signals, *Mol. Cell. Biol.,* 9, 3028—3036, 1989.

443a. **Yamazaki, M., Akaogi, K., Miwa, T., Imai, T., Soeda, E., and Yokoyama, K.,** Nucleotide sequence of a full-legth cDNA for 90 kDa heat-shock protein from human peripheral blood lymphocytes, *Nucl. Acids Res.,* 17, 7108, 1989.

443b. **Young, K. D., Anderson, R. J., and Hafner, R. J.,** Lysis of *Escherichia coli* by the bacteriophage phiX174 protein: Inhibition of lysis by heat shock proteins, *J. Bacteriol.,* 171, 4334—4341, 1989.

444. **Yura, T.,** Genetic control of heat-shock proteins, *Jpn. J. Genet.,* 61, 227—290, 1986.

445. **Yura, T., Tobe, T., Ito, K., and Osawa, T.,** Heat shock regulatory gene (htpR) of *Escherichia coli* is required for growth at high temperature but is dispensable at low temperature, *Proc. Natl. Acad. Sci. U.S.A.,* 81, 6803—6807, 1984.

446. **Zagris, N. and Matthopoulos, D.,** Differential heat shock gene expression in chick blastula, *Roux's Arch. Dev. Biol.,* 195, 403—407, 1986.

446a. **Zakeri, Z. F., Wolgemuth, D. J., and Hunt, C. R.,** Identification and sequence analysis of a new member of the mouse HSP70 gene family and characterization of its unique cellular and developmental pattern of expression in the male germ line, *Mol. Cell. Biol.,* 8, 2925—2932, 1988.

446b. **Zeuthen, M. L., and Howard, D. H.,** Thermotolerance and heat-shock response in *Candida albicans, J. Gen. Microbiol.,* 135, 2509—2518, 1989.

447. **Zimmerman, J. L., Petri, W., and Meselson, M.,** Accumulation of a specific subset of *D. melanogaster* heat shock mRNAs in normal development without heat shock, *Cell,* 32, 1161—1170, 1983.

447a. **Zimmermann, R., Sagstetter, M., Lewis, M. J., and Pelham, H. R. B.,** Seventy-kilodalton heat shock proteins and an additional component from reticulocyte lysate stimulate import of M13 procoat protein into microsomes, *EMBO J,* 7, 2875—2880, 1988.

448. **Zivy, M.,** Genetic variability for heat shock protein in common wheat, *Theor. Appl. Genet.,* 4, 209—213, 1987.

449. **Zuker, C., Cappello, J., Chisholm, R. L., and Lodish, H. F.,** A repetitive *Dictyostelium* gene family that is induced during differentiation and by heat shock, *Cell,* 34, 997—1005, 1983.

450. **Zuker, C., Cappello, J., Lodish, H. F., George, P., and Chung, S.,** *Dictyostelium* transposable element DIRS-1 has 350-base-pair inverted terminal repeats that contain a heat shock promoter, *Proc. Natl. Acad. Sci. U.S.A.,* 81, 2660—2664, 1984.

451. **Zylicz, M., Lebowitz, J. H., McMacken, R., and Georgopoulos, C.,** The dnaK protein of *Escherichia coli* process an ATPase and autophosphorylating activity and is essential in an *in vitro* DNA replication system, *Proc. Natl. Acad. Sci. U.S.A.,* 80, 6431—6435, 1984.

452. **Zylicz, M. and Georgopoulos, C.,** Purification and properties of the *Escherichia coli* dnaK replication protein, *J. Biol. Chem.,* 259, 8820—8825, 1984.

Chapter 3

STRUCTURE OF EUKARYOTIC HEAT SHOCK GENES

L. Nover

TABLE OF CONTENTS

3.1. SURVEY OF HEAT SHOCK GENES

Analysis of heat shock (hs) genes started in 1978/1979 with the description of the first specific clones for *Drosophila melanogaster*.[50,75,79,116] In fact, due to the peculiarities of the system, mapping of the hs genes by visual inspection of the altered puffing pattern of larval polytene chromosomes marked in 1962 the onset of the molecular biological era of the hs response.[110] This led to the discovery of the newly formed HSPs[133] and to the *in situ* hybridization of defined mRNA fractions to the activated hs genes.[46,83,84,130,131] Thus, the actual cloning work was greatly facilitated by the preceding identification of the hs genes, of their mRNAs as well as their protein products.

Basically, the procedure for gene analysis involves five main parts: (1) isolation and fractionation of hs mRNA, generation of cDNA and genomic libraries, selection, and identification of hs clones by hybrid-release or hybrid-arrest translation; (2) mapping of the hs gene loci by *in situ* hybridization with mRNA or cDNA clones and, eventually, by characterization of deletion mutants; (3) generation of physical maps and analysis of the genomic organization by R-loop mapping and Southern blotting; (4) sequencing of coding and flanking regions leading to the identification of putative regulatory elements; and (5) functional analysis of coding and regulatory parts of a gene by expression in homologous and heterologous systems (see Chapter 6).

A survey of >60 hs and related genes, whose sequences are known, is given in Table 3.1. Starting from this, detailed information on the amino acid sequences is included in Chapter 2 (Figures 2.8 to 2.13), whereas results of the structural and functional analyses of the promoter regions are dealt with in Chapter 6 (Figure 6.6 and Table 6.2). Several genes listed in Table 3.1 contain introns. Their physical organization with respect to exon and intron lengths and distribution is summarized in Figure 3.1. In the following, we only briefly compile the references on cloning and sequencing of hs genes in different systems and comment on aspects of their genomic organization. A considerable number of earlier and recent reviews are concerned with this particular part of the hs response.[6,7,16,71,74,74a,91a,93,94,94a,98a,120-122]

3.2. VERTEBRATES

Members of the four major gene families were sequenced (Table 3.1). With the exception of the two hsp70 genes of chicken and man, all other vertebrate genes so far analyzed in detail contain many introns. Partial sequencing of 12 independent hsp90 cDNA clones from a human cDNA library proved them to be identical, i.e., besides the grp94 gene, there are evidently only a few expressed hsp90 genes.[108,144] Identification of two hs-inducible but otherwise independently regulated HSP90 isoforms in mammals [4a,125a] was confirmed by sequencing two genes coding for HSP90α and HSP90β (Figure 2.8).[48a,108a]

More information is available on the hsp70 family. In mammalian cells, four types of genes were sequenced (Figure 2.10). These are the grp78, the constitutively expressed hsc-71-73 (clathrin-uncoating ATPase), the hs-induced hsp70, and a developmentally regulated hsc70 of sperm cells. In addition, there are at least two other genes of this family expressed sequentially during spermatozoan development[1,68,81a,141a] and a number of pseudogenes.[37,90] Complementary to the data given in Figure 2.10, a number of partial sequences of hsp70 cDNA clones were reported by Lowe and Moran[80] and O'Malley et al.[97] An interesting, but so far unexplained feature of the mammalian hsp70 gene family was pointed out by Munn and Mues.[90c] When comparing the Hs-hsc71 and the Rn-hsc73 genes, they found highly conserved 90 bp motifs in introns 5, 6, and 8.

Mapping of the human hsp70 genes using somatic cell hybrids and *in situ* hybridization as well as restriction fragment length polymorphism analyses[39,44,115a] revealed 2 to 3 loci.

TABLE 3.1
Structure of Genes of Eukaryotic hsp Families

Gene	Mol Wt (protein)	TU (nucl.)	Intron	mRNA (nucl.)	5′ L (nucl.)	ORF (nucl.)	3′ UT (nucl.)	Ref.
Vertebrates								
Hs-hsp90α	84,564	5,998	10	3,100	60	2,196	663	48a
Hs-hsp90β	83,303	6,933	11	2,650	99	2,112	260	108, 108a
Hs-hsp70A	70,008	2,440	—	2,600	213	1,986	242	25, 53
Hs-hsc71	70,899	4,600	8	2,400	77	1,938	215	27
Hs-hsp60+	61,049	n.d.	n.d.	2,400	>45	1,719	460	62
Hs-hsp27	22,327	1,800	2	950	49, 90	597	139	48
Mm-grp94+	92,475	n.d.	n.d.	3,200	>89	2,406	261	82
Mm-hsp90+	84,000	n.d.	n.d.	2,650	94	2,172	243	86
Mm-hsc72+	70,830	n.d.	n.d.	2,700	>65	1,938	83	37a
Mm-hsc70	69,734	n.d.	—	2,700	121	1,899	n.d.	145
Mm-hsp25+	22,882	n.d.	n.d.	~800	n.d.	624	127	34b
Rn-grp78	72,341	n.d.	+	2,700	206	1,962	350	13, 91
Rn-hsc73	70,865	3,850	8	2,300	80	1,938	70	97, 128
Rn-hsc70	69,500	2,497	—	2,700	112	1,899	485	141a
Gd-grp94	91,555	9,900	17	2,900	101	2,385	244	66, 69
Gd-hsp90	84,112	?	+	2,900	82	2,184	630	7a, 137a
Gd-hsp70	70,000	2,370	—	2,600	112	1,905	360	88
Xenopus laevis								
X1-hsp70	70,914	2,200	—	2,400	124	1941	140	5
Nematode *(Caenorhabditis elegans)*								
Ce-hsp70A	69,800	2,450	3	2,300	80	1,920	150	127a
Ce-hsp70C =grp78	73,339	3,000	3	2,800	~45	1,983	~620	46a
Ce-hsp16-1	16,301	620	1	750	42	435	80	112
Ce-hsp16-2	16,220	620	1	770	38	435	100	62b
Ce-hsp16-48	16,299	590	1	750	51	429	45	112
Ce-hsp16-41	16,018	630	1	770	42	429	100	62b
Drosophila melanogaster								
Dm-hsp83 (63BC)	81,858	3,700	1	2,900	149	2,151	400	8, 43, 51
Dm-hsp70 (87A7)	70,000	2,400	—	2,600	254	1,929	216	64, 134
Dm-hsp70 (87C1)	70,270	2,400	—	2,600	242	1.923	n.d.	57

TABLE 3.1 (continued)
Structure of Genes of Eukaryotic hsp Families

Gene	Mol Wt (protein)	TU (nucl.)	Intron	mRNA (nucl.)	5′ L (nucl.)	ORF (nucl.)	3′ UT (nucl.)	Ref.
Dm-hsp27 (67B)	23,616	1,250	—	1,250	118	636	n.d.	52, 54, 129
Dm-hsp26 (67B)	22,994	930	—	1,100	182	624	110	52, 54, 129
Dm-hsp-23 (67B)	20,399	860	—	1,050	112	558	180	52, 54, 129
Dm-hsp22 (67B)	19,733	940	—	1,050	251	522	160	52, 54, 129
Dm-gene 1 (67B)	26,560	1,300	—	1,400	93	714	400	2
Dm-gene 2 (67B)	12,600	600	2	560	60	333	52—86	100
		800	3	780	183	333	52—86	
		2,000	2	2,000	60	333	52—86	
Dm-gene 3 (67B)	18,792	975	—	1,100	168	507	300	99
Fungi								
Sc-STI1	66,246	n.d.	—	2,000	n.d.	1,767	n.d.	92a
Sc-hsc82	80,885	n.d.	—	n.d.	n.d.	2,118	n.d.	10a
Sc-hsp82	81,419	2,314	—	2,500	59	2,127	128	30
Sc-KAR2 (grp78)	74,462	n.d.	—	2,300	38	2,046	?	92b, 111a
Sc-hsp70 (SSA1)	70,065	n.d.	—	n.d.	n.d.	1,929	n.d.	126 126a
Sc-SSB1	66,601	n.d.	—	n.d.	n.d.	1,839	n.d.	92b, 126b
Sc-SSC1	70,622	n.d.	—	n.d.	n.d.	1,962	n.d.	23a
Sc-hsp26	23,877	n.d.	—	900	n.d.	642	n.d.	10b, 132
Sc-hsp60	60,830	1,900	—	1,900	n.d.	1,716	n.d.	107
Nc-hsp30	25,262	1,300	—	~1,450	121	684	475	105a
Bl-hsp70	73,981	2,241	—	2,300	60	2,025	150	63a
Plants								
Zm-hsp70	70,361	2,900	1	2,300	107	1,938	66	111, 123
Ph-hsp70	71,238	2,800	1	2,200	~95	1,956	n.d.	141
Le-hsc70$^+$	73,495	n.d.	n.d.	n.d.	>37	1,998	>81	26a
Ps-hsp21$^+$ (IV)	26,276	n.d.	n.d.	1,000	>49	696	>240	136
Gm-hsp26A	25,946	1,335	1	1,110	72	675	190	24a
Cr-hsp20$^+$	16,800	n.d.	n.d.	1,700	>85	471	>600	41
Chr-hsp25$^+$	23,284	n.d.	n.d.	>950	>35	612	55	66a
Gm-hsp17.5E (I)	17,534	~740	—	940	82	462	~190	24, 92, 119

Gm-hsp17.6L (I)	17,572	>720	—	940	93, 96	462	160, 184	24, 92, 119
Gm-hsp17.5M (I)	17,545	>720	—	940	88, 93	459	183, 193	24, 92, 119
Gm-hsp17.3B (6871) (I)	17,346	~690	—	890	103	459	~126	24, 92, 119
Gm-hsp17.9D (VI)	17,878	>820	—	≥1,000	72	477	270—360	106
Gm-hsp18.5C (I)	18,503	>680	—	≥950	76	483	120—260	106
Ta-hsp17+	16,877	>740	n.d.	>940	>75	453	>210	82b
At-hsp17.4 (I)	17,410	680	—	~850	55	471	>160	45a, 132b
At-hsp18.2 (I)	18,174	685	—	~850	42	483	>160	132b
Protozoa								
Tc-p85	80,500	n.d.	—	2,700	n.d.	2,112	n.d.	26
Tb-hsp70	71,370	2,200	—	2,500	53	1,980	160	38
Tc-hsp70 (mt)	71,187	n.d.	—	2,500 3,000	n.d.	1,968	n.d.	28a
Lm-hsp70	70,000	n.d.	—	3,400	172	1,974	1,088	72b

Note: Organisms are abbreviated by a two-letter code as indicated in Abbreviations and Special Terms. The length of the transcription units (TU), 5′ untranslated leaders (5′L), open reading frames (ORF), and 3′ untranslated sequences are given in nucleotides. The size of the mRNAs is estimated including a poly(A)+ tail of ~150 nucleotides. Occasionally two figures or a range is given for the 5′ or 3′ untranslated sequences, because the precise sites of transcription start and stop are variable or not well defined. If present and known, the number of introns is indicated (see Figure 3.1 for details).

+ Date were derived from cDNA clones. For sequences, see Figures 2.8, 2.10 to 2.12, and 2.15.

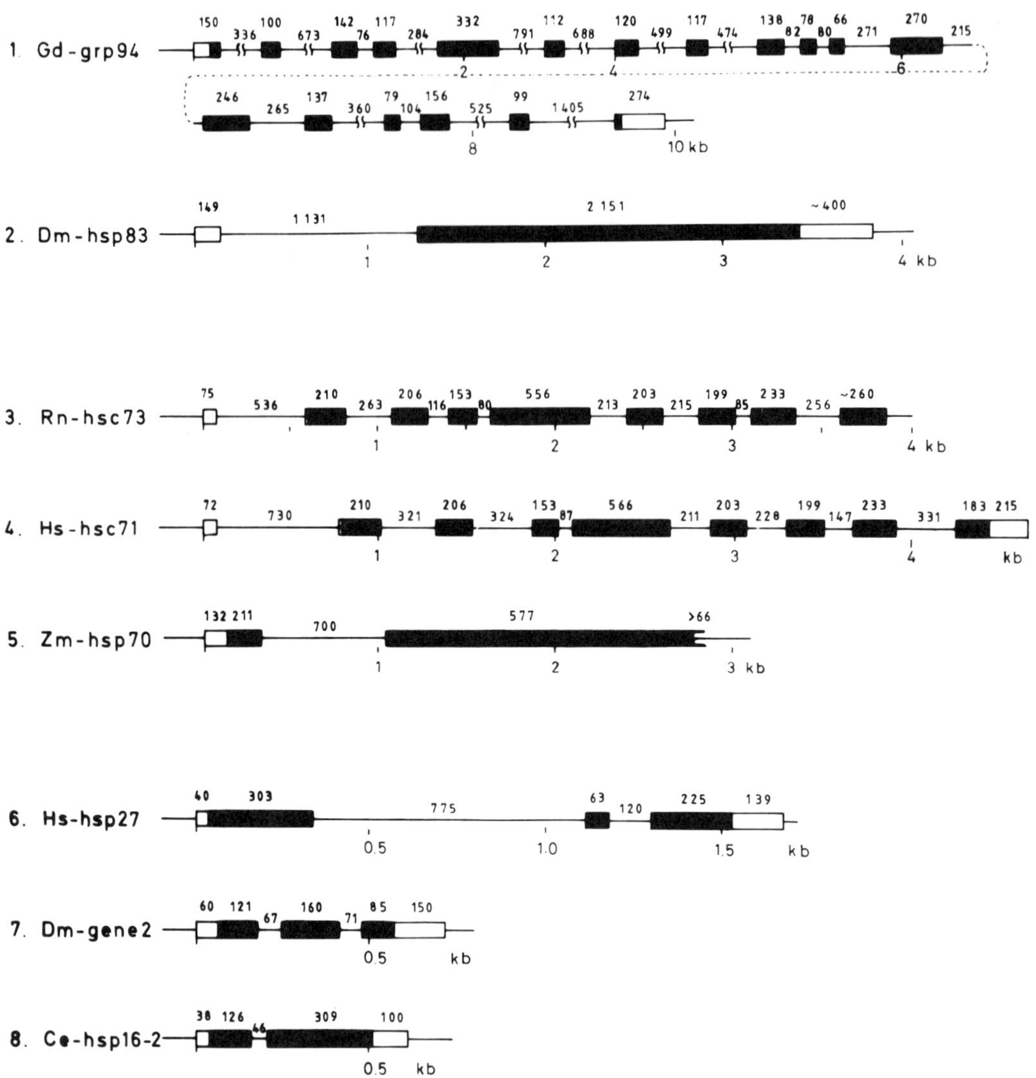

FIGURE 3.1. Basic structure of some intron-containing members of the heat shock (hs) gene families. Genes 1 and 2 belong to the hsp90, genes 3 to 5 to the hsp70, and genes 6 to 8 to the hsp20 family. 1, Chicken grp94;[66] 2, *Drosophila melanogaster* hsp83;[8,43] 3, rat hsc73;[128] 4, human hsc71;[27] 5, maize and *Petunia*[111,141] hsp70; 6, human hsp27;[48] 7, *Drosophila melanogaster* gene 2, hs-induced transcript;[100] and 8, nematode hsp16.[63,112] Length of exons and introns are indicated in base pairs. Parts of the open reading frame are marked in full, 5′ and 3′ untranslated sequences in open boxes. References to other intron-containing genes are in Table 3.1.

One was identified on chromosome 6 as band 6p21, in the class III region of the MHC complex. Two others are presumably on chromosomes 14 and 21, respectively. The mapping of an hsp70 gene inside the major histocompatibility gene complex was also reported for the rat.[143]

Sequences for the mammalian hsp25 to 27 were derived from a mouse cDNA clone[34b] and from genomic clones of the human hsp27 gene.[47,48] The latter was identified in an 18-kb genomic fragment together with a pseudogene and two other homologous sequences. The Hs-hsp27 gene contains two introns. In view of the 18-kb joint genomic fragment, it is surprising that the chromosomal assignment of the hsp27 gene revealed three independent loci on chromosomes 3, 9, and X, respectively.[82c] The functional significance of these loci remains to be established.

Two further hs-induced genes of vertebrates were described. Because they do not belong to any of the hsp gene families, they were not included in Table 3.1, but their promoters contain the usual HSE motifs. These are the rat heme oxygenase gene (see Figure 6.6 and Table 6.2, Müller et al.,[89] Shibahara et al.[124]), a human gene of the MHC complex on the short arm of chromosome 6 coding for the plasma glycoprotein B with 90 kDa[142] (see Table 6.2) as well as the chicken and human polyubiquitin genes[4,9,10,34a] (see Figure 6.6 and Table 6.2).

3.3. NEMATODE *(CAENORHABDITIS ELEGANS)*

Among the HSPs induced in the nematode *Caenorhabditis elegans* are HSP70 and the prominent small HSPs with M_r 16 to 18 kDa.[113,127,127a] There is one hsp70, one hsc70, and evidently one pseudogene.[46a,46b,127] The hsp70A gene is abundantly transcribed under control conditions and induced two- to sixfold by hs.[127a] In contrast, the hsp70C gene codes for a constitutively expressed ER protein of the GRP78 type. Both genes contain three introns (Table 3.1). There are in addition another GRP78 gene, a gene coding for a DnaK-like protein in mitochondria (HSP70F), and a pseudogene derived from the sequence of hsp70A.[46a,46b]

Synthesis of cDNAs containing parts of the hsp16 genes led to the isolation of two types of genomic clones. A pair of two hsp16 genes (16-48 and 16-1) are duplicated in a perfect 1.9 kb inverted repeat, separated by 416 bp of unique DNA.[112] The second type comprises another pair of hsp16 genes (16-41 and 16-2) flanked by low abundance repetitive elements of about 200 bp.[62b] Physical details of both types are given in Figure 3.2. In both cases the adjacent genes are transcribed in divergent directions. The intercistronic HSE elements act on both promoters. Details of the gene structure derived from sequencing of the genomic clones are included in Figure 6.6 and Table 6.2.

Remarkable is the occurrence of short A/T-rich introns in all four genes (Figure 3.2). The intron boundaries match well the consensus sequences described for most eukaryotic genes. However, an intron-internal consensus sequence for lariat formation (PyTPuAPy) is lacking. This may be the reason for the defective splicing of hsp16-1/16-48 pre-mRNA if the corresponding gene pair is introduced into mouse cells[65] (see B17b in Table 6.1). Unspliced pre-mRNAs are usually not detectable in heat-shocked nematode cells.

Despite the virtual identity of the regulatory regions, hs-induced expression of the 16-2/16-41 gene pair in embryos is at least 14-fold higher than that of the two 16-1/16-48 pairs.[62b,63] This effect is much less pronounced in adult tissues and presumably results from differences in the kinetics of induction and decline of hsp16 gene activity.[63]

Comparison of the amino acid sequences (Figure 2.12) shows almost no conservation in the N-terminal part (mainly exon 1) but moderate conservation of the C-terminal part (mainly exon 2).[94a] The latter domain is closely related to corresponding parts of other small HSPs and of vertebrate α-crystalline (Figure 2.13). Nucleotide sequence comparison of the four gene types reveals almost no homology of the intron but high conservation of the 5′ leader sequences and intercistronic regions.[62b] These are evidently important for the coordinate transcription and translation during hs.

In contrast to many other organisms (see Section 2.3.5), the polyubiquitin gene of *Caenorhabditis elegans* is not hs-inducible.[135a] But together with other mRNA of the nematode, including hsp70A mRNA, it is characterized by a 22-nucleotide leader sequence added posttranscriptionally by *trans*-splicing. This leader sequence is coded within the 5S rDNA repeat and transcribed as a 100-nucleotide preleader sequence.[4b,135a]

Another peculiarity of the nematode genome is the existence of a novel repetitive element formed from overlapping tandem repeats of the hs element: –GAA2nTTC2nGAA–, etc. Such elements of up to 240 bp were detected at more than 20 different loci. Though hs

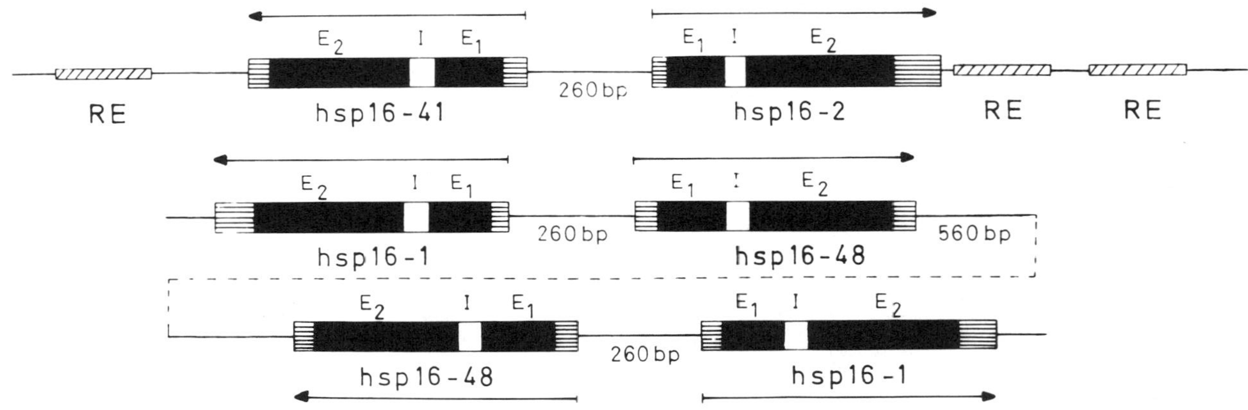

FIGURE 3.2. Physical map of *Caenorhabditis elegans* hsp16 gene clusters. A shows the 16-41/16-2 cluster which contains the two genes in head to head arrangement and is flanked by 200-bp repetitive elements (RE).[63] B shows the 16-48/16-1 cluster which contains two inverted repeats of a head to head arrangement of the two genes.[112] In all genes, the intron (I) separates a small variable exon (E_1) at the 5′ end from a conserved exon (E_2) at the 3′ end. Introns of 58 and 55 bp, respectively, are inserted between codons 46 and 47 in genes 16-41 and 16-48. In genes 16-1 and 16-2 introns of 52 and 46 bp, respectively, separate codons 42 and 43.

activation of adjacent sequences in the genome were not observed, a 242-bp repetitive element inserted 260 bp upstream of the yeast his3 gene conferred hs-inducibility in the heterologous yeast system.[72]

3.4. *DROSOPHILA*

Many papers dealing with the concise analysis of the *Drosophila* hs genes are compiled in Table 3.2. The results for 11 major and minor genes are summarized as follows (Table 3.1 and Figure 3.3):

1. All major hs-induced loci of *Drosophila* are found on the right and left arms of chromosome number three.
2. Two loci (93D and α,β-genes in 87C1) code for RNAs which, though found in the cytoplasmic fraction of poly(A) $^+$RNA, are evidently not translated into heat shock proteins (HSPs) (see Chapter 5).
3. The gene in locus 63BC (hsp83), containing an intron, is the only constitutively expressed HSP, whose synthesis is stimulated under mild but repressed under severe hs conditions. The latter effect is due to a block in pre-mRNP splicing (see Section 9.5). Blackman and Meselson[8] reported on a comparative analyses of the physical organization, of the 5′ flanking and coding sequences of hsp83 genes of *D. melanogaster* and *D. simulans* on the one side and *D. virilis* and *D. pseudoobscura* on the other side. The first two species are separated from each of the latter two by 30 to 40 million years of evolution. Except for some highly conserved regions in the promoter (see Table 6.2), there is almost no similarity of sequences in the leader (exon I) or in the intron. In contrast, coding regions (exon II) of these species are 90% homologous at the DNA and > 97% at the protein level. With the exception of *D. pseudoobscura* the hsp83 locus of the three other species contain a small non-hs inducible transcription unit 0.6 to 0.8 kb upstream of the hsp83 gene (see Figure 3.3).
4. All HSP-coding genes are unique with the exception of the five hsp70 genes found in loci 87A7 and 87C1. Interspersed between the hsp70 genes in 87C1 are the repetitive α,β-elements. Closely related to the hsp70 genes is the unique hsp68 gene of locus 95D, whose analysis has yet to be completed.

   As in all other eukaryotic systems, there is a need for proteins of the HSP70 family also under nonstress conditions. In *Drosophila* these are coded by three hsc70 genes mapping in loci 70C, 87D, and 88E (see summary in Table 2.3). Two of the genes contain introns (hsc1 and hsc2). Whereas the strongly expressed hsc4 gene at locus 88E is intron free. Cellular concentration of its product (HSC70) always surpasses that of HSP70 even under long-term stress conditions.[21,55,98]

   It is an interesting aspect of regulation that the relative activities of hsp70 puffs 87A7 and 87C1 are highly variable.[70] A role of locus 93D for the fine tuning of transcription from both puffs is discussed. Hochstrasser[49] confirmed these peculiarities when analyzing hs-induced gene expression in other polytene tissues of the *Drosophila* larvae (prothoracic gland, hind gut, middle midgut).

   Identification and functional analysis of the two hsp70 loci was facilitated by a number of deletion mutants. Ish-Horowicz et al.[60-62] created a series of overlapping deletions removing parts of or the entire 87A7 and/or 87C1 loci. Deletion of both loci generates HSP70-free mutants, which, unfortunately, were not investigated with respect to their thermotolerance behavior (see Section 17.3). A 2.1 kb 5′ deletion in locus 87A7[135] removed most of the intercistronic region, including the repetitive elements X_a, X_b, and X_c plus part of the proximal hsp70 gene. Expression of the distal hsp70 gene contained in the same locus was not affected, i.e., except for the directly adjacent promoter sequence no other parts of the intercistronic region are required.

TABLE 3.2
Analyses of *Drosophila melanogaster* Heat Shock-Induced Genes

Genes (locus)	hsp70 (87A7,87C1)	hsp68 (95D)	hsp83 (63BC)	Small hsps (67B)	α,β (87C7)	(93D)
A. Selection and characterization of cDNA and genomic clones	20, 40, 79, 116	50	50, 95	15, 17, 137, 138	20, 75, 79	35, 36, 114, 115, 139
B. Chromosomal localization	20, 46, 59—62, 79, 85, 116, 131	50	50, 95	2, 15, 17, 59, 99, 100, 131, 137, 138	20, 46, 62, 75, 78, 79	same
C. Genomic organization, physical maps	3, 20, 40, 59—62, 77, 79, 85, 87, 116	—	50, 95	2, 15, 99, 100, 137, 138	62, 75, 77—79	same
D. Sequencing	51, 57, 64, 77, 81c, 134	—	8, 43, 51	2, 51, 54, 56, 99, 100, 129, 137	42, 77	same

Note: Numbers indicate references.

A: 87A7

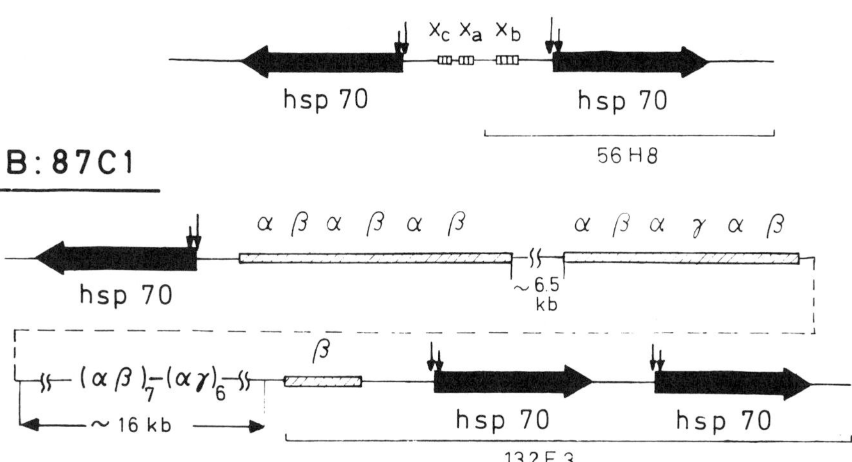

C: 63BC

D: 67B

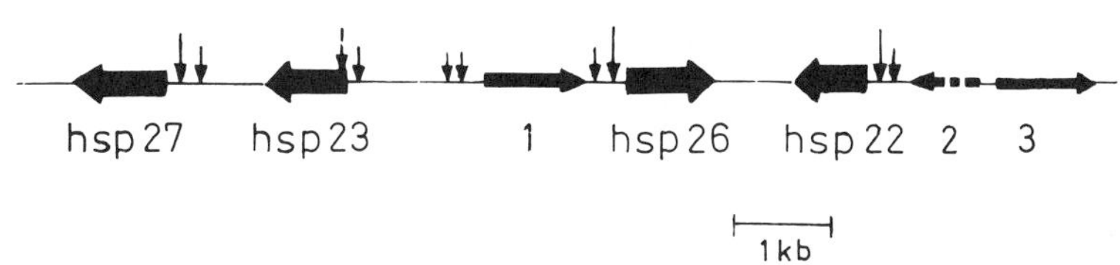

FIGURE 3.3. Genomic organization of *Drosophila* HSP-coding genes. Thick horizontal arrows indicate hs transcription units, and vertical arrows mark DNase I-hypersensitive sites[28a,141] (see Chapter 9). References to the individual genes are given in Table 3.2. Five hsp70 genes are organized in loci 3-87A7 (2 genes) and 3-87C1 (three genes). In the latter a pair of two genes is separated from the third one by ~40 kb of repetitive elements (α,β), which are also transcribed in a hs-dependent manner, if they are connected with the hsp70 promoter element (γ) (see Chapter 5). Brackets denote two clones (56H8 and 132E3) originally isolated by Schedl et al.[116] The intron-containing hsp83 gene in locus 3-63BC is flanked at the 3′ end by a non-hs-induced but constitutively expressed transcription unit T2.[95] The locus 3-67B contains 7 hs-inducible genes. Besides the major hsp27, 26, 23, and 22 genes, there are three minor (genes 1 to 3), whose protein products were not yet identified. Gene 2 contains two small introns. (From Nover, L., Ed., *Heat Shock Response of Eukaryotic Cells,* G. Thieme Verlag, Berlin, 1984. With permission.)

Finally, Caggese et al.[12] isolated an X-ray-induced deletion mutant with a 3′ truncated hsp70 gene coding for a variant "HSP40". This was later used as a convenient reporter gene for promoter analysis (see Section 6.1). Interestingly, this HSP40 mRNA is not affected by the autorepressive destabilization of the mRNA typical for the full-length HSP70 mRNA.[125]

5. The four major small HSP-coding genes (hsp22, 23, 26, and 27) are found clustered with three minor genes, whose transcription is also stimulated under hs (genes 1, 2, and 3), but whose protein products were not yet identified. Expression of all seven inhabitants of locus 67B is also under developmental control. The promoter regions contain HSE and other control elements, e.g., responsive to the ecdysterone receptor (see Figure 6.6 and Chapter 8). Sequence data for all genes were recently completed.[2,99,100] Transcription of the intron-containing gene 2 starts at two different sites, leading to two transcripts with two and three introns, respectively. In addition, during hs a large read-through transcripts of 2000 bp with two introns is formed which extends into the adjacent hsp22 gene.[100]
6. Several minor hs genes were identified by cDNA cloning and *in situ* hybridization. Such hs-induced minipuffs map at loci 63F, 10F, 73D, 54E, and 88B of the *D. melanogaster* polytene chromosomes.[76]
7. *In situ* hybridization and analysis of stress inducibility led to the identification of hs loci also in other *Drosophila* species. With exception of the sibling species of *D. melanogaster,* i.e., *D. simulans,* map positions of hs genes are highly variable:

   Locus 3R-93D of *D. melanogaster* is homologous to 2-48B/C of *D. hydei,*[36,67,71,102] to 2-58C of *D. psudoobscura,*[11] to 2-48A of *D. nasuta,*[71] and to 2-2C of *D. ananassae.*[71]

   Loci 3R-87A7/C1 (hsp70 genes) are homologous to 2-32A of *D. hydei,*[81,101] to 2-29C of *D. virilis,*[29] to 3-89A and 3-94A of *D. subobscura,*[98b] and to 2-53 and 2-58C of *D. pseudoobscura.*[11,105] The insertion of α,β-elements into 87C1 is evidently a very recent event of evolution. In the closely related species *D. simulans* and *D. mauritiana,* the hsp70 genes are found in identical loci as in *D. melanogaster,* but α,β-elements are lacking. Instead they are found only in the chromocentric DNA and are not induced by hs.[73,79]

   Locus 3R-95D (hsp68 gene) corresponds to 2-36A of *D. hydei.*[81,101]

   Locus 3L-63BC (hsp83 gene) corresponds to 4-81B of *D. hydei,*[81,101] to the locus 23 on the X-chromosome of *D. pseudoobscura*[8,28] and to section J on chromosome 3 of *D. virilis.*[8]
8. Similar to most other eukaryotes, an hs-inducible polyubiquitin gene was detected also in *Drosophila.*[72a]

3.5. YEAST AND OTHER FUNGI

Cloning of yeast hs-induced genes started with the isolation of hsp90- and hsp70-specific clones.[33,58] Two genomic clones of hsp90 were later sequenced (Figure 2.8), and the promoter function was analyzed by transformation.[10a,30,31,32,82a] Of the complex hsp70 gene family, nine different members have been characterized by promoter fusion and site-directed insertion mutagenesis[18,19,22,23,23b,140] (see data in Tables 2.4 and 3.1). Five genes of the hsp70 gene family were sequenced (Table 3.1), including sequences for the HSP70 (SSA1), the mitochondrial HSC70 (SSC1), and the GRP78 (KAR2) genes (see Figure 2.10). In addition, there is a mitochondrial GroEL-type HSP60[107] (see Figure 2.11). The only hsp26 gene of yeast was identified by insertion-deletion mutagenesis[104] and sequenced (see Figure 2.12).[132] Finally, similar to the situation in vertebrates, the yeast polyubiquitin gene belongs to the group of hs-induced genes.[34,96,132c]

An unusual HSP of 66.2 kDa was detected by screening for methotrexate resistance in yeast cells transformed with a fusion construct of the SSA4 promoter X mouse dhfr gene. The STI1 gene evidently codes for a regulatory protein which itself is hs-inducible and stimulates expression of the SSA4 gene.[92a] The protein is unrelated to any other yeast HSP so far analyzed.

Heat shock gene analysis of other fungi is just beginning. Judelson and Michelmore[63a] isolated and sequenced an hsp70 gene of the oomycete *Bremia lactucae*. The 2.3 kb mRNA is expressed constitutively, but its abundance increases after heat or cold shock and during sporulation. Four different hsp70 genes (ums 1 to 4) were characterized in the basidiomycete *Ustilago maydis*.[49a,139a] During hs levels of mRNAs increased 5-fold for ums 1 and 2 genes, whereas ums 3 mRNA decreased and ums 4 mRNA remained unchanged. For promoter structures of the *B. lactucae* and *U. maydis* hs genes see Table 6.2.

3.6. PLANTS

The typical, highly complex pattern of small HSPs in plants (HSP20 family) results from at least three major subfamilies of genes[94a] originally defined by hybrid release translation using cDNA clones of soybean.[117,118] Each of the subfamilies I and VI contain several genes coding for 9 to 12 HSPs in the 17 to 18 kDa region. Genes of the subfamily IV code for hs-induced proteins of 20 to 23 kDa which are localized in chloroplasts. As representatives of the latter subfamily, sequences of complete cDNA clones for a *Chlamydomonas reinhardi* HSP22[46] and for a pea HSP21[136] plus an almost complete sequence for the soybean HSP22[136] are available (Figure 2.15). The two proteins of pea and soybean are synthesized as preproteins with a 5 to 6 kDa N-terminal transit peptide, whereas no such transit peptide is contained in the Cr-hsp22 sequence.

So far only five members of subfamily I and one member of subfamily VI (Gm-hsp-17,9D) of soybean were sequenced.[24,92,107,118a,119] In addition, there are sequences of HSP17 proteins of *Arabidopsis*,[45a,132b] pea,[94a] and wheat.[82b] The latter proteins all belong to subfamily I (for sequences, see Figure 2.12).

The complexity of the plant hsp70 family is slowly emerging. Hsp 70 genes from maize and *Petunia* and an hsc70 cDNA clone from tomato were sequenced (Figure 2.10 and Table 3.1). The two hsp70 genes contain introns of 700 and 618 bp, respectively, inserted in codon 71 (Figure 3.1). The same intron position was found for an hsc70 gene and a pseudogene arranged in head-to-tail position in the *Arabidopsis* genome.[142a] *In situ* hybridization and restriction-length polymorphisms allowed the localization of the maize hsp70 gene on chromosome 8.[45]

In addition to the at least two cytoplasmic proteins of the HSP70 family, there are three to four organelle-associated members. Marshall et al.[81b] described one HSC75 at the inner surface of the outer chloroplast membrane and two others in the stroma. By immunological properties, one of the latter (HSC78) was identified as a DnaK-like protein, i.e., it represents the prokaryotic subfamily of the HSP70 proteins. Similar proteins were also reported for spinach and maize chloroplasts and for mitochondria (see Section 2.1.5).

Members of the other HSP families defined in Sections 2.3.1 to 2.3.5 are the HSP80 (HSP90 family, Figure 2.8), the HSC60 (HSP60 family, Figure 2.11)[45b] and the polyubiquitin gene (HSP8.5 family).[14,136a]

3.7. PARASITIC PROTOZOA

In parasitic protozoa of the *Trypanosoma* and *Leishmania* species, expression of HSPs and related proteins is part of their particular life cycle, with transitions from insect vectors into warm-blooded mammals (see Section 18.4). Among them is a constitutive member of the HSP90 family[26] and several HSP70 in *T. brucei*,[38] and *T. cruzi*,[109] and *Leishmania major*.[72b,122a] The HSP83 of *Leishmania mexicana* is an abundant cytoplasmic protein in amastigotes and promastigotes. Four genes arranged in head-to-tail tandem repeats of 4 kb code for a major transcript of 3.5 kb.[123a]

Most analyses were concerned with the hsp70 genes. In all cases, a group of four to seven clustered hsp70 genes is found in a head to tail arrangement. In *T. brucei* one gene is separated by 5.2 kb of spacer DNA from a cluster of five others each with only 200 and 234 bp of the intercistronic region. One of the hsp70 genes was sequenced[38] (Figure 2.9). In *T. cruzi* seven hsp70 genes of 2.5 kb each are arranged in tandem (head to tail) with ~400 bp of spacer between each of two members of the cluster.[109]

In *Leishmania major,* four different hsp70-type genes were mapped and sequenced.[72b,122a] A cluster of four identical, hs-inducible genes with 380 bp of intercistronic spacer, is found on chromosome 17. Three isolated hsc70 genes were detected by pulse-field electrophoresis on chromosomes 19 (mitochondrial HSC70) and 15 and 16, respectively. A similar DnaK-like HSC70 of mitochondria was described by Engman et al.[28a] for *Trypanosoma cruzi.* It is nuclear encoded and synthesized as a precursor with a 25 amino acid residue transit peptide. The levels of the two mt-hsc70 mRNAs of 2.5 and 2.0 kb decrease under hs conditions. Similar to DnaK, this mitochondrial HSC70 is bound to the replication origin of the kinetoplast DNA.

Like other mRNAs of trypanosomes, the hsp70 and hsp85 mRNAs contain a capped 39-nucleotide leader sequence at their 5′ ends, coded in separate 140-nucleotide miniexons and added posttranscriptionally by *trans*-splicing. It is an interesting observation that functional hs mRNAs including the obligate leader sequence are formed, though *trans*-splicing to other non-hs mRNAs is disrupted by heat shock.[90a,90b]

Similar to the findings in vertebrates (Section 3.2) and yeast (Section 3.5), some ubiquitin genes are also hs regulated.[132a] In *T. cruzi* about 100 ubiquitin genes are clustered in a 27 kb genomic fragment. Five of them are fusion genes with a C-terminal extension of their open reading frames by a 52-amino acid residue basic domain. They are transcribed during hs (26°→41°C). A palindromic promoter element 5′-CTGAAT-2n-ATGCAG-3′ is found 16 nucleotides upstream of the putative TATA box (-CTATT-). The function of this *Trypanosoma* variant of the classical HSE (see Table 6.2) was not tested, but a similar element is contained also in the 5′-upstream region of the *T. brucei* hsp70 genes.[38]

REFERENCES

1. **Allen, R. L., O'Brien, D. A., and Eddy, E. M.,** A novel hsp70-like protein (p70) is present in mouse spermatogenic cells, *Mol. Cell. Biol.*, 8, 828—832, 1988.
2. **Ayme, A. and Tissieres, A.,** Locus 67B of *Drosophila melanogaster* contains seven, not four, closely related heat shock genes, *EMBO J.*, 4, 2949—2954, 1985.
3. **Artavanis-Tsakonas, S., Schedl, R., Mirault, M. E., Moran, L., and Lis, J.,** Genes for the 70,000 dalton heat shock protein in two cloned *D. melanogaster* DNA segments, *Cell,* 17, 9—18, 1979.
4 **Baker, Z. T. and Board, P. G.,** The human ubiquitin gene family. Structure of a gene and pseudogenes from the UbB subfamily, *Nucl. Acids Res.*, 15, 443—463, 1987.
4a. **Barner, J. V., Bensaude, O., Moranges, M., and Babinet, C.,** Mouse 89 kD heat shock protein. Two peptides with distinct developmental regulation, *Exp. Cell Res.*, 170, 186—194, 1987.
4b. **Bektesh, S., van Doren, K., and Hirsh, D.,** Presence of the *Caenorhabditis elegans* spliced leader on different mRNAs and in different genera of nematodes, *Genes Dev.*, 2, 1277—1283, 1988.
5. **Bienz, M.,** *Xenopus* hsp70 genes are constitutively expressed in injected oocytes, *EMBO J.*, 3, 2477—2483, 1984.
6. **Bienz, M.,** Transient and developmental activation of heat-shock genes, *Trends Biochem. Sci.*, 10, 157—161, 1985.
7. **Bienz, M. and Pelham, H. R. B.,** Mechanisms of heat-shock gene activation in higher eucaryotes, *Adv. Genet.*, 24, 31—72, 1987.
7a. **Binart, N., Chambraud, B., Dumas, B., Rowlands, D. A., Bigogne, C., Levin, J. M., Garnier, J., Baulieu, E.-E., and Catelli, M.-G.,** The cDNA-derived amino acid sequence of chick heat shock protein Mr 90,000 (HSP90) reveals a "DNA-like" structure: Potential site of interaction with steroid receptors, *Biochem. Biophys. Res. Commun.*,159, 140—147, 1989.

8. **Blackman, R. K. and Meselson, M.,** Interspecific nucleotide sequence comparisons used to identify regulatory and structural features of the *Drosophila* hsp82 gene, *J. Mol. Biol.,* 188, 499—516, 1986.
9. **Bond, U. and Schlesinger, M. J.,** Ubiquitin is a heat shock protein in chicken embryo fibroblasts, *Mol. Cell. Biol.,* 5, 949—956, 1985.
10. **Bond, U. and Schlesinger, M. J.,** The chicken ubiquitin gene contains a heat shock promoter and expresses an unstable mRNA in heat-shocked cells, *Mol. Cell. Biol.,* 6, 4602—4610, 1986.
10a. **Borkovich, K. A., Farrelly, F. W., Finkelstein, D. B., Taulieu, J., and Lindquist, S.,** Hsp82 is an essential protein that is required in higher concentrations for growth of cells at higher temperatures, *Mol. Cell. Biol.,* 9, 3913—3930, 1989.
10b. **Bossier, P., Fitch, I. T., Boucherie, H., and Tuite, M. F.,** Structure and expression of a yeast gene encoding the small heat-shock protein hsp26, *Gene,* 78, 323—330, 1989.
11. **Burma, P. K. and Lakhotia, S. C.,** Cytological identity of 93D-like and 87C-like heat shock loci in *Drosophila pseudoobscura, Indian J. Exp. Biol.,* 22, 577—580, 1984.
12. **Caggese, C., Caizzi, R., Morea, M., Scalenghe, F., and Ritossa, F.,** Mutation generating a fragment of the major heat shock-inducible polypeptide in *Drosophila melanogaster, Proc. Natl. Acad. Sci. U.S.A.,* 76, 2385—2389, 1979.
13. **Chang, S. C., Wooden, S. K., Nakaki, T., Kim, Y. K., Lin, A. Y., Kung, L., Attenello, J. W., and Lee, A. S.,** Rat gene encoding the 78-kDa glucose-regulated protein GRP78: Its regulatory sequences and the effect of protein glucosylation on its expression, *Proc. Natl. Acad. Sci. U.S.A.,* 84, 680—684, 1987.
14. **Christensen, A. H. and Quail, P. H.,** Sequence analysis and transcriptional regulation by heat shock of polyubiquitin transcripts from maize, *Plant Mol. Biol.,* 12, 619—632, 1989.
15. **Corces, V., Holmgren, R., Freund, R., Morimoto, R., and Meselson, M.,** Four heat shock proteins of *Drosophila melanogaster* coded within a 12-kilobase region in chromosome subdivision 67B, *Proc. Natl. Acad. Sci. U.S.A.,* 77, 5390—5393, 1980.
16. **Craig, E. A.,** The heat shock response, *CRC Crit. Rev. Biochem.,* 18, 239—280, 1985.
17. **Craig, E. A. and McCarthy, B. J.,** Four *Drosophila* heat shock genes at 67B: Characterization of recombinant plasmids, *Nucl. Acids Res.,* 8, 4441—4457, 1980.
18. **Craig, E. A. and Jacobsen, K.,** Mutations of the heat inducible 70 kilodalton genes of yeast confer temperature sensitive growth, *Cell,* 38, 841—849, 1984.
19. **Craig, E. A. and Jacobsen, K.,** Mutations in cognate genes of *Saccharomyces cerevisiae* hsp70 result in reduced growth rates at low temperatures, *Mol. Cell. Biol.,* 5, 3517—3524, 1985.
20. **Craig, E. A., McCarthy, B. J., and Wadsworth, S. C.,** Sequence organization of two recombinant plasmids containing genes for the major heat shock-induced protein of *D. melanogaster, Cell,* 16, 575—588, 1979.
21. **Craig, E. A., Ingolia, T. D., and Manseau, L. J.,** Expression of *Drosophila* heat-shock cognate genes during heat-shock and development, *Dev. Biol.,* 99, 418—426, 1983.
22. **Craig, E. A., Slater, M. R., Boorstein, W. R., and Palter, K.,** Expression of the *S. cerevisiae* hsp70 multigene family, *UCLA Symp. Mol. Cell. Biol.,* 30, 659—667, 1985.
23. **Craig, E. A., Kramer, T., and Kosic-Smithers, J.,** SSC1, a member of the 70-kDa heat shock protein multigene family of *Saccharomyces cerevisiae,* is essential for growth, *Proc. Natl. Acad. Sci. U.S.A.,* 84, 4156—4160, 1987.
23a. **Craig, E. A., Kramer, J., Shilling, J., Werner-Washburne, M., Holmes, S., Kosic-Smithers, J., and Nicolet, C. M.,** SSC1, an essential member of the yeast HSP70 multigene family, encodes a mitochondrial protein, *Mol. Cell. Biol.,* 9, 3000—3008, 1989.
23b. **Craig, E., Boorstein, W., Park, H.-O., Stone, D., and Nicolet, C.,** Complex regulation of three heat inducible HSP70 related genes in *Saccharomyces cerevisiae,* in *Stress-Induced Proteins,* Pardue, M. L., Feramisco, J. R., and Lindquist, S., Eds., Alan R. Liss, New York, 1989, 51—61.
24. **Czarnecka, E., Gurley, W. B., Nagao, R. T., Mosquera, L. A., and Key, J. L.,** DNA sequence and transcript mapping of a soybean gene encoding a small heat shock protein, *Proc. Natl. Acad. Sci. U.S.A.,* 82, 3726—3730, 1985.
24a. **Czarnecka, E., Nagao, R. T., Key, J. L., and Gurley, W. B.,** Characterization of Gm-hsp26-A, a stress gene encoding a divergent heat shock protein of soybean: Heavy-metal-induced inhibition of intron processing, *Mol. Cell. Biol.,* 8, 1113—1122, 1988.
25. **Drabent, B., Genthe, A., and Benecke, B.-J.,** In vitro transcription of a human hsp70 heat shock gene by extracts prepared from heat-shocked and non-heat-shocked human cells, *Nucl. Acids Res.,* 14, 8933—8948, 1986.
26. **Dragon, E. A., Sias, S. R., Kato, E. A., and Gabe, J. D.,** The genome of *Trypanosoma cruzi* contains a constitutively expressed, tandemly arranged multicopy gene homologous to a major heat shock protein, *Mol. Cell. Biol.,* 7, 1271—1275, 1987.
26a. **Duck N. and Winter, J.,** personal communication.

27. **Dworniczak, B. and Mirault, M.-E.,** Structure and expression of a human gene coding for a 71 kd heat shock 'cognate' protein, *Nucl. Acids Res.*, 15, 5181—5197, 1987.
28. **Eissenberg, J. C. and Lucchesi, J. C.,** Chromatin structure and transcriptional activity of an X-linked heat shock gene in *Drosophila pseudoobscura, J. Biol. Chem.*, 258, 13986—13991, 1983.
28a. **Engman, D. M., Kirchhoff, L. V., and Donelson, J. E.,** Molecular cloning of mtp70, a mitochondrial member of the hsp70 family, *Mol. Cell. Biol.*, 9, 5163—5168, 1989.
29. **Evgen'ev, M. B., Kolchinski, A., Levin, A., Preobrazenskaya, O., and Sarkisova, E.,** Heat-shock DNA homology in distantly related species of *Drosophila, Chromosoma*, 68, 357—365, 1978.
30. **Farrelly, F. W. and Finkelstein, D. B.,** Complete sequence of the heat shock-inducible HSP90 gene of *Saccharomyces cerevisiae, J. Biol. Chem.*, 259, 5745—5751, 1984.
31. **Finkelstein, D. B. and Strausberg, S.,** Identification and expression of a cloned yeast heat shock gene, *J. Biol. Chem.*, 258, 1908—1913, 1983.
32. **Finkelstein, D. B. and Strausberg, S.,** Heat shock-regulated production of *Escherichia coli* β-galactosidase in *Saccharomyces cerevisiae, Mol. Cell. Biol.*, 3, 1625—1633, 1983.
33. **Finkelstein, D. B., Strausberg, S., and McAlister, L.,** Alterations of transcription during heat shock of *Saccharomyces cerevisiae, J. Biol. Chem.*, 257, 8405—8411, 1982.
34. **Finley, D., Özkaynak, E., and Varshavsky, A.,** The yeast polyubiquitin gene is essential for resistance to high temperatures, starvation and other stresses, *Cell*, 48, 1035—1046, 1987.
34a. **Fornace, A. J., Alamo, I., Hollander, M. C., and Lamoreaux, E.,** Ubiquitin mRNA is a major stress-induced transcript in mammalian cells, *Nucl. Acids Res.*, 17, 1215—1230, 1989.
34b. **Gaestel, M., Gross, B., Benndorf, R., Strauss, M., Schunk, W.-H, Kraft, R., Otto, A., Boehm, H., Stahl, J., Drabsch, H., and Bielka, H.,** Molecular cloning, sequencing and expression in *Escherichia coli* of the 25-kDa growth-related protein of Ehrlich ascites tumor and its homology to mammalian stress proteins, *Eur. J. Biochem.*, 179, 209—213, 1989.
35. **Garbe, J. C. and Pardue, M. L.,** Heat shock locus 93D of *Drosophila melanogaster:* A spliced RNA most strongly conserved in the intron sequence, *Proc. Natl. Acad. Sci. U.S.A.*, 83, 1812—1816, 1986.
36. **Garbe, J. C., Bendena, W. G., Alfano, M., and Pardue, M. L.,** A *Drosophila* heat shock locus with a rapidly diverging sequence but a conserved structure, *J. Biol. Chem.*, 261, 16889—16895, 1986.
37. **Giebel, J. C., Dworniczak, B. P., and Bautz, E. K. F.,** Nucleotide sequence of a processed human hsc70 pseudogene, *Nucl. Acids Res.*, 15, 9605, 1987.
37a. **Giebel, L. B., Dworniczak, B. P., and Bautz, E. K. F.,** Developmental regulation of a constitutively expressed mouse mRNA encoding a 72-kDa heat shock-like protein, *Dev. Biol.*, 125, 200—207, 1988.
38. **Glass, D. J., Polvere, R. J., and Van Der Ploeg, L. H. T.,** Conserved sequences and transcription of the hsp 70 gene family in *Trypanosoma brucei, Mol. Cell. Biol.*, 6, 4657—4666, 1986.
39. **Goate, A. M., Cooper, D. N., Hall, C., Leung, Th. K. C., Solomon, E., and Lim, L.,** Localization of a human heat-shock HSP 70 gene sequence to chromosome 6 and detection of two other loci by somatic cell hybrid and restriction fragment length polymorphism analysis, *Hum. Genet.*, 75, 123—128, 1987.
40. **Goldschmidt-Clermont, M.,** Two genes for the major heat-shock protein of *Drosophila melanogaster* arranged as an inverted repeat, *Nucl. Acids Res.*, 8, 235—252, 1980.
41. **Grimm, B., Ish-Shalom, D., Evan, D., Glaczinski, H., Ottersbach, P., Ohad, I., and Kloppstech, K.,** The nuclear-coded chloroplast 22-kDa heat shock protein of *Chlamydomonas*. Evidence for translocation into the organelle without a processing step, *Eur. J. Biochem.*, 182, 539—546, 1989.
42. **Hackett, R. W. and Lis, J. T.,** DNA sequence analysis reveals extensive homologies of regions preceding hsp70 and alphabeta heat shock genes in *Drosophila melanogaster, Proc. Natl. Acad. Sci. U.S.A.*, 78, 6196—6200, 1981.
43. **Hackett, R. W. and Lis, J. T.,** Localization of the hsp83 transcript within a 3292 nucleotide-sequence from the 63B heat shock locus of *Drosophila melanogaster, Nucl. Acids Res.*, 11, 7011—7030, 1983.
44. **Harrison, G. S., Drabkin, H. A., Kao, F.-T., Hartz, J., Hart, I. M., Chu, E. H. Y., Wu, B. J., and Morimoto, R. I.,** Chromosomal location of human genes encoding major heat-shock protein hsp70, *Somatic Cell Mol. Genet.*, 13, 119—130, 1987.
45. **Helentjaris, T., Weber, D. F., and Wright, S.,** Use of monosomics to map cloned DNA fragments in maize, *Proc. Natl. Acad. Sci. U.S.A.*, 83, 6035—6039, 1986.
45a. **Helm, K. W. and Vierling, E.,** An *Arabidopsis thaliana* cDNA clone encoding a low molecular weight heat shock protein, *Nucl. Acids Res.*, 17, 7995, 1989.
45b. **Hemmingsen, S. M., Woolford, C., van der Vies, S. M., Tilly, K., Dennis, D. T., Georgopoulos, C. P., Hendrix, R. W., and Ellis, R. J.,** Homologous plant and bacterial proteins chaperone oligomeric protein assembly, *Nature*, 333, 330—335, 1988.
46. **Henikoff, S. and Meselson, M.,** Transcription of two heat shock loci in *Drosophila, Cell*, 12, 441—451, 1977.
46a. **Heschl, M. F. P. and Baillie, D. L.,** Characterization of the hsp70 multigene family of *Caenorhabditis elegans, DNA, 8, 233—243, 1989.*

46b. **Heschl, M. F. P. and Baillie, D. L.,** Identification of a heat-shock pseudogene from *Caenorhabditis elegans, Genome,* 32, 190—195, 1989.

47. **Hickey, E., Brandon, S. E., Sadis, S., Smale, G., and Weber, L. A.,** Molecular cloning of sequences encoding the human heat-shock proteins and their expression during hyperthermia, *Gene,* 43, 147—154, 1986.

48. **Hickey, E., Brandon, S. E., Potter, R., Stein, G., Stein, J., and Weber, L. A.,** Sequence and organization of genes encoding the human 27 kDa heat shock protein, *Nucl. Acids Res.,* 14, 4127—4146, 1986.

48a. **Hickey, E., Brandon, S. E., Smale, G., Lloyd, D., and Weber, L. A.,** Sequence and regulation of a gene encoding a human 89-kilodalton heat shock protein, *Mol. Cell. Biol.,* 9, 2615—2626, 1989.

49. **Hochstrasser, M.,** Chromosome structure in four wild-type polytene tissues of *Drosophila melanogaster.* The 87A and 87C heat shock loci are induced unequally in the midgut in a manner dependent on growth temperature, *Chromosoma,* 95, 197—208, 87.

49a. **Holden, D. W., Kronstad, J. W., and Leong, S. A.,** Mutation in a heat-regulated hsp70 gene of *Ustilago maydis, EMBO J.,* 8, 1927—1934, 1989.

50. **Holmgren, R., Livak, K., Morimoto, R., Freund, R., and Meselson, M.,** Studies of cloned sequences from four *Drosophila* heat shock loci, *Cell,* 18, 1359—1370, 1979.

51. **Holmgren, K., Corces, V., Morimoto, R., Blackman, R., and Meselson, M.,** Sequence homologies in the 5′ regions of four *Drosophila* heat-shock genes, *Proc. Natl. Acad. Sci. U.S.A.,* 78, 3775—3778, 1981.

52. **Hultmark, D., Klemenz, R., and Gehring, W. J.,** Translational and transcriptional control elements in the untranslated leader of the heat-shock gene hsp22, *Cell,* 44, 429—438, 1986.

53. **Hunt, C. and Morimoto, R. J.,** Conserved features of eukaryotic hsp70 genes revealed by comparison with the nucleotide sequence of human hsp70, *Proc. Natl. Acad. Sci. U.S.A.,* 82, 6455—6459, 1985.

54. **Ingolia, T. D. and Craig, E. A.,** Primary sequence of the 5′ flanking regions of the *Drosophila* heat shock genes in chromosome subdivision 67 B, *Nucl. Acids Res.,* 9, 1627—1642, 1981.

55. **Ingolia, T. D. and Craig, E. A.,** *Drosophila* gene related to the major heat shock-induced gene is transcribed at normal temperatures and not induced by heat shock, *Proc. Natl. Acad. Sci. U.S.A.,* 79, 525—529, 1982.

56. **Ingolia, T. D. and Craig, E. A.,** Four small *Drosophila* heat shock proteins are related to each other and to mammalian α-crystallin, *Proc. Natl. Acad. Sci. U.S.A.,* 79, 2360—2364, 1982.

57. **Ingolia, T. D., Craig, E. A., and McCarthy, B. J.,** Sequence of three copies of the gene for the major *Drosophila* heat shock induced protein and their flanking regions, *Cell,* 21, 669—679, 1980.

58. **Ingola, T. D., Slater, M. R., and Craig, E. A.,** *Saccharomyces cerevisiae* contains a complex multigene family related to the major heat shock-inducible gene of *Drosophila, Mol. Cell. Biol.,* 2, 1388—1398, 1982.

59. **Ish-Horowicz, D. and Pinchin, S. M.,** Genomic organization of the 87A7 and 87C1 heat-induced loci of *Drosophila melanogaster, J. Mol. Biol.,* 142, 231—245, 1980.

60. **Ish-Horowicz, D., Holden, J. J., and Gehring, W.,** Deletions of two heat-activated loci in *Drosophila melanogaster* and their effects on heat-induced protein synthesis, *Cell,* 12, 643—652, 1977.

61. **Ish-Horowicz, D., Gausz, J., Gyurkovicz, H., Bencze, G., Goldschmidt-Clermont, M., and Holden, J. J.,** Deletion mapping of two *D. melanogaster* loci that code for the 70,000 Dalton heat-induced protein, *Cell,* 17, 565—571, 1979.

62. **Ish-Horowicz, D., Schedl, P., Artavanis-Tsakonas, S., and Mirault, M. E.,** Genetic and molecular analysis of the 87A7 and 87C1 heat-inducible loci of *D. melanogaster, Cell,* 18, 1351—1358, 1979.

62a. **Jindal, S., Dudani, A. K., Singh, B., Harley, C. B., and Gupta, R. S.,** Primary structure of a human mitochondrial protein homologous to the bacterial and plant chaperonins and to the 65-kd mycobacterial antigen, *Mol. Cell. Biol.,* 9, 2279—2283, 1989.

62b. **Jones, D., Russnak, R. H., Kay, R. J., and Candido, E. P. M.,** Structure expression and evolution of a heat shock gene locus in *Caenorhabditis elegans* that is flanked by repetitive elements, *J. Biol. Chem.,* 261, 12006—12015, 1986.

63. **Jones, D., Dixon, D. K., Graham, R. W., and Candido, E. P. M.,** Differential regulation of closely related members of the HSP16 gene family in *Caenorhabditis elegans, DNA,* 8, 481—490, 1989.

63a. **Judelson, H. S. and Michelmore, R. W.,** Structure and expression of a gene encoding heat-shock protein hsp70 from the oomycete fungus *Bremia lactucae, Gene,* 79, 207—217, 1989.

64. **Karch, F., Török, I., and Tissieres, A.,** Extensive regions of homology in front of the two hsp 70 heat shock variant genes in *Drosophila melanogaster, J. Mol. Biol.,* 148, 219—230, 1981.

65. **Kay, R. J., Russnak, R. H., Jones, D., Mathias, C., and Candido, E. P. M.,** Expression of intron-containing *C. elegans* heat shock genes in mouse cells demonstrates divergence of 3′ splice site recognition sequences between nematodes and vertebrates, and an inhibitory effect of heat shock on the mammalian splicing apparatus, *Nucl. Acids Res.,* 15, 3723—3741, 1987.

66. **Kleinsek, D. A., Beattie, W. G., Tsai, M. J., and O'Malley, B. W.,** Molecular cloning of a steroid-regulated 108 K heat shock protein gene from hen oviduct, *Nucl. Acids Res.,* 14, 10053—10071, 1986.

66a. **Knack, G. and Kloppstech, K.**, cDNA sequence of a heat-inducible protein of *Chenopodium* sharing little homology with other heat shock proteins, *Nucl. Acids Res.*, 17, 5380, 1989.
67. **Koninkx, J. F. J. G.**, Protein synthesis in salivary glands of *Drosophila hydei* after experimental gene induction, *Biochem. J.*, 158, 623—628, 1976.
68. **Krawczyk, Z., Szymik, N., and Wisniewski, J.**, Expression of hsp70-regulated gene in developing and degenerating rat testis, *Mol. Biol. Rep.*, 12, 35—41, 1987.
69. **Kulomaa, M. S., Weigel, N. L., Kleinsek, O. A., Beattie, W. G., Connely, O. M., March, C., Jarucki-Schulz, T., Schrader, W. T., and O'Malley, B. W.**, Amino acid sequence of a chicken heat shock protein derived from the complementary DNA nucleotide sequence., *Biochemistry*, 25, 6244—6252, 1986.
70. **Lakhotia, S. C.**, The 93D heat shock locus in *Drosophila:* A review, *J. Genet.*, 66, 139—158, 1987.
71. **Lakhotia, S. C. and Singh, A. K.**, Conservation of the 93D puff of *Drosophila melanogaster* in different species of *Drosophila, Chromosoma*, 86, 265—278, 1982.
72. **LaVolpe, A., Ciaramella, M., and Bazzicalupo, P.**, Structure evolution and properties of a novel repetitive DNA family in *Caenorhabditis elegans, Nucl. Acids Res.*, 16, 8213—8232, 1988.
72a. **Lee, H., Simon, J. A., and Lis, J. T.**, Structure and expression of ubiquitin genes of *Drosophila melanogaster, Mol. Cell. Biol.*, 8, 4727—4735, 1988.
72b. **Lee, M. G. S., Atkinson, B. L., Giannini, S. H., and van der Ploeg, L. H. T.**, Structure and expression of the hsp70 gene family of *Leishmania major, Nucl. Acids Res.*, 16, 9567—9585, 1988.
73. **Leigh Brown, A. J. and Ish-Horowicz, D.**, Evolution of the 87A and 87C heat-shock loci in *Drosophila, Nature*, 290, 677—682, 1981.
74. **Lindquist, S.**, The heat-shock response (review), *Annu. Rev. Biochem.*, 55, 1151—1191, 1986.
74a. **Lindquist, S. and Craig, E. A.**, The heat-shock proteins, *Annu. Rev. Genet.*, 22, 631-677, 1988.
75. **Lis, J., Pretidge, L., and Hogness, D. S.**, A novel arrangement of tandemly repeated genes at a major heat shock site in *D. melanogaster, Cell*, 14, 901—919, 1978.
76. **Lis, J. R., Neckameyer, W., Dubensky, R., and Costlow, N.**, Cloning and characterization of nine heat-shock-induced mRNAs of *Drosophila melanogaster, Gene*, 15, 67—80, 1981.
77. **Lis, J., Neckameyer, W., Mirault, M.-E., Artavanis-Tsakonas, S., Lall, P., Martin, G., and Schedl, P.**, DNA sequences flanking the starts of the hsp70 and alpha, beta heat shock genes are homologous, *Dev. Biol.*, 83, 291—300, 1981.
78. **Lis, J. T., Ish-Horowicz, D., and Pinchin, S. M.**, Genomic organization and transcription of the alpha,beta heat shock DNA in *Drosophila melanogaster, Nucl. Acids Res.*, 9, 5297—5310, 1981.
79. **Livak, K. J., Freund, R., Schweber, M., Wensink, P. C., and Meselson, M.**, Sequence organization and transcription at two heat shock loci in *Drosophila, Proc. Natl. Acad. Sci. U.S.A.*, 75, 5613—5617, 1978.
80. **Lowe, D. G. and Moran, L. A.**, Molecular cloning and analysis of DNA complementary to three mouse 68,000-molecular-weight heat shock protein messenger RNA species, *J. Biol. Chem.*, 261, 2102—2112, 1986.
81. **Lubsen, N. H. and Sondermeijer, P. J. A.**, The products of the 'heat-shock' loci of *Drosophila hydei.* Correlation between locus 2-36A and the 70,000 MW 'heat-shock' peptide, *Chromosoma*, 66, 115—125, 1978.
81a. **Maekawa, M., O'Brien, D. A., Allen, R. L., and Eddy, E. M.**, Heat-shock cognate protein (Hsc71) and releated proteins in mouse spermatogenic cells, *Biol. Rep.*, 40, 843—852, 1989.
81b. **Marshall, J. S., Derocher, A. E., Keegstra, K., and Vierling, E.**, Identification of hsp70 homologues in chloroplasts, *Proc. Natl. Acad. Sci. U.S.A.*, 87, 374—378, 1990.
81c. **Mason, P. J., Töroek, I., Kiss, I., Karch, F., and Udvardy, A.**, Evolutionary implications of a complex pattern of DNA sequence homology extending far upstream of the hsp70 genes at loci 87A7 and 87C1 in *Drosophila melanogaster, J. Mol. Biol.*, 156, 21—35, 1982.
82. **Mazzarella, R. A. and Green, M.**, ERp99, an abundant, conserved glycoprotein of the endoplasmic reticulum, is homologous to the 90-kDa heat shock protein (hsp90) and the 94-kDa glucose regulated protein (GRP94), *J. Biol. Chem.*, 262, 8875—8883, 1987.
82a. **McDaniel, D.-A., Caplan, A. J., Lee, M.-S., Adams, C. C., Fishel, B. R., Gross, D. S., and Garrard, W. T.**, Basal-level expression of the yeast HSP82 gene requires a heat shock regulatory element, *Mol. Cell. Biol.*, 9, 4789—4798, 1989.
82b. **McElwain, E. F. and Spiker, S.**, A wheat cDNA clone which is homologous to the 17kd heat-shock protein gene family of soybean, *Nucl. Acids Res.*, 17, 1764, 1989.
82c. **McGuire, S. E., Fuqua, S. A. W., Naylor, S. L., Helin-Davis, D. A., and McGuire, W. L.**, Chromosomal assignments of human 27-kDa heat shock protein gene family, *Somat. Cell Mol. Genet.*, 15, 167—171, 1989.
83. **McKenzie, S. L. and Meselson, M.**, Translation in vitro of *Drosophila* heat-shock messages, *J. Mol. Biol.*, 117, 279—283, 1977.

84. **McKenzie, S. L., Henikoff, S., and Meselson, M.,** Localization of RNA from heat-induced polysomes at puff sites in *Drosophila melanogaster, Proc. Natl. Acad. Sci. U.S.A.*, 72, 1117—1121, 1975.

85. **Mirault, M. E., Goldschmidt-Clermont, M., Artavanis-Tsakonas, S., and Schedl, P.,** Organization of the multiple genes for the 70,000-dalton heat shock protein in *Drosophila melanogaster, Proc. Natl. Acad. Sci. U.S.A.*, 76, 5254—5258, 1979.

86. **Moore, M., Schaack, J., Baim, S. B., Morimoto, R. I., and Schenk, T.,** Induced heat-shock mRNAs escape the nucleo-cytoplasmic transport block in adenovirus-infected cells, *Mol. Cell. Biol.*, 7, 4505—4512, 1987.

87. **Moran, L., Mirault, M. E., Tissieres, A., Lis, J., Schedl, P., Artavanis-Tsakonas, S., and Gehring, W. J.,** Physical map of two *D. melanogaster* DNA segments containing sequences coding for the 70,000 dalton heat shock protein, *Cell,* 17, 1—8, 1979.

88. **Morimoto, R., Hunt, C., Huang, S.-Y., Berg, K. L., and Banerji, S. S.,** Organization, nucleotide sequence and transcription of the chicken HSP70 gene, *J. Biol. Chem.*, 261, 12692—12699, 1986.

89. **Müller, R. M., Taguchi, H., and Shibahara, S.,** Nucleotide sequence and organization of the rate heme oxygenase gene, *J. Biol. Chem.*, 262, 6795—6802, 1987.

90. **Mues, G. I., Munn, T. Z., and Raese, J. D.,** A human gene family with sequence homology to *Drosophila melanogaster* Hsp70 heat shock genes, *J. Biol. Chem.*, 261, 874—877, 1986.

90a. **Muhich, M. L., Hsu, M. P., and Boothroyd, J. C.,** Heat-shock disruption of *trans*-splicing in trypanosomes—Effect on hsp70, hsp85 and tubulin messenger RNA synthesis, *Gene,* 82, 169—175, 1989.

90b. **Muhich, M. L. and Boothroyd, J. C.,** Synthesis of trypanosome hsp70 mRNA is resistant to disruption of *trans*-splicing by heat shock, *J. Biol. Chem.*, 264, 7107—7110, 1989.

90c. **Munn, T. Z. and Mues, G. I.,** Highly conserved repeats in heat-shock introns, *Nature,* 332, 789, 1988.

91. **Munro, S. and Pelham, H. R. B.,** An hsp70-like protein in the ER: Identity with the 78 kd glucose-regulated protein and immunoglobulin heavy chain binding protein, *Cell,* 46, 291—300, 1986.

91a. **Nagao, R. T. and Key, J. L.,** Heat shock protein genes of plants, in *Cell Cultural and Somatic Cell Genetics of Plants,* Vol. 6, Schell, J. and Vasil, T. K., Eds., Academic Press, Orlando, FL, 1989, 297—328.

92. **Nagao, R. T., Czarnecka, E., Gurley, W. B., Schoeffl, F., and Key, J. L.,** Genes for low-molecular-weight heat shock proteins of soybeans: Sequence analysis of a multigene family, *Mol. Cell. Biol.*, 5, 3417—3428, 1985.

92a. **Nicolet, C. M. and Craig, E. A.,** Isolation and characterization of STI1, a stress-inducible gene from *Saccharomyces cerevisiae, Mol. Cell. Biol.*, 9, 3638—3646, 1989.

92b. **Normington, K., Kohno, K., Kozutsumi, Y., Gething, M.-J., and Sambrook, J.,** *S. cerevisiae* encodes an essential protein homologous to mammalian BiP, *Cell,* 57, 1223—1236, 1989.

93. **Nover, L., Ed.,** *Heat Shock Response of Eukaryotic Cells,* Springer-Verlag, Berlin, 1984.

94. **Nover, L.,** Expression of heat shock genes in homologous and heterologous systems, *Enz. Microb. Technol.*, 9, 130—144, 1987.

94a. **Nover, L., Neumann, D., and Scharf, K. D., Eds.,** *Heat Shock and Other Stress Response Systems of Plants,* Springer-Verlag, Berlin, 1990.

95. **O'Conner, D. and Lis, J. T.,** Two closely linked transcription units within the 63B heat shock puff locus of *D. melanogaster* display strikingly different regulation, *Nucl. Acids Res.*, 9, 5075—5092, 1981.

96. **Özkaynak, E., Finley, D., Solomon, M. J., and Varshavsky, A.,** The yeast ubiquitin genes: A family of natural gene fusions, *EMBO J.*, 6, 1429—1439, 1987.

97. **O'Malley, K., Mauron, A., Barchas, J. D., and Kedes, L.,** Constitutively expressed rat mRNA encoding a 70-kilodalton heat-shock-like protein, *Mol. Cell. Biol.*, 5, 3476—3483, 1985.

98. **Palter, K. B., Watanabe, M., Stinson, L., Mahowald, A. P., and Craig, E. A.,** Expression and localization of *Drosophila melanogaster* hsp70 cognate proteins, *Mol. Cell. Biol.*, 6, 1187—1203, 1986.

98a. **Pardue, M. L., Feramisco, J. R., and Lindquist, S., Eds.,** *Stress-Induced Proteins,* Alan R. Liss, New York, 1989.

98b. **Pascual, L. and de Frutos, R.,** Stress response in *Drosophila subobscura* IV. Differential gene activity induced by heat shock, *Chromosoma,* 97, 164—170, 1988.

99. **Pauli, D. and Tonka, C.-H.,** A new *Drosophila* heat shock gene from locus 67B is expressed during embryogenesis and pupation, *J. Mol. Biol.*, 198, 233—240, 1987.

100. **Pauli, D., Tonka, C.-H., and Ayme-Southgate, A.,** An unusual split *Drosophila* heat shock gene expressed during embryogenesis, pupation and in testis, *J. Mol. Biol.*, 200, 47—53, 1988.

101. **Peters, F. P. A. M. N., Lubsen, N. H., and Sondermeijer, P. J. A.,** Rapid sequence divergence in a heat shock locus of *Drosophila, Chromosoma,* 81, 271—280, 1980.

102. **Peters, F. P. A. M. N., Lubsen, N. H., Walldorf, U., Moormann, R. J. M., and Hovemann, B.,** The unusual structure of heat-shock locus 2-48B in *Drosophila hydei, Mol. Gen. Genet.*, 197, 392—398, 1984.

103. **Petersen, N. S., Moller, G., and Mitchell, H. K.,** Genetic mapping of the coding regions for three heat-shock proteins in *Drosophila melanogaster, Genetics,* 92, 891—902, 1979.
104. **Petko, L. and Lindquist, S.,** Hsp 26 is not required for growth at high temperatures, nor for thermotolerance, spore development, or germination, *Cell,* 45, 885—894, 1986.
105. **Pierce, D. A. and Lucchesi, L. C.,** Dosage compensation of X-linked heat shock puffs in *D. pseudoobscura, Chromosoma,* 76, 245—254, 1980.
105a. **Plessofsky-Vig, N. and Brambl, R.,** in preparation.
106. **Raschke, E., Baumann, G., and Schöffl, F.,** Nucleotide sequence analysis of soybean small heat shock protein genes belonging to two different multigene families, *J. Mol. Biol.,* 199, 549—557, 1988.
107. **Reading, D. S., Hallberg, R., and Myers, A. M.,** Characterization of the yeast HSP60 gene coding for a mitochondrial assembly factor, *Nature,* 337, 655—659, 1989.
108. **Rebbe, N. F., Ware, J., Bectina, R. M., Modrich, P., and Stafford, D. W.,** Nucleotide sequence of a cDNA for a member of the human 90kDa heat-shock protein family, *Gene,* 53, 235—245, 1987.
108a. **Rebbe, N. F., Hickman, W. S., Ley, T. J., Stafford, D. W., and Hickman, S.,** Nucleotide sequence and regulation of a human 90-kDa heat shock protein gene, *J. Biol. Chem.,* 264, 15006—15011, 1989.
109. **Requena, J. M., Lopez, M. C., Jimenez-Ruez, A., De La Torve, J. C., and Alonso, C.,** A head-to-tail tandem organization of hsp70 in *Trypanozoma cruzi, Nucl. Acids Res.,* 16, 1393—1406, 1988.
110. **Ritossa, F.,** A new puffing pattern induced by heat shock and DNP in *Drosophila, Experientia,* 18, 571—573, 1962.
111. **Rochester, D. E., Winter, J. A., and Shah, D. M.,** The structure and expression of maize genes encoding the major heat shock protein, hsp70, *EMBO J.,* 5, 451—458, 1986.
111a. **Rose, M. D., Misra, L. M., Vogel, J. P.,** KAR2, a karyogamy gene, is the yeast homolog of the mammalian BiP/GRP78 genes, *Cell,* 57, 1211—1221, 1989.
112. **Russnak, R. H. and Candido, E. P. M.,** Locus encoding a family of small heat shock genes in *Caenorhabditis elegans:* Two genes duplicated to form a 3.8-kilobase inverted repeat, *Mol. Cell. Biol.,* 5, 1268—1278, 1985.
113. **Russnak, R. H., Jones, D., and Candido, E. P. M.,** Cloning and analysis of cDNA sequences coding for two 16-kilodalton heat shock proteins (hsps) in *Caenorhabditis elegans:* Homology with the small hsps of *Drosophila, Nucl. Acids Res.,* 11, 3187—3205, 1983.
114. **Ryseck, R.-P., Walldorf, U., and Hovemann, B.,** Two major RNA products are transcribed from heat-shock locus 93D of *Drosophila melanogaster, Chromosoma,* 93, 17—20, 1985.
115. **Ryseck, R.-P., Walldorf, U., Hoffman, T., and Hovemann, B.,** Heat shock loci 93D of *Drosophila melanogaster* and 48B of *Drosophila hydei* exhibit a common structural and transcriptional pattern, *Nucl. Acids Res.,* 15, 3317—3333, 1987.
115a. **Sargent, C. A., Dunham, I., Trowsdale, J., and Campbell, R. D.,** Human major histocompatibility complex contains genes for the major heat shock protein HSP70, *Proc. Natl. Acad. Sci. U.S.A.,* 86, 1968—1972, 1989.
116. **Schedl, P., Artavanis-Tsakonas, S., Steward, R., Gehring, W. J., Mirault, M.-E., Goldschmidt-Clermont, M., Moran, L., and Tissieres, A.,** Two hybrid plasmids with *D. melanogaster* DNA sequences complementary to mRNA coding for the major heat shock protein, *Cell,* 14, 921—929, 1978.
117. **Schöffl, F. and Baumann, G.,** Thermo-induced transcripts of a soybean heat shock gene after transfer into sunflower using a Ti plasmid vector, *EMBO J.,* 4, 1119—1124, 1985.
118. **Schöffl, F. and Key, J. L.,** An analysis of mRNAs for a group of heat shock proteins of soybean using cloned cDNAs, *J. Mol. Appl. Genet.,* 1, 301—314, 1982.
118a. **Schöffl, F. and Key, J. L.,** Identification of a multigene family for small heat shock proteins in soybean and physical characterization of one individual gene coding region, *Plant Mol. Biol.,* 2, 269—278, 1983.
119. **Schöffl, F., Raschke, E., and Nagao, R. T.,** The DNA sequence analysis of soybean heat shock genes and identification of possible regulatory promoter elements, *EMBO J.,* 3, 2491—2497, 1984.
120. **Schöffl, F., Lin, C.-Y., Key, J. L.,** Soybean heat shock proteins: Temperature regulated gene expression and the development of thermotolerance, in *Genetic Manipulation of Plants and its Application to Agriculture,* Stewart, G. R. and Lea, P. J., Eds., Oxford University Press, 1984, 129—140.
121. **Schöffl, F., Baumann, G., Raschke, E., and Bevan, M.,** The expression of heat-shock genes in higher plants, *Philos. Trans. R. Soc. London Ser. B,* 314, 453—468, 1986.
122. **Schöffl, F., Baumann, G., and Raschke, E.,** The expression heat shock genes — a model for environmental stress response, in *Plant Gene Research,* Vol. 5, Goldberg, B. and Verma, D. P. S., Eds., Springer Verlag, Vienna, 1988, 253—273.
122a. **Searle, S., Campos, A. J. R., Coulson, R. M. R., Spithill, T. W., and Smith, D. F.,** A family of heat shock protein 70-related genes are expressed in the promastigotes of *Leishmania major, Nucl. Acids Res.,* 17, 5081—5095, 1989.
123. **Shah, D. M., Rochester, D. E., Krivi, G. G., Hironaka, C. M., Mozer, T. J., Fraley, R. T., and Tiemeier, D. C.,** Structure and expression of maize hsp70 gene, in *Cellular and Molecular Biology of Plant Stress,* Key, J. L. and Kosuge, T., Eds., Alan R. Liss, New York, 1985, 181—200.

123a. **Shapira, M. and Pinelli, E.**, Heat-shock protein 83 of *Leishmania mexicana amazonensis* is an abundant cytoplasmatic protein with a tandemly repeated genomic arrangement, *Eur. J. Biochem.*, 185, 231—236, 1989.

124. **Shibahara, S., Müller, R. M., and Taguchi, H.**, Transcriptional control of rat heme oxygenase by heat shock, *J. Biol. Chem.*, 262, 12889—12892, 1987.

125. **Simcox, A. A., Cheney, C. M., Hoffman, E. P., and Shearn, A.**, A deletion of the 3′ end of the *Drosophila melanogaster* hsp70 gene increases stability of mutant mRNA during recovery from heat shock, *Mol. Cell. Biol.*, 5, 3397—3402, 1985.

125a. **Simon, M. C., Kitchener, K., Kao, H.-T., Hickey, E., Weber, L., Voellmy, R., Heintz, N., and Nevins, J. R.**, Selective induction of human heat shock gene transcription by the adenovirus E1A gene products, including the 12S E1A product, *Mol. Cell. Biol.*, 7, 2884—2890, 1987.

126. **Slater, M. R. and Craig, E. A.**, Transcriptional regulation of an hsp70 heat shock gene in the yeast *Saccharomyces cerevisiae, Mol. Cell. Biol.*, 7, 1906—1916, 1987.

126a. **Slater, M. R. and Craig, E. A.**, The SSA1 and SSA2 genes of yeast *Saccharomyces cerevisiae, Nucl. Acids Res.*, 17, 805—806, 1989.

126b. **Slater, M. R. and Craig, E. A.**, The SSB1 heat shock cognate gene of the yeast *Saccharomyces cerevisiae, Nucl. Acids Res.*, 17, 4891, 1989.

127. **Snutch, T. P. and Baillie, D. L.**, A high degree of DNA strain polymorphism associated with the major shock gene in *Caenorhabditis elegans, Mol. Gen. Genet.*, 195, 329—335, 1984.

127a. **Snutch, T. P., Heschl, M. F. P., and Baillie, D. L.**, The *Caenorhabditis elegans* hsp70 gene family: A molecular genetic characterization, *Gene*, 64, 241—255, 1988.

128. **Sorger, P. K. and Pelham, H. R. B.**, Cloning and expression of a gene becoding hsc 73, the major hsp 70-like protein in unstressed rat cells, *EMBO J.*, 6, 993—998, 1987.

129. **Southgate, R., Ayme, A., and Voellmy, R.**, Nucleotide sequence analysis of the *Drosophila* small heat shock gene cluster at locus 67B, *J. Mol. Biol.*, 165, 35—57, 1983.

130. **Spradling, A., Penman, S., and Pardue, M. L.**, Analysis of *Drosophila* mRNA by in situ hybridization: Sequences transcribed in normal and heat shocked cultured cells, *Cell*, 4, 395—404, 1975.

131. **Spradling, A., Pardue, M. L., and Penman, S.**, Messenger RNA in heat-shocked *Drosophila* cells, *J. Mol. Biol.*, 109, 559—587, 1977.

132. **Susek, R. E. and Lindquist, S. L.**, Hsp26 of *Saccharomyces cerevisiae* is homologous to the superfamily of small heat shock proteins, but is without a demonstrable function, *Mol. Cell. Biol.*, 9, 5265—5271, 1989.

132a. **Swindle, J., Ajioka, J., Eisen, H., Sanwal, B., Jacquemot, C., Browder, Z., and Buck, G.**, The genomic organization and transcription of the ubiquitin genes of *Trypanosoma cruzi, EMBO J.*, 7, 1121—1127, 1988.

132b. **Takahashi, T. and Komeda, Y.**, Characterization of two genes encoding small heat-shock proteins in *Arabidopsis thaliana, Mol. Gen. Genet.*, 219, 365—372, 1989.

132c. **Tanaka, K., Matsumoto, K., and Toh-e, A.**, Dual regulation of the expression of the polyubiquitin gene by cyclic AMP and heat shock in yeast, *EMBO J.*, 7, 495—502, 1988.

133. **Tissieres, A., Mitchell, H. K., and Tracy, U. M.**, Protein synthesis in salivary glands of *D. melanogaster*. Relation to chromosome puffs, *J. Mol. Biol.*, 84, 389—398, 1974.

134. **Török, I. and Karch, R.**, Nucleotide sequences of heat shock activated genes in *Drosophila melanogaster*. I. Sequences in the regions of the 5′ and 3′ ends of the hsp70 gene in the hybrid plasmid 56H8, *Nucl. Acids Res.*, 8, 3105—3123, 1980.

135. **Udvardy, A., Sumegi, J., Toth, E. C., Gausz, J., Gyurkovics, H., Schedl, P., and Ish-Horowicz, D.**, Genomic organization and functional analysis of a deletion variant of the 87A7 heat-shock locus of *Drosophila melanogaster, J. Mol. Biol.*, 155, 267—280, 1982.

136. **Vierling, E., Nagao, R. T., De Rocher, A. E., and Harris, L. M.**, A heat shock protein localized to chloroplasts is a member of a eukaryotic superfamily of heat shock proteins, *EMBO J.*, 7, 575—582, 1988.

136a. **Vierstra, R. D., Burke, T. J., Callis, J., Hatfield, P. M., Jabben, M., Shanklin, J., and Sullivan, M. L.**, Characterization of the ubiquitin-dependent proteolytic pathway in higher plants in, *The Ubiquitin System*, Schlesinger, M. and Hershko, A., Eds., Cold Spring Harbor Press, Cold Spring Harbor, NY, 1988, 119—125.

137. **Voellmy, R., Goldschmidt-Clermont, M., Southgate, R., Tissieres, A., Levis, R., and Gehring, W.**, A DNA segment isolated from chromosomal site 67B in *D. melanogaster* contains four closely linked heat-shock genes, *Cell*, 23, 261—270, 1981.

137a. **Vourc'h, C., Binart, N., Chambraud, B., David, J. P., Jerome, V., Baulieu, E. E., and Catelli, M. G.**, Isolation and functional analysis of chicken 90-kDa heat shock protein gene promotor, *Nucl. Acids Res.*, 17, 5259—5272, 1989.

138. **Wadsworth, S. C., Craig, E. A., and McCarthy, B. J.**, Genes for three *Drosophila* heat-shock-induced proteins at a single locus, *Proc. Natl. Acad. Sci. U.S.A.*, 77, 2134—2137, 1980.

139. **Walldorf, U., Richter, S., Ryseck, R.-P., Steller, H., Edstroem, J. E., Bautz, E. K. F., and Hovemann, B.,** Cloning of heat-shock locus 93D from *Drosophila melanogaster, EMBO J.,* 3, 2499—2504, 1984.

139a. **Wang, J., Holden, D. W., and Leong, S. A.,** Gene transfer system for the phytopathogenic fungus *Ustilago maydis, Proc. Natl. Acad. Sci. U.S.A.,* 85, 865-869, 1988.

140. **Werner-Washburne, M., Stone, D. E., and Craig, E. A.,** Complex interactions among members of an essential subfamily of hsp70 genes in *Saccharomyces cerevisiae, Mol. Cell. Biol.,* 7, 2568—2577, 1987.

141. **Winter, J., Wright, R., Duck, N., Gasser, C., Fraley, R., and Shah, D.,** The inhibition of *Petunia* hsp 70 mRNA processing during $CdCl_2$ stress, *Mol. Gen. Genet.,* 211, 315—319, 1988.

141a. **Wisniewski, J., Kordula T., and Krawczyk, Z.,** Isolation and nucleotide sequence analysis of the rat testis-specific major heat-shock protein (HSP70)-related gene, *Biochim. Biophys. Acta,* 1048, 93—99, 1990.

142. **Wu, L., Morley, B. J., and Campbell, R. D.,** Cell-specific expression of the human complement protein factor B gene: Evidence for the role of two distinct 5′-flanking elements, *Cell,* 48, 331—342, 1987.

142a. **Wu, C. H., Caspar, T., Browse, J., Lindquist, S., and Somerville, C.,** Characterization of an HSP70 cognate gene family in *Arabidopsis, Plant Physiol.,* 88, 731—740, 1988.

143. **Wurst, W., Benesch, C., Drabent, B., Rothermel, E., Benecke, B. J., and Gunther, E.,** Localization of heat shock protein 70 genes inside the rate major histocompatibility complex close to class III genes, *Immunogenetics,* 30, 46—49, 1989.

144. **Yamazaki, M., Akaogi, K., Miwa, T., Imai, T., Soeda, E., and Yokoyama, K.,** Nucleotide sequence of a full-length cDNA for 90 kDA heat-shock protein from human peripheral blood lymphocytes, *Nucl. Acids Res.,* 17, 7108, 1989.

145. **Zakeri, Z. F., Wolgemuth, D. J., and Hunt, C. R.** Identification and sequence analysis of a new member of the mouse HSP70 gene family and characterization of its unique cellular and developmental pattern of expression in the male germ line, *Mol. Cell. Biol.,* 8, 2925—2932, 1988.

Chapter 4

THE GLUCOSE-REGULATED PROTEINS (GRP)

L. Nover

TABLE OF CONTENTS

4.1. INDUCTION AND INTRACELLULAR LOCALIZATION

In vertebrate cells, two abundant constitutive proteins with M_r 78,000 and 94,000 are minor heat shock proteins (HSPs), whose synthesis in some cases is moderately stimulated by relatively mild heat shock (hs). However, under conditions of glucose deprivation their synthesis is greatly enhanced. Hence, they were called glucose-regulated proteins (GRPs). Properties of GRPs, regulation of synthesis, and characterization of their coding genes are summarized by Lee.[25]

Both proteins were originally described as retrovirus-induced, but it was shown subsequently that retroviral infection leads to acute glucose deficiency in the host cell.[38] Meanwhile, the list of GRP inducers increased (Table 4.1) including agents interfering with glucose utilization (insulin, 2-deoxyglucose, glucosamine, tunicamycin, and long-term anoxia) or causing Ca^{2+}-deprivation (ionophores A23187 and ionomycin, EGTA, and low pH), but some of them are typical inducers of HSPs (hs, canavanine, and azetidine carboxylic acid see also Table 1.1). In contrast to the latter group of GRP inducers which can induce both HSP and GRP synthesis in some organisms, the former two do not stimulate synthesis of the two major HSPs. Vice versa, there are experimental conditions characterized by increased HSPs 70 and 90 synthesis with little or no change in GRP synthesis, e.g., high extracellular pH. Finally, there are conditions where HSPs are induced while GRPs are simultaneously deinduced (recovery from anoxia, refeeding cells with glucose or 1 m*M* uridine after glucose deprivation). These interesting details of the interaction between the two metabolic stress systems are illustrated in Figure 4.1 (see also Table 4.1 and Kim and Lee,[15] Sciandra and Subjeck,[42] Whelan and Hightower,[51] and Watowich and Morimoto[48a]).

In search for a common target reaction in the signal transformation chain leading to the activation of GRP-coding genes, aberrant proteins in the ER, such as those lacking appropriate glycosylation, may be of prime importance.[4,16b,25] Evidently, it is the presence of malfolded proteins in the ER themselves rather than an aberrant glycosylation which triggers GRP induction. This was demonstrated by accumulation of mutant influenze virus heamaglutinin.[16b] The putative role of abnormal proteins reminds of the signal transduction chain discussed for HSP induction (see Section 1.4). Similar to some chemical stressors included in Table 1.1 GRP inducers act slowly, with lag phases of 3 to more than 24 h. They need ongoing protein synthesis, i.e., their effect is abolished by cycloheximide.

A remarkable finding is the reversion of glucose starvation effect in mouse L929 cells by addition of millimolar amounts of uridine or other ribonucleosides.[8a,22] Induction of GRPs is abolished at the same time as the repression of HSP synthesis and the cytotoxic effects of glucose starvation are released. It is worth noticing that GRP synthesis induced by tunicamycin, glucosamine, EGTA, A23187, or hypoxia cannot be reversed by uridine. Evidently, glucose deprivation acts via the ribonucleoside pool needed for generation of UDP-Gls and similar intermediates of glycosylation reactions, whereas other inducers act on later steps in the single transformation chain. For the calcium ionophore A23187, there is evidence that perturbations of the intracellular Ca^{2+} storage compartments rather than variations of the cytoplasmic Ca^{2+} content are responsible for the inducing effect.[7b]

Another aspect of GRP induction was pointed out by Whelan and Hightower.[51] The characterization of low pH, long term anaerobiosis and mercaptoethanol as inducers led to speculations about the role of reduced disulfide bond formation. The enzyme involved is protein disulfide isomerase, a 58-kDa highly acidic glyoprotein of the ER. Interestingly, a minor GRP58 (pI 5.0) is occasionally observed together with the two major GRPs from chick cells. This protein is probably different from a mere basic protein of similar apparent size identified as a GRP in hamster cells.[24] At any rate, the implication of protein glycosylation in GRP induction is supported by two ts mutant cell lines of mouse[38] (fibroblast line AD6) and of hamster[4,27,28] (fibroblast line K 12), which, under nonpermissive conditions, exhibit increased GRP synthesis together with defective glycoprotein synthesis.

TABLE 4.1
Inducers of Glucose-Regulated Proteins in Vertebrates

	Ref.
Heat shock	9, 28, 49
Canavanine	6, 9—11
Azetidine carboxylic acid	28, 48—50
Glucose deprivation	13, 19, 28, 30, 33, 45, 47, 49
Insulin (5 μg/ml)	14
2-Deoxyglucose (1—10 m*M*)	6, 38
Glucosamine (1—10 n*M*)	24, 33, 38, 47
Tunicamycin (50 n*M*)	33, 35, 47
A23187 (7 μ*M*)	7, 33, 47, 49
Ionomycin (4 μ*M*)	52
Long-term anoxia	43
β-Mercaptoethanol	51
pH 5.8, extracellular	51
EGTA (1 m*M*, Ca^{2+}-deprivation)	18
Lead glutamate (1 m*M*)	44a
>40°C in the ts mutant hamster cell line K 12	10
Infection with lytic and transforming RNA viruses	
Newcastle disease v.	6
Rous sarcoma v.	12, 38, 45
Sendai v.	37
Sindbis v.	8
Vesicular stomatitis v.	8
Hypertonic shock (5*M* NaCl)	47a

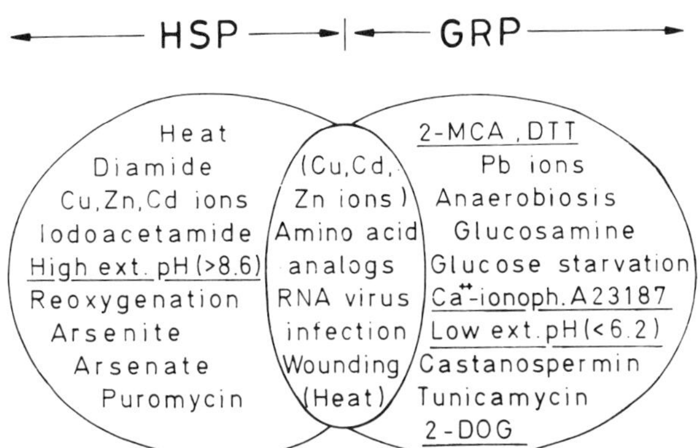

FIGURE 4.1. Three sets of inducers/conditions controlling heat shock protein (HSP) and glucose regulated protein (GRP) synthesis, respectively, in vertebrate cells. The left part shows inducers of HSP; the right part shows inducers of GRP synthesis, whereas in the central part inducers/conditions are listed which stimulate synthesis of both groups of proteins. For details and references see Tables 1.1 and 4.1. Heat shock (hs) and heavy metals induce GRP synthesis only in mammalian cells but not in chicken. Compounds/conditions underlined inhibit synthesis of the other set of proteins. 2-Deoxyglucose (2-DOG) and A23187 decrease basal level of HSP70 but do not affect its hs-inducibility.[48a] MCA, mercaptoethanol; DTT, dithiothreitol. Main difference between both classes may reside in the fact that inducers of HSP synthesis generate aberrant proteins in the cytoplasm, whereas GRP inducers act more specifically on proteins in the ER. (Scheme of Hightower with modifications from Watowich, S. S. and Morimoto, R. I., *Mol. Cell. Biol.*, 8, 393, 1988. With permission).

Despite some structural relations (see below) GRPs are clearly distinguished from the major HSPs. The discrimination is based on selective induction and on electrophoretic separation as shown in Figure 4.2. GRPs 78 and 94 were originally described as membrane-

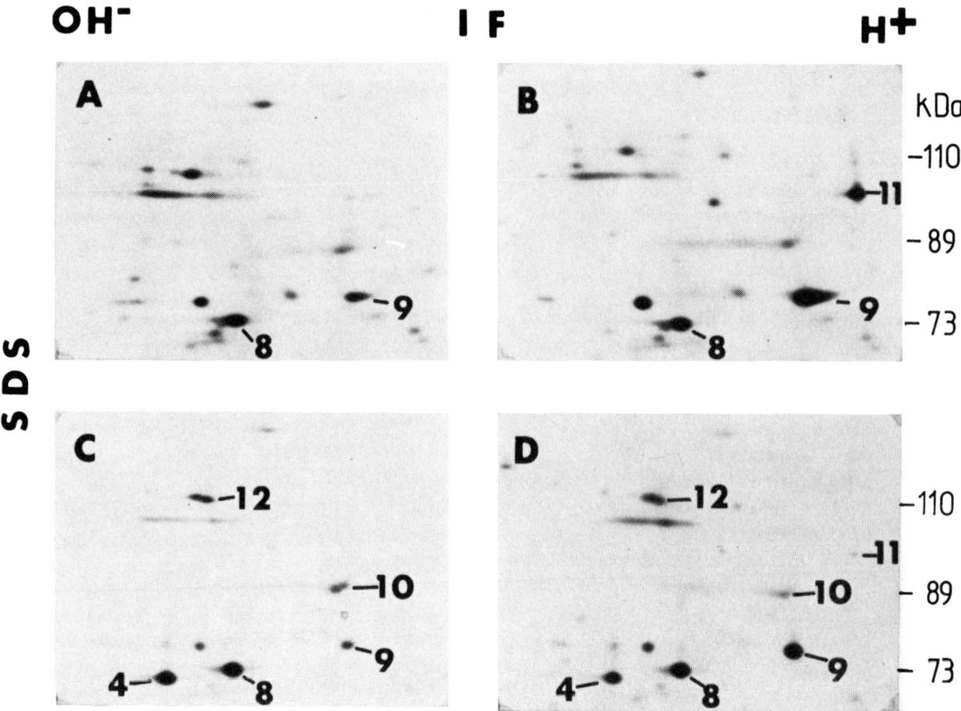

FIGURE 4.2. Survey of heat shock proteins (HSPs) and glucose-regulated proteins (GRPs) in rat embryo fibroblasts. Proteins of rat embryo fibroblasts were labeled for 30 min with ^{35}S-methionine and separated by two-dimensional electrophoresis with isoelectric focusing in the first (IF) and SDS-PAGE (SDS) in the second dimension. Figures represent parts of the autofluorographs (65 to 200-kDa region). A, control cells (37°C); B, glucose-starved cells; C, cells heat shocked for 10 min at 45°C and recovered for 2 h at 37°C; D, cells induced for 3 h with 0.6 m*M* canavanine. The numbered proteins are 4: HSP71; 8: HSC73; 10: HSP89; 12: HSP110; 9: GRP78; 11: GRP94. (From Whelan, S. A. and Hightower, L. E., unpublished.)

bound proteins.[35,45] This led to the idea that they may be the hexose transport proteins or that GRP94 may be the Ca^{2+} transport ATPase of the sarcoplasmic reticulum. Both suggestions are evidently not valid.[28,52,53] GRPs 78 and 94 are acidic proteins with pI 5.0 to 5.3 (Figure 4.1). The larger one is a glycoprotein[6,45,49] whereas GRP78 is not glycosylated, but modified by phosphorylation and ADP-ribosylation, respectively.[3,9a,28,49] Both are mainly localized in the ER/Golgi system. This was shown by cell fractionation as well as by immune staining.[33,34,47,49] In addition, some of the GRP94 may be associated with plasma membrane,[28] whereas under hs or other stress conditions, GRPs 78 and 94 are also found in nuclear fractions.[28,41,49] The structural features responsible for the ER/Golgi localization of these proteins include their synthesis as preproteins with an N-terminal leader sequence[4,28,33] (see Figures 2.8 and 2.10) and the presence of a conserved tetrapeptide -Lys-Asp-Glu-Leu (KDEL) at the C-terminus, shared with other luminal proteins of the ER.[34,36a,47]

The C-terminal -KDEL evidently functions as a general marker to prevent secretion of ER proteins. Interestingly, retention is dependent on Ca^{2+}. Depletion of the ER by addition of the Ca-ionophore A23 187 leads to a massive secretion of GRPs and other KDEL-labeled proteins, e.g., of protein disulfide isomerase. In the later stage of the A23 187 action increased synthesis restores the normal level of ER proteins. This two-phasic response with an initial loss and then restoration is not observed with tunicamycin as inducer of GRP synthesis.[2a]

4.2. GENE STRUCTURE

The overproduction of GRPs 78 and 94 in the ts mutant hamster fibroblast line K 12 facilitated the generation of corresponding cDNA clones.[26,27] It was demonstrated that GRP induction in all cases coincides with a proportional increase of the mRNA levels.[6,27,31,39] Meanwhile partial and complete cDNA clones for GRP94[17,47] and GRP78[26,33] as well as a partial genomic clone for the rat GRP78[4] were sequenced. GRPs are presumably coded by single copy genes.[27]

The sequence of the 2383 bp complete grp78 cDNA[33] predicts a protein of 72,300 kDa with an 18-aa residue N-terminal hydrophobic leader sequence and the above-mentioned C-terminal KDEL tetrapeptide. A remarkable homology with the *Drosophila* HSP70 is documented in Figure 2.10. A partial genomic sequence[4] comprises the promoter, and the first exon with 206 bp of the untranslated leader and 122 bp coding sequence and a short part of the first intron.

Sequencing of a chick "hsp108"[16,17] gave clear evidence that this is in fact the chick grp94 gene. The open reading frame codes for a 91,555-kDa unglycosylated protein with homology to the HSP90 family (Figure 2.8). An N-terminal 21 amino acid residue hydrophobic leader and the ER-specific C-terminus are also present. The genomic clone[16] shows a highly complex single copy gene of 9.9 kb with 18 exons and 17 introns coding for a 2733-bp mRNA (Figure 3.1). In hormone-responsive chick oviduct cells expression of this gene is stimulated by estrogen or progesterone.[1] The homology of chicken GRP94 alias HSP108, and the hamster GRP94 is evident not only from the described structural elements, but also from comparison with the 411 amino acid residues of the C-terminal part of the hamster GRP94.[47] The C-terminal parts of both are highly enriched in acidic amino acid residues, i.e., 23 Glu/Asp residues out of 41 for chick and 31 Glu/Asp residues out of 50 for hamster.[17,28,33,47]

Characterization and functional testing of the grp78 promoter[4,31,40] revealed a highly active enhancer-like element (bp -85 to -480) to be responsible for the activation by the Ca^{2+}-ionophore A23187, by glucose starvation but also under nonpermissive conditions in the hamster fibroblast cell line K 12.[4] Transformation of grp78 promoter × reporter gene fusion constructs into hamster, mouse, human, or rat cell lines gave high inducible activities in all cases. The promoter is characterized by a relatively distant TATA element at bp -54 to -61 and four upstream CCAAT elements within 300 bp of 5′ flanking sequences. Most important is the promoter region between bp -208 to -130.[40] But no hs box was identified in more than 700 bp of 5′ flanking sequences. Thus, the moderate induction of GRPs 78 and 95 under hs conditions[9,28,49] is presumably a slow, secondary effect of hs, e.g., caused by perturbations of glucose utilization, glycoprotein synthesis and/or energy metabolism.[13,19-23] Alternatively, a hs-increased translation of preformed GRP mRNA must be considered.

Recently, genes coding for GRP-type ER proteins were also cloned and sequenced from yeast and the nematode *Caenorhabditis elegans*. In yeast the KAR2 gene codes for a protein of 682 amino acid residues (see Figure 2.10) localized in the perinuclear ring of ER membranes. Expression is strongly enhanced by tunicamycin, A23 187, 2-deoxyglucose, and hs. Mammalian GRP78 can substitute for KAR2 defects.[34a,41a] The nematode grp78 gene (hsp-3) contains three introns and is not hs-inducible. It codes for a protein of 661 amino acid residues (73,339 Da) including a leader sequence of 34 residues. Similar to the vertebrate GRPs it ends at the C-terminus with-KDEL.[9b] Most remarkable is the conservation of two promoter elements between the nematode and the rat grp genes. One of them is part of the essential region (bp -208 to -130) defined by Resendez et al.[40] for the rat gene.

4.3. FUNCTION OF GRPs

A possible function of GRPs is derived form several facts:

1. GRP94 shows sequence homology with the HSP90 family[32,47] (see Figure 2.8), whereas GRP78 is structurally related to the HSP70 family[4,33] (see Figure 2.10). Especially the N-terminal domain of GRP78 corresponds to the ATP-binding N-terminal domain common to many HSP70 proteins.[5,33,36]
2. The abundance of both proteins in the ER/Golgi system also under noninducing conditions is the reason for their identification under different names:[33,36] GRP94 is the HSP100 of Welch et al.,[49] and Koyashu et al.,[16b] the chicken HSP108 of Kulomaa et al.,[17] endoplasmin of Smith and Koch,[46] ERp99 of Lewis et al.[29a] and Mazzarella and Green[32] and presumably also the p105 detected in SV40-transformed mouse cells.[41] On the other hand, GRP78 was referred to as ADP-ribosylated SP83 by Carlsson and Lazarides,[3] as immunoglobulin heavy chain binding protein BiP by Bole et al.,[2] as p77 by Sharma et al.[44] and, because of the sequence-derived exact molecular size of 72,000, as p72 by Munro and Pelham.[33]
3. An essential part of the GRP78 molecule is the conservation of the typical ATP binding domain of the HSP70 family[5,14a,33] (Figure 2.10). This homology and the close relation between GRP94 and members of the HSP90 family led Pelham to propose that all exert basically similar functions in the cell, though in different compartments and under different conditions. The whole group of large HSPs and related proteins may help to assemble protein complexes, to stabilize proteins not yet assembled and to remove or reassemble aberrant protein complexes. For GRP78 this is best illustrated by its putative role for immunoglobulin assembly,[2,33] by its association with aberrant fusion proteins in the ER[44] and by its possible role to prevent cataract development in isolated rat eye lens.[7]

   In support of this, GRP78 is found to be associated with aberrant proteins in the ER.[7a,9a,14a] Interestingly, the retention of mutant tissue plasminogen activator protein in the ER of CHO cells is diminished if the level of GRP78 is decreased, e.g., by introduction of antisense mRNA. Complex formation evidently prevents processing and secretion of aberrant proteins.[7a]
4. Similar to other members of the mammalian HSP90 family, GRP94 binds to actin filaments, and this interaction is inhibited by Ca^{2+}-calmodulin.[16a] Since HSP90 proteins have a putative membrane-spanning domain, it is conceivable that part of the GRP94 is exposed to the cytoplasm.[28]

REFERENCES

1. **Baez, M., Sargan, D. R., Elbrecht, A., Kulomaa, M. S., Zarucki-Schulz, T., Tsai, M.-J., and O'Malley, B. W.,** Steroid hormone regulation of the gene encoding the chicken heat shock protein Hsp108, *J. Biol. Chem.*, 262, 6582—6588, 1987.
2. **Bole, D. G., Hendershot, L. M., and Kearny, J. F.,** Posttranslational association of immunoglobulin heavy chain binding protein with nascent heavy chains in non-secreting and secreting hybridomes, *J. Cell Biol.*, 102, 1558—1566, 1986.

2a. **Booth, C. and Koch, G. L. E.,** Perturbation of cellular calcium induces secretion of luminal ER proteins, *Cell*, 59, 729—737, 1989.

3. **Carlsson, L. and Lazarides, E.,** ADP-ribosylation of the M_r 83,000 stress-inducible and glucose-regulated protein in avian and mammalian cells: modulation by heat shock and glucose starvation, *Proc. Natl. Acad. Sci. U.S.A.*, 80, 4664—4668, 1983.

4. **Chang, S. C., Wooden, S. K., Nakaki, T., Kim, Y. K., Lin, A. Y., Kung, L., Attenello, J. W., and Lee, A. S.,** Rat gene encoding the 78-kDa glucose-regulated protein GRP78: Its regulatory sequences and the effect of protein glycosylation, *Proc. Natl. Acad. Sci. U.S.A.*, 84, 680—684, 1987.
5. **Chappell, T. G., Welch, W. J., Schlossman, D. M., Palter, K. B., Schlesinger, M. J., and Rothman, J. E.,** Uncoating ATPase is a member of the 70 kilodalton family of stress proteins, *Cell,* 45, 3—13, 1986.
6. **Collins, P. L. and Hightower, L. E.,** Newcastle disease virus stimulates the cellular accumulation of stress (heat shock) mRNAs and proteins, *J. Virol.*, 44, 703—707, 1982.
7. **De Jong, W. W., Hoekman, W. A., Mulders, J. W. M., and Bloemendal, H.,** Heat shock response of the rat lens, *J. Cell Biol.*, 102, 104—111, 1986.

7a. **Dorner, A. J., Krane, M. G., and Kaufman, R. J.,** Reduction of endogenous GRP78 levels improves secretion of a heterologous protein in CHO cells, *Mol. Cell. Biol.*, 8, 4063—4070, 1988.

7b. **Drummond, J. A. S., Lee, A. S., Resendez, E., and Steinhardt, C. R. A.,** Deletion of intracellar calcium stress by calcium ionophore A 23187 induces the genes for glucose-regulated proteins in hamster fibroblasts, *J. Biol. Chem.*, 262, 12801—12805, 1987.

8. **Garry, R. F., Ulug, E. T., and Bose, H. R.,** Induction of stress proteins in Sindbis virus- and vesicular stomatitis virus-infected cells, *Virology,* 129, 319—332, 1983.

8a. **Gstraunthaler, J., Harris, H. W., and Handler, J. S.,** Precursors of ribose 5-phosphate supress expression of glucose-regulated proteins in LLC-PK1 cells, *J. Physiol.*, 252, 239—243, 1978.

9. **Hatayama, T., Honda, K., and Yukioka, M.,** HeLa cells synthesize a specific heat shock protein upon exposure to heat shock at 42°C but not at 45°C, *Biochem. Biophys. Res. Commun.*, 137, 957—963, 1986.

9a. **Hendershot, L. M., Ting, J., and Lee A. S.,** Identity of the immunoglobulin heavy-chain-binding protein with the 78,000-dalton glucose-regulated protein and the role of posttranslational modifications in its binding function, *Mol. Cell. Biol.*, 8, 4250—4256, 1988.

9b. **Heschl, M. F. P. and Baillie, D. L.,** Characterization of the hsp70 multigene family of *Caenorhabditis elegans, DNA,* 8, 233—243, 1989.

10. **Hightower, L. E. and White, F. P.,** Cellular response to stress: Comparison of a family of 71—73 kilodalton proteins rapidly synthesized in rat tissue slices, *J. Cell. Physiol.*, 108, 261—275, 1981.
11. **Hightower, L. E. and White, F. P.,** Preferential synthesis of rat heat-shock and glucose-regulated proteins in stressed cardiovascular cells, in *Heat Shock: from Bacteria to Man,* Schlesinger, M. J., Ashburner, M., and Tissieres, A., Eds., Cold Spring Harbor Laboratory, Cold Spring Harbor, NY, 1982, 369—377.
12. **Isaka, T., Yoshida, M., Owada, M., and Toyoshima, K.,** Alterations in membrane polypeptides of chick embryo fibroblasts induced by transformation with avian sarcoma viruses, *Virology,* 65, 226—237, 1975.
13. **Kasambalides, E. J. and Lanks, K. W.,** Dexamethasone can modulate glucose-regulated and heat shock protein synthesis, *J. Cell. Physiol.*, 114, 93—98, 1983.
14. **Kasambalides, E. J. and Lanks, K. W.,** Antagonistic effects of insulin and dexamethasone on glucose-regulated and heat shock protein synthesis, *J. Cell. Physiol.*, 123, 283—287, 1985.

14a. **Kassenbrock, C. K., Garcia, P. D., Walter, P., and Kelly, R. B.,** Heavy-chain binding protein recognizes aberrant polypeptides translocated in vitro, *Nature,* 333, 90—93, 1988.

15. **Kim, K. S. and Lee, A. S.,** The effect of extracellular Ca^{2+} and temperature on the induction of the heat-shock and glucose-regulated proteins in hamster fibroblasts, *Biochem. Biophys. Res. Commun.*, 140, 881-888, 1986.
16. **Kleinsek, D. A., Beattie, W. G., Tsai, M. J., and O'Malley, B. W.,** Molecular cloning of a steroid-regulated 108 K heat shock protein gene from hen oviduct, *Nucl. Acids Res.*, 14, 10053—10071, 1986.

16a. **Koyashu, S., Nishida, E., Miyata, Y., Sakai, H., and Yahara, I.,** Hsp-100, a 100-kDa heat shock protein, is a Ca2 + -calmodulin-regulated actin-binding protein, *J. Biol. Chem.*, 264, 15083—15087, 1989.

16b. **Kozutsumi, J., Segal, M., Normington, K., Gething, M.-J., and Sambrock, J.,** The presence of malfolded proteins in the endoplasmatic reticulum signals the induction of glucose-regulated proteins, *Nature,* 322, 462—464, 1988.

17. **Kulomaa, M. S., Weigel, N. L., Kleinsek, O. A., Beattie, W. G., Conneely, O. M., March, C., Zarucki-Schulz, T., Schrader, W. T., and O'Malley, B. W.,** Amino acid sequence of a chicken heat shock protein derived from the complementary DNA nucleotide sequence, *Biochem. J.*, 25, 6244—6252, 1986.
18. **Lamarche, S., Chretien, P., and Landry, J.,** Inhibition of the heat shock response and synthesis of glucose-regulated proteins in calcium-deprived rat hepatoma cells, *Biochem. Biophys. Res. Commun.*, 131, 868—876, 1985.
19. **Lanks, K. W.,** Metabolite regulation of heat shock protein levels, *Proc. Natl. Acad. Sci. U.S.A.*, 80, 5325—5329, 1983.
20. **Lanks, K. W.,** Modulators of the eukaryotic heat shock responses, *Exp. Cell Res.*, 165, 1—10, 1986.
21. **Lanks, K. W.,** Studies on the mechanism by which glutamine and heat shock increase lactate synthesis by L929 cells in the presence of insulin, *J. Cell. Physiol.*, 129, 385—389, 1986.

22. **Lanks, K. W., Hitti, I. F., and Chin, N. W.,** Substrate utilization for lactate and energy production by heat-shocked L929 cells, *J. Cell. Physiol.*, 127, 451—456, 1986.
23. **Lanks, K. W., Gao, J.-P., and Kasambalides, E. J.,** Nucleosides restore heat resistance and suppress glucose-regulated protein synthesis by glucose-deprived L929 cells, *Cancer Res.*, 48, 1442—1446, 1987.
24. **Lee, A. S.,** The accumulation of three specific proteins related to glucose-regulated proteins in a temperature-sensitive hamster mutant cell line K 12, *J. Cell. Physiol.*, 106, 119—125, 1981.
25. **Lee, A. S.,** Coordinated regulation of a set of genes by glucose and calcium ionophores in mammalian cells, *Trends Biochem. Sci.*, 12, 20—23, 1987.
26. **Lee, A. S., Delegeane, A., and Scharff, D.,** Highly conserved glucose-regulated protein in hamster and chicken cells: Preliminary characterization of its cDNA clone, *Proc. Natl. Acad. Sci. U.S.A.*, 78, 4922—4925, 1981.
27. **Lee, A. S. Delegeane, A. M., Baker, V., and Chow, P. C.,** Transcriptional regulation of two genes specifically induced by glucose starvation in a hamster mutant fibroblast cell line, *J. Biol. Chem.*, 258, 597—603, 1983.
28. **Lee, A. S., Bell, J., and Ting, J.,** Biochemical characterization of the 94- and 78-kilodalton glucose-regulated proteins in hamster fibroblasts, *J. Biol. Chem.*, 259, 4616—4621, 1984.
29. **Lewis, M. J. and Pelham, H. R. B.,** Involvement of ATP in the nuclear and nucleolar functions of the 70 kd heat shock protein, *EMBO J.*, 4, 3137—3143, 1985.
29a. **Lewis, M. J., Mazzarella, R. A., and Green, M.,** Structure and assembly of the endoplasmic reticulum. The synthesis of three major endoplasmic reticulum proteins during lipopolysaccharide-induced differentiation of murine lymphocytes, *J. Biol. Chem.*, 260, 3050-3057, 1985.
30. **Lin, A. Y. and Lee, A. S.,** Induction of two genes by glucose starvation in hamster fibroblasts, *Proc. Natl. Acad. Sci. U.S.A.*, 81, 988—992, 1984.
31. **Lin, A. Y., Chang, S. C., and Lee, A. S.,** A calcium ionophore-inducible cellular promoter is highly active and has enhancer like properties, *Mol. Cell. Biol.*, 6, 1235—1243, 1986.
32. **Mazzarella, R. A. and Green, M.,** ERp99, an abundant, conserved glycoprotein of the endoplasmic reticulum, is homologous to the 90-kDa heat shock protein (hsp 90) and the 94-kDa glucose regulated protein (GRP 94), *J. Biol. Chem.*, 262, 8875—8883, 1987.
33. **Munro, S. and Pelham, H. R. B.,** An hsp70-like protein in the ER: Identity with the 78 kd glucose-regulated protein and immunoglobulin heavy chain binding protein, *Cell*, 46, 291—300, 1986.
34. **Munro, S. and Pelham, H. R. B.,** A C-terminal signal prevents secretion of luminal ER proteins, *Cell*, 48, 899—907, 1987.
34a. **Normington, K., Kohno, K., Kozutsumi, Y., Gething, M.-J., and Sambrook, J.,** *Saccharomyces cerevisiae* encodes an essential protein homologous to mammalian BiP, *Cell*, 57, 1223—1236, 1989.
35. **Olden, K., Pratt, R. M., Jaworski, C., and Yamada, K. M.,** Evidence for role of glycoprotein in membrane transport: Specific inhibition by tunicamycin, *Proc. Natl. Acad. Sci. U.S.A.*, 76, 791—795, 1979.
36. **Pelham, H. R. B.,** Speculations on the functions of the major heat shock and glucose-regulated proteins, *Cell*, 46, 959—961, 1986.
36a. **Pelham, H. R. B.,** Heat shock and the sorting of luminal ER proteins, *EMBO J.*, 8, 3171—3176, 1989.
37. **Peluso, R. W., Lamb, R. A., and Choppin, P. W.,** Infection with paramyxoviruses stimulates synthesis of cellular polypeptides that are also stimulated in cells transformed by Rous sarcoma virus or deprived of glucose, *Proc. Natl. Acad. Sci. U.S.A.*, 75, 6120—6124, 1978.
38. **Pouyssegur, J., Shiu, R. P. C., and Pastan, I.,** Induction of two transformation-sensitive membrane polypeptides in normal fibroblasts by a block in glycoprotein synthesis or glucose deprivation, *Cell*, 11, 941—947, 1977.
39. **Resendez, E., Attenello, J. W., Grafsky, A., Chang, C. S., and Lee, A. S.,** Calcium ionophore A23187 induces expression of glucose-regulated genes and their heterologous fusion genes, *Mol. Cell. Biol.*, 5, 1212—1219, 1985.
40. **Resendez, E., Wooden, S. K., and Lee, A. S.,** Identification of highly conserved regulatory domains and protein-binding sites in the promoters of the rat and human genes encoding the stress-inducible 78-kilodalton glucose-regulated protein, *Mol. Cell. Biol.*, 8, 4579—4584, 1988.
41. **Rose, T. M. and Khandjian, E. W.,** A 105,000-dalton antigen of transformed mouse cells is a stress protein, *Can. J. Biochem. Cell Biol.*, 63, 1258—1264, 1985.
41a. **Rose, M. D., Misra, L. M., and Vogel, J. P.,** KAR2, a karyogamy gene, is the yeast homolog of the mammalian BiP/GRP78 genes, *Cell*, 57, 1211—1221, 1989.
42. **Sciandra, J. J. and Subjeck, J. R.,** Effects of glucose on protein synthesis and thermosensitivity in Chinese hamster ovary cells, *J. Biol. Chem.*, 258, 12091—12093, 1983.
43. **Sciandra, J. J., Subjeck, J. R., and Hughes, C. S.,** Induction of glucose-regulated proteins during anaerobic exposure and of heat-shock proteins after reoxygenation, *Proc. Natl. Acad. Sci. U.S.A.*, 81, 4843—4847, 1984.

44. **Sharma, S., Rodgers, L., Brandsma, J., Gething, M.-J., and Sambrook, J.,** SV40 T antigen and the exocytotic pathway, *EMBO J.*, 4, 1479—1489, 1985.

44a. **Shelton, K. R., Tood, Y. M., and Egle, P. M.,** The induction of stress-related proteins by lead, *J. Biol. Chem.*, 261, 1935—1940, 1986.

45. **Shiu, R. P. C., Pouyssegur, J., and Pastan, I.,** Glucose depletion accounts for the induction of two transformation-sensitive membrane proteins in Rous sarcoma virus-transformed chick embryo fibroblasts, *Proc. Natl. Acad. Sci. U.S.A.*, 74, 3840—3844, 1977.

46. **Smith, M. J. and Koch, G. L.,** Isolation and identification of partial cDNA clones for endoplasmin, the major glycoprotein of mammalian endoplasmic reticulum, *J. Mol. Biol.*, 194, 345—347, 1987.

47. **Sorger, P. K. and Pelham, H. R. B.,** The glucose-regulated protein grp94 is related to heat shock protein hsp90, *J. Mol. Biol.*, 194, 341—344, 1987.

47a. **Tanaka, K., Jay, G., and Isselbacher, K. J.,** Expression of heat-shock and glucose-regulated genes: differential effects of glucose starvation and hypertonicity, *Biochim. Biophys. Acta*, 950, 138—146, 1988.

48. **Thomas, G. P. and Mathews, M. B.,** Alterations of transcription and translation in HeLa cells exposed to amino acid analogs, *Mol. Cell. Biol.*, 4, 1063—1072, 1984.

48a. **Watowich, S. S. and Morimoto, R. I.,** Complex regulation of heat shock and glucose responsive genes in human cells, *Mol. Cell. Biol.*, 8, 393—405, 1988.

49. **Welch, W. J., Garrels, J. I., Thomas, G. P., Lin, J. J. C., and Feramisco, J. R.,** Biochemical characterization of the mammalian stress proteins and identification of two stress proteins as glucose-and calcium ionophore-regulated proteins, *J. Biol. Chem.*, 258, 7102—7111, 1983.

50. **Welch, W. J. and Suhan, J. P.,** Cellular and biochemical events in mammalian cells during and after recovery from physiological stress, *J. Cell Biol.*, 103, 2035—2052, 1986.

51. **Whelan, S. A. and Hightower, L. E.,** Differential induction of glucose-regulated and heat shock proteins: Effects of pH and sulfhydryl-reducing agents on chicken embryo cells, *J. Cell. Physiol.*, 125, 251—258, 1985.

51a. **Whelan, S. A. and Hightower, L. E.,** unpublished.

52. **Wu, F. S., Park, Y.-C., Roufa, D., and Martonosi, A.,** Selective stimulation of the synthesis of an 80,000-dalton protein by calcium ionophores, *J. Biol. Chem.*, 256, 5309—5312, 1981.

53. **Zala, C. A., Salas-Prato, M., Yan, W.-T., Banjo, B., and Perdue, J. F.,** In cultured chick embryo fibroblasts the hexose transport components are not the 75,000 and 95,000 dalton polypeptides synthesized following glucose deprivation, *Can. J. Biochem.*, 58, 1179—1188, 1980.

Chapter 5

RNAs AS HEAT SHOCK-INDUCED PRODUCTS

L. Nover

TABLE OF CONTENTS

5.1. RNA OF *DROSOPHILA* PUFFS 3-93D AND 2-48B

Among the major heat shock (hs)-induced puffs of *Drosophila* are puffs 3-93D of *Drosophila melanogaster* or 2-48B of *D. hydei* (for homologous puffs of other *Drosophila* species see Section 3.4). They are active under control conditions and induced about fivefold under hs conditions. Extensive analysis over many years revealed a number of peculiarities of this hsr ω locus discriminating them from other hs-induced loci.[1a,18,30]

1. Besides hs, they are induced specifically by inducers ineffective on normal hs loci.[1,4,5,16,19-21,34] These are benzamide (1 mg/ml), colchicine (100 μg/ml), vitamin B6 (1 m*M* to 1 μ*M*) or cold shock (10°C).
2. Giant ribonucleoprotein particles accumulate after hs induction. They are not observed in any other puff and contain special snRNP and hnRNP molecules. Remarkable is the selective association of a specific 10S snRNP with the giant 170-220S RNP material generated during hs in puff 93D.[1b,7,8,33a,37]
3. Much of the giant, hs-induced transcripts of puff 93D and its homologs in other *Drosophila* species remain nuclear-bound, though small cytoplasmic transcripts of 1.2 to 1.3 kb are also observed.[8a,10,11,23]
4. From experiments with 93D-specific inducers it is evident that no heat shock protein (HSP) was encoded by these puffs,[20] i.e., no additional polypeptide could be detected after selective induction of 93D.
5. In agreement with this, extensive loading of the chromosomal region with mutations, including deletions covering the whole 93D locus, had no detectable influence on the HSP synthesis pattern nor on recovery of control protein synthesis after hs.[29] However, these 93D deficient animals are not viable and show morphologic anomalities. Thus, the 93D locus may have vital functions for normal development. But "co-deletion" of another, so far undetected, closely linked gene can not be excluded at present.[29] Furthermore, 93D mutants exhibit altered patterns of hs-induced puffing in the 87A7/87C1 regions.[5,18]
6. Contrasting to the high degree of conservation between the HSP coding genes from different organisms (see Figures 2.8 to 2.11) the sequences of the functionally related 93D-like puffs of *D. melanogaster, D. virilis,* and *D. hydei* have completely diverged,[10,11,13,31] i.e., no cross-hybridization was observed using a 93D-specific cDNA probe. However, the repetitive units within one locus are highly conserved.[11,33,37]

Sequencing of the 93D locus and the corresponding locus of *D. hydei* (2-48B) confirmed earlier findings[23] that two types of RNAs are encoded: a large, heterogenously sized ~ 10 kb transcript forming the giant ribonucleoprotein particles[8] and a pair of small transcripts (1.9 kb and 1.2 kb in *D.m.* and 2.0 kb and 1.35 kb in *D.h.*). Only the smallest transcript is found in the cytoplasm.[10,11,13,31-33] Despite the almost complete sequence divergence, the two loci are very similar with respect to their organization. A single unique sequence of ~2 kb is flanked by 10 kb of small repetitive elements of 280 bp in *D. melanogaster*[11,33,37] and of 115 bp in *D. hydei.*[31,33] The 1.9 and 2.0 kb transcripts contain introns of 710 and 740 nucleotides, respectively,[11,13,32] which are removed to give rise to the cytoplasmic poly(A)$^+$ RNAs of 1.2 and 1.35 kb (ω3 transcripts). The 10-kb nuclear transcript is evidently not used as precursor for generation of the small ω3 transcripts, which are bound to the polysomal fraction and released only by inhibitors of initiation.[1,8a] Maximum length of the only open reading frames is 23 to 27 codons. Hence, translation itself and not the formation of a peptide product may be the function of the ω3 transcripts during hs.[8a]

The 93D locus is active already at 25°C. Transcription starts at the promoter of the unique sequence. The 10-kb transcript results from read-through into the flanking region with repetitive elements. Under hs conditions transcription is increased ~ fivefold, but the relative amount of read-through transcripts and processing of the 93D pre-RNP are reduced.[13] As found with the other typical HSP-coding puffs 93D-like puffs are also transcribed by RNA polymerase II. This is clearly evident from the usual transcription signals. In front of the unique sequence, there are a TATA box and the typical hs promoter elements (see Table 6.2). At the 3′ end, the usual polyadenylation signals are found.[13,33] These results are in good agreement with earlier investigations using RNAP II antibodies. The 93D puff was well stained, although at a somewhat reduced level compared to the major HSP70 coding puffs 87A7/87C1.[2]

The functional significance of all these peculiarities of 93D-like puffs in *Drosophila* is far from clear. The existence of this locus in all *Drosophila* species, the conservation of its basic structure with a single unique transcription unit flanked by ~ 10 kb repetitive elements, and the sequence conservation of the latter within the same locus are remarkable. The striking accumulation of giant RNP during hs[8] may be related to hs-induced changes in nucleolar morphology (see Figure 10.4): continued transcription with reduced or lacking processing of the pre-RNP lead to their visible deposition at the nuclear sites of synthesis. Alternatively, stability and/or specificity of the translation apparatus may be influenced by the ω3 transcripts.[8a]

5.2. RNA ENCODED BY α,β-ELEMENTS IN PUFF 3-87C1 OF *DROSOPHILA MELANOGASTER*

Early papers on the characterization of the two major hs puffs of *D. melanogaster* encoding HSP70 (puff 3-87A7 and 3-87C1) revealed remarkable differences between them. In addition to the three hsp70 genes, the 87C1 puff contains 21 tandemly repeated elements of the type α,β or α,β,γ forming the 40-kb spacer between the proximal hsp70 gene pair and the distal single hsp70 gene within this locus[6,12,25-28] (see Figure 3.3). The α-elements comprise ~ 490 bp, the β-elements ~ 1100 bp.[25] Similar α,β-type repetitive elements exist in a dispersed form in chromocentric DNA.[14,26,27] However, in the closely related *Drosophila* species, *D. simulans* and *D. mauritiana*, the α,β-elements were found in the chromocentric DNA but not in the otherwise identically organised 87C1 locus.[22,28] This led to speculations that the existence and activity of α,β-elements in the 87C1 locus of *D. melanogaster* are not essential for the hs response, but reflect a relatively recent evolutionary event of translocation from chromocentric DNA to the 87C1 locus.

Structural analyses by restriction mapping, hybridization and sequencing[12,15,26,36] showed the presence of a third "repetitive" element of about 0.87 kb, initially called γ. It is found in close vicinity to seven α,β-units and, in fact, represents the promoter region of the hsp70 genes, i.e., about 350 nucleotides immediately upstream of the α,β-elements are virtually identical with the corresponding part of the hsp70 promoter. This includes the TATA box as well as the 5 HSE sequences (see Table 6.2). The association with a hs promoter region explained the isolation of cDNA clones of hs-induced polysomal poly(A)$^+$RNA homologous to the repetitive elements in the 87C1 locus.[24,25] α,β-specific cytoplasmic RNAs are 2.5, 1.8, 1.4, and 1.1 kb in size. There is apparently no protein product coded by them. Only the seven α,β-units connected with the γ (promoter)-element are assumed to be transcribed.

The integration of the repetitive elements in the ensemble of the 87C1 locus together with the three hsp70 genes stimulated investigation with *Drosophila* cell cultures[24] on the temperature and time kinetics of expression of both types of transcription units. Although expression of both RNAs always increased and decreased in parallel, the amount of α,β-RNAs in the cytoplasm is always only 2% of hsp70 mRNA. Remarkable differences were found, when nuclear transcription and processing were analyzed at different temperatures. At 34°C up to 19-fold more hps70-specific RNA was synthesized than α,β-RNA, but this ratio declined to 1.5 at 37°C, i.e., despite identical promoter regions of 342 nucleotides plus 64 nucleotides of untranslated leader, there is an increasing inhibition of transcription of hsp70 genes, whereas activity of the α,β-units is even enhanced at temperatures ≥ 37°C. Another difference concerns the nucleo-cytoplasmic transport which at 34 to 35°C is three to fivefold more rapid for hsp70 transcripts. At 37°C, a marked overall reduction is observed and again export of hsp70-specific RNAs is more affected. Summarizing, the expression of α,β-elements is much more heat resistant than expression of hsp70 genes. This leads to the question of whether additional regulatory information is present outside of the virtually identical promoter/leader regions of hsp70 genes and α,β-elements.

5.3. HEAT SHOCK-INDUCED TRANSCRIPTS OF RNA POLYMERASE III

RNA polymerase III is responsible for transcription of different kinds of small stable RNAs, e.g., tRNAs, ribosomal 5S RNA, and some types of small RNAs of snRNP and scRNP particles. RNAP III-specific promoters are basically different from their counterparts found in vicinity of protein coding genes. Usually part of the regulatory sequence is contained in the coding region of RNAP III-transcribed genes.[35]

Activity of RNAP III is primarily not affected by hs (see Section 10.3). Interestingly, however, levels of some transcripts produced by this enzyme are markedly increased. The hs-responsive regulatory element remains to be characterized.

1. Bouche et al.[3] reported on a maximum eightfold increase of snRNAs K and L and of pre-tRNA in heat-shocked CHO cells. RNAs K and L are part of small nuclear RNP particles involved in the processing of pre-mRNP.
2. Fornace and Mitchell[9] and Fornace et al.[9a] characterized a 100 to 600 nucleotides polyadenylated RNA (B2) in rodent cells, whose transcription from repetitive elements rapidly increased 10- to 20-fold under hs. The 10^5 copies of the coding elements of B2 are scattered throughout the rodent genome. They may be part of introns and of 3′ nontranslated regions of RNAP II-specific transcription units. The level of B2 transcripts is evidently increased in malignant cells. The dramatic stimulation of B2 transcription in CHO and mouse 3T3 cells makes the B2 RNA the most abundant transcript under hs conditions, substantially surpassing all other transcripts including hsp70 mRNA. Although at this state of investigation any speculation on a possible function of B2 RNA in the stress response system is premature, the data present another link between malignant transformation and hs (see Chapter 20).
3. A novel 306 nucleotide heat shock and starvation induced RNA (G8) was reported by Kraus et al.[17] for the ciliate *Tetrahymena thermophila.* It is transcribed by RNAP III and quantitatively associated with polysomes. The coincidence of restoration of control protein synthesis and the accumulation of this RNA leads to speculation about its possible function as a translational control RNA (tcRNA). Sequencing of the G8 clone shows two homology regions known to be part of the internal transcription factor binding site of thc *Tetrahymena* 5S rRNA gene, to the eukaryotic 7SL RNA of the signal recognition particles, as well as to the *E. coli* 4.5S RNA.[12a]
4. A hs-induced scRNA of 320 nucleotides was isolated from *Drosophila* S2 cells. It is evidently a blunt-ended dsRNA, which inhibits *in vitro* translation of *Drosophila* mRNAs in a rabbit reticulocyte lysate. Transcription by RNA polymerase II or III was not investigated.[15a,15b]

REFERENCES

1. **Bendena, W. G., Garbe, J. C., Traverse, K. L., Lakhotia, S. C., and Pardue, M. L.,** Multiple inducers of the *Drosophila* heat shock locus 93D (hsr omega): inducer-specific patterns of the three transcripts, *J. Cell Biol.,* 108, 2017—2028, 1989.

1a. **Bendena, W. G., Fini, M. E., Garbe, J. C., Kidder, G. M., Lakhotia, S. C., and Pardue, M. L.,** hsr omega: a different sort of heat shock locus, in *Stress-Induced Proteins,* Pardue, M. L., Feramisco, J. R., and Lindquist, S., Eds., Alan R. Liss, New York, 1989, 3—14.

1b. **Bisseling, T., Berendes, H. D., and Lubsen, N. H.,** RNA synthesis in puff 2-48 BC after experimental induction in *Drosophila hydei, Cell,* 8, 299—304, 1976.

2. **Bonner, J. J. and Kerby, R. L.,** RNA polymerase II transcribes all of the heat shock induced genes of *Drosophila melanogaster, Chromosoma,* 85, 93—108, 1982.

3. **Bouche, G., Caizergus-Ferrer, M., Amalric, F., Zalta, J. P., Banville, D., and Simard, R.,** Synthesis and behaviour of small RNA species of CHO cells submitted to a heat shock, *Nucl. Acids Res.,* 9, 1615—1625, 1981.

4. **Brady, T. and Belew, K.,** Pyridoxine induced puffing (II-48C) and synthesis of a 40 kD protein in *Drosophila hydei* salivary glands, *Chromosoma,* 82, 89—98, 1981.

5. **Burma, P. K. and Lakhotia, S. C.,** Cytological identity of 93D-like and 87C-like heat shock loci in *Drosophila pseudoobscura, Indian J. Exp. Biol.,* 22, 577—580, 1984.

6. **Craig, E. A., McCarthy, B. J., and Wadsworth, S. C.,** Sequence organization of two recombinant plasmids containing genes for the major heat shock-induced protein of *D. melanogaster, Cell,* 16, 575—588, 1979.

7. **Dangli, A. and Bautz, E. K. F.,** Differential distribution of non-histone proteins from polytene chromosomes of *Drosophila melanogaster* after heat shock, *Chromosoma,* 88, 201—207, 1983.

8. **Dangli, A., Grond, C., Kloetzel, P., and Bautz, E. K. F.,** Heat-shock puff 93D from *Drosophila melanogaster:* Accumulation of a RNP-specific antigen associated with giant particles of possible storage function, *EMBO J.,* 2, 1747—1751, 1983.

8a. **Fini, M. E., Bendena, W. G., and Pardue, M. L.,** Unusual behavior of the cytoplasmic transcript of hsr omega: an abundant, stress-inducible RNA that is translated but yields no detectable protein product, *J. Cell Biol.,* 108, 2045—2057, 1989.

9. **Fornace, A. J. and Mitchell, J. B.,** Induction of B2 RNA polymerase III transcription by heat shock: Enrichment for heat shock induced sequences in rodent cells by hybridization subtraction, *Nucl. Acids Res.,* 14, 5793—5811, 1986.

9a. **Fornace, A. J., Alamo, I., Hollander, M. C., and Lamoreaux, E.,** Induction of heat shock protein transcripts and B2 transcripts by various stresses in Chinese hamster cells, *Exp. Cell Res.,* 182, 61—74, 1989.

10. **Garbe, J. C. and Pardue, M. L.,** Heat shock locus 93D of *Drosophila melanogaster:* A spliced RNA most strongly conserved in the intron sequence, *Proc. Natl. Acad. Sci. U.S.A.,* 83, 1812—1816, 1986.

11. **Garbe, J. C., Bendena, W. G., Alfano, M., and Pardue, M. L.,** A *Drosophila* heat shock locus with a rapidly diverging sequence but a conserved structure, *J. Biol. Chem.,* 261, 16889—16895, 1986.

12. **Hackett, R. W. and Lis, J. T.,** DNA sequence analysis reveals extensive homologies of regions precoding hsp70 and alpha, beta heat shock genes in *Drosophila melanogaster, Proc. Natl. Acad. Sci. U.S.A.,* 78, 6196—6200, 1981.

12a. **Hallberg, R. L. and Hallberg, E. M.,** Heat shock in *Tetrahymena* induces the accumulation of a small RNA homologous to eukaryotic 7SL RNA and *E. coli* 4.5S RNA, in *Stress-Induced Proteins,* Pardue, M. L., Feramisco, J. R., and Lindquist, S., Eds., Alan R. Liss, New York 1989, 107—116.

13. **Hovemann, B., Walldorf, U., and Ryseck, R.-P.,** Heat-shock locus 93D of *Drosophila melanogaster:* An RNA with limited coding capacity accumulates precursor transcripts after heat shock, *Mol. Gen. Genet.,* 204, 334—340, 1986.

14. **Ish-Horowicz, D., Schedl, P., Artavanis-Tsakonas, S., and Mirault, M. E.,** Genetic and molecular analysis of the 87A7 and 87C1 heat-inducible loci of *D. melanogaster, Cell,* 18, 1351—1358, 1979.

15. **Karch, F., Toeroek, I., and Tissieres, A.,** Extensive regions of homology in front of the two hsp 70 heat shock variant genes in *Drosophila melanogaster, J. Mol. Biol.,* 148, 219—230, 1981.

15a. **Kawata, Y., Fujiwara, H., and Ishikawa, H.,** Low molecular weight RNA of *Drosophila* cells which is induced by heat shock. I. Synthesis and its effect on protein synthesis, *Comp. Biochem. Physiol.,* 91B, 149—153, 1988.

15b. **Kawata, Y., Fujiwara, H., Shiba, T., Miyake, T., and Ishikawa, H.,** Low molecular weight RNA of *Drosophila* cells which is induced by heat shock. II. Structural properties, *Comp. Biochem. Physiol.,* 91B, 155-157, 1988.

16. **Koninkx, J. F. J. G.,** Protein synthesis in salivary glands of *Drosophila hydei* after experimental gene induction, *Biochem. J.,* 158, 623—628, 1976.

17. **Kraus, K. W., Good, P. J., and Hallberg, R. L.,** A heat shock-induced, polymerase III-transcribed RNA selectively associates with polysomal ribosomes in *Tetrahymena thermophila, Proc. Natl. Acad. Sci. U.S.A.,* 84, 383—387, 1987.
18. **Lakhotia, S. C.,** The 93D heat shock locus in *Drosophila:* A review, *J. Genet.,* 66, 139—158, 1987.
19. **Lakhotia, S. C. and Mukherjee, T.,** Specific activation of puff 93D of *Drosophila melanogaster* by benzamide and the effect of benzamide treatment on the heat shock induced puffing activity, *Chromosoma,* 81, 125—136, 1980.
20. **Lakhotia, S. C. and Mukherjee, T.,** Absence of novel translation products in relation to induced acivity of the 93D puff in *Drosophila melanogaster, Chromosoma,* 85, 369—374, 1982.
21. **Lakhotia, S. C. and Singh, A. K.,** Conservation of the 93D puff of *Drosophila melanogaster* in different species of *Drosophila, Chromosoma,* 86, 265—278, 1982.
22. **Leigh-Brown, A. J. and Ish-Horowicz, D.,** Evolution of the 87A and 87C heat-shock loci in *Drosophila, Nature,* 290, 677—682, 1981.
23. **Lengyel, J. A., Ransom, L. J., Graham, M. L., and Pardue, M. L.,** Transcription and metabolism of RNA from the *Drosophila melanogaster* heat shock puff site 93D, *Chromosoma,* 80, 237—252, 1980.
24. **Lengyel, J. A. and Graham, M. L.,** Transcription, export and turnover of Hsp70 and alpha, beta, two *Drosophila* heat shock genes sharing a 400 nucleotide 5′ upstream region, *Nucl. Acids Res.,* 12, 5719—5735, 1984.
25. **Lis, J., Pretidge, L., and Hogness, D. S.,** A novel arrangement of tandemly repeated genes at a major heat shock site in *D. melanogaster, Cell,* 14, 901—919, 1978.
26. **Lis, J., Neckameyer, W., Mirault, M.-E., Artavanis-Tsakonas, S., Lall, P., Martin, G., and Schedl, P.,** DNA sequences flanking the starts of the hsp70 and alpha, beta heat shock genes are homologous, *Dev. Biol.,* 83, 291—300, 1981.
27. **Lis, J. T., Neckameyer, W., Dubensky, R., and Costlow, N.,** Cloning and characterization of nine heat-shock-induced mRNAs of *Drosophila melanogaster, Gene,* 15, 67—80, 1981.
28. **Livak, K. J., Freund, R., Schweber, M., Wensink, P. C., and Meselson, M.,** Sequence organization and transcription at two heat shock loci in *Drosophila, Proc. Natl. Acad. Sci. U.S.A.,* 75, 5613—5617, 1978.
29. **Mohler, J. and Pardue, M. L.,** Mutational analysis of the region surrounding the 93D heat shock locus of *Drosophila melanogaster, Genetics,* 106, 249—265, 1984.
30. **Pardue, M. L., Bendena, W. G., and Garbe, J. C.,** Heat shock: Puffs and response to environmental stress, *Res. Problems Cell Diff.,* Springer-Verlag, Berlin, 14, 121—131, 1987.
31. **Peters, F. P. A. M. N., Grond, C. J., Sondermeijer, P. J. A., and Lubsen, N. H.,** Chromosomal arrangement of heat shock locus 2-48B in *Drosophila hydei, Chromosoma,* 85, 237—249, 1982.
32. **Ryseck, R.-P., Walldorf, U., and Hovemann, B.,** Two major RNA products are transcribed from heat-shock locus 93D of *Drosophila melanogaster, Chromosoma,* 93, 17—20, 1985.
33. **Ryseck, R.-P., Walldor, U., Hoffman, T., and Hovemann, B.,** Heat shock loci 93D of *Drosophila melanogaster* and 48B of *Drosophila hydei* exhibit a common structural and transcriptional pattern, *Nucl. Acids Res.,* 15, 3317—3333, 1987.

33a. **Schuldt, C., Kloetzel, P. M., and Bautz, E. K. F.,** Molecular organization of RNP complexes containing P11 antigen in heat-shocked and non-heat-shocked *Drosophila* cells, *Eur. J. Biochem.,* 181, 135—142, 1989.

34. **Singh, A. K. and Lakhotia, S. C.,** Lack of effect of microtubule poisons on the 93D or 93D-like heat shock puffs in *Drosophila, Indian J. Exp. Biol.,* 22, 569—576, 1984.
35. **Sollnerb-Webb, B.,** Surprises in polymerase III transcription, *Cell,* 52, 153—154, 1988.
36. **Török, I. and Karch, R.,** Nucleotide sequences of heat shock activated genes in *Drosophila melanogaster.* I. Sequences in the regions of the 5′ and 3′ ends of the hsp70 gene in the hybrid plasmic 56H8, *Nucl. Acids Res.,* 8, 3105—3123, 1980.
37. **Walldorf, U., Richter, S., Ryseck, R.-P., Steller, H., Edstroem, J. E., Bautz, E. K. F., and Hovemann, B.,** Cloning of heat-shock locus 93D from *Drosophila melanogaster, EMBO J.,* 3, 2499—2504, 1984.

Chapter 6

GENE TECHNOLOGY AND FUNCTIONAL ANALYSES OF HEAT SHOCK GENES

L. Nover

TABLE OF CONTENTS

6.1. EXPRESSION OF HEAT SHOCK GENES IN HOMOLOGOUS AND HETEROLOGOUS SYSTEMS

Cloning and sequencing of numerous hs-induced genes from different types of organisms (see Chapters 2 and 3) was complemented by the functional analysis of promoters and genes by transformation of homologous and heterologous expression systems. Updated material of a previous review[96] is presented in this chapter (Table 6.1). After some general remarks on methods and the organization of Table 6.1, selected results are discussed in more detail.

6.1.1. EXPRESSION SYSTEMS

Table 6.1 classifies the experimental results into homologous (A) vs. heterologous expression systems (B). This is based on the origin of the regulatory elements, i.e., mainly of the promoter regions and not on the particular gene associated with them, which in many cases is a reporter gene derived from another, mostly prokaryotic system.

Three types of expression systems are discriminated:

1. T: transient expression systems rely on the microinjection of DNA into *Xenopus* oocyte nuclei or on the massive uptake of DNA into animal or plant cells mediated by calcium phosphate, PEG, or DEAE-dextran. Alternatively, electroporation can be used. The initially high copy number of the transformed gene rapidly declines. Activity tests are therefore performed at the appropriate time shortly after transformation. Though frequently applied to investigate hs-induced gene activity in *Drosophila* and vertebrate cells, there are only three reports on this topic concerning plant protoplasts and DNA transfer by electroporation[1,23a,108] (see nos, A37b, A37c and A37d in Table 6.1)
2. MP: use of autoreplicating multicopy plasmids for transformation of monkey COS and yeast cells leads to stably transformed cell lines with a high copy number of foreign genes.
3. S: in contrast to the preceding two systems, generation of cells with stable integration of one or few copies of the transforming gene(s) into the chromosomal material usually requires some type of selective procedure. Prominent examples of this are the transformation of *Drosophila* preblastoderm embryos by microinjection of P-element derived vectors or the *Agrobacterium*/Ti-plasmid mediated transformation of plant cells. But stably transformed cell lines with low copy number of genes can also be selected from the first type of expression systems.

The high copy number systems (T, MP) allow a rapid and convenient assay of gene expression. However, studies on the fine tuning of regulation frequently require stably transformed cell lines with a low copy number of genes. Moreover, germ-line transformation of *Drosophila* or mammals, so well as the regeneration of whole plants from Ti-plasmid transformed callus, offer the possibility for investigation of tissue-specific expression of the introduced gene(s).

It should be mentioned that the outcome of functional tests of promoters in transformation experiments can be thoroughly influenced by the experimental conditions applied, i.e., by using homologous or heterologous cell systems or when comparing systems with a low copy number or a high copy number of the transfected gene. Examples are summarized in Figure 6.6. Thus, the decisive role of hs box cascades (see Table 6.2) only became apparent when the corresponding constructs of *Drosophila* genes were tested in homologous cells under low copy number conditions. The results are evidently dependent on four factors: (1) the overall efficiency of the promoter construct in the transfected cell, (2) the basal level transcription (leakiness of the promoter), (3) the inducibility by the cell-specific heat shock (hs) treatment, and (4) the level of the activator protein (see Section 7.5).

TABLE 6.1
Use of hs Genes and Their Promoters for Gene Transfer to Homologous and Heterologous Systems

No.	Constructs	Methods	Results	Ref.
A: Homologous expression systems				
***Drosophila* cells**				
A 1	Dm-hsp70P/L(+65) × cat gene, with 1200 bp of 5′ fls	Transformation of S2 cells (T); S1 nuclease analysis and CAT activity assay	Hs-induced 30-fold increase of CAT activity, 65 nucl. of mRNA leader are sufficient for selective translation	38
A 2	Dm-hsp70P/L(+65) × cat gene, with 1200 bp of 5′ fls	Transformation of S3 cells (T); CAT activity measurement	30-fold hs induction of CAT activity, optimum of transient expression 24 h after transformation	10
A 3	Dm-hsp70P/L(+141) × gpt gene	Transformation of *D. hydei* DH33 cells (T); autoradiographic assay for [3H]guanine	>25-fold increase of xanthine: guanine phosphoribosyltransferase activity	22
A 4	Dm-hsp70P/L(+89) × λ-phage b2 DNA (1051 bp), with 194, 146, 70, 52, 44, and 25 bp of 5′ fls	P element transformation with adh^+ vector (S); Northern and S1 nuclease analyses, puff formation	Optimum hs-induced transcription requires sequences upstream of HSE1 (see Figure 6.6); hs-puffs formed with the −194 construct, cointegration of adh gene and hsp70 × λ hybrid but differential transcription at 25°C and 37°C, respectively	26
A 4a	Dm-hsp83 gene, with 171 bp of 5′ fls	DEAE/dextran transformation of S3 cells (T); Northern analysis, comparison of constitutive, hs-induced and ecdysterone-induced levels, for discrimination all genes tagged with λ-DNA	Constitutive expression +, hs-induced expression + + +, ecdysterone-dependent expression +	71d
	Dm-hsp70 gene, with 146 bp of 5′ fls		Constitutive −, hs + + + +, ecdysterone (+)	
	Dm-hsp28 gene, with 1.35 kb of 5′ fls		Constitutive −, hs + +, ecdysterone + +	
	Dm-hsp26 gene, with 2.0 kb of 5′ fls		Constitutive −, hs + +, ecdysterone +	
	Dm-hsp23 gene, with 440 bp of 5′ fls		Constitutive −, hs + +, ecdysterone + +	
A 5	Truncated Dm-hsp70 gene (“hsp40”)	Cotransformation of *D. hydei* DH33 cells together with gpt vector (S); Southern, Northern and dot blot analyses, PAGE of HSP40	50—200 copies integrated into chromosomes, mRNA synthesis about 50-fold increased after hs, but poor translation of “HSP40”	128
A 6	Tuncated Dm-hsp70 gene (“hsp44”), with 1150 bp of 5′ fls, leader deletions and insertions	Cotransformation of S2 cells together with dhfr vector (S); synthesis of HSP44 analysed by PAGE	Sequences of untranslated leader required for preferential translation under hs-conditions	80

TABLE 6.1 (continued)
Use of hs Genes and Their Promoters for Gene Transfer to Homologous and Heterologous Systems

No.	Constructs	Methods	Results	Ref.
A 7	Dm-hsp70 × adh gene, with fusion points A: between hs box (hsp70) and TATA box (adh) and B: in the hsp70 leader sequence	P element transformation (S); S1 nuclease mapping, ADH activity, protein analysis by PAGE	Hs-induced transcription of both types of hybrid genes, but 95 nucl. of hsp70 leader (B) are required for translation during hs	70
A 8	Dm-hsp70P/L(+200) × adh gene, with 440 bp of 5′ fls	P element transfornation (S); analysis of ADH activity in tissues by *in situ* staining, ADH synthesis by PAGE	ADH induced by hs with small puffs at integr. sites, normal processing of the intron-containing primary transcript, no ADH (hs) in primary spermatocytes	20a
A 8a	Dm-hsp70P/L(+200) × adh gene, with 440 and 68 bp of 5′ fls Dm-hsp26P/L(+186) × adh gene, with 178 bp of 5′ fls	P element transformation (S), ethanol selection; ADH activity by *in situ* staining of tissues and enzyme assay	Selection of 21 mutants with constitutive expression of hs genes, expression HSE-dependent but tissue specific, many different genes affected (see Section 1.4)	97
A 9	Dm-hsp70P/L(+200) × adh gene, with 440 bp of 5′ fls	Cotransformation of S2 cells with dhfr plasmid (S); Northern analysis and protein PAGE	Hs at 38°C without preadaption blocks intron splicing but not transcription, accumulation of pre-mRNA (see Section 9.5)	155
A10	Dm-hsp70P/L(+198) × adh gene, with 186, 130, 97, 68, and 44 bp of 5′ fls	P element transformation with ry^+ plasmid (S); Southern and S1 nuclease analyses, chromosomal integration sites	Strong hs-inducibility of consructs with 186, 130, and 97 bp of 5′ fls, moderate puffing at integration sites; inducibility with −68 construct low, with −44 construct lacking	41
A11	Dm-hsp70P/L(+263) × lacZ gene, with 194, 90, 68, 63, and 50 bp of 5′ fls	Transformation of S3 cells (T); assay of beta-Galase	10—15-fold hs-induced expression of constructs with 194 and 90 bp of 5′ fls; weak, constitutive expression with 68 and 63 bp construct, no expression with 50 bp of 5′ fls, strong differences to expression in COS cells (see B4)	1a, 73
A11a	Dm-hsp70 P/L (+263) × lacZ gene, with 87 bp of 5′ fls, many constructs with variations of HSE1 and HSE2 (5′ deletions, insertion of synthetic oligonucleotides, duplications, inversions)	DEAE-dextran transformation of S3 cells (T); S1 nuclease and enzyme assay	Detailed analysis of role and structural requirements of HSE1 and HSE2 (bp −71 to −84), for selected results see Figure 6.2	2, 2a

A12	Dm-hsp70 × lacZ gene, with 194, 89, 73, 59, and 23 bp of 5′ fls	P element transformation with ry^+ vector (S); Southern analysis, puffing pattern by *in situ* hybridisation, enzyme assay with X-Gal	Strong hs-induced puffing and expression of lacZ gene with −194 and −89 constructs, but not with −73, −59, and −23 constructs, enzyme expressed in different tissues, visible puffing depends on transcript size (see Section 6.6.1)	77, 126, 127
A12a	Dm-hsp70P/L × lacZ gene, with 195 and 43 bp of 5′ fls and yolk protein gene enhancers yp1 and yp2 inserted 5′ upstream	P element transformation of ry^- embryos (S); enzyme assay	Tissue-specific expression of beta-galactosidase (yp-dependent) does not interfer with hs-induced expression	48, 77b
A12b	Dm-hsp70P/L(+88) × winter flounder antifreeze protein gene, with 194 bp of 5′ fls	P element transformation of ry^- embryos (S); protein analysis	Hs(36.5°C)-induced synthesis of antifreeze protein, splicing of pre-mRNA, pre-pro-protein processing and protein secretion works normally during hs	109
A13	Truncated Dm-hsp70 gene "hsp40"	Cotransformation of a *D.m.* ts cell line (shi^{ts}) with wild-type DNA and hsp40 plasmid (S); Northern analysis, PAGE of HSP40	Hs-induced expression of hsp40 gene normal, but destabilisation of mRNA in recovery period, lacking 3′ half of hsp70 mRNA required for autorepression	124
A14	Dm-hsp70P/L(+264) × 5′ segment of hsp26 gene, inserted in antisense direction	Cotransformation of S2 cells together with dhfr vector (S); Northern analysis, PAGE of labeled HSPs	Dose-dependent suppression of internal HSP26 synthesis by antisense RNA, other small HSPs not affected, at high gene doses (>1000/cell) low inducibility of hsp70 gene	81
A15	Dm-hsp70P/L(+206) × P element tnp gene, with 250 bp of 5′ fls Dm-hsp70P/L(+206) × neo^r × P element tnp gene, with 250 bp of 5′ fls	P element transformation (S); Northern analysis, tests for transposition function after crossing with marker recipient strains and by plasmid rescue in *E. coli*	Hs induced expression with increased levels of transpositions, not restricted to germ line: general method for gene tagging with P element	138
A16	Dm-hsp70P/L(+90) × P element tnp gene, with 1100 bp of 5′ fls, intron deletion mutants	Transformation of S2 cells with neo^r gene as selectable marker (S); Northern and protein analysis, plasmid rescue experiments	High levels of hs-induced expression (mRNA and proteins), two proteins (p66 and p87) identified, expression of active transposase (p87) also in somatic cells (see Section 6.6.2)	113, 114
A16a	Dm-hsp70P/L(+90) × tnp gene, with 1100 bp of 5′ fls	P element transformation with white gene as selectible marker (S)	Hs-induced transcription in all tissues, but correct splicing of pre-mRNA only in germ line cells (see Section 6.6.2)	72

TABLE 6.1 (continued)
Use of hs Genes and Their Promoters for Gene Transfer to Homologous and Heterologous Systems

No.	Constructs	Methods	Results	Ref.
A17	Dm-hsp70P/L(+206) × SFV capsid protein gene, with 250 bp of 5′ fls	P element transformation (S); Northern analysis and immune fluorescence of SFV proteins	Hs-induced expression of SFV capsid protein in all embryonal tissues, declining levels with ongoing development	136
A18	Dm-hsp70P/L(+206) × white locus, with 250 bp of 5′ fls	P element transformation of w^- embryos (S); recovery of red eye pigment formation	High transcript abundance of rarely transcribed gene, hs-induced normalization of eye pigmentation (see Section 6.6.2)	137
A18a	Dm-hsp70P/L(+216) × truncated white gene, with 177 bp of 5′ fls, with multicloning sites upstream and downstream, insertion of mutant actin gene	P element transformation of w^- embryos (S); selection for normalization of w^--phenotype at 25°C	Use as hs expression vector, expression of truncated actin in indirect flight muscles leads to HSP synthesis (see Section 1.4)	71
A18b	Dm-hsp70P/L(+95) × Dm-hsp70 trailer + 3′ fls, with 250 bp of 5′ fls in pHT4 Carnegie 20 vector, insertion of antennapedia cDNA at KpnI site between 5′ and 3′ service sequences	P element transformation of ry^- embryos (S); hs-induced morphogenetic effects	Use as hs expression vector, stage-dependent transformation of antennae into legs by hs treatment	117
A18c	Dm-hsp70P/L(+95) × Dm-EF-1α cDNA, with 250bp of 5′ fls	See 18b	Increased lifetime of transformed flies (18% at 25°C vs. 41% at 29.5°C) indicates role of elongation factor for aging	121a
A18d	Dm-hsp70P/L × Dm-tra cDNA	See 18b	Ectopic expression of transformer gene product causes transformation of XY (male) flies into phenotypically female	81a
A18e	Dm-hsp70P/L(+211) × sevenless cDNA, with 250 bp of 5′ fls	See 18b	Development of omatidia in the *Drosophila* eye requires expression of sevenless gene at defined stages, gene codes for membrane receptor mediating cell-cell signaling	20b
A18f	Dm-hsp70P/L(+65) × Dm-deformed cDNA, including construct with ubx homeobox	See 18b	Hs-induced expression of this selector gene causes transformation of thoracic into head segments: with the ubx homeobox replacing the dfd homeobox, transition of head into thoracic segments is observed	71b,c

A18g	Dm-hsp70P/L(+95) × Dm-fasciclin cDNA, with 250 bp of 5′ fls	Ca-phosphate transformation of S2 cells, together with α-amanitin resistance vector (S); Northern analysis, and cell adhesion	After induction (15 min 37°C plus 1 h 25°C), synthesis of fasciclin and Ca^{2+}-independent adhesion of S2 cells	129a
A18h	Dm-hsp70P/L(+94) × Sc-FLP recombinase gene, with 250 bp of 5′ fls	See 18b; combination of the FLP gene with reporter gene (white) flanked by the yeast recombination targets; analysis of eye colors	Hs-induced site-directed recombination of genes flanked by the appropriate target sequences	49a
A18i	Dm-hsp70P/L (+90) × ubx cDNA, with 1.5 kb of 5′ fls	See 18a	Postembryonic heat shocks induce homeotic transformations (cuticle, nervous system, antennae)	78a
A19	Dm-hsp70P/L(+200) × ftz gene(s) with 250 bp of 5′ fls	P element transformation together with adh gene (S); Northern analysis, larval morphology	Hs-induced synthesis, of ftz gene product(s) in blastoderm stage leads to pair rule phenotypes (see Section 6.6.2)	139
A20	Dm-hsp70 × tk gene, with bp -3000 to −8 of hsp70 gene Dm-hsp22 × tk gene with bp −400 to +215 of hsp22 gene Dm-hsp23 × tk gene with bp −900 to +10 and −119 to +10 of hsp23 gene Dm-hsp26 × tk gene with bp -2800 to +11 of hsp26 gene Dm-hsp27 × tk gene, with bp −1300 to +87 of hsp27 gene	Calcium-phosphate transformation of S3 cells (T); slot blot and S1 nuclease analyses	Good hs-induced expression of all five hs promoter constructs in the homologous system contrasts to the results in COS cells (B6, B8), truncated hsp23 promoter (−119 to +10) uninducible, hs and ecdysterone-induced transcription of hsp22 use same start site	87
A20a	Dm-hsp83P/L(codon 110) × lacZ gene, with 870 bp of 5′ fls Dm-hsp70P/L × lacZ gene, with 194 bp of 5′ fls Dm-hsp70P/L(+65) × cat gene, with 1200 bp of 5′ fls Dm-hsp26P/L (+600) × lacZ gene	Cotransformation of S2 cells (calcium phosphate) together with HSE-plasmid containing 40 tandem copies of the hsp70 HSE pair (bp −37 to −108) (T); enzyme assays	Increasing amounts of ''HSE-plasmid'' in transformation mixture inhibits hs-induced expression from hs gene promoter, but does not affect expression from histone and copia promoters (see Figure 7.4)	153
A21	Synthetic hs box × tk gene, inserted 13 bp upstream of TATA box of tk gene	Transformation of S3 cells (T); slot blot and S1 nuclease analysis	Contrasting to COS cells (B9) and *Xenopus* oocytes (B21) plasmid with synthetic hs box is not inducible in *Drosophila* S3 cells	87

TABLE 6.1 (continued)
Use of hs Genes and Their Promoters for Gene Transfer to Homologous and Heterologous Systems

No.	Constructs	Methods	Results	Ref.
A21a	Dm-hsp70P/L(+198) × adh gene, with 186, 130, 97, 68, and 44 bp of 5′ fls	Transformation of S3 cells (T); slot blot hybridization after hs induction (1 h 37°C)	Only constructs with more than 97 bp hs-inducible	10a
	Synthetic HSE-containing oligonucleotides, inserted 14 nucl. upstream of TATA box of tk gene		HSE monomer weekly active, overlapping trimer highest activity, overlapping tetramer moderately active	10a
A21b	Dm-hsp27/23 genes, with 1.35 kb of 5′ fls, hsp27 gene tagged with 1.4 kb adenovirus DNA insert	P element transformation (S); Northern analysis and DNase I hypersensitive sites	Tagged transcript detected after 45 min 37°C, DNase I hypersensitive sites similar to WT gene	42a
A21c	Dm-hsp70P/L (codon 7) × lacZ gene, with 194 bp of 5′ fls; numerous constructs with insertions and substitutions in HSE 1, 2, and flanking sequences	P element transformation; beta-Galase assay	HS-inducibility optimum with multimers of nGAAnnTTCn	154
A22	Truncated Dm-hsp27 ("hsp18.5"), with 2100, 1100, 227, and 124 bp of 5′ fls	P element transformation with ry^+ vector (S); Northern analysis and *in situ* hybridization of polytene chromosome	Full hs control requires sequences 2100 to 1100 bp upstream, total loss of hs inducibility with −124 construct; full developmental control with −2100 and −1100 construct, but still 20—40% with −124 construct	56
A22a	Dm-hsp27P/L(+87) × lacZ or cat genes, with 1200, 782, 579, 540, 523, 516, 480, 455, 405, 378, 368, 348, 323, 319, 293, 227, 129, 46, and 10 bp of 5′ fls, complementation with Dhsp70 bp −37 to −108 element Dm-hsp27P/L(+87) × cat gene, with 1200 bp of 5′ fls and internal del. of promoter proximal sequences (bp −40 to −57, −46 to −57, −46 to −96, −46 to −154, and −111 to −154)	DEAE/dextran transformation of S1 cells (T); CAT and beta-galactosidase assays, S1 nuclease mapping after induction by hs (37°C) and 2 μ*M* ecdysterone	Centers of HSE (bp −370 to −270) and of ecdysterone elements (bp −579 to −455) defined (see Figure 6.6), TATA proximal deletion without effect on induction	111, 112
A23	Dm-hsp26(+600) × lacZ gene, with 2200 bp of 5′ fls	P element transformation with ry^+ vector (S); Southern analysis, puffing of polytene chromosomes, enzyme assay with X-Gal	Hs-induced formation of beta-galactosidase and puffing at integration sites	85, 126, 127

A23a	Dm-hsp26P/L(+600) × lacZ gene, with 2000 and 278 bp of 5′ fls, deletion of bp −350 to −52 and replacement by a Dm-hsp70 51 bp HSE1/2 fragment	P element transformation of ry^- strains; enzyme assay, RNA levels	Hs-induced promoter activity reduced 30-fold by removing HSE6 (Figure 6.6), totally abolished in the deletion construct, but half of the activity can be restituted by insertion of the Dm-hsp70 element	125
A23b	Dm-hsp70P/L(+260) × lacZ gene, with 194 bp of 5′ fls, deletions of HSE1, of nucleotides between HSE1 and HSE2, insertions of 10, 127, and 331 bp between HSE1 and HSE2	P element transformation of ry^- strains; enzyme assay, RNA levels	Delection of HSE1 reduces hs-inducibility to 15%, but deletion or insertions of nucleotides between HSE1 and HSE2 have no or moderate effects	125
A23c	Dm-hsp70 P/L(+89) × cat gene, with 146 bp of 5′ fls, insertions of 4 to 304 bp between HSE1 and HSE2	DEAE dextran transformation of *Drosophila* S2 cells (T); Northern analysis of CAT mRNA	For high hs inducibility distance between HSEs must be (10)n bp with n = 1...8; 50% of max. inducibility with insertions of 291 to 304 bp	27a
A24	Dm-hsp26 gene with 1051 bp of λ phage b2 DNA insert at bp +493, with 9500, 2000, 522, 341, 236 bp of 5′ fls and a bp −351 to −53 deletion	P element transformation with adh+ vector (S); Southern, Northern and primer extension analyses	Developmental (ovarian) and hs-induced expression of λ-RNA dependent on different elements: bp −522 to −352 for ovarian and bp −341 to −53 for hs control	27
A25	Dm-hsp23 × tk gene, with 618, 402, 321, 263, and 149 bp of 5′ fls Dm-hsp26 × tk gene, with 1700, 350, 171, and 52 bp of 5′ fls	P element transformation with ry^+ vector (S); Southern, Northern analyses, analysis of chromosomal integration sites	Several hs boxes upstream of −149 (hsp23) and −52 (hsp26) required for optimum hs-induced expression (see Figure 6.6), differences to results with COS cells (see B8) and *Xenopus* oocytes (B19)	100
A26	Dm-hsp26(+600) × lacZ gene, with 2000 and 278 bp of 5′ fls	P element transformation with ry^+ vector (S); *in situ* assay of beta-Galase with X-Gal, chromosomal integration sites	Developmental regulation of hsp26 promoter in different tissues, at least three different expression states depend on separate promoter elements (see Section 6.5)	49
A27	Dm-hsp22 gene, with 1 kb of 5′ fls	Polyornithine-mediated transformation of Kc and S3 cells (T); Northern analysis and PAGE of HSP22	Efficient, hs-induced expression of transformed hsp22 gene	86
A28	Dm-hsp22 gene, with numerous Bal31 leader deletion mutants and HSV DNA tag in 3′ trailer	P element transformation with ry^+ vector (S) using recipient embryos with hsp22 mutant gene; S1 nuclease analysis and PAGE of HSP22	Only the first 26 nucleotides of leader required for hs translation, these 26 nucl. also required for optimum transcription, definition of a new consensus sequence (see Section 11.5.5)	59

TABLE 6.1 (continued)
Use of hs Genes and Their Promoters for Gene Transfer to Homologous and Heterologous Systems

No.	Constructs	Methods	Results	Ref.
A29	Dm-hsp23P/L(+248) × lacZ gene, with 1500 and 147 bp of 5′ fls	Transformation of S3 cells (T); assay of activity and amount (ELISA) of β-galactosidase	Ecdysterone and hs induction depend on sequences upstream of bp −147	73
	Dm-hsp70P/L(+270) × lacZ gene, with 1100 and 195 bp of 5′ fls		Both constructs induced tenfold by hs, but not by ecdysterone	
	Dm-hsp83 gene × lacZ gene, fused in C-terminal part of hsp83, with 800 bp of 5′ fls		High constitutive level, low hs inducibility, no influence of ecdysterone	
A29a	Dm-hsp23P/L(+112) × lacZ gene, with 1500, 554, 406, 379, 363, 333, 295, 274, 197, 186, and 147 bp of 5′ fls and numerous internal deletions between bp −391 and +1	Transformation of S3 cells (T); ELISA of beta-galactosidase after induction by hs (37°C) or 3μ*M* ecdysterone	Definition of functional HSEs and position of ecdysterone receptor box (see Figure 6.6)	83, 84
	Dm-hsp70P/L × lacZ gene, with 67 bp of 5′ fls, joining with different hsp23 promoter elements between bp −465 and −143	Transformation of S3 cells (T); ELISA of beta-galactosidase after hs-induction (37°C)	Screening for functional HSEs in hsp23 promoter (see Table 6.2), HSE at bp −146 to −133 essential	
A30	Dm-hsp22 gene, with 3200, 382, 273, 262, 244, 209, 194, 176, 175, 134, and 96 bp of 5′ fls	P element transformation with ry^+ vector (S) using recipient embryos with mutant hsp22 gene; Southern and primer extension analysis, PAGE of HSP22	Three hs boxes (Figure 6.6) required for optimum inducibility, developmental control depends on sequences bp −194 to −134	69
A30a	Hs-hsp90P/L(+740) × nptII gene, with 1044 bp of 5′ fls	Electroporation of K562 erythroid leukemia cells, 3 h 42°C, S1 nuclease mapping	Proof that hsp90 promoter confers hs-inducibility	110a
	Hs-hsp90P/L (+38) × cat gene, with 280 bp of 5′ fls	Ca-P transformation of HeLa cells (T); CAT assay	About threefold induction of CAT activity by hs	144b
A30b	Hs-hsp70P/L(+122) × cat gene, with 285, 223, 162, 146, 106, and 69 bp of 5′ fls	Transformation of COS(MP) and HeLa cells (T); enzyme assay	Only 69 bp of 5′ fls required for optimum inducibility in COS cells, but 285 bp required in HeLa cells	116b

A30c	Hs-hsp70P/L (+150) × cat gene, with 188, 105, and 100 bp of 5′ fls	Ca-P transformation of K562 erythroleukemia cells (T); hemin-induced maturation; S1 nuclease and CAT assay	Hemin-induced activity of hsp70 gene needs intact HSE and activation of HSF, −100 construct with defective HSE is inactive	139e
Human (primate) cells				
A31	Hs-hsp70P/L(+491) × lacZ gene, with 3150 bp of 5′ fls	Transformation of COS cells (MP) and human wish cells (T); enzyme assay	Hs-induced synthesis of beta-galactosidase	144
A32	Hs-hsp70P/L(+150) × cat gene, with 2400 bp of 5′ fls	Transformation of 293 cells (human embryonic kidney cells) (T); S1 nuclease mapping	Hs-induced and E1A-dependent transcription of fusion gene	147
		Ca-phosphate transformation of human fibroblasts (T); 4.5 h 42°C, CAT assay	Inducibility by hs decreases with age of fibroblasts, corresponds to decreasing capability for induced HSP synthesis	77a
A33	Hs-hsp70P/L(+150) × cat gene, with 2400, 292, 188, 150, 131, 112, 107, 68, 58, and 47 bp of 5′ fls	Cotransformation of 293 cells together with gpt vector (S); S1 nuclease mapping	Characterization of promoter elements for hs, Cd^{2+} and serum stimulation (see Section 6.5)	146, 148
A33a	Hs-hsp70P/L(+150) × cat gene, with 2400 bp of 5′ fls	Cotransformation of HeLa cells with E1A cDNA-containing plasmids (T); S1 nuclease mapping	Adenovirus E1A activation only by intact 13S product, independent of serum stimulation	149
A33b	Hs-hsp70P/L(+113) × Hepatitis virus B surface antigen gene, with 500 bp of 5′ fls and SV40P × neo^r gene as selectable marker	Transformation of human Wish cells with G418 selection (S); S nuclease and protein analysis	Efficient hs-inducted production and excretion of HVB-SAG, transformed gene stable also after tumor passage in nude mice (see Figure 6.7)	40
A33c	Hs-hsp70P/L(+113) × human GH gene, with 500 bp of 5′ fls	COS cells (see B2); analysis of labeled HGH secretion	Hs-induced growth hormone formation (3 h 43°C + 15 h 37°C) (see Figure 6.7)	39
	Hs-hsp70P/L(+113) × chicken lys cDNA with 1100 bp of 5′ fls	COS cells (see B2); analysis of labeled lysozyme secretion	Hs-induced lysozyme formation (3 h 43°C + 15 h 37°C)	
A33d	Hs-hsp70P/L × cat gene, with 1250 and 84 bp of 5′ fls, numerous linker scan mutations in bp −17 to −69	Ca-phosphate transformation of HeLa and BALB/c3T3 cells (T); S1 nuclease mapping and CAT assay	Full basal level expression in mouse 3T3 cells requires only 84 bp of 5′ fls, in HeLa cells sequences further upstream are also involved (see Figure 6.6)	50a
A33e	Hs-hsp70 P/L(+101) × cat gene, with 107, 68, 58, 43, and 28 bp of 5′ fls; replacement of TATAA by SV40 derived TATT/TAT	Ca-P transformation of HeLa cells (T); induction by pEA1 cotransformation or by adenovirus infection; CAT assay	Only hsp70 TATA box required for induction by EA1 protein, replacement by the SV40-derived TATTTAT blocks induction	127a

TABLE 6.1 (continued)
Use of hs Genes and Their Promoters for Gene Transfer to Homologous and Heterologous Systems

No.	Constructs	Methods	Results	Ref.
Rodent cells				
A33f	Rn-heme oxygenase × gpt fusion gene, with 549 and 210 bp of 5′ fls	Transformation of mouse melanoma cells (T); S1 nuclease mapping	Strong hs (42°C) and heme induced expression with −549 but not with −210 construct	122
Avian cells				
A33g	Gd-hsp90P/L(+99) × cat gene, with 890 and 378 bp of 5′ fls	Ca-phosphate transformation of quail myoblast cells (T); 90 min 45°C + 90 min 37°C, CAT assay	Less than 378 bp of 5′ fls required for optimum hs-inducibility	144a
***Xenopus* oocytes**				
A34	X1-hsp70 gene, X1-hsp30 gene	Microinjection into oocytes (T); S1 nuclease mapping	hsp70 constitutively expressed, hsp30 gene strongly induced by hs (see also B2)	11
A34a	X1-hsp70 gene, with 196, 179, 138, 118, 108, 79, 55, and 37 bp of 5′ fls; deletion of CAT1 (bp −49 to −58)	Microinjection into oocytes (T); S1 nuclease mapping	Constitutive but not hs-induced expression of hsp70, declines in construct with <118 bp of 5′ fls, deletion of CAT1 makes promoter hs-inducible (see Figure 6.4)	14
	X1-hsp70 HSE1 (bp −106 to −129), inserted upstream of TATA box of tk gene		HSE1 (dimer of hs box) fused to tk gene makes the construct hs-inducible	
Plant cells				
A35	Zm-hsp70 gene, with about 1 kb of 5′ and 3′ fls	Agrobacterium/pTi-mediated transformation of *Petunia* leaf discs (S); Northern, primer extension analyses	Faithful hs-induced transcription	115
A36	Gm-hsp17.3b gene, with −1000 bp of 5′ fls	Agrobacterium/pTi-mediated transformation of sunflower hypocotyl (S); Southern, S1 nuclease analyses	0.1 to 0.5 copies of hsp17 gene/cell, correct hs-induced initiation, but very low level of expression	118
A36a	Gm-hsp17.3B gene, with 336, 304, 195, 139, 78, 51, and 12 bp of 5′ fls	Agrobacterium/pTi-mediated transformation of tobacco leaf discs, plant regeneration (S); dot blot, Northern ans S1 nuclease analyses	Optimum expression of hsp17.3 gene after HS (2 h 40°C) requires >300 bp of 5′ fls, role of non-HSE promoter elements (see Figure 6.6)	8

A36b	Truncated CaMV35S P/L × cat gene with synthetic HSE2 oligonucleotide derived from Gm-hsp17.3B gene; Gm-hsp17.3BP/L(+131) × cat gene, with numerous internal deletions of leader sequence	Agrobacterium transformation of tobacco leaf discs(S); cat assay and Northern analysis	2 h at 40°C required for cat expression, leader sequence of Gm-hsp17.3 or of CaMV required for translation under hs conditions	120b
A36c	Gm-hsp17.6L gene fused to cauliflower mosaic virus 35S promoter	See A36a (S); Northern analysis	High levels of constitutive expression of hsp17 mRNA (~ 20,000 mol/cell), decline after hs	120a
A37	Gm-hsp17.5E gene, with 3250, 1175 and 95 bp of 5′ fls, many internal deletions tested	See A36 (S); Northern and S1 nuclease analyses	Hs-induced expression with correct 5′ and 3′ ends, all three constructs hs-inducible but −95 construct with lower efficiency, only −3250 construct also inducible by arsenite and Cd^{2+}, totally 5 promoter regions between bp −244 and −25 occupied by different proteins	37a, 51
A37a	5′ truncated Gm-hsp17 gene fused to the 5′ terminal 68 codons of small subunit gene of ribulose-bisphosphate carboxylase, with adjacent SP6 promoter	*In vitro* transcription (SP6 polymerase), translation (wheat germ system) and transport (isolated pea chloroplasts)	HSP17 hybrid protein taken up into chloroplasts with correct cleavage of the Rubisco transit peptide (55 aa residues)	78
A37b	Gm-hsp17.5EP/L(+77) × gus gene, with 3250 bp of 5′ fls	Electroporation of *Nicotiana plumbaginifolia* leaf protoplasts (T); GUS assay	10-fold increase of GUS activity by 2 h at 39°C, applied 24 h after transformation	1
A37c	Zm-hsp70P/L(+84) × cat gene, with ~650 bp of 5′ fls	Electroporation of maize protoplasts (T); CAT assay	200-fold increase of CAT activity after 6 h hs at 40°C, applied 12 h after transformation	23a
A37d	Truncated CaMV35S P/L × cat gene with synthetic HSE oligonucleotides inserted at EcoRV site	Electroporation of *Orychophragmus violaceus* suspension culture protoplasts (T); CAT assay	Weak stimulation of CAT activity by 3 h hs at 40°C, applied 24 h after transformation	108
A37e	Zm-hsp70P/L (+107) × Ti-plasmid iaaM gene or ipt gene, with ~600 bp of 5′ fls	Ti-plasmid mediated transformation of Petunia leaf discs (S); morphological effects after hs-induced hyperproduction of auxin or cytokinin	Hyper-auxin effects only observed after hs, hyper-cytokinin morphology also without hs (leakiness of promoter?)	82a
A37f	Ps-rbcSP/L(+15) × cat gene, with 410 bp of 5′ fls, HSE2 of Gm-hsp17.3-B gene inserted upstream or at bp −48	Agrobacterium/Ti-plasmid transformation of *N. tabacum* leaf discs, plant regeneration (S); S.1 nuclease mapping	Hs-inducibility only in leaves, because rbcS promoter contains tissue-specific silencer active in roots	138a

TABLE 6.1 (continued)
Use of hs Genes and Their Promoters for Gene Transfer to Homologous and Heterologous Systems

No.	Constructs	Methods	Results	Ref.
Yeast cells				
A38	3′ truncated Sc-hsp90 gene × lacZ gene, with >1 kb of 5′ fls	Transformation with 2 μ multicopy plasmid (MP); X-Gal selection of cells with beta-galactosidase enzyme assay and PAGE	High constitutive expression of fusion gene, two- to threefold induction by hs	45
A39	Sc-hsp90 gene and 3′ truncated hsp90 gene ("hsp69.2"), with >1 kb of 5′ fls	Li-acetate-transformation of leu-cells with 2 μ multicopy plasmid (leu 2 gene) (MP); Northern analysis and protein PAGE	Fourfold increase of hs-induced synthesis of HSP90 and "HSP 69.2"	46
A39a	Sc-hsp90 gene with point mutations in TATA-proximal –TTC2nGAA–group	Transformation of haploid yeast with YIp5 vector (S); Northern and footprint analyses	Mutants are defective in basal level expression of hsp90 but not in hs-inducibility	79a
A39b	Sc-hsp70(SSA4)P × Mm-dhfr gene	Transformation with centromeric vector (S); methotrexate resistance	Hs-induced methotrexate resistance used to select for regulatory genes from a genomic library, isolation of STI1 gene	36a, 94a
A40	Sc-hsp70 × lacZ gene (SSA1), Sc-hsc70 × lacZ gene (SSB1), fusion point in 5′ parts of coding region	See A 39 (MP); beta-galactosidase assay	60-fold induction of SSA1 construct, high basal level expression of SSB1 construct	21
A41	Sc-hsp70 × lacZ gene (SSA1), with bp −499 to +285	See A39 (MP); S1 nuclease and enzyme assay	Low constitutive level and strong hs induced expression	43
	Sc-hsp70 × lacZ gene (SSA2), with bp −755 to +30	See A39 (MP); S1 nuclease and enzyme assay	Medium constitutive level and srong hs induced expression	
	Sc-hsc70 × lacZ gene (SSB1), with bp −650 to +30	See A39 (MP); S1 nuclease and enzyme assay	High constitutive level, synthesis repressed by hs	
A41a	Sc-hsp70P/L(+90) (SSA1 gene) × lacZ gene, with 1200, 339, 280, 223, 181, 173, 148, and 123 bp of 5′ fls	See A39; enzyme assay, S1 nuclease and primer extension analyses	Among the three HSEs only HSE2 required for full hs-induced activity (see Figure 6.6)	129
	Truncated Sc-cyc1P × lacZ gene, with insertion of HSE2 sequences from hsp70 promoter (bp −185 to −199, −182 to −203 and −131 to −262) and of synthetic oligonucleotides mimicking HSE2 and its flanking sequences		HSE2 and its fls also determine basal level expression typical for SSA1 combination of positive and negative regulatory sequences which act also on truncated cyc1 promoter	

A41b	Truncated Sc-cyc1P × lac Z gene, with four tandem repeats of HSE inserted upstream of TATA box	Li-acetate transformation with 2μ multicopy plasmid (MP) of yeast cells containing the HSF1 gene under gal promoter control; enzyme assay with X-Gal	Overexpression of hs transcription factor after galactose activation of HSF1 gene leads to increased levels of β-galactosidase at 30°C	145a
A42	Sc-hsp70 genes SSA1 and SSA2 and hsp70 genes SSB1 and SSB2 each with leu2 gene inserted instead of codons 307 to 386	Site directed insertion mutagenesis of leu^- strains (S); isolation of $SSA1^-/A2^-$ and $SSB1^-/B2^-$cells; Southern analysis, viability tests	SSA1-/A2-cells are deficient in growth and plating at 37°C, whereas SSB1-/B2-cells are deficient at 19°C but grow normally at 37°C (see Table 2.4)	35, 36
A43	Sc-hsp26 mutant gene, with his3 gene inserted at 5′ end or replacing the hsp26 gene	LiCl transformation of $his3^-$yeast cells (S); generation of hsp26-defective strains by gene conversion; Southern, Northern, immunological analyses, viability tests	Functional HSP26 not required for induced thermotolerance, ethanol resistance, sporulation, spore germination or survival after long-term storage	107
B: Heterologous expression systems				
Human and monkey COS cells				
B 1	Dm-hsp70P/L(+89) × histon H3 gene, with 200 bp of 5′ fls	Transformation of HeLa cells (T) and cotransformation of HeLa cells together with neo^r gene in hybrid plasmid (S); S1 nuclease analysis	Hs-induced transcription of hsp70 × H3 hybrid mRNA, not destabilized after block of DNA synthesis, role of leader sequence for H3-mRNA stability	89
B 1a	X1-hsp70 gene with 700, 196, 170, 138, 118, 108, and 79 bp of 5′ fls, and a 215 bp insertion at bp −88 of the −700 construct	Calcium phosphate transformation of HeLa cells (T) and transformation of COS cells (see B2, MP); S1 nuclease mapping	Hs inducibility decreases with deletions to bp −138, lacking in −118, −108, and −79 constructs, improved by interaction with CAT element centered at bp −55 (see Figure 6.4)	16
	X1-hsp70P × Hsβ-globin gene, bp −700 to −89 and −196 to −89 inserted at bp −800 of human beta-globin gene	Calcium phosphate transformation of HeLa cells (T); S1 nuclease mapping	Hs-induced transcription of globin gene independent of distance and orientation of hs boxes from TATA box (see Figure 6.5)	
	Dm-hsp70 P × Hsβ-globin gene, bp −108 to −37 inserted as monomer or dimer at bp −800 of human beta-globin gene	Calcium phosphate transformation of HeLa cells (T); S1 nuclease mapping	Weak hs-induced transcription with monomer, but very strong with dimer (see Section 6.4)	
B 1b	Rn-hsc73P/L(+65) × cat gene, with 2500, 343, 271, 241, and 84 bp of 5′ fls	Transformation of HeLa (T) and COS cells (MP); CAT assay after 45 min 44°C + 15 h recovery	Weak hs-inducibility (twofold) of the 2500 bp construct, increases when upstream sequences are deleted, max. 10- to 15-fold with the 84 bp construct (see Figure 6.6)	130

TABLE 6.1 (continued)
Use of hs Genes and Their Promoters for Gene Transfer to Homologous and Heterologous Systems

No.	Constructs	Methods	Results	Ref.
B 2	X1-hsp70 gene, XL-hsp30 gene	DEAE dextran/chloroquine transformation of COS cells with SV40 ori-containing vector (MP); protein analysis	Hs-induced synthesis of HSP70 and HSP30	11
B 3	Dm-hsp70 gene, with 2000 bp of 5′ fls	See B2; S1 nuclease analysis	Hs-induced expression, first definition of hs box upstream of bp −66, acts positively (no escape)	102
	Dm-hsp70P(−10) × tk gene, with 186, 108, 97, 68, 66, 44, and 28 bp of 5′ fls			
B 3a	Dm-hsp70 gene starting at codon no. 5, with 31 aa residue leader of rat grp78 gene fused to the Adv major late promoter, hsp70 gene tagged at 3′ end by insertion of 10 codons of Hs-c-myc gene	See B2; analysis of expression and intracellular localization using monoclonal antibodies against c-myc tag	Rat GRP78 leader sequence directs Dm-HSP70 into lumen of ER, where it is highly glycosylated	92
B 4	Dm-hsp70P/L(+270) × lacZ gene, with 194, 90, 68, 63, and 50 bp of 5′ fls	See B2; enzyme assay	5- to 35-fold hs induced increase of beta-Galase activity in all constructs, except deletion construct starting at −50 (see A11)	1a, 73
B 5	Dm-hsp70 gene, with 186, 68, and 53 bp of 5′ fls.	See B2; Northern and S1 nuclease analysis	Inducibility by hs and arsenite, abolished in −53 construct	84b
B 6	Dm-hsp70P(−10) × tk gene with 2000, and 186 bp of 5′ fls	See B2; S1 nuclease analysis	20- to 50-fold hs-induced transcription	104, 105
	Dm-hsp70 gene, with 2000, 186, 66, 44, and 28 bp of 5′ fls	See B2; S1 nuclease analysis	20- to 50-fold hs-induced transcription, bp −44 to −66 essential	
B 7	Dm-hsp70 gene, 5′ truncated to bp +269 and fused to adenovirus major late promoter (bp −260 to +33), constructs with oligonucleotide inserted at 3′ end and with N- and C-terminal deletions of HSP70	See B2; protein blots and immunological analysis of intracellular compartmentation	High constitutive expression by means of adenovirus promoter, Dm-HSP70 functions "normally" in COS cells (see Sections 10.6 and 16.2), definition of nuclear targeting domain	91, 103, 106

B 8	Dm-hsp22P(+5) × tk gene, with 520, 197, 99, and 66 bp of 5′ fls	See B2; primer extension analysis	Strong hs-induced transcription two hs boxes defined (Figure 6.6)	6
	Dm-hsp23P(+5) × tk gene, with 950, 450, 400, 264, and 148 bp of 5′ fls	See B2; primer extension analysis	Only weak constitutive expression, one hs box at by −140	
	Dm-hsp26P(+5) × tk gene, with 1700, 351, 172, and 53 bp of 5′ fls	See B2; primer extension analysis	Strong, hs induced transcription two hs boxes defined (Figure 6.6)	
	Dm-hsp27P/L(+100) × tk gene with 420 and 126 bp of 5′ fls	See B2; primer extension analysis	No expression (but see A20 and A22)	
B 8a	Dm-hsp70P/L(+270) × chicken lys cDNA with 194 bp of 5′ fls	COS cells (see B2); analysis of labeled lysozyme secretion	Hs-induced lysozyme secretion (3 h 43°C + 15 h 37°C)	39
B 9	Synthetic hs box and derivatives, inserted upstream of TATA box into tk gene promoter	See B2; S1 nuclease analysis	CTnGAAnnTTCnAG sufficient for hs induction, insertions into palindrome abolish activity (see A21)	104
B 9a	Dm-hsp70P/L × mouse IFN-β gene, with ~1600 bp of 5′ fls	Ca-P transformation of monkey kidney cells and human epithelial cells (T); biological assay of interferon synthesis	Strong induction of interferon synthesis by hs (37°C → 3 h 41°C) and arsenite, assayed after 24 h	4a
Rodent cells				
B10	Dm-hsp70 gene, with 1100 bp of 5′ fls	Stable cotransformation of tk^- mouse L cells together with tk^+ plasmid (S); Southern, Northern and S1 nuclease analyses	First publication on stable integration of hsp70 gene and hs-induced expression in a heterologous system	30
B11	Dm-hsp70 P/L(+198) × human growth hormone gene, with 1100 bp of 5′ fls	Cotransformation of tk^- mouse L cells with tk^+ plasmid (S); Northern analysis	Hs-induced transcription of growth hormone mRNA	31
B11a	Dm-hsp70P/L × hepatitis virus B surface antigen cDNA, with ~750 bp of 5′ fls	Ca-P cotransformation of tk^- cells together with tk vector (S); radioimmunoassay	Hs-induced synthesis of HVB surface antigen is enhanced by interferon treatment; → stabilization of mRNAs	40b
B12	Dm-hsp70 P/L (+85) × cat gene, with 780 bp of 5′ fls	Cotransformation of BALB/c mice 3T3 cells together with c-myc containing plasmid (T); assay of CAT activity	c-myc gene product stimulates transcription from Dm-hsp70 promoter (see also B16b, B17e)	67
B13	Dm-hsp70P/L(+85) × dhfr gene, with 1300, 780, and 200 bp of 5′ fls	Cotransformation of $dhfr^-$ CHO cells with c-myc containing plasmid (S); assay of colony number on selective medium	c-myc gene product stimulates transcription from Dm-hsp70 promoter, sequences bp −340 to −780 required (see also B16b, B17e)	67
B14	Dm-hsp70 genes, from 87A7 locus	Transformation of Rat-1 cells with an hsp70/gpt vector (S); Southern and S1 nuclease analyses	Hs-induced transcription of hsp70 genes, at very low level, no protein products detected	22

TABLE 6.1 (continued)
Use of hs Genes and Their Promoters for Gene Transfer to Homologous and Heterologous Systems

No.	Constructs	Methods	Results	Ref.
B14a	Dm-hsp70 P/L (+88) × truncated mouse c-myc gene, with 780 bp of 5′ fls	Cotransformation of $dhfr^-$ CHO cells together with $pdhfr^+$, methotrexate selection (S); mRNA and protein analysis	About 100-fold increase of c-myc mRNA during hs (43°C), but translation of c-myc protein only at 37°C, overproduction of c-myc has cytotoxic effect	152
B15	Dm-hsp70 gene, with 30 different Bal31 deletion constructs starting at bp −1100 and ending at bp −9	Cotransformation of tk^- mouse cells with neor plasmid (S); S1 nuclease analysis	In addition to the hs box, transcription is influenced by two alternatively acting elements with negative control (see Section 6.5)	29
	Dm-hsp70 P and P/L × tk gene, many different constructs with fusion points in the P and L regions	Transformation of tk^- L cells (S); Northern analysis		
B15a	Dm-hsp70P/L (+171) × cat gene, with 440 bp of 5′ fls	Transformation of murine plasmacytoma cells (T); CAT activity assay	No hs-inducibility in PCC4 and 1009EC cells, because of lacking HSF; EC cells become inducible after differentiation with retinoic acid	84a
B16	Dm-hsp70 gene, with 2000 bp of 5′ fls	Transformation of mouse tk^- L cells with an hsp70/tk^+ plasmid (S); immunological assay of HSP70	Normal, hs-induced synthesis and intracellular translocation and function in mouse cells	103
B16a	Hs-hsp70P/L(+160) × cat gene, with 1250, 131, 107, 68, and 58 bp of 5′ fls	Ca-P cotransformation of mouse 3T3 cells with large polyoma virus T-antigen expression plasmid (T); primer extension analysis	Stimulation of cat transcription by large T-antigen, undiminished even in −58 construct; no direct interaction with promoter	67
B16b	Hs-hsp70P/L (+160) × cat gene, with 1250, 120, and 84 bp of 5′ fls	Ca-P cotransformation of mouse 3T3 cells with SV40P × c-myc vector (T); CAT activity and S1 nuclease mapping	For c-myc-dependent stimulation of hsp70 transcription sequences upstream of bp 120 required, HSE not involved, only exon 3 of c-myc protein required	63a
B16c	Hs-hsp70P/L (+150) × c-Ha-ras gene, with 300 bp of 5′ fls	Ca-P cotransformation of mouse 3T3 cells with neo^+ vector (S)	Hs induction of p21 oncoprotein synthesis causes sequential increase of ODC, stromelysin, and c-jun mRNAs	128a
B16d	X1-hsp70P/L fragment (bp −250 to +470) inserted in dhfr-containing vector	Transformation of $dhfr^-$ CHO cells, methotrexate selection (S); HSP70 synthesis, decay of heat resistance	Gene amplification by methotrexate selection results in scavenging of HSF by increasing copy numbers of X1-hsp70 promoter, decline of induced HSP70 synthesis and of cell survival at 45°C	62a

B17	Hs-hsp70 gene, 6.3 kb genomic clone in pBr322	Transformation of hamster V79 cells (T); assay of HSP70 synthesis, Northern analysis	Hs-induced expression of hsp70 gene (mRNA, protein)	147
B17a	Hs-hsp70 P/L(+113) × Hs-growth hormone gene, with 500 bp of 5′ fls	Transformation of $dhfr^-$ CHO cells with $dhfr^+$/Hs-gh plasmid (S); transformation of mouse 3T3 cells with Hs-gh/neo^r plasmid (S) and cotransformation with Hs-gh and c-ras gene containing plasmids (S); protein analyses	Stable cell lines secrete human growth hormone after hs-induction, optimum production systems for 3T3 cells (see Figure 6.7), increased productivity after passage of transformed cell line as tumor in mouse.	39, 40a
B17b	Ce-hsp16-1/16-48 gene pair, with different HSE constructs in the intercistronic region	Transformation of mouse fibroblasts with disarmed BPV vector containing a HSV tk P × npt II gene construct for selection (MP); S1 nuclease mapping	Optimun hs-induced expr. with overlapping HSE tetramer surpasses transcription from tk promoter 100-fold, splicing of nematode introns in mouse cells defective (see Figure 6.1)	65
B17c	Dm-hsp70P/L (+212) × Hs-tissue plasminogen activator cDNA, with 98 bp of 5′ fls Hs-hsp70P/L(+113) × Hs-tPA cDNA, with 350 bp of 5′ fls.	Constructs in BPV vector with Hs-metallothionein gene; transformation of mouse c127 cells, selection with 20/μM $CdCl_2$/$ZnCl_2$ (S); protein assay with fibrin-agarose overlay, primer extension analysis	40-fold increase of tPA after 2—4 h 42°C, maximum 1.3 mg tPA/1	9
***Caenorhabditis elegans* (nematode)**				
B17d	Dm-hsp70P/L(+240) × lacZ gene, with about 200 bp of 5′ fls	Cotransformation of germ line cells of Ce-tra-3 amber mutant strains with sup-7 tRNA gene as selectable marker (S); enzyme assay by *in situ* staining	New transformation system, expression of beta-galactosidase only after hs (3 h 34°C)	47
Avian cells				
B17e	Dm-hsp70P/L (+80) × cat gene with 190bp of 5′ fls Hs-hsp70P/L (+150) × cat gene, with 120 and 84 bp of 5′fls	Ca-P cotransformation of quail fibroblasts with v-myb vector (T); CAT assay	Activation of *Drosophila* and human hsp70 promoter by v-myb protein even observed with truncated v-myb not able to bind to DNA; → interaction with other nuclear proteins essential	71a
***Xenopus laevis* oocytes**				
B18	Dm-hsp70 gene, with 2500, 186, 138, 130, 97, 66, 44, and 28 bp of 5′ fls Dm-hsp70P(−10) × tk gene, with 2500 bp of 5′ fls	Microinjection into oocyte nuclei (T); S1 nuclease analysis	Strong hs induction, except with −44 and −28 constructs, but results with −55 and −97 constructs not reproducible, proper function requires sequences upstream of bp −66, no translation of hsp70 mRNA during hs (see Section 11.3)	15

TABLE 6.1 (continued)
Use of hs Genes and Their Promoters for Gene Transfer to Homologous and Heterologous Systems

No.	Constructs	Methods	Results	Ref.
B18a	Dm-hsp70 gene with 5′ fls	See B18; primer extension analysis	Induction by hs (34°C or injection of purified *Drosophila* HSF	151
B19	Dm-hsp70 P/L × lacZ gene, with 195 and 50 bp of 5′ fls	See B18; S1 nuclease analysis, enzyme assay	Strong hs-induced transcription, but translation only at 22°C (see Section 11.3), promoter sequence between bp −195 and −50 required	73, 83
	Dm-hsp23P/L × lacZ gene, with 379, 186, 147, 139, 109, 84, and 83 bp of 5′ fls	See B18; S1 nuclease analysis, enzyme assay	Strong hs-induced expression, hs boxes between bp −181 and −133 required, differences to results with COS cells (B8) and *Drosophila* cells (A25)	83
B19a	Dm-hsp70P/L × lacZ gene, with 194, 88, 67, and 50 bp of 5′ fls Hs-hsp70P/L × Dhsp70L × lacZ gene, with 106 bp of 5′ fls	See B18; enzyme assay, induction by injection of abnormal proteins	Induction of beta-galactosidase expression by abnormal proteins operates with both promoters but depends on the presence of HSE, i.e., no induction with −50 construct of Dm-hsp70 promoter (See Figure 1.4)	3
B20	Dm-hsp70 genes (10.5 kb of 87A7 locus with two hsp70 genes)	See B18; S1 nuclease and Northern analyses	Fidele, hs-induced transcription of hsp70 genes, correct 3′ and 5′ ends of mRNA	143
B21	Synthetic hs box, inserted upstream of TATA box of tk gene promoter	See B18; S1 nuclease analysis	CTnGAAnnTTCnAG sufficient for hs-induced transcription of tk mRNA, insertions into palindrome abolish activity (see A21 and B9)	104
B21a	Synthetic oligonucleotides fused to tk gene upstream of TATA box: (a) dimer of hs box, (b) CCAAT element, and (c) hs box dimer + CCAAT	See B18; S1 nuclease mapping	Strong hs-induced transcription with construct (a), but high constitution transcription with constructs (b)and (c)	14
	Dm-hsp70 P/L (+65) × X1-hsp70 gene	See B18; S1 nuclease mapping	Strong hs-induced transcription	
B22	Hs-hsp70 gene, with 3150 bp of 5′ fls	See B18; S1 nuclease analysis	Hs-induced synthesis of hsp70 transcription, correct 5′ end	144

B23	Hs-hsp70 × lacZ gene, fusion point at bp 491 of hsp70 gene, with 3150, 600, and 105 bp of 5′ fls	See B18; enzyme assay	Strong hs-induced synthesis β-galactosidase, hs promoter needs only 105 bp of 5′ fls	144
B23a	Dm-hsp70P/L(+65) × Hs-infl. haemaglutinine cDNA, with about 200 bp of 5′ fls	See B18; protein assay with anti-HA	Hs-induced precursor HA secretion after 90 min 36°C + 90 min 21°C	39
	Dm-hsp70P/L(+89) × Gd-lysozyme cDNA, with 194 bp of 5′ fls	See B18; protein assay with anti-Lys	Hs-induced secretion of lysozyme	
	Hs-hsp70P/L(+113) × Gd-lysozyme cDNA with 1100 bp of 5′ fls	See B18; protein assay with anti-Lys	Hs-induced secretion of lysozyme	
	Hs-hsp70P/L(+113) × Hs-growth hormone gene, with 500 bp of 5′ fls	See B18; protein assay with anti-HGH	Hs-induced secretion of human growth hormone, max. 10 ng/10 oocytes/15 h	
B23b	Hs-hsp27 gene, with about 1 kb of 5′ fls	See B18; S1 nuclease mapping after 90 min hs (36°C)	hsp27 Transcripts only in hs-induced oocytes	54
Sea urchin eggs				
B24	Dm-hsp70P/L(+65) × cat gene, with 1100 bp of 5′ fls	Injection of linearized plasmid into *Strongylocentrotus* eggs, hs at 25°C with pluteus stage embryos (T); enzyme assay	About tenfold hs induction of CAT activity, very low efficiency on a per gene basis	82
Mosquito cells				
B25	Dm-hsp70P/L(+65) × cat gene, with 1100 bp of 5′ fls	Transformation of *Aedes albopictus* cells (T), hs at 41°C; enzyme assay	Effective transformation depends on cell culture medium, hs-induced increase of CAT activity	10
Plant cells				
B26	Dm-hsp70P/L(+199) × nptII gene with 258 bp of 5′ fls	Agrobacterium/pTi-mediated transformation of tobacco protoplasts (S); Northern analysis and neomycin phosphotransferase assay	Faithful, transcription of npt II gene, effective translation only in recovery period, expression also in root, stem, leaves but not in germinating pollen of regenerated plants	134, 135
B26a	Dm-hsp70P/L (+199) × Ti-plasmid ipt gene, with 258 bp of 5′ fls	Agrobacterium/pTi-mediated transformation of tobacco (S); cytokinin requirement of transformed calli	Daily hs-induction (1 h 40°C) of calli confers cytokinin independence, regenerated plants phenotypically normal at 25°C	116c
Yeast cells				
B27	Dm-hsp70 gene	Transformation of $leu2^-$ cells with hsp genes in leu^+ fusion plasmid (S); Southern, dot blot and Northern analyses	Constitutive expression, no hs-induction, 5′-end not exact	32, 76
	Dm-hsp83 gene	Transformation of $leu2^-$ cells with hsp genes in $leu2^+$ fusion plasmid (S); Southern, dot blot and Northern analyses	Constitutive expression with sixfold stimulation by hs, correct transcript size, DNase I hypersensitive sites preserved in yeast	

TABLE 6.1 (continued)
Use of hs Genes and Their Promoters for Gene Transfer to Homologous and Heterologous Systems

No.	Constructs	Methods	Results	Ref.
B28	Dm-hsp70 × lacZ fusion gene, with 194, 145, 74, 43, and 23 bp of 5′ fls	Transformation with episomal centromer containing plasmid (S); digestion of nuclear DNP with DNase I, Southern blots	DNase I hypersensitive sites at −93 and −6 preserved, determined by presence of hs boxes and HSF binding (?)	33
B29	Dm-hsp70 gene	Transformation with 2 μ multicopy plasmid (MP); Northern and S1 nuclease analysis	Fivefold increase of transcription after hs, transcription start incorrect, no translation	94
B29a	Dm-hsp70 gene, with 194 bp of 5′ fls	Transformation of leu$^-$ cells with YEp13 plasmid (S); Northern and 2D protein analysis	Constitutive transcription hsp70 gene, but translation of mRNA only after hs	28
B29b	Synthetic hs boxes inserted upstream of bp −178 of truncated Sc-cyc1P/L(+74) × lacZ gene	Transformation with 2μ plasmid pLG −178 (MP); enzyme assay after 60 min at 39°C	Optimum hs-induced expression requires 3—4 hs boxes in series or overlapping tetramer (see Figure 6.3)	132, 145
B30	*Dictyostelium* promoter with flanking sequences of DIRS	Transformation of ura$^-$ cells with ura$^-$ 2μ multicopy plasmid (MP); Northern and S1 nuclease analysis	381-bp promoter fragment of *Dictyostelium* functions in yeast, tenfold hs-induced transcription, initiation point flexible	24
B31	Ce-repetitive element with tandemly repeated HSE inserted 260 bp 5′ upstream of Sc-his3 gene	Transformation of his$^-$ cells with 2μ multicopy plasmid (MP); restoration of his prototrophy	The 242-bp element confers hs-inducibility to the his3 gene, prototrophy observed by cultivation > 30°C	72a

Note: Fusion constructs using parts of hs genes are characterized by the fusion point in the promoter (P), leader (L), or coding region, respectively, e.g., for construct A1 (Dm-hsp70P/L (+65) × cat gene) the *E. coli* chloramphenicol acetyltransferase gene as reporter gene was fused at bp +65 of the leader region to a *Drosophila melanogaster* hsp70 promoter/leader fragment with about 1200 bp of 5′ flanking sequences (5′ fls). For functional analyses promoter deletions are generated by restriction enzyme cutting eventually followed by Bal31 nuclease digestion. In these cases, the last remaining base of the truncated promoter is indicated, e.g., in A4: with 194 bp of 5′ fls or, shorter, the −194 construct. For further explanation see text (Sections 6.1.1 and 6.1.2).

Extended version of Table 1 from Nover, L., *Enz. Microb. Technol.*, 9, 130, 1987.

6.1.2. METHODS OF ANALYSIS

For convenient discrimination between resident hs genes and those introduced by transformation, several methods were used.

Electrophoresis — Heterologous but in rare cases also homologous expression systems allow a direct electrophoretic or immunological discrimination of the gene products (A27, A28, A30, B5, B7, B10, B14, B15, B16, B17, B18, B20, B22, B27, B29).* The situation is facilitated by using truncated hsp-coding genes, e.g., a 3′ truncated *Drosophila* hsp70 gene (hsp40, see A15, A13), a *Drosophila* hsp70 gene with a deletion of codons 114 to 337 (hsp44, see A6), a *Drosophila* hsp27 gene with deletion of codons 12 to 68 (hsp18.5; see A22), or a yeast 3′ truncated hsp90 gene (hsp69.2, see A39).

RNA or protein tagging — For detection of transformed genes and their products with available probes (nucleic acid probes, antibodies), the genes are tagged by insertion of corresponding synthetic oligonucleotides or DNA fragments. Examples are the insertion of λ DNA into *Drosophila* HSP-coding genes (A4, A4a, A24), of an adenovirus fragment into the Dm-hsp27 gene (A21b), of a herpes simplex virus fragment into the Dm-hsp22 gene (A28), or of synthetic oligonucleotides coding for a P-peptide or part of the c-myc gene product into the 3′ part of the Dm-hsp70 gene (B3a, B7).

Fusion to reporter genes — Related to this is the fusion of hsp genes or their promoter/leader (P/L) regions with reporter genes. In most cases, genes from *Escherichia coli* or one of its plasmids were used, e.g., the lacZ gene (β-galactosidase, see A11, A12, A12a, A18a, A20a, A21c, A23, A23a, A23b, A26, A29, A29a, A31, A38, A40, A41, A41a, B4, B19, B19a, B23, B28, B29b), the cat gene (chloramphenicol acetyltransferase, see A1, A2, A20a, A22a, A30b, A32, A33, A33a, A33d, A33g, A36c, A37c, A37d, B1b, B12, B16b, B17e, B24, B25), the nptII gene (neomycin phosphotransferase, see A15, A17, A30a, A33b, B26), the gpt gene (guanosine phosphoribosyltransferase, see A3), or the gus gene (β-glucuronidase, see A37b). But the list also includes eukaryotic genes, e.g., the mouse interferon gene (B9a), the hGH gene (human growth hormone, see B11), the tk gene (herpes simplex virus thymidine kinase, see A20, A21, A25, A34a, B3, B6, B8, B9, B15, B17a, B18, B21, B21a), the hβ-gl gene (human β-globin, see B1a), the human histone H3 gene (B1), the semliki forest virus capsid protein gene (A17), the dhfr gene (mouse dihydrofolate reductase, see A39b, B13), and a number of *Drosophila* genes. These are the larval adh gene (alcohol dehydrogenase isozyme, see A4, A7, A8, A8a, A9, A10, A37a), the w locus (white locus, see A18, A18a), the ftz gene (segmentation gene fushi tarazu, see A19), and the P-element tnp gene (transposase, see A15, A16).

Selection — For selection of stably transformed cells, suitable genes were either directly fused to hs control regions, inserted on the same plasmid, or introduced on a separate DNA fragment (cotransformation). Leucine auxotrophic yeast cells ($leu2^-$) were transformed with $leu2^+$ plasmids (A39-A42, B27), a ura3 plasmid was used to correct the lack of active orotidine-5′-phosphate decarboxylase in ura^- cells (B30), and a yeast deficiency in histidine biosynthesis was repaired by introduction of a his3 genes (A43, B21). Defective *Drosophila* mutants were normalized by transformation with the corresponding wild-type (WT) genes, DNA or gene constructs: w^- (white eyes, see A16, A16a, A18, A18a, A18i), ry^- (rosy gene, reddish-brown eye color mutant because of lacking xanthine dehydrogenase, see A10, A12, A12a, A12b, A18b—h, A22, A23, A25, A26, A30) and a shi^{ts} cell line which stops to grow at 30°C (A13). Chemical resistance markers are neo^r (resistance of plant and animal cells to kanamycin or the gentianamycin derivative G418 due to the presence of the bacterial nptII gene, see A15, A16, A33b, A35, B15, B17a, B26), adh (resistance to ethanol treatment of *Drosophila,* see A8, A8a, A19, A24), dhfr (resistance of animal cells to methotrexate, see A6, A9, A14, A39b, B13) and gpt, tk (normalization of mammalian tk^- cells to grow on HAT medium, see A5, A33, B10, B11, B14, B16, B17a, B21a). A *Drosophila* gene coding for a subunit of RNA polymerase II, which confers α-amanitin resistance, was used to select stably transformed S2 cells of the fly (see A18g).

* The letter-number codes (A1, A2, B1, B2, etc.) used throughout this chapter refer to entries in Table 6.1.

Analytical techniques for characterization of transformed cell lines are indicated under Methods (Section 6.1.2). They rely on specific probes for testing nucleic acids or proteins. The integration of hs constructs into the chromosomal material of the recipient cell (organism) is demonstrated by means of Southern blots or, with polytene chromosomes of *Drosophila* larvae salivary glands, by *in situ* hybridization. Formation of hs-induced transcripts of the foreign gene is analysed by dot blot, Northern blot, S1 nuclease, or primer extension analyses. The latter two methods can give information about the fidelity of transcription initiation and termination. Protein products of the hs gene constructs under investigation (HSPs, ADH) were analyzed by one- or two-dimensional electrophoresis (PAGE), by immunological techniques, or by *in vitro* or *in situ* assays of their enzymic or biological activities, e.g., products of the lacZ, cat, nptII, tk, adh, w, ftz, or tnp genes.

6.2. THE HEAT SHOCK ELEMENT (HSE)

Sequence comparison of several hsp-coding genes of *Drosophila* revealed homology boxes in the 5′ flanking and mRNA leader regions including the universal TATA box at −30 and a specific 14 bp palindrome 20 to 30 nucleotides further upstream of this.[57] The core of the latter sequence (HSE), essential for hs induction, was defined by Pelham and Bienz:[15,102,104] -CTnGAAnnTTCnAG-. Usually a minimum of seven correct nucleotides out of the ten given in bold face letters is required for activity (see Table 6.2). A synthetic oligonucleotide containing the HSE conferred hs inducibility to the HSV tk gene, if the hybrid was tested in COS cells and *Xenopus* oocytes.[104]

Meanwhile, many more promoters were sequenced and analysed by transformation of homologous and heterologous cells (for reviews see Bienz,[13] Bienz & Pelham,[17] Craig,[34] Lindquist,[75] Lindquist and Craig[75a], Nover[95,96], Nover et al.,[96a] Pelham[1032] Schoeffl et al.,[120] and Voellmy[142]). The experiments included a great number of promoter deletion mutants generated from appropriate restriction sites, mostly in combination with Bal31 exonuclease treatment. The positions and sequences of HSEs relative to the TATA box and transcription start sites are compiled in Table 6.2. It is essential that localization of a HSE by sequence comparison is followed by functional analysis in appropriate expression systems (see last column of Table 6.2 and Figure 6.6). Despite the general role of the HSE for hs inducibility in all eukaryotic cells so far tested, details of its function and interaction with other promoter elements may be highly variable. Many examples in Figure 6.6 illustrate that the outcome of expression experiments in quantitative and qualitative aspects are dependent on the cell system used. It is a matter of curiosity that the synthetic HSE inserted into the promoter of the HSV tk gene was active in COS and *Xenopus* cells,[104] but later proved to be inactive in *Drosophila* cells.[87] Functional analyses of HSEs must be aware of the influence of the sequence and cellular context, e.g., availability of promoter binding proteins, stability and fate of the gene expression products, etc. A number of interesting examples are compiled in the following sections.

The survey of HSEs in Table 6.2 reveals several remarkable features:

1. The extent of conservation (%) is different for the individual nucleotides: 5′-C(80)T(52)A(45)G(91)A(90)A(88)nnT(86)T(81)C(91)T(42)A(73)G(75)-3′. The highly conserved palindrome CnnGAAnnTTCnnG contains two less conserved peripheral TA-dinucleotides and a central nonconserved dinucleotide (nn).
2. The central 10 nucleotides are formed by a 5′ purine cluster (-AGAAn-) and a 3′ pyrimidine cluster (-nTTCT-), which are situated on opposite sites of the DNA helix.
3. The 5′ and 3′ terminal tetranucleotides are identical (-CTAG-). This pecularity gives rise to the frequently encountered direct 10 bp repeats formed of overlapping duplicate or triplicate hs boxes. In these types of constructs the repeating purine and pyrimidine clusters are spaced by 10 bp, i.e., by one turn of the DNA helix. The significance of this is dealt with in the following section.

TABLE 6.2
Heat Shock Promoter Elements

Gene	Heat shock boxes 5′-CTnGAAnnTTCnAG-3′	10	TATA box (•, −30)	TS	5′L (nucl.)	Functional analysis (Table 6.1)	Ref.
Dm-hsp83	(−84) CcaGAAgccTCtAG (−71)	8				—	8, 53, 57
	(−74) CTaGAAgtTTCtAG (−61)	10					
	(−64) CTaGAgacTTCcAG (−51)	9	TATȦAA	-G A	149		
Ds-hsp83	(−84) CcaGAAgccTCtAG (−71)	8				—	8
	(−74) CTaGAAgtTTCtAG (−61)	10					
	(−64) CTaGAgtcTTCgAG (−51)	9	TATȦAA	-G A	149		
Dp-hsp83	(−85) CTcGtAgccTCtAG (−72)	8				—	8
	(−75) CTaGAAacgTCtAG (−62)	9					
	(−65) CTaGAAaaTTCtAc (−52)	9	TATAȦA	-G A	136		
Dv-hsp83	(−85) CgaGAAgtcTCtAG (−72)	8				—	8
	(−75) CTaGAAgtgTCtAG (−62)	9					
	(−65) CTaGAAgtTTCgAG (−52)	10	TATAȦA	-C A	150		
Dm-hsp70 (87C1)	(−262) agcGAAtaTTCtAG (−249)	8				A4, A10, A11, A11a, A12, A21c, B2, B3, B6, B15, B18, B19	52, 62, 64, 79, 98, 141
	(−202) aaaGAAaacTCgAG (−189)	7					
	(−192) CgaGAAttTTCtct (−179)	7					
	←(−84) CtcGttggTTCgAG (−71)	7					
	(−61) CTcGAAtgTTCgcG (−48)→	9	TAṪAAA	-C A	242		
Dm-hsp70 (87A7)	(−253) aaaGAAtaTTCtAG (−240)	8					52, 62, 64, 79, 98, 141
	(−193) aagGAAaacTCgAG (−180)	7					
	(−183) CgaGAAatTTCtct (−170)	7					
	←(−84) CTcGttgcTTCgAG (−71)	8					
	(−61) CTcGAAtgTTCgcG (−48)→	9	TAṪAAA	-C A	242	see 87C1	
Dm-hsp68	(−94) CTgGAAtgTTCtga (−81)	8				—	6, 57
	(−61) CTcGAAttTTCccc (−48)*	8	TATȦAA	-C A			
Dm-hsp27	(−369) CaaGAgaacTCcAG (−356)	7				A22, A22a	25, 61, 56, 133
	(−355) aaaGAAatgTCaAG (−342)	7					
	(−345) CaaGAAgtTTCtgG (−332)	8					
	(−296) CTaGAAagagCcAG (−283)	8					
	(−286) CcaGAAgaTgCgAG (−273)	8					
	(−219) CTtaAA-cTTtaAG (−207)	8	TAṪAAA	-C A	118		

TABLE 6.2 (continued)
Heat Shock Promoter Elements

Gene	Heat shock boxes 5′-CTnGAAnnTTCnAG-3′	10	TATA box (●, −30)	TS	5′L (nucl.)	Functional analysis (Table 6.1)	Ref.
Dm-hsp26	(−350) CTaGAAacTTCggc (−337)	8				A23a, A24, A25, A26	25, 56, 61, 133
	(−281) tTtaAAatTTCtcG (−268)	7					
	(−271) CTcGAAacTcatgG (−258)	7					
	(−188) aaaGAAatTTCtAa (−175)	7					
	←(−72) CTgtcActTTCcgG (−59)	7					
	(−62) CcgGActcTTCtAG (−49)→	8	TATȦAA	-C A	182		
Dm-hsp23	(−413) CTcGAAgtTTCgcG (−400)→	9				A25, A29, A29a	25, 56, 61, 133
	*(−393) CTccAtccTTCgtG (−380)	7					
	*(−372) CTccAAccTTCcta (−359)	7					
	(−213) CTctcActTTCaAt (−200)	7					
	(−181) gcgGcAaaTTCgAG (−168)→	7					
	(−146) CgaGAAgtTTCgtG (−133)	8	TATȦAA	-C A	112		
Dm-hsp22	*(−196) CcaGAAacTTCcAc (−183)*	8				A28, A30	25, 56, 61, 133
	(−95) gaaGAAaaTTCgAG (−82)	8					
	(−75) CcgGtAttTTCtAG (−62)	8	TATȦAA	-C A	251		
Dm-gene1	(−94) CTgGAAgtgaCcAG (−81)	8	CAṪAAG	-C C	93	—	5
Dm-gene2	(−339) tTcGttgaTTCtAG (−326)	7				—	101
	(−316) tggGAAccTTCtgG (−303)	7					
	(−52) CccGAAtcTTtaAt (−39)	7					
	(−38) CTtGAAccTcatAa (−25)	7	:CATAAA	-C A	60		
Dm-gene3	(−404) tTatgAggTTCaAG (−391)	7				—	99
	(−390) aTtaAAgaTTCggG (−377)	7					
	(−127) CcaGAAggTTCcca (−114)	7					
	(−104) CTaGAAtcaaCgAa (−91)	7	TATAȦA	-A T	168		
Dm-93D	(−463) CTgGAtatTTCcAT (−450)	8				—	58a
	(−304) CacGAAatTcttAG (−291)	7					
	(−252) gcaGtAgtTTCcAG (−239)	7	TAṪAAA	-A T	—		
Dh-48B	(−60) CTcGAAtcTTCgAG (−47)→	9	:TATA	-A G	—	—	116a
Hs-hsp90α	(−85) tccGgAagTTCggG (−72)	6				—	54a, 144b
	(−75) CggGAggcTTCtgG (−62)→	7	:TATATA	-T C	60		

Hs-hsp90β	(−650)	CTgGAAacTgCtgG	(−637)	8				—	110a
	(−640)	CTgGAAatgcCgca	(−627)	6	−25 TATATA	-T A	99		
Hs-hsp70A	*(−189)	CTgGAgagTTCtga	(−176)	7				A33, B23	38a, 60
	*(−108)	CTgGAAtaTTCccG	(−95)	9	TATAAA	-G A	213		
Hs-hsp70B	*(−259)	CTaGAAccTTCccc	(−246)*	8				A30b	116b, 144
	(−160)	CccGAAccTTCtcc	(−147)	7					
	(−75)	CggGAAggTgCggG	(−62)	7					
	(−65)	CggGAAggTTCgcG	(−52)	8	AAAAGG	-G A	119		
Hs-hsc71	(−223)	CTtGAAggTTCcAG	(−210)→	10				—	42
	←(−49)	CTgGAAggTTCtAa	(−36)	9	−28 TATAA	-C C	77		
Hs-hsp27	(−185)	agaGAAggTTCcAG	(−172)	8	CATAAA	-C G	40	—	54
Hs-factor B gene	(−757)	CctGAAttTTttAc	(−744)	7				—	150
	(−467)	aTgGAAttTcCcAG	(−454)	8					
	(−98)	CTtGgAtgTTCcgG	(−85)	7	−28 TATAA	-C A	?		
Rn-hsc73	(−216)	CgtGAAagTTCcAG	(−203)	9				—	130
	(−206)	CcaGAAcgcTgcgG	(−193)	6					
	(−63)	CggaAcccTTCtgG	(−50)	6					
	(−53)	CTgGAAggTTCtAa	(−40)	9	TATAAG	-T C	80		
Rn-hsc70 (testis)	(−75)	CTgagAgtTTCcAG	(−62)→	8	TATAA	-G A	112	—	145c
Rn-ho gene	(−287)	CTgGAAccTTCcAG	(−273)	10	ACTTAA	-C A	?	A33e	90, 122
Gd-hsp90	(−73)	agtGAAggTTCccG	(−60)	7				A33g	144a
	(−63)	CccGAtggTTCtgG	(−50)	7	TATATA	-G C	82		
Gd-hsp70	(−65)	CcttAgcgTTCtgG	(−52)	6				—	88
	(−55)	CTgGcAggTTCcAG	(−42)→	9	TATAA	-A A	112		
Gd-ubi gene	(−369)	CTcGAAtcTTCcAG	(−356)	10				—	19, 20
	(−359)	CcaGAgctTTCttt	(−346)	6	ATAAA	-A C	63		
Xl-hsp70	(−252)	CgaGAAagcTCgcG	(−239)	7				B1b, A34a	11, 16
	(−242)	CgcGAAtcTTCcgc	(−229)	7					
	(−196)	CcgGAAacctCgcG	(−183)	7					
	(−186)	CgcGAAagTTCttc	(−173)	7					
	(−129)	aTgGAAgccTCggG	(−116)	7					
	(−119)	CggGAAacTTCggG	(−106)	8	TATAAA	-A A	105		
Xl-hsp30	(−155)	CcaGAAgtTgCtAG	(−142)	7				—	11, 13
	(−70)	aTgGAAgtcTCggG	(−57)	7					
	(−60)	CggGAAcgTcCcAG	(−47)→	8	TATAAA	-C A	237	—	
Ce-hsp70A	(−283)	CgcGAAcaTTCtct	(−270)	7				—	129b
	(−72)	tTcGAAttTTCtAG	(−59)→	9	TAAATT	-C A			

TABLE 6.2 (continued)
Heat Shock Promoter Elements

Gene	Heat shock boxes 5′-CTnGAAnnTTCnAG-3′	10	TATA box (●, −30)	TS	5′L (nucl.)	Functional analysis (Table 6.1)	Ref.
Ce-hsp70F	←(−67) CgaGAAatTTCtcG (−54)	8	ATTAAA	-C A	?	—	53b
Ce-hsp16-48	(−212) CTaGAAcaTTCgAG (−199)	10				B17b	116
	(−202) CgaGctgcTTCttG (−189)	6					
	(−68) CTaGgAccTTCtAG (−55)	9					
	(−58) CTaGAAcaTTCtAa (−45)	9	$^{-25}$TATATAA	-A A	51		
Ce-hsp16-1	(−211) tTaGAAtgTTCtAG (−198)	9				B17b	116
	(−201) CTaGAAggTcCtAG (−188)	9					
	(−67) CaaGAAgcagCtcG (−54)	6					
	(−57) CTcGAAtgTTCtAG (−44)	10	$^{-25}$TATAAA	-C A	42		
Ce-hsp16-41	(−212) CTaGAAcaTTCgAG (−199)	10				—	63
	(−68) CTaGgAccTTCtAG (−55)	9					
	(−58) CTaGAAcaTTCtga (−45)	8	$^{-25}$TATAT	-A A	42		
Ce-hsp16-2	(−211) acaGAAtgTTCtAG (−198)	8				—	63
	(−201) CTaGAAggTcCtAG (−188)	9					
	(−57) CTcGAAtgTTCtAG (−44)	10	$^{-25}$TATAA	-G A	38		
Zm-hsp70	(−109) CcaGAgccTTCcAG (−96)→	8				—	115
	(−63) CccGAAtcTTCtgG (−50)	8	AATAAA	-G T	107		
At-hsp70-1	(−138) atcGAAcaTTCtcG (−125)→	7	TATATA	-T T	106	—	151a
Gm-hsp17.3B	(−173) CccGAAacTTCtAG (−160)	9				A36a	119
	(−142) CcaGAAtgTTtctG (−129)	7					
	(−132) tctGAAagTTtcAG (−119)	7					
	(−122) tcaGAAaaTTCtAG (−109)	8					
	(−70) CaaGgActTTCtcG (−57)	7					
	(−60) CTcGAAagTaCtAt (−47)	8	$^{-25}$TTAAAT	-A C	103		
Gm-hsp17.5E	(−371) CTttAAcaTTCtAa (−358)	8				A37	37
	(−100) gTgGAgaaTTCaAc (−87)	7					
	(−72) tagGAtttTTCtgG (−59)	6					
	(−62) CTgGAAcaTaCaAG (−49)	9	TTAAAT	-C G	82		

Gene	Start	Sequence	End	Matches	TATA		Distance		Ref.
Gm-hsp17.5M	(-342)	aaacAAtaTTCtAG	(-329)	7				—	93, 109a
	(-332)	CTaGAAaaaTattt	(-319)	6					
	(-271)	CaacAAtaTTtcAG	(-258)→	7					
	(-155)	CTaGAAccTTCgta	(-142)	8					
	(-135)	gTgGAgaagTCcAG	(-122)	7					
	(-125)	CcaGAAgtTTttAt	(-112)	7					
	(-72)	CacGAtttTTCtgG	(-59)	7					
	(-62)	CTgGAAcgTaCacG	(-49)	8	ṪTAAAT	-G A	88		
Gm-hsp17.6L	(-182)	CTaGAAggTTgtAG	(-169)	9				—	93, 109a
	(-160)	CTaGAAcgTaCgta	(-147)	7					
	(-141)	gTgGAgaagTCctG	(-128)	6					
	(-131)	CctGAAgtTTatcG	(-118)	7					
	(-121)	aTcGAAtcaTCtAa	(-108)	7					
	←(-62)	CTgGAAcaTaCaAG	(-49)	9	ṪTAAAT	-C A	93		
Gm-hsp18.5C	(-130)	CTgGttttcTgtAG	(-117)	6				—	110
	(-120)	gTaGAAagcTCtAG	(-107)	8					
	(-110)	CTaGAActTgggAt	(-97)	7					
	(-70)	acaGAAttTTCtgG	(-57)	7					
	(-60)	CTgGAAaaacacAG	(-47)	7					
	(-50)	CagGAttcTTCctc	(-37)	6	ȦTATAA		76		
Gm-hsp17.9D	(-218)	gcacAAgaTTCtgG	(-205)	6				—	110
	(-208)	CTgGAcatTaCtAG	(-195)	8					
	(-198)	CTaGAAagaTCcga	(-185)	7					
	(-159)	accGAAccTaCtgG	(-146)	6					
	(-149)	CTgGAAgtTTCaca	(-136)	8					
	(-128)	aTtGctttcTCcAG	(-115)	6					
	(-118)	CcaGAAacTTCcAt	(-105)	8					
	(-74)	tcaGAAacTTCcAt	(-61)	7					
	(-54)	CTcGAAttaTCtAt	(-41)	8	TAṪAA		72		
Dd-IR element	(-87)	tTcGAAtgTTCtAG	(-74)	9				—	24
	(-77)	CTaGAAcaTTCtAa	(-64)	9	TAȦATA				
Sc-grp78	(-169)	---GAAccTTCtgG	(-156)		−60			—	94b
(KAR2)	(-159)	CTgGAAatTTCacc	(-146)	7	TATȦAA	-C A	38		
Sc-hsp70	(-312)	tgtaAActTTCcAG	(-299)	7				—	129
(SSA1)	(-302)	CcaGAAcaTTCtAG	(-289)→	9	−100				
	←(-199)	CcaGAAcgTTCcAt	(-186)	8	TATATAȦ	-C A	60		
Sc-ubi 4	←	CTaGAAcgTTCtAG	→	10	?		?	—	96b

TABLE 6.2 (continued)
Heat Shock Promoter Elements

Gene	Heat shock boxes 5'-CTnGAAnnTTCnAG-3'	10	TATA box (•, −30)	TS	5'L (nucl.)	Functional analysis (Table 6.1)	Ref.
Bl-hsp70	←(−202) CgtGAActTTCggG (−189)	8	TATCṪT (−35)	-T T	60	—	63a
Um-hsp70	(−93) CTgGAAgtTcatAG (−80)	8	AĊATAA (−50)	-T T	126	—	56a
Nc-hsp30	(−340) gTgGAActTTCtAG (−327)	9				—	108b
	(−330) CTaGgctgcTCccG (−317)	6					
	(−169) agtGgccaTTCtAG (−156)	6					
	(−159) CTaGAAtgacCtgG (−146)	7					
	(−149) CTgGcAtcTgCatc (−136)	6	ATȦTAA (−50)	-A G	121		

Note: Positions and sequences of hs boxes were derived from sequence comparison (see references given) and from functional analyses by transformation experiments (see references to Table 6.1). Only 7 out of 10 bp matches with the consensus sequence defined by Pelham and Bienz[102,104] were included. Numbers in parentheses mark positions of the first and last nucleotides, respectively. Downstream of the last hs box, positions of the TATA-box and of the transcription start point (TS) are given. If not indicated otherwise, the dot at the TATA-box marks position −30. The numbering in the promoter region depends on the position of the transcription start, which is not always unambiguous. Length of the untranslated leader (5'L) of mRNA is given in nucleotides. A survey on the position of the hs boxes and other promoter elements is presented in Figure 6.6. In several cases, extensions of the hs boxes are marked. Arrows denote AA at the 3' and TT at the 5' ends, respectively. Stars denote -4n-TTC at the 3' and GAA-4n- at the 5' ends, respectively (see Section 6.3). Organisms are abbreviated by a two letter code as indicated in Abbreviations and Special Terms.

6.3. INTERACTION OF HSEs

The finding of overlapping oligomers of HSEs in promoters of plant,[119] *Drosophila,*[105] *Xenopus,*[11,12] and nematode hs genes[116] led to a reconsideration of the early observation that in monkey COS cells a synthetic overlapping dimer was not more active than the corresponding monomer.[104] Constructs on the basis of the *Caenorhabditis elegans* promoter were introduced into mouse cells and tested in a transient expression assay[65] (Figure 6.1). In addition to the hs construct, the disarmed bovine papilloma virus vector contained a hybrid of the bacterial nptII gene fused to the constitutively active thymidine kinase promoter. Expression of the npt gene was used for internal standardization (= 1). Compared to this hs-induced activity of the promoter with a HSE monomer was 3-fold, that with the overlapping tetramer was even 90-fold higher. These results demonstrate the considerable potentiation of HSE activity by organization into oligomers. The efficiency of such promoters surpasses that of good constitutive promoters by about two orders of magnitude.

Another type of amplification by HSE interaction is exemplified by the *Drosophila* hsp70 promoter. The original deletion analysis using multicopy conditions in monkey, mouse, and *Xenopus* cells as heterologous expression systems[15,30,85,102] led to the definition of one HSE only, about 15 nucleotides upstream of the TATA box. However, expression of similar deletion constructs in the homologous *Drosophila* cells after stable integration[41] demonstrated the interaction with a second HSE localized nine nucleotides 5′-upstream of the first. HSE1 alone gives only 2% of the normal hs-induced activity. Using purified heat shock transcription factor (HSF) Shuey and Parker[123] found a cooperative binding interaction between both HSE regions. First the TATA proximal then the distal HSE is occupied (see Figure 7.3). "Resonance" between both HSEs is only observed, if they are spaced by multiples of a helical turn, i.e., $(10)_n$ bp with n = 1 to 8.[27a]

Meanwhile, the sequence and positional details of the two dominant HSEs of the Dm-hsp70 promoter were characterized.[2,2a,154] So far no contribution of the other three HSEs further upstream was recognized (see Table 6.2). Results with numerous constructs modifying the position and sequence of the distal HSE2 are summarized in Figure 6.2. A transient assay in *Drosophila* S3 cells was used in these experiments. The usual spacing of the two HSEs by one turn of the DNA double helix (D88, wild-type promoter) is superior to a spacing of a $1^1/_2$ turn (IN4),[27a] but even 29 nucleotides of spacer (three turns of the helix) are tolerated (IN20). Increase of the number of upstream HSEs to two (R6) and three (R7) even enhances the promoter activity two to threefold. Remarkable are the findings with a number of HSE2 sequence variants shown in the lower part of Figure 6.2. Despite a good agreement with the consensus sequence, constructs S4, S2, and D78F have only negligible effects on the hs inducibility of these modified hsp70 promoters. In addition to the basic sequence homology, there must be other, so far ill defined, requirements for the effective interaction between both HSEs.

A similar reinvestigation of HSE1 and its flanking sequences [2a,154] led to an interesting analogy with the results on overlapping HSE oligomers. Evidently the wild type HSE1 plus flanking sequences functions as $1^1/_2$ HSE; (−61)ctcGAAtgTTCgcGAA (−46). Removal or modification of the GAA at the 3′ end inactivates HSE1. On the basis of these results, Nover[96] and Xiao and Lis[154] proposed to define the consensus HSE as a repetitive 10 bp element (-nGAAnnTTCn-). Functional variations to this general theme could be (1) nTTCnnGAAnnTTCn, (2) nGAAnnTTCnnGAAn, (3) nGAAnnTTC7nTTCn, and (4) nGAA7nGAAnnTTCn.

So far only variations (1) and (2) with $1^1/_2$ HSE were shown to be functionally significant.[2,154] They reflect the basic structure of the active hs transcription factor, which was found to be a homotrimer in yeast and *Drosophila*. Each subunit interacts with $^1/_2$ HSE, and at least two, but better three half-sites are required for stable binding of HSF[106a,132a]

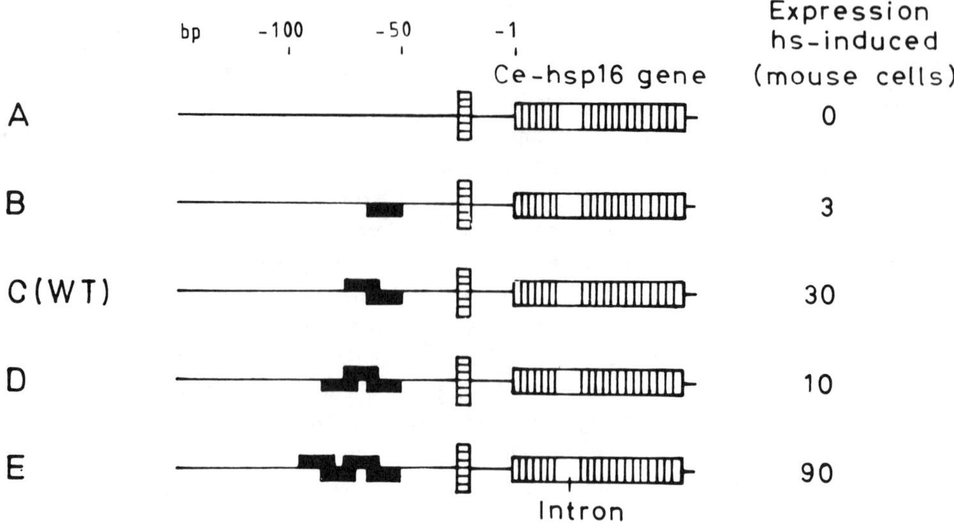

FIGURE 6.1. Effect of overlapping multimers of HSE on the hs-induced expression in mouse cells. A, nematode *(Caenorhabditis elegans)* hsp16 gene pair (see Figure 3.2) was introduced into mouse cells on plasmids containing an HSVtk(P) × nptII gene as selector gene and internal standard. Expression of the hsp16 gene after 60 min at 42.5°C + 30 min at 37°C was analyzed by S1 nuclease mapping. Using the HSE tetramer, maximum transcript levels were about 90-fold higher than the constitutive levels of the nptII gene (= 1). Constructs C and D contain further hs elements upstream which may influence the transcript levels. (Data from Kay, R. J., Boissy, R. J., Russnak, R. H., and Candido, E. P. M., *Mol. Cell. Biol.*, 6, 3134, 1986.)

(see Section 7.6). Variations (3) and (4) with deletion of one trinucleotide motif of the palindrome are also frequently observed, extended forms of the HSE. Both types of extensions are marked in the compilation of HSEs in Table 6.2 (see note at the end of Table 6.2 for explanation).

Consideration of HSE in this more general sense of a repetitious decamer helps the understanding of the results with yeast promoters. On the one hand, the Sc-hsp90 promoter shows poor homology with the classical HSE,[44] but upstrean of a putative TATA motif, centered around bp −79, there is a long HSE-like sequence containing repeats of -TTC- and -GAA- spaced by two to three and six to seven nucleotides, respectively: (−228)GAA3nTTC7nTTC3nTTC3nGAA6nGAA3nTTC6n<u>TTC2nAGA</u>7nGAA(−159). Similar variants of yeast hs promoter elements were reported for the SSA1, 3, and 4 genes,[36a] for the STI1 gene,[94a] for the hsp26 gene,[139a] and for the phosphoglycerate kinase gene.[108a] Mutations of the hsp90 promoter demonstrated that transitions of T⟶G and G⟶C nucleotides in the TATA-proximal –TTC2nGAA– motif (underlined) result in defective basal level expression, whereas hs-inducibility is unchanged.[79a] The only detailed analyses on functional requirements of the yeast hs promoters were reported by Slater and Craig[129] for the SSA gene family (see also Craig et al.[36a]). In the SSA1 promoter, several HSEs with extensions at their 5′ and 3′ ends are present and the proximal two are required for efficient hs-induced activity.

Tests with yeast cells as the expression system and insertion of synthetic HSE into the iso-1-cytochrome c promoter fused to the *E. coli* lacZ gene were reported by Wei et al.[145] and Sorger and Pelham[132] (Figure 6.3). A maximum of four synthetic HSEs with ten matches of the consensus sequence were each spaced by five nucleotides. The hs-induced activity was half of the very active wild-type enhancer usually found in this position of the cyc1 gene. In Figure 6.3B the effects with insertion of overlapping dimers of HSEs are illustrated. The HSE2 with 10/10 matches gives almost the same activity as the four spaced monomers

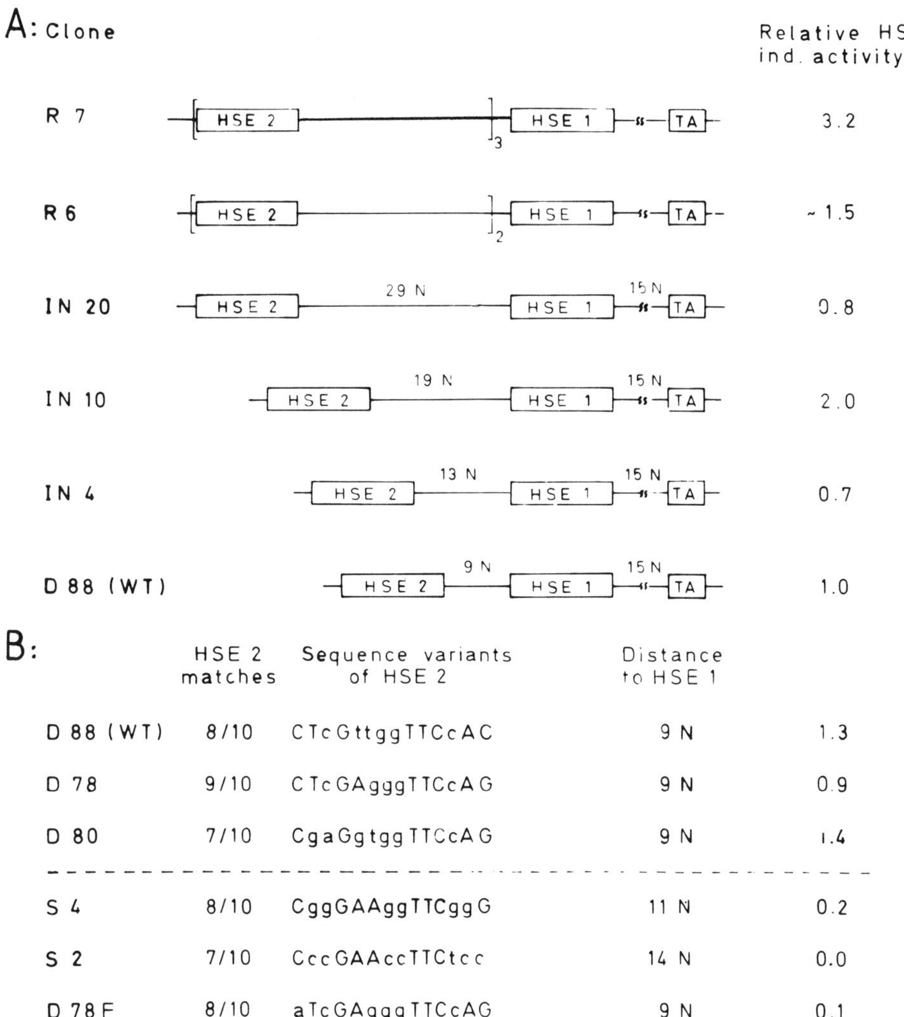

FIGURE 6.2. Influence of structure and distance of the second HS-box (HSE 2) on the activity of the *Drosophila* hsp70 gene promoter. HS-induced transcript levels of Dm-hsp70 (P/L) × lacZ fusion constructs were tested in *Drosophila* S3 cells (transient expression). In the wild type promoter both HSE are spaced by 9 nucleotides and separated from the TATA box by 15 nucleotides (strain D88 in the middle). In part A, the distance between HSE 1 and 2 was increased from 1 to 1.5 (IN 4), 2 (IN 10), and 3 (IN 20) helical turns, and the HSE 2 was duplicated (R6) or triplicated (R7). In part B, sequence of the HSE 2 part was modified. Despite the good matches with the consensus, HSEs of the second group (S4, S2, D78F) have practically no activity (HSE1 alone gives ~ 0.1). (Compiled from Amin, J., Mestril, R., Schiller, P., Dreano, M., and Voellmy, R., *Mol. Cell. Biol.*, 7, 1055, 1987.)

in part A. Remarkable are the drastic differences between the two fairly similar dimers with 8/8 matches each. HSE27 is active whereas HSE12 is not. The essential difference between both is a regular pattern of GAA7nGAA7nGAA in the HSE27, which is distorted to GAg7nGAg7nGAg in the HSE12 construct.

6.4. THE HEAT SHOCK ENHANCER

It is evident from Table 6.2 and Figure 6.6 that maximum hs inducibility of many genes in different kinds of organisms depends on multiple HSEs frequently scattered over several hundred base pairs of the promoter region (see data given in Figure 6.6 for the *Drosophila*

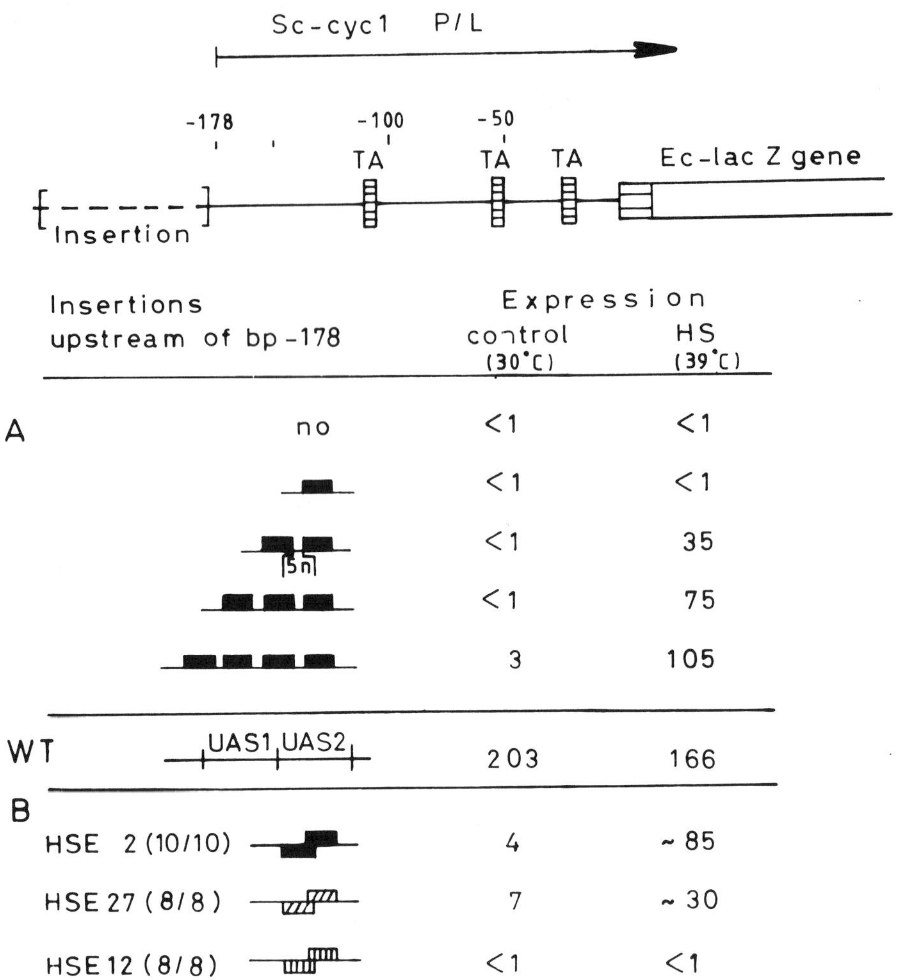

FIGURE 6.3. Activation of the truncated yeast iso-1-cytochrome c gene promoter by synthetic hs box oligonucleotides. Synthetic oligonucleotides with 10/10 matches of the hs box were inserted upstream of bp − 178 of a cycl promoter/leader element fused to the bacterial lacZ gene. Three potential TATA boxes are contained in the cycl promoter. β-Galactosidase activity in control (30°C) and hs cells (39°C) was measured. Hs-induced expression increases with the number of hs boxes. In part A, each hs box is separated by 5 nucleotides from the adjacent. In part B, activity of overlapping dimers with 10/10 (HSE2) and with 8/8 matches (dimer between bp − 296 and − 273 of Dm-hsp27 promoter as well as a synthetic HSE12 of Pelham and Bienz[104]) were tested. For comparison, promoter activity with the wild-type hemin-regulated enhancer elements (UAS) is given. (Data were compiled from Wei, R., Wilkinson, H., Pfeifer, K., Schneider, C., Young, R., and Guarente, L., *Nucl. Acids Res.*, 14, 8183, 1986 (Part A) and (Part B) from Sorger, P. K., Lewis, M. L., and Pelham, H. R. B., *Nature*, 321, 81, 1987.)

genes hsp27, 26, 23, and 22). The cascades of HSEs extend up to bp − 350 for the hsp 26,[27,100] to bp − 620 or more of the hsp23,[74,100] and to bp − 2100 of the hsp27 promoter.[55,56] The results derived from functional analyses with corresponding deletion promoters were confirmed by *in vivo* footprint analysis using methidiumpropyl EDTA-iron(II) as the cleavage reagent.[25]

In optimum combination, HSEs can act as an enhancer, i.e., they act irrespective of their direction of insertion and distance from the transcription start.[121] Two examples from a paper by Bienz and Pelham[16] are given in Figures 6.4 and 6.5. The hs-induced expression of *Xenopus laevis* hsp70 gene constructs in HeLa cells is illustrated in Figure 6.4. The wild-

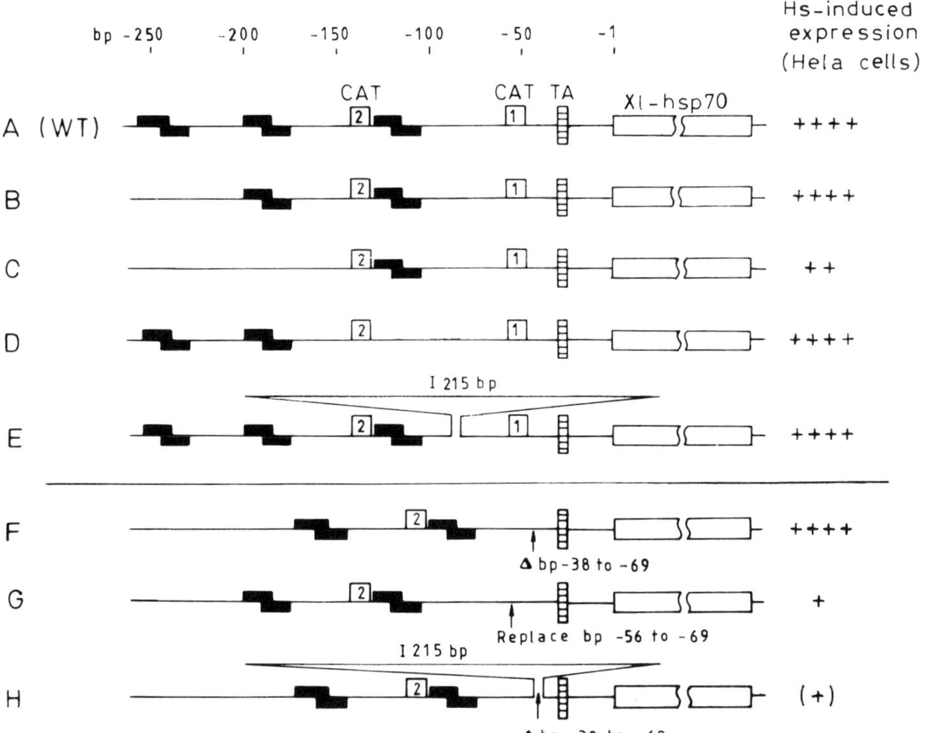

FIGURE 6.4. Expression of the *Xenopus* hsp70 gene with different promoter constructs in HeLa cells. The wild-type promoter (WT) contains three overlapping HSE dimers and two CAT-boxes interspersed. This promoter configuration is engineered as indicated by removing small parts or by insertion of 215 bp of "neutral" DNA. The hs-induced transcription level is indicated on the right side. For experimental details see Bla of Table 6.1. (Data from Bienz, M. and Pelham, H. R. B., *Cell,* 45, 753, 1986.)

type gene (A), containing three overlapping dimers of HSE and two CAT boxes in front of the TATA box, is highly expressed in HeLa cells. Removal of one HSE dimer has no effect, irrespective of whether it is the distal or the TATA-proximal one (constructs B, D). The distance between the HSE containing region of the promoter and the TATA box can be increased by insertion of 215 bp of foreign DNA (construct E) without influence on the hs inducibility, unless the TATA-proximal helper element (CAT-box 1) is eliminated (construct H). It is remarkable that two overlapping HSE dimers in combination function perfectly without CAT-box 1, but only if they are in close vicinity to the TATA box (construct F). One has to conclude that the enhancer-like activity of HSE combinations need complementation by TATA proximal helper elements, e.g., a CAT box or a HSE monomer itself. These findings may explain why the wild-type *Drosophila* hsp27 and hsp23 genes are not expressed in monkey COS cells (see numbers 12 and 14 of Figure 6.6).

These observations were extended (Figure 6.5) by use of the human β-globin gene as reporter gene and enhancer elements from three different sources: (1) the SV40 enhancer, (2) the *Xenopus laevis* hsp70 promoter element between bp −89 and −196, and (3) a duplication of the *Drosophila melanogaster* hsp70 promoter region between bp −37 and −108. All three enhancers were inserted about 750 bp upstream of the TATA/CAT-box combination of the β-globin gene. The SV40 enhancer confers strong constitutive expression, whereas the combinations with the two hs-enhancer constructs (C, D) are effectively transcribed only after hs induction of the HeLa cells. Similar to the results in Figure 6.4, the activity of the hs enhancer to act at a distance is abolished if the TATA-proximal CAT box is removed (construct E).

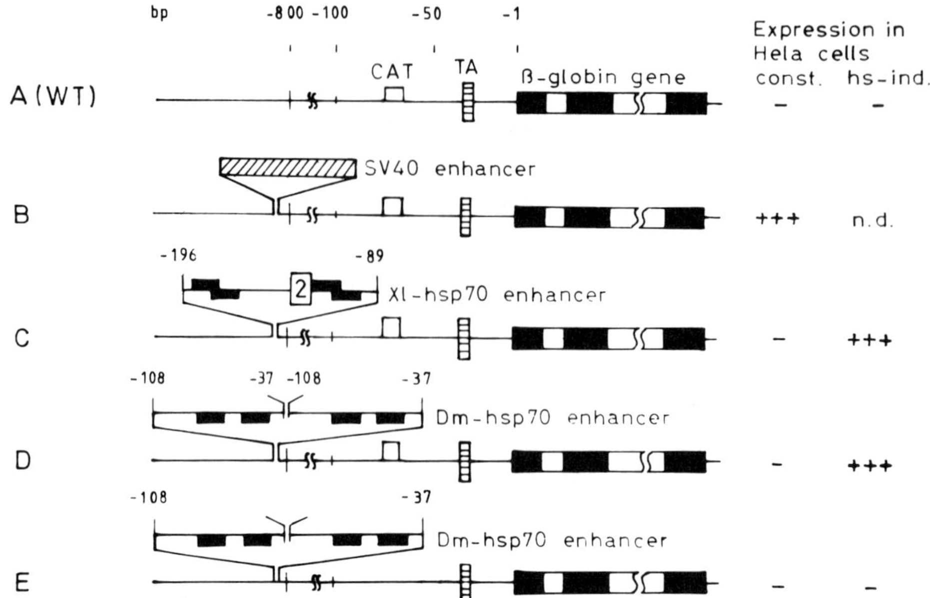

FIGURE 6.5. Effects of enhancers inserted in front of the human β-globin gene. The human β-globin gene was used as reporter for the activity of three types of enhancer element inserted 800 bp upstream of the transcription start. These are the SV40 enhancer as well as corresponding parts of the *Xenopus* and *Drosophila* hsp70 promoters. The extent of transcription in HeLa cells is documented on the right side at 37°C (constitutive) and after hs induction, respectively. (For experimental details see Bla of Table 6.1.) (Data from Bienz, M. and Pelham, H. R. B., *Cell,* 45, 753, 1986.)

A direct combination of the SV40 enhancer with a HSE-containing promoter was generated earlier by insertion of the former at bp −200 of the Dm-hsp70 promoter. The fusion construct gave high constitutive expression in monkey COS cells which could not be further increased after hs induction.[102] At the same position of the *Drosophila* hsp70 promoter Garabedian et al.[48] introduced a developmental enhancer, derived from the *Drosophila* yolk protein gene (ypl), which is expressed selectively in follicle and fat body cells. P-element transformation of *Drosophila* embryos (see A12a of Table 6.1) resulted in a correct developmental, i.e., tissue-specific control of the ypl × hsp70 hybrid promoter at 25°C, combined with a general induction under hs conditions.

The need for basic promoter elements with TATA-box and TATA-proximal helper elements and for their proper interaction explains why the mere existence of heat shock elements does not necessarily indicate hs-inducibility. A particularly striking example was reported for the nematode *C. elegans.* A novel repetitive element formed of multiple perfect matches of the HSE is found at 20 different loci of the genome without detectable hs-induced transcription of adjacent sequences. But transfer of such a 242-bp element to the yeast his3 gene promoter confers hs-inducibility in yeast.[72a]

6.5. COMPLEXITY OF HEAT SHOCK PROMOTERS

In addition to the position and possible contribution of HSEs to the hs inducibility, the survey of hs promoters from different organisms (Figure 6.6) demonstrates the presence of several other consensus elements with known but hs-unrelated functions. Thus, many promoters activated under hs are in fact bi- or even multifunctional. Some interesting examples are briefly described.

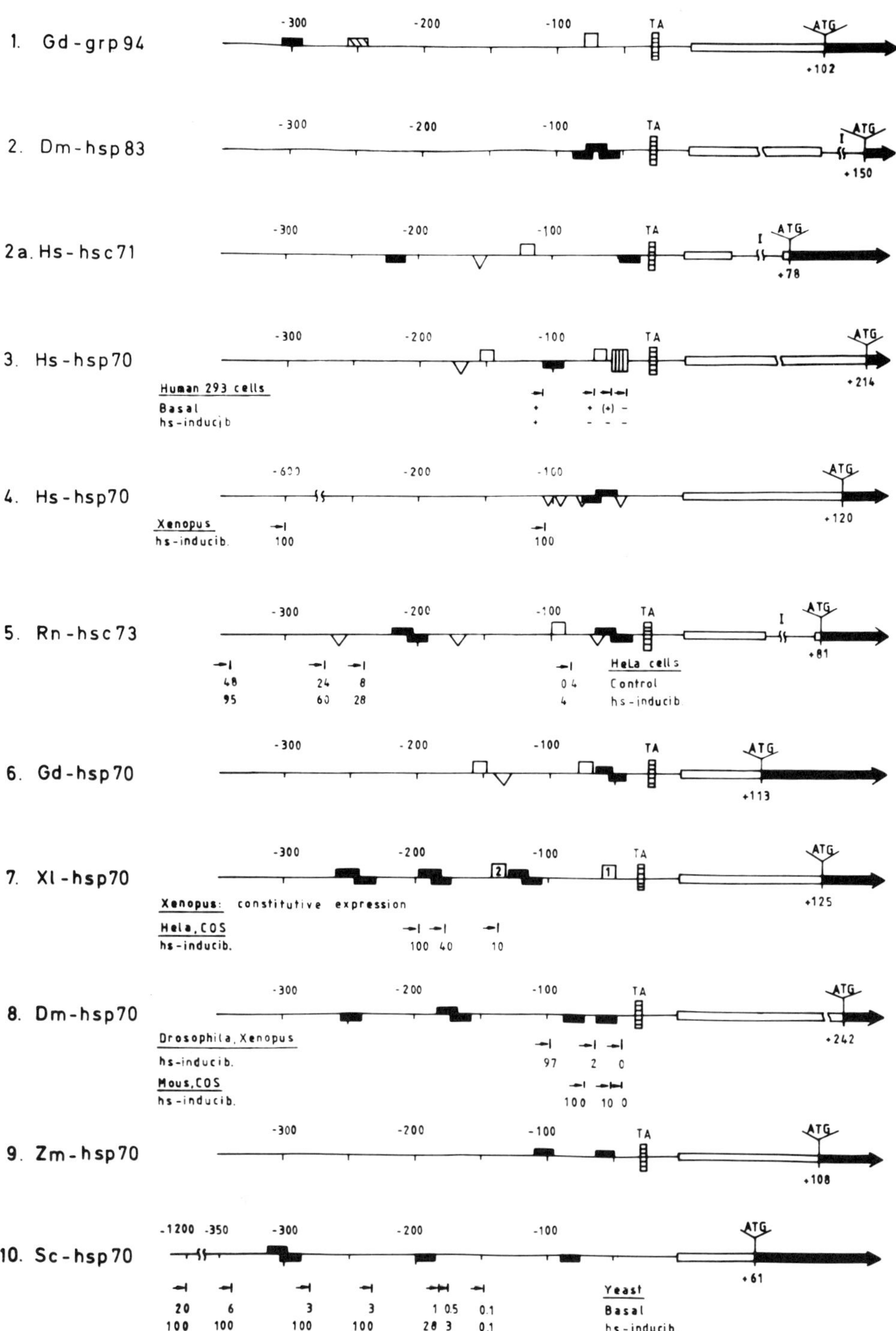

FIGURE 6.6. Survey of promoter elements 5′ upstream of hs and related genes. Symbols used for promoter elements are given below: TATA box, usually TATAAA; CAT box, -ATTGG- or -CCAAT-; Spl box, -GGGGCGGGG-; SRE = serum responsive element -(AA)GGGAAA(AG)-; HRE, hormone receptor element for ecdysterone receptor *(Drosophila)* or progesterone receptor (chicken); A_n or T_n, simple sequences with 10 to 14 A or T residues; HSE, heat shock element (see Table 6.2). Whenever analyzed, expression data obtained by means of deletion promoters are given. End points of deletions are marked by arrows. The numbers indicate hs-inducibility

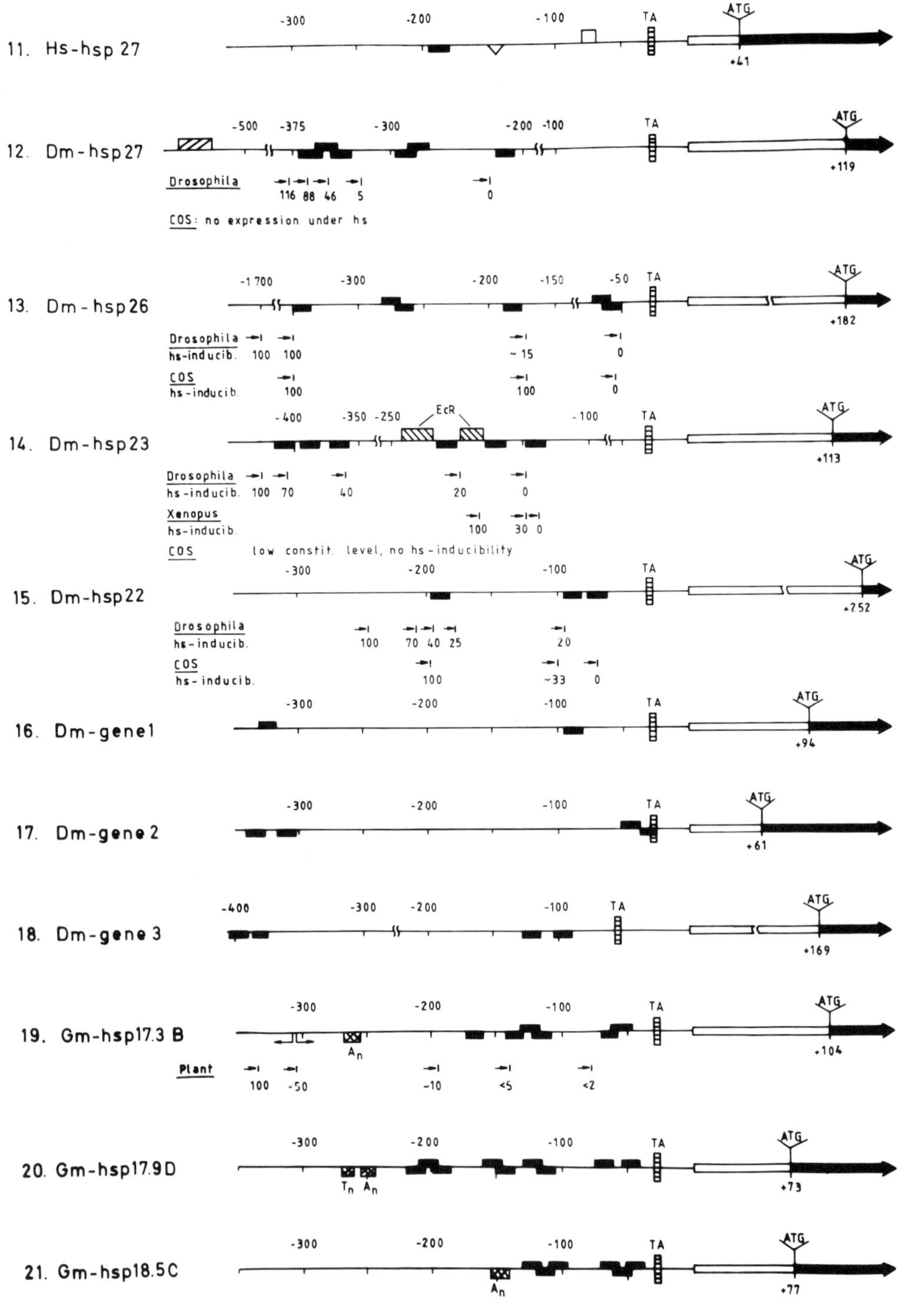

FIGURE 6.6 (continued).

referred to the wild-type promoter (= 100%). Note differences in cases where the same constructs were tested in different expression systems (numbers 7, 8, and 12 to 15). Translation start sites are indicated on the right by ATG. I, intron in the leader sequences of genes 2, 5, 28, and 29. Genes and references for the data are (1) chicken grp94[68]; (2) *Drosophila melanogaster* hsp83;[18,53] (2a) human hsc71;[42] (3) human hsp70;[38a,60,148,149] (4) human hsp70;[116b,144] (5) rat hsc73;[130] (6) chicken hsp70;[88] (7) *Xenopus* hsp70;[12,14,16] (8) *Drosophila* hsp70;[62,64,140] functional

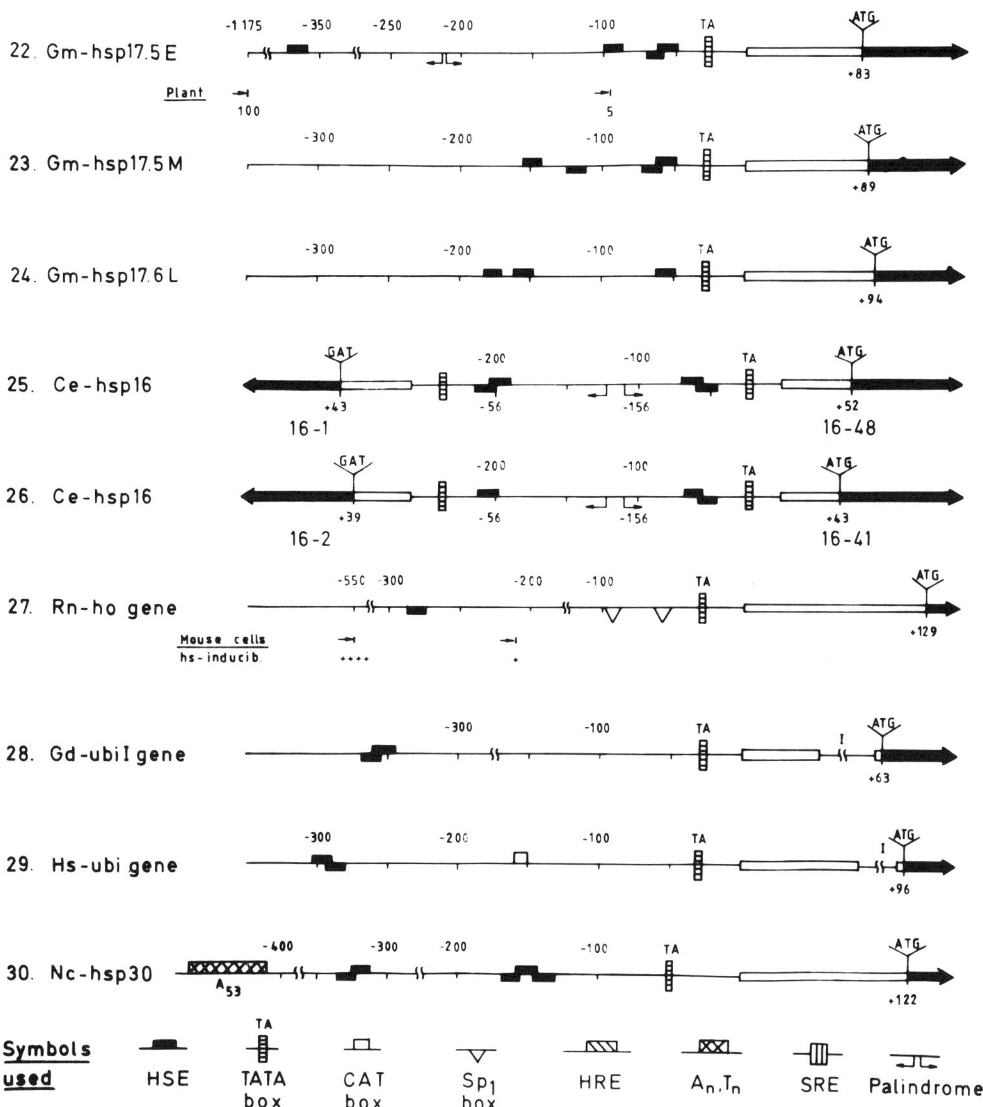

FIGURE 6.6 (continued).

analysis;[15,29,41,102] (9) maize hsp70;[115] (10) yeast hsp70;[129] (11) human hsp27;[54] (12—18) *Drosophila* small HSP-coding genes;[5,133] (12) *Drosophila* hsp27;[6,56,111,112] (13) *Drosophila* hsp26;[6,27,100] (14) *Drosophila* hsp23;[6,83,84,100] (15) *Drosophila* hsp22;[6,69] (16—18) *Drosophila* genes 1 to 3 of locus 67B;[5,99,101] (19—24) *Glycine max* (soybean) small HSP coding genes: (19)[8,119] (20, 21)[110] (22)[51] (23, 24);[93] (25, 26) *Caenorhabditis elegans* hsp16 genes organized in two clusters;[63,116] (27) rat heme oxygenase gene;[90,122] (28) chicken polyubiquitin gene,[19,20] (29) human polyubiquitin gene;[7] (30) *Neurospora crassa* hsp30 gene coding for a mitochondrial HSP.[108b] Additional hs promoters were included into Table 6.2. Hs-hsp 90;[44a,110a] Gd-hsp 70;[53b,129b] At-hsp 70;[151a] Sc-Kar2;[94b] BL-hsp 70;[63a] and Um-hsp 70.[56a].

1. Constitutive expression of hs genes (see Sections 8.4 and 8.5), especially in the vertebrate hsp70 family, is brought about by hybrid promoters containing HSEs and CAT or Sp1 boxes. Characteristic sequences are -CCAAT-, -ATTGG- (CAT), or -GGGGCGGG- (Sp1) in the human hsc71 (number 2a), the human hsp70 (numbers 3 and 4), the rat hsc73 (number 5), and in the *Xenopus* hsp70 (number 7) promoters, respectively. The regulatory behavior of the rat hsc73[132] illustrates a remarkable point. The gene, which, similar to the

Hs-hsc71 gene, codes for the clathrin-uncoating ATPase, is highly expressed under non-hs conditions, this level being only twofold increased during hs. In this case, the two overlapping HSE dimers in the promoter may be considered as elements for maintaining transcription under stress conditions. It can be anticipated that this type of promoter arrangement will be found in many other genes whose continued transcription during hs is essential for the cell. Relevant examples in this direction may be the human hsc71 gene (number 2a),[42] the human heme oxygenase gene[90,122] (number 27), a gene from the human MHC complex coding for the complement plasma factor B (90-kDa glycoprotein[150]), the yeast phosphoglycerate kinase gene,[108a] or the vertebrate hsp90 genes.[110a,144a]

In addition to CAT and Sp1 boxes, one of the human hsp70 genes (number 3) contains a homopurine stretch adjacent to the CAT box: (−59) AGAAGGGAAAAGGCG (−45) with strong homology to the negative control region of the human β-interferon gene promoter[50] but also to bp −48 to −34 of the Dm-hsp70 promoter. This part of the Hs-hsp70 promoter was identified as a serum control element[148,149a] linking the human hsp70 expression to cell cycle control (see Section 8.5), but even the particular sequence of the TATA box (-TATAA-) has evidently more regulatory significance than previously assumed.[127a] Its replacement by the SV40-derived -TATTTAT- abolishes the specific induction of the Hs-hsp70 gene by the adenovirus E1A protein (see Figure 8.5).

Though the direct role of the TATA-box was questioned, Taylor and Kingston[139d] and Williams et al.[145a] confirmed that the transcriptional complex involved in basal level expression of the Hs-hsp70, which occupies the TATA-box, the CCAAT-box, and the homopurine stretch, is responsible for E1A activation of the gene. Similar to the stimulation of hsp70 transcription by the SV40 large T-antigen[139b] or by the c-myc and v-myb onc-gene products,[63b,71a] an interaction of the adenovirus protein with the preexisting DNP complex at the hsp70 promoter can be discussed. Alternatively, the E1A protein may act as a general DNA-binding protein, which facilitates the entry of other transcription factors necessary for transcriptosome formation.[47b]

The multiplicity of promoter elements is typical for many genes, at least those of vertebrates and plants. Gel retardation and footprint analyses demonstrate the sequence-specific association with a series of distinct promoter-binding proteins.[37a,53a,85a,b,145b] Disregarding the complexity of RNA polymerase II itself, formation of a multiprotein/DNA complex (transcriptosome), whose promoter specificity and efficiency for initiation depend on protein/DNA and protein/protein interactions, is evidently a normal prerequisite for transcription and its regulation.[139b,c]

The particular role of the CAT box as TATA-proximal helper element was already discussed, e.g., when describing the expression of the Xl-hsp70 gene constructs in HeLa cells (Figure 6.4). If injected into the homologous system (*Xenopus* oocytes) this gene is transcribed constitutively, independent of hs control. Only removal of the TATA-proximal CAT box (see construct F of Figure 6.4) makes the gene hs inducible.[14] Thus the same gene, in its wild type composition shows three types of control: (1) no transcription of the resident gene in oocytes (see Section 8.3), (2) high constitutive transcription, if injected into oocytes, and (3) hs-induced transcription in somatic cells of *Xenopus*[11,12,14,58] but also in the heterologous COS and HeLa cells (Figure 6.4). It is tempting to speculate that a negative control system, somehow connected with the CAT box or flanking sequences, is active in oocytes but not in somatic cells and that injection of a high copy number of genes from outside outcompetes the putative repressor protein usually silencing the resident hsp70 gene.

2. Negative control elements, acting on the *Drosophila* hsp70 promoter, were described by Corces and Pellicer[29] when investigating many different constructs in stably transformed mouse L cells as the expression system (see B15 of Table 6.1). They defined two alternative acting sequences (bp −200 to −70, and −11 to +88), only one of which

is required to suppress constitutive transcription. Hence, the extent of hs inducibility, although defined primarily by the presence and strength of the HSE, also depends on the effective suppression of the constitutive level. However, these results need reinvestigation by using a homologous expression system.

In this respect, the situation is much clearer with the yeast hsp70 promoter (SSA1 gene[36a,129]). The strong basal level expression is mostly due to sequences far upstream between bp −233 and −1200, whereas hs inducibility depends on the two HSEs proximal to the transcription start (see number 10 of Figure 6.6). Interestingly the 3′ flanking sequences of HSE2 (bp −199 to −186) exert also a strong negative control on basal level activity but not after hs induction. Insertion of a 137 bp fragment containing this region in the heterologous cyc1 promoter reduced the expression about 17-fold. Induction of gene activity by hs is largely due to the binding of an activator protein (see Chapter 7). However, the simultaneous release of negative control may be involved as well (see also discussion on the regulation of X1-hsp70 expression in *Xenopus* oocytes).

An interesting example of a tissue-specific silencer was revealed by Strittmatter and Chua.[138a] Combination of the light-inducible ribulosebisphosphate carboxylase small subunit promoter with the highly active HSE2 oligonucleotide derived from the soybean hsp-17.3-B promoter conferred hs-inducibility to the hybrid construct. But this was restricted to leaves of the transgenic tobacco plants, whereas constructs with the HSE2 sequence inserted upstream of the truncated CaMV 35S promoter were induced equally well in roots and leaves. These results suggest that a root-specific silencer is responsible for the specificity of light-induction of the rbcS gene in leaves. The existence of this promoter region became apparent by combination with the hs promoter element.

3. Two types of developmental control are superimposed to the hs inductibility of small HSPs in *Drosophila* (see Section 8.4). The major developmental signal, acting at distinct larval and pupal stages, is the steroid hormone ecdysterone. Ecdysterone induction was tested in transformed *Drosophila* cell lines[86,87] and by P element transformation of recipient eggs. The functional analysis of promoters revealed the ecdysterone receptor element (HRE) of the hsp27 gene in the region between bp −1100 to −455.[56,111] By gel retardation assays and DNaseI footprinting a 23 bp palindromic element was identified in this region:[112] (−536) GACAAGgGttcaatgCaCTTGTC- (−524). For the hsp26 promoter, the developmental control depends on bp −522 to −352 and possible −52 to +14.[27,49] In contrast to the element defined for hsp27, Mestril et al.[84] mapped a putative palindromic hormone receptor sequence in the hsp23 promoter at about bp −200: (−228)ATtTtCCAT-19n-ATGGcAgAT(−193). Similar sequences were also found around bp −148 of the hsp27, bp −151 of hsp26 and bp −293 of hsp22 genes. But at least for hsps26 and 27, these positions were not identified by the functional analysis of the promoter regions as detailed above. Thus, the exact structure and positions of the ecdysterone receptor elements remain to be established.

4. Usually the gene-specific combination of HSEs (Table 6.2) provides both aspects of a promoter function, i.e. the qualitative of hs inducibility and the quantitative of bf high efficiency, but this need not be true in all cases. Extensive deletion analyses with the soybean hsp17.3 promoter[8] (see number 19 of Figure 6.6) showed that 90% of the transcription activity was lost in a construct with 195 bp of 5′ flanking sequences. However, all HSEs in this promoter map between bp −47 and −173. The only prominent elements in the deleted part of the promoter are a palindromic sequence centered around bp −304 and a stretch of 14 A-residues, which may function as an attachment site to the nuclear matrix (see Section 9.2). Transfer of individual parts of the promoter to appropriate reporter genes[117a] revealed that the sequences upstream of bp −200 have a quantitative enhancing effect. But this is only observed in combination with the HSEs further downstream, which are fully responsible for hs inducibility. "Simple sequences", i.e., stretches of 10 to 15 A

or T, are also found in the upstream parts of two other plant hs promoters (see numbers 20 and 21 of Figure 6.6). Their functional significance remains to be shown.

A search for corresponding quantitative enhancer proteins was started.[117a] A relevant example may be the auxiliary activator protein GRF2 isolated from yeast.[25a] It is a 127-kDa DNA-binding protein that generates a 230-bp nucleosome-free region. It potentiates the activity of other weak promoter elements in the close vicinity, e.g., if connected with the UAS_G of the yeast GAL1-GAL10 gene complex. But GRF2 has practically no stimulatory activity on its own.[25a]

6.6. ANALYSIS OF GENE FUNCTION

In addition to the detailed analysis of promoter elements, the far advanced methods for dissection of genes into functional parts and testing these in connection with appropriate reporter genes were used to study other aspects of gene expression under hs conditions. The relevant papers are included in Table 6.1, but details are discussed in the topically related parts of this book, e.g., signal systems (Section 1.2), pre-mRNP splicing (Section 9.5), the role of the untranslated leader (Section 11.5.5), the influence of the 3′ end of the gene for mRNA stability (Section 8.7), and the function of hs genes (Sections 10.6 and 16.2). In the following, we will briefly summarize special aspects of gene function in *Drosophila,* which is by far the dominant object in this area.

6.6.1. PUFFING OF *DROSOPHILA* POLYTENE CHROMOSOMES

Germ line transformation of *Drosophila* using P element derived cloning vectors and different fusion constructs with hs promoters provided new insights into the puffing phenomenon in general.

1. *In situ* hybridization demonstrates that integration proceeds almost randomly, mostly without effects on the hs inducibility.[20a,26,41,55,77,126,127]
2. Large puffs are formed only if the transcript size exceeds 4 kb.[126] With single hsp70 genes or hsp70 P/L × adh gene hybrids only moderate puffs are produced.[20a,41] No puffs are observed after integration of single small hsp coding genes.[55]
3. Hs induced puffing reflects the promoter strength and activity state, i.e., loci containing gene constructs with truncated promoters, lacking the appropriate combination of hs boxes, do not respond.[41,126]
4. Similar to the natural situation in locus 63BC cointegration of two genes in the same chromosomal site does not necessarily lead to coexpression, i.e., the *Drosophila* adh gene and hsp70 gene, if transformed by the same P element vector, retain their regulatory independence.[26]

In summary, visible puffing is neither automatically coupled to gene activity nor is it a prerequisite for it. Large puffs result from the combination of highly active promoter(s) with large amounts of transcript(s), either from a single large transcription unit or from several smaller ones, e.g., encountered in locus 67B.

6.6.2. FUSION OF *DROSOPHILA* GENES TO THE hsp70 PROMOTER

Larval adh gene (A7—A10) — Bonner er al.[20a] introduced an hsp70 P/L × adh gene construct into ADH-deficient *Drosophila* embryos. Except in primary spermatocytes, hs-dependent expression of alcohol dehydrogenase occurred in all tissues without significant deleterious effects on behavior, fertility, or development. The negative outcome with primary spermatocytes may indicate lacking competence for hsp synthesis, e.g., because of lacking HSF. The hsp70 × adh gene transformed flies can be selected by ethanol treatment. This procedure also allowed the selection of regulatory mutants of the hs response with a considerably lower threshold of hs induction (28°C instead of 35°C). The genetic and molecular analysis of these mutants were used to search for elements of the signal transformation chain[97] (see A8 and Section 1.4).

White locus (A18, A18a) — Steller and Pirrotta[137] made an hsp70 (P/L) × white locus fusion and transformed w^- mutants of *Drosophila* characterized by white eyes instead of the normally red. Hs induced transcription of the otherwise rarely transcribed gene allowed the exact determination of the transcription start site and the definition of the critical period, i.e., the first 2 d of pupal stage, in which the white locus determines the development of red eyes. Because of the leakiness of the hsp70 promoter at 25°C, hs-induced normalization of eye pigmentation was only observed with a transformed line containing the P element inserted in close vicinity to heterochromatin. Based on the white gene as the appropriate selection marker, Klemenz et al.[71] reported on a new P element vector with multicloning sites upstream or downstream of a Dm-hsp70 P/L (−177 to +216) × white gene fusion construct.

Segmentation gene ftz (A19) — To overcome the cell type-specific expression of the fushi tarazu (ftz) gene, an hsp70 P/L × ftz fusion construct was introduced into the *Drosophila* germ line.[139] Hs treatment in the blastoderm stage led to the overexpression of ftz gene products in all cells. The resulting developmental abnormalities (pair rule phenotypes) were very similar to those observed in embryos with a defective ftz gene. They were not observed, if the hs was applied in later stages of the embryo development.

P element transposase function (A15, A16, A16a) — Similar to the situation with the former genes, the P element transposase gene is expressed cell specifically (germ line cells) and at a very low level. Use of a fusion construct with the hsp70 P/L region[72,113,114,138] allowed the hs-induced transcription, increase of transcription frequency, and the identification of two gene products[72,114] with 87 kDa (p87, active transposase) and 66 kDa (p66, inactive, 3′ truncated protein). The germ line-restricted formation of active transposase (p87) is evidently brought about by a specific splicing activity in these cells, joining the 3′ terminal exon 3 to the body of exons 0, 1, and 2. Deletion of the intron between exons 2 and 3 makes the mutant P element transposase active in all cells with corresponding high levels of P element instability.[72,114]

The homeotic antennapedia locus (A18B) — Two types of mutants of this gene complex are known which affect the development of all three thoracic elements of *Drosophila*. These are loss-of-function and gain-of-function mutants. To study the latter type in a similar way as done by Struhl[139] for the ftz gene, Schneuwly et al.[117] used a P-element vector with hsp70 promoter/leader and 3′ flanking sequences to insert a 2.8-kb antennapedia cDNA at a unique KpnI site. Hs-induced overexpression altered the body plan of developing larvae in a predictable manner, depending on the time point of induction.

This type of analysis of developmental genes, put under the control of the inducible hsp70 promoter, was extended to several other genes:

1. The gene sevenless codes for a membrane receptor tyrosine kinase (A18e), which at distinct developmental stages is required for the differentiation of the photoreceptors of omatidia.[20b]
2. A *Drosophila* cDNA clone of the elongation factor EF-1α gene (A18c) can markedly elongate the average lifetime of transformed flies, if maintained at 29.5°C.[121a]
3. Hs-induced expression of a homeotic selector gene cDNA (deformed gene, A18f) leads to transformation of heterologous segments into mandibular or maxillary segments.[71b] The specificity in determination of head segments depends on the particular homeobox. Replacing the deformed homeobox by the ultrabithorax homeobox in a similar gene construct with hsp70 promoter results in a complete failure of head involution and in formation of only thoracic segments.[71c]

4. Transformer gene cDNA under the control of the hsp70 promoter (A18d) allows sexual reprogramming of XY males into females.[81a]
5. An 80-kDa cell adhesion protein (fasciclin III), expressed from an hs vector in S2 cells (A18g), causes hs-dependent aggregation of transformed cells.[129a]
6. Introduction of the *yeast FLP recombinase* gene under control of the hsp70 promoter (A18h) allows hs-induction of site specific recombination, if corresponding genes are inserted between the 599-bp inverted repeats characterizing the yeast recombinase targets.[49a]
7. A fusion construct of the hsp70 promoter/leader with an ultrabithorax (ubx) cDNA was introduced by P-element transformation (A18i). Postembryonic heat shocks induced homeotic transformations of the cuticle, the peripheral nervous system, and the antennae.[78a]

6.7. HEAT SHOCK EXPRESSION VECTORS FOR BIOTECHNOLOGY

The convenient use of 5′ and 3′ regulatory elements of hs genes for specific and highly efficient expression of heterologous genes inserted between them, is documented in the numerous entries compiled in Table 6.1. Yet the application of such constructs for biotechnological purposes is just appearing. The critical analysis of the HSE structure and function (Sections 6.3 and 6.4) illustrates that, despite its conserved basic sequence in all eukaryotic cells, the optimum HSE combination has to be elucidated for each expression system. In addition, whenever stable multicopy systems are considered, attempts must be made to overcome the limiting cellular concentration of the HSF (see Chapter 7). Finally, so far as experimental data are available, the efficient translation of mRNA under hs conditions depends on the presence of a homologous leader sequence (see Section 11.5.5). The latter drawback can be overcome by hs pulse induction of transcription followed by translation of the preformed mRNA in the recovery period (see Figure 6.7).

Two examples of production systems, based on vectors with the human hsp70 promoter/leader sequence and the 3′ flanking sequences of SV40,[39,40,40a] are illustrated in Figure 6.7. The genes inserted between the regulatory elements are the human growth hormone (HGH) gene (A) and the hepatitis virus B surface antigen (SAG) gene (B). Two resistance markers serve the selection of the shuttle vectors in bacteria (amp^r) or mammalian cells (neo^r). Figure 6.7 documents the optimization of the production system with respect to the length of the hs and the following recovery period as well as the application of repeated production cycles. In addition to the HGH and SAG genes, cDNA clones for chicken lysozyme and human influenza haemaglutinin were also used in a similar way.[39] Transformed tumor-cell lines were passed as tumors through mice, and the cells were reisolated by dissociation of solid tumors, yielding about 10^9 cells per 20-mm tumor. After repeated hs induction, 1 to 4 mg/l HGH were obtained when these tumor cells were cultivated in a fermentor on spherical microcarriers.[40a]

Another example in this direction is the incorporation of the human tissue plasminogen activator (PA) cDNA downstream of the *Drosophila* or human hsp70 promoter/leader regions. The fusion constructs were inserted into a bovine papilloma virus vector with a mouse metallothionein gene as a selectable marker. Stable transformants of mouse C127 cells were selected in the presence of 20 μM $CdCl_2$. Hs induced a massive induction of the PA gene and led to an accumulation of maximum 1.3 mg PA/l within 5 h.[9]

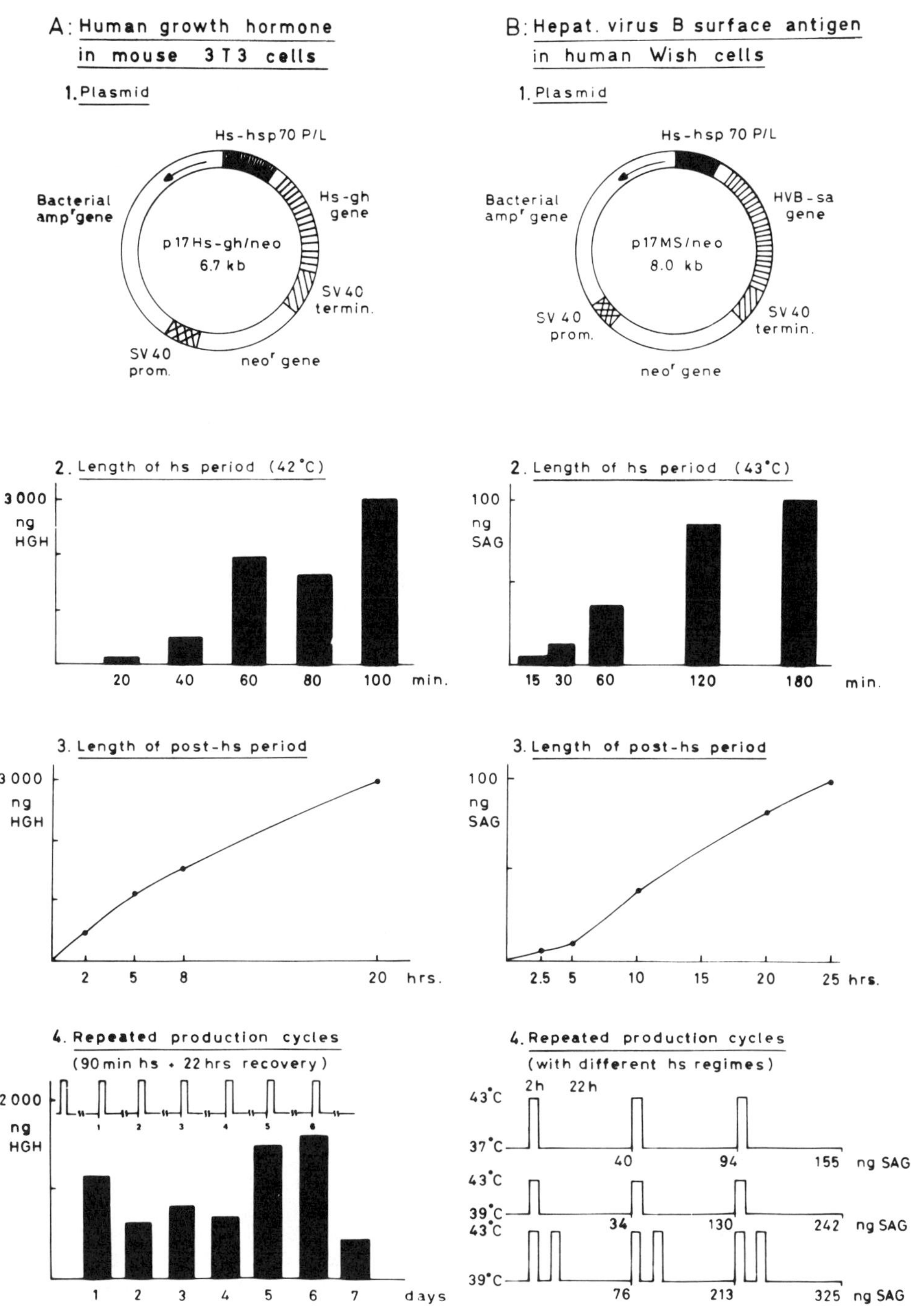

FIGURE 6.7. Two stable protein production systems under the control of the human hsp70 promoter. A 0.6 kb fragment of the human hsp70 promoter[144] with 500 bp 5′ flanking sequence and 113 bp of nontranslated leader were fused to the human growth hormone gene (A) or to the hepatitis virus B surface antigen gene (B). In addition, the plasmids (1) contained resistance genes for selection in bacterial and animal cells (amp^r, neo^r). The optimum length for the hs induction (2) and post-hs production periods (3) were tested and combined to design repeated production cycles (4). Amounts of HGH and SAG secreted to the culture medium are expressed in ng/10^6 cells. (Data were compiled from Dreano, M. et al., *Gene,* 49, 1—8, 1986; and Dreano, M. et al., *Virus Res.*, 8, 43—59, 1987.)

REFERENCES

1. **Ainley, W. M. and Key, J. L.,** Development of a heat shock inducible expression cassette for plants: characterization of parameters for its use in transient expression assays and transgenic plants, *Plant Mol. Biol.,* 14, 949—967, 1990.

1a. **Amin, J., Mestril, R., Lawson, R., Klapper, H., and Voellmy, R.,** The heat shock consensus sequence is not sufficient for hsp70 gene expression in *Drosophila melanogaster, Mol. Cell. Biol.,* 5, 197—203, 1985.

2. **Amin, J., Mestril, R., Schiller, P., Dreano, M., and Voellmy, R.,** Organization of the *Drosophila melanogaster* hsp70 heat shock regulation unit, *Mol. Cell. Biol.,* 7, 1055—1062, 1987.

2a. **Amin, J., Ananthan, J., and Voellmy, R.,** Key features of heat shock regulatory elements, *Mol. Cell. Biol.,* 8, 3761—3769, 1988.

3. **Ananthan, J., Goldberg, A. L., and Voellmy, R.,** Abnormal proteins serve as eukaryotic stress signals and trigger the activation of heat shock genes, *Science,* 232, 522—524, 1986.

4. **Artavanis-Tsakonas, S., Schedl, P., Mirault, M. E., Moran, L., and Lis, J.,** Genes for the 70,000 dalton heat shock protein in two cloned *D. melanogaster* DNA sgements, *Cell,* 17, 9—18, 1979.

4a. **Asano, M., Nagashima, H., Iwakura, Y., and Kawade, Y.,** Interferon production under the control of heterologous inducible enhancers and promoters, *Microbiol. Immunol.,* 32, 589—596, 1988.

5. **Ayme, A. and Tissieres, A.,** Locus 67B of *Drosophila melanogaster* contains seven, not four, closely related heat shock genes, *EMBO J.,* 4, 2949—2954, 1985.

6. **Ayme, A., Southgate, R., and Tissieres, A.,** Nucleotide sequences responsible for the thermal inducibility of the *Drosophila* small heat-shock protein genes in monkey COS cells, *J. Mol. Biol.,* 182, 469—475, 1985.

6a. **Blackmann, R. K. and Meselson, M.,** Interspecific nucleotide sequence comparisons used to identify regulatory and structural features of the *Drosophila* hsp82 gene, *J. Mol. Biol.,* 188, 499—516, 1986.

7. **Baker, Z. T. and Board, P. G.,** The human ubiquitin gene family structure of a gene and pseudogenes from the UbB subfamily, *Nucl. Acids Res.,* 15, 443—463, 1987.

8. **Baumann, G., Raschke, E., Bevan, M., and Schoeffl, F.,** Functional analysis of sequences required for transcriptional activation of a soybean heat shock gene in transgenic tobacco plants, *EMBO J.,* 6, 1161—1166, 1987.

9. **Bending, M. M., Stephens, P. E., Cockett, M. I., and Hentschel, C. C.,** Mouse cell lines that use heat shock promoters to regulate the expression of tissue plasminogen activator, *J. Mol. Biol.,* 6, 343—352, 1987.

10. **Berger, E. M., Marino, G., and Torrey, D.,** Expression of *Drosophila* hsp70-cat hybrid gene in Aedes cells induced by heat shock, *Somat. Cell Mol. Genet.,* 11, 371—378, 1985.

10a. **Berger, E. M., Torrey, D., and Morganelli, C.,** Natural and synthetic heat shock protein gene promoters assayed in *Drosophila melanogaster, Somat. Cell Mol. Genet.,* 12, 433—441, 1986.

11. **Bienz, M.,** *Xenopus* hsp70 genes are constitutively expressed in injected oocytes, *EMBO J.,* 3, 2477—2483, 1984.

12. **Bienz, M.,** Developmental control of the heat shock response in *Xenopus, Proc. Natl. Acad. Sci. U.S.A.,* 81, 3138—3142, 1984.

13. **Bienz, M.,** Transient and developmental activation of heat-shock genes, *Trends Biochem. Sci.,* 10, 157—161, 1985.

14. **Bienz, M.,** A CCAAT box confers cell-type-specific regulation on the *Xenopus* hsp70 gene in oocytes, *Cell,* 46, 1037—1042, 1986.

15. **Bienz, M. and Pelham, H. R. B.,** Expression of a *Drosophila* heat-shock protein in *Xenopus* oocytes: Conserved and divergent regulatory signals, *EMBO J.,* 1, 1583—1588, 1982.

16. **Bienz, M. and Pelham, H. R. B.,** Heat shock regulatory elements function as an inducible enhancer in the *Xenopus* hsp70 gene and when linked to a heterologous promoter, *Cell,* 45, 753—760, 1986.

17. **Bienz, M. and Pelham, H. R. B.,** Mechanisms of heat-shock gene activation in higher eucaryotes, *Adv. Genet.,* 24, 31-72, 1987.

18. **Blackman, R. K. and Meselson, M.,** Interspecific nucleotide sequence comparisons used to identify regulatory and structural features of the *Drosophila* hsp82 gene, *J. Mol. Biol.,* 188, 499—516, 1986.

19. **Bond, U. and Schlesinger, M. J.,** Ubiquitin is a heat shock protein in chicken embryo fibroblasts, *Mol. Cell. Biol.,* 5, 949—956, 1985.

20. **Bond, U. and Schlesinger, M. J.,** The chicken ubiquiten gene contains a heat shock promoter and express an unstable mRNA in heat-shocked cells, *Mol. Cell. Biol.,* 6, 4602—4610, 1986.

20a. **Bonner, J. J., Parks, C., Parker-Thornburg, J., Mortin, M. A., and Pelham, H. R.,** The use of promoter fusion in *Drosophila* genetics: Isolation of mutations affecting the heat shock response, *Cell,* 37, 979—991, 1984.

20b. **Bowtell, D. D., Simon, M. A., and Rubin, G. M.,** Ommatidia in the developing *Drosophila* eye require and can respond to sevenless for only a restricted period, *Cell,* 56, 931—936, 1989.

21. **Brazzell, C. and Ingolia, T. D.**, Stimuli that induce a yeast heat shock gene fused to β-galactosidase, *Mol. Cell. Biol.*, 4, 2573—2579, 1984.
22. **Burke, J. F. and Ish-Horowicz, D.**, Expression of Drosophila heat shock genes is regulated in Rat-1 cells, *Nucl. Acids Res.*, 10, 3821—3830, 1982.
23. **Burke, J. F., Sinclair, J. H., Sang, J. H., and Ish-Horowicz, D.**, An assay for transient gene expression in transfected *Drosophila* cells, using /3H/guanine incorporation, *EMBO J.*, 3, 2549—2554, 1984.
23a. **Callis, J., Fromm, M., and Walbot, O.**, Heat inducible expression of a chimeric maize hsp70 CAT gene in maize protoplasts, *Plant Physiol.*, 88, 965—968, 1988.
24. **Cappello, J., Zuker, C., and Lodish, H. F.**, Repetitive Dictyostelium heat-shock promoter functions in *Saccharomyces cerevisiae, Mol. Cell. Biol.*, 4, 591—598, 1984.
25. **Cartwright, I. L. and Elgin, S. C. R.**, Nucleosomal instability and induction of new upstream protein-DNA associations accompany activation of four small heat shock protein genes in *Drosophila melanogaster, Mol. Cell. Biol.*, 6, 779—791, 1986.
25a. **Chasman, D. I., Lue, N. F., Buchman, A. R., LaPointe, J. W., Lorch, Y., and Kornberg, R. D.**, A yeast protein that influences the chromatin structure of UAS_G and functions as a powerful auxiliary gene activator, *Genes Devel.*, 4, 503—514, 1990.
26. **Cohen, R. S. and Meselson, M.**, Inducible transcription and puffing in *Drosophila melanogaster* transformed with hsp70-phage hybrid heat shock genes, *Proc. Natl. Acad. Sci. U.S.A.*, 81, 5509—5513, 1984.
27. **Cohen, R. S. and Meselson, M.**, Separate regulatory elements for the heat-inducible and ovarian expression of the *Drosophila* hsp26 gene, *Cell*, 43, 737—746, 1985.
27a. **Cohen, R. S. and Meselson, M.**, Periodic interactions of heat shock transcriptional elements, *Nature*, 332, 856—858, 1988.
28. **Collatz, E., Plesset, J., Foy, J., and McLaughlin, C. S.**, Expression of the *Drosophila* 70,000-dalton heat-shock protein is translationally controlled in yeast, *Yeast*, 1, 49—56, 1985.
29. **Corces, V. and Pellicer, A.**, Identification of sequences involved in the transcriptional control of a *Drosophila* heat-shock gene, *J. Biol. Chem.*, 259, 14812—14817, 1984.
30. **Corces, V., Pellicer, A., Axel, R., and Meselson, M.**, Integration, transcription, and control of a *Drosophila* heat shock gene in mouse cells, *Proc. Natl. Acad. Sci. U.S.A.*, 78, 7038—7042, 1981.
31. **Corces, V., Pellicer, A., Axel, R., Mei, S.-Y., and Meselson, M.**, Approximate localization of sequences controlling transcription of a *Drosophila* heat-shock gene, in *Heat Shock from Bacteria to Man*, Schlesinger, M. J., Ashburner, M., and Tissieres, A., Eds., Cold Spring Harbor Laboratory, Cold Spring Harbor, NY, 1982, 27—34.
32. **Costlow, N. and Lis, J. T.**, High-resolution of DNase I-hypersensitive sites of *Drosophila melanogaster* and *Sarcharomyces cerevisiae* heat shock genes, *Mol. Cell. Biol.*, 4, 1853—1863, 1984.
33. **Costlow, N. A., Simon, J. A., and Lis, J. R.**, A hypersensitive site in hsp70 chromatin requires adjacent not internal DNA sequence, *Nature*, 313, 147—149, 1985.
34. **Craig, E. A.**, The heat shock response, *CRC Crit. Rev. Biochem.*, 18, 239—280, 1985.
35. **Craig, E. A. and Jacobsen, K.**, Mutations of the heat inducible 70 kilodalton genes of yeast confer temperature sensitive growth, *Cell*, 38, 841—849, 1984.
36. **Craig, E. A. and Jacobsen, K.**, Mutations in cognate genes of *Saccharomyces cerevisiae* hsp70 result in reduced growth rates at low temperatures, *Mol. Cell. Biol.*, 5, 3517—3524, 1985.
36a. **Craig, E., Boorstein, W., Park, H.-O., Stone, D., and Nicolet, C.**, Complex regulation of three heat inducible HSP70 related genes in *Saccharomyces cerevisiae*, in *Stress-Induced Proteins*, Pardue, M. L., Feramisco, J. R., and Lindquist, S., Eds., Alan R. Liss, New York, 1989, 51-61
37. **Czarnecka, E., Gurley, W. B., Nagao, R. T., Mosquera, L. A., and Key, J. L.**, DNA sequence and transcript mapping of a soybean gene encoding a small heat shock protein, *Proc. Natl. Acad. Sci. U.S.A.*, 82, 3726—3730, 1985.
37a. **Czarnecka, E., Key, J. L., and Gurley, W. B.**, Regulatory domains of the Gmhsp17.5-E heat shock promoter of soybean, *Mol. Cell. Biol.*, 9, 3457—3463, 1989.
38. **Di Nocera, P. P. and Dawid, I. B.**, Transient expression of genes introduced into cultured cells of *Drosophila, Proc. Natl. Acad. Sci. U.S.A.*, 80, 7095—7098, 1983.
38a. **Drabent, B., Genthe, A., and Benecke, B.-J.**, In vitro transcription of a human hsp 70 heat shock gene by extracts prepared from heat-shocked and non-heat-shocked human cells, *Nucl. Acids Res.*, 14, 8933—8948, 1986.
39. **Dreano, M., Brochot, J., Myers, A., Cheng-Meyer, C., Rungger, D., Voellmy, R., and Bromley, R.**, High-level, heat regulated synthesis of proteins in eucaryotic cells, *Gene*, 49, 1—8, 1986.
40. **Dreano, M., Fouillet, X., Brochot, J., Vallet, J.-M., Michel, M.-L., Rungger, D., and Bromley, P.**, Heat-regulated expression of the hepatitis B virus surface antigen in the human Wish cell line, *Virus Res.*, 8, 43—59, 1987.
40a. **Dreano, M., Fischbach, M., Montandon, F., Salina, C., Padieu, P., and Bromley, P.**, Production of secretable proteins using the passage *in vivo* as tumours of cells carrying heat-inducible expression constructs, *Biotechnology*, 6, 953—958, 1988.

40b. **Dubois, M. F., Mezger, V., Morange, M., Ferrieux, C., Lebon, P., and Bensaude, O.,** Regulation of the heat-shock response by interferon in mouse L cells, *J. Cell. Physiol.*, 137, 102—109, 1988.

41. **Dudler, R. and Travers, A. A.,** Upstream elements necessary for optimal function of the hsp70 promoter transformed flies, *Cell,* 38, 391—398, 1984.

42. **Dworniczak, B. and Mirault, M.-E.,** Structure and expression of a human gene coding for a 71 kd heat shock 'cognate' protein, *Nucl. Acids Res.*, 15, 5181—5197, 1987.

42a. **Eissenberg, J. C. and Elgin, S. C. R.,** Chromatin structure of a P-element-transduced hsp 28 gene in *Drosophila melanogaster, Mol. Cell. Biol.*, 6, 4126—4129, 1986.

43. **Ellwood, M. S. and Craig, E. A.,** Differential regulation of the 70K heat shock gene and related genes in *Saccharomyces cerevisiae, Mol. Cell. Biol.*, 4, 1454—1459, 1984.

44. **Farrelly, F. W. and Finkelstein, D. B.,** Complete sequence of the heat shock-inducible HSP90 gene of *Saccharomyces cerevisiae, J. Biol. Chem.*, 259, 5745—5751, 1984.

45. **Finkelstein, D. B. and Strausberg, S.,** Heat shock-regulated production of *Escherichia coli* β-galactosidase in *Saccharomyces cerevisiae, Mol. Cell. Biol.*, 3, 1625—1633, 1983.

46. **Finkelstein, D. B. and Strausberg, S.,** Identification and expression of a cloned yeast heat shock gene, *J. Biol. Chem.*, 258, 1908—1913, 1983.

47. **Fire, A.,** Integrative transformation of *Caenorhabditis elegans, EMBO J.*, 5, 2673—2680, 1986.

47a. **Fischer, J. A., Giniger, E., Maniatis, T., and Ptashne, M.,** Gal4 activates transcription in *Drosophila, Nature,* 332, 853—856, 1988.

47b. **Flint, J. and Shenk, T.,** Adenovirus E1A protein: Paradigm viral transactivator, *Annu. Rev. Genet.*, 23, 141—161, 1989.

48. **Garabedian, M. J., Shepherd, B. M., and Wensink, P. C.,** A tissue-specific transcription enhancer from the *Drosophila* yolk protein 1 gene, *Cell,* 45, 859—867, 1986.

49. **Glaser, R. L., Wolfner, M. F., and Lis, J. T.,** Spatial and temporal pattern of hsp 26 expression during normal development, *EMBO J.*, 5, 747—754, 1986.

49a. **Golic, K. G. and Lindquist, S.,** The FLP recombinase of yeast catalyzes site-specific recombination in the *Drosophila* genome, *Cell,* 59, 499—509, 1989.

50. **Goodbourn, S., Burstein, H., and Maniatis, T.,** The human β-interferon gene enhancer is under negative control, *Cell,* 45, 601—610, 1986.

50a. **Greene, J. M., Larin, Z., Taylor, I. C. A., Prentice, H., Gwinn, K. A., and Kingston, R. E.,** Multiple basal elements of a human hsp 70 promoter function differently in human and rodent cell lines, *Mol. Cell. Biol.*, 7, 3646—3655, 1987.

51. **Gurley, W. B., Czarnecka, E., Nagao, R. T., and Key, J. L.,** Upstream sequences required for efficient expression of a soybean heat shock gene, *Mol. Cell. Biol.*, 6, 559—565, 1986.

52. **Hackett, R. W. and Lis, J. T.,** DNA sequence analysis reveals extensive homologies of regions preceding hsp70 and alpha, beta heat shock genes in *Drosophila melanogaster, Proc. Natl. Acad. Sci. U.S.A.*, 78, 6196—6200, 1981.

53. **Hacket, R.W. and Lis, J. T.,** Localization of the hsp83 transcript within a 3292 nucleotide-sequence from the 63B heat shock locus of *Drosophila melanogaster, Nucl. Acids Res.*, 11, 7011—7030, 1983.

53a. **Herbomel, P., Rollier, A., Tronche, F., Ott, M. -O., Yaniv, M., and Weiss, M. C.,** The rat albumin promoter is composed of six distinct positive elements within 130 nucleotides, *Mol. Cell. Biol.*, 9, 4750—4758, 1989.

53b. **Heschl, M. F. P. and Baillie, D. L.,** Characterization of the hsp70 multigene family of *Caenorhabditis elegans, DNA,* 8, 233—243, 1989.

54. **Hickey, E., Brandon, S. E., Potter, R., Stein, G., Stein, J., and Weber, L. A.,** Sequence and organization of genes encoding the human 27 kDa heat shock protein, *Nucl. Acids Res.*, 14, 4127—4146, 1986.

54a. **Hickey, E., Brandon, S. E., Smale, G., Lloyd, D., and Weber, L. A.,** Sequence and regulation of a gene encoding a human 89-kilo-dalton heat shock protein, *Mol. Cell. Biol.*, 9, 2615—2626, 1989.

55. **Hoffman, E. P. and Corces, V. G.,** Correct temperature induction and developmental regulation of a cloned heat shock gene transformed into *Drosophila* germ line, *Mol. Cell. Biol.*, 4, 2883—2889, 1984.

56. **Hoffman, E. and Corces, V.,** Sequences involved in temperature and ecdysterone-induced transcription are located in separate regions of a *Drosophila melanogaster* heat shock gene, *Mol. Cell. Biol.*, 6, 663—673, 1986.

56a. **Holden, D. W., Kronstad, J. W., and Leong, S. A.,** Mutation in a heat-regulated hsp70 gene of *Ustilago maydis, EMBO J.*, 8, 1927—1934, 1989.

57. **Holmgren, K., Corces, V., Morimoto, R., Blackman, R., and Meselson, M.,** Sequence homologies in the 5′ regions of four *Drosophila* heat-shock genes, *Proc. Natl. Acad. Sci. U.S.A.*, 78, 3775—3778, 1981.

58. **Horrell, A., Shuttleworth, J., and Colman, A.,** Transcript levels and translational control of hsp70 synthesis in *Xenopus* oocytes, *Genes Devel.*, 1, 433—444, 1987.

58a. **Hovemann, B., Walldorf, U., and Ryseck, R.-P.,** Heat shock locus 93D of *Drosophila melanogaster:* An RNA with limited coding capacity accumulates precursor transcripts after heat shock, *Mol. Gen. Genet.*, 204, 334—340, 1986.

59. **Hultmark, D., Klemenz, R., and Gehring, W. R.,** Translational and trancriptional control elements in the untranslated leader of the heat-shock gene hsp22, *Cell,* 44, 429—438, 1986.

60. **Hunt, C. and Morimoto, R. J.,** Conserved features of eukaryotic hsp70 genes revealed by comparison with the nucleotide sequence of human hsp70, *Proc. Natl. Acad. Sci. U.S.A.*, 82, 6455—6459, 1985.

61. **Ingolia, T. D. and Craig, E. A.,** Primary sequence of the 5′ flanking regions of the *Drosophila* heat shock genes in chromosome subdivision 67B, *Nucl. Acids. Res.*, 9, 1627—1642, 1981.

62. **Ingolia, T. D., Craig, E. A., and McCarthy, B. J.,** Sequence of three copies of the gene for the major *Drosophila* heat shock induced protein and their flanking regions, *Cell,* 21, 669—679, 1980.

62a. **Johnston, R. N. and Kucey, B. L.,** Competitive inhibition of hsp70 gene expression causes thermosensitivity, *Science,* 242, 1551—1554, 1988.

63. **Jones, D., Russnak, R. H., Kay, R. J., and Candido, E. P. M.,** Structure, expression and evolution of a heat shock gene locus in *Caenorhabditis elegans* that is flanked by repetitive elements, *J. Biol. Chem.*, 261, 12006—12015, 1986.

63a. **Judelson, H. S. and Michelmore, R. W.,** Structure and expression of a gene encoding heat-shock protein hsp70 from the oomycete fungus *Bremia lactucae, Gene,* 79, 207—217, 1989.

63b. **Kaddurah-Daouk, R., Greene, J. M., Baldwin, A. S., and Kingston, R. E.,** Activation and repression of mammalian gene expression by the c-myc protein, *Genes Dev.*, 1, 347—357, 1987.

64. **Karch, F., Toeroek, I., and Tissieres, A.,** Extensive regions of homology in front of the two hsp 70 heat shock variant genes in *Drosophila melanogaster, J. Mol. Biol.*, 148, 219—230, 1981.

65. **Kay, R. J., Boissy, R. J., Russnak, R. H., and Candido, E. P. M.,** Efficient transcription of a *Caenorhabditis elegans* heat shock gene pair in mouse fibroblasts is dependent on multiple promoter elements which can function bidirectionally, *Mol. Cell. Biol.*, 6, 3134—3143, 1986.

66. **Kay, R. J., Russnak, R. H., Jones, D., Mathias, C., and Candido, E. P. M.,** Expression of intron-containing *C. elegans* heat shock genes in mouse cells demonstrates divergence of 3′ splice site recognition sequence between nematodes and vertebrates and inhibitory effect of heat shock on the mammalian splicing apparatus, *Nucl. Acids Res.*, 15, 3723, 1987.

67. **Kingston, R. E., Baldwin, A. S., and Sharp, P. A.,** Regulation of heat shock protein 70 gene expression by c-myc, *Nature,* 312, 280—282, 1984.

68. **Kleinsek, D. A., Beattie, W. G., Tsai, M. J., and O'Malley, B. W.,** Molecular cloning of a steroid-regulated 108 K heat shock protein gene from hen oviduct, *Nucl. Acids Res.*, 14, 10053—10071, 1986.

69. **Klemenz, R. and Gehring, W. J.,** Sequence requirement for expression of the *Drosophila melanogaster* heat shock protein hsp22 gene during heat shock and normal development, *Mol. Cell. Biol.*, 6, 2011—2019, 1986.

70. **Klemenz, R., Hultmark, D., and Gehring, W. J.,** Selective translation of heat shock mRNA in *Drosophila melanogaster* depends on sequence information in the leader, *EMBO J.*, 4, 2053—2060, 1985.

71. **Klemenz, R., Weber, U., and Gehring, W. J.,** The white gene as a marker in a new P-element vector for gene transfer in *Drosophila, Nucl. Acids Res.*, 15, 3947—3959, 1987.

71a. **Klempnauer, K. -H, Arnold, H., and Biedenkapp, H.,** Activation of transcription by v-myb: evidence for two different mechanisms, *Genes Dev.*, 3, 1582—1589, 1989.

71b. **Kuziora, M. A. and McGinnis, W.,** Autoregulation of a *Drosophila* homeotic selector gene, *Cell,* 55, 477—485, 1988.

71c. **Kuziora, M. A. and McGinnis, W.,** A homeodomain substitution changes the regulatory specificity of the deformed protein in *Drosophila* embryos, *Cell,* 59, 563—571, 1989.

71d. **Larocca, D.,** Ecdysterone and heat shock induction of transfecting and endogenous heat shock genes in cultured *Drosophila* cells, *J. Mol. Biol.*, 191, 563—567, 1986.

72. **Laski, F. A., Rio, D. C., and Rubin, G. M.,** Tissue specificity of *Drosophila* P element transposition is regulated at the level of mRNA splicing, *Cell,* 44, 7—19, 1986.

72a. **LaVolpe, A., Ciaramella, M., and Bazzicalupo, P.,** Structure evolution and properties of a novel repetitive DNA family in *Caenorhabditis elegans, Nucl. Acids Res.*, 16, 8213—8232, 1988.

73. **Lawson, R., Mestril, R., Schiller, P., and Voellmy, R.,** Expression of heat shock-beta-galactosidase hybrid genes in cultured *Drosophila* cell, *Mol. Gen. Genet.*, 198, 116—124, 1984.

74. **Lawson, R., Mestril, R., Luo, Y., and Voellmy, R.,** Ecdysterone selectively stimulates the expression of a 23000-Da heat-shock protein-beta-galactosidase hybrid gene in cultured *Drosophila* cells, *Dev. Biol.*, 110, 321—330, 1985.

75. **Lindquist, S.,** The heat-shock response (review), *Annu. Rev. Biochem.*, 55, 1151—1191, 1986.

75a. **Lindquist, S. and Craig, E. A.,** The heat-shock proteins, *Annu, Rev. Genet.*, 22, 631—677, 1988.

76. **Lis, J., Costlow, N., De Banzie, J., Knipple, D., O'Connor, D., and Sinclair, S.,** Transcription and chromatin structure of *Drosophila* heat shock genes in yeast, in *Heat Shock from Bacteria to Man,* Schlesinger, M. J., Ashburner, M., and Tissieres, A., Eds., Cold Spring Harbor Laboratory, Cold Spring Harbor, NY, 1982, 57—62.
77. **Lis, J. T., Simon, J. A., and Sutton, C. A.,** New heat shock puffs and beta-galactosidase activity resulting from transformation of *Drosophila* with an hsp70-lacZ hybrid gene, *Cell,* 35, 403—410, 1983.
77a. **Liu, A. Y. C., Lin, Z., Choi, H. S., Sorhage, F., and Li, B. S.,** Attenuated induction of heat shock gene expression in aging diploid fibroblasts, *J. Biol. Chem.,* 264, 12037—12045, 1989.
77b. **Logan, S. K., Garabedian, M. J., and Wensink, P. C.,** DNA regions that regulate the ovarian transcriptional specificity of *Drosophila* yolk protein genes, *Genes Dev.,* 3, 1453—1461, 1989.
78. **Lubben, T. H. and Keegstra, K.,** Efficient in vitro import of a crytosolic heat shock protein into pea chloroplasts, *Proc. Natl. Acad. Sci. U.S.A.,* 83, 5502—5506, 1986.
78a. **Mann, R. S. and Hogness, D. S.,** Functional dissection of ultrabithorax proteins in *D. melanogaster, Cell,* 60, 597—610, 1990.
79. **Mason, P. J., Toeroek, I., Kiss, I., Karch, F., and Udvardy, A.,** Evolutionary implications of a complex pattern of DNA sequence homology extending far upstream of the hsp70 genes at loci 87A7 and 87C1 in *Drosophila melanogaster, J. Mol. Biol.,* 156, 21—35, 1982.
79a. **McDaniel, D.-A., Caplan, A. J., Lee, M.-S., Adams, C. C., Fishel, B. R., Gross, D. S., and Garrard, W. T.,** Basal-level expression of the yeast HSP82 gene requires a heat shock regulatory element, *Mol. Cell. Biol.,* 9, 4789—4798, 1989.
80. **McGarry, T. J. and Lindquist, S.,** The preferential translation of *Drosophila* hsp70 mRNA requires sequences in the untranslated leader, *Cell,* 42, 903—911, 1986.
81. **McGarry, T. J. and Lindquist, S.,** Inhibition of heat shock protein synthesis by heat-inducible antisense RNA, *Proc. Natl. Acad. Sci. U.S.A.,* 83, 399—403, 1986.
81a. **McKeown, M., Belote, J. M., and Boggs, R. T.,** Ectopic expression of the female transformer gene product leads to female differnetiation of chromosomally male *Drosophila, Cell,* 53, 887—895, 1988.
82. **McMahon, A. P., Novak, T. J., Britten, R. J., and Davidson, E. H.,** Inducible expression of a cloned heat shock fusion gene in sea urchin embryos, *Proc. Natl. Acad. Sci. U.S.A.,* 81, 7490—7494, 1984.
82a. **Medford, J. and Klee, H.,** Manipulation of endogenous auxin and cytokinin levels in transgenic plants, in *Molecular Basis of Plant Development,* Goldberg, R., Ed., Alan R. Liss, New York, 1989, 211—222.
83. **Mestril, R., Rungger, D., Schiller, P., and Voellmy, R.,** Identification of a sequence element in the promoter of the *Drosophila melanogaster* hsp23 gene that is required for its heat activation, *EMBO J.,* 4, 2971—2976, 1985.
84. **Mestril, R., Schiller, P., Amin, J., Klapper, H., Ananthan, J., and Voellmy, R.,** Heat shock and ecdysterone activation of the *Drosophila melanogaster* hsp 23 gene: A sequence element implied in developmental regulation, *EMBO J.,* 5, 1667—1673, 1986.
84a. **Mezger, V., Bensaude, O., and Morange, M.,** Deficient activation of heat shock gene transcription in embryonal Carcinoma cells, *Dev. Biol.,* 124, 544, 1987.
84b. **Mirault, M. E., Southgate, R., and Delwart, E.,** Regulation of heat-shock genes: A DNA sequence upstream of *Drosophila* hsp70 genes is essential for their induction in monkey cells, *EMBO J.,* 1, 1279—1285, 1982.
85. **Monsma, S. A., Ard, R., Lis, J. T., and Wolfner, M. F.,** Localized heat-shock inducton in *Drosophila melanogaster, J. Exp. Zool.,* 247, 279—284, 1988.
85a. **Morgan, W. D.,** Transcription factor Sp1 binds to and activates a human hsp70 gene promoter, *Mol. Cell. Biol.,* 9, 4099—4104, 1989.
85b. **Morgan, W. D., Williams, G. T., Morimoto, R. I., Greene, J., Kingston, R. E., and Tjian, R.,** Two transcriptional activators, CCAAT-box-binding transcription factor and heat shock transcription factor, interact with a human hsp70 gene promoter, *Mol. Cell. Biol.,* 7, 1129—1138, 1987.
86. **Morganelli, C. M. and Berger, E. M.,** Transient expression of homologous genes in *Drosophila* cells, *Science,* 224, 1004—1006, 1984.
87. **Morganelli, C. M., Berger, E. M., and Pelham, H. R. B.,** Transcription of *Drosophila* small hsp-tk hybrid genes is induced by heat shock and by ecdysterone in transfected *Drosophila* cells, *Proc. Natl. Acad. Sci. U.S.A.,* 82, 5865—5869, 1985.
88. **Morimoto, R., Hunt, C., Huang, S.-Y., Berg, K. L., and Banerji, S. S.,** Organization, nucleotide sequence and transcription of the chicken HSP70 gene, *J. Biol. Chem.,* 261, 12692—12699, 1986.
89. **Morris, T., Marashi, F., Weber, L., Hickey, E., Greespan, D., Bonner, J., Stein, J., and Stein, G.,** Involvement of the 5′-leader sequence in coupling the stability of a human H3 histone mRNA with DNA replication, *Proc. Natl. Acad. Sci. U.S.A.,* 83, 981—985, 1986.
90. **Müller, R. M., Taguchi, H., and Shibahara, S.,** Nucleotide sequence and organization of the rat heme oxygenase gene, *J. Biol. Chem.,* 262, 6795—6802, 1987.
91. **Munro, S. and Pelham, H. R. B.,** Use of peptide tagging to detect proteins expressed from cloned genes: Deletion mapping functional domains of *Drosophila* hsp70, *EMBO J.,* 3, 3087—3093, 1984.

92. **Munro, S. and Pelham, H. R. B.,** An hsp70-like protein in the ER: Identity with the 78 kd glucose-regulated protein and immunoglobulin heavy chain binding protein, *Cell,* 46, 291—300, 1986.
93. **Nagao, R. T., Czarnecka, E., Gurley, W. B., Schoeffl, F., and Key, J. L.,** Genes for low-molecular-weight heat shock proteins of soybeans: Sequence analysis of a multigene family, *Mol. Cell. Biol.,* 5, 3417—3428, 1985.
94. **Nicholson, R. C. and Moran, L. A.,** Expression of a *Drosophila* heat-shock gene in cells of the yeast *Saccharomyces cerevisiae, Biosci. Rep.,* 4, 963—972, 1984.
94a. **Nicolet, C. M. and Craig, E. A.,** Isolation and characterization of STI1, a stress-inducible gene from *Saccharomyces cerevisiae, Mol. Cell. Biol.,* 9, 3638—3646, 1989.
94b. **Normington, K., Kohno, K., Kozutsumi, Y., Gething, M.-J., and Sambrook, J.,** *S. cerevisiae* encodes an essential protein homologous to mammalian BiP, *Cell,* 57, 1223—1236, 1989.
95. **Nover, L., Ed.,** *Heat Shock Response of Eukaryotic Cells,* Springer-Verlag, Berlin, 1984.
96. **Nover, L.,** Expression of heat shock genes in homologous and heterologous systems, *Enz. Microb. Technol.,* 9, 130—144, 1987.
96a. **Nover, L., Neumann, D., and Scharf, K.-D., Eds.,** *Heat Shock and other Stress Response Systems of Plants,* Springer-Verlag, Berlin, 1990.
96b. **Oezkaynak, E., Finlay, D., Salomon, M. J., and Varshavsky, A.,** The yeast ubiquitin genes: A family of natural gene fusions, *EMBO J.,* 6, 1429—1439, 1987.
97. **Parker-Thornburg, J. and Bonner, J. J.,** Mutations that induce the heat shock response of *Drosophila, Cell,* 51, 763, 1987.
98. **Parker, C. S. and Topol, J.,** A *Drosophila* RNA polymerase II transcription factor binds to the regulatory site of an hsp 70 gene, *Cell,* 37, 273—283, 1984.
99. **Pauli, D. and Tonka, C.-H.,** A new *Drosophila* heat shock gene from locus 67B is expressed during embryogenesis and pupation., *J. Mol. Biol.,* 198, 233—240, 1987.
100. **Pauli, D., Spierer, A., and Tissieres, A.,** Several hundred base pairs upstream of *Drosophila* hsp23 and 26 genes are required for their heat induction in transfomed flies, *EMBO J.,* 5, 755—761, 1986.
101. **Pauli, D., Tonka, C.-H., and Ayme-Southgate, A.,** An unusual split *Drosophila* heat shock gene expressed during embryogenesis, pupation and in testis, *EMBO J.,* 200, 47—53, 1988.
102. **Pelham, H. R. B.,** A regulatory upstream promoter element in the *Drosophila* hsp 70 heat-shock gene, *Cell,* 30, 517—528, 1982.
103. **Pelham, H. R. B.,** Hsp 70 accelerates the recovery of nucleolar morphology after heat shock, *EMBO J.,* 3, 3095—3100, 1984.
103a. **Pelham, H. R. B.,** Heat shock and the sorting of luminal ER proteins, *EMBO J.,* 8, 3171—3176, 1989.
104. **Pelham, H. R. B. and Bienz, M.,** A synthetic heat-shock promoter element confers heat-inducibility on the herpes simplex virus thymidine kinase gene, *EMBO J.,* 1, 1473—1477, 1982.
105. **Pelham, H. and Bienz, M.,** DNA sequences required for transcriptional regulation of the *Drosophila* hsp 70 heat-shock gene in monkey cells and *Xenopus* oocytes in *Heat Shock from Bacteria to Man,* Schlesinger, M. J., Ashburner, M., and Tissieres, A., Eds., Cold Spring Harbor Laboratory, Cold Spring Harbor, NY, 1982, 43—48.
106. **Pelham, H., Lewis, M., and Lindquist, S.,** Expression of a *Drosophila* heat shock protein in mammalian cells: Transient association with nucleoli after heat shock, *Philos. Trans. R. Soc. London Ser. B.,* 307, 301—307, 1984.
106a. **Perisic, O., Xiao, H., and Lis, J. T.,** Stable binding of *Drosophila* heat shock factor to head-to-head and tail-to-tail repeats of a conserved 5 bp recognition unit, *Cell,* 59, 797—806, 1989.
107. **Petko, L. and Lindquist, S.,** Hsp 26 is not required for growth at high temperatures, nor for thermotolerance, sport development, or germination, *Cell,* 45, 885—894, 1986.
108. **Pietrzak, M., Burri, M., Herrero, J. J., and Mosbach, K.,** Transcriptional activity is inducible in the cauliflower mosaic virus 35S promoter engineered with the heat shock consensus sequence, *FEBS Lett.,* 249, 311—315, 1989.
108a. **Piper, P. W., Curran, B., Davies, M. W., Hirst, K., Lockheart, A., Ogden, J. E., Stanway, C. A., Kingsman, A. J., and Kingsman, S. M.,** A heat shock element in the phosphoglycerate kinase gene promoter of yeast, *Nucl. Acids Res.,* 16, 1333—1348, 1988.
108b. **Plesofsky-Vig, N. and Brambl, R.,** personal communication.
109. **Rancourt, D. E., Walker, V. K., and Davies, P. L.,** Flounder antifreeze protein synthesis under heat shock control in transgenic *Drosophila melanogaster, Mol. Cell. Biol.,* 7, 2188—2195, 1987.
109a. **Raschke, E.,** Molekulare Analyse verschiedener Gene für kleine Hitzeschock Poteine der Sojabohne *(Glycine Max* (L.) Merrill), Thesis, University of Bielefeld (FRG), 1987.
110. **Raschke, E., Baumann, G., and Schöffl, F.,** Nucleotide sequence analysis of soybean small heat shock protein genes belonging to two different multigene families, *J. Mol. Biol.,* 199, 549—557, 1988.
110a. **Rebbe, N. F., Hickman, W. S., Ley, T. J., Stafford, D. W., and Hickman, S.,** Nucleotide sequence and regulation of a human 90-kDa heat shock protein gene, *J. Biol. Chem.,* 264, 15006—15011, 1989.

111. **Riddihough, G. and Pelham, H. R. B.,** Activation of the *Drosophila* hsp 27 promoter by heat shock and ecdysone involves independent and remote regulatory sequences, *EMBO J.*, 5, 1653—1658, 1986.
112. **Riddihough, G. and Pelham, H. R. B.,** An ecdysone response element in the *Drosophila* hsp27 promoter, *EMBO J.*, 6, 3729—3734, 1987.
113. **Rio, D. C. and Rubin, G. M.,** Transformation of cultured *Drosophila melanogaster* cells with a dominant selectable marker, *Mol. Cell. Biol.*, 5, 1833—1838, 1985.
114. **Rio, D. C., Laski, F. A., and Rubin, G. M.,** Identification and immuno-chemical analysis of biologically active *Drosophila* P element transposase, *Cell*, 44, 21—32, 1986.
115. **Rochester, D. E., Winter, J. A., and Shah, D. M.,** The structure and expression of maize genes encoding the major heat shock protein, hsp70, *EMBO J.*, 5, 451—458, 1986.
116. **Russnak, R. H. and Candido, E. P. M.,** Locus encoding a family of small heat shock genes in *Caenorhabditis elegans:* Two genes duplicated to form a 3.8-kilobase inverted repeat, *Mol. Cell, Biol.*, 5, 1268—1278, 1985.
116a. **Ryseck, R.-P., Walldorf, U., Hoffmann, T., and Hovemann, B.,** Heat shock loci 93D of *Drosophila melanogaster* and 48B of *Drosophila hydei* exhibit a common structural and transcriptional pattern, *Nucl. Acids Res.*, 15, 3317—3333, 1987.
116b. **Schiller, P., Amin, J., Ananthan, J., Brown, M. E., Scott, W. A., and Voellmy, R.,** Cis-acting elements involved in the regulated expression of a hunam HSP70 gene, *J. Mol. Biol.*, 203, 97—105, 1988.
116c. **Schmülling, T., Beinsberger, S., Degreef, J., Schell, J., van Onckelen, H., and Spena, A.,** Construction of a heat-inducible chimaeric gene to increase the cytokinin content in transgenic plant tissue, *FEBS Lett.*, 249, 401—406, 1989.
117. **Schneuwly, S., Klemenz, R., and Gehring, W. J.,** Redesigning the body plan of *Drosophila* by ectopic expression of the homoeotic gene *Antennapedia, Nature,* 325, 816—818, 1987.
117a. **Schöffl. F.,** personal communication.
118. **Schöffl, F. and Baumann, G.,** Thermo-induced transcripts of a soybean heat shock gene after transfer into sunflower using a Ti plasmid vector, *EMBO J.*, 4, 1119—1124, 1985.
119. **Schöffl, F., Raschke, E., and Nagao, R. T.,** The DNA sequence analysis of soybean heat shock genes and identification of possible regulatory promoter elements, *EMBO J.*, 3, 2491—2497, 1984.
120. **Schöffl, F., Baumann, G., Raschke, E., and Bevan, M.,** The expression of heat-shock genes in higher plants, *Philos. Trans. R. Soc. London Ser. B,* 314, 453—468, 1986.
120a. **Schöffl, F., Rieping, M., and Baumann, G.,** Constitutive transcription of a soybean heat-shock gene by a cauliflower mosaic virus promoter in transgenic tobacco plants, *Dev. Genet.*, 8, 365—374, 1987.
120b. **Schöffl, F., Rieping, M., Baumann, G., Bevan, M., and Angermueller, S.,** The function of plant heat shock promoter elements in the regulated expression of chimaeric genes in transgenic tobacco, *Mol. Gen. Genet.*, 217, 246—253, 1989.
121. **Serfling, E., Jasin, M., and Schaffner, W.,** Enhancers and eukaryotic gene transcription, *Trends Genet.*, 1, 224—230, 1985.
121a. **Shepherd, J. C. W., Walldorf, U., Hug, P., and Gehring, W. J.,** Fruit flies with additional expression of the elongation factor EF-1alpha live longer, *Proc. Natl. Acad. Sci. U.S.A.*, 86, 7520—7521, 1989.
122. **Shibahara, S., Müller, R. M., and Taguchi, H.,** Transcriptional control of rat heme oxygenase by heat shock, *J. Biol. Chem.*, 262, 12889—12892, 1987.
123. **Shuey, D. J. and Parker, C. S.,** Binding of *Drosophila* heat-shock gene transcription factor to the hsp70 promoter, *J. Biol. Chem.*, 261, 7934—7940, 1986.
124. **Simcox, A. A., Cheney, C. M., Hoffman, E. P., and Shearn, A.,** A deletion of the 3′ end of the *Drosophila melanogaster* hsp70 gene increases stability of mutant mRNA during recovery from heat shock, *Mol. Cell. Biol.*, 5, 3397—3402, 1985.
125. **Simon, J. A. and Lis, J. T.,** A germline transformation analysis reveals flexibility in the organization of heat shock consensus elements, *Nucl. Acids Res.*, 15, 2971—2988, 1987.
126. **Simon, J. A., Sutton, C. A., Lobell, R. B., Glaser, R. L., and Lis, J. T.,** Determinants of heat shock-induced chromosome puffing, *Cell*, 40, 805—817, 1985.
127. **Simon, J. A., Sutton, C. A., and Lis, J. T.,** Localization and expression of transformed DNA sequences within heat shock puffs of *Drosopila melanogaster, Chromosoma,* 93, 26—30, 1985.
127a. **Simon, M. C., Fisch, T. M., Benecke, B. J., Nevins, J. R., and Heintz, N.,** Definition of multiple functionally distinct TATA elements, one of which is a target in the hsp 70 promoter for E1A regulation, *Cell,* 52, 723—729, 1988.
128. **Sinclair, J. H., Saunders, S. E., Burke, J. F., and Sang, J. H.,** Regulated expression of a *Drosophila melanogaster* heat shock locus after stable intergration in a *Drosophila hydei* cell line, *Mol. Cell. Biol.*, 5, 3208—3213, 1985.
128a. **Sistonen, L., Holttä, E., Mäkelä, T. P., Keski-Oja, J., and Alitalo, K.,** The cellular response to induction of the p21$^{c\text{-}Ha\text{-}ras}$ oncoprotein includes stimulation of jun gene expression, *EMBO J.*, 8, 815—822, 1989.

129. **Slater, M. R. and Craig, E. A.,** Transcriptional regulation of an hsp70 heat shock gene in the yeast *Saccharomyces cerevisiae, Mol. Cell. Biol.,* 7, 1906—1916, 1987.

129a. **Snow, P. M., Bieber, A. J., and Goodman, C. S.,** Fasciclin III: a novel homophilic adhesion molecule in *Drosophila, Cell,* 59, 313—323, 1989.

129b. **Snutch, T. P., Heschl, M. F. P., and Baillie, D. L.,** The *Caenorhabditis elegans* hsp70 gene family: a molecular genetic characterization, *Gene,* 64, 241—255, 1988.

130. **Sorger, P. K. and Pelham, H. R. B.,** Cloning and expression of a gene encoding hsc 73, the major hsp 70-like protein in unstressed rat cells, *EMBO J.,* 6, 993—998, 1987.

131. **Sorger, P. K., Lewis, M. J., and Pelham, H. R. B.,** Heat shock factor is regulated differently in yeast and HeLa cells, *Nature,* 329, 81—84, 1987.

132. **Sorger, P. K. and Pelham, H. R. B.,** Purification and characterization of a heat-shock element binding protein from yeast, *EMBO J.,* 6, 3035—3042, 1987.

132a. **Sorger, P. K. and Nelson, H. C. M.,** Trimerization of a yeast transcriptional activator via a coiled-coil motif, *Cell,* 59, 807—813, 1989.

133. **Southgate, R., Ayme, A., and Voellmy, R.,** Nucleotide sequence analysis of the *Drosophila* small heat shock gene cluster at locus 67B, *J. Mol. Biol.,* 165, 35—57, 1983.

134. **Spena, A. and Schell, J.,** The expression of a heat-inducible chimeric gene in transgenic tobacco plants, *Mol. Gen. Genet.,* 206, 436—440, 1987.

135. **Spena, A., Hain, R., Ziervogel, U., Saedler, H., and Schell, J.,** Construction of a heat-inducible gene for plants. Demonstration of heat-inducible activity of the *Drosophila* hsp70 promoter in plants, *EMBO J.,* 4, 2739—2743, 1985.

136. **Steller, H. and Pirrotta, V.,** Regulated expression of genes injected into early *Drosophila* embryos, *EMBO J.,* 3, 165—173, 1984.

137. **Steller, H. and Pirrotta, V.,** Expression of the *Drosophila* white gene under the control of the hsp70 heat shock promoter, *EMBO J.,* 4, 3765—3772, 1985.

138. **Steller, H. and Pirrotta, V.,** P transposons controlled by the heat shock promoter, *Mol. Cell. Biol.,* 6, 1640—1649, 1986.

138a. **Strittmatter, G. and Chua, N. H.,** Artificial combination of two cis-regulatory elements generates a unique attern of expression in transgenic plants, *Proc. Natl. Acad. Sci. U.S.A.,* 84, 8986—8990, 1987.

139. **Struhl, G.,** Near-reciprocal phenotypes caused by inactivation or indiscriminate expression of the *Drosophila* segmentation gene ftz, *Nature,* 318, 677—680, 1986.

139a. **Susek, R. E. and Lindquist, S. L.,** Hsp26 of *Saccharomyces cerevisiae* is homologous to the superfamily of small heat shock proteins, but is without a demonstrable function, *Mol. Cell. Biol.,* 9, 5265—5271, 1989.

139b. **Taylor, I. C. A., Solomon, W., Weiner, B. M., Paucha, E., Bradley, M., and Kingston, R. E.,** Stimulation of the human heat shock protein 70 promotor *in vitro* by Simian virus 40 large T-antigen, *J. Biol. Chem.,* 264, 16160—16164, 1989.

139c. **Taylor, I. C. A. and Kingston, R. E.,** Factor substitution in a human HSP70 gene promoter: TATA-dependent and TATA-independent interactions, *Mol. Cell. Biol.,* 10, 165—175, 1990.

139d. **Taylor, I. C. A. and Kingston, R. E.,** E1a transactivation of human HSP70 gene promoter substitution mutants is independent of the composition of upstream and TATA elements, *Mol. Cell. Biol.,* 10, 176—183, 1990.

139e. **Theodorakis, N. G., Zand, D. J., Kotzbauer, P. T., Williams, G. T., and Morimoto, R. I.,** Hemin-induced transcriptional activation of the hsp70 gene during erythroid maturation in K526 cells is due to a heat shock factor-mediated stress response, *Mol. Cell. Biol.,* 9, 3166—3173, 1989.

140. **Toeroek, I. and Karch, R.,** Nucleotide sequences of heat shock activated genes in *Drosophila melanogaster.* I. Sequences in the regions of the 5′ and 3′ ends of the hsp70 gene in the hybrid plasmid 56H8, *Nucl. Acids Res.,* 8, 3105—3123, 1980.

141. **Topol, J., Ruden, D. M., and Parker, C. S.,** Sequences required for in vitro transcriptional activation of a *Drosophila* hsp 70 gene, *Cell,* 42, 527—537, 1985.

142. **Voellmy, R.,** The heat shock genes: a family of highly conserved genes with a superbly complex expression pattern, *Bio-Essays,* 1, 213—217, 1985.

143. **Voellmy, R. and Rungger, D.,** Transcription of a *Drosophila* heat shock gene is heat-induced in *Xenopus* oocytes, *Proc. Natl. Acad. Sci. U.S.A.,* 79, 1776—1780, 1982.

144. **Voellmy, R., Ahmed, A., Schiller, P., Bromley, P., and Rungger, D.,** Isolation and functional analysis of a human 70,000-dalton heat shock protein gene segment, *Proc. Natl. Acad. Sci. U.S.A.,* 82, 4949—4953, 1985.

144a. **Vourc'h, C., Binart, N., Chambraud, B., David, J. P., Jerome, V., Baulieu, E. E., and Catelli, M. G.,** Isolation and functional analysis of chicken 90-kDa heat shock protein gene promoter, *Nucl. Acids Res.,* 17, 5259—5272, 1989.

144b. **Walter, T., Drabent, B., Krebs, H., Tomalak, M., Heiss, S., and Beneke, B.-J.,** Cloning and analysis of a human 86-kDa heat-shock protein-encoding gene, *Gene,* 83, 105—116, 1989.

145. **Wei, R., Wilkinson, H., Pfeifer, K., Schneider, C., Young, R., and Guarente, L.,** Two or more copies of *Drosophila* heat shock consensus sequence serve to activate transcription in yeast, *Nucl. Acids Res.*, 14, 8183—8189, 1986.

145a. **Wiederrecht, G., Seto, D., and Parker, C. S.,** Isolation of the gene encoding the *S. cerevisiae* heat shock transcription factor, *Cell,* 54, 841—853, 1988.

145b. **Williams, G. T., McClanahan, T. K., and Morimoto, R. I.,** E1a transactivation of the human HSP70 promoter is mediated through the basal transcription complex, *Mol. Cell. Biol.*, 9, 2574—2587, 1989.

145c. **Wisniewski, J., Kordula, T., and Krawczyk, Z.,** Isolation and nucleotide sequence analysis of the rat testis-specific major heat-shock protein (HSP70)-related gene, *Biochim. Biophys. Acta,* 1048, 93—99, 1990.

146. **Wu, B. J. and Morimoto, R. I.,** Transcription of the human hsp70 gene is induced by serum stimulation, *Proc. Natl. Acad. Sci. U.S.A.*, 82, 6070—6074, 1985.

147. **Wu, B., Hunt, C., and Morimoto, R.,** Structure and expression of the human gene encoding major heat shock protein HSP70, *Mol. Cell. Biol.*, 5, 330—341, 1985.

148. **Wu, B. J., Kingston, R., and Morimoto, R. I.,** Human HSP70 promoter contains at least two distinct regulatory domains, *Proc. Natl. Acad. Sci. U.S.A.*, 83, 629—633, 1986.

149. **Wu, B. J., Hurst, H. C., Jones, N. C., and Morimoto, R. I.,** The E1A 13S product of adenovirus S activates transcription of the cellular human HSP70 gene, *Mol. Cell. Biol.*, 6, 2994—2999, 1986.

149a. **Wu, B. J., Gregg, T. W., and Morimoto, R. J.,** Detection of three protein binding sites in the serum-regulated promoter of human gene encoding the 70-kDa heat shock protein, *Proc. Natl. Acad. Sci. U.S.A.*, 84, 2203—2207, 1987.

150. **Wu, L., Morley, B. J., and Campbell, R. D.,** Cell-specific expression of the human complement protein factor B gene: Evidence for the role of two distinct 5′-flanking elements, *Cell,* 48, 331—342, 1987.

151. **Wu, C., Wilson, S., Walker, B., David, I., Paisley, T., Zimarino, V., and Ueda, H.,** Purification and properties of *Drosophila* heat shock activator protein, *Science,* 238, 1247—1253, 1987.

151a. **Wu, C. H., Caspar, T., Browsc, J., Lindquist, S., and Somerville, C.,** Characterization of an HSP70 cognate gene family in *Arabidopsis, Plant Physiol.*, 88, 731—740, 1988.

152. **Wurm, F. M., Gwinn, K. A., and Kingston, R. E.,** Inducible overproduction of the mouse c-myc protein in mammalian cells, *Proc. Natl. Acad. Sci. U.S.A.*, 83, 5414—5418, 1986.

153. **Xiao, H. and Lis, J. T.,** A consensus sequence polymer inhibits in vivo expression of heat shock genes, *Mol. Cell. Biol.*, 6, 3200—3206, 1986.

154. **Xiao, H. and Lis, J. T.,** Germline transformation used to define key feature of heat shock response elements, *Science,* 239, 1139—1142, 1988.

155. **Yost, H. J. and Lindquist, S.,** RNA splicing is interrupted by heat shock and is rescued by heat shock protein synthesis, *Cell,* 45, 185—193, 1986.

Chapter 7

HEAT SHOCK TRANSCRIPTION FACTORS

K.-D. Scharf and L. Nover

TABLE OF CONTENTS

7.1. THE HEAT SHOCK REGULON OF *ESCHERICHIA COLI*

Contrasting with the situation in eukaryotes, heat shock proteins (HSP) of *E. coli* (see Table 2.2) were defined as members of a regulatory entity (regulon) because their synthesis was blocked in a pleiotropically acting amber mutant (Tsn-K165). This mutant, originally isolated by Cooper and Ruettinger,[12] was shown to be defective in the regulatory gene (rpoH = htpR = hin) of the high-temperature production (HTP) regulon.[41,70a] The product of the rpoH gene was identified as a hs specific σ factor of RNA polymerase (M_r 32 kDa). This led to a renaming of the corresponding gene from originally htpR or hin to rpoH.[42,72]

Cloning and sequencing of the rpoH gene[34,43,57,73] revealed an open reading frame of 852 nucleotides coding for a protein with regions of pronounced sequence homology to the σ factor of RNA polymerase (σ 70 coded by the rpoD gene). This includes also regions of homology to the *E. coli* nusA protein[73] as well as to two *Bacillus subtilus* proteins (σ^{43} and SpoI G).[52] The initial evidence for a σ-like factor coded by the rpoH gene[71] was supported by the phenotypical suppression of temperature-sensitive mutants of the rpoD gene by mutations in the rpoH gene[26] so well as by the isolation and *in vitro* characterization of the corresponding protein.[25,49c] An *in vitro* mixture of RNA polymerase core enzyme and the purifed RpoH protein (σ^{32}) correctly initiated transcription from the hs promoters. The control σ^{70}-subunit was not required. These results were later confirmed by purification of the σ^{32}-containing holoenzyme and by the demonstration that both σ-factors are exchangeable *in vitro*.[21a] The effects are summarized in Figure 7.1 using an *in vitro* transcription/translation system in the absence (lanes 1 to 4) and in the presence (lanes 5 to 9) of a rpoH gene containing template.[6] β-Lactamase synthesis is unaffected, but expression of templates with hs promoters (groEL, dnaK, σ^{70}) is considerably stimulated, if the system is substituted with a high expression plasmid for σ^{32}. The *in vitro* synthesis and correct activity of the heat shock (hs) σ factor suggests that the relative concentrations of σ-factors may be decisive for the shift of transcription specificity during hs *in vivo* (see Section 7.3).

The observation of an overlapping promoter specificity between the *E. coli* σ^{32} and the *Bacillus subtilis* σ^{28} [9] led to the idea that both factors may function similarly, but closer investigation of the hs response in *Bacillus subtilis* showed that levels of σ^{28} and σ^{43} declined[2] and that σ^{28} is evidently not involved.

7.2. SEQUENCE ELEMENTS OF *E. COLI* HEAT SHOCK PROMOTERS

Promoter recognition in *E. coli* is mainly based on two regions upstream of the transcription start. For the control RNA polymerase, containing the σ^{70} initiation factor, these are a pentanucleotide -TTGAC- in the −35 region and -TATAAT- in the −10 region (see bottom line of Figure 7.2). Replacement of σ^{70} by the hs-specific σ^{32} shifts the promoter specificity. All hs promoters so far sequenced (Figure 7.2) contain -CnCcc$\underline{\text{TTGAA}}$- in the −35 region and -CCCCATnT- in the −10 region. In addition, there are blocks of 7 to 16 uninterrupted A/T pairs between bp −40 and −71.[35a,42,72] Travers[60] pointed out that there may be an evolutionary link between the hs promoter elements of *E. coli* and eukaryotes, because the underlined part of the −35 region is virtually identical to half of the eukaryotic HSE: cTnGAA- (see Table 6.2).

Special interest deserves the demonstration that the control σ factor is itself a hs-induced protein.[24,54] The gene rpoD is part of a polycistronic operon, which may be crucial for the control of macromolecule synthesis in *E. coli*.[36] It contains, from 5′ to 3′ end, the rpsU gene coding for ribosomal protein S21, the dnaG gene for DNA primase (synthesis of RNA primer as starter for DNA replication), and rpoD. Multiple promoters evidently ensure the necessary versatility of the regulatory response. *Inter alia,* an internal hs promoter (see

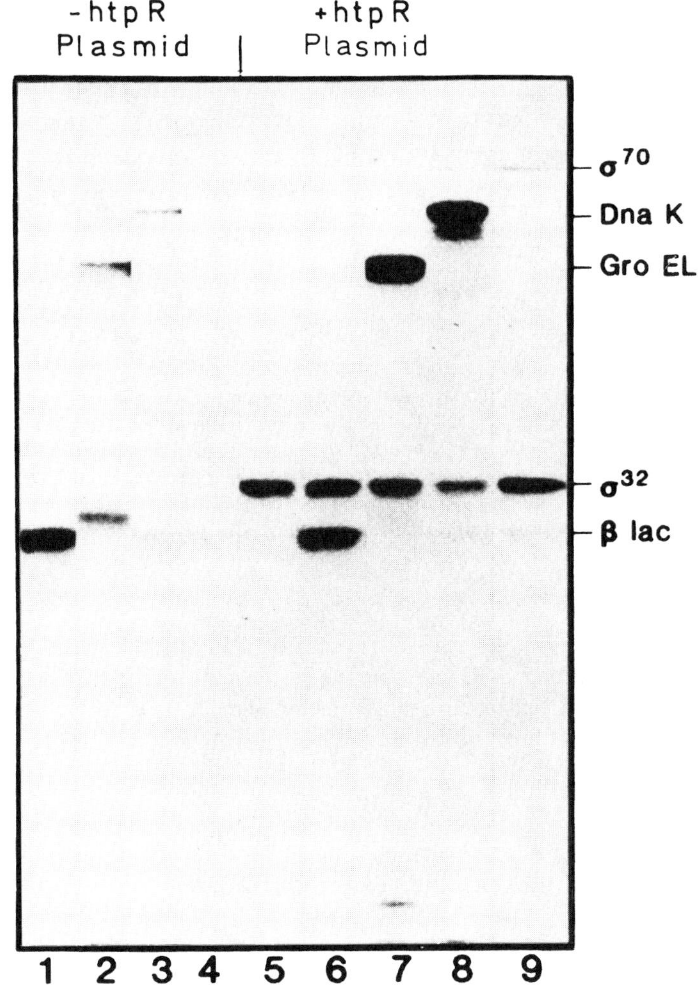

FIGURE 7.1. Effect of htpR containing plasmid on the *in vitro* transcription/translation of *E. coli* heat shock proteins (HSPs). Plasmids containing the indicated genes were used as templates for an *E. coli in vitro* transcription/translation system. Lanes 1 to 4 in the absence of htpR (= rpoH) plasmid coding for the hs σ^{32} subunit, lanes 5 to 9 in presence of htpR plasmid. Note the marked increase of HSP synthesis (σ^{70}, GroE, DnaK) while expression of the β-lactamase gene is not influenced. The figure shows an autofluorograph of labeled proteins synthesized *in vitro* from the templates indicated on the right margin. (From Bloom, M., et al., *J. Bacteriol.*, 166, 380—384, 1986. With permission.)

Figure 7.2) allows the third gene (rpoD) to be selectively transcribed under hs, when initiation in the "normal" promoter region upstream of rpsU, is blocked.[37,54]

7.3. CHANGING LEVELS OF SIGMA FACTORS IN *E. COLI*

The transient reprogramming of transcription by hs is brought about by the changing levels of three σ subunits of RNA polymerase. Initially, the level of σ^{32} rises rapidly up to 17-fold.[20,21,26a,35,49c,53,56] Several factors are responsible for this: (1) the level of mRNA increases and a new rpoH transcript is formed.[20,56] Analysis of the rpoH promoter revealed four overlapping, different sequence combinations in the −35 and −10 regions giving rise to three different transcripts. Promoters 1 and 4 are constitutively active and σ^{70}-depen-

```
               -35 region                  -10 region
P   (groE)   CCCCCTTGAAGGGGCG--AAGCCATCCCCATTTTCTCTG
P1  (dnaK)   CCCCCTTGATGACGTG-GTTTACGACCCCATTTAGTAGT
P2  (dnaK)   GGCAGTTGAAACCAGA--CGTTTCGCCCC-TATTACAGA
P   (lon)    CGGCGTTGAATGTGGG-GGAAACATCCCCATATACTGAC
Phs(rpoD)    CACCCTTGAAAAACTGTCGATGTGGGACGATATAGCAGA
P   (hptG)   CTCGCTTGAAATTATTCTCCCTTGTCCCCATCTCTCCCA
Cons. σ32    CnCccTTGAA- 13-15 nucl. -CCCCATnT-6 nucl.

Cons. σ70    ---tcTTGACaatt- 14 nucl.  -   TAtAaT---
```

FIGURE 7.2. Sequences of heat shock-induced promoters of *Escherichia coli* (bp −1 to −39) of the noncoding strands. (For summaries, see Neidhardt et al.[42] and Yura.[72]) Sequences are from Taylor et al.[54] (rpoD), Cowing et al.[13] and Hemmingson et al.[27b] (groE, dnaK), Gayda et al.[22] (lon), and Bardwell and Craig[4] (hptG). At the bottom, the consensus sequences derived for σ^{32} and that published for σ^{70} are compared. Transcription start is on the right. (From Neidhardt, F. C., Van Bogelen, R. A., and Vaughn, V., *Annu. Rev. Genet.*, 18, 259, 1984; and Yura, T., *Jpn. J. Genet.*, 61, 277, 1986.)

dent.[20,21] In addition, there is a gene proximal promoter P5, which is sensitive to catabolite repression.[40a] Most interesting is P3 interspersed between P1 and P4. Its activity depends on the presence of a high temperature sigma factor with 24 kDa (σ^{24}), which mediates transcription at 42 to 50°C.
The promoter sequence:

$$\overset{-35}{\overline{\text{GAACTT}}}\text{4nATAAAA6n}\overset{-10}{\overline{\text{TCTGA}}}$$

may be characteristic of a separate high temperature regulon, which includes also the htrA gene coding for a 48-kDa periplasmic protease.[19b,35b,61b] Remarkably, at 50°C transcription of hs mRNAs and their translation is not subject to autorepression (see Section 8.7). (2) There is an enhanced translation of the rpoH mRNA.[20,53] (3) The stability of the otherwise very short-lived σ^{32} is markedly improved from a half-life of 1 min to 8 min.[3,53] The latter effect also explains the induction of HSP synthesis in λ-phage infected cells. The phage cIII protein stabilizes the σ^{32} protein, i.e., it acts like a specific proteinase inhibitor.[3]

Similar to the *in vitro* situation (Figure 7.1), σ^{32} replaces σ^{70} in the RNA polymerase and, thus, shifts promoter recognition toward the hs genes. The situation is reversed by the increasing σ^{70} level as a consequence of the internal hs promoter of the rpsU-dnaG-rpoD operon. Simultaneously, accumulation of the Lon and DnaK proteins contributes to a decrease of the σ^{32} level. Autorepression of HSP synthesis in *E. coli* by DnaK is presumably mediated by an inhibition of rpoH mRNA translation,[26a,56a] by complex formation between DnaK and sigma 32[49d] and by Lon-dependent degradation of σ^{32} (see review by Gottesman[23b]). The extent of hs gene transcription always parallels the σ^{32} level except after shift down from 42°→30°C. In this case, a transient block of hs transcription precedes the decrease of σ^{32} levels, indicating a direct, additional control mechanism for its activity.[53a]

The transient stabilization of the σ^{32}-factor so well as the mode of action of the λ-phage cIII protein are in good agreement with the hypothesis of Munro and Pelham[40] on the signal transformation events during hs (see Section 1.4): an overload of the cell with aberrant proteins may protect the constitutively expressed, but short-lived transcription factor from degradation, probably by direct interaction of unfolded proteins with σ^{32}.[45a] Indeed, the overproduction of aberrant or foreign proteins in *E. coli* results in an increased HSP synthesis even at control temperatures,[23,28,45a] but the same effect is also brought about by direct rise of the internal σ^{32} levels after transformation of *E. coli* cells with a tac promoter × rpoH gene fusion plasmid followed by induction with IPTG.[56,56a,61]

Finally, it is worth noticing that the interaction between the two σ-factors may by far exceed the "classical" hs situation. Extensive mutation analysis of the rpoH gene including deletions[23c] showed that these mutations stopped growing at >20°C. Evidently the effective transcription of dnaK, groE, rpoD, and other genes even at non-hs temperatures needs the presence of low amounts of σ^{32}. Multifunctional promoter combinations, as illustrated in Figure 6.6 for the eukaryotic hs genes, are well known for *E. coli* and other bacteria.[20,37]

7.4. DETECTION OF EUKARYOTIC PROMOTER BINDING PROTEINS

The regular organization of DNA into nucleosomes held together by histone octamers and, starting from this, into higher order structures is the dominant feature of eukaryotic chromatin. However, at regulatory sites, e.g., in promoter regions, it is interrupted by nonnucleosomal DNP.[10a,19a,46c] This is brought about by the sequence or region-specific binding of nonhistone proteins. Such are RNA polymerase and its helper proteins so well as other transcription factors, whose nature depends on the gene investigated. In most cases of plant and mammalian promoters, transcription initiation evidently depends on multifactorial DNP complexes ("transcriptosomes") composed of RNA polymerase II, several promoter-binding proteins, and additional transcription factors, which exert their effects by protein-protein interaction.[9a,15a,20a,27,27d,38c,39,46a,62b,68]

The rapid progress in the isolation and characterization of eukaryotic promoter binding proteins was extensively reviewed.[29b,38b,62a] It was facilitated by a number of methods derived from studies on promoters of *E. coli* and its phages. Thus, the occupancy of a given DNA region by a protein is revealed by protection from degradation by nucleases, e.g., DNase I or exonuclease III, or by chemical cleavage, e.g., by methidium propyl-EDTA·Fe(II). Alternatively, accessibility of base residues or phosphate groups in the DNA to chemical alkylation is impaired. In both cases the protein leaves a "footprint" of a protected area, which can be detected in a sequencing gel (Figure 7.3B). Complementary to this footprint analysis is the gel retardation assay.[46b] To test for the presence and/or activity of a DNA-binding transcription factor, an appropriate promoter fragment or synthetic oligonucleotide is labeled and incubated *in vitro* with crude or purified fractions of the binding protein. Formation of the DNP complex can be demonstrated by retardation of the DNA probe in an acrylamide or agarose gel (Figure 7.3A). Both, footprint and gel retardation assays are prerequisites to monitor procedures for purification of promoter binding proteins. They are complemented by functional tests performed *in vitro* and *in vivo*.

A decisive step for successful separation of the usually small amounts of promoter binding factors from the mass of unrelated proteins is the affinity chromatography through columns of sepharose modified with an appropriate oligonucleotide. In addition to heat shock transcription factor (HSF), other promoter binding proteins, which play a role for the non-hs control of expression of vertebrate hs genes, were purified. These are the Sp1-specific protein,[19,31,32] different CAT-box binding factors (CTF)[16,20a,30,39] and TATA-binding proteins.[9a,29b,38b,62c] Purification of promoter-binding proteins to homogeneity allowed the preparation of corresponding antibodies, followed by immunological screening of expression libraries for specific cDNA clones and isolation and sequencing of the regulatory genes. This procedure was exemplified for the yeast HSF.[50a,62a] A much more rapid and direct access to cDNA clones coding for the promoter binding domain of a given transcription factor is possible, if the corresponding promoter fragment can be used as probe to screen for binding proteins blotted on nitrocellulose membranes (Southwestern screening).[34b,47a,49b,51a,61b]

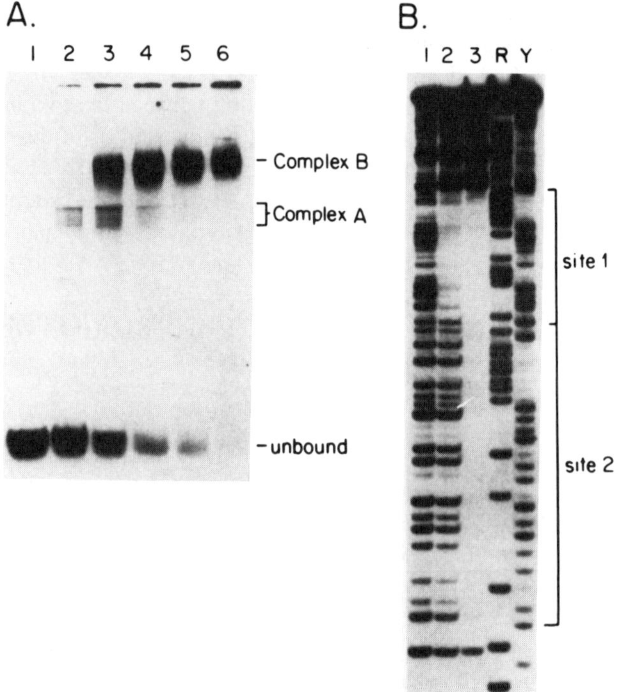

FIGURE 7.3. Binding of *Drosophila* HSF to the hsp70 promoter. (A), Gel retardation assay. A labeled hsp70 promoter fragment of 101 base pairs was incubated with increasing amounts of partially purified HSF. Binding results in the retardation of the DNA probe in an agarose gel. Complex A contains HSF bound to HSE1; in complex B both adjacent HSE1 and 2 are occupied by HSF. (B) DNaseI footprinting: unbound DNA (lane 1) so well as DNP complexes A (lane 2) and B (lane 3) were digested with DNase I, denatured and rerun in a sequencing gel. Binding of HSF protects cleavage by DNaseI in the HSE 1 region alone (complex A) or over the whole distance of HSE 1 plus 2 (complex B). R and Y are chemical cleavages at purine and pyrimidine residues for control. (From Shuey, D. J. and Parker, C. S., *J. Biol. Chem.*, 261, 7934, 1986. With permission.)

7.5. THE HEAT SHOCK TRANSCRIPTION FACTOR (HSF) OF *DROSOPHILA*

Early attempts to characterize the hs promoter binding protein of *Drosophila* relied on four different methods:

1. Jack et al.[29] used a nitrocellulose filter binding assay to detect three proteins of M_r 35, 34, and 29 kDa attached to the 5′ upstream regions of 87A7 and 87C1 clones.
2. Cytoplasmic factors from heat-shocked *Drosophila* cells were demonstrated to induce hs puffs in isolated polytene nuclei.[7,11] *In situ* hybridization of the *in vitro* transcribed RNA to polytene chromosomes revealed homology to sequences of hsp70 genes in puffs 87A7/C1.[8]
3. Craine and Kornberg[14,15] followed a similar line with isolated nuclei from control cell cultures of *Drosophila* activated by cytoplasmic factors from hs cells. Exogenous *E. coli* RNA polymerase was used for the *in vitro* transcription assay. Transcription of hsp70 species RNA was found selectively enhanced, whereas transcription of histone genes remained constant. The putative HSF is maximum active after 10 min of a hs treatment.
4. Wu[63-66] applied footprint analyses to the Dm-hsp83 and hsp70 promoters. By DNaseI and exonuclease III protection assay, he detected two proteins associated with these

promoters in a sequence-specific manner. One was identified as the TATA-box binding protein, the other as the HSF (heat shock activator protein [HAP] in Wu's nomenclature). Exonuclease III protection could be used to monitor the *in vitro* binding of HSF to a hsp83 promoter fragment[65,66] and to confirm earlier findings that under non-hs conditions the transcription factor exists in preformed, but inactive state.[74] Activation is probably brought about by hs-dependent phosphorylation of the homotrimeric form.[45b] It is detectable after 30 s of hs, is optimum after 5 min, and is rapidly reversed in the recovery period. Several cycles of activation and deactivation were performed in the presence of cycloheximide. Evidently no new synthesis of protein is required. This contrasts to the situation in bacteria (see Section 7.3) and to the hypothesis of Munro and Pelham[40] on the hs signal transformation chain based on the short-lived transcription factor. Interestingly, specific binding of HSF could also be observed after 2,4-DNP or salicylate treatment. Both compounds were originally used by Ritossa[47] as examples for chemical stressors. In a similar way, new promoter-protein complexes were also found upstream of the small HSP coding genes after cleavage with DNaseI and methidiumpropyl-EDTA-Fe(II). All are characterized by the presence of HSEs.[10]

The observation on the hs-induced activation of HSF in the presence of cycloheximide was extended and modified: (1) Similar results were obtained with chicken MSB cells and human HeLa cells[33,75] but also with cell cultures of tomato.[47a] (2) The independence of protein synthesis is only valid for severe hs conditions. Using *Drosophila,* chicken, and human cells, Zimarino et al.[75] demonstrated that the noticeable HSF activation by heat shock at intermediate temperatures is abolished in the presence of cycloheximide. Evidently two different mechanisms for signal transduction are operative under mild vs. severe hs conditions.

There is evidently a limiting amount of active HSF available in *Drosophila* but also in vertebrate cells. This may explain why in expression systems based on a multicopy situation (see Section 6.1.1), the promoter activity calculated on a per gene dosis may be as low as 1% of the wild-type context with one or few genes. Using competitor plasmids with tandem insertions of a *Drosophila* hsp70 promoter fragment (bp −89 to −38), Xiao and Lis[69] directly demonstrated the inhibition of transcription from hs promoters, when increasing amounts of competitor plasmid were added to the transformation mixture (Figure 7.4). In agreement with findings of McGarry and Lindquist,[38] the hsp83 promoter is least sensitive. It can be speculated that the overlapping HSE trimer of this promoter (Table 6.2) provides it with the highest affinity for HSF. This may explain the enhanced transcription of hsp83 at rather low hs temperatures when the rise of HSF activity is barely detectable and other hs genes of *Drosophila* do not yet respond (see Figure 1.1). Even the constitutive level of hsp83 expression may depend on this particular quality of its promoter.

7.6. PURIFICATION OF HSF FROM *DROSOPHILA* AND YEAST-CLONING OF HSF GENES

Partial purification of HSF and TATA-box binding protein from *Drosophila* cells[44,45] was achieved by chromatography through DEAE cellulose, DEAE sephadex, Biorex70, and, finally, through a DNA cellulose column. The enriched fraction, containing about 10% of HSF, was analyzed by *in vitro* transcription, footprint, and gel retardation assays.[49,59] On the one side, binding activity from control cells was much lower than that from hs cells. On the other hand, careful studies with the Dm-hsp70 promoter (Figure 7.3) supported the results from promoter deletion experiments:[1,18,59] there is a cooperative interaction between the two TATA-proximal HSE motifs. At low concentrations of HSF only HSE 1 (complex A in the retardation assay Figure 7.3A), and at higher concentrations both HSEs (complex B) are occupied, resulting in a continuous footprint of resistance to DNase I (Figure 7.3B,

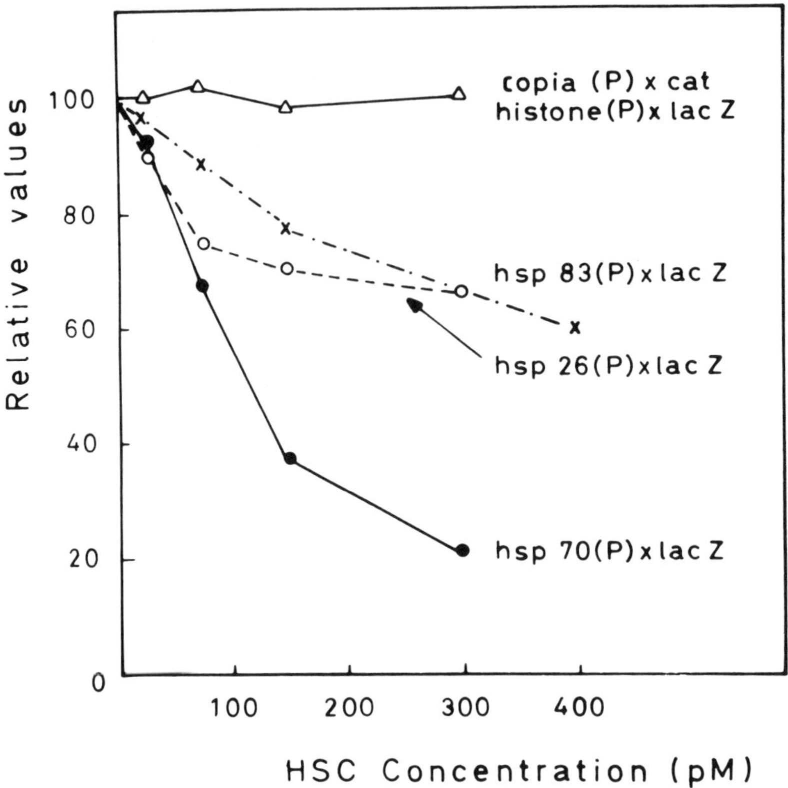

FIGURE 7.4. Competition for hs transcription factor (HSF) in a transient expression assay. Different promoter fusion constructs with the *E. coli* cat or lacZ as reporter genes were used to monitor the hs inducibility in the presence of increasing amounts of a competitor plasmid (HSC) containing 40 tandem copies of the Dm-hsp70 promoter region (bp −89 to −38). *Drosophila* S2 cells were cotransformed with corresponding mixtures of the two types of plasmids (for details see Chapter 6 and Table 6.1). Transcription from the two constitutive *Drosophila* promoters of the copia element and a histone gene is unaffected. Inducible activity of the hs promoters is inhibited, with strongest effects on the homologous hsp70 promoter. (Modified from Xiao, H. and Lis, J. T., *Mol. Cell. Biol.*, 6, 3200, 1986.)

lane 3). This second type of DNP complex is evidently required for the high transcription activity of the hsp70 promoter.

The purification of HSF was thoroughly improved in a three step procedure. It includes a combination of phosphocellulose, heparin agarose, and affinity chromatography using sepharose covalently modified with a concatemer of a hsp83-like HSE. By this procedure, the *Drosophila* and the yeast HSF were enriched 250,000- and 86,000-fold, respectively. Both exhibited apparent molecular weights of 70 kDa, and both gave essentially identical footprints, irrespective of whether the yeast (SSA1 gene) or the *Drosophila* hsp70 promoters were used. The same result was obtained, if the *Drosophila* HSF was separated on SDS-polyacrylamide gels and the 70-kDa band was isolated, renatured, and used for footprint analysis. Furthermore, in an *in vitro* transcription assay, the pure *Drosophila,* but not the yeast, HSF was active.[62]

Meanwhile, it is apparent that the 70-kDa proteins isolated by Wiederrecht et al.[62] are in fact active HSF fragments generated by limited proteolytic degradation during the purification procedure. The 210,000-fold purification of the yeast HSF (Figure 7.5)[50,51] revealed a 130-kDa protein with specific affinity for synthetic HSE-containing oligonucleotides,

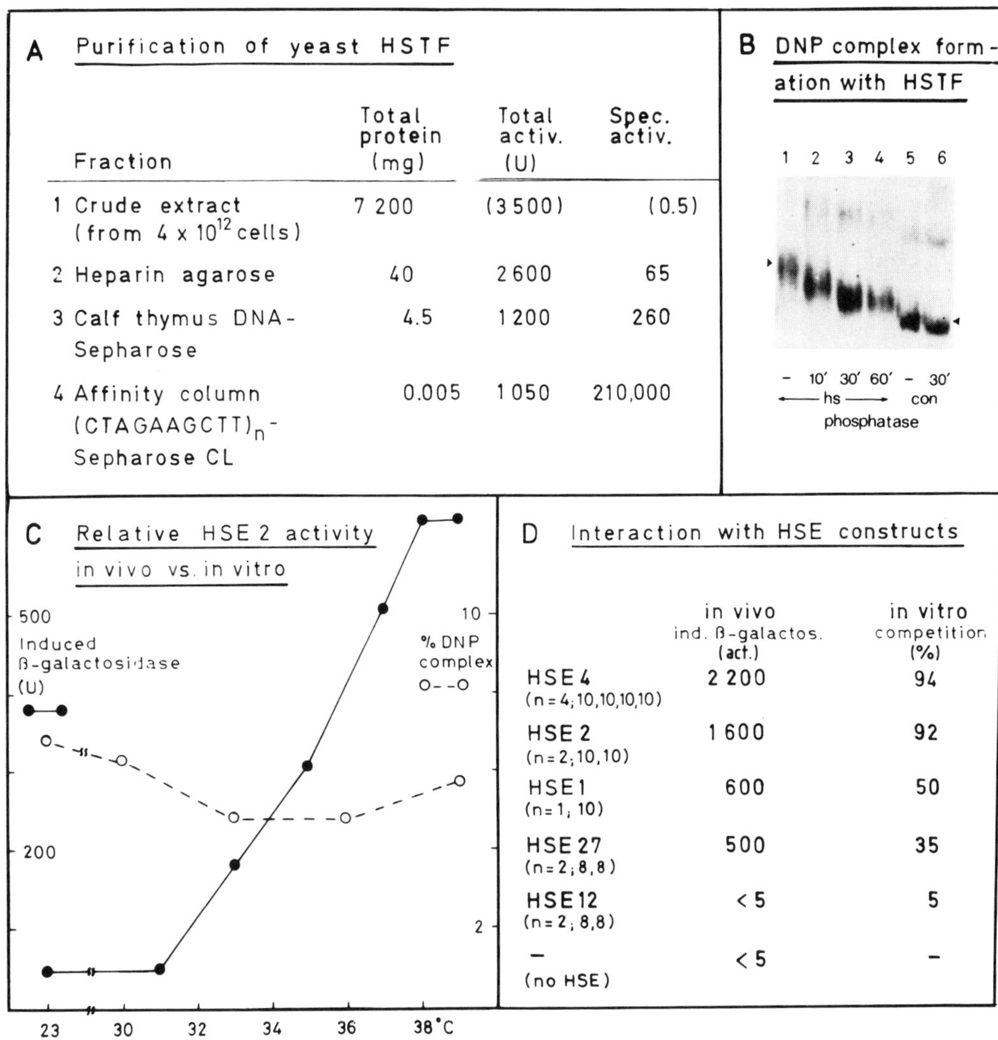

FIGURE 7.5. "Fingerprint" of the yeast heat shock transcription factor (HSF). (A) Purification of the 130-kDa HSF by a three step procedure. (B) *In vitro* DNP-complex formation (gel retardation assay) using a synthetic HSE 2 oligonucleotide and a step 3 HSF preparation from control (lanes 5 and 6) and heat-shocked cells (lanes 1 to 4). The slower migration of the heat shock (hs)-DNP complex (lane 1) is reversed by the action of phosphatase (lanes 2 to 4). (C) Temperature dependent induction of a HSE 2 × cyc1(P)-lac Z fusion gene measured by the β-galactosidase activity is not paralleled by similar changes of the HSE-binding activity measured by *in vitro* complex formation with labeled HSE2. (D) Relative activities of different HSE constructs for HSF1 binding. Comparison of the *in vivo* induction of β-galactosidase, using HSE × cyc1(P)-lac Z fusion constructs (see A41a in Table 6.1), with the extent of *in vitro* competition of a 250-fold excess of the indicated HSE in the DNP complex formation with labeled HSE2 (see B). HSE constructs used are a synthetic monomer (HSE1), two overlapping dimers with 10 (HSE2) and 8 matches of the consensus sequence (HSE12), a tetramer (HSE4), and an overlapping dimer (HSE27) from the Dm-hsp27 promoter (bp −296 to −273) with 8 matches. (Compiled from Sorger, P. K., Lewis, M. J., and Pelham, H. R. B., *Nature,* 329, 81, 1987; and Sorger, P. K. and Pelham, H. R. B., *EMBO J.*, 6, 3035, 1987. With permission.)

irrespective of its isolation from control or heat-shocked cells (Figure 7.5B and C). However, the migration of the DNP complex formed with hs-HSF and control cell-HSF was different (Figure 7.5B), and this difference could be abolished by incubating the former with phosphatase. These results indicate a covalent phosphorylation of the preformed HSF, which is connected with the temperature-dependent transition in the *in vivo* observed activity state (see curve with closed symbols in Figure 7.5C).

The biological significance of the isolated yeast HSF was underlined by comparing the relative *in vivo* activity of hs promoter constructs with different HSE motifs with the *in vitro* competition in the DNP complex formation assay (Figure 7.5D). The *in vivo* data almost perfectly correspond to the *in vitro* data. There is a strongly decreasing activity when changing from the overlapping tetramer to the dimer and monomer HSE and the structural peculiarities of the two distorted dimers can also be clearly discriminated *in vivo* as well as *in vitro* (see Section 6.3).

Another three-step purification of the *Drosophila* HSF was reported by Wu et al.[67] starting with a 0.35 *M* NaCl nuclear extract of *Drosophila* cells. They used chromatography through heparin sepharose coupled with affinity chromatography and a Mono S column. The 7000-fold enriched HSF of 110 kDa constituted 95% of the isolated protein fraction. Its identity was analyzed by DNase I and exonuclease III footprints and by its promoter-specific inducing activity revealed by injection into *Xenopus* oocytes together with a complete Dm-hsp70 gene. Contrasting to the situation in yeast, the *Drosophila* HSF exhibits affinity for binding to hs promoters only after hs-induced modification.[67]

The yeast gene (HSF1) was cloned and sequenced using corresponding antibodies and a λgt11 expression library.[50a,62a] A protein of 833 amino acid residues with an actual molecular weight of 93,218 Da was derived. The deviation of the apparent M_r of 130 kDa from this value is probably due to peculiarities of the protein structure. Sequences from both laboratories are almost identical except a frame shift changing a 24 amino acid region (pos. 557 to 580). It is strongly negatively charged (11 E/D vs. 2 K/R residues) in the sequence of Wiederrecht et al.,[62a] but positively charged in the sequence given by Sorger and Pelham[50a] (no E/D but 7K residues).

The single gene in yeast is essential under all growth conditions. HSF is a constitutive protein bound to the hs genes irrespective of its activity state.[29a] Activation of this preformed DNP complex is connected with hs-induced phosphorylation.[50a,62a] Expression of deletion constructs in *E. coli* coupled to Southwestern screening, gave evidence for the localization of the DNA binding domain between amino acid residues 167 and 284. In addition, a protein domain between residues 327 and 424 is essential for interaction of three monomers to form a homotrimer, which exists irrespective of the activity state or association with DNA.[50b] Inspection of the amino acid sequence in this part reveals a large α-helical region with the typical interspersed pattern of hydrophobic amino acid residues (isoleucine, leucine) characteristic of a leucine zipper. Other well-known motifs of DNA-binding proteins (Zn-finger or helix-turn-helix) are lacking.[62a] Interestingly, sequence comparison between the HSF of yeast and the closely related *Kluyveromyces lactis* showed homology only in the two regions defined for DNA binding and trimerization, respectively (B. Jakobsen and H. Pelham, *EMBO J.*, 10, 1991, in press).

Recently, Southwestern screening led to the isolation of three HSF-coding genes of tomato. Sequencing of the DNA-binding domains revealed a remarkable homology with the yeast and *Kluyveromyces* HSF in this region. This allows definition of a general HSE recognition domain of about 100 amino acid residues, probably present also in other eukaryotic HSF[47a] (X, any amino acid; Ψ, hydrophobic amino acid; invariant residues underlined):

$$-\underline{P}AX\underline{F}{}^{L}_{V}X\underline{K}X^{Y}_{W}XMVXDDXXT^{D}_{E}X\Psi IX\underline{W}X_5SF^{I}_{V}\underline{V}X_5\underline{F}X_3{}^{I}_{L}\underline{LPK}Y\underline{FKH}X\underline{NF}{}^{S}_{A}\underline{SF}V\underline{RQLN}X\underline{YG}{}^{F}_{W}{}^{R}_{H}\underline{K}VX_{1-12}{}^{P}_{D}D^{R}_{K}\underline{W}E\underline{F}X\underline{NE}X\underline{F}XR\underline{G}{}^{Q}_{R}XX\underline{LL}XX\underline{I}X\underline{R}{}^{RK}_{QR}-$$

7.7. MAMMALIAN HSF AND THE INTERACTION BETWEEN PROMOTER BINDING PROTEINS

Of particular interest is the interaction between different regulatory elements and their binding proteins in complex promoters. Results with the human hsp70 promoter show that downstream of the HSE three different types of DNP complexes are encountered, dependent on the specific interaction of proteins with TATA (bp −36 to −22), CAT (bp −76 to −52), and serum-responsive elements (bp −68 to 32). They were detected by exonuclease III protection assay after *in vitro* DNP complex formation with nuclear proteins from HeLa cells.[68] In agreement with the promoter structure (number 3 in Figure 6.6) another CAT-specific DNP complex was detected around bp − 147, whereas protection by HSF is centered around bp −92.[39] Interestingly, the TATA- so well as the CAT-box binding proteins exhibit an unexpected sequence specificity, i.e., in each case there is evidently a family of activator proteins discriminating between different promoters.[10b,16,49a,53b]

HSF binding activity from nonheat-shocked HeLa cells is very low. Induction by hs is rapid, and, similar to the results of Zimarino and Wu[74] with *Drosophila* cell cultures, it is not inhibited in the presence of cycloheximide.[33,51] Evidently, not increased synthesis of HSF but modification of a preexisting inactive form by phosphorylation triggers the transcription of hs genes.[34a] Crude and highly purified preparations of HSF from HeLa cells were characterized by gel retardation, footprint and *in vitro* transcription assays.[17,23a,34a,39a] The protein, purified 14,000-fold in a three-step procedure, reveals an M_r of 83 kDa.[23a] Activation *in vivo* is brought about by hs, but also by Cd^{2+} or azetidine-carboxylic acid.[39a] Formation of the hs DNP complex was even demonstrated *in situ* after introduction of labeled HSE oligonucleotides into human T-cells by electroporation.[27a] A kind of titration of HSF in Chinese hamster ovary cells was achieved by amplification of a dhfr gene connected with a *Xenopus* hsp70 promoter/leader fragment. Almost complete inhibition of hs-induction of the resident hsp70 gene was observed at 10^4 promoter copies per cell, i.e., the number of HSF should be in the range of 10^3 copies per CHO cell.[29c] Results on the HSE-binding activity of the non-activated mammalian HSF are equivocal, if the results of Larson et al.[34a] and Sorger et al.[51] are compared with those of Mosser et al.[39a] and Harel-Bellan et al.[27a]

Part of the evident discrepancies may be due to remarkable cell-specific differences in the hs signal transduction chain. (1) Mezger et al.[38a] reported on changes of binding activities in different mouse cells if analyzed by gel retardation assays. In fibroblasts the usual hs activation of HSF binding is observed, whereas in embryonal carcinoma (EC) cells high constitutive levels of active HSF are connected with constitutive expression of hs genes. Moreover, some EC lines exhibit defects in hs induction of HSP synthesis because the constitutively active HSF is evidently inactivated by hs. But this abnormal behavior disappears, if embryonal carcinoma cells are allowed to differentiate in the presence of retinoic acid, i.e., normal hs induction of HSP synthesis and HSF binding reappears. (2) Despite formation of a HSE-DNP complex revealed by gel retardation assay, murine erythroleukemia cells are defective in hs induction of the hsp70 gene. Evidently, promoter binding of HSF is independent of a second step (HSF modification?) required for transcription activation.[27c] (3) Liu et al.[35c] observed a decay of the hs signal mechanism with aging of human fibroblasts.

The functional interaction between different promoter binding proteins is not yet clear. On the one hand, a mutual substitution may be envisaged, e.g., in promoters with a high constitutive level of expression maintained during hs. On the other hand, there is good evidence (see Figures 6.4 and 6.5) that a TATA-proximal CAT-box may function as helper element for distantly positioned HSE.[5] Protein-protein interaction between remote DNP complexes was explained in a model of Ptashne,[46] which is based on experiments with λ phage promoters. The bending of the DNA and eventually formation of loops can bring about a direct contact between distantly positioned promoter regions with proteins attached. Indeed, such a DNA bending was observed by Shuey and Parker,[48] when the two TATA-proximal HSEs of the Dm-hsp70 promoter were loaded with HSF.

The looping model was strongly supported by linking the SV40 enhancer to the β-globin gene noncovalently, via a biotin-streptavidin bridge. Transcription activation was not disturbed by the protein bridge in this unusual promoter construct.[39b] In the *Drosophila* hsp26 gene the interaction of two essential, but relatively distant HSE centered at bp −342 and −60 respectively is brought about by looping the linker DNA around a nucleosome. This particular mechanism is apparent from the DNase I cleavage patterns.[15b,54a]

REFERENCES

1. **Amin, J., Mestril, R., Schiller, P., Dreano, M., and Voellmy, R.,** Organization of the *Drosophila melanogaster* hsp70 heat shock regulation unit, *Mol. Cell. Biol.*, 7, 1055—1062, 1987.
2. **Arnosti, D. N., Singer, V. L., and Chamberlin, M. J.,** Characterization of heat shock in *Bacillus subtilis, J. Bacteriol.*, 168, 1243—1249, 1986.
3. **Bahl, H., Echols, H., Straus, D. B., Court, D., Crowl, R., and Georgopoulos, C. P.,** Induction of the heat shock response of *E. coli* through stabilization of sigma-32 by the phage lambda cIII protein, *Genes Devel.*, 1, 57—64, 1987.
4. **Bardwell, J. C. A. and Craig, E. A.,** Eukaryotic Mr 83,000 heat shock protein has a homologue in *Escherichia coli, Proc. Natl. Acad. Sci. U.S.A.*, 84, 5177—5181, 1987.
5. **Bienz, M. and Pelham, H. R. B.,** Mechanisms of heat-shock gene activation in higher eucaryotes, *Adv. Genet.*, 24, 31—72, 1987.
6. **Bloom, M., Skelly, S., Van Bogelen, R., Neidhardt, F., Brot, N., and Weissbach, H.,** In vitro effect of the *Escherichia coli* heat-shock regulatory protein on expression of heat shock genes, *J. Bacteriol.*, 166, 380—384, 1986.
7. **Bonner, J. J.,** Induction of *Drosophila* heat shock puffs in isolated polytene nuclei, *Dev. Biol.*, 86, 409—418, 1981.
8. **Bonner, J.,** Mechanism of transcriptional control during heat shock, in *Changes in Eukaryotic Gene Expression in Response to Environmental Stress,* Atkinson, B. G. and Walden, D. B., Eds., Academic Press, Orlando, FL, 1985, 31—51.
9. **Briat, J.-F., Gilman, M. Z., and Chamberlin, M. J.,** *Bacillus subtilis* sigma 28 and *Escherichia coli* sigma 32(htpR) are minor sigma factors that display an overlapping promoter specificity, *J. Biol. Chem.*, 260, 2038—2041, 1985.

9a. **Buratowski, S. T., Hahn, S., Guarante, L., and Sharp, P. A.,** Five intermediate complexes in transcription initiation by RNA polymerase II, *Cell,* 56, 549—561, 1989.

10. **Cartwright, I. L. and Elgin, S. C. R.,** Nucleosomal instability and induction of new upstream protein-DNA associations accompany activation of four small heat shock protein genes in *Drosophila melanogaster, Mol. Cell. Biol.*, 6, 779—791, 1986.

10a. **Chasman, D. I., Lue, N. F., Buchman, A. R., LaPointe, J. W., Lorch, Y., and Kornberg, R. D.,** A yeast protein that influences the chromatin structure of UAS_G and functions as a powerful auxiliary gene activator, *Genes Devel.*, 4, 503—514, 1990.

10b. **Chodosh, L. A., Baldwin, A. S., Carthew, R. W., and Sharp, P. A.,** Human CCAAT-binding proteins have heterologous subunits, *Cell,* 53, 11—24, 1988.

11. **Compton, J. L. and McCarthy, J. B.,** Induction of the *Drosophila* heat shock response in isolated polytene nuclei, *Cell,* 14, 191—201, 1978.
12. **Cooper, S. and Ruettinger, T.,** A temperature sensitive nonsense mutation affecting the synthesis of a major protein of *Escherichia coli* K12, *Mol. Gen. Genet.*, 139, 167—176, 1975.
13. **Cowing, D. W., Bardwell, J. C. A., Craig, E. A., Woolford, C., Hendrix, R. W., and Gross, C. A.,** Consensus sequence for *Escherichia coli* heat shock gene promoters, *Proc. Natl. Acad. Sci. U.S.A.*, 82, 2679—2683, 1985.
14. **Craine, B. L. and Kornberg, T.,** Activation of the major *Drosophila* heat-shock genes in vitro, *Cell,* 25, 671—681, 1981.
15. **Craine, B. L. and Kornberg, T.,** Transcription of the major *Drosophila* heat-shock gene in vitro, *Biochemistry,* 20, 6584—6588, 1981.

15a. **Czarnecka, E., Key, J. L., and Gurley, W. B.,** Regulatory domains of the Gmhsp17.5-E heat shock promoter of soybean, *Mol. Cell. Biol.*, 9, 3457—3463, 1989.

15b. **Dietz, T. J., Cartwright, I. L., Gilmour, D. S., Siegfried, E., Thomas, G. H., and Elgin, S. C. R.,** The chromatin structure of hsp26, in *Stress-Induced Proteins,* Pardue, M. L., Feramisco, J. R., and Lindquist, S., Eds., Alan R. Liss, New York, 1989, 15—24.

16. **Dorn, A., Bollekens, J., Straub, A., Benvist, C., and Mathis, D.,** A multiplicity of CCAAT box-binding proteins, *Cell,* 50, 863—872, 1987.
17. **Drabent, B., Genthe, A., and Benecke, B.-J.,** In vitro transcription of a human hsp 70 heat shock gene by extracts prepared from heat-shocked and non-heat-shocked human cells, *Nucl. Acids Res.,* 14, 8933—8948, 1986.
18. **Dudler, R. and Travers, A. A.,** Upstream elements necessary for optimal function of the hsp70 promoter transformed flies, *Cell,* 38, 391—398, 1984.
19. **Dynan, W. S.,** Promoters for house keeping genes, *Trends Genet.,* 2, 196—197, 1986.
19a. **Eissenberg, J. C., Cartwright, I. L., Thomas, G. H., and Elgin, S. C. R.,** Selected topics in chromatin structure, *Annu. Rev. Genet.,* 19, 485—536, 1985.
19b. **Erickson, J. W. and Gross, C. A.,** Identification of the sigmaE subunit of *Escherichia coli* RNA polymerase: a second alternate sigma factor involved in high-temperature gene expression, *Genes Devel.,* 3, 1462—1471, 1989.
20. **Erickson, J. W., Vaughn, V., Walter, W. A., Neidhardt, F. C., and Gross, C. A.,** Regulation of the promoters and transcripts of rpoH, the *Escherichia coli* heat shock regulatory gene, *Genes Devel.,* 1, 419—432, 1987.
20a. **Forsburg, S. L. and Guarente, L.,** Identification and characterization of HAP4: a third component of the CCAAT-bound HAP2/HAP3 heteromer, *Genes Devel.,* 3, 1166—1178, 1989.
20b. **Fujita, N. and Ishihama, A.,** Heat-shock induction of RNA polymerase sigma-32 synthesis in *Escherichia coli:* Transcriptional control and a multiple promoter system, *Mol. Gener. Genet.,* 210, 10—15, 1987.
21. **Fujita, N., Nomura, T., and Ishihama, A.,** Promoter selectivity of *Escherichia coli* RNA polymerase. Purification and properties of holoenzyme containing the heat shock sigma subunit, *J. Biol. Chem.,* 262, 1855—1859, 1987.
22. **Gayda, R. C., Stephens, P. E., Hewick, R., Schoemaker, J. M., Dreyer, W. J., and Markovitz, A.,** Regulatory region of the heat shock-inducible capR(lon) gene: DNA and protein sequences, *J. Bacteriol.,* 162, 271—275, 1985.
23. **Goff, S. A. and Goldberg, A. L.,** Production of abnormal proteins in *E. coli* stimulates transcription of lon and other heat shock genes, *Cell,* 41, 587—595, 1985.
23a. **Goldenberg, C. J., Luo, Y., Fenna, M., Baler, R., Weinmann, R., and Voellmy, R.,** Purified human factor activates heat shock promoter in a HeLa cell-free transcription system, *J. Biol. Chem.,* 263, 19734—19739, 1988.
23b. **Gottesman, S.,** Genetics of proteolysis in *Escherichia coli, Annu. Rev. Genet.,* 23, 163—198, 1989.
24. **Gross, C. A., Grossman, A. D., Liebke, H., Walther, W., and Burgess, R. R.,** Effects of the mutant sigma allele rpoD800 on the synthesis of specific macromolecular components of the *E. coli* K12 cell, *J. Mol. Biol.,* 172, 283—300, 1984.
25. **Grossman, A. D., Erickson, J. W., and Gross, C. A.,** The htpR gene product of *E. coli* is a sigma factor for heat-shock promoters, *Cell,* 38, 383—390, 1984.
26. **Grossman, A. D., Zhou, Y.-N., Gross, C., Heilig, J., Christie, G. E., and Calendar, R.,** Mutations in the rpoH (htpR) gene of *Escherichia coli* K-12 phenotypically suppress a temperature-sensitive mutant defective in the sigma 70 subunit of RNA polymerase, *J. Bacteriol.,* 161, 939—943, 1985.
26a. **Grossman, A. D., Straus, D. B., Walter, W. A., and Gross, C. A.,** Sigma 32 synthesis can regulate the synthesis of heat shock proteins in *Escherichia coli, Genes Devel.,* 1, 179—184, 1987.
27. **Ha, S.-B. and An, G.,** Cis-acting regulatory elements controlling temporal and organ-specific activity of nopaline synthase promoter, *Nucl. Acids Res.,* 17, 215—223, 1989.
27a. **Harel-Bellan, A., Brini, A. T., Ferris, D. K., Robin, P., and Farrar, W. L.,** In situ detection of a heat-shock regulatory element binding protein using a soluble short synthetic enhancer sequence, *Nucl. Acids Res.,* 17, 4077—4087, 1989.
27b. **Hemmingsen, S. M., Woolford, C., van der Vies, S. M., Tilly, K., Dennis, D. T., Georgopoulos, C. P., Hendrix, R. W., and Ellis, R. J.,** Homologous plant and bacterial proteins chaperone oligomeric protein assembly, *Nature,* 333, 330—335, 1988.
27c. **Hensold, J. O., Hunt, C. R., Calderwood, S. K., Housman, D. E., and Kingston, R. E.,** DNA binding of heat shock factor to the heat shock element is insufficient for transcriptional activation in murine erythroleukemia cells, *Mol. Cell. Biol.,* 10, 1600—1608, 1990.
27d. **Herbomel, P., Rollier, A., Tronche, F., Ott, M.-O., Yaniv, M., and Weiss, M. C.,** The rat albumin promoter is composed of six distinct positive elements within 130 nucleotides, *Mol. Cell. Biol.,* 9, 4750—4758, 1989.
28. **Ito, K., Akiyama, Y., Yura, T., and Shiba, K.,** Diverse effects of the MalE-LacZ hybrid protein on *Escherichia coli* cell physiology, *J. Bacteriol.,* 167, 201—204, 1986.
29. **Jack, R. S., Gehring, W. J., and Brack, C.,** Protein component from *Drosophila* larval nuclei showing sequence specificity for a short region near a major heat-shock protein gene, *Cell,* 24, 321—331, 1981.
29a. **Jakobsen, B. K. and Pelham, H. R. B.,** Constitutive binding of yeast heat shock factor to DNA in vivo, *Mol. Cell. Biol.,* 8, 5040—5042, 1988.

29b. **Johnson, P. F. and McKnight, S. L.,** Eukaryotic transcriptional regulatory proteins, *Annu. Rev. Biochem.*, 58, 799—839, 1989.

29c. **Johnston, R. N. and Kucey, B. L.,** Competitive inhibition of hsp70 gene expression causes thermosensitivity, *Science*, 242, 1551—1554, 1988.

30. **Jones, K. A., Kadonaga, J. T., Rosenfeld, P. J., Kelly, T. J., and Tjian, R.,** A cellular DNA-binding protein that activates eukaryotic transcription and DNA replication, *Cell*, 48, 79—89, 1987.

31. **Kadonaga, J. T., Jones, K. A., and Tjian, R.,** Promoter-specific activation of RNA polymerase II transcription by Sp1, *Trends Biochem. Sci.*, 11, 20—23, 1986.

32. **Kadonaga, J. T., Carner, K. R., Masiarz, F. R., and Tjian, R.,** Isolation of cDNA encoding transcription factor Sp1 and functional analysis of the DNA binding domain, *Cell*, 51, 1079—1090, 1987.

33. **Kingston, R. E., Schuetz, T. J., and Larin, Z.,** Heat-inducible human factor that binds to a human hsp70 promoter, *Mol. Cell. Biol.*, 7, 1530—1534, 1987.

34. **Landick, R., Vaughn, V., Lau, E. T., Van Bogelen, R. A., Erickson, J. W., and Neidhardt, F. C.,** Nucleotide sequence of the heat shock regulatory gene of *E. coli* suggests its protein product may be a transcription factor, *Cell*, 38, 175—182, 1984.

34a. **Larson, J. S., Schuetz, T., and Kingston, R. E.,** Activation in vitro of sequence specific DNA binding by a human regulatory factor, *Nature*, 335, 372—375, 1988.

34b. **Lelong, J. C., Prevost, G., Lee, K. I., and Crepin, M.,** South-western blot mapping. A procedure for simultaneous characterization of DNA binding proteins and their specific genomic DNA target sites, *Anal. Biochem.*, 179, 299—303, 1989.

35. **Lesley, S. A., Thompson, N. E., and Burgess, R. R.,** Studies of the role of the *Escherichia coli* heat shock regulatory protein sigma 32 by the use of monoclonal antibodies, *J. Biol. Chem.*, 262, 5404—5407, 1987.

35a. **Lipinska, B., King, J., Ang, D., and Georgopoulos, C.,** Sequence analysis and transcriptional regulation of the *Escherichia coli* grpE gene encoding a heat shock protein, *Nucl. Acids Res.*, 16, 7545—7562, 1988.

35b. **Lipinska, B., Zylicz, M., and Georgopoulos, C.,** The HtrA (DegP) protein, essential for *Escherichia coli* survival at high temperatures, is an endopeptidase, *J. Bacteriol.*, 172, 1791—1797, 1990.

35c. **Liu, A. Y. C., Lin, Z., Choi, H. S., Sorhage, F., and Li, B. S.,** Attenuated induction of heat shock gene expression in aging diploid fibroblasts, *J. Biol. Chem.*, 264, 12037—12045, 1989.

36. **Lupski, J. R. and Gosdon, G. N.,** The rspU-dnaG-rpoD macromolecular synthesis operon of *E. coli*, *Cell*, 39, 251—252, 1984.

37. **Lupski, J. R., Ruiz, A. A., and Godson, G. N.,** Promotion, termination, and antitermination in the rpsU-dnaG-rpoD macromolecular synthesis operon of *E. coli* K-12, *Mol. Gen. Genet.*, 195, 391—401, 1984.

38. **McGarry, T. J. and Lindquist, S.,** Inhibition of heat shock protein synthesis by heat-inducible antisense RNA, *Proc. Natl. Acad. Sci. U.S.A.*, 83, 399—403, 1986.

38a. **Mezger, V., Bensaude, O., and Morange, M.,** Unusual levels of heat shock element-binding activity in embryonal carcinoma cells, *Mol. Cell. Biol.*, 9, 3888—3896, 1989.

38b. **Mitchell, P. J. and Tjian, R.,** Transcriptional regulation in mammalian cells by sequence-specific DNA binding proteins, *Science*, 245, 371—378, 1989.

38c. **Morgan, W. D.,** Transcription factor Sp1 binds to and activates a human hsp70 gene promoter, *Mol. Cell. Biol.*, 9, 4099—4104, 1989.

39. **Morgan, W. D., Williams, G. T., Morimoto, R. I., Greene, J., Kingston, R. E., and Tjian, R.,** Two transcriptional activators, CCAAT-box-binding transcription factor and heat shock transcription factor, interact with a human hsp70 gene promoter, *Mol. Cell. Biol.*, 7, 1129—1138, 1987.

39a. **Mosser, D. D., Theodorakis, N. G., and Morimoto, R. I.,** Coordinate changes in heat shock element-binding activity and hsp70 gene transcription rates in human cells, *Mol. Cell. Biol.*, 8, 4736—4744, 1988.

39b. **Mueller-Storm, H.-P., Sogo, J. M., and Schaffner, W.,** An enhancer stimulates transcription in *trans* when attached to the promoter via a protein bridge, *Cell*, 58, 767—777, 1989.

40. **Munro, S. and Pelham, H.,** What turns on heat shock genes?, *Nature*, 317, 477—478, 1985.

40a. **Nagai, H., Yano, R., Erickson, J. W., and Yura, T.,** Transcriptional regulation of the heat shock regulatory gene rpoH in *Escherichia coli:* involvement of a novel catabolite-sensitive promoter, *J. Bacteriol.*, 172, 2710—2715, 1990.

41. **Neidhardt, F. C. and Van Bogelen, R. A.,** Positive regulatory gene for temperature-controlled proteins in *Escherichia coli, Biochem. Biophys. Res. Commun.*, 100, 894—900, 1981.

42. **Neidhardt, F. C., Van Bogelen, R. A., and Vaughn, V.,** The genetics and regulation of heat-shock proteins, *Annu. Rev. Genet.*, 18, 295—329, 1984.

43. **Neidhardt, F. C., Van Bogelen, R. A., and Lau, E. T.,** Molecular cloning and expression of a gene that controls the high-temperature regulon of *Escherichia coli, J. Bacteriol.*, 153, 597—603, 1983.

44. **Parker, C. S. and Topol, J.,** A *Drosophila* RNA polymerase II transcription factor binds to the regulatory site of an hsp 70 gene, *Cell*, 37, 273—283, 1984.

45. **Parker, C. S. and Topol, J.,**. A *Drosophila* RNA polymerase II transcription factor contains a promoter-region-specific DNA-binding activity, *Cell*, 36, 357—369, 1984.

45a. **Parsell, D. A. and Sauer, R. T.,** Induction of a heat shock-like response by unfolded protein in *Escherichia coli*: dependence on protein level not protein degradation, *Genes Devel.*, 3, 1226—1232, 1989.

45b. **Perisic, O., Xiao, H., and Lis, J. T.,** Stable binding of *Drosophila* heat shock factor to head-to-head and tail-to-tail repeats of a conserved 5 bp recognition unit, *Cell,* 59, 797—806, 1989.

46. **Ptashne, M.,** Gene regulation by proteins acting nearby and at a distance, *Nature,* 322, 697—701, 1986.

46a. **Ransone, L. J., Visvader, J., Sassone-Corsi, P., and Verma, I. M.,** Fos-jun interaction: mutational analysis of the leucine zipper domain of both proteins, *Genes Devel.*, 3, 770—781, 1989.

46b. **Revzin, A.,** Gel electrophoresis assay for DNA-protein interactions, *Biotechniques,* 7, 346—355, 1989.

46c. **Rindt, K. P. and Nover, L.,** Chromatin structure and function, *Biol. Zentralbl.*, 99, 641—673, 1980.

47. **Ritossa, F.,** A new puffing pattern induced by heat shock and DNP in *Drosophila, Experientia,* 18, 571—573, 1962.

47a. **Scharf, K.-D., Rose, R., Zott, W., Schöffl, F., and Nover, L.,** Three tomato genes code for heat stress transcription factors with a region of remarkable homology to the DNA-binding domain of the yeast HSF, *EMBO J.*, 9, 1990, in press.

48. **Shuey, D. J. and Parker, C. S.,** Bending of promoter DNA on binding of heat shock transcription factor, *Nature,* 323, 459—461, 1986.

49. **Shuey, D. J. and Parker, C. S.,** Binding of *Drosophila* heat-shock gene transcription factor to the hsp70 promoter, *J. Biol. Chem.*, 261, 7934—7940, 1986.

49a. **Simon, M. C., Fisch, T. M., Benecke, B. J., Nevins, J. R., and Heintz, N.,** Definition of multiple functionally distinct TATA elements, one of which is a target in the hsp70 promoter for E1A regulation, *Cell,* 52, 723—729, 1988.

49b. **Singh, H., Lebowitz, J. H., Baldwin, A. S., and Sharp, P. A.,** Molecular cloning of an enhancer binding protein: isolation by screening of an expression library with a recognition site DNA, *Cell,* 52, 415—423, 1988.

49c. **Skelly, S., Coleman, T., Fu, C. F., Brot, N., and Weissbach, H.,** Correlation between the 32-kDa factor levels and in vitro expression of *Escherichia coli* heat shock genes, *Proc. Natl. Acad. Sci. U.S.A.*, 84, 8365—8369, 1987.

49d. **Skelly, S., Fu, C.-F., Dalie, B., Redfield, B., Coleman, T., Brot, N., and Weissbach, H.,** Antibody to sigma 32 cross-reacts with DnaK: Association of DnaK protein with *Escherichia coli* RNA polymerase, *Proc. Natl. Acad. Sci. U.S.A.*, 85, 5497—5501, 1988.

50. **Sorger, P. K. and Pelham, H. R. B.,** Purification and characterization of a heat-shock element binding protein from yeast, *EMBO J.*, 6, 3035—3042, 1987.

50a. **Sorger, P. K. and Pelham, H. R. B.,** Yeast heat shock factor is an essential DNA-binding protein that exhibits temperature-dependent phosphorylation, *Cell,* 54, 855—864, 1988.

50b. **Sorger, P. K. and Nelson, H. C. M.,** Trimerization of a yeast transcriptional activator via a coiled-coil motif, *Cell,* 59, 807—813, 1989.

51. **Sorger, P. K., Lewis, M. J., and Pelham, H. R. B.,** Heat shock factor is regulated differently in yeast and HeLa cells, *Nature,* 329, 81—84, 1987.

51a. **Staudt, L. M., Clerc, R. G., Singh, H., Lebowitz, J. H., Sharp, P. A., and Baltimore, D.,** Cloning of a lymphoid-specific cDNA encoding a protein binding the regulatory octamer DNA motive, *Science,* 241, 577—580, 1988.

52. **Stragier, P., Parsol, C., and Bouvier, J.,** Two functional domains conserved in major and alternate bacterial sigma factors, *FEBS Lett.*, 187, 11—15, 1985.

53. **Straus, D. B., Walter, W. A., and Gross, C. A.,** The heat shock response of *Escherichia coli* is regulated by changes in the concentration of sigma 32, *Nature,* 329, 348—350, 1987.

53a. **Straus, D. B, Walter, W. A., and Gross, C. A.,** The activity of sigma32 is reduced under conditions of excess heat shock protein production in *Escherichia coli, Genes Devel.*, 3, 2003—2010, 1989.

53b. **Taylor, I. C. A. and Kingston, R. E.,** E1a transactivation of human HSP70 gene promoter substitution mutants is independent of the composition of upstream and TATA elements, *Mol. Cell. Biol.*, 10, 176—183, 1990.

54. **Taylor, W. E., Straus, D. B., Grossman, A. D., Burton, Z. F., Gross, C. A., and Burgess, R. R.,** Transcription from heat-inducible promoter causes heat shock regulation of the sigma subunit of *E. coli* RNA polymerase, *Cell,* 38, 371—381, 1984.

54a. **Thomas, G. H. and Elgin, S. C. R.,** Protein/DNA architecture of the DNase I hypersensitive region of the *Drosophila* hsp 26 promoter, *EMBO J.*, 7, 2191—2201, 1988.

55. **Tilly, K., McKittrick, N., Zylicz, M., and Georgopoulos, C.,** The dnaK protein modulates the heat-shock response of *Escherichia coli, Cell,* 34, 641—646, 1983.

56. **Tilly, K., Erickson, J., Sharma, S., and Georgopoulos, C.,** Heat shock regulatory gene rpoH mRNA level increases after heat shock in *Escherichia coli, J. Bacteriol.*, 168, 1155—1158, 1986.

56a. **Tilly, K., Spence, J., and Georgopoulos, C.,** Modulation of stability of the *Escherichia coli* heat shock regulatory factor sigma32, *J. Bacteriol.*, 171, 1585—1589, 1989.

57. **Tobe, T., Ito, K., and Yura, T.,** Isolation and physical mapping of temperature-sensitive mutants defective in heat-shock induction of proteins in *Escherichia coli, Mol. Gen. Genet.*, 195, 10—16, 1984.
58. **Tobe, T., Kusukawa, N., and Yura, T.,** Suppression of rpoH (htpR) mutations of *Escherichia coli:* Heat shock response in suhA revertants, *J. Bacteriol.*, 169, 4128—4134, 1987.
59. **Topol, J., Ruden, D. M., and Parker, C. S.,** Sequences required for in vitro transcriptional activation of a *Drosophila* hsp 70 gene, *Cell*, 42, 527—537, 1985.
60. **Travers, A.,** Sigma factors in multitude, *Nature*, 313, 15—16, 1985.
61. **Van Bogelen, R. A., Acton, M. A., and Neidhardt, F. C.,** Induction of the heat shock regulon does not produce thermotolerance in *Escherichia coli, Genes Devel.*, 1, 525—531, 1987.
61a. **Vinson, C. R., Lamarco, K. L., Johnson, P. F., Landschulz, W. H., and McKnight, S. L.,** In situ detection of sequence-specific DNA binding activity specified by a recombinant bacteriophage, *Genes Devel.*, 2, 801—806, 1988.
61b. **Wang, Q. and Kaguni, J. M.,** A novel sigma factor is involved in expression of the rpoH gene of *Escherichia coli, J. Bacteriol.*, 171, 4248—4253, 1989.
62. **Wiederrecht, G., Shuey, D. J., Kibbe, W. A., and Parker, C. S.,** The *Saccharomyces* and *Drosophila* heat shock transcription factors are identical in size and DNA binding properties, *Cell*, 48, 507—515, 1987.
62a. **Wiederrecht, G., Seto, D., and Parker, C. S.,** Isolation of the gene encoding the *S. cerevisiae* heat shock transcription factor, *Cell*, 54, 841—853, 1988.
62b. **Williams, G. T., McClanahan, T. K., and Morimoto, R. I.,** E1a transactivation of the human HSP70 promoter is mediated through the basal transcription complex, *Mol. Cell. Biol.*, 9, 2574—2587, 1989.
62c. **Wingender, E.,** Compilation of transcription regulatory proteins, *Nucl. Acids Res.*, 16, 1879—1902, 1988.
63. **Wu, C.,** The 5′ends of *Drosophila* heat shock genes in chromatin are hypersensitive to DNase I, *Nature*, 286, 854—860, 1980.
64. **Wu, C.,** Activating protein factor binds in vitro to upstream control sequences in heat shock gene chromatin, *Nature*, 311, 81—84, 1984.
65. **Wu, C.,** Two protein-binding sites in chromatin implicated in the activation of heat-shock genes, *Nature*, 309, 229—234, 1984.
66. **Wu, C.,** An exonuclease protection assay reveals heat-shock element and TATA box DNA-binding proteins in crude nuclear extracts, *Nature*, 317, 84—87, 1985.
67. **Wu, C., Wilson, S., Walker, B., David, I., Paisley, T., Zimarino, V., and Ueda, H.,** Purification and properties of *Drosophila* heat shock activator protein, *Science*, 238, 1247—1253, 1987.
68. **Wu, B. J., Gregg, T. W., and Morimoto, R. J.,** Detection of three protein binding sites in the serum-regulated promoter of human gene encoding the 70-kDa heat shock protein, *Proc. Natl. Acad. Sci. U.S.A.*, 84, 2203—2207, 1987.
69. **Xiao, H. and Lis, J. T.,** A consensus sequence polymer inhibits in vivo expression of heat shock genes, *Mol. Cell. Biol.*, 6, 3200—3206, 1986.
70. **Xiao, H. and Lis, J. T.,** Germline transformation used to define key features of heat-shock response elements, *Science*, 239, 1139—1142, 1988.
70a. **Yamamori, T. and Yura, T.,** Genetic control of heat shock protein synthesis and its bearing on growth and thermal resistance in *Escherichia coli* K-12, *Proc. Natl. Acad. Sci. U.S.A.*, 79, 860—864, 1982.
71. **Yamamori, T., Osawa, T., Tobe, T., Ito, K., and Yura, T.,** *Escherichia coli* gene (hin) controls transcription of heat-shock operons and cell growth at high temperatures, in *Heat Shock from Bacteria to Man*, Schlesinger, M. J., Ashburner, M., and Tissieres, A., Eds., Cold Spring Harbor Laboratory, Cold Spring Harbor, NY, 1982, 131—137.
72. **Yura, T.,** Genetic control of heat-shock proteins, *Jpn. J. Genet.*, 61, 277—290, 1986.
73. **Yura, T., Tobe, T., Ito, K., and Osawa, T.,** Heat shock regulatory gene (htpR) of *Escherichia coli* is required for growth at high temperature but is dispensable at low temperature, *Proc. Natl. Acad. Sci. U.S.A.*, 81, 6803—6807, 1984.
74. **Zimarino, V. and Wu, C.,** Induction of sequence-specific binding of *Drosophila* heat shock activator protein without protein synthesis, *Nature*, 327, 727—730, 1987.
75. **Zimarino, V., Tsai, C., and Wu, C.,** Complex modes of heat shock factor activation, *Mol. Cell. Biol.*, 10, 752—759, 1990.

Chapter 8

CONTROL OF HSP SYNTHESIS

L. Nover

TABLE OF CONTENTS

8.1. COORDINATE VS. NONCOORDINATE CONTROL

With the discovery of induced heat shock protein (HSP) synthesis in *Drosophila* and its extension to other types of organisms[7,8a,106] (see Table 2.1), there was the general impression of a rapid and highly coordinate response of prokaryotic and eukaryotic cells to heat shock (hs) or chemical stressors. Due to the particular induction mechanism operative at the HSE-containing promoter regions, this is basically true. However, with the advancing fine analyses of hs-induced protein patterns, with our increasing knowledge about hs genes, their regulatory sequences so well as the multiplicity of temperature effects observed at all levels of gene expression, the picture became much more diversified. Each hs gene exhibits individual traits with respect to the quantitative and qualitative aspects of its expression, and considerable organism-specific variation emerged.

Similar to other aspects of the response (see Sections 8.7, 9.5, and 11.5.5), the first concise analysis came from the laboratory of Susan Lindquist,[77] who investigated temperature and time-dependent changes of HSP synthesis in cell cultures of *Drosophila*. The main results were the following (see data given in Figure 1.1): (1) the coordinate induction of HSP synthesis at intermediate hs temperatures constrasts with a selective synthesis of HSP83 at low (⩾33°C) and HSP70 at high (⩾37°C) temperatures; (2) the maximum temperature for induced HSP synthesis can be considerably extended, if cell cultures are heated gradually rather than immediately (see also Figure 2.1); (3) there is a strong influence of the cell culture medium on the extent of induction and the temperature optimum; and (4) peculiarities in the regulation of HSP70 synthesis at 33°C were recognized later as autorepression (see Section 8.7). Her conclusion, relevant for practically all other hs systems as well, was: "Several heat-shock proteins . . . have very individual induction characteristics with respect to the temperatures at which they are maximally induced, the range of temperatures over which they are synthesized, and the kinetics of their induction."[77] The results were used as a starting point for analysis of different regulatory mechanisms, which form the basis for this individuality. They are dealt with in different parts of this book (see Sections 8.7, 9.5, and 11.5.5).

Essential aspects of the investigations by Lindquist were reproduced in our laboratory using tomato cell cultures (see Figure 2.1). The outcome was similar. What may be an interesting extension is demonstrated in Figure 8.1, which gives a summary of labeling data derived from two-dimensional electrophoresis of corresponding total protein fractions labeled with ^{35}S-methionine.[93] Three classes of HSPs can be discerned: (1) HSPs 22 and 95 with transient synthesis in the early phase of the hs period never accumulate to significant levels; (2) the bulk of HSPs including HSPs 70 and 17, whose synthesis starts rapidly and is maintained for several hours form stable proteins accumulating to Coomassie-stainable amounts; and (3) HSPs18/19 with delayed onset of synthesis become prominent HSPs after 8 to 12 h, when the synthesis of the bulk of HSPs has declined.

Whenever analyzed appropriately, these three types of HSPs can also be detected in other systems, e.g., in plants[36,81,88] or in rainbow trout fibroblasts.[87b] In the latter, the hs-induced increase of HSP70 and HSP27 precedes that of HSP87, whereas a late HSP19 appears after only 6 h. Above a certain threshold level, the amount of heat stress applied to a system is defined by the temperature and the duration (see Chapter 1). The actual pattern of HSP synthesis is markedly influenced by this. Similar to the HSP83 of *Drosophila,* a low stress protein (HSP90) was detected in HeLa cells after 2 h 42°C but not after 15 min 45°C induction.[53] On the other hand, in most cases, high-stress conditions are marked by the striking prevalence of HSP70 synthesis until an upper limit of its cellular level is reached (see Section 8.7).

Under natural conditions, high stress frequently results from a combination of factors. Kimpel and Key[65] documented this for field-grown soybeans subjected to daily temperature

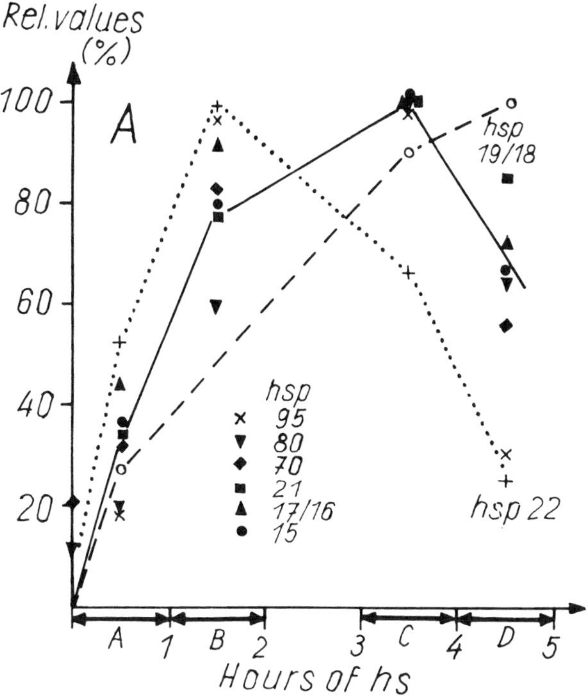

FIGURE 8.1. Time course of HSP labeling in tomato cell cultures during a 5 h period of heat shock (hs) at 39°C. The labeling periods with ^{35}S-methionine are indicated by arrows at the bottom. Proteins were separated by two-dimensional electrophoresis and radioactivity in a given heat shock protein (HSP) was determined after excision of the corresponding spot from the gel. To give an estimate of the radioactivity distribution between different HSPs the following values (cpm $\times$ 10^{-3}) apply to the 4 h sample (C): HSP95, 8.5; HSP80, 16.5; HSP70, 64.2; HSP22, 3.2; HSP21, 7.3; HSPs18/19, 7.1, and HSP17, 31.0. (From Nover, L. and Scharf, K.-D., *Eur. J. Biochem.*, 139, 303, 1984. With permission.)

variations with or without sufficient water supply, i.e., a model system with the usual combination of stress factors encountered by plants in their natural surrounding. They found that induction of HSP-specific mRNAs under a given regime of ambient temperature was 5 to 9 times higher in the water-deficient soybean.

8.2. PROTOTYPES OF HEAT SHOCK-INDUCED GENES

Although many of the deviations from the regulatory norm of coordinate HSP synthesis are so far unexplained, the rapid achievements in isolation and functional characterization of hs genes have provided important insights in quite a number of cases. Thus, despite a concise treatment of this topic in Chapter 6, it is important to summarize relevant data as a basis for the following discussion. Four types of hs genes were selected for this purpose (Figure 8.2). The soybean hsp18.5C gene (number 1) is a typical representative of the hsp20 gene family of plants.[99] It is a gene with almost nonexistent basal level expression. Its high inducibility is determined by the two overlapping trimers of the HSE and presumably by an uninterrupted stretch of 13A-residues further upstream, which are found also in other hs genes. They may help in the effective stabilization of the transcribed gene by interaction with the nuclear matrix (see Section 9.2).

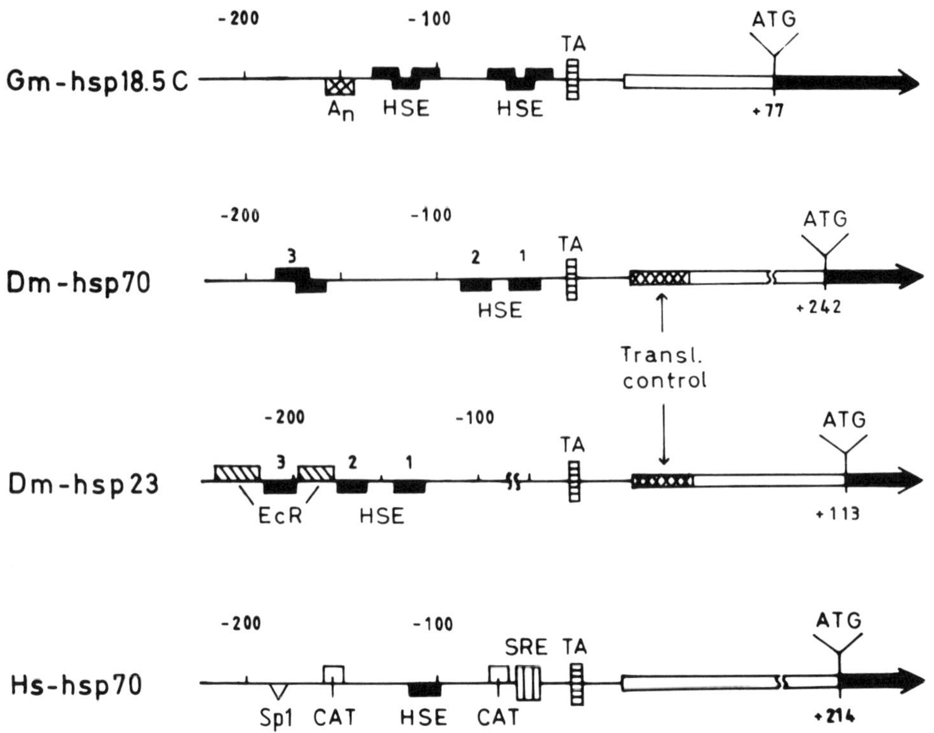

FIGURE 8.2. Four types of promoters connected with heat shock (hs)-inducible genes. 1, 2: Promoters of the hsp18,5 gene of soybean and of the hsp70 gene of *Drosophila;* both genes have a very low basal level but a high level of hs-induced expression. 3: *Drosophila* hsp23 gene which is bifunctional. The very low basal level expression can be increased by hs or the steroid hormone ecdysterone. 4: Multifunctional human hsp70 gene has a fairly high basal level expression, is under cell cycle control, and moderately hs-inducible. Promoter elements are characterized as in Figure 6.6. HSE, hs activator element; EcR, ecdysterone receptor element; SRE, serum responsive element; CAT, CCAAT box; TA, TATA box; ATG, start of translation. The untranslated leader sequences of the genes are drawn with open bars, the coding sequences with full bars, and an arrow to indicate direction of translation (for References see legend to Figure 6.6).

The Dm-hsp83 gene (number 2) is the only hs gene in *Drosophila* with significant expression under control conditions. So far, no other regulatory element besides the overlapping HSE trimer was identified.[26] It is very likely that this particular promoter structure, not found in other *Drosophila* hs genes, provides the basis for sufficient binding of the heat shock transcription factor (HSF) even at 25°C. At any rate, hsp83 is a low temperature hs gene, whose expression is heavily limited at ⩾37°C by the presence of the intron (see Section 9.5). A similar, dual function of the HSE for constitutive as well as hs-induced promoter activity was also documented for the corresponding yeast gene (hsp82). Point mutations in the TATA-proximal –TTC2nGAA– motif causes defects in basal level expression with reduced binding of HSF at 25°C (see Section 6.2).[79a]

A step up in promoter complexity is documented for the *Drosophila* hsp23 gene (number 3). In addition to the scattered HSE monomers, it contains an ecdysterone receptor element responsible for developmental control of its expression, which is typical for all small HSPs of this organism[80,114] (see Section 8.4.). It is worth noticing that there is practically no basal level expression of these genes, and that maximum hs inducibility requires interaction between several HSE monomers. The three HSE of Dm-hsp23 shown in Figure 8.2 provide only 40% of the maximum. Two additional HSEs further upstream complement the set. An important accessory is indicated at the 5′ end of the untranslated leader: the cross-hatched

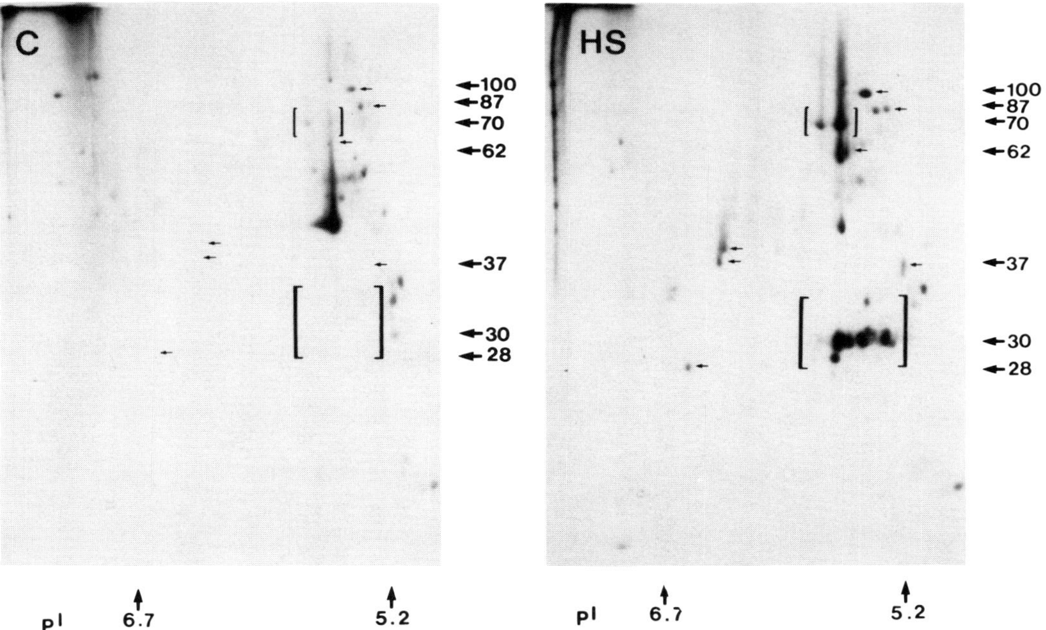

FIGURE 8.3. Heat shock (hs) proteins from *Xenopus laevis* kidney epithelial cells. Figures represent autoradiographs of two-dimensional electropherograms of control (C) and heat shocked cultures (2 h 35°C) after labeling of proteins with ^{35}S-methionin. HSPs are indicated by arrows and by their M_r on the margin. The 70-kDa region comprises 5 heat shock proteins (HSPs), the 28- to 30-kDa region comprises 16 HSPs.[37a] (From Darasch, S., Mosser, D. D., Bols, N. C., and Heikkila, J. J., *Biochem. Cell Biol.*, 66, 862, 1988. With permission.)

region of 26 nucleotides, also found in a similar position in other *Drosophila* hs genes, is not only involved in the highly efficient transcription but also in selective translation of the mRNA (see Section 11.5.5).

The ultimate type of a multifunctional promoter is represented by one of the two human hsp70 genes (number 4), which serves many regulatory purposes. A relatively high constitutive expression, presumably controlled by the two CCAAT elements, is connected with a cell cycle-dependent induction-repression mechanism (serum factor responsive element, [SRE]) and with a moderate induction by heat or metal ion stress (HSE monomer). Four different binding factors were found to interact in a sequence-specific manner with TATA, CCAAT, SRE, and HSE, respectively.[86,131] Variation of the factor concentration or activity depending on the cell type, cell cycle state, or environmental conditions determines the regulatory pattern of this hsp70 gene.

This type of promoter may be the rule rather than the exception, at least for the hsp70 gene family. Other interesting examples are the *Xenopus* hsp70 and the rat hsc73 promoters (see Figure 6.6 and references therein). The coexistence of constitutive and hs-inducible expression in one gene leads to an important reconsideration of the definition of hs genes and the function of HSE in particular (see Section 2.2). It may serve not only a dramatic increase of gene expression as observed with promoters of type 1 to 3, but also maintenance of the transcription level of constitutively active genes during the stress period (see Section 6.5).

8.3. CELL-TYPE-SPECIFIC VARIATION OF HEAT SHOCK INDUCIBILITY AND HSP PATTERNS

With the advancing analysis of HSP patterns in different tissues of the same organism more and more subtle details were detected, which are modulations or even striking exceptions of the general theme: coordinate induction of HSP synthesis in all types of living cells. A few examples will be given to illustrate this point.

1. In complex multicellular organisms, cell-specific variation of HSP patterns is common. This is true for *Drosophila*,[113] the butterfly *Calpodes ethlius*,[39] for vertebrates,[8,87,124] the aquatic fungus *Blastocladiella emersonii*,[27] and plants.[98] An impressive survey emerges from the summary of data on *Xenopus*,[30a] derived from a broad range of analyses of different developmental stages (see Figure 8.3, Table 8.1 and references given therein). Not only is the pattern of HSPs induced in different cell types greatly variable but also the mode of hs-induction. Additional factors causing variability of HSP patterns became apparent when spawned eggs or artificially activated eggs (see Table 8.1C,D) were heat-shocked *in vitro* in different culture solutions (summarized by Browder et al.[30a]). The Cl^- concentration, pH value, and extracellular Ca^{2+} levels thoroughly influence inducible synthesis of HSPs 48, 62, and 66. However, all three factors may be interrelated by effects on ion balance and/or intracellular pH.

   Remarkable is the regulatory behavior of the X1-hsp70 gene. It is constitutively transcribed at a high level, if injected into *Xenopus* oocytes, and hsp70 mRNA is also synthesized under hs-independent control during oocyte maturation. Translation of the preformed mRNA proceeds at 20°C, but becomes prominent at 35°C after hs-induced changes of the translation apparatus[56] (see Section 11.5). In contrast to this, the same hsp70 gene is hs inducible in all somatic tissues of tadpoles and adult frogs.[23] Sequencing and functional analysis of the promoter region revealed three overlapping dimers of HSE with two CAT boxes interspersed (see Figure 6.4). Removal of the TATA-proximal CAT box abolishes the constitutive type of expression in oocytes. The deletion promoter only responds after hs induction.[24] This example demonstrates that the same promoter exhibits two types of transcription control (constitutive vs. hs inducible) in two different cell types of the same organism depending on the cell-specific pattern of promoter binding proteins.

2. A new type of analysis was introduced by Brown and coworkers[31,115] by *in situ* hybridization of a hsp70 antisense RNA to rabbit brain tissue. Hs was done by lysergic acid diethylamide (LSD) administration, which resulted in a 2 to 3°C increase of the body temperature. Not only the constitutive level of hsp70-specific mRNA, but also the inducibility by LSD treatment shows striking regional differences. Thus, neurons in the cerebellum and fiber tracts throughout the brain are well inducible, whereas neurons of the hippocampus are not. The same method applied to sea urchin gastrulae revealed similar cell-specific differences of hs inducibility, with much more hsp70 mRNA detected in the ectoderm than in the intestine.[107]
3. The nonresponsiveness to hs induction of fertilized eggs and early cleavage stage embryos and its reappearance at the transition from blastula to gastrula stage correlates with acquisition of the capability to become thermotolerant. This is particularly striking for *Drosophila, Xenopus,* molluscs, and sea urchin embryos,[21,29a,44,48,55,91,104] (see also Figure 17.1 and the survey by Browder et al.[30a]) but it also holds true for early embryonic stages of mammals[51,54,70,85,121,128] and for ripe and germinating plant pollen.[37,45c,132]
4. Circadian control of hs-inducibility was detected in green plants kept in a 12 h/12 h day/night regimen.[93b] Using pea plants and 2-h induction at 42°C, the authors found a three- to fivefold variation of the hs mRNA levels during a 24-h cycle. The maximum was between midnight and early morning and the minimum was afternoon. Other RNAs, e.g., of light-regulated genes, showed similar variations.

Experimental models to study the switch from the early embryonic state without induced HSP synthesis to the late embryonic/adult state may be found in the form of certain mouse embryonal carcinoma cell types, which can be triggered by retinoic acid to undergo differentiation with concomitant transition to stress inducibility of HSP synthesis.[9,10,85] In this case, transformation of plasmacytoma (PC) cell lines with a Dm-hsp70 P/L × cat gene construct revealed that peculiarities of the HSF are responsible for the deficiency of PC C4 and 1009 cells and the normalization of the hs response after differentiation (see Section 7.7).[80a,80b]

Similar techniques were applied to charcterize the remarkable decline of HSP inducibility by hs or canavanine in aging human fibroblasts. Analysis of mRNA levels, nuclear runoff assays, and tests in fibroblasts, transformed with a Hs-hsp70 promoter × cat gene fusion construct, demonstrated that a specific deficiency of the hs signal transformation system is responsible for this effect.[77a] This may also be true for the retardation of hsp70 transcription in peripheral blood mononuclear cells of aged vs. young human beings (Deguchi et al.[39a]). Contrasting to this, changes of HSP synthesis observed in aging *Drosophila* flies, are connected with a dramatic increase in the complexity of the induced HSP pattern. Two-dimensional gel electrophoresis shows 14 major spots in 10-d-old flies but 50 hs-induced polypeptides in 45-d-old flies.[45b]

In other cases, however, the lack of stress-induced HSP synthesis is presumably caused by the transient general inactivity of the transcription apparatus, i.e., no regulatory peculiarities to the hs genes are involved.[3,44] It is interesting to notice that in two cell types with similar ''problems'' (*Xenopus* oocytes and chicken reticulocytes) special mechanisms for hs activation of HSP70 synthesis at the translation level may be operative (see Section 8.6). Two other modes of substitution are worth mentioning: (1) storage or constitutive synthesis of HSP70 isoforms in germ cells and early developmental stages,[4,18,19,71,77b,134] but also during sporulation in yeast[74,125a] and (2) the massive cell-cell transfer of stress-induced HSP70 (traversin) from glial cells to the axon in neural tissues of squid[75,119a,120] and possibly also in the sea mollusc *Aplysia californica.*[49]

TABLE 8.1
Changing Patterns of Heat Shock-Induced Protein Synthesis during *Xenopus* Development

	A	B	C	D	E	F	G	H	I	J	K	L	M	N
Ref.	(23,30b,56)	(30b)	(30b)	(30b)	(30b)	(55)	(55,91)	(55,91)	(91)	(91a)	(55b)	(127)	(127)	(127)
Stages	Oocyte	Ovulating oocyte	Spawned egg	Egg, artificial activation	Egg, fertilized	Cleavage stage embryo to large cell blastula	Midcell blastula	Fine cell blastula	Gastrula	Neurula	Kidney epithelial cells	Basophil. erythroblast	Orthochr. erythroblast	Mature erythrocyte
HSPs induced	83	83	(83)	—	—	—	87	87	87	87	87	87	87	—
(M_r, kDa)	76	76	76	—	—	—	—	—	—	76	—	—	—	—
	70[a]	70[a]	70[a]	(70)	(70)[a]	—	68—70[b]	68—70[b]	68—70[b]	70[b]	70—73[b]	70[b]	70[b]	(70)
	—	—	—	66	—	—	—	—	—	66	—	—	66	—
	62[c]	62[c]	62[c]	—	—	—	—	62[c]	62[c]	62[c]	59[c]	62[c]	—	61[c]
	—	—	—	48	—	—	—	48[c]	48	48	51, 54	—	—	48
	—	—	—	—	—	—	—	35[d]	35[d]	35[d]	—	—	—	—
	—	—	—	—	—	—	—	—	—	—	30—32[e]	30—32[e]	—	30—32[e]
	—	—	—	—	—	—	—	—	—	25	—	—	—	—
	—	22	(22)	—	—	—	—	—	—	—	—	—	—	—
	—	16	16	—	—	—	—	—	—	—	—	—	—	—
	—	—	—	—	—	—	—	ubi	ubi	ubi	ubi	?	?	?
Remarks	Heat shock response controversial[f]				Low levels of HSP 83, and 70 mRNAs; constitutive synthesis of both HSPs (55a)		Hs-induced HSP synthesis controlled at transcription level[g]						HSPs 87 and 70 synthesis also constitutive	

[a] Oocytes, fertilized eggs and early cleavage stage embryos contain maternal hsp70 mRNA, but their translation is not hs-inducible.[53a,56]

[b] The HSP68-73 complex contains 5 proteins in the pI range 6.0 to 5.8[37a] (Figure 8.3).

[c] HSPs62, 48, and 35 are major HSPs of the vegetal half only. HSP62 is a complex of several proteins. Identity of HSP62 between different developmental stages was not analyzed.[30a]

[d] HSP35 represents a minor isozyme of glyceraldehyde phosphate dehydrogenase.[91a]

[e] HSPs30 to 32 are the major HSPs of mature tissues, especially under conditions of severe hs. The HSP30 complex contains about 16 proteins of pI 6.0 to 5.2[37a] (Figure 8.3). There is no relation between the embryonic HSPs25 and 35 and members of the adult HSP30 complex.

[f] Oocytes, were reported to induce HSP synthesis from preformed mRNAs.[25,30b] However, recent experimental evidence is in conflict with these results.[56,66] For details see Section 8.6.

[g] Together with the inducibility of HSPs 68-70 from the midblastula stages on, ubiquitin mRNAs are detected as well. But HSP30 appears only much later, i.e., in the tailbud stage following the neurula stage. In contrast to these differential effects, all three genes are coordinately regulated by hs in tadpoles and adult frogs.[71a]

Data compiled from References 53a and 56. For a summary see Browder et al.[30a]

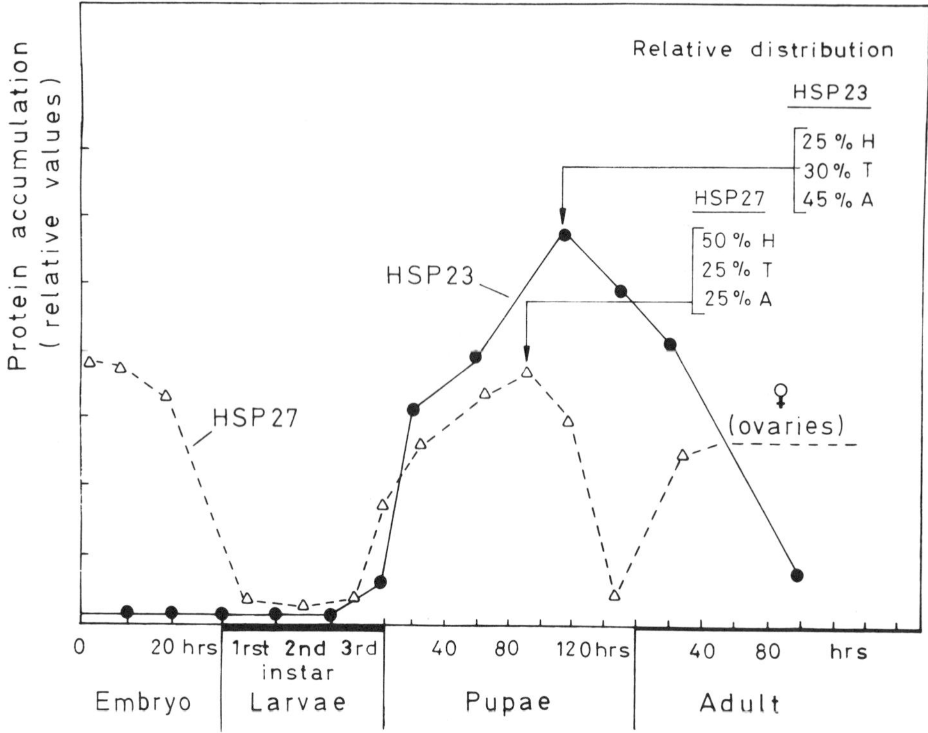

FIGURE 8.4. Developmental synthesis of HSPs 23 and 27 in *Drosophila*. Amount of the two proteins were analyzed immunologically. The basic pattern of expression described in the text is modulated by tissue-specific factors as indicated for the maximum in late pupal stages: H, head; T, thorax; A, abdomen. (Arrigo, P. and Pauli, D., personal communication.)

8.4. HSP SYNTHESIS INDUCED BY DEVELOPMENTAL SIGNALS

With the advancing gene analysis, a 15-kb chromatin domain of *Drosophila* (locus 67B) was characterized, which contains the four small HSP-coding genes (hsp27, 26, 23, and 22), and three additional small transcription units under hs control (genes 1, 2, and 3)[11,96,97,114] (see Figure 3.3). Interest in this particular locus was enhanced when it was recognized that these genes exhibit a highly complex pattern of developmentally controlled expression. The initial observation was based on the ecdysterone induction of small *Drosophila* HSPs in cell cultures and imaginal discs.[32,60,61]

Nowadays, different types of hs-independent expression patterns can be discriminated[11,78,125] (Figure 8.4). (1) In ovarian nurse cells of adult females high amounts of hsp83, 27, and 26 mRNAs are produced and transferred to the maturing oocytes, which are connected with the nurse cells by plasma bridges.[2,33,47,135] Recent immunological analyses showed, at least for HSP27, that considerable levels of the corresponding protein are found in oocytes[5] (Figure 8.4). The mRNAs rapidly disappear in early embryonic stages.[118] (2) In mid-embryonic stages a transient rise in the ecdysterone levels leads to an increase of transcripts coded by hsps83, and 26 genes, but mainly by genes 2 and 3.[96,97,118] (3) A major and more durable increase of expression of all small HSP-coding genes plus gene 1 is connected with another increase of ecdysterone titer at the transition between larval and pupal stages.[112,118,135] (4) Finally, a low level of HSP26 and a high level expression of gene 2 was observed in adult testis[46b,47,97] (spermatocytes).

Sequence analyses of the multifunctional promoters are not yet completed. Generally, hs elements, scattered over a long range of 5′ flanking sequences contribute to optimum hs inducibility of these genes (see summary in Figure 6.6). Signals for developmental control are separated from them. By sequence comparison and functional analysis, three putative ecdysterone receptor elements were defined in the hsp23 promoter[80] (see Figure 6.6). Functionally similar elements are localized between bp −579 to −455 of the hsp27 promoter[102,102a] and between bp −194 to −134 of hsp22.[68] The situation with the hsp26 promoter and the other three genes of locus 67B is unclear. But for the former, two separate regions were shown to be important for the ovarian type (bp −522 to −352[33]) and for the testis type of hsp26 expression, respectively, which depends on two types of sequences downstream of bp −351.[46b,47]

By use of a hsp26 P/L × lac Z fusion construct introduced into the *Drosophila* germ line by P-element transformation, Glaser et al.[47] analyzed the tissue specificity of developmental control using a histological assay. Main sites of expression in the third instar larval and prepupal stages are epithelial cells, testis, gut, and imaginal discs. Other parts of the body are virtually free of staining. This may reflect the tissue-specific competence to respond to ecdysterone, e.g., dependent on the presence or absence of the corresponding receptor protein. An interesting modulation of the developmental control pattern is also indicated by the unequal distribution of HSPs 27 and 23 between parts of the late pupa (Figure 8.4).

An interesting contribution to the identification of developmental control elements was reported by Xiao and Lis.[132a] By introducing minor changes in the HSE1/HSE2 region of the Dm-hsp70 promoter, they generated constructs with a remarkable cell-specific expression in the brain and ejaculatory duct, respectively. These results demonstrate that even a few base exchanges in this promoter region may abolish the typical hs-inducibility and create binding sites for the putative developmental control proteins. Though completely artificial, the constructs may help to identify these proteins.

Though investigations with other organisms are not so extended, they indicate similar complex expression patterns of members of the HSP families under the control of developmental signals.

1. Lindquist and co-workers[73,74] reported on the sporulation-dependent accumulation of hsps 84 and 26, but not of hsp70 mRNAs in yeast. They discuss this phenomenon as an ancient developmental pattern closely related to the ovarian type of control in *Drosophila* (see above). In fact, careful analysis of the whole yeast hsp70 family gave a much more detailed pattern. The individuality of the eight genes is already evident from their differential regulatory behavior under control vs. hs conditions (Table 2.4),[125a] During sporulation, mRNA levels of SSB1/2, SSC1, and SSD genes decrease continuously, whereas levels of SSA1/2 mRNAs initially increase until 6 h of sporulation and then slowly decrease. A third pattern is observed for the SSA3 gene, which is usually inactive except under hs conditions. But from the onset of sporulation to late stages, there is a constant increase of SSA3 mRNA. This type of regulation is apparently connected with decreasing cAMP levels, i.e., it is blocked by addition of glucose, and constitutive activity of the SSA3 genes is observed in an adenylate cyclase mutant (cyr1). If corresponding analyses are performed during the growth cycle of yeast cultures, similar complex patterns are noticed.[125a] SSB1/2 and SSC1 mRNA levels are high during growth, but rapidly decline during stationary phase. In contrast to this, the relatively low level of SSA1/2 mRNA in the early growth phase markedly increases just before transition to stationary phase and then decreases. Again, SSA3 gene regulation is totally different. The gene is inactive during growth. But with decreasing cAMP levels in the stationary phase, it is induced, and mRNA accumulates to fairly high levels.

   To summarize, the four physiological situations described for yeast, i.e., exponential growth, stationary phase, sporulation, and heat shock, can each be characterized by a typical pattern of activity of the hsp70 gene family. Changes of mRNA synthesis and/or turnover may contribute to the overall effects. Though details of the regulatory mechanisms remain to be elucidated, five factors are already evident (see summary by Craig et al.[32b]): (1) multiple overlapping HSEs in the promoter regions of SSA genes are responsible for hs-inducibility and extent of basal level expression; (2) a negative control element partially overlaps with the TATA-proximal HSE of the SSA1 gene; (3) activity of the SSA1 promoter under hs conditions is blocked in the presence of excess SSA1 protein, and this autorepression is not dependent on the HSE (see Section 8.7); (4) an activator protein (STI1) was characterized that acts specifically on the SSA4 gene (see Section 2.1.3); and (5) the hs-independent expression levels of SSA3 and SSA1 genes are inversely correlated with the cAMP content of the cell.[32b,116b]

2. With the ongoing analysis of the plant HSP70 family, specific gene probes became available for the constitutively expressed members. Duck et al.[43b] used a tomato hsc70 cDNA to screen expression levels of the corresponding gene in different tissues by Northern and *in situ* hybridization. They detected high levels of mRNA in the vascular systems of the ovary, in dividing cells of the root tips, and in the inner integument of seeds, moderate levels in immature anthers and embryos, and no hsc70 mRNA in mature pollen, xylem, and ovules. Investigations on the HSC70 protein itself were not done in this case.

3. Examples for developmental control of HSP70 synthesis in vertebrates were reported for early embryonic stages[19,51,70,85] during maturation of the mouse brain[38] and during hemin induced maturation of human crythroid cell lines.[111] In the latter case, increased levels of GRP78 were detected as well. Surprisingly, the hemin-induced transcriptional activation of the hsp70 gene depends on the intact heat shock element and involves activation of the heat shock transcription factor. This was shown by corresponding gel retardation assays and transfection of a human K562 cell line with an hsp70 promoter × cat gene fusion constructs.[127a] Heat shock induction (3 h 42°C) and hemin induction (24 h) always gave similar results. Because hemin itself is not cytotoxic and does not induce HSP70 synthesis in nonerythroid cells, the effect may be the result of the massive reconstruction of these erythroleukemia cells during maturation.

   A particularly interesting example of developmental control is the multiplicity of HSP70-type proteins expressed differentially during mammalian spermatogenesis.[1,1a,71,77b,78a,134,134a] Experiments were done mostly with mouse testis. Besides the hs-inducible HSP70,[1,1a] there are the abundant constitutively expressed clathrin uncoating ATPase (HSC71), three sperm-specific HSC70,[1,1a,78a,134a] and an HSC74. All are ATP-binding proteins and cross-react with the family-specific monoclonal antibody 7.10.

   Though data between the different groups are not fully comparable, there is evidence for three sperm-specific HSC70. Zakeri et al.[134a] isolated and sequenced an HSC70 gene (see Figure 2.10) expressed early in the meiotic prophase of spermatocytes. A second clone of this group characterizes a related protein found in postmeiotic spermatids. Its sequence was reported by Matsumoto and Fujimoto.[78a] Other groups[1a,77b] defined a third protein of this family (P70) synthesized primarily in pachytene-stage spermatocytes. It is the dominant protein of this family in spermatids. Finally, there is a P69 (pI 7.0) detected only in spermatozoa.[77b] Because of the particular role of mitochondria for the mobility of sperm, it can be speculated that P69 represents the mitochondrial member of the HSP70 family. Remarkably, each of the two hsp90 genes of mammals likewise exhibits its particular pattern of developmental expression during germ cell formation.[75a]

4. There is an increasing number of examples for steroid hormones stimulating synthesis of one or few HSPs, dependent on the cell type. This was extensively discussed for the ecdysterone control of small HSP synthesis in *Drosophila*. In addition there are

reports on steroid hormones increasing levels of HSP90-type proteins, known to be associated with the hormone receptor complexes (see Section 16.8). This is true for HSP85 in the water mold *Achlya ambisexualis* induced by antheridiol[31a,31b] but also for the murine mammary gland with the highest levels of HSP90 in pregnant and lactating animals.[31c] Estrogen stimulation of murine uterus results in a 15-fold increase of HSP90 mRNA levels, whereas hsp70 mRNA is not affected.[98a,108a] Probably connected with the receptor-HSP90 complex of mammary tumor cells is the HSP25, whose synthesis is likewise increased by estrogen.[45d,79b] Finally, estrogen-induced increase of HSC70 levels in the nucleus of the hypothalamus of female rats may indicate a participation of this protein in the organismic signal systems controlling mating behavior.[83a]

5. An interesting example of tissue-specific and developmental regulation emerged when Giebel et al.[46a] analyzed the expression HSC72 of mouse, which is presumably the clathrin uncoating ATPase. The hsc72 mRNA levels are especially high in brain, correlated with particularly abundant clathrin protein and in F9 teratocarcinoma cells, but they are relatively low in other tissues (heart, lung, liver, ovaries, 14-d embryos). Intermediate levels are found in tongue, kidney, pancreas, and adrenal gland. Remarkable is the high expression in teratocarcinoma cells, which are considered as model cells for embryonic differentiation. After stimulation with retinoic acid and dibutyryl-cAMP, they differentiate into nontumorigenic cells of the parietal endoderm with expression of cell-type specific genes, e.g., of collagen type IV, plasminogen activator and laminin. Simultaneously, expression of cellular oncogens (p53, myc) is depressed.[43a] Additionally, this reduction is also observed for the hsc72 gene.[46a] Unlike the mouse plasmacytoma cell lines mentioned in Section 8.3, the F9 teratocarcinoma cells can be hs-induced to synthesize the HSP70, which is discriminated by size and immunological data from the constitutively expressed HSC72.

8.5. CELL CYCLE CONTROL OF HSP SYNTHESIS

Investigations of DNA viruses as inducers of HSP synthesis in mammalian cells[63,90] led to the discovery of a transient stimulation of HSP70 synthesis in adenovirus-infected HeLa cells (see Figure 8.5). Induction is dependent on a functional viral E1A gene, whose 13S-mRNA codes for a 289 amino acid residue phosphoprotein. The effect on the hsp70 promoter is indirect, i.e., it is mediated by an endogenous regulatory system of the host cell.[59,64,130] Remarkably, the induction-deinduction pattern of the hsp70 gene during the lytic cycle of the adenovirus (Figure 8.5) exactly mimicks the behavior of early viral gene products.

The description of a cellular E1A-like function[59,64] and reports on the cell cycle-dependent expression of this hsp70 gene in the S phase[82,129] make it very likely that the viral induction is related to this mode of control and not to the hs induction mechanism.[110] Other immortalizing factors, e.g., the polyoma virus large tumor antigen may act in a similar manner.[67] This led Imperiale et al.[59] to the investigation of many human tumor cell lines. Most of them had increased levels of HSP70 combined with high activity of the endogenous E1A-like system. An essential part of the recognition mechanism may be the TATA box, which evidently exhibits more specificity than hitherto assumed. Replacement of the hsp70-specific TATAA box (see Figure 8.2, promoter number 4) by the SV40-derived TATA-box variant (-TATTTAT-) abolishes the induction by the adenovirus without affecting hs induction of the hsp70 gene.[110a] However, this result was not supported by analyses of the same promoter by Morimoto and co-workers.[87a,126a] They found the CCAAT-box and its flanking sequences (bp -74 to -63) to be essential for E1A stimulation of the hsp70 promoter. At any rate, two facts are established: (1) increased transcription of hsp70 by nuclear onc-proteins coded by DNA and RNA tumor viruses (see Section 19.1) does not involve the hs transcription

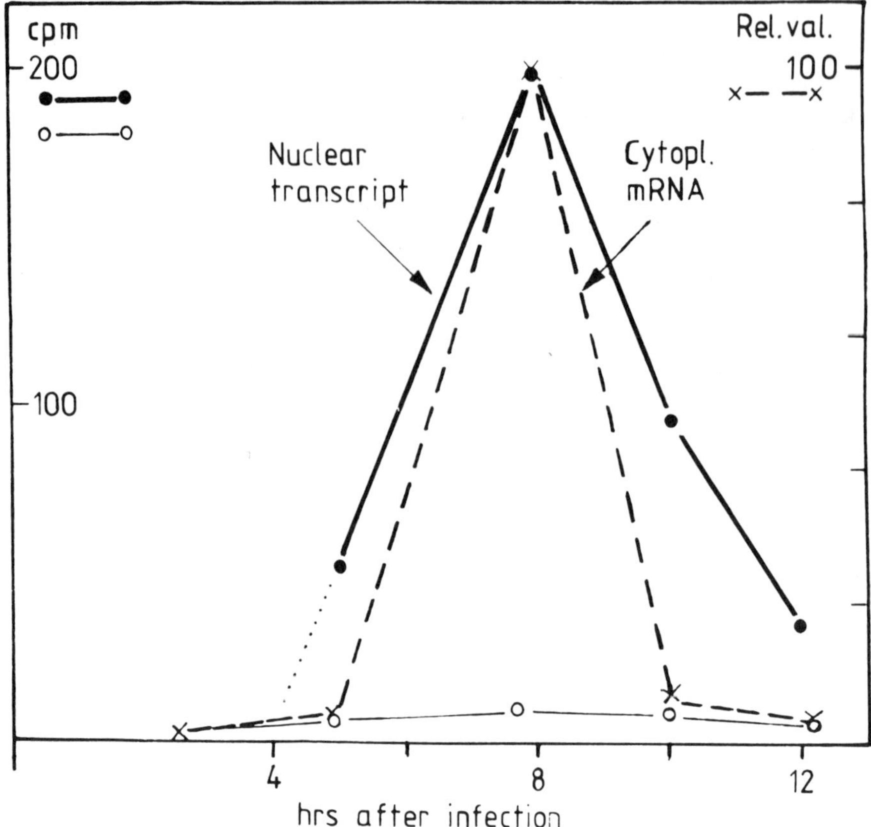

FIGURE 8.5. Transcription and accumulation of hsp70-specific mRNA in adenovirus infected HeLa cells. Transcription of the hsp70 gene (full lines) was investigated by pulse labeling of isolated HeLa cell nuclei with ^{32}P-UTP. Dots represent cells infected with wild-type adenovirus; circles represent cells infected with an E1A gene deletion mutant. Cytoplasmic mRNA (dashed line) was determined by Northern analysis. (Data from Kao, H.-T., and Nevins, J. R., *Mol. Cell. Biol.*, 3, 2058, 1983; reprinted from Nover, L., Ed., *Heat Shock Response of Eukaryotic Cells,* G. Thieme Verlag, 1984. With permission.)

factor; and (2) the effect is indirect and mediated by interaction with the promoter DNP complex, active in the basal level or cell-cycle-controlled expression of the hsp70 gene.

A different type of cell cycle control was reported by Iida and Yahara.[57,58] They found hs-independent synthesis of high molecular weight HSPs in resting cells of yeast and other eukaryotes (*Drosophila,* chicken, mouse). Their description of certain HSPs as G_0-specific proteins rely on careful two-dimensional electrophoresis, investigations with mutant yeast strains and peptide mapping. The increase of HSP synthesis at the transition of cells from growing to resting state is significant but much less than the hs-induced stimulation. Analysis of the expression of the yeast polyubiquitin gene showed that the G_0-type of induction is dependent on decreasing levels of cAMP required for G_0/G_1 arrest in *Saccharomyces cerevisiae.* This negative control by cAMP is independent of the hs induction.[116a] In support of this, increased hsp70 gene activity was found after prostaglandin-induced cell cycle block in human erythroleukemia and HeLa cells.[93a,105]

Experiments with mitogen stimulation of quiescent G_0 cells are equivocal. On the one hand, Kaczmarek et al.[62] reported on a decrease of the hsp70 mRNA after phytohemagglutinin (PHA) treatment of peripheral blood mononuclear cells. On the other hand, Haire et al.[51a]

and Ferris et al.[45a] observed increased HSP70 and HSP90 synthesis 4 to 6 h after stimulation of human lymphocytes with PHA or interleukin 2. Possible explanations for these discrepancies are (1) cell type-specific variations in the hs-independent modes of HSP synthesis, (2) the carry-over of the mRNA into the G_0 state from previous S/G_2 phases of synthesis, as found by Milarski and Morimoto[82] and Wu and Morimoto,[129] and, (3) the human peripheral blood monocytes are naturally occurring G_0 cells, whereas the cells used in the cell culture studied were made artificially quiescent by serum deprivation. It will be necessary to study exactly the transcription phases of these mRNAs found in G_0 cells so well as the promoter sequences required for this type of control. At any rate, it is very likely that HSP70 and probably others have an essential function for cell cycle activities in eukaryotes. In keeping with this, Milarski and Morimoto[82] observed a transient accumulation of the HSP70 in HeLa cell nuclei during the S phase, and the finding of increased hsp70 mRNA levels after prolactin treatment of rat lymphoma cells[40a] may also depend on the mitogenic effect of this hormone.

8.6. TRANSLATION CONTROL OF HEAT SHOCK INDUCTION?

The discovery of hs-induced changes of gene expression[103] was based on cytological observations of an altered pattern of gene activity. The primary effects at the transcription level were later confirmed for all kinds of systems using nuclear run off transcription assays, appropriate methods for characterization, and quantification of the newly formed mRNAs or by inhibiting induction by actinomycin D. However, fine analyses of the HSP patterns observed in the presence or absence of actinomycin D, cordycepin, or other drugs led to the description of minor proteins, whose induction was not affected by these inhibitors of mRNA synthesis. This is true for HeLa cells,[100] for rat thymic lymphocytes,[34,79] and *Tetrahymena pyriformis*.[1b] In the latter two cases, the hs activation of preexisting mRNAs was confirmed by *in vitro* translation and by Northern analysis, respectively. The same may be true for HSP25 mRNA in chicken embryo cells. HSP induction by 600 μM canavanine evidently involves activation of a stored or incompletely processed mRNA for this stress protein.[125b]

A most surprising result in this respect was reported by Bienz and Gurdon.[25] They found that in *Xenopus* oocytes, but not in somatic cells of the bullfrog, hs-induced HSP70 synthesis is exclusively controlled at the translation level. It is brought about by activation of maternal mRNA.[22,23] In keeping with this, transformation of oocytes by microinjection of the homologous X1-hsp70 gene is connected with constitutive transcription of the corresponding mRNA, which is translated at 20°C and is highly selective at 35°C.[24,56] These findings were confirmed later by developmental study of Browder et al.[30b] (see Table 8.1).

However, careful reinvestigations by King and Davis,[66] and Horrell et al.[56] led to contradictory results. It turns out that oocytes are not able to respond to hs. The induced synthesis of HSP70 observed by Bienz and Gurdon,[25] is presumably due to adhering follicle cells. Although oocytes contain a low amount of maternal hsp70 mRNA, it is never translated either under control or under hs conditions. Remarkably, this does not hold true for mRNA transcribed from the microinjected hsp70 gene.[22,56]

Clearly, this part of the hs story needs further investigation, including the question of other HSPs observed in oocytes by Browder et al[30b]. and the intriguing differences in the state of the hsp70 mRNAs produced from the endogenous gene during oocyte maturation or from microinjected genes.[56] The situation appears similar to sea urchin eggs and zygotes.[16] In this case, the maternal hsp90 mRNA is stored in the pool of free mRNP. It is in a poly$(A)^-$ form, which is reactivated in the morula stage under developmental, but not hs, control. This transition into the polysomal fraction is connected with polyadenylation. Hs-induced synthesis of HSP90 starts much later during embryogenesis in the gastrula stage.

Another example of translation control was analyzed by Morimoto and co-workers. In chicken lymphocytes, three major HSPs are induced by hs: HSPs 89, 70A/B, and 25, whereas in reticulocytes only HSP70B is expressed constitutively and further enhanced by hs. It is a special isoform of HSP70 found in addition to HSP70A also in lymphocytes.[14,87] In reticulocytes, the hsp70 mRNA is moderately abundant at 1 to 2% of the globin mRNA level. The 20-fold increase of HSP70 synthesis under hs conditions is not inhibited by actinomycin D or 5,6-dichloro-1-,3-D-ribofuranosylbenzimidazole, and, correspondingly, the levels of hsp70 mRNA do not change significantly, i.e., maximum two-fold. Repression of globin and induction of HSP70 synthesis are exclusively due to changes at the translation level.[13,117a] This regulatory pattern is established late in the developmental pathway of erythroid cells. Primitive red cells exhibit constitutive expression of HSP70 without hs inducibility.[15] There is good evidence that the hsp70 mRNA in its repressed state exists in polysomes, i.e., the major control point for hs induction is the elongation rate of nascent HSP70 protein.[117a] This, however, may be a reticulocyte-specific mode of hs control.

In apparent contrast to these results, Atkinson et al.[8b] found that in reticulocytes of quail and chicken, recovering from experimental anemia, hs-induced synthesis of HSPs 90, 70, and 25 were inhibited by actinomycin D, but the inductive effect was delayed by 30 to 60 min. Although in *Xenopus laevis* there is a progressive decrease of HSP expression during maturation of erythroid cells,[1,27] (see Table 8.1), the hs-induced HSP70 synthesis in mature erythrocytes is controlled in the "normal" way, i.e., by transcription of new mRNA. Thus, the quotation mark behind the heading of this section remains. There is no doubt that the pattern of protein synthesis during the hs response is thoroughly influenced by translational control (see Chapter 11), but the reports on an exclusive induction of HSP synthesis by activation of preformed mRNA in *Xenopus* oocytes and chicken reticulocytes are controversial.

8.7. AUTOREPRESSION OF HSP SYNTHESIS

The early cytological investigations on hs-induced gene activity patterns using *Drosophila* salivary glands led to a number of conclusions, which later were confirmed with cell cultures and biochemical methods.[6,7,17,29,44,45,76,77,101] First, induction is very rapid and does not require protein synthesis (see also Section 1.4). Second, the extent of puffing is dependent on the hs dose. Higher temperatures lead to larger puffs. Third, puffing is transient; recovery requires prior protein synthesis, and its velocity is reversed proportional to the stress dose.

A concise analysis of the response in *Drosophila* Schneider cells led Lindquist and coworkers to the concept of a self-regulated system of HSP70 synthesis.[42] Essential results are presented in Figure 8.6. In all experiments, 1 h of hs at 36.5°C was followed by several hours of recovery. Usually synthesis of hsp70-specific mRNA rapidly declines during recovery (Figure 8.6A) unless HSP70 synthesis was blocked by addition of cycloheximide (broken curve in Figure 8.6A) or, even more intriguingly, if HSP70 synthesis proceeds in the presence of canavanine. This arginine analog is incorporated into HSP70, but makes it nonfunctional. The hs protein neither binds to the nuclear compartment (see Section 16.2.) nor is it self-regulating, i.e., the corresponding mRNA continues being formed in the recovery period. It is stable, and huge amounts of aberrant HSP70 accumulate. A second level of control is illustrated in Figure 8.6B. HSP70 synthesis is also self-limiting at the translation level. If the amount of induced mRNA is fixed at a certain time point of the induction period by adding actinomycin D, HSP70 synthesis is prolonged during the recovery period. This kind of message dosage experiment shows that a certain amount of HSP70 is required before its translation begins to decline. It is worth noticing that restoration of control protein synthesis was delayed in proportion to the delay of autorepression.[42,43] This inverse correlation between repression of HSP70 expression and restoration of control protein translation is documented in Figure 13.3.

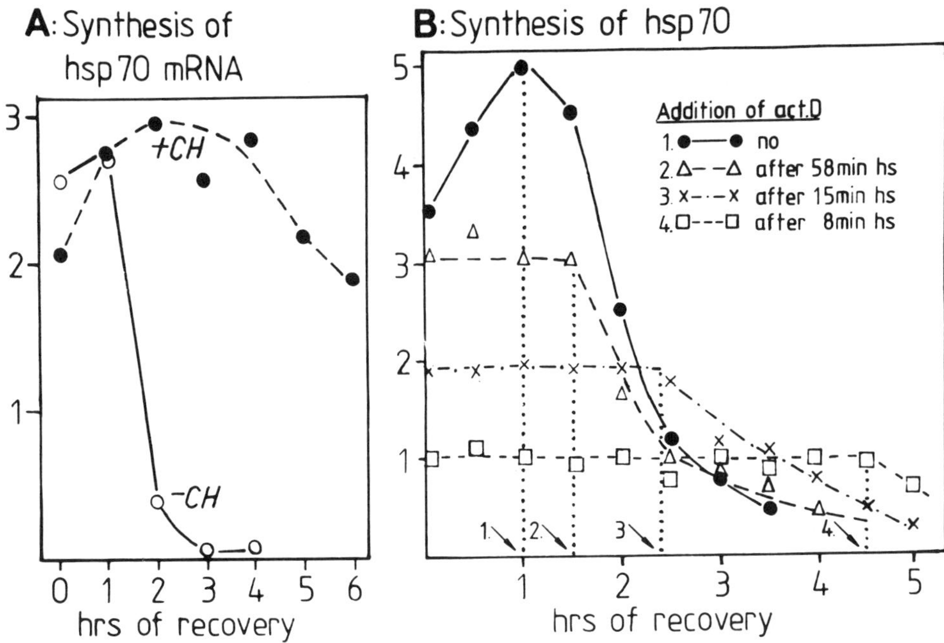

FIGURE 8.6. Autorepression of hsp70 synthesis at the transcription and translation level. A, transcription level: *Drosophila* cell cultures were heat-shocked for 1 h at 36.5°C, then returned to 25°C for the indicated time and pulse labeled with ^{3}H-uridine. After electrophoretic separation synthesis of hsp70-specific mRNA was quantified. Full line, without cycloheximide; dashed line, cycloheximide added before the hs. The inhibition of HSP synthesis leads to an undiminished synthesis of hsp70 mRNA in the recovery period. B, translation level: same procedure as in A, but actinomycin added 8 min, 15 min, or 58 min after the onset of the heat shock (hs). Incorporation of ^{3}H-isoleucine into HSP70 was determined during the recovery period. Arrows at the time scale indicate the points where HSP70 accumulation reaches the level triggering the autorepression of translation. The decreasing amount of hsp70 mRNA in the four samples proportionally extends the time period required to reach this autorepressor level. (Modified from DiDomenico, B. J., Bugaisky, G. E., and Lindquist, S., *Cell,* 31, 593, 1982.)

The molecular mechanisms involved in the autorepressive control of the *Drosophila* HSP70 level are not yet clear, but it is remarkable that similar results were obtained with rat embryo fibroblasts and other mammalian cell lines.[83] The stress dosis applied is reflected in the amount of HSP72/73 accumulated, at least to an upper level. Other HSPs (M_r 110, 90, 27) are not controlled in the same way. Similar to the results with *Drosophila,* the velocity of restoration of control synthesis is directly proportional to the HSP72/73 levels. Moreover, if HSP synthesis is induced by the amino acid analog azetidine carboxylic acid, this particular thermotolerance effect is only observed after the cells were allowed to recover from the AzC stress and to synthesize sufficient amounts of functional HSP72/73.

In support of these results, HSP70 was repeatedly found to be an RNA-binding protein.[43,95,116] Furthermore, self-regulation in *Drosophila* is prominent under conditions of recovery or moderate stress, e.g., 33°C, when the hsp70 mRNA is particularly short lived, and its synthesis rapidly stops.[42,72,77] In contrast, during severe hs, HSP70 synthesis proceeds at a high level for a long time. Its mRNA is much more stable.[42] An essential aspect of self-regulation and restoration of control protein synthesis is evidently the release of HSP70 from its high affinity nuclear binding sites.[122] On the other hand, experiments with a truncated hsp70 gene lacking the whole 3′ half demonstrated that the capability for self-regulation at the translational level was lost, despite the presence of a normal hsp70 gene and its protein in the same cell.[109] A direct destabilization of hsp70 mRNA by interaction of its 3′ half with the cognate protein must be considered. The mechanism may be basically similar to the self-regulation of tubulin synthesis in vertebrate cells and *Tetrahymena*.[33a,46] Evidently, free tubulin subunits bind to the N-terminal MREI-peptide of nascent tubulin chains. This leads to activation of a short-lived, ribosome-associated nuclease that acts on the 3′ ends of the polysome-bound mRNA (see reviews by Ben-Zeev,[20] Brawerman,[30] and Cleveland[32a]).

The expression of other *Drosophila* hs genes are not affected to the same extent. It is well documented that mRNA levels of small HSPs are maintained over several hours, and their translation proceeds much longer in the recovery period.[44,95,109] This peculiarity may be related to their organization in high molecular weight scRNP complexes (pre-HSG, see Section 16.7). On the other hand, evidence points to a self-regulation mechanism also for HSP83. In trisomies of *Drosophila*, the increased gene dosage for hsp83 resulted in an increased mRNA level, but the protein level was normal, i.e., gene dosage compensation was at the posttranscriptional level.[41] More convincing are experiments by Yost and Lindquist[133] involving addition of cycloheximide prior to the hs induction. Similar to the results for hsp70 (see Figure 8.6A), hsp83 mRNA accumulated to high levels under these conditions. It is remarkable to recall that Bonner[28] reported on an *in vitro* autorepression of puffing with isolated polytene nuclei of *Drosophila* larvae by adding partially purified fractions of HSP83. Puffs 93D, 63C, and 87C were repressed, but not 67B coding for the small HSPs. This type of experiment was not developed further.

It is important to notice that self-regulation in *Drosophila* is evidently dependent solely on the hs-inducible form of HSP70 which is virtually absent in control cells.[123] The much more abundant, constitutively expressed form (HSC70), coded by gene hsc4 (see Table 2.3), is not active, although both proteins are highly homologous (see Figure 2.10). They bind equally well to the network of intermediate filaments, and HSC70 may even substitute for its hs-inducible counterpart in the nuclear compartment.[95]

What is the situation in other eukaryotic organisms with respect to self-regulation of HSP synthesis? Although the results are circumstantial, it is useful to briefly summarize them. Destabilization of hsp70 mRNA in the recovery period or under conditions of moderate stress was reported for vertebrates,[14,35,40,126] for fibroblast cell cultures of rainbow trout,[69] plants,[65,105a] and *Tetrahymena*.[52] Theodorakis and Morimoto[117] found a tenfold difference for the half-life of hsp70 mRNA in HeLa cells with about 50 min at 37°C and more than 500 min at 43°C. On the basis of the results of Simcox et al.[109] for the *Drosophila* hsp70 mRNA they discuss the role of a AUUUA signal in the 3′ untranslated part of the human hsp70 mRNA for the destabilization in the recovery period.[108]

The two levels of control of HSP70 expression, described for *Drosophila*, can also be discriminated in chicken lymphoblastoid cells.[14] During long term hs at 45°C, increase of hsp70 transcription is transient, but the newly formed mRNA is stable leading to continued accumulation of HSP70. Translation and mRNA levels only decline in the recovery period. Interestingly, synthesis of the chicken HSP25 even increases in the recovery period, i.e., comparable to the situation in *Drosophila*, self-regulation seems to be limited to the hsp70 system, and this also holds true for tomato cell cultures[105a] and chicken myotubes.[12]

However, despite many similarities of the phenomena in *Drosophila* and vertebrate cells, there may be essential differences in the mechanism. When White and Hightower[125a] investigated HSP synthesis in canavanine-induced chicken embryo fibroblasts, they found rapidly increasing levels of hsp90 and hsp70 mRNAs, but unlike the situation in *Drosophila* (Figure 8.6), the half-lives were always very short with 89 min for the hsp90 and 46 min for the hsp70 mRNAs. There is no stabilization of hsp70 mRNA during the stress period. Under these conditions deinduction in the recovery period could simply reflect the shut-off of transcription due to the lacking stress signal.[55c]

Remarkably clear are results on self-regulation of SSA1-coded HSP70 synthesis in yeast. Craig and co-workers[32b,115a] prepared fusion constructs of the SSA1 promoter with lacZ as reporter gene and transformed yeast cells together with a plasmid containing the SSA1 gene itself under the control of a gal promoter. Hence, HSP70 levels in these strains can be modulated by the presence or absence of galactose. They found that hs-induced activity of the SSA1 promoter is almost completely repressed by the SSA1 protein. The autorepression depends on flanking sequences 5′ upstream or 3′ downstream of the TATA-proximal HSE (see Table 6.2). The HSE itself is not sufficient. But the mechanism may involve interaction of SSA1 with HSF to prevent its activation. At least in ssal/ssa2 double mutants, the constitutive overexpression of SSA3 depends on its functional hs promoter element.[29b]

In contrast to the situation in most eukaryotic cells, induced HSP synthesis in bacteria is exclusively regulated by the interplay among three σ subunits of RNA polymerase required for promoter recognition (see Sections 7.3 and 13.1). The transient increase of hs gene transcription coincides with a rise of hs-specific σ^{32}, whereas restoration of the control transcription pattern is brought about by production of the normal σ^{70} as an integral part of the hs response. At high temperatures (≥50°C), the continuous production of σ^{32} by transcription of the rpoH gene requires an additional sigma factor (σ^{24}), which is probably part of a special high-temperature regulon.[44a] Interestingly, this finely tuned system is complemented by an autorepression mechanism dependent on the bacterial HSP70 analog, the DnaK protein.[119] The results are summarized by Neidhardt et al.[89] The mechanism is not much clearer than in eukaryotic systems. In DnaK-defective mutants, HSP synthesis proceeds for a long time,[94,119] and this is connected with an undiminished high cellular concentration of σ^{32}. Thus, it is very likely that DnaK exerts its effect by modulation of the σ^{32} level, presumably by repression of σ^{32} mRNA translation.[50] Support for the direct role of σ^{32} levels for autorepression comes also from analyses of high-temperature stress. At 50°C, the expression of all parts of the rpoH-controlled hs regulon proceeds undiminished because high levels of σ^{32} are maintained.[44a]

REFERENCES

1. **Allen, R. L., O'Brien, D. A., and Eddy, E. M.,** A novel hsp70-like protein (P70) is present in mouse spermatogenic cells, *Mol. Cell. Biol.*, 8, 828—832, 1988.

1a. **Allen, R. L., O'Brien, D. A., Jones, C. C., Rockett, D. L., and Eddy, E. M.,** Expression of heat shock proteins by isolated mouse spermatogenic cells, *Mol. Cell. Biol.*, 8, 3260—3266, 1988.

1b. **Amaral, M. D., Galego, L., and Rodrigues-Pousada, C.,** Stress response of *Tetrahymena pyriformis* to arsenite and heat shock: Differences and similarities, *Eur. J. Biochem.*, 171, 463—470, 1988.

2. **Ambrosio, L. and Schedl, P.,** Gene expression during *Drosophila melanogaster* oogenesis: Analysis by in situ hybridization to tissue sections, *Dev. Biol.*, 105, 80—92, 1984.

3. **Anderson, K.-V. and Lengyel, J. A.,** Rates of synthesis of major class of RNA in *Drosophila* embryos, *Dev. Biol.*, 70, 217—231, 1979.

4. **Anderson, N. L., Giometti, C. S., Gemmell, M. A., Nance, S. L., and Anderson, N. G.,** A two-dimensional electrophoretic analysis of the heat-shock-induced proteins of human cells, *Clin. Chem.*, 28, 1084—1092, 1982.

5. **Arrigo, A.-P. and Pauli, D.,** Characterization of HSP27 and three immunologically related polypeptides during *Drosophila* development, *Exp. Cell Res.*, 175, 169—183, 1988.

5a. **Arrigo, P. and Pauli, D.,** personal communication.

6. **Ashburner, M.,** Patterns of puffing activity in salivary gland chromosomes of *Drosophila*. V. Responses to environmental treatments, *Chromosoma*, 31, 356—376, 1970.

7. **Ashburner, M. and Bonner, J. J.,** The induction of gene activity in *Drosophila* by heat shock, *Cell*, 17, 241—254, 1979.

8. **Atkinson, B. G.,** Synthesis of heat-shock proteins by cells undergoing myogenesis, *J. Cell Biol.*, 89, 666—671, 1981.

8a. **Atkinson, B. G. and Walden, D. W., Eds.,** *Changes in Eukaryotic Gene Expression in Response to Environmental Stress*, Academic Press, Orlando, FL, 1985.

8b. **Atkinson, B. G., Dean, R. L., and Blaker, T. W.,** Heat shock induced changes in the gene expression of terminally differentiating avian red blood cells, *Can. J. Genet. Cytol.*, 28, 1053—1063, 1987.

9. **Aujame, L.,** Murine plasmacytoma constitute a class of natural heat-shock variants in which the major inducible hsp-68 gene is not expressed, *Can. J. Genet. Cytol.,* 28, 1064—1075, 1987.
10. **Aujame, L. and Morgan, C.,** Nonexpression of major heat shock gene in mouse plasmacytoma MPC-11, *Mol. Cell. Biol.,* 5, 1780—1783, 1985.
11. **Ayme, A. and Tissieres, A.,** Locus 67B of *Drosophila melanogaster* contains seven, not four, closely related heat shock genes, *EMBO J.,* 4, 2949—2954, 1985.
12. **Bag, J.,** Regulation of heat-shock protein synthesis in chicken muscle culture during recovery from heat shock, *Eur. J. Biochem.,* 135, 373—378, 1983.
13. **Banerji, S. S., Theodorakis, N. G., and Morimoto, R. I.,** Heat shock-induced translational control of HSP70 and globin synthesis in chicken reticulocytes, *Mol. Cell. Biol.,* 4, 2437—2448, 1984.
14. **Banerji, S. S., Berg, L., and Morimoto, R. I.,** Transcription and post-transcriptional regulation of avian HSP70 gene expression, *J. Biol. Chem.,* 261, 15740—15745, 1986.
15. **Banerji, S. S., Laing, K., and Morimoto, R. I.,** Erythroid lineage specific expression and inducibility of the major heat shock protein HSP 70 during avian embryogenesis, *Genes Devel.,* 1, 946—953, 1987.
16. **Bedard, P. A. and Brandhorst, B. P.,** Translational activation of maternal mRNA encoding the heat-shock protein hsp90 during sea urchin embryogenesis, *Dev. Biol.,* 117, 286—293, 1986.
17. **Behnel, H. J. and Wekbart, G.,** Induced stabilization of the transmembrane potential of *Drosophila* cells by heat shock and periodic applications of chloramphenicol, *J. Cell Sci.,* 87, 197—201, 1987.
18. **Bensaude, O. and Morange, M.,** Spontaneous high expression of heat-shock proteins in mouse embryonal carcinoma cells and ectoderm from day 8 mouse embryo, *EMBO J.,* 2, 173—177, 1983.
19. **Bensaude, O., Babinet, C., Morange, M., and Jacob, F.,** Heat shock proteins, first major products of cygotic gene activity in mouse embryo, *Nature,* 305, 331—333, 1983.
20. **Ben-Zeev, A.,** The relationship between cytoplasmic organization, gene expression and morphogenesis, *Trends Biochem. Sci.,* 11, 478—481, 1986.
21. **Bergh, S. and Arking, R.,** Developmental profile of the heat shock response in early embryos of *Drosophila, J. Exp. Zool.,* 231, 379—391, 1984.
22. **Bienz, M.,** *Xenopus* hsp70 genes are constitutively expressed in injected oocytes, *EMBO J.,* 3, 2477—2483, 1984.
23. **Bienz, M.,** Developmental control of the heat shock response in *Xenopus, Proc. Natl. Acad. Sci. U.S.A.,* 81, 3138—3142, 1984.
24. **Bienz, M.,** A CCAAT box confers cell-type-specific regulation on the *Xenopus* hsp70 gene in oocytes, *Cell,* 46, 1037—1042, 1986.
25. **Bienz, M. and Gurdon, J. B.,** The heat shock response in *Xenopus* oocytes is controlled at the translational level, *Cell,* 29, 811—819, 1982.
26. **Blackman, R. K. and Meselson, M.,** Interspecific nucleotide sequence comparisons used to identify regulatory and structural features of the *Drosophila* hsp83, *J. Mol. Biol.,* 188, 499—516, 1986.
27. **Bonato, M. C. M., Silva, A. M., Gomes, S. L., Maia, J. C. C., and Juliani, M. H.,** Differential expression of heat-shock proteins and spontaneous synthesis of HSP70 during the life cycle of *Blastocladiella emersonii, Eur. J. Biochem.,* 163, 211—220, 1987.
28. **Bonner, J. J.,** Regulation of the *Drosophila* heat-shock response, in *Heat Shock from Bacteria to Man,* Schlesinger, M. J., Ashburner, M., and Tissieres, A., Eds., Cold Spring Harbor Laboratory, Cold Spring Harbor, NY, 1982, 147—153.
29. **Bonner, J.,** Mechanism of transcriptional control during heat shock, in *Changes in Eukaryotic Gene Expression in Response to Environmental Stress,* Atkinson, B. G. and Walden, D. W., Eds., Academic Press, Orlando, FL, 1985.

29a. **Boon-Niermeijer, E. K., De Waal, A. M., Souren, J. E. M., and Van Wijk, R.,** Heat-induced changes in thermosensitivity and gene expression during development, *Devel. Growth Differ.,* 30, 705—715, 1988.

29b. **Boorstein, W. R. and Craig, E. A.,** Transcriptional regulation of SSA3, an HSP70 gene from *Saccharomyces cerevisiae, Mol. Cell. Biol.,* 10, 3262—3267, 1990.

30. **Brawerman, G.,** mRNA decay: Finding the right targets, *Cell,* 57, 9—10, 1989.

30a. **Browder, L. W., Pollock, M., Nickells, R. W., Heikkila, J. J., and Winning, R. S.,** Developmental regulation of the heat shock response, in *Developmental Biology: A Comprehensive Synthesis,* Vol 6, Di Berardina, M. A. and Etkin, L., Eds., Plenum Press, New York, 1988.

30b. **Browder, L. W., Pollock, M., Heikkila, J. J., Wilkes, J., Wang, T., Krone, P., Ovsenek, N., and Kloc, M.,** Decay of the oocyte-type heat shock response of *Xenopus laevis, Dev. Biol.,* 124, 191—199, 1987.

31. **Brown, I. R., Rush, S., and Ivy, G. O.,** Induction of a heat shock gene at the site of tissue injury in the rate brain, *Neuron,* 2, 1559—1564, 1989.

31a. **Brunt, S. A. and Silver, J. C.,** Cellular localization of steroid hormone-regulated proteins during sexual development in Achlya, *Exp. Cell Res.,* 165, 306—319, 1986.

31b. **Brunt, S. A., Riehl, R., and Silver, J. C.,** Steroid hormone regulation of the *Achlya ambisexualis* 85 kDa heat shock protein, a component of the *Achlya* steroid receptor complex, *Mol. Cell. Biol.,* 10, 273—281, 1990.

31c. **Catelli, M. G., Ramachandran, C., Gauthier, Y., Legagneux, V., Quelard, C., Baulieu, E.-E., and Shyamala, G.,** Developmental regulation of murine mammary gland 90 kDa heat shock proteins, *Biochem. J.*, 258, 895—901, 1981.

32. **Cheney, C. M. and Shearn, A.,** Developmental regulation of *Drosophila* imaginal disk proteins: Synthesis of a heat-shock protein under nonheat-shock conditions, *Dev. Biol.*, 95, 325—330, 1983.

32a. **Cleveland, D. W.,** Autoregulated instability of tubulin mRNAs: a novel eukaryotic regulatory mechanism, *Trends Biochem. Sci.*, 13, 339—343, 1988.

32b. **Craig, E., Boorstein, W., Park, H.-O., Stone, D., and Nicolet, C.,** Complex regulation of three heat inducible HSP70 related genes in *Saccharomyces cerevisiae*, in *Stress-Induced Proteins*, Pardue, M. L., Feramisco, J. R., and Lindquist, S., Eds., Alan R. Liss, NY, 1989, 55—61.

33. **Cohen, R. S. and Meselson, M.,** Separate regulatory elements for the heat-inducible and ovarian expression of the *Drosophila* hsp 26 gene, *Cell*, 43, 737—746, 1985.

33a. **Coias, R., Galego, L., Barchona, I., and Rodrigues-Pousada, C.,** Destabilization of tubulin mRNA during heat-shock in *Tetrahymena pyriformis*, *Eur. J. Biochem.*, 175, 467—474, 1988.

34. **Colbert, R. A. and Young, D. A.,** Detection of mRNAs coding for translationally regulated heat-shock proteins in non-heat-shocked thymic lymphocytes, *J. Biol. Chem.*, 262, 9939—9941, 1987.

35. **Collier, N. C. and Schlesinger, M. J.,** Induction of heat-shock proteins in the embryonic chicken lens, *Exp. Eye Res.*, 43, 103—117, 1986.

36. **Cooper, P. and Ho, T.-H. D.,** Heat shock proteins in maize, *Plant Physiol.*, 71, 215—222, 1983.

37. **Cooper, P., Ho, T.-H. D., and Hauptmann, R. M.,** Tissue specificity of the heat-shock response in maize, *Plant Physiol.*, 75, 431—441, 1984.

37a. **Darasch, S., Mosser, D. D., Bols, N. C., and Heikkila, J. J.,** Heat shock gene expression in *Xenopus laevis* A6 cells in response to heat shock and sodium arsenite, *Biochem. Cell Biol.*, 66, 862—870, 1988.

38. **Darmon, M. C. and Paulin, D. J.,** Translational activity of mRNA coding for cytoskeletal brain proteins in newborn and adult mice: A comparative study, *J. Neurochem.*, 44, 1672—1678, 1985.

39. **Dean, R. L. and Atkinson, B. G.,** The acquisition of thermal tolerance in larvae of *Calpodes ethlius* (*Lepidoptera*) and the in situ and in vitro synthesis, *Can. J. Biochem. Cell Biol.*, 61, 472—479, 1983.

39a. **Deguchi, Y., Negoro, S., and Kishimoto, S.,** Age-related changes of heat shock protein gene transcription in human peripheral blood mononuclear cells, *Biochem. Biophys. Res. Commun.*, 157, 580—584, 1988.

40. **Delpino, A., Mileo, A. M., Mattei, E., and Ferrini, U.,** Characterization of the heat shock response in M-14 human melanoma cells continuously exposed to supranormal temperature, *Exp. Mol. Pathol.*, 45, 128—141, 1986.

40a. **De Toledo, S. M., Murphy, L. J., Hatton, T. H., and Friesen, H. G.,** Regulation of 70-Kilodalton heat-shock like messenger ribonucleic acid in vitro and in vivo by prolactin, *Mol. Endocrinol.*, 1, 430—434, 1987.

41. **Devlin, R. H., Grigliatti, T. A., and Holm, D. G.,** Gene dosage compensation in trisomies of *Drosophila melanogaster*, *Dev. Genet.*, 6, 39—58, 1985.

42. **Di Domenico, B. J., Bugaisky, G. E., and Lindquist, S.,** The heat shock response is self-regulated at both the transcriptional and posttranscriptional levels, *Cell*, 31, 593—603, 1982.

43. **Di Domenico, B. J., Bugaisky, G. E., and Lindquist, S.,** Heat shock and recovery are mediated by different translational mechanisms, *Proc. Natl. Acad. Sci. U.S.A.*, 79, 6181—6185, 1982.

43a. **Dony, C., Kessel, M., and Gruss, P.,** Post-transcriptional control of myc and p53 expression during differentiation of the embryonal carcinoma cell line F9, *Nature*, 317, 636—639, 1985.

43b. **Duck, N., McCormick, S., and Winter, J.,** Heat shock protein hsp70 cognate gene expression in vegetative and reproductive organs of *Lycopersicon esculentum*, *Proc. Natl. Acad. Sci. U.S.A.*, 86, 3674—3678, 1989.

44. **Dura, J. M.,** Stage dependent synthesis of heat shock induced proteins in early embryos of *Drosophila melanogaster*, *Mol. Gen. Genet.*, 184, 381—385, 1981.

44a. **Erickson, J. W. and Gross, C. A.,** Identification of the sigmaE subunit of *Escherichia coli* RNA polymerase: a second alternate sigma factor involved in high-temperature gene expression, *Genes Dev.*, 3, 1462—1471, 1989.

45. **Evgen'Ev, M. B., Kolchinski, A., Levin, A., Preobrazenskaya, O., and Sarkisova, E.,** Heat-shock DNA homology in distantly related species of *Drosophila*, *Chromosoma*, 68, 357—365, 1978.

45a. **Ferris, D. K., Harel-Bellan, A., Morimoto, R. I., Welch, W. J., and Farrar, W. L.,** Mitogen and lymphokine stimulation of heat shock proteins in T lymphocytes, *Proc. Natl. Acad. Sci. U.S.A.*, 85, 3850—3854, 1988.

45b. **Fleming, J. E., Walton, J. K., Dubitsky, R., and Bensch, K. G.,** Aging results in an unusual expression of *Drosophila* heat shock proteins, *Proc. Natl. Acad. Sci. U.S.A.*, 85, 4099—4103, 1988.

45c. **Frova, C., Taramino, G., and Binelli, G.,** Heat-shock proteins during pollen development in maize, *Dev. Genet.*, 10, 324—332, 1989.

45d. **Fuqua, S. A. W., Blumsalingaros, M., and McGuire, W. L.,** Induction of the estrogen-related 24K protein by heat shock, *Cancer Res.*, 49, 4126—4129, 1989.

46. **Gay, D. A., Yen, T. J., Lau, T. Y., and Cleveland, D. W.,** Sequences that confer beta-tubulin autoregulation through modulated mRNA stability reside within exon 1 of a beta-tubulin mRNA, *Cell,* 50, 671—679, 1987.

46a. **Giebel, L. B., Dworniczak, B. P., and Bautz, E. K. F.,** Developmental regulation of a constitutively expressed mouse mRNA encoding a 72-kDa heat shock-like protein, *Dev. Biol.,* 125, 200—207, 1988.

46b. **Glaser, R. L. and Lis, J. T.,** Multiple, compensatory regulatory elements specify spermatocyte-specific expression of the *Drosophila melanogaster* hsp26 gene, *Mol. Cell. Biol.,* 10, 131—137, 1990.

47. **Glaser, R. L., Wolfner, M. F., and Lis, J. T.,** Spatial and temporal pattern of hsp 26 expression during normal development, *EMBO J.,* 5, 747—754, 1986.

48. **Graziosi, G., Decristini, F., Di Marcotullio, A., Marzari, R., Micali, F., and Savoini, A.,** Morphological and molecular modifications induced by heat shock in *Drosophila melanogaster* embryos, *J. Embryol. Exp. Morphol.,* 77, 167—182, 1983.

49. **Greenberg, S. G. and Lasek, R. L.,** Comparison of labelled heat-shock proteins in neuronal and non-neuronal cells of *Aplysia californica, J. Neurosci.,* 5, 1239—1245, 1985.

50. **Grossman, A. D., Straus, D. B., Walter, W. A., and Gross, C. A.,** Sigma 32 synthesis can regulate the synthesis of heat shock proteins in *Escherichia coli, Genes Devel.,* 1, 179—184, 1987.

51. **Hahnel, A. C., Gifford, D. J., Heikkila, J. J., and Schultz, G. A.,** Expression of the major heat shock protein (hsp70) family during early mouse development, *Teratog. Cancerog. Mutag.,* 6, 493—510, 1986.

51a. **Haire, R. N., Peterson, M. S., and O'Leary, J. J.,** Mitogen activation induces the enhanced synthesis of two heat-shock proteins in human lymphocytes, *J. Cell Biol.,* 106, 883—891, 1988.

52. **Hallberg, R. L., Kraus, K. W., and Findly, R. C.,** Starved *Tetrahymena thermophila* cells that are unable to mount an effective heat shock response selectively degrade their r-RNA, *Mol. Cell. Biol.,* 4, 2170—2179, 1984.

53. **Hatayama, T., Honda, K., and Yukioka, M.,** HeLa cells synthesize a specific heat shock protein upon exposure to heat shock at 42°C but not at 45°C, *Biochem. Biophys. Res. Commun.,* 137, 957—963, 1986.

53a. **Heikkila, J. J.,** personal communication.

54. **Heikkila, J. J. and Schultz, G. A.,** Different environmental stresses can activate the expression of a heat shock gene in rabbit blastocysts, *Gamete Res.,* 10, 45—56, 1984.

55. **Heikkila, J. J., Kloc, M., Bury, J., Schultz, G. A., and Browder, L. W.,** Acquisition of the heat shock response and thermotolerance during early development of *Xenopus laevis, Dev. Biol.,* 107, 483—489, 1985.

55a. **Heikkila, J. J., Ovsenek, N., and Krone, P.,** Examination of heat shock protein mRNA accumulation in early *Xenopus laevis* embryos, *Biochem. Cell Biol.,* 65, 87—94, 1987.

55b. **Heikkila, J. J., Darasch, S. P., Mosser, D. D., and Bols, N. C.,** Heat and sodium arsenite act synergistically on the induction of heat shock gene expression in *Xenopus laevis* A6 cells, *Biochem. Cell Biol.,* 65, 310—316, 1987.

55c. **Hightower, L.,** personal communication.

56. **Horrell, A., Shuttleworth, J., and Colman, A.,** Transcript levels and translational control of hsp 70 synthesis in *Xenopus* oocytes, *Genes Devel.,* 1, 433—444, 1987.

57. **Iida, H. and Yahara, I.,** Durable synthesis of high molecular weight heat shock proteins in GO cells of the yeast and other eukaryotes, *J. Cell Biol.,* 99, 199—207, 1984.

58. **Iida, H. and Yahara, I.,** A heat shock-resistant mutant of Saccharomyces cerevisiae shows constitutive synthesis of two heat shock proteins and altered growth, *J. Cell Biol.,* 99, 1441—1450, 1984.

59. **Imperiale, M. J., Kao, H.-T., Feldman, L. T., Nevins, J. R., and Strickland, S.,** Common control of the heat shock gene and early adenovirus genes: Evidence for a cellular E1A-like activity, *Mol. Cell. Biol.,* 4, 867—874, 1984.

60. **Ireland, R. C. and Berger, E. M.,.** Synthesis of low molecular weight heat shock peptides stimulated by ecdysterone in a cultured *Drosophila* cell line, *Proc. Natl. Acad. Sci. U.S.A.,* 79, 855—859, 1982.

61. **Ireland, R. C., Berger, E., Sirotkin, K., Yund, M. A., Osterbur, D., and Fristrom, J.,** Ecdysterone induces the transcription of four heat-shock genes in *Drosophila* S3 cells and imaginal disks, *Dev. Biol.,* 93, 498—507, 1982.

62. **Kaczmarek, L., Calabretta, B., Kao, H.-T., Heintz, N., Nevins, J., and Baserga, R.,** Control of hsp 70 RNA levels in human lymphocytes, *J. Cell Biol.,* 107, 183—189, 1987.

62a. **Kaddurah-Daouk, R., Greene, J. M., Baldwin, A. S., and Kingston, R. E.,** Activation and repression of mammalian gene expression by the c-myc protein, *Genes Dev.,* 1, 347—357, 1987.

63. **Kao, H.-T. and Nevins, J. R.,** Transcriptional activation and subsequent control of the human heat shock gene during adenovirus infection, *Mol. Cell. Biol.,* 3, 2058—2065, 1983.

64. **Kao, H.-T., Capasso, O., Heintz, N., and Nevins, J. R.,** Cell cycle control of the human HSP70 gene: Implications for the role of a cellular E1A-like function, *Mol. Cell. Biol.,* 5, 628—633, 1985.

65. **Kimpel, J. A. and Key, J. L.,** Presence of heat shock messenger RNA species in field grown soybeans *Glycine max, Plant Physiol.,* 79, 672—678, 1985.

66. **King, M. L. and Davis, R.,** Do *Xenopus* oocytes have a heat shock response?, *Dev. Biol.,* 119, 532—539, 1987.
66a. **Kingston, R. E., Baldwin, A. S., and Sharp, P. A.,** Regulation of heat shock protein 70 gene expression by c-myc, *Nature,* 312, 280—282, 1984.
67. **Kingston, R. E., Cowie, A., Morimoto, R. I., and Gwinn, K. A.,** Binding of polyomavirus large T antigen to the human hsp 70 promoter is not required for transactivation, *Mol. Cell. Biol.,* 6, 3180—3190, 1986.
68. **Klemenz, R. and Gehring, W. J.,** Sequence requirement for expression of the *Drosophila melanogaster* heat shock protein hsp22 gene during heat shock and normal development, *Mol. Cell. Biol.,* 6, 2011—2019, 1986.
69. **Kothary, R. K., Jones, D., and Candido, E. P. M.,** 70-Kilodalton heat shock polypeptides from rainbow trout: Characterization of cDNA sequences, *Mol. Cell. Biol.,* 4, 1785—1791, 1984.
70. **Kothary, R., Perry, M. D., Moran, L. A., and Rossant, J.,** Cell-lineage-specific expression of the mouse hsp 68 gene during embryogenesis, *Dev. Biol.,* 121, 342—348, 1987.
71. **Krawczyk, Z., Szymik, N., and Wisniewski, J.,** Expression of hsp70-regulated gene in developing and degenerating rat testis, *Mol. Biol. Rep.,* 12, 35—41, 1987.
71a. **Krone, P. H. and Heikkila, J. J.,** Analysis of hsp30, hsp70 and ubiquitin gene expression in *Xenopus laevis* tadpoles, *Development,* 103, 59—67, 1988.
72. **Krueger, C. and Benecke, B. J.,** Translation and turnover of Drosophila heat-shock and non-heat-shock mRNAs, in *Heat Shock from Bacteria to Man,* Schlesinger, M. J., Ashburner, M., and Tissieres, A., Eds., Cold Spring Harbor Laboratory, Cold Spring Harbor, NY, 1982, 191—197.
73. **Kurtz, S. and Lindquist, S.,** Changing patterns of gene expression during sporulation in yeast, *Proc. Natl. Acad. Sci. U.S.A.,* 81, 7323—7327, 1984.
74. **Kurtz, S., Rossi, J., Petko, L., and Lindquist, S.,.** An ancient developmental induction: Heat-shock proteins induced in sporulation and oogenesis, *Science,* 231, 1154—1157, 1986.
75. **Lasek, R. J. and Tytell, M.,** Macromolecular transfer from glia to axon, *J. Exp. Biol.,* 95, 153—165, 1981.
75a. **Lee, S.-J.,** Expression of HSP86 in male germ cells, *Mol. Cell. Biol.,* 10, 3239—3242, 1990.
76. **Lewis, M., Helmsing, P. J., and Ashburner, M.,** Parallel changes in puffing activity and patterns of protein synthesis in salivary glands of *Drosophila, Proc. Natl. Acad. Sci. U.S.A.,* 72, 3604—3608, 1975.
77. **Lindquist, S.,** Varying patterns of protein synthesis in *Drosophila* during heat shock: Implications for regulation, *Dev. Biol.,* 77, 463—479, 1980.
77a. **Liu, A. Y. C., Lin, Z., Choi, H. S., Sorhage, F., and Li, B. S.,** Attenuated induction of heat shock gene expression in aging diploid fibroblasts, *J. Biol. Chem.,* 264, 12037—12045, 1989.
77b. **Maekawa, M., O'Brien, D. A., Allen, R. L., and Eddy, E. M.,** Heat-shock cognate protein (Hsc71) and related proteins in mouse spermatogenic cells, *Biol. Reprod.,* 40, 843—852, 1989.
78. **Mason, P. J., Hall, L. M. C., and Gausz, J.,** The expression of heat shock genes during normal development in *Drosophila melanogaster, Mol. Gen. Genet.,* 194, 73—78, 1984.
78a. **Matsumoto, M. and Fujimoto, H.,** Cloning of an HSP70-related gene expressed in mouse spermatids, *Biochem. Biophys. Res. Commun.,* 166, 43—49, 1990.
79. **Maytin, E. V., Colbert, R. A., and Young, D. A.,** Early heat shock proteins in primary thymocytes. Evidence for transcriptional and translational regulation, *J. Biol. Chem.,* 260, 2384—2392, 1985.
79a. **McDaniel, D.-A., Caplan, A. J., Lee, M.-S., Adams, C. C., Fishel, B. R., Gross, D. S., and Garrard, W. T.,** Basal-level expression of the yeast HSP82 gene requires a heat shock regulatory element, *Mol. Cell. Biol.,* 9, 4789—4798, 1989.
79b. **McGuire, S. E., Fugua, S. A. W., Naylor, S. L., Helin-Davis, D. A., and McGuire, W. L.,** Chromosomal assignments of human 27-kDa heat shock protein gene family, *Somat. Cell Mol. Genet.,* 15, 167—171, 1989.
80. **Mestril, R., Schiller, P., Amin, J., Klapper, H., Ananthan, J., and Voellmy, R.,** Heat shock and ecdysterone activation of the *Drosophila melanogaster* hsp 23 gene. A sequence element implicated developmental regulation, *EMBO J.,* 5, 1667—1673, 1986.
80a. **Mezger, V., Bensaude, O., and Morange, M.,** Deficient activation of heat shock gene transcription in embryonal carcinoma cells, *Dev. Biol.,* 124, 544—550, 1987.
80b. **Mezger, V., Bensaude, O., and Morange, M.,** Unusual levels of heat shock element-binding activity in embryonal carcinoma cells, *Mol. Cell. Biol.,* 9, 3888—3896, 1989.
81. **Meyer, J. and Chartier, Y.,** Long-lived and short-lived heat shock proteins in tobacco mesophyll protoplasts, *Plant Physiol.,* 72, 26—32, 1983.
82. **Milarski, K. L. and Morimoto, R. I.,** Expression of human HSP70 during the synthetic phase of the cell cycle, *Proc. Natl. Acad. Sci. U.S.A.,* 83, 9517—9522, 1986.
83. **Mizzen, L. A. and Welch, W. J.,** Characterization of the thermotolerant cell. I. Effects on protein synthesis activity and the regulation of HSP70 expression, *J. Cell Biol.,* 106, 1105—1116, 1988.

84. **Mobbs, C. V., Romano, G. J., Schwartz-Giblin, S., and Pfaff, D. W.,** Biochemistry of a steroid-regulated mammalian mating behavior; heat shock proteins and secretion, enkephalin and GABA, in *Neural Control of Reproductive Function,* Alan R. Liss, New York, 1989, 95—116.
85. **Morange, M., Diu, A., Bensaude, O., and Babinet, C.,** Altered expression of heat shock proteins in embryonal carcinoma and mouse early embryonic cells, *Mol. Cell. Biol.,* 4, 730—735, 1984.
86. **Morgan, W. D., Williams, G. T., Morimoto, R. I., Greene, J., Kingston, R. E., and Tjian, R.,** Two transcriptional activators, CCAAT-box-binding transcription factor and heat shock transcription factor, interact with a human hsp 70 gene promoter, *Mol. Cell. Biol.,* 7, 1129—1138, 1987.
87. **Morimoto, R. and Fodor, E.,** Cell-specific expression of heat shock proteins in chicken reticulocytes and lymphocytes, *J. Cell Biol.,* 99, 1316—1323, 1984.
87a. **Morimoto, R., Mosser, D.,, McClanahan, T. K., Theodorakis, N. G., and Williams, G.,** Transcriptional regulation of the human hsp70 gene, in *Stress-Induced Proteins,* Pardue, M. L., Feramisco, J. R., and Lindquist, S., Alan R. Liss, New York, 1989, 83—94.
87b. **Mosser, D. D., Heikkila, J. J., and Bols, N. C.,** Temperature ranges over which rainbow trout fibroblasts survive and synthesize heat-shock proteins, *J. Cell. Physiol.,* 128, 432—440, 1986.
88. **Necchi, A., Pogna, N. E., and Mapelli, S.,** Early and late heat shock proteins in wheats and other cereal species, *Plant Physiol.,* 84, 1378—1384, 1987.
89. **Neidhardt, F. C., Van Bogelen, R. A., and Vaughn, V.,** The genetics and regulation of heat-shock proteins, *Annu. Rev. Genet.,* 18, 295—329, 1984.
90. **Nevins, J. R.,.** Induction of the synthesis of a 70,000 dalton mammalian heat shock proteins by the adenovirus E1A gene product, *Cell,* 29, 913—919, 1982.
91. **Nickells, R. W. and Browder, L. W.,** Region-specific heat-shock protein synthesis correlates with a biphasic acquisition of thermotolerance in *Xenopus laevis, Dev. Biol.,* 112, 391—395, 1985.
91a. **Nickells, R. W. and Browder, L. W.,** A role for glyceraldehyde-3-phosphate dehydrogenase in the development of thermotolerance in *Xenopus laevis* embryos, *J. Cell Biol.,* 107, 1901—1910, 1988.
92. **Nover, L., Ed.,** *Heat Shock Response of Eukaryotic Cells,* Springer-Verlag, New York, 1984.
93. **Nover, L. and Scharf, K.-D.,** Synthesis, modification and structural binding of heat shock proteins in tomato cell cultures, *Eur. J. Biochem.,* 139, 303—313, 1984.
93a. **Ohno, J., Fukushima, M., Fujiwara, M., and Narumiya, S.,** Induction of 68,000-dalton heat shock proteins by cyclopentenone prostaglandins. Its association with prostaglandin-induced G-1 block in cell cycle progression, *J. Biol. Chem.,* 263, 19764—19770, 1988.
93b. **Otto, B., Grimm, B., Ottersbach, P., and Kloppstech, K.,** Circadian control of the accumulation of messenger RNAs for light-inducible and heat-inducible chloroplast proteins in pea (*Pisum sativum* L.), *Planta Physiol,* 88, 21—25, 1988.
94. **Paek, K.-H. and Walker, G. C.,** *Escherichia coli* dnaK null mutants are inviable at high temperature, *J. Bacteriol.,* 169, 283—290, 1987.
95. **Palter, K. B., Watanabe, M., Stinson, L., Mahowald, A. P., and Craig, E. A.,** Expression and localization of *Drosophila melanogaster* hsp70 cognate proteins, *Mol. Cell. Biol.,* 6, 1187—1203, 1986.
96. **Pauli, D. and Tonka, C.-H.,** A new *Drosophila* heat shock gene from locus 67B is expressed during embryogenesis and pupation, *J. Mol. Biol.,* 198, 253—240, 1987.
97. **Pauli, D., Tonka, C.-H., and Ayme-Southgate, A.,** An unusual split *Drosophila* heat shock gene expressed during embryogenesis, pupation and in testis, *J. Mol. Biol.,* 200, 47—53, 1988.
98. **Pitto, L., Loschiavo, F., Gioliano, G., and Terzi, T.,** Analysis of heat-shock protein pattern during somatic embryogenesis of carrot, *Plant Mol. Biol.,* 2, 231—237, 1983.
98a. **Ramachandran, C., Catelli, M. G., Schneider, W., and Shyamala, G.,** Estrogenic regulation of uterine 90-kilodalton heat shock protein, *Endocrinology,* 123, 956—961, 1988.
99. **Raschke, E., Baumann, G., and Schoeffl, F.,** Nucleotide sequence analysis of soybean small heat shock protein genes belonging to two different multigene families, *J. Mol. Biol.,* 199, 549—557, 1988.
100. **Reiter, T. and Penman, S.,** "Prompt" heat shock proteins: Translationally regulated synthesis of new proteins associated with nuclear matrix-intermediate filaments as an early response to heat shock, *Proc. Natl. Acad. Sci. U.S.A.,* 80, 4737—4741, 1983.
101. **Rensing, L., Olomski, R., and Drescher, K.,** Kinetics and models of the *Drosophila* heat shock system, *Biosystems,* 15, 341—356, 1982.
102. **Riddihough, G. and Pelham, H. R. B.,** Activation of the *Drosophila* hsp 27 promoter by heat shock and ecdysone involves independent and remote regulatory sequences, *EMBO J.,* 5, 1653—1658, 1986.
102a. **Riddihough, G. and Pelham, H. R. B.,** An ecdysone response element in the *Drosophila* hsp 27 promoter, *EMBO J.,* 6, 3729—3734, 1987.
103. **Ritossa, F.,** A new puffing pattern induced by heat shock and DNP in *Drosophila, Experientia,* 18, 571—573, 1962.
104. **Roccheri, M. C., Sconzo, G., Larosa, M., Oliva, D., Abrignani, A., and Giudice, G.,** Response to heat shock of different sea urchin species, *Cell Differ.,* 18, 131—135, 1986.

105. **Santoro, M. G., Garaci, E., and Amici, C.,** Prostaglandins with anti-proliferative activity induce the synthesis of a heat shock protein in human cells, *Proc. Natl. Acad. Sci. U.S.A.*, 86, 8407—8411, 1989.

105a. **Scharf, K. D., et al.,** unpublished.

106. **Schlesinger, M. J., Ashburner, M., and Tissieres, A., Eds.,** *Heat Shock from Bacteria to Man,* Cold Spring Harbor Laboratory, Cold Spring Harbor, NY, 1982.

107. **Sconzo, G., Roccheri, M. C., Oliva, D., La Rosa, M., and Giudice, G.,** Territorial localization of heat shock mRNA production in sea urchin gastrulae, *Cell Biol. Int. Rep.*, 9, 877—881, 1985.

108. **Shaw, G. and Kamen, R.,** A conserved AU sequence from the 3′ untranslated region of GM-CSF mRNA mediates selectivity in RNA degradation, *Cell,* 46, 659—667, 1986.

108a. **Shyamala, G., Gauthier, Y., Moore, S. K., Catelli, M. G., and Ullrich, S. J.,** Estrogenic regulation of murine uterine 90-kD heat shock protein gene expression, *Mol. Cell. Biol.*, 9, 3567—3570, 1989.

109. **Simcox, A. A., Cheney, C. M., Hoffman, E. P., and Shearn, A.,** A deletion of the 3′end of the *Drosophila melanogaster* hsp70 gene increases stability of mutant mRNA during recovery from heat shock, *Mol. Cell. Biol.*, 5, 3397—3402, 1985.

110. **Simon, M. C., Kitchener, K., Kao, H.-T., Hickey, E., Weber, L., Voellmy, R., Heintz, N., and Nevins, J. R.,** Selective induction of human heat shock gene transcription by the adenovirus E1A gene products, including the 12S E1A product, *Mol. Cell. Biol.*, 7, 2884—2890, 1987.

110a. **Simon, M. C., Fisch, T. M., Benecke, B. J., Nevins, J. R., and Heintz, N.,** Definition of multiple functionally distinct TATA elements, one of which is a target in the hsp70 promoter for E1A regulation, *Cell,* 52, 723—729, 1988.

111. **Singh, M. K. and Yu, J.,** Accumulation of heat shock-like protein during differentiation of human erythroid cell line K562, *Nature,* 309, 631—633, 1984.

112. **Sirotkin, K. and Davidson, N.,** Developmentally regulated transcription from *Drosophila melanogaster* chromosomal site 67B, *Dev. Biol.*, 89, 196—210, 1982.

113. **Sondermeijer, P. J. A. and Lubsen, N. H.,** Heat-shock peptides in *Drosophila hydei* and their synthesis in vitro, *Eur. J. Biochem.*, 88, 331—339, 1978.

114. **Southgate, R., Ayme, A., and Voellmy, R.,** Nucleotide sequence analysis of the *Drosophila* small heat shock gene cluster at locus 67B, *J. Mol. Biol.*, 165, 35—57, 1983.

115. **Sprang, G. K. and Brown, I. R.,** Selective induction of a heat shock gene in fibre tracts and cerebellar neurons of the rabbit brain detected by in situ hybridization, *Mol. Brain Res.*, 3, 89—93, 1987.

115a. **Stone, D. E. and Craig, E. A.,** Self-regulation of 70-kilodalton heat shock proteins in *Saccharomyces cerevisiae, Mol. Cell. Biol.*, 10, 1622—1632, 1990.

116. **Storti, R. V., Scott, M. P., Rich, A., and Pardue, M. L.,** Translational control of protein synthesis in response to heat shock in *D. melanogaster* cells, *Cell,* 22, 825—834, 1980.

116a. **Tanaka, K., Matsumoto, K., and Tohe, A.,** Dual regulation of the expression of the polyubiquitin gene by cyclic AMP and heat shock in yeast, *EMBO J.*, 7, 495—502, 1988.

116b. **Tanaka, K., Yatomi, T., Matsumoto, K., and Toh-E, A.,** Transcriptional regulation of the heat shock genes by cyclic AMP and heat shock in yeast, in *Stress-Induced Proteins,* Pardue, M. L., Feramisco, J. R., and Lindquist, S., Eds., Alan R. Liss, New York, 1989, 63—72.

117. **Theodorakis, N. G. and Morimoto, R. I.,** Post-transcriptional regulation of HSP 70 expression in human cells: Effects of heat shock, inhibition of protein synthesis, and adenovirus infection on translation and mRNA stability, *Mol. Cell. Biol.*, 7, 4357—4368, 1987.

117a. **Theodorakis, N. G., Banerji, S. S., and Morimoto, R. I.,** HSP70 mRNA translation in chicken reticulocytes is regulated at the level of elongation, *J. Biol. Chem.*, 263, 14579—14585, 1988.

117b. **Theodorakis, N. G., Zand, D. J., Kotzbauer, P. T., Williams, G. T., and Morimoto, R. I.,** Hemin-induced transcriptional activation of the hsp70 gene during erythroid maturation in K526 cells is due to a heat shock factor-mediated stress response, *Mol. Cell. Biol.*, 9, 3166—3173, 1989.

118. **Thomas, S. R. and Lengyel, J. A.,** Ecdysteroid-regulated heat-shock gene expression during *Drosophila melanogaster* development, *Dev. Biol.*, 115, 434—438, 1986.

119. **Tilly, K., McKittrick, M., Zylicz, M., and Georgopoulos, C.,** The dnaK protein modulates the heat-shock response of *Escherichia coli, Cell,* 34, 641—646, 1983.

119a. **Tytell, M. and Lasek, R. J.,** Glia polypeptides transferred into the squid giant axon, *Brain Res.*, 324, 223—232, 1984.

120. **Tytell, M., Greenberg, S. G., and Lasek, R. J.,** Heat shock-like protein is transferred from glia to axon, *Brain Res.*, 363, 161—164, 1986.

121. **Ulberg, L. C. and Sheean, L. A.,** Early development of mammalian embryo in elevated temperatures, *J. Reprod. Fertil. Suppl.*, 19, 155—161, 1973.

122. **Velazquez, J. M. and Lindquist, S.,** Hsp70: Nuclear concentration during environmental stress and cytoplasmic storage during recovery, *Cell,* 36, 655—662, 1984.

123. **Velazquez, J., Sonoda, S., Bugaisky, G. E., and Lindquist, S.,** Is the major *Drosophila* heat shock protein present in cells that have not been heat shocked?, *J. Cell Biol.*, 96, 286—290, 1983.

124. **Voellmy, R. and Bromley, P. A.,** Massive heat-shock polypeptide synthesis in late chicken embryos: convenient system for study of protein synthesis in highly differentiated organisms, *Mol. Cell. Biol.*, 2, 479—483, 1982.

125. **Voellmy, R.,** The heat shock genes: a family of highly conserved genes with a superbly complex expression pattern, *Bio Essays,* 1, 213—217, 1985.

125a. **Werner-Washburne, M., Becker, J., Kosic-Smithers, J., and Craig, E. A.,** Yeast Hsp70 RNA levels vary in response to the physiological status of the cell, *J. Bacteriol.*, 171, 2680—2688, 1989.

125b. **White, C. N. and Hightower, L. E.,** Stress mRNA metabolism in canavanine-treated chicken embryo cells, *Mol. Cell. Biol.*, 4, 1534—1541, 1984.

126. **Widelitz, R. B., Duffy, J. J., and Gerner, E. W.,** Accumulation of heat shock protein 70 RNA and its relationship to protein synthesis after heat shock in mammalian cells, *Exp. Cell Res.*, 168, 539—545, 1987.

126a. **Williams, G. T., McClanahan, T. K., and Morimoto, R. I.,** E1a transactivation of the human HSP70 promoter is mediated through the basal transcription complex, *Mol. Cell. Biol.*, 9, 2574—2587, 1989.

127. **Winning, R. S. and Browder, L. W.,** Changes in heat shock protein synthesis and hsp70 gene transcription during erythropoiesis of *Xenopus laevis, Dev. Biol.*, 128, 111—120, 1988.

128. **Wittig, S., Hensse, S., Keitel, C., Elsner, C., and Wittig, B.,** Heatshock gene expression is regulated during teratocarcinoma cell differentiation and early embryonic development, *Dev. Biol.*, 96, 507—514, 1983.

129. **Wu, B. J. and Morimoto, R. I.,** Transcription of the human hsp70 gene is induced by serum stimulation, *Proc. Natl. Acad. Sci. U.S.A.*, 82, 6070—6074, 1985.

130. **Wu, B. J., Hurst, H. C., Jones, N. C., and Morimoto, R. I.,** The E1A 13S product of adenovirus 5 activates transcription of the cellular human HSP70 gene, *Mol. Cell. Biol.*, 6, 2994—2999, 1986.

131. **Wu, B. J., Gregg, T. W., and Morimoto, R. J.,** Detection of three protein binding sites in the serum-regulated promoter of human gene encoding the 70-kDa heat shock protein, *Proc. Natl. Acad. Sci. U.S.A.*, 84, 2203—2207, 1987.

132. **Xiao, C. M. and Mascarenhas, J. P.,** High temperature-induced thermotolerance in pollen tubes of *Tradescantia* and heat-shock proteins, *Plant Physiol.*, 78, 887—890, 1985.

132a. **Xiao, H. and Lis, J. T.,** Closely related DNA sequences specify distinct patterns of developmental expression in *Drosophila melanogaster, Mol. Cell. Biol.*, 10, 3272—3276, 1990.

133. **Yost, H. J. and Lindquist, S.,** RNA splicing is interrupted by heat shock and is rescued by heat shock protein synthesis, *Cell,* 45, 185—193, 1986.

134. **Zakeri, Z. F. and Wolgemuth, D. J.,** Developmental-stage-specific expression of the hsp70 gene family during differentiation of the mammalian male germ line, *Mol. Cell. Biol.*, 7, 1791—1796, 1987.

134a. **Zakeri, Z. F., Wolgemuth, D. J., and Hunt, C. R.,** Identification and sequence analysis of a new member of the mouse HSP70 gene family and characterization of its unique cellular and developmental pattern of expression in the male germ line, *Mol. Cell. Biol.*, 8, 2925—2932, 1988.

135. **Zimmerman, J. L., Petri, W., and Meselson, M.,** Accumulation of a specific subset of *D. melanogaster* heat shock mRNAs in normal development without heat shock, *Cell,* 32, 1161—1170, 1983.

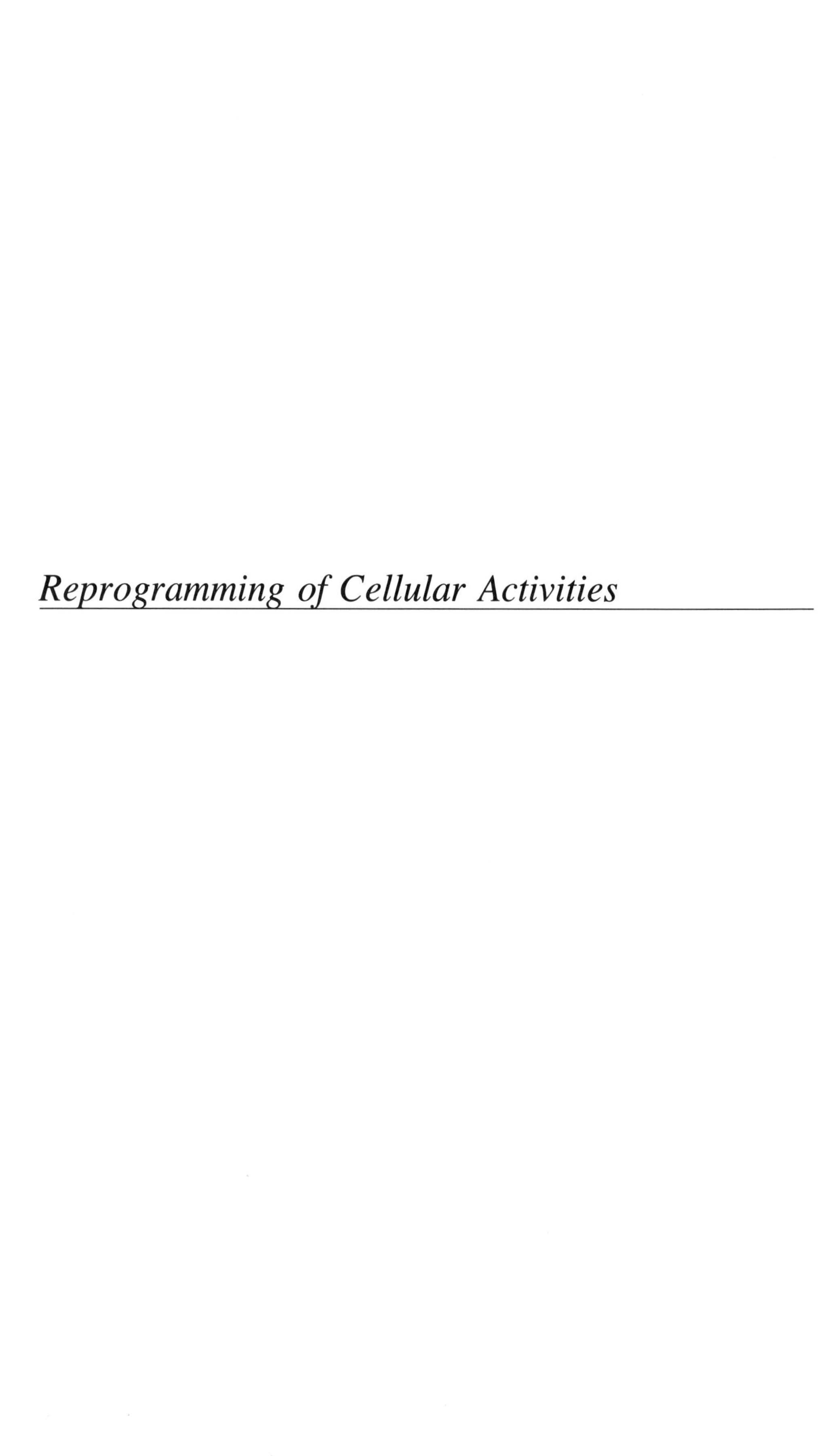

Reprogramming of Cellular Activities

Chapter 9

CHROMATIN STRUCTURE, TRANSCRIPTION, AND PRE-mRNP PROCESSING

L. Nover

TABLE OF CONTENTS

9.1. REPROGRAMMING OF TRANSCRIPTION

At the onset of a fascinating 25-year development of the molecular biology of cellular stress response was the observation, by light microscopy, of the heat shock (hs)-induced reprogramming of transcription in *Drosophila* larvae.[72] With extraordinary rapidity, sites of gene activity (puffs of polytene chromosomes) disappeared and new hs puffs appeared instead. Incorporation of radioactively labeled uridine at the hs-activated gene loci was noticeable after less than 1 min.[3,4] This visual analysis of transcriptional reprogramming was confirmed later by testing nuclear RNA from *Drosophila* cell cultures. Using appropriate gene-specific probes, i.e., hsp70, as representative of hs-induced genes and actin as representative of control genes, Findly and Pederson,[21] demonstrated that the hs-induced increase of hsp70 transcription coincides with the rapid repression of the actin gene (Figure 9.1). Effects are measurable as early as 60 s after the onset of the hs and are optimum after 5 to 10 min. The data given in the legend to Figure 9.1 indicate that under optimum hs conditions the abundance of hs gene transcripts reaches or even surpasses the abundance of actin transcripts at 25°C. Note that there is one hsp26 gene as opposed to five hsp70 genes. Similar results were also reported for *Drosophila* Kc cells,[85] but much lower levels of hs gene transcription at 25°C were observed in this case.

With the exception of *Escherichia coli* (see Section 7.1), data from other hs systems are not exactly comparable, but whenever tested, it is evident that the transcription pattern of RNA polymerase II and not its activity is affected by hs. For evident reasons, this is the prerequisite for efficient expression of hsp-coding genes, but it is also true for RNA polymerases I and III. As outlined in Section 10.2, synthesis of preribosomes is continued for a considerable period of time, and only the increasing accumulation of ribosomal wastage may finally limit the activity of RNA polymerase I.[63,80] Transcripts of RNA polymerase III (tRNA, 5S RNA) continue to be formed (see Section 10.3). In addition, hs-specific new RNAs were detected in CHO and mouse cells[23] and in *Tetrahymena*.[46] They are described in more detail in Section 5.3. The characteristic pattern of transcription by all three forms of RNA polymerase was reproduced with runoff assays using nuclei from heat-shocked cells of *Drosophila*[57] or plants.[81]

The simple method of observing selective gene activation in *Drosophila* salivary glands also provided the basis for immunofluorescence studies on the site-specific redistribution of RNA polymerase II. In correspondence with the biochemical and morphological data, the enzyme was found to "move" from sites with actively transcribed control genes to the new hs puffs.[7,30,79] This is true for all hs puffs, including the nonprotein-coding puff 93D (see Chapter 5).

A different type of analysis was reported by Gilmour and Lis.[26,27] They used *in vivo* cross-linking of DNA with associated proteins by UV irradiation followed by immunoprecipitation of the DNP complexes and analysis with specific probes. The data compiled in Table 9.1 give an estimate of the relative densities of the enzyme under control (22°C) and hs conditions (37°C). Four classes can be discriminated. First, ribosomal RNA genes are not complexed with RNAPII. Second, copia genes and the cytoplasmic actin genes are repressed, i.e., the density of RNAPII decreases under hs. Third, in agreement with results by Spradling et al.[90] and Spadoro et al.,[88] transcription of histone genes is unaffected. Fourth, density of RNAPII increases on hs genes. Changes are moderate for the constitutively expressed hsp83 and 10- to 100-fold for the other hs genes tested.

Because of region-specific differences within one gene and the unequal quality of restriction fragments used for the assay, a direct comparison of the absolute numbers may be misleading. Thus, values for the relative distribution of RNAPII along the hsp70 gene is (37°/22°C): 7/2 for the 5′ flanking sequences, 13/0.5 for the 5′ part and 20/0.2 for the 3′ part of the coding region, and 3/not detectable for the 3′ flanking region.[26] It is evident that

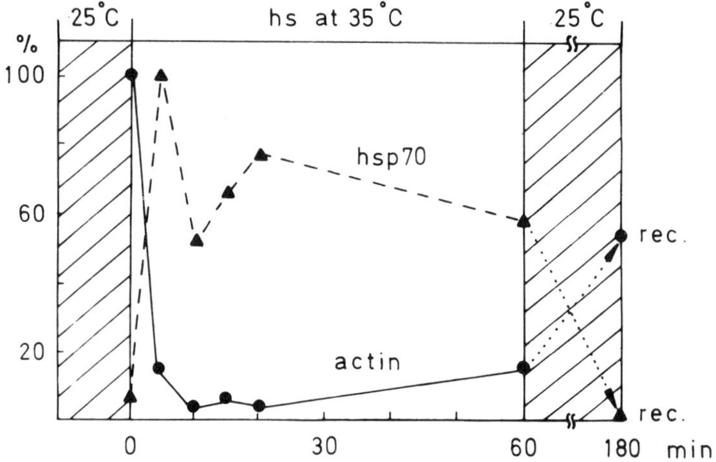

FIGURE 9.1. Transcription of hsp70 and actin genes in *Drosophila* S2 cell cultures under conditions of hs and recovery. After 5 min pulse labeling with ^{3}H-uridine, the relative amounts of gene-specific products in nuclear RNA were analyzed by means of corresponding cDNA clones. For maximum hybridization values (100%) see table. The rapid, hs-induced reprogramming of transcription of RNA polymerase II is evident from the increase of hsp70 gene and coincident decrease of actin gene transcription.

cDNA clone	**Nuclear transcripts (% of total radioactivity)**	
	25°C	**hs**
Actin	0.1	0.0075
hsp70	0.045	1.3
hsp26	0.008	0.085

(Modified from Findly, R. C. and Pederson, T., *J. Cell Biol.,* 88, 323, 1981; *Biol. Zbl.*, 103, 1984. With permission.)

inclusion of varying parts of the 5′ or 3′ flanking sequences into the gene clones, e.g., in the case of the cluster of histone or small hsp genes, have great influence on the absolute numbers. For the sake of simplicity, these details were omitted from Table 9.1. Nevertheless, it is worth noticing that 20 RNAPII molecules per 1 kb in the hsp70 gene is apparently the upper limit in *Drosophila.* This value considerably surpasses the maximum density of RNAPII found for comparable coding regions of the actin gene (3.0) and copia (0.5), respectively. The modest repression ratios for the actin gene of 3 to 1 and 2 to 1, respectively, do not agree with the results of Findly and Pederson[21] on a 15-fold decrease of the level of nuclear actin-specific RNAs (see Figure 9.1). Despite differences in the methods used for analysis and for the hs regime (37°C vs. 35°C) the discrepancy must be taken seriously because the data for the copia element (Table 9.1) are almost identical with earlier experiments on a sixfold repression of copia RNA synthesis during hs.[64]

The relatively high density of RNAPII in the 550 bp 5′ flanking region of the hsp70 genes in noninduced cells (2 compared to 7 in heat-shocked cells) deserves special interest. The enzyme is confined to the 5′ end of the gene covering nucleotides −12 to +65.[27] Detailed analysis of the complex showed that it is a stalled RNA polymerase II with 30 to 50 nucleotides of the mRNA 5′ end attached to it.[76a] These findings must be considered when the mechanism of hs activation of hsp coding genes is discussed (Section 1.4).

Similar immunofluorescence[22,36] and UV cross-linking studies[28] revealed that DNA topoisomerase I co-localizes with RNA polymerase II, i.e., it is likewise highly enriched in hs puffs. This contrasts to topoisomerase II which, independent of hs or control cells, is

TABLE 9.1
Gene-Specific Variation of RNA Polymerase II Binding in Heat-Shocked and Nonheat-Shocked *Drosophila* Cells

Genes	Class of genes	Relative densities of RNA polymerase II[a] (37°C/22°C)	
rDNA	A: not transcribed by RNAPII	0.03/0.03;	0.03/0.04
Copia	B: repressed	3.0/0.2;	0.5/0.09
Actin[b]		3.0/1.0;	2.0/1.0
Histone cluster	C: unaffected	0.4/0.3;	0.5/0.5
hsp83	D: induced	3.0/2.0;	
hsp70		13/0.5;	20/0.2
hsp26		5.0/0.2;	
hsp23/27		3.0/0.2;	

Note: RNA polymerase molecules were cross-linked to DNA by UV irradiation of *Drosophila* S2 cells. DNP complexes were isolated by CsCl centrifugation and immunoprecipitation using antibodies against the 215-kDa subunit of RNA polymerase II. The RNAPII-associated DNA was analyzed by means of dot or Southern blots. Numbers represent the specific density of RNAPII molecules per 1 kb corrected for the copy number of the gene and size of the restriction fragment used for analysis. Control cells were cultivated at 22°C, heat-shocked cells were kept for 30 min at 37°C.

[a] If two data are given, these represent different parts of the gene plus flanking regions.
[b] Only one of the two cytoplasmic actin genes in locus 5C is bound to RNAPII.

Compiled from Gilmour, D. S. and Lis, J. T., *Mol. Cell. Biol.*, 5, 2009, 1985.

generally distributed over the whole length of the *Drosophila* polytene chromosomes (see following section). Interestingly, the nucleoli of hs cells are intensively stained also by antibodies against topoisomerase I,[22] indicative of the ongoing transcription of pre-rRNA during hs (20 min 37°C, see Chapter 10). Similar to RNAPII, topoisomerase I is mainly localized in the coding regions and much less or not at all in the spacer regions, e.g., between the small hsp coding genes of locus 67B,[28] but for unknown reasons, the interaction of both enzymes with the transcribed regions is independent. Relatively low densities of topoisomerase II are found in the highly transcribed hsp70 gene, whereas the value for the moderately transcribed copia element is much higher.

9.2. CHROMATIN STRUCTURE OF HEAT SHOCK GENES

A sensitive probe for chromatin structure is the localization of cleavage sites after treatment with nucleases, such as DNase I, micrococcal nuclease, S1 nuclease, *Neurospora crassa* nuclease or exonuclease III.[17] DNA covered by nonhistone proteins or organized in nucleosomes is relatively resistant, whereas sequences between such DNP regions may be highly sensitive to nuclease attack. The resolution of this type of footprint analysis can be improved by use of chemical reagents, e.g., methidium propyl-EDTA-iron (II), instead of the bulky enzymes.[10] Except for a few reports on the chromatin structure surrounding the yeast hsp90 gene,[92b] the human hsp70B gene,[8a] and the nematode hsp16 genes,[16c] most papers deal with the *Drosophila* hs genes i.e., with the hsp83 gene,[33,102,103] the hsp70 genes[13,32,34,41,51,52,95,102,103,105] and the small hsp-coding genes of locus 67B.[10,16b,43,93a]

The results are basically similar for all genes. The coding regions as well as the spacer regions between genes are covered by nucleosomes generating the typical pattern of cleavage sites spaced by about 180 bp. In contrast, a defined region upstream of the transcription start sites so well as a short region downstream of it are free of nucleosomes, but covered by nonhistone proteins. In the uninduced state these are (1) RNA polymerase II covering the immediate vicinity of the transcription start site, e.g., bp −12 to +65 of the hsp70 gene[27] and (2) the TATA-box binding protein, which protects bp −12 to −40 in the hsp70 and bp −17 to −39 in the hsp83 gene[103] of *Drosophila*.

In the hs-induced state, the whole gene regions become much more nuclease sensitive. Evidently disruption of the regular nucleosome structure is combined with the high rate of transcription.[10,16b,41,49,51,52,104] Using crosslinking methods, Karpov et al.[41] reported on a total histone depletion of the actively transcribed hsp70 gene, whereas Cartwright and Elgin[10] reached their conclusions from DNase I cleavage patterns on the existence of half nucleosomes. The extent of histone depletion may depend on the actual transcription state. The rapid loss of ubiquitinated forms of histones H2A and H2B during hs (see Table 12.1) may be related to these rearrangements of the chromatin structure.[49]

The protein footprints in the promoter regions also change. Binding of the hs transcription factor (HSF) results in protection of previously sensitive regions containing the hs elements.[10,103] By using *Drosophila* genes for transformation of yeast[12,13] or P-element mediated integration at different sites of the *Drosophila* genome,[18] it became apparent that the characteristic footprint of a hs promoter with nuclease protected regions interrupted by hypersensitive sites is maintained even in the heterologous cell or at unusual sites of the *Drosophila* genome. As to be expected, the flanking sequences are less important for the particular positioning of promoter binding proteins.

A peculiarity of the hsp26 promoter structure became apparent when DNase I and exonuclease III cleavage patterns of induced and noninduced cells of *Drosophila* were compared.[16b,93a] Between two DNase I hypersensitive sites in the promoter region, there is an ordered array covered by a nucleosome. The folding back of the DNA around the nucleosome is discussed as the essential means to bring into close vicinity hs transcription factors bound to the two distant HSE motifs centered at bp −61 and −343, respectively (see promoter no. 13 in Figure 6.6).

Brown et al.[8a] reported on similar studies with the human hsp70B gene using different promoter deletion constructs and monkey COS, HeLa, and mouse cells as test systems. A DNase I hypersensitive region (bp −50 to −260) containing the four HSE (see Table 6.2) is defined by a sequence element between bp −223 and −162. Removing this internal part of the promoter abolishes DNase I hypersensitivity but does not affect hs inducibility in COS cells. Evidently, the presence of abundant, tightly bound proteins, but not binding of the hs transcription factor, determines the sites of DNase I hypersensitivity.

Similar to the results with other genes, the typical pattern of DNase I hypersensitive sites around the two loci with the *Caenorhabditis elegans* hsp16 genes disappears upon hs induction. Instead the whole gene regions become more nuclease sensitive, indicating active transcription.[16c] Interestingly, despite the almost identical sequence elements of their promoter regions and the coordinate induction (see Section 3.3 and Figure 3.2), the two hsp16 loci are differentially regulated in the course of a hs treatment. Locus hsp16-48/16-1 is shut down relatively soon, whereas the second locus (hsp16-41/16-2) continues to be active. The differences in the extent of mRNA accumulation are also detected at the level of chromatin structure.[16c]

An interesting aspect of transcription is the organization of genes in chromatin domains attached to the nuclear matrix by scaffold associated region (SAR).[25,62a,105] A dominant protein of the nuclear scaffold is topoisomerase II. SAR are found in the 3′ and 5′ flanking regions of hs genes or gene groups. They are preferred cleavage sites of topoisomerase II.[77,96] Potential SARs are so-called simple sequences, i.e., stretches of oligo T or oligo A residues found, e.g., in close vicinity of plant hs genes (see Figure 6.6). In addition to the function of such T-boxes and A-boxes as SARs, Gasser and Laemmli[25] defined a topo II box of 15 nucleotides GTn(A/T)A(T/C)ATTnATnn(G/A).

Most details of the intricate role of the nuclear matrix in transcription are unknown. However, a few facts are worth noticing. (1) In *Drosophila* Kc cells, hs genes are associated with the nuclear matrix irrespective of their transcription state, i.e., whether induced or noninduced cells were investigated. In contrast, the chorion gene, which is under developmental control and not inducible in these cells, is not bound to the nuclear matrix. Hence, binding of a chromatin domain to the nuclear matrix means preactivation, whereas the actual transcription state is controlled by additional factors. This explains why genes within the same chromatin domain may be regulated independently of each other (see Sections 3.4 and 6.6.1). In artificial constructs with hs genes in close vicinity to non-hs genes used for selection, both retain their individuality with respect to their control patterns.[11,83] (2) Potential SARs were localized in the 5′ and 3′-flanking regions of hsp70 and hsp83 genes mostly by detection of unusual nuclease cleavage sites or by use of topoisomerase II.[34,59,77,96,97]

Induction does not basically change the number and type of anchorage sites, but the yield of cleavage products varied considerably, indicating changes in the interaction with the nuclear matrix complex. Moreover, inhibitors of topoisomerase II have profound influence on the expression of hs genes. Rowe et al.[77] found that low doses of the antitumor drugs, epipodophyllotoxin (teniposide) and novobiocin, slightly induced hsp70 expression at 25°C, but strongly reduced transcription during hs. The latter effect is particularly striking for novobiocin. Detailed analyses of Han et al.[34] defined three topoisomerase II-dependent parts of gene expression: hs-induction with the typical disruption of the regular nucleosome pattern and binding of HSF to the promoter region, ongoing transcription and, finally, regeneration of the pre-hs chromatin organization in the recovery period.

9.3. TIGHTLY BOUND NUCLEAR MATRIX PROTEINS

A striking peculiarity of the hs response in vertebrate cells is the rapid increase of the nuclear matrix protein content.[73,74,94] Most of the proteins are apparently of cytoplasmic origin and not synthesized in the hs period. In HeLa or CHO cells, a specific protein set with M_r 28.5, 38.5, 60, 66, 81, 88, and 100 kDa was detected.[9,99,99a] They are not removed by 2 *M* NaCl or by treatment with nonionic detergent. The implications of the drastically changed protein: DNA ratio during hs are unclear. It is not unlikely that the increasing inhibition of DNA repair, e.g., of radiation-induced DNA lesions (see Section 15.3), results from a decreasing accessibility of the DNA to repair enzymes.[58,69,98] Binding of cytoplasmic proteins to the nuclear matrix is not dependent on HSP synthesis, nor is there any influence on the inducibility of hs genes. The slow removal of excess proteins from the nuclear compartment in the recovery period precedes the resumption of cell cycle activity.[75,76]

The situation in other organisms is not clear. Berrios and Fisher[4a] described the accumulation of tightly bound proteins in nuclei of heat-shocked yeast. Topoisomerase II was detected exclusively in a detergent- and high salt-resistent form in cultures heated to 37 to 42°C. In addition, McConnell et al.[56a] reported on the formation of a stable nuclear matrix-protein complex in *Drosophila melanogaster* embryos after 15 min hs at 37°C. Moreover, in an early paper on HSP localization in *Drosophila* larvae, Mitchell and Lipps[60] reported that a "substantial quantity of normal cellular protein labeled at 25°C or during the first few minutes of heat treatment becomes specifically and rapidly localized to chromosomes." It should be stressed at this point that the collapse of the intermediate size cytoskeleton in the perinuclear region (see Section 16.7) leads to a detergent- and high salt-resistant structure,[64b] which may entrap other cytoplasmic proteins and which is very difficult to remove from nuclear preparations. This problem is best illustrated by the erroneous localization of the small HSPs in the nuclear fractions or even explicitly in the nuclear matrix[49,84] of heat shocked cells. In fact, most of them are part of a perinuclear aggregate of cytoskeletal elements with hs granules (HSG) (see Section 16.7).

Only the generation of antibodies and their use for a fine localization of the proteins with electronmicroscopic techniques (see Figure 16.5) can help overcome this ambiguity.

In support of marked changes of nuclear protein composition during hs, the following results are worth noticing: in different mammalian cell lines the disappearance of the cytoplasmic microfilament system is connected with crystal-like actin depositions (actin rods) in the nucleus.[38,62b,100a] Such intranuclear actin bundles were also observed in heat-shocked cells of the water mold *Achlya*.[68] Though a transport of cytoplasmic actin to the nucleus under DMSO stress was demonstrated,[78a] actin is also a major component of the nucleus under normal conditions. An actin network is essential for the spatial orientation of transcription loops,[11a] which are destroyed by antibodies to actin or under stress conditions.[80a] A 21-kDa actin-modulating protein (cofilin), found in the cytoplasm of fibroblasts, is dephosphorylated during hs, migrates to the nucleus, and participates in the formation of actin rods.[62b,64a]

On the other hand, the rapid formation of high salt insoluble nuclear protein complexes including nuclear oncoproteins was described by Staufenbiel and Deppert,[91] Evan and Hancock,[19] and Littlewood et al.[54] Although a much lower threshold temperature is required (35 to 37°C instead of 41 to 43°C for whole cells), such complexes are also formed in isolated nuclei. Oncoproteins detected in these complexes are p53, $p62^{c\text{-}myc}$, $p66^{N\text{-}myc}$, $p58^{v\text{-}myc}$, $p45^{v\text{-}myb}$, $p75^{c\text{-}myb}$, Ela protein and the SV40 large T-antigen.[54] Formation of the insoluble complexes is fast, but restoration of the original state with simultaneous normalization of the nuclear morphology takes several hours. It is an intriguing possibility that complexation transiently inactivates and stabilizes the short-lived oncoproteins and that the resolubilization in the recovery period depends on HSP70 and its capability to act as ATPase[53] (see Section 2.3.2). Such a mechanism would allow a rapid and reversible reprogramming of gene expression and cell cycle activities. At any rate, proteins of the HSP70 family were found associated with different types of nuclear tumor antigens, e.g., p53 or mutant polyoma virus T-antigen,[68a,97a] and the p53/HSC70 complex can be dissociated *in vitro* by addition of ATP.[10a] In support of this mechanism are also findings on a hs-induced complexation and five- to eightfold increase of the half-lives of c-myc and c-myb proteins in chicken bursal lymphoma cells. Resolubilization and restoration of the normal, rapid turnover in the recovery period is enhanced in preconditioned cells.[54a]

9.4. MODIFICATION OF NUCLEAR PROTEINS

The deep influence of hs on practically all nuclear functions coincides with various changes in the modification pattern of nuclear proteins (see Table 12.1). Unfortunately, the multiplicity of reported examples contrasts to our almost complete ignorance about the causal relationship between both parts of the nuclear hs response. Three types of changes of histone modifications may contribute to the reprogramming of transcription in *Drosophila:*

1. The striking, reversible deubiquitination of H2A[29,50] is presumably connected with the transition of the regular nucleosomal structure of hs genes into the highly transcribed form without recognizable nucleosomes.[50] It is remarkable that a similar loss of ubiquitin-H2A was also reported for vertebrate cells.[6,65]
2. Tanguay and Desrosiers[93] summarized the experimental evidence for a correlation between altered patterns of H3 and H2B methylation and the repression of control gene transcription. At 37°C changes are noticeable after 5 min, whereas under conditions of mild stress (32°C or 8% ethanol), induction of hs genes proceeds without repression of other genes, and, consequently, changes of histone methylation are not observed.[14-16] Changes of histone modification, similar to those observed under hs, were reported for treatments with transcriptional inhibitors affecting DNA topisomerase II (ethidium bromide, novobiocin, teniposide). Chromatin conformation may be an

important factor controlling accessibility to histone methyltransferases.[16a] Novobiocin and teniposide were also shown to induce transcription of hsp70 genes in *Drosophila*.[77]

3. These results corroborate earlier findings of Arrigo.[1] In addition, he reported on the rapid deacetylation of all four core histones (H2A, H2B, H3, H4) by heat or arsenite stress. Restoration of normal modification patterns in the recovery period requires preceding HSP synthesis, i.e., it is inhibited, if cycloheximide or emetine are added prior to the hs treatment.

9.5. PROCESSING OF PRE-mRNP

Interruption of the coding parts of a gene by intervening sequences (introns) is common to all eukaryotic cells. It has great evolutionary and regulatory implications. The exact removal of the introns requires an elaborate processing machinery, whose composition and function has been partially elucidated in recent years, mainly using mammalian cells and yeast.[46,55,61a,82,85a] Five types of small nuclear RNPs (snRNPs) with an abundance of more than 10^5 to 10^6 particles per cell are involved. snRNPs are composed of small RNAs of 56 to 217 nucleotides (U1, U2, U4, U5, U6 RNAs) and 5 to 9 polypeptides.[65b] The core polypeptides of all snRNP are similar. Splicing proceeds in close association with the nuclear matrix after formation of a spliceosome (50 to 60S), in which the pre-mRNP is complexed and appropriately folded by binding to U1, U2, U4, U5, and U6 snRNPs. U1 and U5 snRNPs participate in the early steps of spliceosome assembly by binding to the 5′ and 3′ splice sites, respectively.[82,85a]

As is true for synthesis and processing of rRNP precursors (see Section 10.2), synthesis of pre-mRNP by RNA polymerase II is significantly more heat resistant than its processing and export to the cytoplasm. Although intron splicing is the most elaborate part of the co- and postsynthetic modifications of the primary transcript, others are equally important, e.g., capping of the 5′ end, methylations and poly(A) addition at the 3′ end, which is probably connected with the termination mechanism. With improved experimental techniques, more and more examples for hs-induced alterations at this stage of gene expression emerge. However, which part of the processing pathway is primarily affected in the particular case is mostly unknown. Spinelli et al.[89] reported on a block in the processing of histone mRNA precursors in sea urchin blastula mesenchyme. The nuclear accumulation of relatively large pre-mRNP suggests formation of polycistronic transcripts, i.e., of a read-through by RNA polymerase II under hs conditions. Such hs-induced "problems" with the recognition of normal termination signals were confirmed for *Drosophila* cells by the observations of Garbe et al.[24a] and Pauli et al.[66] The former found an increase of the 10 kb read-through transcript of the hs locus 93D (see Section 5.1), whereas the latter observed an unusually large transcript of gene 2 (locus 67B) whose 3′ end extends into the adjacent hsp22 gene. But defective termination with production of giant transcripts was also found in *Xenopus* oocytes and HeLa cells, when transcription of rRNA genes by RNA polymerase I was analyzed.[47,65a]

So far, interest has concentrated mainly on hs effects on the splicing machinery because the growing list of data on the structure of hs genes show that many are intron free (see Table 3.1). This led Pederson[67] to the concept of the coexistence of two different processing pathways: type I for intron-containing pre-mRNP with a relatively high protein: RNA ratio and type II for intron-free pre-mRNP with a lower protein: RNA ratio. Under hs conditions, only the type II pathway should be operative and thus, expression of many control genes is blocked at this stage. Similar to the situation in the nucleolus (Section 10.5), accumulation of perichromatin granules in hs mammalian cells[20,35] may reflect this discrepancy between ongoing transcription and blocked processing. In keeping with this concept, Mayrand and Pederson[56] described a rapid and pronounced shift of the isopycnic density of hnRNP in both heat-shocked *Drosophila* and mammalian cells, indicating a loss of protein. They concluded that hs leads to an immediate block in the assembly of hnRNP and to a disassembly of preexisting hnRNP.

For a number of reasons, this stimulating concept must be modified:

1. Kloetzel and Schuldt[45] reinvestigated the postulated changes in the protein composition of hnRNPs in *Drosophila* cells, using cesium sulfate gradients and recombinant clones to detect individual gene products (rRNA, hsp70, tubulin, and U1 RNAs). Ribonucleoprotein material containing rRNA is found at rho = 1.56 g/cm^3, whereas that specific for the three other clones is found in the range of 1.31 to 1.34. No change after hs was observed. Thus, the marked increase of RNP material with isopycnic density of 1.58[56] is presumably due to the hs-induced accumulation of pre-rRNP (see Section 10.5).
2. Intron-containing reporter genes, e.g., the *Drosophila* alcohol dehydrogenase gene,[8,44] the flounder antifreeze protein gene,[70] or the P-element transposase gene of *Drosophila*[71,92] were combined with the Dm-hsp 70 promoter/leader element and correctly expressed in hs *Drosophila* cells (see Table 6.1, numbers A8, A8a, A12b, A16a). These results show that splicing is not principally blocked in hs cells.
3. Although it is still true that most hs genes are intron free, the number of exceptions to this rule is constantly increasing (see Table 3.1, Figure 3.1). Examples are the genes of the vertebrate hsp90 family,[36b,70a] the human hsp27 gene,[36a] the *Drosophila* hsp83 gene,[5,31] the nematode hsp16 and hsp70 genes,[39,78,85b] the plant hsp70 genes,[72a,101a,101c] the soybean hsp26 gene,[13a] and the constitutively expressed mammalian hsc72 genes.[16d,87]

Detailed investigations with *Drosophila* S2 cells[106] gave new insights into the temperature dependence of the splicing system and thus clarified part of the seemingly controversial results. The authors studied expression of the hsp83 gene as well as of a hybrid construct of the adh gene fused to the hsp70 promoter/leader region (Table 6.1, number A9). Results with both genes were identical. With increasing hs temperature (≥36.5°C), expression of intron-containing genes decreased due to an increasing deficiency of the splicing apparatus. Pre-mRNP began to accumulate in the nucleus. At 38°C, no splicing was observed, i.e., the *Drosophila* HSP83 is a low temperature HSP with optimum expression between 33 and 35°C. In earlier experiments with Dm-hsp 70 (P/L) × adh fusion constructs[8,44] using 36.5 to 37°C as hs temperature, the splicing deficiency escaped notice because it was not complete. Some hs-induced ADH was formed and other hs temperatures, as well as nuclear transcripts, were not analyzed. Interestingly, the unspliced transcripts of hsp83 accumulated during hs are unable to reenter the processing pathway in the recovery period. However, they are transported to the cytoplasm and translated into abnormal protein fragments.[107]

Recovery of splicing activity after a severe hs (30 min 38°C) is markedly improved by a preceding conditioning treatment at 35°C (Figure 9.2). This thermotolerance effect evidently depends on HSP synthesis. It is inhibited by addition of cycloheximide prior to the conditioning treatment (samples 5 and 6). It is tempting to speculate that HSP70 synthesis and its intranuclear localization are important for this protection. Support for this assumption comes from two observations. Kloetzel and Bautz[44a] described cross-linking by UV irradiation of the nuclear HSP70 to hnRNP material of *Drosophila* cells. Unfortunately, these authors also found cross-linking with small HSPs, which are certainly not localized in the nuclear compartment. Evidently their analysis in cesium chloride gradients could not discriminate between nuclear hnRNP and perinuclear mRNP aggregates (see Section 16.7).

Indicators of the profound perturbations of the splicing apparatus, probably connected with the disruption of the nuclear actin network (see Section 9.3), are alterations of the distribution and immune recognition of snRNP particles. This is true for *Drosophila*,[101b] HeLa cells,[5,54b] and rat embryo fibroblasts.[101] The usually finely speckled immunostaining

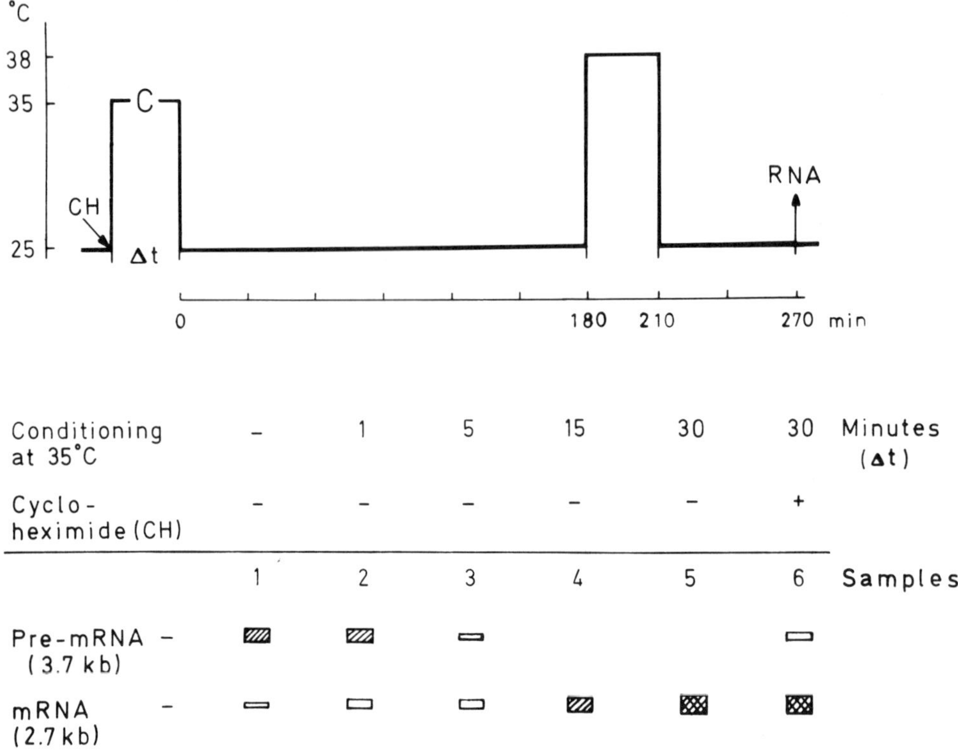

FIGURE 9.2. Protection of *Drosophila* hsp83 pre-mRNA splicing from damage by severe heat shock (hs). The pictograph on top indicates the hs regime with a conditioning treatment of *Drosophila* S2 cells at 35°C (0 to 30 min), followed by 180 min recovery at 25°C, a severe hs (30 min 38°C) and again 60 min recovery at 25°C. Content of pre-hsp83-mRNA (3.7 kb) and mature hsp83-mRNA (2.7 kb) was analyzed by Northern hybridization. The persistent block in pre-mRNA splicing (samples 1, 2) can be relieved by a 5 to 30 min conditioning treatment at 35°C (samples 3 to 5). (Data compiled from Yost, H. J. and Lindquist, S., *Cell,* 45, 185, 1986.)

with antisera to snRNP disappears more and more and gives rise to a coarsely speckled pattern of tightly bound aggregates, typical of stored, inactive forms of snRNP.[2a] Similar to the observations on the splicing activities, the extent and reversibility of aggregation are influenced by a prior conditioning treatment.[5,101]

Hs effects on the splicing efficiency also extend to another transcript of *Drosophila.* Locus 93D of *D. melanogaster* and related loci of other *Drosophila* species give rise to an abundant cytoplasmic poly(A)$^+$-RNA of unknown function (see Section 5.1). This 1.2-kb RNA derives from a 1.9-kb precursor by excision of a 0.7-kb intron. Despite marked hs induction of transcription at the 93D locus, formation of the cytoplasmic RNA decreases at ⩾37°C with simultaneous accumulation of the nuclear precursor RNA and a 10-kb read-through product.[24,37]

First direct experimental evidence on this subject concerning mammalian cells came from experiments of Kay et al.[42] introducing the intron-containing hsp16 genes of the nematode *Caenorhabditis elegans* into mouse cells. Due to a divergence of the 3′ splice sites between donor and recipient organism, splicing was aberrant. However, under hs conditions, the efficiency of splicing was greatly reduced. The unspliced transcription products, accumulating in the nucleus, were only processed and exported to the cytoplasm in the following recovery period. It is worth noticing that the same transient block of splicing can be achieved by arsenite stress[42] known to induce HSP synthesis in vertebrates and many other organisms (see Chapter 1). In an extension of these results to other organisms, inhibition of splicing activity was also reported in hs cells of chicken[5b] and the cellular slime mold *Dictyostelium*.[54c] In HeLa cells splicing of the hsp27 pre-mRNA is blocked under conditions of severe hs, unless cells were made thermotolerant by a prior conditioning treatment. Evidently changes in the U2, U4, U5, and U6 snRNP lead to formation of aberrant splicing complexes.[5]

So far no splicing defects were detected in heat-shocked plant cells. Instead, Cd^{2+} ions, in concentrations inducing HSP synthesis, block splicing of the hsp70 and hsp26 mRNAs. Other heavy metals (Zn^{2+}, Cu^{2+}), though active as inducers, have no effect.[13a,31a,101a]

A peculiarity was reported for trypanosomes. As already mentioned (Section 3.7), all mRNAs of parasitic protozoa are modified at their 5′ ends by *trans*-splicing of a short capped leader sequence formed from a separate miniexon which codes for a precursor of this leader. In *Trypanosoma brucei* part of the processing machinery is blocked under hs conditions (42°C). *Trans*-splicing to the dicistronic pre-mRNA of α- and β-tubulin with cleavage into the two monocistronic parts is defective, whereas addition of the leader to the hsp83 and hsp70 mRNAs proceeds normally.[61b-d]

In addition to the interference with intron-splicing, there is evidence for peculiarities of nuclear-cytoplasmic transport in connection with the hs response. This is vividly illustrated by a concise analysis of hs-induced gene expression in the *D. melanogaster* locus 87Cl,[48] which contains 3 hsp70 genes plus 7 α,β-transcription units (see Section 3.4 and Figure 3.3). All share the same 389 nucleotides of 5′ flanking sequences, i.e., all essential elements of the hsp70 promoter. Correspondingly, their expression is highly coordinate with respect to time and temperature dependence, but three remarkable and so far unexplained exceptions were noticed: (1) synthesis and cytoplasmic accumulation of hsp70 RNA was usually much higher than α,β-specific RNA. (2) At ⩾37°C transcription starting at the hsp70 promoters rapidly declines, whereas at the (identical!) α,β-promoters transcription continues. (3) Under these conditions of severe hs, the nuclear-cytoplasmic transport becomes selective. Transcripts from hsp70 genes are rapidly exported to the cytoplasm, but α,β-transcripts are largely retained in the nucleus.

Interestingly, a preferential transport of hsp70 transcripts was also observed in HeLa cells after adenovirus infection.[61] It is well known that the drastic reduction of host cell gene expression is brought about by two effects, a greatly reduced nuclear-cytoplasmic transport of newly formed mRNA and, late in the infectious cycle, a shut-off of translational initiation of host cell mRNAs.[2] In contrast to most other host cell mRNAs, hsp70 mRNAs are not affected by the Adv-induced block on nuclear-cytoplasmic transport, which is not relieved by hs.[61] The molecular basis and biological meaning for this selective escape, shared also by the β-tubulin mRNA, is unclear. Both gene products may play some role in the middle phase of the Adv infectious cycle since expression of hsp70 and β-tubulin genes are stimulated by the early Adv E1A gene product[40,61,62] (see also Section 8.5). Similar effects with cessation of normal processing and retention of actin pre-mRNP bound to the nuclear matrix were also observed by Denome et al.[13b] after adenovirus infection or heat stress, but in agreement with the results of Moore et al.,[61] hsp70 mRNA was not affected.

REFERENCES

1. **Arrigo, A. P.,** Acetylation and methylation patterns of core histones are modified after heat or arsenite treatment of *Drosophila* tissue culture cells, *Nucl. Acids Res.*, 11, 1389—1404, 1983.
2. **Babich, A., Feldman, L. T., Nevins, J. R., Darnell, J. E., and Weinberger, C.,** Effects of adenovirus on metabolism of specific host mRNAs: Transport control and specific translational discrimination, *Mol. Cell. Biol.*, 3, 1212—1221, 1983.

2a. **Bachmann, M., Schroeder, H. C., Falke, D., and Mueller, W. E. G.,** Alteration of the intracellular localization of the La protein compared with the localization of UsnRNPs, *Cell Biol. Intern. Rep.,* 12, 101—125, 1988.

3. **Belyaeva, E. S. and Zimulev, J. F.,** RNA synthesis in *Drosophila melanogaster* puffs, *Cell Differ.,* 4, 415—427, 1976.

4. **Berendes, H. D.,** Factors involved in the expression of gene activity in polytene chromosomes, *Chromosoma,* 24, 418—437, 1968.

4a. **Berrios, S. and Fisher, P. A.,** Thermal stabilization of putative karyoskeletal protein-enriched fractions from *Saccharomyces cerevisiae, Mol. Cell. Biol.,* 8, 4573—4575, 1988.

4b. **Birnstiel, M. L., Ed.,** *Structure and Function of Major and Minor Small Nuclear Ribonucleoprotein Particles,* Springer-Verlag, Berlin, 1988.

5. **Blackman, R. K. and Meselson, M.,** Interspecific nucleotide sequence comparisons used to identify regulatory and structural features of the *Drosophila* hsp82 gene, *J. Mol. Biol.,* 188, 499—516, 1986.

5a. **Bond, U.,** Heat shock but not other stress inducers leads to the disruption of a sub-set of snRNPs and inhibition of in vitro splicing in HeLa cells, *EMBO J.,* 7, 3509—3518, 1988.

5b. **Bond, U. and Schlesinger, M. J.,** The chicken ubiquiten gene contains a heat shock promotor and express an unstable mRNA in heat-shocked cells, *Mol. Cell. Biol.,* 6, 4602—4610, 1986.

6. **Bond, U., Agell, N., Haas, A. L., Redman, K., and Schlesinger, M. J.,** Ubiquitin in stressed chicken embryo fibroblasts, *J. Biol. Chem.,* 263, 2384—2388, 1988.

7. **Bonner, J. J. and Kerby, R. L.,** RNA polymerase II transcribes all of the heat shock induced genes of *Drosophila melanogaster, Chromosoma,* 85, 93—108, 1982.

8. **Bonner, J. J., Parks, C., Parker-Thornburg, J., Mortin, M. A., and Pelham, H. R. B.,** The use of promoter fusion in *Drosophila* genetics: Isolation of mutations affecting the heat shock response, *Cell,* 37, 979—991, 1984.

8a. **Brown, M. F., Amin, J., Schiller, P., Voellmy, R., and Scott, K. A.,** Determinants for the DNaseI-hypersensitive chromatin structure 5′ to a human HSP70 gene, *J. Mol. Biol.,* 203, 107—117, 1988.

9. **Caizergues-Ferrer, M., Dousseau, F., Gas, N., Bouche, G., Stevens, B., and Amalric, F.,** Induction of new proteins in the nuclear matrix of CHO cells by heat shock: detection of a specific set in the nucleolar matrix, *Biochem. Biophys. Res. Commun.,* 118, 444—450, 1984.

10. **Cartwright, I. L. and Elgin, S. C. R.,** Nucleosomal instability and induction of new upstream protein-DNA associations accompany activation of four small heat shock protein genes in *Drosophila melanogaster, Mol. Cell. Biol.,* 6, 779—791, 1986.

10a. **Clarke, C. F., Cheng, K., Frey, A. B., Stein, R., Hinds, P. W., and Levien, A. J.,** Purification of complexes of nuclear oncogene p53 with rat and *Escherichia coli* heat shock proteins. In vitro dissociation of hsc 70 and dnak from murine p53 by ATP, *Mol. Cell. Biol.,* 8, 1206—1215, 1988.

11. **Cohen, R. S. and Meselson, M.,** Inducible transcription and puffing in *Drosophila melanogaster* transformed with hsp70-phage hybrid heat shock genes, *Proc. Natl. Acad. Sci. U.S.A.,* 81, 5509—5513, 1984.

11a. **Cook, P. R.,** The nucleoskeleton and the topology of transcription, *Eur. J. Biochem.,* 185, 487—501, 1989.

12. **Costlow, N. and Lis, J. T.,** High-resolution of DNase I-hypersensitive sites of *Drosophila melanogaster* and *Saccharomyces cerevisiae* heat shock genes, *Mol. Cell. Biol.,* 4, 1853—1863, 1984.

13. **Costlow, N. A., Simon, J. A., and Lis, J. T.,** A hypersensitive site in hsp70 chromatin requires adjacent not internal DNA sequence, *Nature,* 313, 147—149, 1985.

13a. **Czarnecka, E., Nagao, R. T., Key, J. L., and Gurley, W. B.,** Characterization of Gm-hsp26-A, a stress gene encoding a divergent heat shock protein of soybean: heavy-metal-induced inhibition of intron processing, *Mol. Cell. Biol.,* 8, 1113—1122, 1988.

13b. **Denome, R. M., Werner, E. A., and Patterson, R. J.,** RNA metabolism in nuclei: adenovirus and heat shock alter intranuclear RNA compartmentalization, *Nucl. Acids Res.,* 17, 2081—2090, 1989.

14. **Desrosiers, R. and Tanguay, R. M.,** The modification in the methylation patterns of H2B and H3 after heat shock can be correlated with the inactivation of normal gene expression, *Biochem. Biophys. Res. Commun.,* 133, 823—829, 1985.

15. **Desrosiers, R. and Tanguay, R. M.,** Further characterization of the posttranslational modifications of core histones in response to heat and arsenite stress in *Drosophila, Biochem. Cell Biol.,* 64, 750—758, 1986.

16. **Desrosiers, R. and Tanguay, R. M.,** Methylation of *Drosophila* histones at proline, lysine and arginine residues during heat shock, *J. Biol. Chem.,* 263, 4686—4692, 1988.

16a. **Desrosiers, R. and Tanguay, R. M.,** Transcriptional inhibitors affecting topoisomerase II induce changes in histone methylation patterns similar to those induced by heat shock, *Biochem. Biophys. Res. Commun.,* 162, 1037—1043, 1989.

16b. **Dietz, T. J., Cartwright, I. L., Gilmour, D. S., Siegfried, E., Thomas, G. H., and Elgin, S. C. R.,** The chromatin structure of hsp26, in *Stress-Induced Proteins,* Pardue, M. L., Feramisco, J. R., and Lindquist, S., Eds., Alan R. Liss, New York, 1989, 15—24.

16c. **Dixon, D. K., Jones, D., and Candido, E. P. M.,** The differentially expressed 16-Kd heat shock genes of *Caenorhabditis elegans* exhibit differential changes in chromatin structure during heat shock, *DNA Cell Biol.*, 9, 177—191, 1990.

16d. **Dworniczak, B. and Mirault, M.-E.,** Structure and expression of a human gene coding for a 71 kd heat shock 'cognate' protein, *Nucl. Acids Res.*, 15, 5181-5197, 1987.

17. **Eissenberg, J. C., Cartwright, I. L., Thomas, G. H., and Elgin, S. C. R.,** Selected topics in chromatin structure, *Annu. Rev. Genet.*, 19, 485—536, 1985.

18. **Eissenberg, J. C. and Elgin, S. C. R.,** Chromatin structure of a P-element-transduced hsp 28 gene in *Drosophila melanogaster, Mol. Cell. Biol.*, 6, 4126—4129, 1986.

19. **Evan, G. I. and Hancock, D. C.,** Studies on the interaction of the human c-myc protein with cell nuclei: $p62^{c-myc}$ as a member of a discrete subset of nuclear proteins, *Cell,* 43, 253—261, 1985.

20. **Fakan, S. and Puvion, E.,** The ultrastructural visualization of nucleolar and extranucleolar RNA synthesis and distribution, *Int. Rev. Cytol.*, 65, 255—299, 1980.

21. **Findly, R. C. and Pederson, T.,** Regulated transcription of the genes for actin and heat-shock proteins in *Drosophila* cells, *J. Cell Biol.*, 88, 323—328, 1981.

22. **Fleischmann, G., Pflugfelder, G., Steiner, E. K., Javaherian, K., Howard, G. C., Wang, J. C., and Elgin, S. C. R.,** *Drosophila* DNA topoisomerase I is associated with transcriptionally active regions of the genome, *Proc. Natl. Acad. Sci. U.S.A.*, 81, 6958—6962, 1984.

23. **Fornace, A. J. and Mitchell, J. B.,** Induction of B2 RNA polymerase III transcription by heat shock: Enrichment for heat shock induced sequences in rodent cells by hybridization subtraction, *Nucl. Acids Res.*, 14, 5793—5811, 1986.

24. **Garbe, J. C. and Pardue, M. L.,** Heat shock locus 93D of *Drosophila melanogaster:* A spliced RNA most strongly conserved in the intron sequence, *Proc. Natl. Acad. Sci. U.S.A.*, 83, 1812—1816, 1986.

24a. **Garbe, J. C., Bendena, W. G., Alfano, M., and Pardue, M. L.,** A *Drosophila* heat shock locus with a rapidly diverging sequence but a conserved structure, *J. Biol. Chem.*, 261, 16889—16895, 1986.

25. **Gasser, S. M. and Laemmli, U. K.,** A glimpse at chromosomal order, *Trends Genet.*, 3, 16—22, 1987.

26. **Gilmour, D. S. and Lis, J. T.,** In vivo interaction of RNA polymerase II with genes of *Drosophila melanogaster, Mol. Cell. Biol.*, 5, 2009—2018, 1985.

27. **Gilmour, D. S. and Lis, J. T.,** RNA polymerase II interacts with the promoter region of the noninduced hsp70 gene in *Drosophila melanogaster* cells, *Mol. Cell. Biol.*, 6, 3984—3989, 1986.

28. **Gilmour, D. S., Pflugfelder, G., Wang, J. C., and Lis, J. T.,** Topoisomerase I interacts with transcribed regions in *Drosophila* cells, *Cell,* 44, 401—407, 1986.

29. **Glover, C. V. C.,** Heat-shock effects on protein phosphorylation in *Drosophila* in *Heat Shock from Bacteria to Man,* Schlesinger, M. J., Ashburner, M., and Tissieres, A., Eds., Cold Spring Harbor Laboratory, Cold Spring Harbor, NY, 1982, 227—234.

30. **Greenleaf, A., Plagens, U., Jamrich, M., and Bautz, E. K. F.,** RNA polymerase B (or II) in heat-induced puffs on *Drosophila* polytene chromosomes, *Chromosoma,* 65, 127—136, 1978.

31. **Hacket, R. W. and Lis, J. T.,** Localization of the hsp83 transcript within a 3292 nucleotide-sequence from the 63B heat shock locus of *Drosophila melanogaster, Nucl. Acids Res.*, 11, 7011—7030, 1983.

31a. **Hagen, G., Uhrhammer, N., and Guilfoyle, T. J.,** Regulation of expression of an auxin-induced soybean sequence by cadmium, *J. Biol. Chem.*, 263, 6442—6446, 1988.

32. **Han, S., Udvardy, A., and Schedl, P.,** Transcriptionally active chromatin is sensitive to *Neurospora crassa* and S1 nucleases, *J. Mol. Biol.*, 179, 469—496, 1984.

33. **Han, S., Udvardy, A., and Schedl, P.,** *Neurospora crassa* and S1 nuclease cleavage in hsp83 gene chromatin, *J. Mol. Biol.*, 184, 657—666, 1985.

34. **Han, S., Udvardy, A., and Schedl, P.,** Chromatin structure of the 87A7 heat-shock locus during heat induction and recovery from heat shock, *Biochim. Biophys. Acta,* 825, 154—160, 1985.

35. **Heine, U., Sverak, L., Kondratick, J., and Bonar, R. A.,** The behaviour of HeLa-S3 cells under the influence of supranormal temperatures, *J. Ultrastruct. Res.*, 34, 375—396, 1971.

36. **Heller, R. A., Shelton, E. R., Dietrich, V., Elgin, S. C. R., and Brutlag, D. L.,** Multiple forms and cellular localization of *Drosophila* DNA topoisomerase II, *J. Biol. Chem.*, 261, 8063—8069, 1986.

36a. **Hickey, E., Brandon, S. E., Potter, R., Stein, G., Stein, J., and Weber, L. A.,** Sequence and organization of genes encoding the human 27 kDa heat shock protein, *Nucl. Acids Res.*, 14, 4127—4146, 1986.

36b. **Hickey, E. Brandon, S. E., Smale, G., Lloyd, D., and Weber, L. A.,** Sequence and regulation of a gene encoding a human 89-kilo-dalton heat shock protein, *Mol. Cell. Biol.*, 9, 2615—2626, 1989.

37. **Hovemann, B., Walldorf, U., and Ryseck, R.-P.,** Heat-shock locus 93D of *Drosophila melanogaster:* An RNA with limited coding capacity accumulates precursor transcripts after heat shock, *Mol. Gen. Genet.*, 204, 334—340, 1986.

38. **Iida, K., Iida, H., and Yahara, I.,** Heat shock induction of intranuclear actin rods in cultured mammalian cells, *Exp. Cell Res.*, 165, 207—215, 1986.

39. **Jones, D., Russnak, R. H., Kay, R. J., and Candido, E. P. M.,** Structure expression and evolution of a heat shock gene locus in *Caenorhabditis elegans* that is flanked by repetitive elements, *J. Biol. Chem.*, 261, 12006—12015, 1986.
40. **Kao, H.-T. and Nevins, J. R.,** Transcriptional activation and subsequent control of the human heat shock gene during adenovirus infection, *Mol. Cell. Biol.*, 3, 2058—2065, 1983.
41. **Karpov, V. L., Preobrazhenskaya, O. V., and Mirzabekov, A. D.,** Chromatin structure of hsp 70 genes, activated by heat shock: Selective removal of histones from the coding region and their absence from the 5′ region, *Cell,* 36, 423—431, 1984.
42. **Kay, R. J., Russnak, R. H., Jones, D., Mathias, C., and Candido, E. P. M.,** Expression of intron-containing *C. elegans* heat shock genes in mouse cells demonstrates divergence of 3′ splice site recognition sequence between nematodes and vertebrates, and an inhibitory effect of heat shock on the mammalian splicing apparatus, *Nucl. Acids Res.*, 15, 3723—3741, 1987.
43. **Keene, M. A., Corces, V., Lowenhaupt, K., and Elgin, S. C. R.,** DNase I hypersensitive sites in *Drosophila* chromatin occur at the 5′ ends of regions of transcription, *Proc. Natl. Acad. Sci. U.S.A.*, 78, 143—146, 1981.
44. **Klemenz, R., Hultmark, D., and Gehring, W. J.,** Selective translation of heat shock mRNA in *Drosophila melanogaster* depends on sequence information in the leader, *EMBO J.*, 4, 2053—2060, 1985.
44a. **Kloetzel, P. and Bautz, E. K. F.,** Heat-shock proteins are associated with hnRNP in *Drosophila melanogaster* tissue culture cells, *EMBO J.*, 2, 705—710, 1983.
45. **Kloetzel, P.-M. and Schuldt, C.,** The packaging of nuclear ribonucleoprotein in heat-shocked *Drosophila* cells is unaltered, *Biochim. Biophys. Acta,* 867, 9—15, 1986.
46. **Kraus, K. W., Good, P. J., and Hallberg, R. L.,** A heat shock-induced, polymerase III-transcribed RNA selectively associates with polysomal ribosomes in *Tetrahymena thermophila, Proc. Natl. Acad. Sci. U.S.A.*, 84, 383—387, 1987.
46a. **Kroeger, P. E. and Rowe, T. C.,** Interaction of topoisomerase I with the transcribed region of the *Drosophila* HSP70 heat shock gene, *Nucl. Acids Res.*, 17, 8495—8509, 1989.
47. **Labhart, P. and Reeder, R. H.,** Heat shock stabilizes highly unstable transcripts of the *Xenopus* ribosomal gene spacer, *Proc. Natl. Acad. Sci. U.S.A.*, 84, 56—60, 1987.
48. **Lengyel, J. A. and Graham, M. L.,** Transcription, export and turnover of Hsp70 and alpha,beta, two *Drosophila* heat shock genes sharing a 400 nucleotide 5′ upstream region, *Nucl. Acids Res.*, 12, 5719—5735, 1984.
49. **Levinger, L. and Varshavsky, A.,** Heat-shock proteins of *Drosophila* are associated with nuclease-resistant, high-salt resistant nuclear structures, *J. Cell Biol.*, 90, 793—796, 1981.
50. **Levinger, L. and Varshavsky A.,** Selective arrangement of ubiquitinated and D1 protein-containing nucleosomes with the *Drosophila* genome, *Cell,* 28, 375—385, 1982.
51. **Levy, A. and Noll, M.,** Multiple phases of nucleosomes in the hsp 70 genes of *Drosophila melanogaster, Nucl. Acids Res.*, 8, 6059—6068, 1980.
52. **Levy, A. and Noll, M.,** Chromatin fine structure of active and repressed genes, *Nature,* 289, 198—203, 1981.
53. **Lewis, M. J. and Pelham, H. R. B.,** Involvement of ATP in the nuclear and nucleolar functions of the 70 kd heat shock protein, *EMBO J.*, 4, 3137—3143, 1985.
54. **Littlewood, T. D., Hancock, D. C., and Evan, G. I.,** Characterization of a heat shock-induced insoluble complex in the nuclei of cells, *J. Cell Sci.*, 88, 65—72, 1987.
54a. **Lüscher, B. and Eisenman, R. N.,** c-myc and c-myb protein degradation: Effect of metabolic inhibitors and heat shock, *Mol. Cell. Biol.*, 8, 2504—2512, 1988.
54b. **Lutz, Y., Jacob, M., and Fuchs, J. P.,** The distribution of two hnRNP-associated proteins defined by a monoclonal antibody is altered in heat-shocked HeLa cells, *Exp. Cell Res.*, 175, 109—124, 1988.
54c. **Maniak, M. and Nellen, W.,** A developmentally regulated membrane protein gene in *Dictyostelium discoideum* is also induced by heat shock and cold shock, *Mol. Cell. Biol.*, 8, 153—159, 1988.
55. **Maniatis, T. and Reed, R.,** The role of small nuclear ribonucleoprotein particles in pre-mRNA splicing, *Nature,* 325, 673—678, 1987.
56. **Mayrand, S. and Pederson, T.,** Heat shock alters nuclear ribonucleoprotein assembly in *Drosophila* cells, *Mol. Cell. Biol.*, 3, 161—171, 1983.
56a. **McConnel, M., Whalen, A. M., Smith, D. E., and Fisher, P. A.,** Heat shock-induced changes in the structural stability of proteinaceous karyoskeletal elements in vivo and morphologic effects in situ, *J. Cell Biol.*, 105, 1087—1098, 1987.
57. **Miller, D. W. and Elgin, S. C. R.,** Transcription of heat shock loci of *Drosophila* in a nuclear system, *Biochemistry,* 20, 5033—5042, 1981.
58. **Mills, M. D. and Meyn, R. E.,** Effects of hyperthermia on repair of radiation induced DNA strand breaks, *Radiat. Res.*, 87, 314—328, 1981.
59. **Mirkovitch, J., Mirault, M. E., and Laemmli, U. K.,** Organization of the higher-order chromatin loop: Specific DNA attachment sites on nuclear scaffold, *Cell,* 39, 223—232, 1984.

60. **Mitchell, H. and Lipps, L.,** Rapidly labeled proteins on the salivary gland chromosomes of *Drosophila melanogaster, Biochem. Genet.,* 13, 585—627, 1975.

61. **Moore, M., Schaack, J., Baim, S. B., Morimoto, R. I., and Schenk, T.,** Induced heat-shock mRNAs escape the nucleo-cytoplasmic transport block in adenovirus-infected cells, *Mol. Cell. Biol.,* 7, 4505—4512, 1987.

61a. **Mowry, K. L. and Steitz, J. A.,** snRNP mediators of 3′ end processing: functional fossils?, *Trends Biochem. Sci.,* 13, 447—451, 1988.

61b. **Muhich, M. L. and Boothroyd, J. C.,** Polycistronic transcripts in trypanosomes and their accumulation during heat shock: evidence for a precursor role in mRNA synthesis, *Mol. Cell. Biol.,* 8, 3837—3846, 1988.

61c. **Muhich, M. L., Hsu, M. P., and Boothroyd, J. C.,** Heat-shock disruption of trans-splicing in trypanosomes — effect on hsp70, hsp85 and tubulin messenger RNA synthesis, *Gene,* 82, 169—175, 1989.

61d. **Muhich, M. L. and Boothroyd, J. C.,** Synthesis of trypanosome hsp70 mRNA is resistant to disruption of trans-splicing by heat shock, *J. Biol. Chem.,* 264, 7107—7110, 1989.

62. **Nevins, J. R.,** Induction of the synthesis of a 70,000 dalton mammalian heat shock protein by the adenovirus E1A gene product, *Cell,* 29, 913—919, 1982.

62a. **Newport, J. W. and Forbes, D. J.,** The nucleus: Structure, function, and dynamics, *Annu. Rev. Biochem.,* 56, 535—565, 1987.

62b. **Nishida, E., Iida, K., Yonezawa, N., Koyasu, S., Yahara, I., and Sakai, H.,** Cofilin is a component of intranuclear and cytoplasmic actin rods induced in cultured cells, *Proc. Natl. Acad. Sci. U.S.A.,* 84, 5262—5266, 1987.

63. **Nover, L., Munsche, D., Ohme, K., and Scharf, K.-D.,** Ribosome biosynthesis in heat shocked tomato cell cultures. I. Ribosomal RNA, *Eur. J. Biochem.,* 160, 297—304, 1986.

64. **O'Connor, D. and Lis, J. T.,** Two closely linked transcription units within the 63B heat shock puff locus of *D. melanogaster* display strikingly different regulation, *Nucl. Acids Res.,* 9, 5075—5092, 1981.

64a. **Ohta, Y., Nishida, E., Sakai, H., and Miyamoto, E.,** Dephosphorylation of cofilin accompanies heat shock-induced nuclear accumulation of cofilin, *J. Biol. Chem.,* 264, 16143—16148, 1989.

64b. **Ornelles, D. A. and Penman, S.,** Prompt heat shock and heat-shifted proteins associated with the nuclear matrix intermediate filament scaffold in *Drosophila melanogaster* cells, *J. Cell Sci.,* 95, 393—404, 1990.

65. **Parag, H. A., Raboy, B., and Kulka, R. G.,** Effect of heat shock on protein degradation in mammalian cells: Involvement of the ubiquitin system, *EMBO J.,* 6, 55—63, 1987.

65a. **Parker, K. A. and Bond, U.,** Analysis of pre-rRNAs in heat shocked HeLa cells allows identification of the upstream termination site of human polymerase I transcription, *Mol. Cell. Biol.,* 9, 2500—2512, 1989.

65b. **Parry, H. D., Scherly, D., and Mattaj, I. W.,** 'Snurpogenesis': the transcription and assembly of UsnRNP components, *Trends Biochem. Sci.,* 14, 15—19, 1989.

66. **Pauli, D., Tonka, C.-H., and Ayme-Southgate, A.,** An unusual split *Drosophila* heat shock gene expressed during embryogenesis, pupation and in testis, *J. Mol. Biol.,* 200, 47—53, 1988.

67. **Pederson, T.,** Nuclear RNA-protein interactions and messenger RNA processing, *J. Cell Biol.,* 97, 1321—1326, 1983.

68. **Pekkala, D., Heath, I. B., and Silver, J. C.,** Changes in chromatin and the phosphorylation of nuclear proteins during heat shock of *Achlya ambisexualis, Mol. Cell. Biol.,* 4, 1198—1205, 1984.

68a. **Pinhashi-Kimhi, O., Michalovitz, D., Ben-Zeev, A., and Oren, M.,** Specific interaction between the p53 cellular tumor antigen and major heat shock proteins, *Nature,* 320, 182—185, 1986.

69. **Radford, I. B.,** Effects of hyperthermia on the repair of X-ray induced DNA double strand breaks in mouse L. cells, *Int. J. Radiat. Biol.,* 5, 551—557, 1983.

70. **Rancourt, D. E., Walker, V. K., and Davies, P. L.,** Flounder antifreeze protein synthesis under heat shock control in transgenic *Drosophila melanogaster, Mol. Cell. Biol.,* 7, 2188—2195, 1987.

70a. **Rebbe, N. F., Hickman, W. S., Ley, T. J., Stafford, D. W., and Hickman, S.,** Nucleotide sequence and regulation of a human 90-kDa heat shock protein gene, *J. Biol. Chem.,* 264, 15006—15011, 1989.

71. **Rio, D. C., Laski, F. A., and Rubin, G. M.,** Identification and immunochemical analysis of biologically active *Drosophila* P element transposase, *Cell,* 44, 21—32, 1986.

72. **Ritossa, F.,** A new puffing pattern induced by heat shock and DNP in *Drosophila, Experientia,* 18, 571—573, 1962.

72a. **Rochester, D. E., Winter, J. A., and Shah, D. M.,** The structure and expression of maize genes encoding the major heat shock protein, hsp70, *EMBO J.,* 5, 451—458, 1986.

73. **Roti-Roti, J. L. and Winward, R. T.,** The effects of hyperthermia on the protein-to-DNA ratio of isolated HeLa cell chromatin, *Radiat. Res.,* 74, 159—169, 1978.

74. **Roti-Roti, J. L., Henle, K. J., and Winward, R. T.,** The kinetics of increase in chromatin protein content in heated cells. A possible role in cell killing, *Radiat. Res.,* 78, 522—531, 1979.

75. **Roti-Roti, J. L. and Winward, R. T.,** Factors affecting the heat-induced increase in protein content of chromatin, *Radiat. Res.,* 81, 138—144, 1980.

76. **Roti-Roti, J. L., Uygur, N., and Higashikubo, R.,** Nuclear protein following heat shock: Protein removal kinetics and cell cycle rearrangements, *Radiat. Res.*, 107, 250—261, 1986.
76a. **Rougvie, A. E. and Lis, J. T.,** The RNA polymerase II molecule at the 5′ end of the uninduced hsp70 gene of *D. melanogaster* is transcriptionally engaged, *Cell,* 54, 795—804, 1988.
77. **Rowe, T. C., Wang, J. C., and Liu, L. F.,** In vivo localization of DNA topoisomerase cleavage sites on *Drosophila* heat shock chromatin, *Mol. Cell. Biol.*, 6, 985—992, 1986.
78. **Russnak, R. H. and Candido, E. P. M.,** Locus encoding a family of small heat shock genes in *Caenorhabditis elegans:* Two genes duplicated to form a 3.8-kilobase inverted repeat, *Mol. Cell. Biol.*, 5, 1268—1278, 1985.
78a. **Sanger, J. W., Sanger, J. M., Kreis, T. E., and Jokusch, B. M.,** Reversible translocation of cytoplasmic actin into the nucleus caused by dimethylsulfoxide, *Proc. Natl. Acad. Sci. U.S.A.*, 77, 5268—5272, 1980.
79. **Sass, H.,** RNA polymerase B in polytene chromosomes: Immunofluorescent and autoradiographic analysis during stimulated and repressed RNA synthesis, *Cell,* 28, 269—278, 1982.
80. **Scharf, K.-D. and Nover, L.,** Control of ribosome biosynthesis in plant cell cultures under heat shock conditions. II. Ribosomal proteins, *Biochim. Biophys. Acta,* 909, 44—57, 1987.
80a. **Scheer, U., Hinssen, H., Franke, W. W., and Jokusch, B. M.,** Microinjection of actin-binding proteins and actin antibodies demonstrates involvement of nuclear actin in transcription of lampbrush chromosomes, *Cell,* 39, 111—122, 1984.
81. **Schöffl, F., Rossol, I., and Angermueller, S.,** Regulation of the transcription of heat shock genes in nuclei from soybean *(Glycine max)* seedlings, *Plant. Cell Environ.*, 10, 113—119, 1987.
82. **Sharp, P. A.,** Splicing of messenger RNA precursors, *Science,* 235, 766—771, 1987.
83. **Simon, J. A., Sutton, C. A., and Lis, J. T.,** Localization and expression of transformed DNA sequences within heat shock puffs of *Drosophila melanogaster, Chromosoma,* 93, 26—30, 1985.
84. **Sinibaldi, R. M. and Morris, P. W.,** Putative function of *Drosophila melanogaster* heat shock proteins in the nucleoskeleton, *J. Biol. Chem.*, 256, 10735—10738, 1981.
85. **Small, D., Nelkin, B., and Vogelstein, B.,** The association of transcribed genes with the nuclear matrix of *Drosophila* cells during heat shock, *Nucl. Acids Res.*, 13, 2413—2431, 1985.
85a. **Smith, C. W. J., Patton, J. G., and Nadel-Ginard, B.,** Alternative splicing in the control of gene expression, *Annu. Rev. Genet.*, 23, 527—577, 1989.
85b. **Snutch, T. P., Heschl, M. F. P., and Baillie, D. L.,** The *Caenorhabditis elegans* hsp70 gene family: a molecular genetic characterization, *Gene,* 64, 241—255, 1988.
86. **Soltyk, A., Tropak, M., and Friesen, J. D.,** Isolation and characterization of the $RNA2^+$, $RNA4^+$, and $RNA11^+$ genes of *Saccharomyces cerevisiae, J. Bacteriol.*, 160, 1093—1100, 1984.
87. **Sorger, P. K. and Pelham, H. R. B.,** Purification and characterization of a heat shock element binding protein from yeast, *EMBO J.*, 6, 3035—3042, 1987.
88. **Spadoro, J. P., Copertino, D. W., and Strausbaugh, L. D.,** Differential expression of histone sequences in *Drosophila* following heat shock, *Dev. Genet.*, 7, 133—148, 1986.
89. **Spinelli, G., Casano, C., Gianguzza, F., Ciaccio, M., and Palla, F.,** Transcription of sea urchin mesenchyme blastula histone genes after heat shock, *Eur. J. Biochem.*, 128, 509—513, 1982.
90. **Spradling, A., Pardue, M. L., and Penman, S.,** Messenger RNA in heat shocked *Drosophila* cells, *J. Mol. Biol.*, 109, 559—587, 1977.
91. **Staufenbiel, M. and Deppert, W.,** Different structural systems of the nucleus are targets for SV40 large T antigen, *Cell,* 33, 173—181, 1983.
92. **Steller, H. and Pirrotta, V.,** P transposons controlled by the heat shock promoter, *Mol. Cell. Biol.*, 6, 1640—1649, 1986.
92a. **Stuerzbecher, H.-W., Chumakov, P., Welch, W. J., and Jenkins, J. R.,** Mutant p53 proteins bind hsp 72/73 cellular heat shock-related proteins in SV40-transformed monkey cells, *Oncogene,* 1, 201—211, 1987.
92b. **Szent-Györgyi, C., Finkelstein, D. B., and Garrard, W. T.,** Sharp boundaries demarcate the chromatin structure of a yeast heat-shock gene, *J. Mol. Biol.*, 143, 71—80, 1987.
93. **Tanguay, R. M. and Desrosiers, R.,** Histone methylation and modulation of gene expression in response to heat shock and chemical stress in *Drosophila*, in *Advances in Post-Translational Modifications of Proteins and Ageing,* Zappia, V., Galletti, P., Porta, R., and Wold, F., Eds., Plenum Press, NY, 1988, 353—362.
93a. **Thomas, G. H. and Elgin, S. C. R.,** Protein/DNA architecture of the DNaseI-hypersensitive region of the *Drosophila* hsp26 promoter, *EMBO J.*, 7, 2191—2201, 1988.
94. **Tomasovic, S. P., Turner, G. N., and Dewey, W. C.,** Effect of hyperthermia on nonhistone proteins isolated with DNA, *Radiat. Res.*, 73, 535—552, 1978.
95. **Udvardy, A. and Schedl, P.,** Chromatin organization of the 87A7 heat shock locus of *Drosophila melanogaster, J. Mol. Biol.*, 172, 385—403, 1984.
96. **Udvardy, A., Schedl, P., Sander, M., and Hsieh, T.-S.,** Novel partitioning of DNA cleavage sites for *Drosophila* topoisomerase II, *Cell,* 40, 933—941, 1985.

97. **Udvardy, A., Maine, E., and Schedl, P.,** The 87A7 chromomere. Identification of novel chromatin structures flanking the heat-shock locus that may define the boundaries of higher order domains, *J. Mol. Biol.*, 185, 341—358, 1985.

97a. **Walter, G., Carbone, A., and Welch, W. J.,** Medium tumor antigen of polyomavirus transformation-defective mutant NG59 is associated with 73-kilodalton heat shock protein, *J. Virol.*, 61, 405—410, 1987.

98. **Warters, R. L. and Roti-Roti, J. L.,** Excision of X-ray-induced thymine damage in chromatin from heated cells, *Radiat. Res.*, 79, 113—121, 1979.

99. **Warters, R. L., Brizgys, L. M., Sharma, R., and Roti Roti, J. L.,** Heat shock (45°C) results in an increase of nuclear matrix protein mass in HeLa cells, *Int. J. Radiat. Biol.*, 50, 253—268, 1986.

99a. **Warters, R. L., Brizgys, L. M., and Lyons, B. W.,** Alteration in the nuclear matrix protein mass correlates with heat-induced inhibition of DNA single-strand-break repair, *Int. J. Radiat. Biol.*, 52, 299—314, 1987.

100. **Warters, R. L., Brizgys, L. M., and Lyons, B. W.,** Alteration in the nuclear matrix protein mass correlate with heat-induced inhibition of DNA single-strand-break repair, *Int. J. Radiat. Biol.*, 52, 299—314, 1987.

100a. **Welch, W. J. and Suhan, J. P.,** Morphological study of the mammalian stress response: characterization of changes in cytoplasmic organelles, cytoskeleton, and nucleoli, and appearance of intranuclear actin filaments in rat fibroblasts, *J. Cell Biol.*, 101, 1198—1211, 1985.

101. **Welch, W. J. and Mizzen, L. A.,** Characterization of the thermotolerant cell. II. Effects on the intracellular distribution of HSP70, intermediate filaments, and snRNP's, *J. Cell Biol.*, 106, 1117—1130, 1988.

101a. **Winter, J., Wright, R., Duck, N., Gasser, C., Fraley, R., and Shah, D.,** The inhibition of *Petunia* hsp 70 mRNA processing during $CdCl_2$ stress, *Mol. Gen. Genet.*, 211, 315—319, 1988.

101b. **Wright-Sandor, L. G., Reichlin, M., and Tobin, S. L.,** Alteration by heat shock and immunological characterization of *Drosophila* small nuclear ribonucleoproteins, *J. Cell Biol.*, 108, 2007—2016, 1989.

101c. **Wu, C. H., Caspar, T., Browse, J., Lindquist, S., and Somerville, C.,** Characterization of an HSP70 cognate gene family in *Arabidopsis, Plant Physiol.*, 88, 731—740, 1988.

102. **Wu, C.,** The 5′ ends of *Drosophila* heat shock genes in chromatin are hypersensitive to DNase I, *Nature,* 286, 854—860, 1980.

103. **Wu, C.,** Two protein-binding sites in chromatin implicated in the activation of heat-shock genes, *Nature,* 309, 229—234, 1984.

104. **Wu, C., Wong, Y.-C., and Elgin, S. C. R.,** The chromatin structure of specific genes. II. Disruption of chromatin structure during gene activity, *Cell,* 16, 807—814, 1979.

105. **Wu, C., Pelham, P. M., Livak, K. J., Holmgren, R., and Elgin, S. C. R.,** The chromatin structure of specific genes. I. Evidence for higher order domains of defined DNA sequence, *Cell,* 16, 797—806, 1979.

106. **Yost, H. J. and Lindquist, S.,** RNA splicing is interrupted by heat shock and is rescued by heat shock protein synthesis, *Cell,* 45, 185—193, 1986.

107. **Yost, H. J. and Lindquist, S.,** Translation of unspliced transcripts after heat shock, *Science,* 242, 1544—1548, 1988.

Chapter 10

SYNTHESIS AND DEGRADATION OF RIBOSOMES

K.-D. Scharf and L. Nover

TABLE OF CONTENTS

10.1. THE NORMOTHERMIC STATE

Ribosome biosynthesis is a highly complex, ordered assembly of 3 to 4 rRNA species with 70 to 80 ribosomal proteins. To facilitate understanding of heat shock (hs) effects on this system, we briefly summarize general traits of ribosome formation in eukaryotic cells. Investigations on hs-induced changes were mainly done with this group of organisms.[4,21,24,49,56,56a,64,69]

Three rRNAs are generated in the nucleolus from a joint primary transcript of greatly varying length. Different amounts of intercistronic spacer RNA, not the size of the rRNAs, are the main reason for this variation. The primary transcript is 4.2×10^6 Da for HeLa cells and many other vertebrates (sedimentation in sucrose gradients with about 45S), 2.4 to 2.8×10^6 Da (38- to 40S) for plants[21] and 2.5×10^6 Da (38S) for yeast.[64] Processing involves site-specific base modifications (mainly methylation) and an ordered sequence of endonucleolytic cleavages. The principle, valid for all eukaryotic cells, is exemplified for HeLa cells:

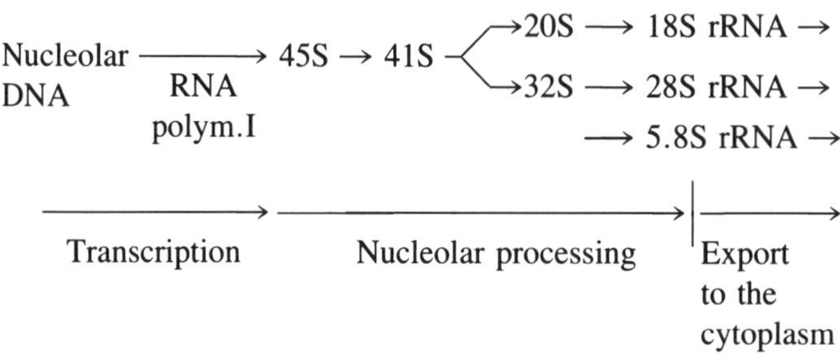

Besides the 3 mature rRNAs transcribed by RNA polymerase I from nucleolar rRNA genes, the fourth rRNA of eukaryotic ribosomes (5S rRNA) is transcribed outside of the nucleoli by RNA polymerase III. For hs-induced peculiarities of 5S rRNA synthesis see Section 10.3.

Accumulation and processing of pre-rRNA is dependent on the addition of ribosomal proteins (rproteins) already *in statu nascendi.* The primary product, released from the nucleolar transcriptional machinery, is not polycistronic pre-rRNA but rather an 80S preribosome, and only this can be further processed. 80S preribosomes contain a distinct subset of early binding rproteins. Further rproteins are added during later steps of maturation in the nucleus or even after export of the preribosomal subunits to the cytoplasm.[4,24] Differences in rprotein composition of precursor vs. mature ribosomal particles, the basis of discrimination between rproteins added early and late are clearly visible from appropriate pulse-labeling experiments[53] (see Figure 10.3).

Usually there is a strict coordination of rRNA and rprotein synthesis, i.e., no significant nucleolar or cytoplasmic pools of free rproteins are detected. Inhibition of cytoplasmic protein synthesis, e.g., by cycloheximide, rapidly blocks also preribosome assembly and processing. Pre-rRNA transcribed under these conditions is degraded. The same is true for an excess of rproteins formed in the presence of actinomycin D or other inhibitors of RNA synthesis. Details of these control mechanisms were summarized by Hadjiolov.[24] His "autogenous regulation model" of ribosome biosynthesis is based on the "excess" synthesis of nonnucleolar components of the ribosomal assembly line, i.e., of 5S rRNA and rproteins, whose continuous supply is required for effective transcription of pre-rRNA. Accumulation of unprocessed or partially processed preribosomal material leads to autorepression of the nucleolar transcription machinery.

The morphology of the nucleolus as the cellular site of ribosome assembly is characterized by electron dense material of fibrillar and granular appearance. Pulse chase experiments using labeled uridine revealed that the fibrillar regions very likely contain the rRNA genes (rDNP), the transcription apparatus, and the preribosomes *in statu nascendi,*[18,55,56,63] whereas the granular parts contain the processed intermediates. Dynamic changes of fibrillar and granular components can be observed under different states of ribosome biosynthesis.[24]

Studies of hs effects on ribosome biosynthesis gave major new insights into details of this complex process and of the regulatory mechanisms involved. This is especially true for tomato cell cultures where the four major facets, i.e., synthesis of rRNA and rproteins, nucleolar ultrastructure and the role of HSP70, were investigated in the same experimental system under carefully optimized conditions of hs and recovery.[43,44,46a,53] These results are the main bases for the following sections and for the comparison with data reported for animals and yeast. In most cases, the outcome was very similar. Discrepancies may be more apparent than real and presumably result from two factors: (1) lack of variation of the experimental conditions, i.e., only a single or few points of the continuum of hs-induced changes were analyzed and (2) restriction of the analyses to one or two of the four main aspects cited above.

10.2. rRNA SYNTHESIS UNDER HEAT SHOCK CONDITIONS

Almost immediately after the onset of even a moderate hs, formation of mature rRNA as assayed by labeling of cytoplasmic ribosomes after feeding appropriate precursors of nucleic acid synthesis, is blocked. Ribosome biosynthesis as analyzed by this simple method is among the most heat-sensitive functions of cells. This is true for vertebrate cells,[1,5,6,14,52a,63] for *Drosophila,*[2,3a,17,57] *Tetrahymena,*[9,29,70] yeast,[22,32,61b,64] and plants.[16,20,46a,53]

The effect is illustrated in Figure 10.1 using plant cell cultures and separation of nucleic acids by polyacrylamide gel electrophoresis. Labeling of the two major rRNA species of mature ribosomes (A, C, E) is discontinued during hs (C), but restored in the recovery period (E). However, inspection of labeled RNA from the nuclear compartment (B, D, F) shows that synthesis of the pre-rRNA is not affected to the same extent. Compared to the control labeling at 25°C (B), there is an almost unchanged amount of the polycistronic pre-rRNA (arrows in B, D, F), i.e., the primary effect of hs is a block of processing. Synthesis of pre-rRNA goes on, though at a reduced rate. Careful studies of the influence of hs temperature and duration on rRNA synthesis[46a] revealed a continuum of changes with undiminished transcription and reduced processing at mild hs (36 to 37°C), reduced transcription and blocked processing at intermediate hs (39°C) and total absence of all measurable pre-rRNA labeling at 40°C. In contrast to processing of pre-rRNA, inhibition of synthesis is not immediate and increases with the extension of the hs. On the other hand, full recovery of the processing activity is evidently also faster.[5-7,17,46a]

A main factor contributing to the variability of effects on the state of ribosome biosynthesis under hs conditions is illustrated in Figure 10.2 by comparing data from tomato cell cultures, which were either preinduced or not. A 15 min pulse hs followed by 3 to 4 h of expression of the thermotolerant state at 25°C (see Chapter 17) leads to a marked improvement of synthesis and processing of pre-rRNA in a subsequent hs period. At 37°C both parts of ribosome biosynthesis proceed almost normally, whereas at 40°C a complete separation of pre-rRNA synthesis and its effective processing in the recovery period can be achieved. Cytoplasmic ribosomes generated from their hs precursor particles (samples 5 → 6 of Figure 10.2) are normal by all criteria of structure and function.[46a] It is very likely that the accumulation of HSP70 in the nucleolus is an important factor in the protection and/or repair of structure and function of the machinery engaged in ribosome biosynthesis (see Section 10.6).

Major conclusions from analyses using other hs systems (animals, yeast, and plants) are in agreement with the picture just outlined: (1) the primary effect of hs is a block of pre-rRNA processing and (2) transcription of pre-rRNA by RNA polymerase I only slowly

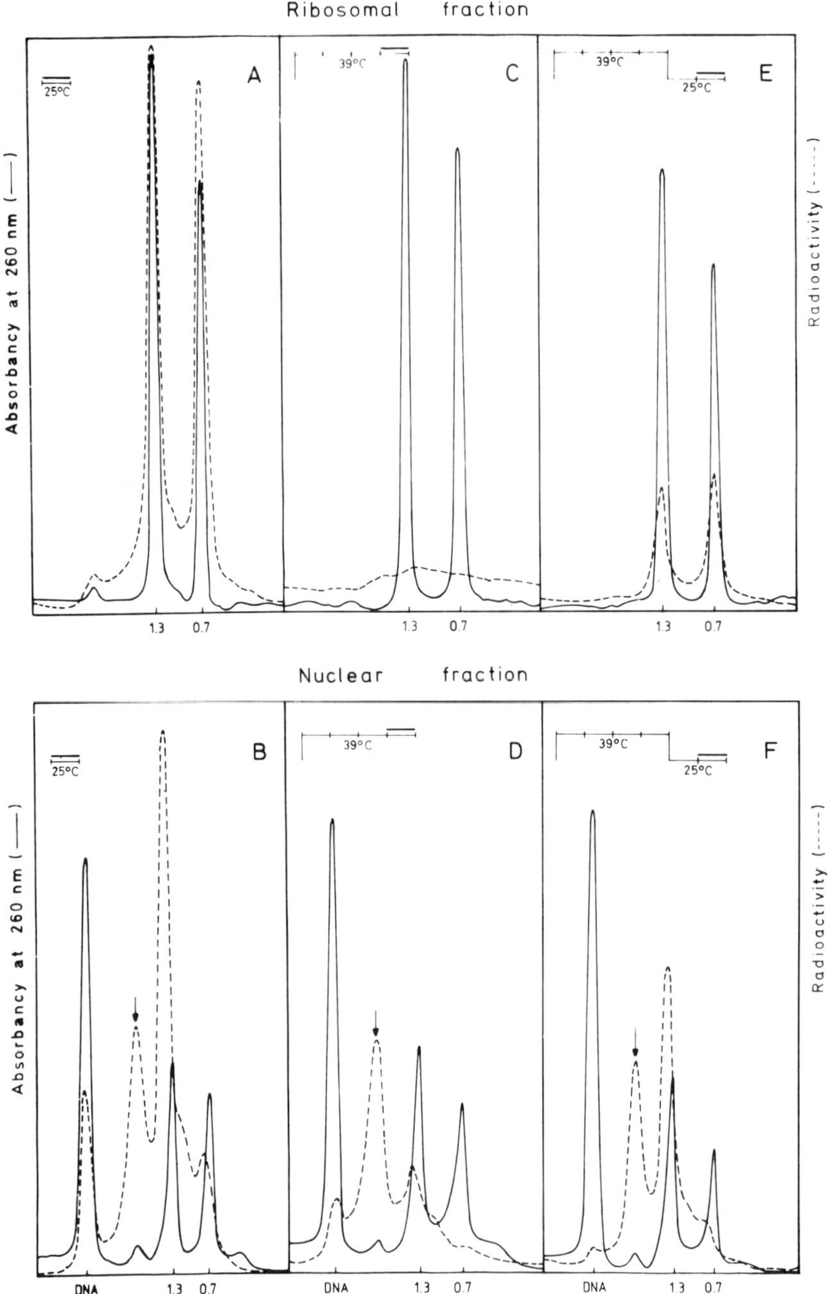

FIGURE 10.1. Synthesis and processing of ribosomal RNA in heat-shocked tomato cell cultures. Heat shock (hs) conditions and labeling periods (1 h) with ^{32}P-phosphate are indicated by the pictographs. Total nucleic acids from the cytoplasmic and from crude nuclear fractions were separated on 2.4% polyacrylamide/agarose gels. Full lines represent the absorbancy scanned at 260 nm. Positions of the DNA and of the two major rRNAs are indicated. The rRNA of 1.3×10^6 Da corresponds to the 28S and the rRNA of 0.7×10^6 Da corresponds to the 18S rRNA of the processing scheme given in the text. The dashed lines represent the distribution of the radioactivity derived from autoradiographs. Contrasting to A and E, peaks of radioactivity in the nuclear fractions (B, D, F) are not coincident with the positions of the two marker rRNAs because they represent the slightly larger precursors. The polycistronic precursor of about 2.5×10^6 Da is marked by arrows. (From Nover, L., Munsche, D., Ohme, K., and Scharf, K.-D., *Eur. J. Biochem.*, 160, 297, 1986. With permission.)

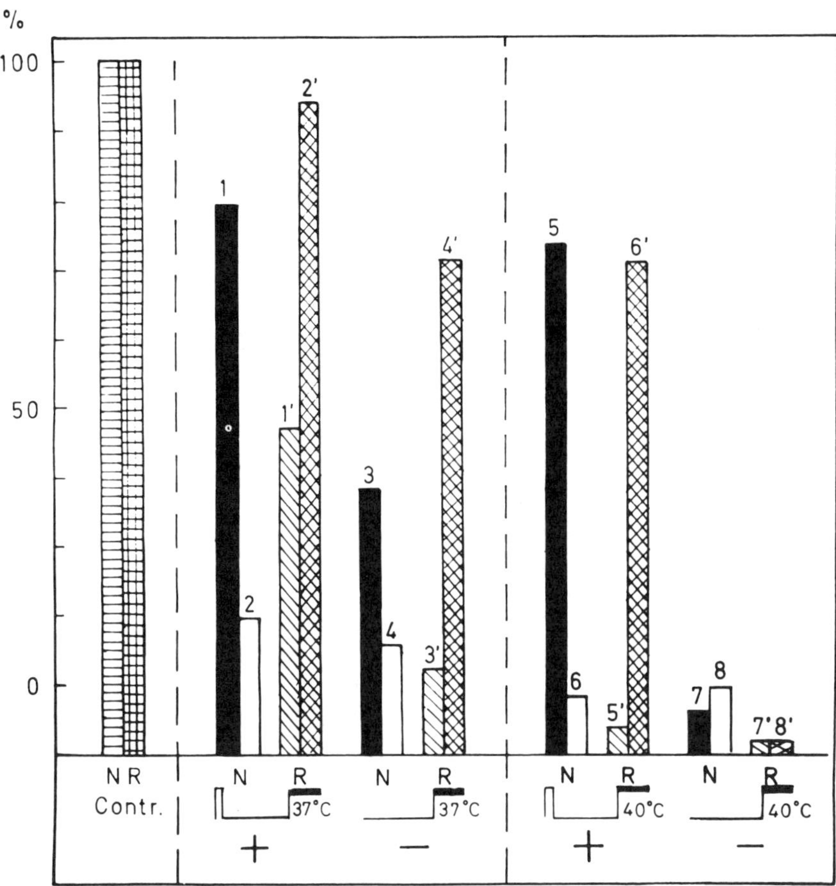

FIGURE 10.2. Protection of synthesis and processing of preribosomes by prior heat shock (hs) conditioning. The figure is based on a similar kind of analysis to that shown in Figure 10.1. The full data are contained in Nover et al.[46a] Radioactivity in mature ribosomal RNAs (R) and in the pre-rRNAs of nuclear fractions (N) was estimated by integration of corresponding densitometer scans of autoradiographs and referred to the values obtained under control conditions as 100%. The hs regimes, indicated by the pictographs at the bottom, are characterized by the use of preinduced vs. nonpreinduced cultures and by the 1 h ^{32}P-phosphate labeling at 37°C (left part) and 40°C (right part), respectively. Analysis of radioactivity distribution was done immediately after the hs labeling (samples 1, 3, 5, 7) or after 1 h recovery at 25°C in the presence of actinomycin D (sampels 2, 4, 6, 8). The results, especially at 40°C, show the profound protective influence of a conditioning pulse hs on the synthesis and subsequent processing of pre-rRNA. (From Nover, L., Ed., *Heat Shock Response of Eukaroytic Cells,* G. Thieme Verlag, 1984. With permission.)

declines.[1,2,3a,5-7,16,17,52a,63] Studies of the nucleolar ultrastructure in animal cells including labeling of pre-rRNA with ^{3}H-uridine in connection with microautoradiographic methods confirmed these results. Precursor labeling immediately before or after the onset of the hs treatment is retained in distinct areas of the nucleolus, whereas labeling during later periods of the hs was not observed.[14,55] Generation of mature rRNA is clearly more heat sensitive than that of mRNA and small cytoplasmic RNAs as demonstrated by Spradling et al.[57] for *Drosophila* S2 cells. This statement although almost trivial, reflects the basic reprogramming of gene expression during hs, from ''housekeeping functions'' to stress-specific functions connected with mass synthesis of new mRNAs.

An exciting variation to the theme was reported by Labhart and Reeder[34] for *Xenopus laevis* oocytes. Within minutes after the onset of a hs at 34°C, two types of changes in the processing of pre-rRNA are observed. First, in agreement with results in other systems preexisting 40S pre-rRNA is retained and completely stabilized in the nucleolus. Second, initiation and elongation by RNA polymerase I is not impaired, but abnormally large precursor molecules are formed (10 to 20 kb) which span at least several rDNA transcription units. Defective termination and processing at the 3′ end of the transcription unit so well as an unusual stability of the spacer transcript are the main factors contributing to this result. The giant hs pre-rRNA is slowly degraded, but not to the 40S pre-rRNA normally formed as the primary product of the transcription process. Similar results with deficiency of termination and partial read-through into the next rDNA transcription unit were also reported for HeLa cells if incubated under mild hs conditions (42 to 43°C).[46c]

Under conditions of severe hs, the differences of temperature sensitivity between synthesis and processing of pre-rRNA may disappear. Synthesis itself is severely repressed. This is evident from samples 7 and 8 in Figure 10.2. Similar results have also been reported for yeast[32] and vertebrate cells.[1,5,14,54] The molecular basis for this immediate halt of RNA polymerase I activity is not clear. Recovery requires several hours, despite the fact that the activity of the enzyme measurable *in vitro* is essentially unchanged by the hs treatment. These results obtained with Chinese hamster ovary cells[5,6] need reinvestigation and confirmation using other hs systems.

The fate of the pre-rRNA synthesized under hs conditions was generally neglected. Our results with tomato cell cultures prove that a complete processing with generation of functional ribosomal subunits in the recovery period is possible but contingent on a hs preconditioning (Figure 10.2). The only system that has been investigated in more detail are Chinese hamster ovary cells.[1,5-7] In this case, the block of pre-rRNA processing is connected with defective methylation. Restoration of processing during the recovery period is rapid for the nucleolytic activities, but delayed for methylation, i.e., in CHO cells the whole process leads to defective rRNAs. Part of an aberrant 4.6×10^6 Da pre-rRNA is maintained in the nucleolus even after 7 h of recovery. It remains to be shown whether real differences exist between plant and animal cells in this respect or whether they are due rather to peculiarities of the experimental conditions. We show in the following sections that, also in plant cell cultures, the hs pre-rRNA is functionally "intact" only for a short time (less than 1 h) at the transition between completed synthesis and aberrant partial processing giving rise to ribosomal wastage.

10.3. GENERATION OF 5S rRNA

One of the two small ribosomal RNAs (5S rRNA) is not synthesized in the nucleolar complex, but is transcribed outside by RNA polymerase III. Few papers deal with this part of the hs response.[7,31,40,52,52a] Pre-5S rRNA is formed as a 135-nucleotide precursor to the mature 120 nucleotide species. In *Drosophila* cells, the precursor form accumulates during hs and can be processed in a recovery period.[31,52] Investigations of the *in vitro* transcription patterns of isolated nuclei from control and hs *Drosophila* embryos confirmed that the activity of the 5S rRNA genes is unchanged.[40]

10.4. RIBOSOMAL PROTEIN SYNTHESIS AND PRERIBOSOME ASSEMBLY

Even the transient existence but, in particular, the faithful processing of hs pre-rRNA (Figure 10.2) is dependent on the cotranscriptional assembly of preribosomal RNP, i.e., on ongoing translation and nuclear import of ribosomal proteins. Experimental proof for this was achieved, first, by use of thermotolerant cell cultures allowing the separation of synthesis and processing of preribosomes under conditions of severe hs and, second, by special extraction procedures for preribosomal proteins from nuclear fractions.[53] The results, summarized in Figure 10.3, gave new insights into ribosome biosynthesis in general:

1. Under control conditions (Figure 10.3A and B) early binding (nucleolar) rproteins (A) can be clearly distinguished from late binding (cytoplasmic) rproteins. The latter group is marked in Figure 10.3A by open arrowheads. As was outlined in Section 10.1, this result for plant cell cultures confirmed the elaborate investigations with animal systems summarized by Hadjiolov.[24] The normothermic state in plants is "normal". An interesting completion of our knowledge is given in the inserts. A special protein of the small ribosomal subunit (S6) may play a prominent role for the hs response because of its rapid dephosphorylation (see Section 11.5.2). rProtein S6 belongs to the primary binding proteins, but it is unphosphorylated in the nuclear fraction. The usual pattern of phosphorylated isoforms (a to d in the insert of Figure 10.3B) is established only in the cytoplasm, i.e., late in ribosome maturation.
2. With few exceptions (see below), most of the newly synthesized rproteins detected at 25°C (B) disappear from the cytoplasmic fraction under hs conditions. However, this is not true for the rproteins in the nuclear fraction (C), confirming the results of the rRNA analysis: processing is blocked but synthesis of preribosomes goes on. Moreover, preribosome-associated proteins can be chased into cytoplasmic ribosomes, if cells are allowed to recover at 25°C (Figure 10.3E and F).
3. A remarkable but hitherto unexplained feature of ribosome biosynthesis and function in all types of organisms[24,64] is the existence of "exchange proteins", which can be incorporated even into mature ribosomal subunits. They evidently constitute parts of the ribosome with a considerably shorter half life than the bulk of rproteins. A convenient means to characterize exchange rproteins is to block ribosome synthesis by actinomycin D or by hs (Figure 10.3D). Using either type of inhibition in plant cell cultures, our comparison gave very similar but not identical results,[53] i.e., there are special hs exchange proteins.
4. The rprotein composition of hs preribosomes is not identical with control preribosomes. Open arrowheads in Figure 10.3C point to proteins absent or in reduced amount in comparison to Figure 10.3A. The most likely explanation is a reduced supply of these proteins because of inefficient translation of the corresponding mRNAs (see Section 11.3). Alternatively, they may constitute semi-early binding proteins added after the initial cleavage of the preribosome to the preribosomal subunits. This and many other problems can only be solved by isolation and *in vitro* processing of the pre-rRNP.
5. An important contribution to our understanding about altered nucleolar functions emerged when we investigated the fate of labeled pre-rRNP after extending the hs period (Figure 10.3C to F) by one more hour at 40°C under chase conditions. The rprotein distribution was virtually identical to that seen in Figure 10.3C and D. However, the pre-rRNP material had a new quality. If these cultures were allowed to recover at 25°C, the labeled rproteins from the nuclear fraction disappeared without formation of mature ribosomes. These results indicate two fundamentally different states of the nucleolar pre-rRNP under hs conditions: first, freshly synthesized and intact for eventual processing in a recovery period and, second, senescent, presumably with aberrantly cleaved rRNA. The second type cannot be processed to mature ribosomal subunits, but is rapidly degraded in the recovery period (ribosomal wastage). Evidently disposal of the ribosomal wastage is also blocked under hs conditions with the result of an abnormal structural reorganization of the nucleolus (see following Section). The two states of hs pre-rRNP were also characterized by rRNA labeling under similar conditions (data not shown). These results give a possible explanation for the defective processing of hs pre-rRNA in CHO cells mentioned above.[6,7]

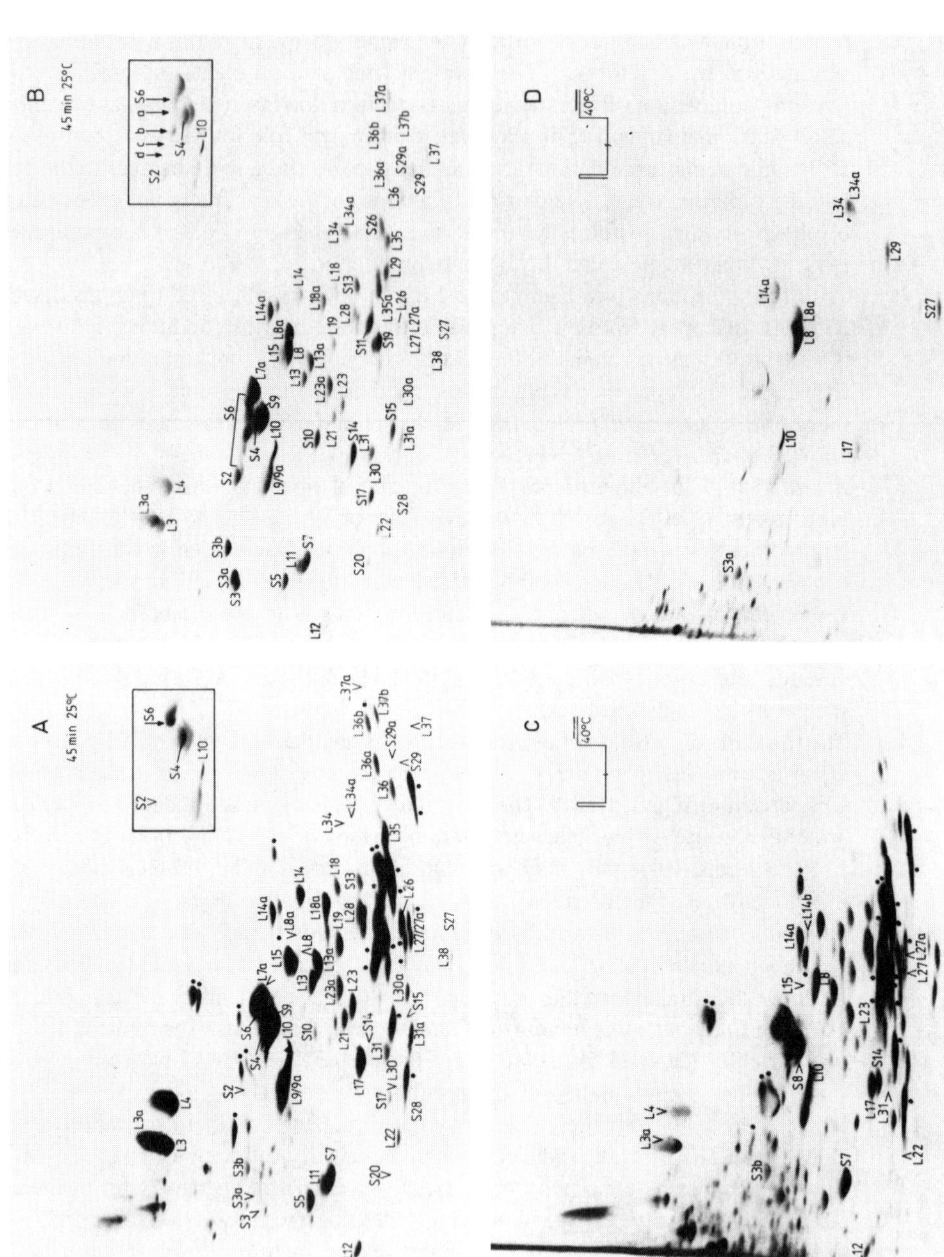
Nuclear fraction
Ribosomal fraction
A
45 min 25°C
B
45 min 25°C
C
40°C
D
40°C

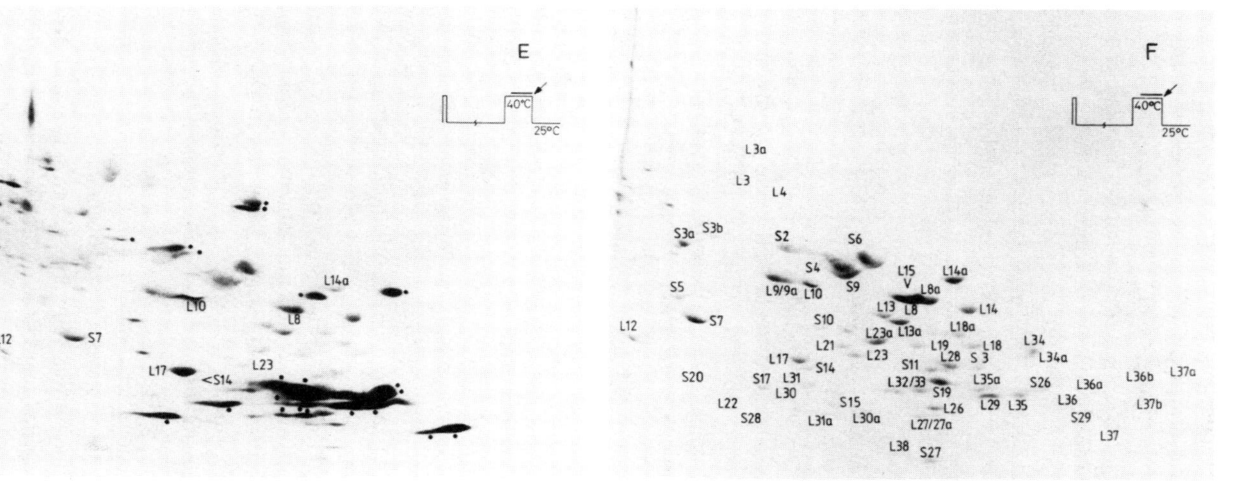

FIGURE 10.3. Synthesis of ribosomal proteins and distribution between nuclear and cytoplasmic rRNP. Ribosomal proteins of tomato cell cultures were labeled for 45 min with ^{14}C-leucine under control (A, B) or heat shock (hs) conditions (C and D). In samples E and F the pattern of rprotein labeling was analyzed after 1 h hs at 40°C followed by 1 h recovery at 25°C under chase conditions (washing of cells and addition of 100-μg/ml cold leucine + 10 μg/ml actinomycin). The principles of labeling and hs treatments are indicated by the pictographs. The two-dimensional separation of the basic rproteins is based on a basic-urea polyacrylamide gel in the first and an SDS-polyacrylamide gradient gel in the second dimension. Dots in A, C, and E (nuclear fractions) mark nonribosomal basic proteins (histones, HMG proteins), whose synthesis is barely affected by hs. Open arrowheads were used in A to indicate late-adding rproteins mainly found as parts of cytoplasmic ribosomal subunits (B). In contrast, open arrowheads in C denote primary binding rproteins with strongly diminished label under hs conditions. (From Scharf, K.-D. and Nover, L., *Biochim. Biophys. Acta,* 909, 44, 1987. With permission.)

Examples of similar analyses integrating results on both parts of ribosome biosynthesis were reported by Warner and co-workers for yeast[64] and by Bell et al.[3a] for *Drosophila*. In yeast, the accumulation of unprocessed pre-rRNA is primarily due to a coordinate, transient depletion of rprotein mRNAs. They are short-lived, and their synthesis is blocked even under mild hs conditions.[22,33,51] In fact, several factors contribute to the overall effect. First, investigations with a ts mutant, defective in pre-mRNA processing at the nonpermissive temperature, make it very likely that not transcription but processing of the mostly intron-containing pre-mRNAs may be rate limiting for the supply of mature rprotein mRNAs in yeast[33] (see also Section 9.5). Second, a general, approximately twofold reduction of transcription is combined with a specific nucleolytic breakdown of rprotein mRNAs induced by hs at 36°C. Use of transcription inhibitor *o*-phenanthroline, added prior to the temperature shift, indicates that the 3- to 5-fold reduction of the halflife of rprotein mRNAs requires preceding synthesis of an unknown hs-induced factor.[29a] In *Drosophila* S2 cells[3a] rprotein synthesis is greatly reduced during severe hs at 37°C. Pre-rRNA formed under these conditions is rapidly degraded. Unfortunately, ribosome biosynthesis in preconditioned cells was not investigated.

10.5. CHANGING NUCLEOLAR ULTRASTRUCTURE

Biosynthesis of ribosomes is a particularly elaborate, material- and energy-intensive process of cells. The need for new ribosomes, adjusted to the cell cycle activity and/or metabolic situation, varies over a broad range with profound influences on nucleolar ultrastructure.[18,24] It is not surprising that the rapid breakdown of ribosome biosynthesis during hs evokes similar changes of nucleolar morphology, e.g., observed in *Neurospora*[60,61] the slime mold *Physarum*,[10,37] in *Tetrahymena*,[13] plants,[20,43,44a,50] *Drosophila*,[3] and vertebrates.[14,28,38,47,54,55,61a,63] Although there is some variation in the details as a result of different hs regimes, results are basically similar for all types of eukaryotes. They are documented in Figure 10.4 for tomato cell cultures. The usually fine granular material of preribosomal intermediates disappears. This process is frequently described as ''degranulation'' of the nucleolus. Instead, with the ongoing hs, larger RNP granules are formed (Figure 10.4A to C), which were called perichromatin granules in animal cells.[28,54,55,63] The extent of nucleolar disintegration is particularly striking in these rapidly dividing tomato cells. The final stage of the hs nucleolus (Figure 10.4C) is characterized by several fibrillar centers of rDNP surrounded by circles of granular pre-rRNP (arrowheads) and flanked by an increasing mass of granular RNP material.[43,46] Our extensive documentation of electron microscopic data show that this type of nucleolar disintegration with mass accumulation of preribosomal granules is particularly prominent in rapidly dividing cells with a high rate of ribosome synthesis before the onset of hs. In resting cultures these pictures are rare. Instead, hs induces a condensation of nucleoli, which may be comparable to the shrinkage reported for animal systems[38,47] or for *Physarum*.[37]

The biochemical analyses, outlined in Sections 10.2 and 10.4, led us to a new interpretation of the morphological data, defining three states of preribosomal material in hs nucleoli: (1) preribosomes *in statu nascendi* in the fibrillar remnants, (2) freshly synthesized preribosomal granules in close vicinity of the fibrillar centers (arrowheads in Figure 10.4C), and (3) the mass of granular material, which is defectively processed ribosomal wastage. Only the small fraction of type 2 pre-rRNP can be eventually processed to give mature ribosomal subunits. In short-term labeling experiments, used for the analyses shown in Figures 10.3 and 10.4, this small part of pre-rRNP is predominantly labeled. Processing of intact preribosomal material and also disposal of the ribosomal wastage is only possible in the recovery period. According to the ''autogenous regulation model'' of Hadjiolov,[24] it may be the accumulation of this ribosomal wastage, which leads to the increasing inhibition of RNA polymerase I activity during long-term hs periods and, vice versa, recovery of the full transcription activity may be contingent on the complete removal of the defective rRNP.

10.6. ROLE OF HSPs 70 AND 110 IN THE NUCLEOLUS

A remarkable property of HSP70 is its reversible recompartmentation from a predominantly cytoplasmic localization under control to a nuclear/nucleolar localization under hs conditions (see the summary in Sections 2.3.2 and 16.2). Immunological analyses, exemplified in Figures 16.1 to 16.3 indicated that in plant and vertebrate cells, most of the nuclear HSP70 is in the nucleolus.[15,44,46b,47,65-68] In hs *Drosophila* cells this unequal distribution between the nucleolar and nucleoplasmic compartments is not so pronounced.[3,62]

The hs-induced changes of nucleolar structure and function and the rapid reversibility in the recovery period are evidently connected to the presence of HSP70:

1. The nucleolar HSP70 in plants and vertebrates is tightly bound and not removed by treatment with nonionic detergents, whereas most of the nucleoplasmic HSP70 is extracted under these conditions.[15,35] This simple test reflects different functional states dependent on different parts of the HSP70 molecule. The highly conserved N-terminal region is mainly responsible for nucleolar binding.[42]
2. Electronmicroscopic fine localization using microautoradiographic (Figure 10.4D) and immunological techniques (Figure 10.4E) shows that HSP70 is almost exclusively found in the region with granular preribosomal material. The fibrillar centers of pre-rRNP synthesis are largely free.[44,46a,68]
3. The state of acquired thermotolerance, as illustrated in the preceding sections for the nucleolar machinery of plant cells, is decisively influenced by the HSP70 level. The recovery of normal nucleolar morphology coincides with the release of HSP70 from its tight binding sites in the organelle. Pelham and co-workers[42,47] introduced a modified ("tagged") *Drosophila* hsp70 gene into mouse and monkey COS cells (see Table 6.1, constructs B7, B16). Two results are worth noticing: the foreign HSP70 functions normally in the heterologous cells, i.e., it is mainly localized in the nucleolus after hs. On the other hand, overproduction of the Dm-HSP70 improves the protection and recovery of the nucleolar apparatus.
4. If HSP70-containing nuclei are isolated from monkey COS and chicken embryo fibroblasts, most of the HSP70 can be released in an energy-dependent process by addition of adenosinetriphosphate (ATP) but not of other nucleotides or nonhydrolyzable ATP analogs.[15,35] This led to the hypothesis that HSP70 may function in the stabilization of preribosomal intermediates and their rapid export in the recovery period.[47,48] According to our data this shuttle function mainly concerns the removal of the ribosomal wastage. It was discussed in Section 2.3.2 that all members of the HSP70 family are ATP-binding proteins, frequently involved in the stabilization and ordered assembly of protein complexes and/or in the disposal of aberrant intermediates.

In vertebrate cells there is another type of nucleolus-associated HSP (HSP110), but contrasting to HSP70, it is a constitutive, dominant phosphoprotein also of nucleoli of nonstressed cells.[58] Immune electronmicroscopical localization showed that HSP110 is mainly bound to the fibrillar part of the nucleoli.[53a] Although definite proof is lacking, the prominent nucleolar phosphoproteins pp95 or pp100 described by Caizergues-Ferrer et al.[11,12] and Bourbon et al.[8] may be related to or identical with HSP110.[58] The function of HSP110 for preribosomal processing may be connected with an autoproteolytic cleavage of the protein.[8] The existence of an analogous protein in other systems is unclear.

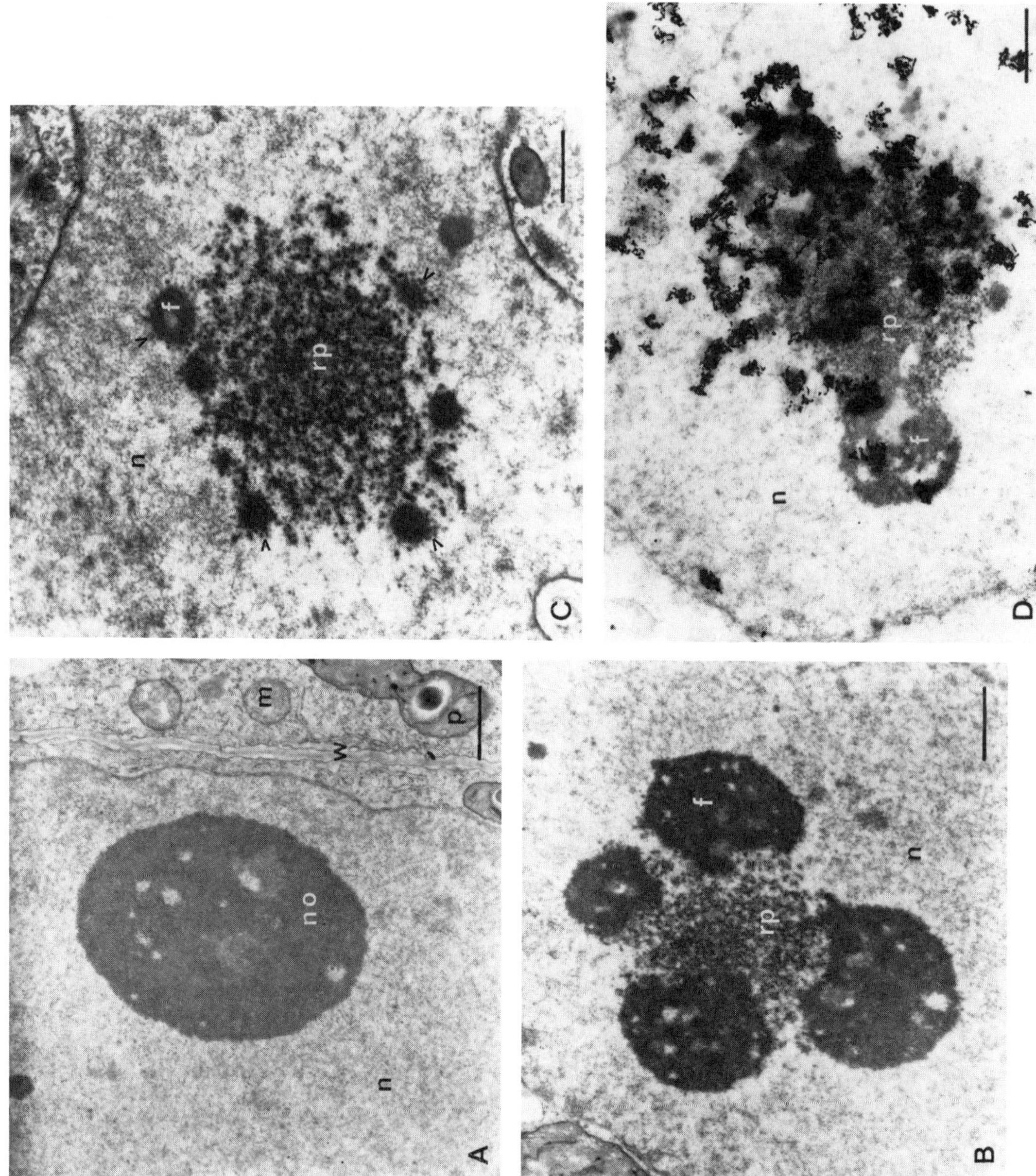
A
n
no
w
m
p
B
n
rp
f
C
n
f
rp
D
n
f
rp

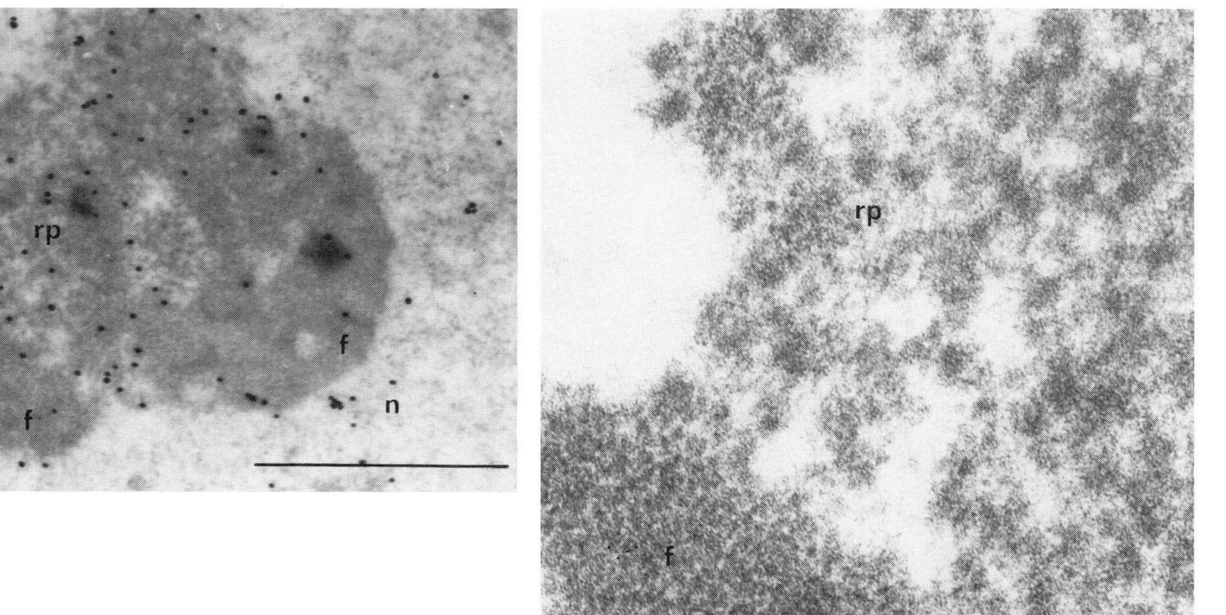

FIGURE 10.4. Heat shock (hs) induced changes of nucleolar ultrastructure and storage of pre-rRNP. Tomato cell cultures were fixed with glutaraldehyde and stained with lead. m, mitochondria; n, nucleus; no, nucleolus; p, proplastid; rp, ribonucleoprotein granules; f, fibrillar part of nucleolus; w, cell wall; Bars in A to E are 1 μm, in F 0.1 μm. A is from a control cell. B and C show hs cells after 1 h at 40°C and 3 h at 39°C. The nucleolus disaggregates and large amounts of preribosomal RNP accumulate (rp). Open arrowheads in C mark freshly synthesized pre-rRNP surrounding the fibrillar centers. D shows hs for 3 h at 39°C with ^{3}H-leucin labeling in the last 90 min, microautoradiography. The newly formed HSPs (black star-like filaments) are mainly bound to the preribosomal granules. E shows the staining of nucleolus from heat-shocked cells with HSP70 antibodies followed by protein A-gold. The gold particles (black dots) in the optically dense granular parts indicate that HSP70 is the hs protein accumulated in the nucleolus and is mainly connected with the pre-rRNA material. F is a high resolution picture of the disaggregated nucleolus.

10.7. MODIFICATION AND DEGRADATION OF RIBOSOMES

Usually the velocity and extent of HSP synthesis observed under hs conditions require a highly active transcription and translation apparatus. Despite some remarkable changes in the phosphorylation pattern of rproteins and of translation factors (see Section 11.5) the translation efficiency is maintained,[36] and ribosomes are stable during the stress period. There are two notable exceptions to this rule.

10.7.1. *TETRAHYMENA THERMOPHILA*

Starvation of this unicellular ciliate in 60 m*M* Tris pH 7.5 prior to a sublethal hs treatment (41°C) prevents a normal hs response. The cells finally lose their viability. The lack of HSP synthesis is not due to the lack of corresponding mRNAs but rather to an inhibition of all protein synthesis because of ribosome degradation.[26] Protection of the translation machinery against hs damage can be achieved by either pretreatment with a short hs under growth conditions or with low doses of cycloheximide,[26,27] i.e., evidently no HSP synthesis is involved. On the other hand, a mutant strain (MC-3) is unable to acquire this special type of induced thermotolerance.[33a] A search for the molecular basis of this effect revealed two types of ribosomal modifications. (1) Phosphorylation of a small subunit rprotein[25] is incomplete in the MC-3 mutant. The extent of this modification correlates with the resistance of starved cells to the otherwise lethal hs.[33a] (2) A new basic protein with M_r 22,000 (p22) tightly binds to ribosomes of starved or heat-shocked *Tetrahymena* cells. It exists in a preformed state and is not synthesized during hs. All conditioning treatments found to protect ribosomes against hs-induced degradation also cause association of p22 with ribosomes.[39] p22, which can be considered to be a special type of hs exchange protein, may help to stabilize ribosomes. In addition, suppression of control protein synthesis in hs cells containing p22-ribosomes is less severe. Thus, in addition to its role in ribosomal integrity, p22 should be considered as a possible factor for restoration of normal protein synthesis.

10.7.2. BACTERIA

In a wide range of bacteria, a massive degradation of mainly the 30S small ribosomal subunit is observed during hs.[19,23,41,59] The preferential nuclease attack on the 16S rRNA may be due to conformational changes of the small ribosomal subunit evoked by hs and Mg^{2+} loss from the cell. In heat-treated cells of *Escherichia coli* or *Staphylococcus aureus* no intact 30S subunits were found after 25 min at 48°C and 20 min at 52°C, respectively.[19,59]

When rRNA labeling was studied during the recovery period precursor particles of 49S, 36S, and 30S were detected. The restoration of the normal state proceeds under unusual conditions. Similar to the situation in eukaryotic cells (see above), the generation of all parts of the ribosomal apparatus is usually tightly coordinated by a set of regulatory interactions.[44b] In *E. coli,* the selective decay of the 30S subunit under hs proceeds with the destruction of the 16S rRNA, but with the conservation of most of the rproteins bound to it.[59] In the recovery period, synthesis of rRNA is restored almost immediately, whereas protein synthesis remains blocked for about 90 min. The newly assembled 30S subunits are built from new 16S rRNA but old rproteins. However, a few proteins are underrepresented or even absent[59] (S1, S2, S10, S21). Whether this has functional consequences and when the 30S subunits are completed is unclear. Interestingly, the imbalance between preserved 50S subunits and the newly assembled 30S subunits leads to an increased turnover of the 50S subunits during the recovery period.[59]

10.7.3. PLANT CHLOROPLASTS

An interesting peculiarity of the hs effects on ribosome biosynthesis is observed in developing plants.[18a] Seedlings grown under very mild hyperthermic conditions (32 to 34°C) become deficient of chloroplastic ribosomes.[18a,18b] The leaves are largely free of chlorophyll and other components required for chloroplast function. Normal chloroplast development is dependent on the coordinate activities of both gene expression systems of the cell, i.e, of the nuclear-cytoplasmic and of the organellar system. Lack of chloroplast ribosomes blocks the whole process. The temperature-sensitive step could not be defined. At least the organellar genome and the corresponding RNA polymerase remain intact.[8a,28a] Thus, assembly and processing of preribosomes are the most likely candidates. In support of this assumption, even a short interruption of the long-term hs at 32°C, e.g., by 1 h at 22°C, was found sufficient to restore a considerable part of ribosome formation and greening in developing oat seedlings. Interestingly, a number of ribosomal proteins, destined for assembly into chloroplast ribosomes, are found in the cytoplasm of heat-treated rye leaves.[18c]

REFERENCES

1. **Amalric, F., Simard, R., and Zalta, J. P.,** Effet de la temperature supra-optimale sur les ribonucleoproteines et le RNA nucleolaire. II. Etude biochimique, *Exp. Cell Res.*, 55, 370—377, 1969.
2. **Arrigo, A. P.,** Investigation of the function of the heat shock proteins in *Drosophila melanogaster* tissue culture cells, *Mol. Gen. Genet.*, 178, 517—524, 1980.
3. **Arrigo, A. P., Fakan, S., and Tissieres, A.,** Localization of the heat shock-induced proteins in *Drosophila melanogaster* tissue culture cells, *Dev. Biol.*, 78, 86—103, 1980.

3a. **Bell, J., Neilson, L., and Pellegrini, M.,** Effect of heat shock on ribosome synthesis in *Drosophila melanogaster, Mol. Cell. Biol.*, 8, 91—95, 1988.

4. **Bielka, H., Ed.,** *The Eukaryotic Ribosome,* Akademie Verlag, Berlin, 1982.
5. **Bouche, G., Amalric, F., Caizergues-Ferrer, M., and Zalta, J. P.,** Effects of heat shock on gene expression and subcellular protein distribution in Chinese hamster ovary cells, *Nucl. Acids Res.*, 7, 1739—1747, 1979.
6. **Bouche, G., Raynal, T., Amalric, F., and Zalta, J. P.,** Unusual processing of nucleolar RNA synthesized during a heat shock in CHO cells, *Mol. Biol. Rep.*, 7, 253—258, 1981.
7. **Bouche, G., Caizergues-Ferrer, M., Amalric, F., Zalta, J. P., Banville, D., and Simard, R.,** Synthesis and behaviour of small RNA species of CHO cells submitted to a heat shock, *Nucl. Acids Res.*, 9, 1615—1625, 1981.
8. **Bourbon, H. M., Bugler, B., Caizergues-Ferrer, M., Amalric, F., and Zalta, J. P.,** Maturation of a 100 kDa protein associated with preribosome in Chinese hamster ovary cells, *Mol. Biol. Rep.*, 9, 39—47, 1983.

8a. **Bünger, W. and Feierabend, J.,** Capacity for RNA synthesis in 70S ribosome-deficient plastids of heat-bleached rey leaves, *Planta,* 149, 163—169, 1980.

9. **Byfield, J. E. and Scherbaum, O. H.,** Temperature-dependent decay of RNA and of protein synthesis in a heat-synchronized protozoan, *Proc. Natl. Acad. Sci. U.S.A.*, 57, 602—606, 1967.
10. **Cadrin, M., Ashraf, M., and Lafontaine, J. G.,** Effects of heat shock on nuclear organization in plasmodia of *Physarum polycephalum, Can. J. Bot.*, 62, 1550—1560, 1984.
11. **Caizergues-Ferrer, M., Bouche, G., and Amalric, F.,** Phosphorylated protein involved in the regulation of rRNA synthesis in CHO cells recovering from heat shock, *FEBS Lett.*, 116, 261—264, 1980.
12. **Caizergues-Ferrer, M., Bouche, G., Amalric, F., and Zalta, J. P.,** Effects of heat shock on nuclear and nucleolar protein phosphorylation in Chinese hamster ovary cells, *Eur. J. Biochem.*, 108, 399—407, 1980.
13. **Cameron, I. L., Padilla, G. M., and Miller, O. L.,** Macronuclear cytology of synchronized *Tetrahymena pyriformis, J. Protozool.*, 13, 336—341, 1966.
14. **Cervera, J.,** Effects of thermic shock on HEp-2 cells. An ultrastructural and high-resolution autoradiographic study, *J. Ultrastruct. Res.*, 63, 51—63, 1978.
15. **Collier, N. C. and Schlesinger, M. J.,** The dynamic state of heat shock proteins in chicken embryo fibroblasts, *J. Cell Biol.*, 103, 1495—1507, 1986.
16. **Dawson, W. O. and Grantham, G. L.,** Inhibition of stable RNA synthesis and production of a noval RNA in heat-stressed plants, *Biochem. Biophys. Res. Commun.*, 100, 23—30, 1981.
17. **Ellgaard, E. G. and Clever, U.,** RNA metabolism during puff induction in *Drosophila melanogaster, Chromosoma,* 36, 60—78, 1971.
18. **Fakan, S. and Puvion, E.,** The ultrastructural visualization of nucleolar and extranucleolar RNA synthesis and distribution, *Int. Rev. Cytol.*, 65, 255—299, 1980.

18a. **Feierabend, J.,** Role of cytoplasmic protein synthesis and its coordination with the plastidic protein synthesis in the biogenesis of chloroplasts, *Ber. Dtsch. Bot. Ges.*, 92, 553—574, 1979.

18b. **Feierabend, J. and Mikus, M.,** Occurrence of a high temperature sensitivity of chloroplast ribosome formation in several higher plants, *Plant Physiol.*, 59, 863—867, 1977.

18c. **Feierabend, J., Schlüter, W., and Tebartz, K.,** Unassembled polypeptides of the plastidic ribosomes in heat-treated 70S ribosome-deficient rye leaves, *Planta,* 174, 542—550, 1988.

19. **Flowers, R. S. and Martin, S. E.,** Ribosome assembly during recovery of heat-injured *Staphylococcus aureus* cells, *J. Bacteriol.*, 141, 645—651, 1980.

20. **Fransolet, S., Deltour, R., Bronchart, R., and Van De Walle, C.,** Changes in ultrastructure and transcription by elevated temperatures in *Zea mays* embryonic root cells, *Planta,* 146, 7—18, 1979.

21. **Grierson, D.,** RNA processing and other posttranscriptional modifications, *Encycl. Plant Physiol. New Ser.*, Vol. 14B, Springer-Verlag, Berlin, 1982, 192—223.

22. **Gorenstein, C. and Warner, J. R.,** Coordinate regulation of the synthesis of eukaryotic ribosomal proteins, *Proc. Natl. Acad. Sci. U.S.A.*, 73, 1547—1551, 1976.

23. **Gray, R. J. H., Witter, L. D., and Ordal, Z. J.,** Characterization of mild thermal stress in *Pseudomonas fluorescens* and its repair, *Appl. Microbiol.*, 26, 78—85, 1973.

24. **Hadjiolov, A. A.,** *The Nucleolus and Ribosome Biogenesis. Cell. Biol. Monographs,* Vol. 12, Springer-Verlag, Wien, 1985.

25. **Hallberg, R. L., Wilson, P. G., and Sutton, C. A.,** Regulation of ribosome phosphorylation and antibiotic sensitivity in *Tetrahymena thermophila:* A correlation, *Cell,* 26, 47—56, 1981.

26. **Hallberg, R. L., Kraus, K. W., and Findly, R. C.,** Starved *Tetrahymena thermophila* cells that are unable to mount an effective heat shock response selectively degrade their rRNA, *Mol. Cell. Biol.*, 4, 2170—2179, 1984.

27. **Hallberg, R. L.,** No heat shock protein synthesis is required for induced thermostabilization of translational machinery, *Mol. Cell. Biol.*, 6, 2267—2270, 1986.

28. **Heine, U., Sverak, L., Kondratick, J., and Bonar, R. A.,** The behaviour of HeLa-S3 cells under the influence of supranormal temperatures, *J. Ultrastruct. Res.*, 34, 375—396, 1971.

28a. **Hermann, R. G. and Feierabend, J.,** The presence of DNA in ribosome-deficient plastids of heat-bleached rye leaves, *Eur. J. Biochem.*, 104, 603—609, 1980.

29. **Hermolin, J. and Zimmerman, A. M.,** RNA synthesis in division synchronized *Tetrahymena:* Macronuclear and cytoplasmic RNA, *J. Protozool.*, 23, 594—600, 1976.

29a. **Herruer, M. H., Mager, W. H., Raue, H. A., Vreken, P., Wilms, E., and Planta, R. J.,** Mild temperature shock affects transcription of yeast ribosomal protein genes as well as the stability of their messenger RNA, *Nucl. Acids Res.*, 16, 7917—7930, 1988.

30. **Jackson, P. J., Naranjo, C. M., McClure, P. R., and Roth, E. J.,** The molecular response of cadmium resistant *Datura innoxia* cells to heavy metal stress, in *Cellular and Molecular Biology of Plant Stress,* Key, J. L. and Kossuge, T., Eds., Alan R. Liss, New York, 1985, 145—160.

31. **Jacq, B., Jourdan, R., and Jourdan, B. R.,** Structure and processing of precursor 5S RNA in *Drosophila melanogaster, J. Mol. Biol.*, 117, 785—795, 1977.

32. **Johnston, G. C. and Singer, R. A.,** Ribosomal precursor RNA metabolism and cell division in the yeast *Saccharomyces cerevisiae, Mol. Gen. Genet.*, 178, 357—360, 1980.

33. **Kim, C. H. and Warner, J. R.,** Mild temperature shock alters transcription of a discrete class of *Saccharomyces cerevisiae* genes, *Mol. Cell. Biol.*, 3, 457—465, 1983.

33a. **Kraus, K. W., Hallberg, E. M., and Hallberg, R.,** Characterization of a *Tetrahymena thermophila* mutant strain unable to develop normal thermotolerance, *Mol. Cell. Biol.*, 6, 3854—3861, 1986.

34. **Labhart, P. and Reeder, R. H.,** Heat shock stabilizes highly unstable transcripts of the *Xenopus* ribosomal gene spacer, *Proc. Natl. Acad. Sci. U.S.A.*, 84, 56—60, 1987.

35. **Lewis, M. J. and Pelham, H. R. B.,** Involvement of ATP in the nuclear and nucleolar functions of the 70 kd heat shock protein, *EMBO J.*, 4, 3137—3143, 1985.

36. **Lindquist, S.,** Translational efficiency of heat-induced messages in *Drosophila melanogaster* cells, *J. Mol. Biol.*, 137, 151—158, 1980.

37. **Lomagin, A. G.,** Repair of functional and ultrastructural alterations after thermal injury of *Physarum polycephalum, Planta,* 142, 123—134, 1978.

38. **Love, R., Soriano, R. Z., and Walsh, R. J.,** Effect of hyperthermia on normal and neoplastic cells in vitro, *Cancer Res.*, 30, 1525—1533, 1970.

39. **McMullin, T. W. and Hallberg, R. L.,** Effect of heat shock on ribosome structure: Appearance of a new ribosome-associated protein, *Mol. Cell. Biol.*, 6, 2527—2535, 1986.

40. **Miller, D. W. and Elgin, S. C. R.,** Transcription of heat shock loci of *Drosophila* in a nuclear system, *Biochemistry,* 20, 5033—5042, 1981.

41. **Miller, D. W. and Ordal, J. Z.,** Thermal injury and recovery of *Bacillus subtilis, Appl. Microbiol.*, 24, 878—884, 1972.

42. **Munro, S. and Pelham, H. R. B.,** Use of peptide tagging to detect proteins expressed from cloned genes: Deletion mapping functional domains of *Drosophila* hsp70, *EMBO J.*, 3, 3087—3093, 1984.
43. **Neumann, D., Scharf, K.-D., and Nover, L.,** Heat shock induced changes of plant cell ultrastructure and autoradiographic localization of heat shock proteins, *Eur. J. Cell Biol.*, 34, 254—264, 1984.
44. **Neumann, D., Zur Nieden, U., Manteuffel, R., Walter, G., Scharf, K.-D., and Nover, L.,** Intracellular localization of heat shock proteins in tomato cells cultures, *Eur. J. Cell Biol.*, 43, 71—81, 1987.
44a. **Neumann, D., Nover, L., Parthier, B., Rieger, R., Scharf, K.-D., Wollgiehn, R., and Zur Nieden, U.,** Heat shock and other stress response systems of plants, *Biol. Zbl.*, 108, 1—155, 1989.
44b. **Nomura, M., Gouse, R., and Baughman, G.,** Regulation of the synthesis of ribosomes and ribosomal components, *Annu. Rev. Biochem.*, 53, 75—117, 1984.
45. **Nover, L., Ed.,** *Heat Shock Response of Eukaryotic Cells*, Springer-Verlag, Berlin, 1984.
46. **Nover, L., Scharf, K.-D., and Neumann, D.,** Formation of cytoplasmic heat shock granules in tomato cell cultures and leaves, *Mol. Cell. Biol.*, 3, 1648—1655, 1983.
46a. **Nover, L., Munsche, D., Ohme, K., and Scharf, K.-D.,** Ribosome biosynthesis in heat shocked tomato cell cultures. I. Ribosomal RNA, *Eur. J. Biochem.*, 160, 297—304, 1986.
46b. **Ohtsuka, K., Nakamura, H., and Sato, C.,** Intracellular distribution of 73000 and 72000 dalton heat shock proteins in HeLa cells, *Int. J. Hyperthermia*, 2, 267—276, 1986.
46c. **Parker, K. A. and Bond, U.,** Analysis of pre-rRNAs in heat shocked HeLa cells allows identification of the upstream termination site of human polymerase I transcription, *Mol. Cell. Biol.*, 9, 2500—2512, 1989.
47. **Pelham, H. R. B.,** Hsp 70 accelerates the recovery of nucleolar morphology after heat shock, *EMBO J.*, 3, 3095—3100, 1984.
48. **Pelham, H. R. B.,** Speculations on the functions of the major heat shock and glucose-regulated proteins, *Cell*, 46, 959—961, 1986.
49. **Perry, R. P.,** Processing of RNA, *Annu. Rev. Biochem.*, 45, 605—629, 1976.
50. **Risueno, M. C., Stockert, J. C., Gimenez-Martin, G., and Diez, J. L.,** Effect of supraoptimal temperatures on meristematic cell nucleoli, *J. Microscopie*, 16, 87—94, 1973.
51. **Rosbash, M., Harris, P. K. W., Woolford, J. L., and Teem, J. L.,** The effect of temperature-sensitive RNA mutants on the transcription products from cloned ribosomal protein genes of yeast, *Cell*, 24, 679—686, 1981.
52. **Rubin, G. M. and Hogness, D. S.,** Effect of heat shock on the synthesis of low molecular weight RNAs in *Drosophila:* Accumulation of a novel form of 5S RNA, *Cell*, 6, 207—213, 1975.
52a. **Sadis, S., Hickey, E., and Weber, L. A.,** Effect of heat shock on RNA metabolism in HeLa cells, *J. Cell. Physiol.*, 135, 377—386, 1988.
53. **Scharf, K.-D. and Nover, L.,** Control of ribosome biosynthesis in plant cell cultures under heat shock conditions. II. Ribosomal proteins, *Biochim. Biophys. Acta*, 909, 44—57, 1987.
53a. **Shyy, T. T., Subjeck, J. R., Heinaman, R., and Anderson, G.,** Effect of growth state and heat shock on nucleolar localization of the 110,000-Da heat shock protein in mouse embryo fibroblasts, *Cancer Res.*, 46, 4738—4745, 1986.
54. **Simard, R. and Bernhard, W.,** A heat-sensitive cellular function located in the nucleolus, *J. Cell Biol.*, 34, 61—76, 1967.
55. **Simard, R., Amalric, F., and Zalta, J. P.,** Effet de la temperature supra-optimale sur les ribonucleoproteines et le RNA nucleolaire. I. Etude ultrastructurale, *Exp. Cell Res.*, 55, 359—369, 1969.
56. **Sommerville, J.,** Nucleolar structure and ribosome biogenesis, *Trends Biochem. Sci.*, 11, 438—442, 1986.
56a. **Sollner-Webb, B. and Tower, J.,** Transcription of cloned eukaryotic ribosomal RNA genes, *Annu. Rev. Biochem.*, 55, 801—830, 1986.
57. **Spradling, A., Pardue, M. L., and Penman, S.,** Messenger RNA in heat-shocked *Drosophila* cells, *J. Mol. Biol.*, 109, 559—587, 1977.
58. **Subjeck, J. R., Shyy, T., Shen, J., and Johnson, R. J.,** Association between the mammalian 110,000-dalton heat-shock protein and nucleoli, *J. Cell Biol.*, 97, 1389—1395, 1983.
59. **Tal, M., Silberstein, R., and Myner, K.,** In vivo reassembly of 30S ribosomal subunits following their specific destruction by thermal shock, *Biochim. Biophys. Acta*, 479, 479—496, 1977.
60. **Ton-That, T. C. and Turian, G.,** High-resolution autoradiography of nuclear modifications during and after heat treatment of *Neurospora crassa, Protoplasma*, 120, 165—171, 1984.
61. **Ton-That, T. C., Turian, G., Fakan, J., and Gautier, A.,** Ultrastructural cytochemistry of perinucleolar dense spots in heat-treated macroconidia of *Neurospora crassa, Eur. J. Cell Biol.*, 24, 317—319, 1981.
61a. **Van Bergen en Henegouwen, P. M. P. and Linnemans, W. A. M.,** Heat shock gene expression and cytoskeletal alterations in mouse neuroblastoma cells, *Exp. Cell Res.*, 71, 367—375, 1988.
61b. **Veinot-Drebot, L. M., Singer, R. A., and Johnston, G. C.,** Heat shock causes transient inhibition of yeast ribosomal RNA gene transcription, *J. Biol. Chem.*, 264, 19473—19474, 1989.
62. **Velazquez, J. M., DiDomenico, B. J., and Lindquist, S.,** Intracellular localization of heat shock proteins in *Drosophila, Cell*, 20, 679—689, 1980.
63. **Warocquier, R. and Scherrer, K.,** RNA metabolism in mammalian cells at elevated temperature, *Eur. J. Biochem.*, 10, 362—370, 1969.

64. **Warner, J. R.,** The yeast ribosome: Structure, function, and synthesis, in *The Molecular Biology of the Yeast Saccharomyces. Metabolism and Gene Expression,* Strathern, J. N., Jones, E. W., and Broach, J. R., Eds., Cold Spring Harbor Laboratory, Cold Spring Harbor, NY, 1982, 529—560.
65. **Welch, W. J. and Feramisco, J. R.,** Nuclear and nucleolar localization of the 72,000 dalton heat shock protein in heat-shocked mammalian cells, *J. Biol. Chem.,* 259, 4501—4513, 1984.
66. **Welch, W. J. and Feramisco, J. R.,** Disruption of the three cytoskeletal networks in mammalian cells does not affect transcription, translation, or protein translocation changes induced by heat shock, *Mol. Cell. Biol.,* 5, 1571—1581, 1985.
67. **Welch, W. J. and Mizzen, L. A.,** Characterization of the thermotolerant cell. II. Effects on the intracellular distribution of HSP70, intermediate filaments, and snRNPs, *J. Cell Biol.,* 106, 1117—1130, 1988.
68. **Welch, W. J. and Suhan, J. P.,** Cellular and biochemical events in mammalian cells during and after recovery from physiological stress, *J. Cell Biol.,* 103, 2035—2052, 1986.
69. **Wool, I. G.,** The structure and function of eucaryotic ribosomes, *Annu. Rev. Biochem.,* 48, 719—754, 1979.
70. **Yuyama, S. and Zimmermann, A. M.,** RNA synthesis in *Tetrahymena:* Temperature-pressure studies, *Exp. Cell Res.,* 71, 193—203, 1972.

Chapter 11

TRANSLATIONAL CONTROL

L. Nover

TABLE OF CONTENTS

The detection of hs-induced proteins (HSPs) in *Drosophila* and the characterization of the newly formed mRNAs including their *in vitro* translation and their specific hybridization to the chromosomal sites known to be activated by hs,[72,74,110,111,122] led to the discovery of similar responses in other organisms and gave the signal for broad range investigations of this field.[79,83] The striking changes in translation patterns, i.e., inhibition of synthesis of most proteins formed prior to the hs ("control proteins") and the rapid appearance of the new heat shock proteins (HSPs), was observed in all kinds of prokaryotic and eukaryotic cells (Table 2.1). Important analytical tools developed by Laemmli[56] (SDS-polyacrylamide gel electrophoresis) and O'Farrell[84] (two-dimensional separation of proteins using isoelectric focusing in the first and the Laemmli method in the second dimension) were essential for this development. Induced HSP synthesis was almost considered synonymous with hs response. In fact, data in this book and earlier reviews[64,79,83,113,122a] emphasize that formation of the new proteins, though central, is only part of the complex reprogramming of many cellular activities in response to hs. Moreover, detailed investigations of the translational level revealed a multiplicity of factors contributing to a dynamic continuum of changes in the pattern of protein synthesis. Such are synthesis and selective decay of mRNAs, structural peculiarities of hs mRNAs, modifications of the translation machinery, reorganization of the intermediate size filament system so well as increasing levels and compartmentation of HSPs.

11.1. POLYSOMAL REORGANIZATION

Similar to other stress situations (starvation of cells, intoxication with chemical stressors, cold shock)[83] an immediate effect of hs is a transient halt of all protein synthesis connected with a breakdown of polysomes. Years before the detection of HSPs, McCormick and Penman[70] and Goldstein and Penman[36] provided important insights into the underlying mechanisms in HeLa cells by analyzing the changing levels of labeled amino acid incorporation and the disruption and reformation of polysomes. Similar results obtained with tomato cell cultures are illustrated in Figures 11.1 and 11.2. Polysomal reorganization is considered to be an essential part of the efficient reorientation of the gene expression machinery to the synthesis of stress proteins, e.g., HSPs. It was observed in electronmicroscopic investigations of heat-shocked mammalian cells[39,65,102] and analyzed in more detail in HeLa cells,[28,70] rabbit brain,[17c,20b,38a] *Drosophila*,[32,54,63,72] plants[47] (see Figures 11.1 and 11.2), and the slime mold *Physarum*.[104]

The following results are worth noticing.

1. Main factor for the rapid decay of polysomes is a block of translation initiation. Ribosomes engaged in protein synthesis run off and do not reinitiate. Cycloheximide, added prior to the hs treatment, inhibits the decay of polysomes. The initiation defect was characterized in HeLa and *Drosophila* cells by using low doses of cycloheximide or NaCl[44,44a] and by analyzing initiation factor modification under hs conditions (see Section 11.5.3). An important event may be the collapse of the intermediate size cytoskeletal system (see Section 11.5.1) known to be tightly connected with polysomes.[18,33,43,76]
2. Reformation of polysomes under hs conditions is dependent on RNA synthesis, i.e., it is inhibited by actinomycin[7,36,70] (see Figures 11.1 and 11.2). Synthesis of new hs-specific mRNAs is not sufficient to explain this effect. The need for an unidentified translation control RNA (tcRNA) was discussed by Goldstein and Penman[36] because of the low dose of actinomycin D (0.1 to 1 μg/ml) required for inhibition of polysome reformation in HeLa cells. The results are particularly striking in chicken reticulocytes[7] where the hs-induced synthesis of HSP70 is exclusively regulated at the translational

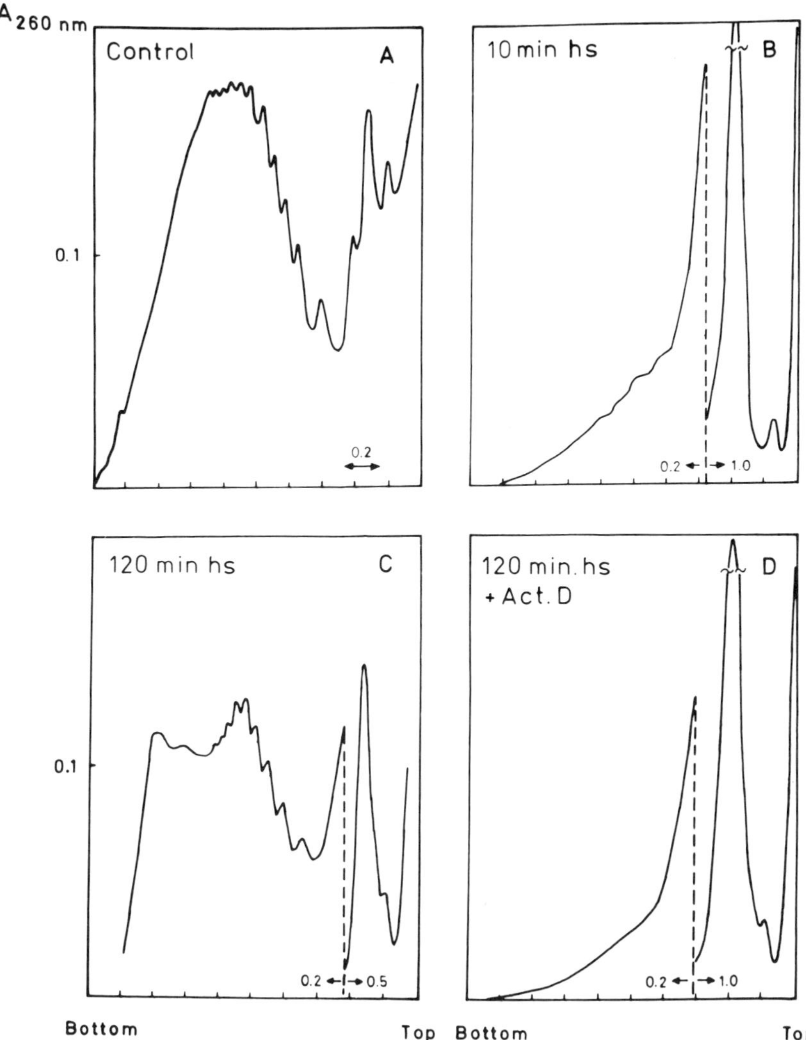

FIGURE 11.1. Changing polysome profiles in tomato cell cultures under heat shock (hs) conditions. Polysomes were separated on a linear sucrose gradient (10→40%) by 3 h centrifugation in a Spinco ultracentrifuge (SW 40 rotor, 35,000 rpm). The optical density at 260 nm was analyzed with a 1 mm flow cell. A, control cells; B, 10 min hs at 39°C; C, 120 min hs at 39°C; D, same as C, but 50 μg/ml actinomycin D added 15 min before the hs. Dashed lines indicate points where the measuring range of the spectrophotometer was extended to monitor the monosomal peak, e.g., from O.D. 0.2 → O.D. 1.0. The complete breakdown of control polysomes (B) is followed by formation of new hs polysomes (C), but not in the presence of actinomycin (D). (From Nover, L., Ed., *Heat Shock Response of Eukaryotic Cells,* G. Thieme Verlag, Berlin, 1984. With permission.)

level.[119a] Although the HSP70 mRNA exists in a preformed state in association with polysomes, its hs-induced translation is inhibited by actinomycin D or 5,6-dichloro-l-β-D-ribofuranosylbenzimidazole. The existence of hs-induced tcRNAs of about 300 nucleotides were reported for *Tetrahymena*[53] and *Drosophila*[45a,b] (see Section 5.3).

3. An important argument in favor of the tcRNA hypothesis is the conservation of most mRNAs existing prior to the hs treatment (see Section 11.4). This is also evident from the reinitiation of protein synthesis in the presence of actinomycin if the normothermic state is reestablished (Figure 11.2).

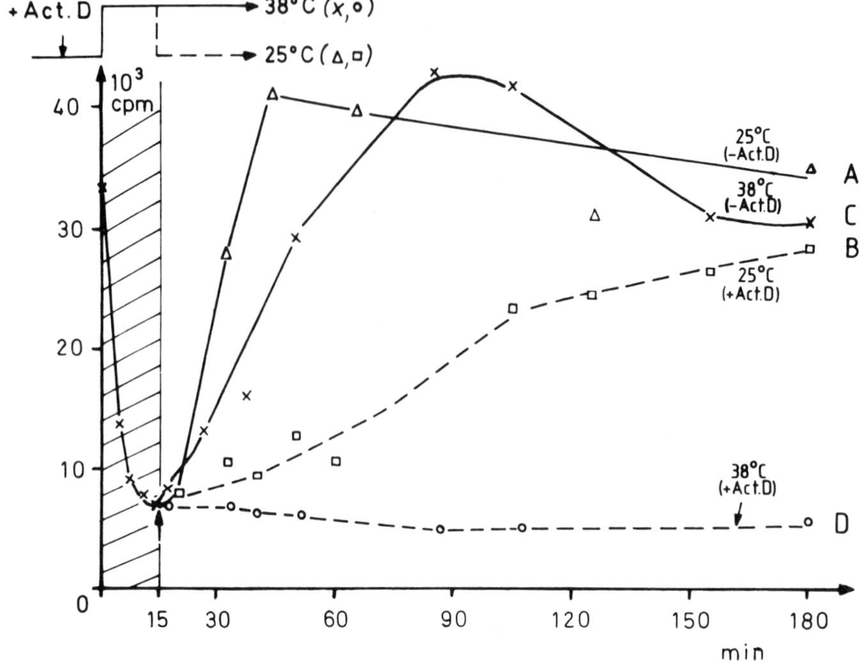

FIGURE 11.2. Methionine incorporation during heat shock (hs) and recovery and influence of actinomycin D. Tomato cell cultures were subjected to hs treatments as illustrated in the pictogram on top of the figure. Actinomycin D (25 μg/ml) was added 10 min prior to the hs. At the indicated time points ^{35}S-methionine incorporation into TCA soluble material was determined. The decay and reformation of polysomes (Figure 11.1) is reflected in the overall methionine incorporation. The inhibitory effect of actinomycin D (sample D) is relieved if cultures are returned to 25°C (sample B) because of the reinitiation of control mRNA species. (From Nover, L., Ed., *Heat Shock Response of Eukaryotic Cells,* G. Thieme Verlag, Berlin, 1984. With permission.)

4. Due to the two prominent size classes of HSPs (see Chapter 2) the polysomal profile of long-term hs cells in noticeably different from that of control cells compare Figure 11.1, parts A and C. Using *Drosophila* cell cultures, the elegant analyses of McKenzie et al.[72] and Mirault et al.[74] demonstrated that small polysomes with 4 to 8 ribosomes per message program the synthesis of the small HSPs (M_r 22 to 27 kDa), whereas the peak of large polysomes with 20 to 30 ribosomes per message is responsible for synthesis of the large HSPs (M_r 68 to 83 kDa).
5. An interesting functional detail of the translational machinery under hs conditions is the maintenance of the overall efficiency. Despite a temperature increase of about 10°C, the rate of protein synthesis, as deduced from incorporation of labeled amino acids, is essentially unchanged in *Drosophila* and plants (see Figure 11.2). These observations are supported by quantitative estimates given by Lindquist[63] for *Drosophila* cells.

11.2. CHANGING mRNA LEVELS: SYNTHESIS AND DEGRADATION

Perhaps with a few exceptions (see Section 8.6), induction of HSP synthesis is dependent on the transcription of corresponding mRNAs. As indicated above, experiments with actinomycin D or other inhibitors of transcription cannot give unequivocal proof for this, but the early investigations with *Drosophila*[72,74,110,111] directly analyzed the newly formed mRNAs by *in vitro* translation and *in situ* hybridization to the hs-activated chromosomal loci. Following this, similar data were provided for practically all other types of hs systems, mainly using *in vitro* translation or Northern and dot blot analyses, respectively. The picture emerging for most eukaryotic cells involves the rapid accumulation and effective translation of hs mRNAs and the conservation of many control protein mRNAs in a nontranslated state.

It is well documented for *Drosophila,*[19,25,62,93] vertebrates,[75,102] and plants[80] that recovery of control protein synthesis after severe hs is drastically improved in cells made thermotolerant by a hs conditioning treatment. Decisive for this effect is the preservation of mRNAs in an intact state, probably in connection with HSPs (see Section 16.7). Earlier reports on bulk degradation of poly(A^+)-RNA in *Drosophila,*[111] if not caused by the particular hs treatment, may in fact reflect the hs-induced loss of the poly(A)-tail.[6,17] At least the results from *in vitro* translation experiments (Figure 11.5) are in contrast to a massive degradation of control mRNAs also in *Drosophila.*

However, with accumulating data for different organisms and with careful investigations on the fate of individual mRNAs, a much more diversified picture emerged. In yeast and probably related organisms the changing translation pattern is largely controlled at the transcriptional level. Polysome breakdown is not observed, and control protein mRNAs decay.[37,48,59,61,69] Herrick et al.[39a] studied half-lives of 20 different mRNAs under hs conditions (36°C). There were essentially no changes independent of the type of mRNA studied, i.e., of short-lived (< 7 min) or long-lived mRNAs (half-life > 25 min). In both cases degradation depends on ongoing translation and is independent of the presence or absence of a poly(A)-tail. Only in the case of the hs-enhanced degradation of rprotein mRNAs is there weak evidence for an association with an hs-induced nuclease.[39b]

Usually, different classes of control protein mRNAs can be defined with respect to their changing levels and rates of translation during hs. This is illustrated by three examples based on data obtained with the ciliate *Tetrahymena thermophila,* with plants (tomato) and *Drosophila,* respectively. The outcome is very similar, but, due to different experimental techniques and peculiarities of the system, the results are not identical.

1. During a permanent hs mRNAs for the two prominent HSPs (M_r 80 and 73) are rapidly induced and accumulate in fairly large amounts in *Tetrahymena*[38] (Figure 11.3A). Using a collection of cDNA probes derived from abundant polysomal RNAs of nonheat-shocked cells, different patterns emerged (Figure 11.3B). First, clone pC8 characterizes an mRNA, whose abundance in the polysomal fraction is essentially unchanged. Its protein product is unknown. Second, clones pC5 and pC3 hybridize to RNAs with a delayed increase during hs. Both RNAs are also found in starved or stationary phase cells, presumably in connection with a cell cycle block. Finally, pC4 and pC6 describe RNAs immediately broken down during hs, but pC4 RNA levels are restored, whereas pC6 is not. The latter mRNA codes for a surface protein of *Tetrahymena* (SerH3).[37a] Selective degradation of tubulin mRNAs in heat-shocked *Tetrahymena*[19a] probably reflects a special type of autorepression by free tubulin subunits as demonstrated for tubulin synthesis in mammalian cells (see reviews by Brawerman[17b] and Cleveland[18a]). Similar to the situation in yeast, an hs-induced factor (nuclease?) increases degradation of tubulin mRNAs.[19a]
2. On the basis of temperature shift experiments in the presence of actinomycin D we distinguished three classes of control proteins in tomato cell cultures: [80] (1) few constant proteins with continued synthesis under hs, (2) many proteins affected by translation discrimination, i.e., with mRNAs being conserved in an untranslated state, and (3) a few proteins, whose synthesis is discontinued and not immediately resumed during the recovery period. This classification is corroborated by *in vitro* translation data.[82] It is important to notice that even within a group of functionally closely related proteins,

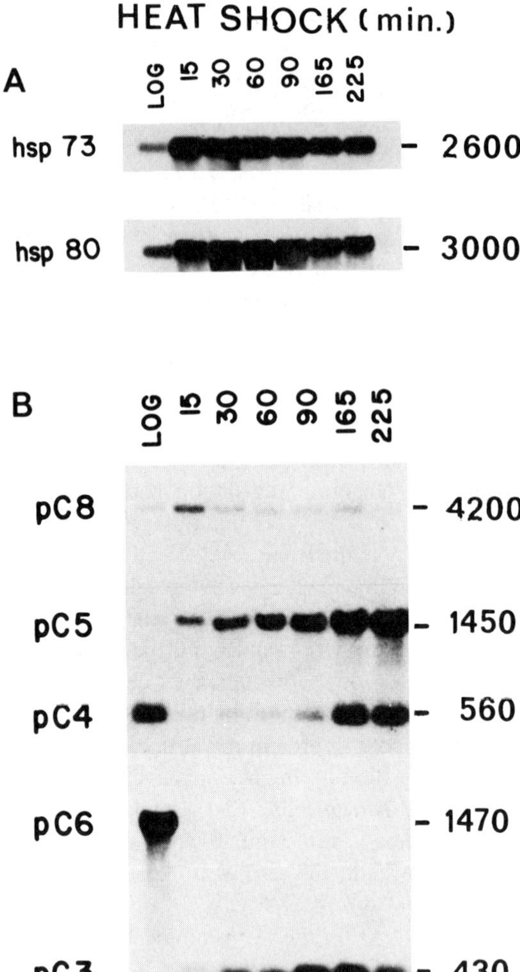

FIGURE 11.3. Changing levels of different RNA-species during long-term heat shock (hs) in *Tetrahymena thermophila*. After separation in agarose gels levels of the different RNA species were analyzed by Northern hybridization. Probes were two genomic clones for HSPs 80 and 73, respectively, (A) and cDNA clones for abundant cytoplasmic *Tetrahymena* RNAs (B). Sizes of the RNAs in bases are indicated on the margin. (From Hallberg, R. L., Kraus, K. W., and Findly, R. C., *Mol. Cell. Biol.*, 4, 2170, 1984. With permission.)

e.g., the ribosomal proteins of plant cells, these different classes can be identified.[102a] Most ribosomal proteins are synthesized almost undiminished and become part of the hs preribosomal RNP accumulating in the nucleolus (class 1, see Section 10.4). However, few of them usually found in preribosomes are lacking under hs, i.e., they belong to classes (2) or (3).

3. Even more complex is the situation with the expression pattern of *Drosophila* histone genes. Under conditions of moderate hs (≤35°C), all four core histones (H2A, H2B, H3, H4) and H1 continue to be synthesized, but tranferring cell cultures to severe hs (37°C) leads to noncoordinate expression of histone genes.[34,108] H2B expression is increased, whereas all others are repressed. This led to the definition of H2B as a hs protein.[100,117] However, this classification is certainly not correct. On the one hand, H2B synthesis is not enhanced by other hs-like inducers like arsenite[117] or Cd^{2+},[21] and the preferential synthesis of H2B after hs is not observed in different larval tissues nor in cell cultures at low density.[108] On the other hand, careful investigations by

Farrell-Towt and Sanders[34] demonstrate that the effect is essentially controlled at the translational level. Synthesis of H2B mRNA is unchanged, but its half-life is increased, and it is found predominantly in polysomes. In contrast, transcription of H2A, H3, and H4 mRNAs is slightly reduced. Their stability increases as well, but most of them are not associated with polysomes. Finally, regulation of Hl is completely different. Transcription is stopped and preexisting mRNA is broken down. Even within this small group of proteins, whose synthesis is usually coordinate and tightly tuned to DNA synthesis, we find three different types of regulatory behavior under given stress conditions (severe stress, high cell density).

Massive degradation of mRNAs in animal cells has been reported occasionally. This may reflect system-specific or experimental peculiarities precluding an efficient protection during hs. Thus, Panniers et al.[91] found an increased requirement of Ehrlich ascites tumor cell lysates for external mRNAs if prepared from cells stressed for 20 min at ≥44°C. On the other hand, recovery of control protein synthesis in chicken myotubes after 2 h at 45°C was completely dependent on *de novo* synthesis of mRNA and could be stimulated by liposome-mediated transfer of isolated mRNAs.[5] In contrast to this, Sadis et al.[98a] reported on increased stability of c-myc mRNA in HeLa cells under relatively mild hs conditions (42°C).

Other examples concern the degradation of single, mostly abundant mRNAs in highly specialized cells. In heat-shocked *Xenopus laevis* hepatocytes, the estrogen-induced transcription and accumulation of vitellogenin mRNA are blocked, and preformed mRNA is rapidly degraded.[126] Related to these findings is the decay of mouse tumor virus RNA observed in mammary carcinoma cells shifted for 1 to 2 h to 42°C. At the same time, the level of actin mRNA is unchanged.[95] Vierling and Key[123] presented a detailed analysis of protein synthesis in green suspension cultures of soybean (*Glycine max*). Synthesis of the most prominent chloroplast constituent, ribulose 1,5-bisphosphate carboxylase, declines with increasing temperatures from 33 to 40°C. The small subunit of this enzyme is produced in the cytoplasm. Its mRNA is completely degraded. In contrast, the large subunit mRNA, coded and translated within the chloroplast compartment, is stable (see Figure 13.1).

A particularly striking example from plants is illustrated in Figure 11.4.[8] In response to the phytohormone gibberellic acid, large amounts of different hydrolytic enzymes, e.g., α-amylase, are produced and secreted by aleurone cells of germinating barley seeds. If heat shocked at 40°C, α-amylase synthesis stops within 1 to 2 h due to a decay of the corresponding mRNA (Figure 11.4A, lanes 2 and 3). This was shown by *in vitro* translation (Figure 11.4A, lanes 4 and 5) but also by dot blot quantification. The dramatic shift in the pattern of protein synthesis from secreted proteins made by ER-bound polysomes to cytoplasmic HSPs can be visualized also at the electronmicroscopic level (Figure 11.4B). The areas with abundant rER disappear. Both, degradation of amylase in mRNA and the disruption of the rER are not dependent on concommitant HSP synthesis.[17c]

11.3. PREFERENTIAL TRANSLATION OF HEAT SHOCK mRNAs

In agreement with the model of Lodish[64a] on translational specificity resulting from the differential affinity of mRNAs for rate-limiting components of the translational machinery, hs mRNAs must be classified as relatively strong mRNAs. Peculiarities of the unusually long leader sequences are important for this effect (see Section 11.5.5). They ensure high frequencies of initiation even under conditions when, by depletion or modification of initiation factors the overall activity of the translational apparatus is reduced. A telling example is the resistance of HSP70 mRNA translation to the shutoff of protein synthesis after poliovirus infection.[78] Thus, selectivity of hs mRNA translation is the result of a multifactorial interaction, whose components slowly begin to emerge (see Section 11.5).

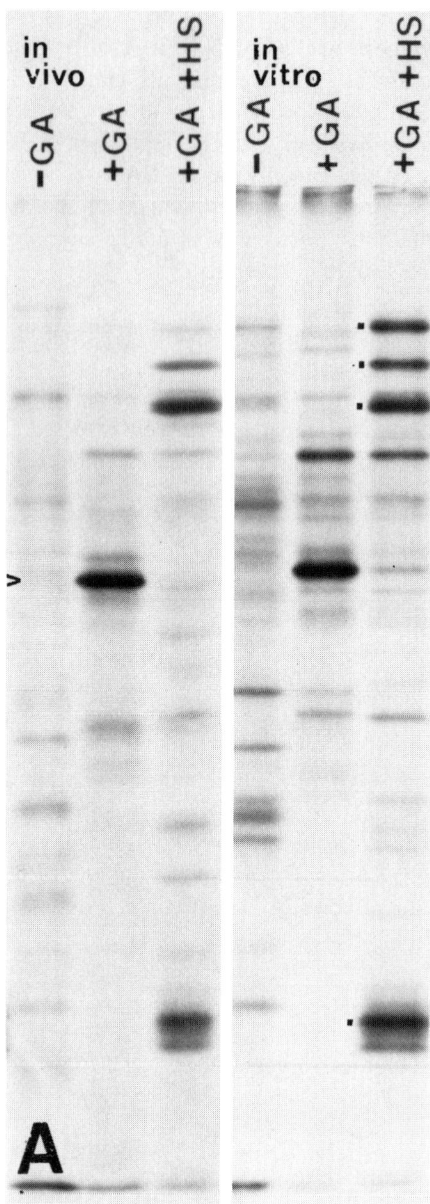

FIGURE 11.4. Heat shock (hs) effects on ultrastructure and protein synthesis in aleurone layers of germinating barley seeds. Isolated aleurone layers of barley seeds were incubated 19 h in the presence or absence of 1 μM gibberellic acid (+GA, −GA). Part of the +GA sample was shifted in the last 3 h to 40°C (+GA+HS). A shows the autoradiographs of ^{35}S-methionine labeled proteins separated by SDS-PAGE. Lanes 1 to 3, *in vivo* labeled, lanes 4 to 6 *in vitro* labeled proteins synthesized from total RNA in a wheat germ system. Addition of GA or hs treatments are indicated. Dots mark position of HSPs, 95, 80, 70, and 17, open arrowhead points to α-amylase (44 kDa). B and C are electron micrographs of GA-treated aleurone cells at 25°C (B) and after 3 h hs at 40°C (C). The prominent rER region (brace in B) almost disappeared in C. g, Golgi; p, proplastid; s, spherosome. (From Belanger, F. C., Brodl, M. R., and Ho, T.-H. D., *Proc. Natl. Acad. Sci. U.S.A.*, 83, 1354, 1986. With permission.)

An important achievement for investigations of hs-specific translation control was the establishment of homologous *in vitro* translation systems of *Drosophila* culture cells mimicking this effect.[54,101,112] The results in Figure 11.5 show autoradiographs of two-dimensional

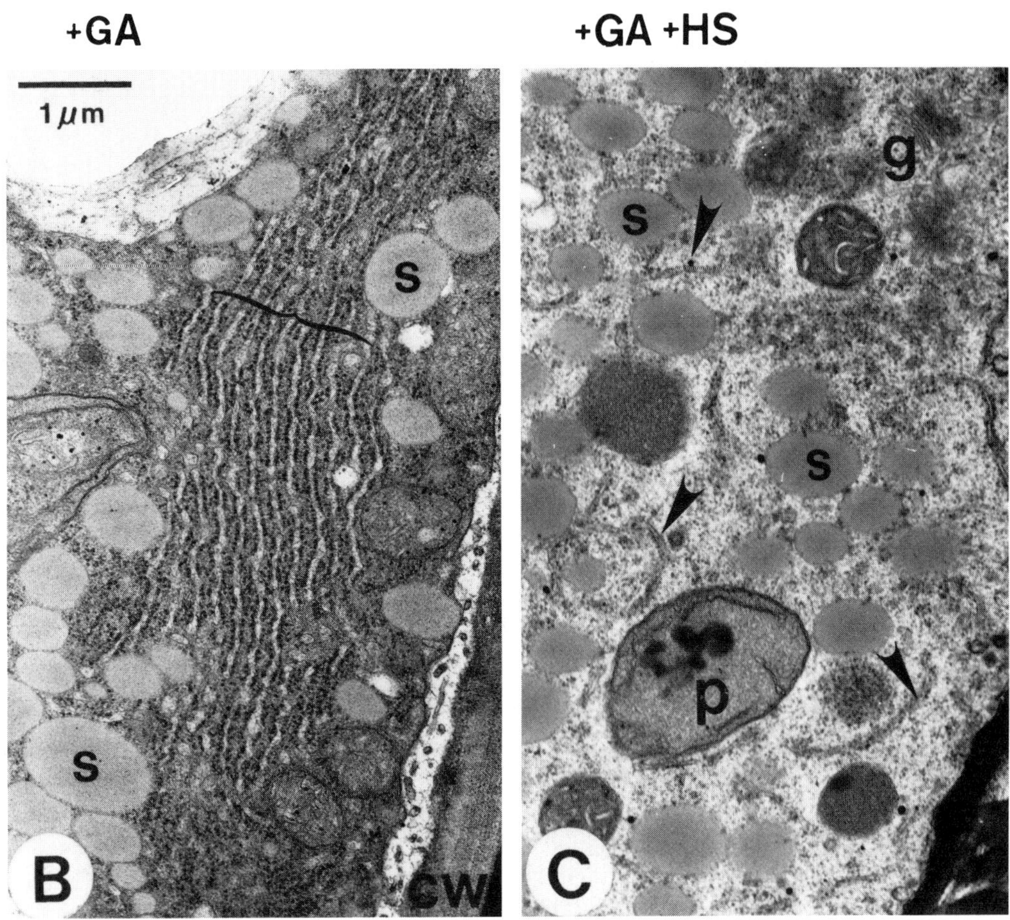

FIGURE 11.4 (continued).

product analyses using cell lysates from control (A, C) and hs cells (B, D). Similar to the *in vivo* situation, the control lysate translates hs and control mRNAs equally well (see position of HSPs indicated on the margin in A), whereas the hs cell lysate is highly discriminating. It operates efficiently only if hs mRNAs are supplied (B), but is unable to use control protein mRNAs (D). Selectivity of *in vitro* translation systems from hs cells was also reported for wheat germ[52] and HeLa cells.[24,67,91]

Despite some controversial reports, which may simply reflect inadequate experimental conditions, preferential translation seems to be a general trait of all hs systems, possibly with the exception of yeast, *Neurospora* and related fungi.[21a,37,59,61,69] However, in contrast to the transcription control system of hs (see Chapters 6 and 7), selectivity of translation is species-specific, i.e., hs mRNAs are only preferred in their homologous cell types. This became already apparent in the early phase of gene transfer introducing the *Drosophila* hsp70 gene into *Xenopus* oocytes.[12] In oocytes, transformed with the heterologous Dm-hsp70 gene, hs triggers correct transcription of the *Drosophila* gene, but translation is repressed, i.e., the *Drosophila* leader sequence is not accepted by the translation machinery of *Xenopus* oocytes altered by hs. As expected, this discrimination disappears after return to normothermic conditions.[12] When thoroughly tested, similar results were found by other groups studying the transfer of the human hsp70 gene into *Xenopus* oocytes,[125] of a Dm-hsp70 (P/L) × lac Z fusion construct into *Xenopus* oocytes,[57] or of a Dm-hsp70 (P/L) ×

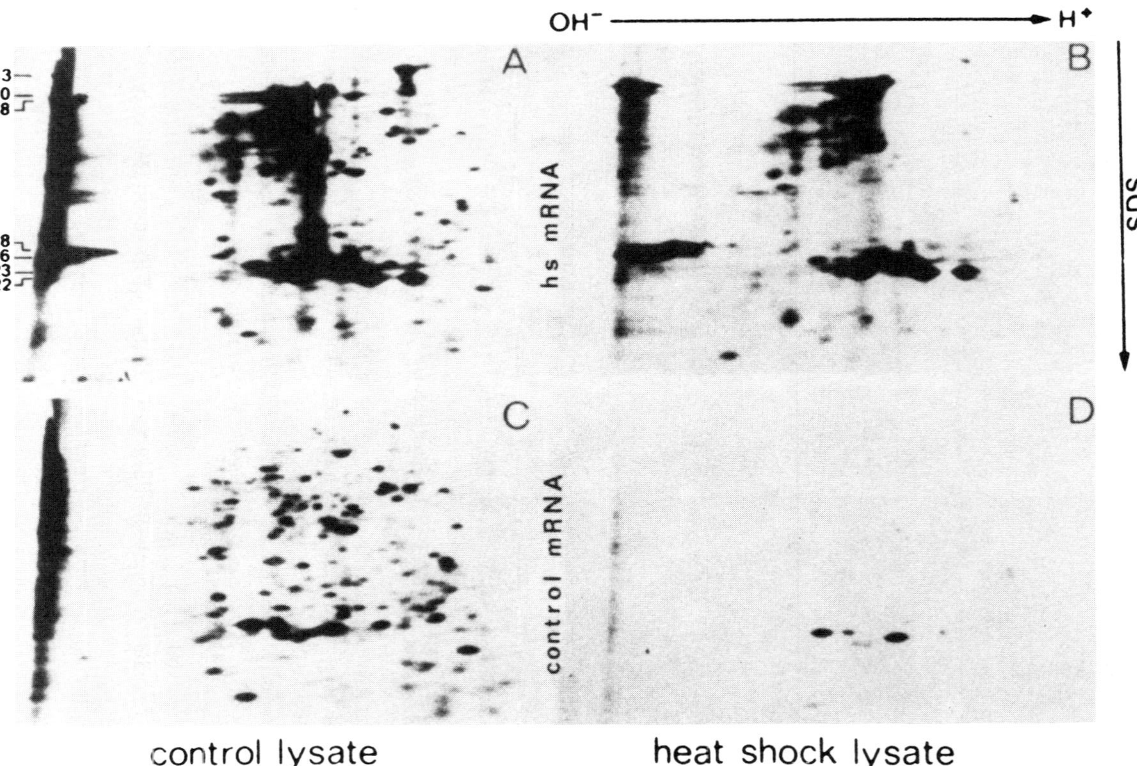

FIGURE 11.5. Discrimination between control and heat shock (hs) mRNAs in a homologous *in vitro* translation system from *Drosophila* cell cultures. Polysomal mRNA was translated in cell lysates from heat shocked (B and D) and control (A and C) *Drosophila* cell cultures. ^{35}S-Methionine-labeled products were separated by two-dimensional electrophoresis analogous to the procedure described in the legend to Figure 2.2. A and B show mRNA prepared from hs cells; C and D show mRNA from control cells. Positions of hsps are indicated on the left margin of A. The strong discrimination of the hs lysate against control mRNAs (see B and D) and the preservation of these mRNAs in hs cells (see A) are clearly visible. (Modified from Krüger, C. and Benecke, B.-J., *Cell,* 23, 595, 1981. With permission.)

neo^r fusion construct into plant cells.[109] These results provide an unusual but strong argument for the universal existence of translation discrimination in heat-shocked eukaryotic cells.

Though certainly not an important factor in the overall control of gene expression patterns during hs, discrimination of mRNAs was even observed in yeast and *E. coli.* Due to an hs element in the promoter region, the level of the yeast phosphoglycerate kinase mRNA increases at least fivefold during hs at 42°C.[93a,b] But at the same time, PGK synthesis is reduced to about 25% of its normal level. In *E. coli,* translation of β-lactamase mRNA ceases transiently after a temperature shift from 30 to 42°C. The effect, which is even observed in an *in vitro* transcription/translation system incubated at 45°C, is due to inactivation of an initiation factor. The deficiency can be restored by addition of a 160,000 × g supernatant prepared from *E. coli* cells grown at 30°C. The specificity for the β-lactamase mRNA is evidently bound to a short nucleotide motif in its leader sequence with the Shine-Dalgarno sequence and the ATG codon (–ATTGAAAAAGGAAGAGTATGAG–). If transferred to the 5′ end of the lacZ mRNA, this oligonucleotide confers hs repression to this heterologous mRNA.[55,55a]

11.4. CONSERVATION OF CONTROL mRNAs AND RECOVERY

It is a general experience with different types of hs systems that the translation capacity and the capability for control protein synthesis during heat stress but especially in the recovery period are markedly influenced by prior heat conditioning. This thermotolerance effect at the translation level was demonstrated for *Drosophila,*[19,62,93] plants,[80,82,103] and vertebrates.[75] Central to this is the protection of the bulk of control mRNAs from being degraded during hs. Two types of investigations in *Drosophila* and plants point to two different mechanism, which must not be mutually exclusive, but may rather reflect early and late stages of the hs response.

1. In *Drosophila* cell cultures after 1 h at 36°C many poorly translated control mRNAs (class II of Figure 11.6) are still connected with ribosomes, but in smaller polysomes then under control conditions. This led Ballinger and Pardue[6] to the suggestion that stalled polysomes are main sites of mRNA storage and that translational specificity, at least in part, is due to elongation control. However, the arguments for this are not really conclusive since rare initiation would give essentially the same mRNA distribution in the polysomal profile.
2. In plant cells, many of the untranslated mRNAs are found in close association with large cytoplasmic HSP aggregates, i.e., with the hs granules,[81,82] which are free of polysomes, but presumably attached to elements of the cytoskeleton.[82] The existence and biochemical characterization of these aggregates (see Section 16.7) give a direct hint on a causal relationship between mRNA preservation and synthesis of HSPs. The function of the HSG in this respect can be directly visualized in the electronmicroscope. If tomato cells, after a period of severe hs with ample accumulation of heat shock granules (HSG), are allowed to recover at 25°C, the slowly decomposing HSG are surrounded by all types of cellular organelles actively involved in protein synthesis: many polysomes, rough ER, and stacks of Golgi membranes[81,82,83] (see Figure 13.2).

The discovery of similar HSP-aggregates also in *Drosophila*[3,27,58] and vertebrate cells[4,20,20a] makes a related mechanism of mRNA storage in other organisms very likely, but so far experimental proof is lacking, because isolation of HSG from other organisms proves to be difficult. One step in this direction would be a reinvestigation of the *Drosophila* system using preconditioned cells and higher hs temperatures: are there forms of preserved control mRNAs not associated with ribosomes? It is remarkable to notice that in the early analysis

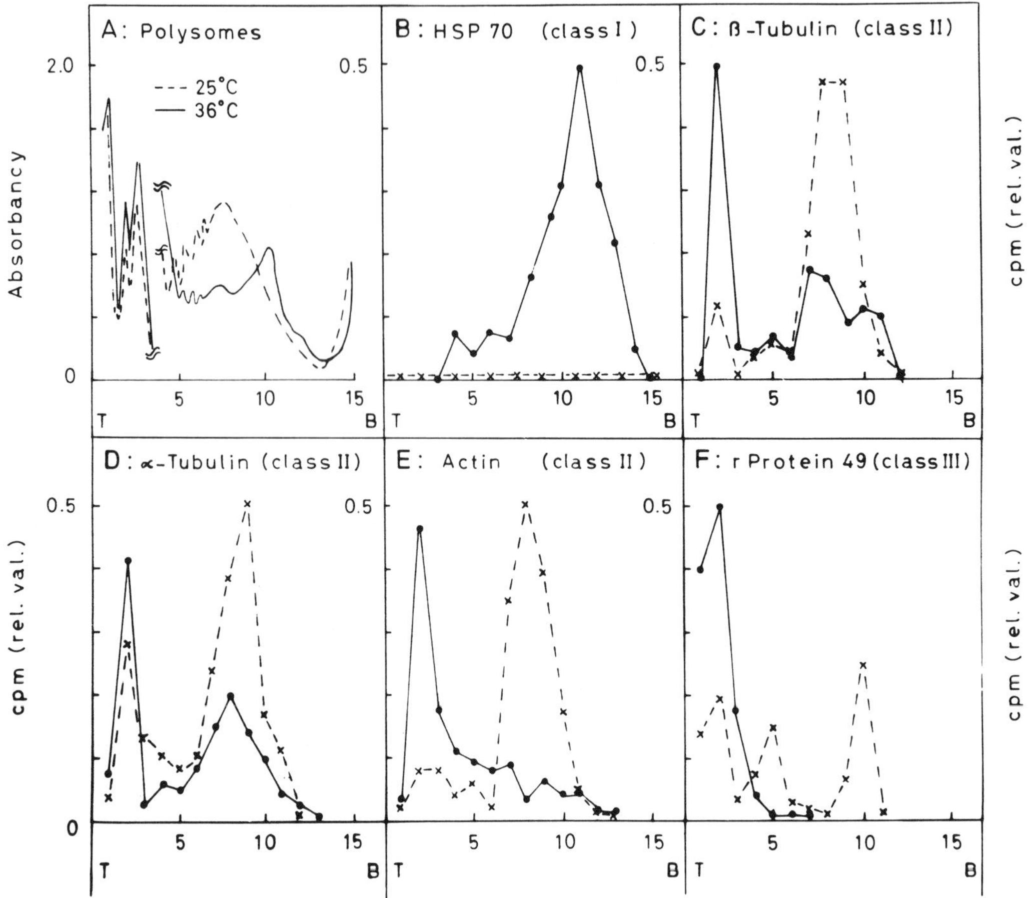

FIGURE 11.6. Polysomal profiles and distribution of mRNAs in *Drosophila* S2 cells under control (25°C) and heat shock (hs) conditions (36°C). Polysomes were separated on sucrose gradients, RNAs were prepared from each fraction and analyzed by Northern hybridization using appropriate clones. Three classes of mRNAs are distinguished: Class I shows that hs mRNA is essentially absent under control conditions, but is actively transcribed and translated under hs conditions; class II shows control protein mRNAs, whose translation is strongly reduced in hs cells but at least part of which remains in the polysomal fraction, and class III shows that rprotein 49 mRNA is found exclusively in the post-polysomal supernatant. (Modified from Ballinger, D. G. and Pardue, M. L., *Cell,* 33, 103, 1983.)

by *in vitro* translation of mRNAs from hs *Drosophila* Kc cells control protein mRNAs were found to be preserved, but not in the polysomal fraction.[74]

11.5. SEARCH FOR MECHANISMS OF TRANSLATION CONTROL

11.5.1. CYTOSKELETAL REARRANGEMENT

When investigating the hs-induced disintegration of polysomes in *Drosophila* cells, Falkner and Biessmann[32] observed the simultaneous translocation of two prominent cytoplasmic proteins with M_r 40 and 46 kDa to the nuclear fraction. Later on, this recompartmentation, associated with phosphorylation of the 46-kDa protein,[124] was characterized as the collapse of the *Drosophila* intermediate size cytoskeletal system (IF) in the perinuclear region. The two proteins represent vimentin-like components and, by use of corresponding antibodies, similar rearrangements of the IF system were also detected in baby hamster kidney cells.[13,32] These results were confirmed by other workers using other vertebrate cells (see Section 14.3).

The essential role of cytoskeletal elements for the organization of translation in animal cells (sea urchin, BHK, and HeLa cells) is well documented.[18,60,76,87,129] Howe and Hershey[43] demonstrated the intimate connection of mRNAs with the cytoskeletal system of HeLa cells. They are part of polysomes which can be released by decomposition of the IF network in a low salt buffer in the presence of detergents (Tween, deoxycholate). Hs causes a release of ribosomes and initiation factors from the cytoskeletal framework, but the mRNAs remain attached. These findings agree with our model on the protection of mRNAs under hs conditions in the complex of HSP aggregates bound to the perinuclear cytoskeletal system. Although at this time point essential experimental data are still lacking, the cytoskeletal system may play a dual role in the reprogramming of translation:

1. It is the cytoplasmic framework for the efficient translation of control protein mRNAs. Its collapse in the early phase of the hs response is an essential part of the transient initiation block.[33] Its restoration during long-term hs or in the recovery period is a prerequisite for normalization of the control protein pattern.
2. The perinuclear aggregates of the cytoskeleton with the control mRNAs attached are the sites of accumulation of mainly small HSPs in the hs granules (see Section 16.7) which help to preserve the untranslated control mRNAs in intact form. The HSP70 associated with these aggregates may assist in the rapid release of mRNAs in the recovery period and the final decay of the hs granules.[82,83]

11.5.2. RIBOSOMAL MODIFICATIONS

In all kinds of eukaryotic organisms, a few ribosomal proteins are modified by phosphorylation (Figure 11.7). Most prominent are a basic 30 kDa rprotein of the small subunit (rprotein S6) and a few acidic proteins of the large subunit (38 and 15 to 16 kDa). Ribosomal protein S6 undergoes rapid changes in the phosphorylation state in response to the cell cycle activity or hormonal stimulation. It is weakly or nonphosphorylated in quiescent, nonstimulated animal cells, whereas the addition of serum, mitogenic factors, hormones, or transformation with oncogenes results in a rapid increase of the phosphorylation level.[10,14,31,116,120,127] Ribosomal protein S6 is localized in the close vicinity of the small subunit site binding initiator tRNA and mRNA.[15,119] This led to the hypothesis that its rapidly changing levels of phosphorylation may be involved in cell cycle control, e.g., by altered rates of translational initiation with distinct mRNA species,[89,120] but so far there is poor experimental evidence for this. Only the transition of ribosomal subunits into polysomes may be enhanced in the highly phosphorylated state.

The hs response is also connected with dephosphorylation of the small subunit protein S6. This is true for plants,[103] *Drosophila,*[35,86] mammalian cells,[46,96,118] and fungi.[21a,92] The results with tomato cell cultures (Figure 11.7) confirm the complete dephosphorylation and rephosphorylation of S6 in the course of a hs and recovery treatment[103] (for details see legend to Figure 11.7).

In search of the underlying mechanism and possible function of this process, the following additional findings may be relevant.[103a]

1. The hs-induced block of the putative rprotein S6 kinase and its reversal is rapid and independent of protein synthesis. It is also observed in cells treated with cycloheximide. In the latter case, even the stalled polysomes are dephosphorylated.
2. In a polysomal gradient, the phosphorylation level is identical in the monosomal and in the polysomal region, i.e., under a given experimental condition the whole ribosomal population is in a uniform phosphorylation state. There is no evidence for a preferential engagement of highly phosphorylated ribosomes in translation.
3. Recovery of S6 phosphorylation usually leads to a higher level of phosphate modifi-

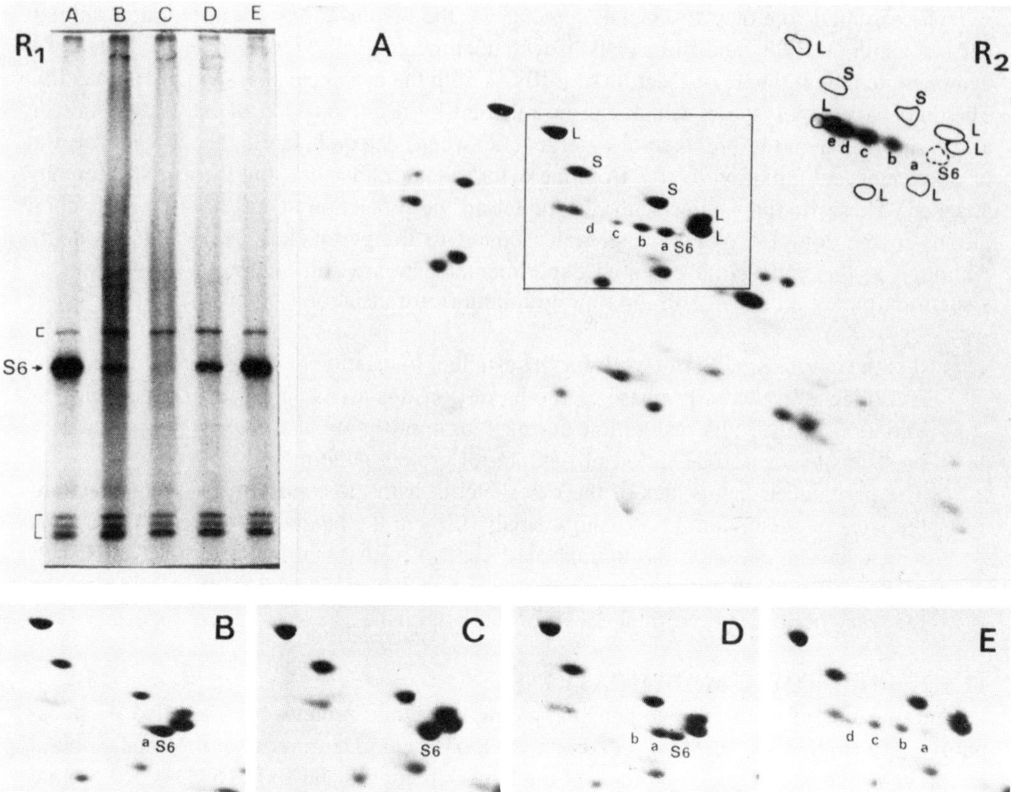

FIGURE 11.7. Reversible changes of phosphorylation levels of ribosomal protein S6 under conditions of heat shock (hs) and recovery. Tomato cell cultures were labeled for 2 h at 28°C with radioactive phosphate. Part of it was removed (sample A, control), and the remainder subjected to a hs at 40°C. Further samples were taken after 30 min (B) and 120 min (C) of hs, respectively. The rest of the culture was allowed to recover at 28°C for 60 min (D) and 120 min (E), respectively. Ribosomal proteins were prepared and separated by electrophoresis.[103] R_1 (upper left) shows the autoradiograph of a one-dimensional polyacrylamide gel of ribosomal proteins with a prominent phosphoprotein (M_r 30 kD, S6) of the small ribosomal subunit. The strong labeling in sample A disappears during the hs (B and C) and reappears in the recovery period (D and E). Labeling of other ribosomal phosphoproteins residing in the large subunit (brackets) does not change. The main part of the figure represents coomassie-stained two-dimensional electrophoretic separations of the ribosomal proteins. The total pattern is only shown for sample A, whereas samples B to E are represented by the boxed region indicated in A. The series of 4 to 5 phosphorylated isoforms of protein S6 (a to e, see also autoradiograph R_2 inserted in the upper right corner) disappears during hs (samples B and C) and reappears during hs recovery (samples D and E). (From Scharf, K.-D. and Nover, L., *Cell,* 30, 421, 1982. With permission.)

cation than before the hs treatment (compare the two-dimensional separations in Figure 11.7A and E). After severe hs (>40°C for tomato cell cultures), recovery is delayed unless cells were made thermotolerant by hs preconditioning.[103]

The kinetics of the hs-induced dephosphorylation and the cycloheximide experiment preclude a direct connection with the much more rapid decay of the polysomes.[118] A contribution to the translation selectivity, possibly in connection with altered initiation-elongation factors and peculiarities of the hs mRNAs (see Sections 11.5.3 and 11.5.5), is an intriguing possibility, but experimental proof is lacking. Finally, it is worth noticing that, similar to the situation in quiescent animal cells, the unphosphorylated state of the small ribosomal subunit under hs coincides with a block of mitotic activity and, vice versa, recovery of S6 phosphorylation and an overshoot of the mitotic index are characteristic events after reestablishment of normothermic conditions.

Another type of hs-induced modification of the ribosomal apparatus was described by Hallberg and co-workers in *Tetrahymena thermophila.* They found a new basic protein of 22 kDa firmly associated with the small ribosomal subunit[73] and a new 270 nucleotides hs-RNA transcribed by RNA polymerase III.[53] Both additions to the translation machinery are roughly stochiometric, i.e., one molecule of each per ribosome. They are presumably involved in the recovery of control protein translation. The exchange rprotein p22 exists in a preformed state and is not synthesized during hs. Association with ribosomes, which is also observed in nutrient-starved cells, may also contribute to the protection of ribosomes against heat damage (see Section 10.7). Most remarkably, a stress-induced association of a preformed 57-kDa protein with the polysomal apparatus was also reported for plants, e.g., if maize seedlings are exposed to heat shock or oxidative stress by the herbicide paraquat or to biogenic stress by pathogen infection.[127a] Isolated P57 was shown to inhibit *in vitro* translation of maize polysomes.

11.5.3. MODIFICATION OF INITIATION FACTORS

Long before the first detection of induced HSP synthesis,[122] Penman and co-workers reported on detailed investigations on the translation machinery of heat-shocked HeLa cells.[36,70] They showed that a transient initiation block with ongoing elongation is responsible for the decay of polysomes. These results were later confirmed for other systems and other stress-induced changes of the translation patterns.[85,90,121]

A deficiency of the initiation step is evidently also maintained during hs and may contribute to the preferential translation of hs mRNAs.[24,40,44a] In cells made defective in initiation by other treatments, e.g., in a ts mutant of yeast at 36°C,[94] in *Drosophila* or mammalian cells under hypertonic stress,[40,44a,52a,b] or after virus infection,[78,128] translation of the relatively strong hs mRNAs is more resistant. Similar effects are also observed in an initiation deficient *in vitro* translation system from wheat germs[51,52] or Ehrlich ascites tumor cells.[91] Interestingly, other strong mRNAs, e.g., viral mRNAs in *Drosophila,*[106] tobacco leaves,[22,104a] and HeLa cells[78] also escape the hs-induced discrimination. Their products become especially prominent in early phases of the hs response.

Careful investigations with HeLa cells under mild (41 to 42°C) and severe (45°C) hs conditions demonstrated that the typical depression of translation with polysomal breakdown at 42°C is not connected with corresponding modifications of initiation factors.[29b] Adaption to long-term incubation at 41 to 42°C requires RNA synthesis. It is tempting to speculate that reorganization of the intermediate-size cytoskeleton is the dominant factor for translation reprogramming at 42°C.

Analyses of the putative factors involved in the translational control mechanism are based on the use of *in vitro* translation systems prepared from hs cells (see Section 11.3). Fractionation of the *Drosophila* cell lysate proved that not the altered ribosomes *per se* but rather factors missing from the high salt wash fraction are responsible for the nontranslation of control mRNAs: the defect in the hs lysate could be repaired by adding soluble proteins from a control cell lysate.[101,107] A likely explanation of these results is the inactivation of the cap recognition factor observed in *Drosophila* embryos by hs.[67a] Evidently cap-independent, internal initiation of translation is a characteristic feature of the hs response.

So far the analysis of initiation factors has produced equivocal results. On the one hand, phosphorylation of the eIF-2α subunit, possibly by hs activation of hemin-regulated protein kinase (HRI), was reported in HeLa cells and rabbit reticulocytes.[16,23,28,29b,30] Interestingly, the inactive HRI is detected in complex with HSP90 in rabbit reticulocytes. The complex is dissociated after activation of the protein kinase under conditions of hemin deficiency, heat shock, and oxidation stress.[67b] Even low levels of phosphorylated eIF-2α can exert dramatic effects on translational initiation by trapping the low amounts of the catalytically active RF factor required for regeneration of active eIF2-GTP from the inactive eIF2-GDP complex.[68,99] In fact, addition of hemin to HeLa cells prior to a 5 min 42.5°C hs reduced the initial decrease of protein synthesis and enhanced the beginning of HSP synthesis.[23] However, the role of eIF-2α phosphorylation on the primary defect of initiation leading to polysome breakdown could not be verified in the studies of Mariano and Siekierka[67] using lysates of HeLa cells. In keeping with this, Morley et al.[77] purified a rabbit reticulocyte factor different from eIF-2α which was able to rescue initiation in lysates from heat-shocked mouse cells. At any rate, the α-subunit of eIF-2 was found to be a delayed HSP in rat thymus cells[19b] with slowly increasing synthesis during hs, reaching 15-fold higher levels of synthesis in the recovery period.

Results with *Drosophila* embryos[67a] and with the water mold *Achlya ambisexualis*[92] indicate that other soluble factors are also involved in the alterations of the translation machinery. In support of this, Duncan and Hershey [28,29b] found an inhibition of activities of initiation factors eIF-2, eIF-3 + 4F, and eIF-4B in HeLa cells. Both subunits of eIF-2 became modified, whereas some forms of eIF-4B disappear. Altered modification in the case of 4B and 2α means phosphorylation/dephosphorylation. Investigations on mouse Ehrlich ascites tumor cells led to the purification of eIF-4F, which can rescue control protein synthesis in an *in vitro* translation assay.[91] This factor is composed of 3 subunits (220, 45, and 28 kDa) and helps in the recognition of capped mRNAs. In HeLa cells the small subunit of eIF-4F (p28) was found to become dephosphorylated under hs.[29a] Unfortunately, the unphosphorylated and the phosphorylated forms exhibit the same cap recognition activity in an *in vitro* assay. But the complexation of p28 with p220 is disturbed. At least the eIF-4F preparation, active in the restoration of *in vitro* translation, contains highly phosphorylated p28 associated with p220.[56a] Preferential translation of hsp70 mRNA in poliovirus-infected cells is also connected with a loss of 4F activity, presumably caused by degradation of the 220-kDa subunit.[78]

11.5.4. POLY(A) METABOLISM

The role of the posttranscriptionally added poly(A)-tail of eukaryotic mRNAs is slowly emerging. Earlier hypotheses considered influences on mRNA stability or on translational, control.[9,45,88] Causal analyses are hindered by the fact that in different organisms many mRNAs also exist without or with very short poly(A)-tails, and these are active in *in vitro* translation systems as well. But recently experimental proof in support of these hypotheses was provided for yeast[97] and human erythroleukemia cells.[9a] The poly(A) tail is required for stability of mRNAs and/or for translation initiation. The poly(A)-binding protein protects the 3′-region from nucleolytic attack. There are several examples, including the *Drosophila* hsp70 mRNA (see Section 8.7), suggesting that determinants for mRNA stability reside in the 3′-untranslated part. Evidently there is an effective "cross-talk" between 5′ and 3′ regulatory sequences in the adjustment of the overall effect.[17a,17b,18a] Coexistence of poly(A)$^+$ and poly(A)$^-$ forms of mRNAs is also true for hsp70 mRNA in HeLa cells[49] and for hsp83 mRNA in *Drosophila* cells.[112] In the latter, the constitutively expressed hsp83 mRNA of control cells is exclusively of the poly(A)$^-$ type, whereas under hs conditions hsp83 mRNA is also found in the poly(A)$^+$ fraction. A simple, but hypothetic explanation is that there are two forms of hsp83 mRNA, a hs-induced, polyadenylated form and nonpolyadenylated form synthesized at 25°C. Alternatively, cytoplasmic polyadenylation of poly(A)$^-$ mRNAs, as observed during maturation of *Xenopus* oocytes,[71a] cannot be excluded.

Alterations of poly(A) metabolism in *Drosophila* cells were considered by Storti et al.[112] to account for the apparent loss of mRNA observed in earlier experiments of this group,[111] i.e., reduction of the poly(A)-tail and not the physical degradation of mRNAs was responsible. This agrees with results of thorough investigations with heat-shocked *Drosophila* S2 cells, showing a strong decrease of cytoplasmic and an increase of nuclear poly(A) content. The effects are observed under conditions of severe hs (30 to 60 min ≥37°C). Activation of nuclear poly(A) synthesis is independent of protein and RNA synthesis, i.e., it proceeds also in the presence of cycloheximide and actinomycin D. Evidently poly(A) tails are added to preexisting RNAs.

So far alterations of poly(A) metabolism have been analyzed only in *Drosophila* cells. Instead, special interest concentrated on the poly(A)$^+$-associated protein of 78 kDa, when Schönfelder et al.[105] claimed after purification, immune coprecipitation, and peptide analysis that the HeLa poly(A)-binding protein is identical or closely related to HSP70. Although very suggestive, this result is in conflict with a report by Sunitha and Slobin:[115] (1) the 78-kDa poly(A)-binding protein is basic, whereas HSP70 is an acidic protein (see Figure 2.2) and (2) its synthesis is strongly inhibited and very slowly recovers in heat-shocked Friend erythroleukemia cells. Other strong arguments against a role of HSP70 as the poly(A)-binding protein are the following:[66] on the one hand, sequence comparison of the yeast genes coding for the poly(A)-binding protein[1,98] and HSP70 (see Figure 2.9) shows no homology. On the other hand, the poly(A)-associated proteins in *Dictyostelium* (31 and 31.5 kDa) are not synthesized during hs but rather degraded. Antibodies raised against both proteins do not crossreact with the *Dictyostelium* HSP70, which however, is structurally related to the HSP70 of higher eukaryotes. Interestingly, the decay of p31/31.5 in these slime molds during hs coincides with a reduction of translational initiation. This again supports the conclusion that the poly(A) tail may be an important factor in the translational control mechanism.

11.5.5. ROLE OF HEAT SHOCK LEADER SEQUENCE

Messenger RNAs coding for HSP synthesis in eukaryotic cells are "normal" with respect ot their 5′ cap and the 3′ poly(A) tail. However, there is no doubt that the striking translational specificity in hs cells requires peculiarities of the hs mRNAs recognized by protein factors of the altered ribosomal machinery. With the onset of hs gene isolation and sequencing, two structural features were recognized in *Drosophila*: (1) mRNAs have unusually long untranslated leaders (see Table 3.1), and (2) there are essentially no elements for secondary structure formation.[41] This agrees with the finding that insertion of oligonucleotides into the leader of herpes simplex virus thymidine kinase mRNA decreased translational efficiency if the inserted sequence tends to form stable hairpin loops.[92a] Evidently, excessive formation of secondary structure at the 5′ end of eukaryotic mRNAs impedes translation under stress conditions.

Direct analyses of the role of the mRNA leader sequence was performed by gene technology (see Table 6.1). The following results are worth noticing:

1. The hs leader sequence is only preferred in the homologous hs translation system. Transfer of the *Drosophila* hsp70 gene or the promoter/leader region as part of a fusion construct into *Xenopus*[57] or plants[109] abolishes the effect.
2. Only 65 nucleotides of the hsp70 leader sequence are required to confer preferential translation to heterologous reporter genes, i.e., to the *Drosophila* adh gene[50] or the *Eschericha coli* cat gene.[26]
3. Extensive deletion and insertion analyses with the *Drosophila* hsp70 and hsp22 leaders[44,71] and sequence comparison[42] led to the definition of a 26 nucleotide consensus sequence immediately following the cap structure: (5′)-AgTTnAAaTnaAAnAanCnAAgn-GanAACA-(3′). It represents the minimum information for effective translation under hs conditions (Figure 11.8). In the 242 nucleotide hsp70 leader there is another element further upstream which may replace the 5′ consensus sequence.[71] Similar results were reported by Schoeffl et al.[104a] for the soybean hsp17.3B leader.

It is worth summarizing once more the evidence that certain control mRNAs are able to escape the hs-induced discrimination. In *Drosophila* these are histone and certain viral

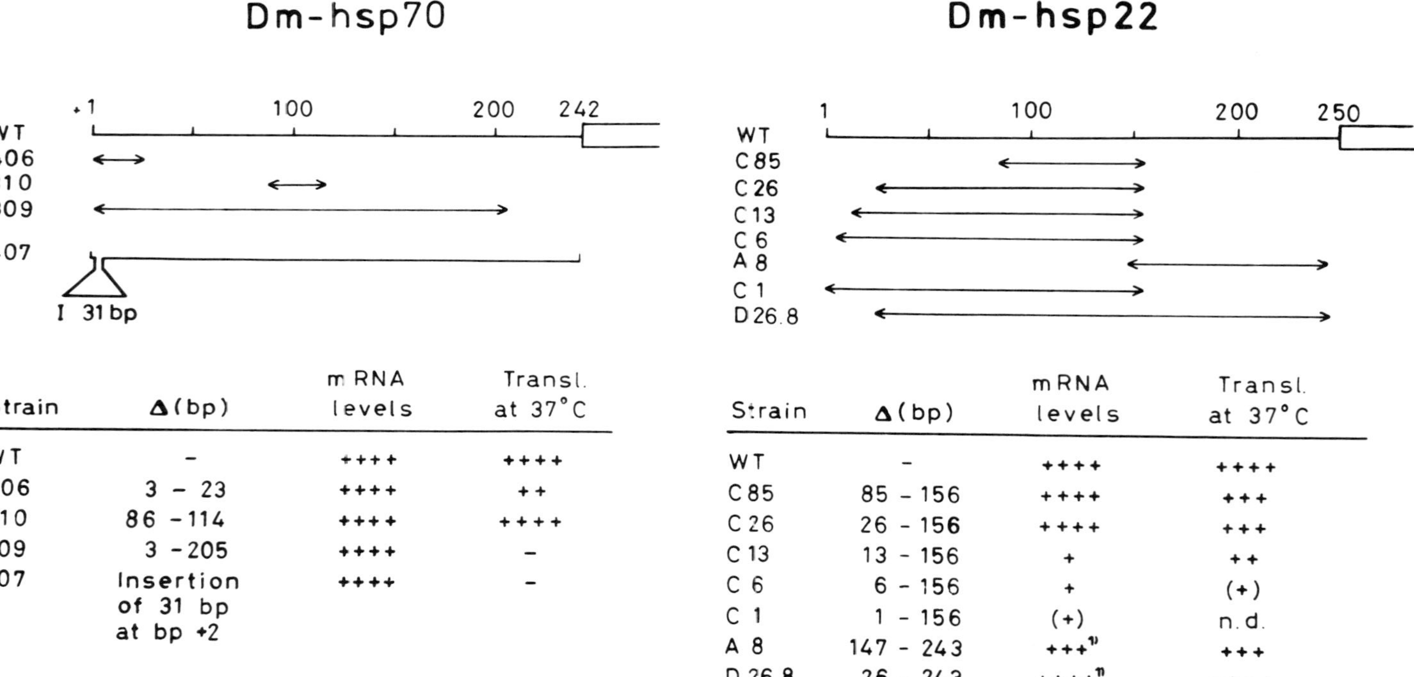

Strain	Δ (bp)	mRNA levels	Transl. at 37°C
WT	-	++++	++++
406	3 - 23	++++	++
310	86 - 114	++++	++++
309	3 - 205	++++	-
407	Insertion of 31 bp at bp +2	++++	-

Strain	Δ (bp)	mRNA levels	Transl. at 37°C
WT	-	++++	++++
C 85	85 - 156	++++	+++
C 26	26 - 156	++++	+++
C 13	13 - 156	+	++
C 6	6 - 156	+	(+)
C 1	1 - 156	(+)	n.d.
A 8	147 - 243	+++[1]	+++
D 26.8	26 - 243	++++[1]	++++

FIGURE 11.8. Functional analysis of leader sequences of *Drosophila melanogaster* hsp70 and hsp22 genes. Constructs with hs genes modified by deletion (hsp70) or insertion (hsp22) were combined with the indicated leader sequences containing various deletions (arrows) or an insertion at the 5′ end (hsp70, construct 407). mRNA levels and translation under hs conditions were estimated and referred to corresponding data of the wild-type gene (WT, + + + +). Main element of translation discrimination is a 26 nucleotide consensus element at the 5′ end (see text). This is particularly clear in the hsp22 construct D 26.8. It is worth noticing that deletion of this 26 nucleotide part of the hsp22 leader leads to a strong reduction of transcription (see Section 6.5). (Data from McGarry, T. J. and Lindquist, S., *Cell*, 42, 903, 1986; and Hultmark, D., Klemenz, R., and Gehring, W. J., *Cell*, 44, 429, 1986.)

mRNAs.[100,106,108,117] But undiminished translation of some control proteins was also reported for plant cells[80] and *Tetrahymena*[2] as well as for proteins coded by viral mRNAs in tobacco leaves.[22,104a] In HeLa cells, different viral mRNAs were unaffected by the translational block, e.g., reovirus, vesicular stomatitis virus, and encephalomyocarditis virus.[78]

A surprising aspect of the discrimination between hs and control mRNAs was elaborated by Denisenko and Yarchuk.[24b] Regarding the remarkable conservation of the C-terminus of practically all members of the HSP90 and HSP70 families (–EEVD; see Figures 2.8 and 2.10), they transferred codons for these four amino acid residues to a bacterial lacZ gene and generated corresponding mRNAs by an *in vitro* transcription system. Compared to the unmodified lacZ mRNA, this hybrid mRNA was preferentially translated *in vitro* in reticulocyte lysates if incubated at physiological ionic strength and 43°C. If reproducible with homologous constructs in other *in vivo* and *in vitro* translation systems, this gives another example for the regulatory "cross-talk" between 5′ and 3′ ends of mRNAs.

11.6. MODEL OF TRANSLATIONAL CONTROL

Though still rather fragmentary, the experimental data from different organisms provide a sufficient basis for a general model of translational reprogramming under hs conditions.

1. The primary effect of heat stress is the polysomal breakdown presumably connected with alterations of the cytoskeleton.
2. Alterations of the ribosomes (dephosphorylation of S6 protein) and initiation factors (cap recognition factor) make the translation system ineffective in the usual cap recognition and the ATP-dependent scanning to the initiator AUG.[52a,52b,95a]
3. The long leader sequences of hs mRNAs are essentially without secondary structures and allow a cap-independent internal initiation of translation.

   The inhibitory effects of hairpin-loop structures in the mRNA leader sequences, especially under conditions of stress, and the possibility for internal initiation were repeatedly documented.[5a,29d,52a,b] Interestingly, additional cytoplasmic and nuclear proteins were shown by gel retardation assays to be associated with stem-loop structures of viral mRNAs capable of internal initiation.[24a,29c,73a] Unwinding of double-stranded regions in the leader region can be envisaged. This might also explain the striking restriction of preferential translation in a given system to homologous hs mRNAs (see Section 11.3).
4. Protection of untranslated control mRNAs during the stress period is brought about by close association with HSPs, probably attached to the cytoskeleton.

REFERENCES

1. **Adam, S. A., Nakagawa, T., Swanson, M. S., Woodruff, T. K., and Dreifuss, G.,** mRNA polyadenylate-binding protein: Gene isolation and sequencing and identification of a ribonucleoprotein consensus sequence, *Mol. Cell. Biol.*, 6, 2932—2943, 1986.
2. **Amaral, M. D., Galego, L., and Rodrigues-Pousada, C.,** Stress response of *Tetrahymena pyriformis* to arsenite and heat shock: Differences and similarities, *Eur. J. Biochem.*, 171, 463—470, 1988.
3. **Arrigo, A.-P.,** Cellular localization of HSP23 during *Drosophila* development and following subsequent heat shock, *Dev. Biol.*, 122, 39—48, 1987.
4. **Arrigo, A.-P., and Welch, W. J.,** Characterization and purification of the small 28,000 dalton mammalian heat shock protein, *J. Biol. Chem.*, 262, 15359—15369, 1988.
5. **Bag, J.,** Recovery of normal protein synthesis in heat-shocked chicken myotubes by liposome-mediated transfer of mRNAs, *Can. J. Biochem. Cell Biol.*, 63, 231—235, 1985.

5a. **Baim, St. B. and Sherman, F.,** mRNA structures influencing translation in the yeast *Saccharomyces cerevisiae, Mol. Cell. Biol.,* 8, 1591—1601, 1988.
6. **Ballinger, D. G. and Pardue, M. L.,** The control of protein synthesis during heat shock in *Drosophila* cells involves altered polypeptide elongation rates, *Cell,* 33, 103—114, 1983.
7. **Banerji, S. S., Theodorakis, N. G., and Morimoto, R. I.,** Heat shock-induced translational control of HSP70 and globin synthesis in chicken reticulocytes, *Mol. Cell. Biol.,* 4, 2437—2448, 1984.
8. **Belanger, F. C., Brodl, M. R., and Ho, T.-H. D.,** Heat shock causes destabilization of specific mRNAs and destruction of endoplasmic reticulum in barley aleurone cells, *Proc. Natl. Acad. Sci. U.S.A.,* 83, 1354—1358, 1986.
9. **Bergman, I. E. and Brawerman, G.,** Loss of the polyadenylate segment from mammalian messenger RNA. Selective cleavage of the sequence from polyribosomes, *J. Mol. Biol.,* 139, 439—454, 1980.
9a. **Bernstein, P., Peltz, S. W., and Ross, J.,** The poly(A)-poly(A)-binding protein complex is a major determinant of mRNA stability in vitro, *Mol. Cell. Biol.,* 9, 659—670, 1989.
10. **Bielka, H., Ed.,** *The Eukaryotic Ribosomes,* Akademie Verlag, Berlin, 1982.
11. **Bienz, M. and Gurdon, J. B.,** The heat-shock response in *Xenopus* oocytes is controlled at the translational level, *Cell,* 29, 811-819, 1982.
12. **Bienz, M. and Pelham, H. R. B.,** Expression of a *Drosophila* heat-shock protein in *Xenopus* oocytes: Conserved and divergent regulatory signals, *EMBO J.,* 1, 1583—1588, 1982.
13. **Biessmann, H., Falkner, F. G., Saumweber, H., and Walter, M. F.,** Disruption of the vimentin cytoskeleton may play a role in heat-shock response, in *Heat Shock from Bacteria to Man,* Schlesinger, M. J., Ashburner, M., and Tissieres, A., Eds., Cold Spring Harbor Laboratory, Cold Spring Harbor, NY, 1982, 275—281.
14. **Blenis, J. and Erickson, R. L.,** Regulation of a ribosomal protein S6 kinase activity by the Rous sarcoma virus transforming protein, serum, or phorbol ester, *Proc. Natl. Acad. Sci. U.S.A.,* 82, 7621—7625, 1985.
15. **Bommer, U. A., Noll, F., Lutsch, G., and Bielka, H.,** Immunochemical detection of protein in the small subunit of rat liver ribosomes involved in binding of the ternary initiation complex, *FEBS Lett.,* 111, 171-174, 1980.
16. **Bonanou-Tzedaki, S., Sohl, M. K., and Arnstein, H. R. V.,** Regulation of protein synthesis in reticulocyte lysates. Characterization of the inhibitor generated in the post-ribosomal supernatant by heating at 44°C, *Eur. J. Biochem.,* 114, 69-77, 1981.
17. **Brandt, C. and Milcarek, C.,** Heat shock induced alterations in polyadenylate metabolism in *Drosophila melanogaster, Biochemistry,* 19, 6152—6158, 1980.
17a. **Braun, R. E., Peschon, J. J., Behringer, R. R., Brinster, R. L., and Palmiter, R. D.,** Protamine 3′-untranslated sequences regulate temporal translation control and subcellular localization of growth hormone in spermatids of transgenic mice, *Genes Dev.,* 3, 793—802, 1989.
17b. **Brawerman, G.,** mRNA decay: Finding the right targets, *Cell,* 57, 9—10, 1989.
17c. **Brodl, M. R., Belanger, F. C., and Ho, T. H. D.,** Heat shock proteins are not required for the degradation of α-amylase messenger RNA and the delamellation of endoplasmic reticulum in heat-stressed barley aleurone cells, *Plant Physiol.,* 92, 1133—1141, 1990.
17d. **Brown, I. R.,** Modification of gene expression in the mammalian brain after hyperthermia in *Gene Expression in Brain,* Zomzely-Neurath, C. and Walker, W. A., Eds., John Wiley & Sons, New York, 1985, 157—171.
18. **Cervera, M., Dreyfuss, G., and Penman, Sh.,** Messenger RNA is translated when associated with the cytoskeletal framework in normal and VSV-infected HeLa cells, *Cell,* 23, 113—120, 1981.
18a. **Cleveland, D. W.,** Autoregulated instability of tubulin mRNAs: a novel eukaryotic regulatory mechanism, *Trends Biochem. Sci.,* 13, 339—343, 1988.
19. **Chomyn, A., Moller, G., and Mitchell, H. K.,** Patterns of protein synthesis following heat shock in pupae of *Drosophila melanogaster, Dev. Genet.,* 1, 77—95, 1979.
19a. **Coias, R., Galego, L., Barchona, I., and Rodrigues-Pousada, C.,** Destabilization of tubulin mRNA during heat-shock in *Tetrahymena pyriformis, Eur. J. Biochem.,* 175, 467—474, 1988.
19b. **Colbert, R. A., Hucul, J. A., Scorsone, K. A., and Young, D. A.,** Alpha-subunit of eukaryotic translational initiation factor-2 is a heat shock protein, *J. Biol. Chem.,* 262, 16763—16766, 1987.
20. **Collier, N. C. and Schlesinger, M. J.,** The dynamic state of heat shock proteins in chicken embryo fibroblasts, *J. Cell Biol.,* 103, 1495—1507, 1986.
20a. **Collier, N. C., Heuser, J., Levy, M. A., and Schlesinger, M. J.,** Ultra-structural and biochemical analysis of the stress granule in chicken embryo fibroblasts, *J. Cell Biol.,* 106, 1131—1139, 1988.
20b. **Cosgrove, J. W., Clark, B. D., and Brown, I. R.,** Effect of intravenous administration of d-lysergic acid diethylamide on subsequent protein synthesis in a cell-free system derived from brain, *J. Neurochem.,* 36, 1037—1045, 1981.
21. **Courgeon, A. M., Maisonhaute, C., and Best-Belpomme, M.,** Heat-shock proteins are induced by cadmium in *Drosophila* cells, *Exp. Cell Res.,* 153, 515—521, 1984.

21a. **Curle, C. A. and Kapoor, M.,** A *Neurospora crassa* heat-shocked cell lysate translates homologous and heterologous messenger RNA efficiently without preference for heat shock messages, *Curr. Genet.,* 13, 401—409, 1988.

21b. **Da Silva, A. M., Juliani, M. H., and Bonato, M. C. M.,** Effect of heat shock on S6 phosphorylation during the development of *Blastocladiella emersonii, Mol. Cell. Biochem.,* 78, 27—36, 1987.

22. **Dawson, W. O. and Boyd, C.,** TMV protein synthesis is not translationally regulated by heat shock, *Plant Mol. Biol.,* 8, 145-149, 1987.

23. **de Benedetti, A. and Baglioni, C.,** Activation of hemin-regulated initiation factor-2 kinase in heat-shocked HeLa cells, *J. Biol. Chem.,* 261, 338—342, 1986.

24. **de Benedetti, A. and Baglioni, C.,** Translational regulation of the synthesis of a major heat shock protein in HeLa cells, *J. Biol. Chem.,* 261, 15800—15805, 1986.

24a. **Del Angel, R. M., Papavassilou, A. G., Fernandez-Tomas, C., Silverstein, S. J., and Racaniello, V. R.,** Cell proteins bind to multiple sites within the 5′ untranslated region of poliovirus RNA, *Proc. Natl. Acad. Sci. U.S.A.,* 86, 8299—8303, 1989.

24b. **Denisenko, O. N. and Yarchuk, O. B.,** Regulation of LacZ mRNA translatability in a cell free system at heat shock by the last four sense codons, *FEBS Lett.,* 247, 251—254, 1989.

25. **Di Domenico, B. J., Bugaisky, G. E., and Lindquist, S.,** Heat shock and recovery are mediated by different translational mechanisms, *Proc. Natl. Acad. Sci. U.S.A.,* 79, 6181—6185, 1982.

26. **DiNocera, P. P. and Dawid, I. B.,** Transient expression of genes introduced into cultured cells of *Drosophila, Proc. Natl. Acad. Sci. U.S.A.,* 80, 7095—7098, 1983.

27. **Duband, J. L., Lettre, F., Arrigo, A. P., and Tanguay, R. M.,** Expression and localization of hsp-23 in unstressed and heat-shocked *Drosophila* cultured cells, *Can. J. Genet. Cytol.,* 28, 1088—1092, 1987.

28. **Duncan, R. and Hershey, J. W. B.,** Heat shock-induced translational alterations in HeLa cells, *J. Biol. Chem.,* 259, 11882—11889, 1984.

29. **Duncan, R. and Mc Conkey, E. H.,** S6 phosphorylation accompanies recruitment of ribosomes and mRNA into polysomes in response to dichlororibofuranosyl benzimidazole, *Exp. Cell Res.,* 152, 520—527, 1984.

29a. **Duncan, R., Milburn, S. C., and Hershey, J. W. B.,** Regulated phosphorylation and low abundance of HeLa cell initiation factor eIF-4F suggest a role in translational control. Heat shock effects on eIF-4F, *J. Biol. Chem.,* 262, 380—389, 1987.

29b. **Duncan, R. F. and Hershey, J. W. B.,** Protein synthesis and protein phosphorylation during heat stress, recovery, and adaptation, *J. Cell Biol.,* 109, 1467—1481, 1989.

29c. **Edery, I., Petryshyn, R., and Sonenberg, A.,** Activation of double-stranded RNA-dependent kinase (dsI) by the TAR region of HIV-1 mRNA: A novel translational mechanism, *Cell,* 56, 303—312, 1989.

29d. **Elroy-Stein, O., Fuerst, T. R., and Moss, B.,** Cap-independent translation of mRNA conferred by encephalomyocarditis virus 5′ sequence improves the performance of the vaccinia virus/bacteriophage T7 hybrid expression system, *Proc. Natl. Acad. Sci. U.S.A.,* 86, 6126—6130, 1989.

30. **Ernst, V., Baum, E. Z., and Reddy, P.,** Heat shock protein phosphorylation, and the control of translation in rabbit reticulocytes, reticulocyte lysates, and HeLa cells, in *Heat Shock from Bacteria to Man,* Schlesinger, M. J., Ashburner, M., and Tissieres, A., Eds., Cold Spring Harbor Laboratory, Cold Spring Harbor, NY 1982, 215—225.

31. **Evans, S. W. and Farrer, W. I.,** Interleukin 2 and diacylglycerol stimulate phosphorylation of 40S ribosomal S6 protein. Correlation with increased protein synthesis and S6 kinase activation, *J. Biol. Chem.,* 262, 4624—4630, 1987.

32. **Falkner, F. G. and Biessmann, H.,** Nuclear proteins in *Drosophila melanogaster* cells after heat-shock and their binding to homologous DNA, *Nucl. Acids Res.,* 8, 943—955, 1980.

33. **Falkner, F. G., Saumweber, H., and Biessmann, H.,** Two *Drosophila melanogaster* proteins related to intermediate filament proteins of vertebrate cells, *J. Cell Biol.,* 91, 175—183, 1981.

34. **Farrell-Towt, J. and Sanders, M. M.,** Noncoordinate histone synthesis in heat-shocked *Drosophila* cells is regulated at multiple levels, *Mol. Cell. Biol.,* 4, 2676—2685, 1984.

35. **Glover, C. V. C.,** Heat shock induces rapid dephosphorylation of a ribosomal protein in *Drosophila, Proc. Natl. Acad. Sci. U.S.A.,* 79, 1781—1785, 1982.

36. **Goldstein, E. and Penman, S.,** Regulation of protein synthesis in mammalian cells. Further studies on the effect of actinomycin D on translation control in HeLa cells, *J. Mol. Biol.,* 80, 243—254, 1973.

37. **Gwynne, D. I. and Brandhorst, B. P.,** Alterations in gene expression during heat shock of *Achlya ambisexualis, J. Bacteriol.,* 149, 488—493, 1982.

37a. **Hallberg, R. L.,** personal communication.

38. **Hallberg, R. L., Kraus, K. W., and Findly, R. C.,** Starved *Tetrahymena thermophila* cells that are unable to mount an effective heat shock response selectively degrade their rRNA, *Mol. Cell. Biol.,* 4, 2170—2179, 1984.

38a. **Heikkila, J. J. and Brown, I. R.,** Hyperthermia and disaggregation of brain polysomes induced by bacterial pyrogen, *Life Sci.,* 25, 347—351, 1979.

39. **Heine, U., Sverak, L., Kondratick, J., and Bonar, R. A.,** The behaviour of HeLa-S3 cells under the influence of supranormal temperatures, *J. Ultrastruct. Res.*, 34, 375—396, 1971.
39a. **Herrick, D., Parker, R., and Jacobson, A.,** Identification and comparison of stable and unstable mRNAs in *Saccharomyces cerevisiae*, *Mol. Cell. Biol.*, 10, 2269—2284, 1990.
39b. **Herruer, M. H., Mager, W. H., Raue, H. A., Vreken, P., Wilms, E., and Planta, R. J.,** Mild temperature shock affects transcription of yeast ribosomal protein genes as well as the stability of their messenger RNA, *Nucl. Acids Res.*, 16, 7917—7930, 1988.
40. **Hickey, E. D. and Weber, L. A.,** Modulation of heat shock polypeptide synthesis in HeLa cells during hyperthermia and recovery, *Biochemistry*, 21, 1513—1521, 1982.
41. **Hoffman, E. P. and Corces, V. G.,** Correct temperature induction and developmental regulation of a cloned heat shock gene transformed into *Drosophila* germ line, *Mol. Cell. Biol.*, 4, 2883—2889, 1984.
42. **Holmgren, K., Corces, V., Morimoto, R., Blackman, R., and Meselson, M.,** Sequence homologies in the 5′ regions of four *Drosophila* heat-shock genes, *Proc. Natl. Acad. Sci. U.S.A.*, 78, 3775—3778, 1981.
43. **Howe, J. G. and Hershey, J. W. B.,** Translation initiation factor and ribosome association with the cytoskeletal frame work fraction from HeLa cells, *Cell*, 37, 85—93, 1984.
44. **Hultmark, D., Klemenz, R., and Gehring, W. J.,** Translational and transcriptional control elements in the untranslated leader of the heat-shock gene hsp22, *Cell*, 44, 429—438, 1986.
44a. **Jackson, R. J.,** The heat-shock response in *Drosophila* KC 161 cells, mRNA competition is the main explanation for reduction of normal protein synthesis, *Eur. J. Biochem.*, 158, 623—634, 1986.
45. **Jacobson, A. and Favreau, M.,** Possible involvement of poly(A) in protein synthesis, *Nucl. Acids Res.*, 11, 6353—6368, 1983.
45a. **Kawata, Y., Fujiwara, H., and Ishikawa, H.,** Low molecular weight RNA of *Drosophila* cells which is induced by heat shock. I. Synthesis and its effect on protein synthesis, *Comp. Biochem. Physiol.*, 91B, 149—153, 1988.
45b. **Kawata, Y., Fujiwara, H., Shiba, T., Miyake, T., and Ishikawa, H.,** Low molecular weight RNA of *Drosophila* cells which is induced by heat shock. II. Structural properties, *Comp. Biochem. Physiol.*, 91B, 155—157, 1988.
46. **Kennedy, I. M., Burdon, R. H., and Leader, D. P.,** Heat shock causes diverse changes in the phosphorylation of the ribosomal proteins of mammalian cells, *FEBS Lett.*, 169, 267—273, 1984.
47. **Key, J. L., Lin, C. Y., and Chen, Y. M.,** Heat shock proteins of higher plants, *Proc. Natl. Acad. Sci. U.S.A.*, 78, 3526—3530, 1981.
48. **Kim, C. H. and Warner, J. R.,** Mild temperature shock alters transcription of a discrete class of *Saccharomyces cerevisiae* genes, *Mol. Cell. Biol.*, 3, 457—465, 1983.
49. **Kioussis, J., Cato, A. C. B., Slater, A., and Burdon, R. H.,** Polypeptides encoded by polyadenylated and non-polyadenylated messenger RNAs from normal and heat shocked HeLa cells, *Nucl. Acids Res.*, 9, 5203—5214, 1981.
50. **Klemenz, R., Hultmark, D., and Gehring, W. J.,** Selective translation of heat shock mRNA in *Drosophila melanogaster* depends on sequence information in the leader, *EMBO J.*, 4, 2053—2060, 1985.
51. **Kloppstech, K. and Ohad, I.,** Heat-shock protein synthesis in *Chlamydomonas reinhardi*. Translational control at the level of initiation of a poly(A)-rich-RNA coded 22-kDa protein in a cell-free system, *Eur. J. Biochem.*, 154, 63—68, 1986.
52. **Kloppstech, K., Lorberboum. H., Degroot, N., and Hochberg, A. A.,** Translational control at the level of initiation between mRNAs for pre-existing proteins and a 22-kDa heat-shock protein of *Chlamydomonas* by small cytosolic RNAs, *Eur. J. Biochem.*, 167, 501—505, 1987.
52a. **Kozak, M.,** Leader length and secondary structure modulate mRNA function under conditions of stress, *Mol. Cell. Biol.*, 8, 2737—2744, 1988.
52b. **Kozak, M.,** Circumstances and mechanisms of inhibition of translation by secondary structure in eucaryotic mRNAs, *Mol. Cell. Biol.*, 9, 5134—5142, 1989.
53. **Kraus, K. W., Good, P. J., and Hallberg, R. L.,** A heat shock-induced, polymerase III-transcribed RNA selectively associates with polysomal ribosomes in *Tetrahymena thermophila*, *Proc. Natl. Acad. Sci. U.S.A.*, 84, 383—387, 1987.
54. **Krüger, C. and Benecke, B.-J.,** In vitro translation of *Drosophila* heat-shock and non-heat-shock mRNAs in heterologous and homologous cell-free systems, *Cell*, 23, 595—603, 1981.
55. **Kuriki, Y.,** Translational repression of TEM beta-lactamase synthesis as a response of *Escherichia coli* to heat shock, *Mol. Microbiol.*, 3, 1131—1140, 1989.
55a. **Kuriki, Y.,** The translation start signal region of TEM beta-lactamase mRNA is responsible for heat shock-induced repression of amp gene expression in *Escherichia coli*, *J. Bacteriol.*, 171, 5452—5457, 1989.
56. **Laemmli, U. K.,** Cleavage of structural proteins during the assembly of the head of bacteriophage T4, *Nature*, 227, 680—685, 1970.
56a. **Lamphear, B. J. and Panniers, R.,** Cap binding protein complex that restores protein synthesis in heat-shocked Ehrlich cell lysates contains highly phosphorylated eIF-4E, *J. Biol. Chem.*, 265, 5333—5336, 1990.

57. **Lawson, R., Mestril, R., Schiller, P., and Voellmy, R.,** Expression of heat shock-beta-galactosidase hybrid genes in cultured *Drosophila* cells, *Mol. Gen. Genet.,* 198, 116—124, 1984.
58. **Leicht, B. G., Biessmann, H., Palter, K. B., and Bonner, J. J.,** Small heat shock proteins of *Drosophila* associate with the cytoskeleton, *Proc. Natl. Acad. Sci. U.S.A.,* 83, 90—94, 1986.
59. **LeJohn, H. B. and Braithwaite, E. E.,** Heat and nutritional shock-induced proteins of the fungus *Achlya* are different and under independent transcriptional control, *Can. J. Biochem. Cell Biol.,* 62, 837—846, 1984.
60. **Lenk, R. and Penman, S.,** The cytoskeletal framework and poliovirus metabolism, *Cell,* 16, 289—301, 1979.
61. **Lindquist, S.,** Regulation of protein synthesis during heat shock, *Nature,* 293, 311—314, 1981.
62. **Lindquist, S.,** Varying patterns of protein synthesis in *Drosophila* during heat shock: implications for regulation, *Dev. Biol.,* 77, 463—479, 1980.
63. **Lindquist, S.,** Translational efficiency of heat-induced messages in *Drosophila melanogaster* cells, *J. Mol. Biol.,* 137, 151—158, 1980.
64. **Lindquist, S.,** The heat-shock response (review), *Annu. Rev. Biochem.,* 55, 1151—1191, 1986.
64a. **Lodish, H. F.,** Translational control of protein synthesis, *Annu. Rev. Biochem.,* 45, 39—72, 1976.
65. **Love, R., Soriano, R. Z., and Walsh, R. J.,** Effect of hyperthermia on normal and neoplastic cells in vitro, *Cancer Res.,* 30, 1525—1533, 1970.
66. **Manrow, R. E. and Jacobson, A.,** Increased rates of decay and reduced levels of accumulation of the major poly (A)-associated proteins of *Dictyostelium* during heat shock and development, *Proc. Natl. Acad. Sci. U.S.A.,* 84, 1858—1862, 1987.
67. **Mariano, T. M. and Siekierka, J.,** Inhibition of HeLa cell protein synthesis under heat shock conditions in the absence of initiation factor eIF-2alpha phosphorylation, *Biochem. Biophys. Res. Commun.,* 138, 519—525, 1986.
67a. **Maroto, F. G. and Sierra, J. M.,** Translational control in heat-shocked *Drosophila* embryos. Evidence for the inactivation of initiation factor(s) involved in the recognition of mRNA cap structure, *J. Biol. Chem.,* 263, 15720—15725, 1988.
67b. **Matts, R. L. and Hurst, R.,** Evidence for the association of the heme-regulated eIF-kinase with the 90-kDa heat shock protein in rabbit reticulocyte lysate in situ, *J. Biol. Chem.,* 264, 15542—15547, 1989.
68. **Matts, R. L., Levin, D. H., and London, I. M.,** Effect of phosphorylation of the alpha-subunit of eukaryotic initiation factor 2 on the function of reversing factor in the initiation of protein synthesis, *Proc. Natl. Acad. Sci. U.S.A.,* 80, 2559—2563, 1983.
69. **McAlister, L. and Finkelstein, D. B.,** Alterations in translatable ribonucleic acid after heat shock of *Saccharomyces cerevisiae, J. Bacteriol.,* 143, 603—612, 1980.
70. **McCormick, W. and Penman, S.,** Regulation of protein synthesis in HeLa cells: translation at elevated temperatures, *J. Mol. Biol.,* 39, 315—333, 1969.
71. **McGarry, T. J. and Lindquist, S.,** The preferential translation of *Drosophila* hsp70 mRNA requires sequences in the untranslated leader, *Cell,* 42, 903—911, 1986.
71a. **McGrew, L. L., Dworkin-Rastl, E., Dworkin, M. B., and Richter, J. D.,** Poly(A) elongation during *Xenopus* oocyte maturation is required for translational recruitment and is mediated by a short sequence element, *Genes Dev.,* 3, 803—815, 1989.
72. **McKenzie, S. L., Henikoff, S., and Meselson, M.,** Localization of RNA from heat-induced polysomes at puff sites in *Drosophila melanogaster, Proc. Natl. Acad. Sci. U.S.A.,* 72, 1117—1121, 1975.
73. **McMullin, T. W. and Hallberg, R. L.,** Effect of heat shock on ribosome structure: Appearance of a new ribosome-associated protein, *Mol. Cell. Biol.,* 6, 2527—2535, 1986.
73a. **Meerovitch, K., Pelletier, J., and Sonenberg, N.,** A cellular protein that binds to the 5′-noncoding region of polio-virus RNA. Implications for internal translation initiation, *Genes Dev.,* 3, 1026—1034, 1989.
74. **Mirault, M. E., Goldschmidt-Clermont, M., Moran, L., Arrigo, A. P., and Tissieres, A.,** The effect of heat shock on gene expression in *Drosophila melanogaster, Cold Spring Harb. Symp. Quant. Biol.,* 42, 819—827, 1978.
75. **Mizzen, L. A. and Welch, W. J.,** Characterization of the thermotolerant cell. I. Effects on protein synthesis activity and the regulation of HSP70 expression, *J. Cell Biol.,* 106, 1105—1116, 1988.
76. **Moon, R. T., Nicosia, R. F., Olsen, C., Hille, M. B., and Jeffery, W. R.,** The cytoskeletal framework of sea urchin eggs and embryos: Developmental changes in the association of messenger RNA, *Dev. Biol.,* 95, 447—458, 1983.
77. **Morley, S. J., Buhl, W.-J., and Jackson, R. J.,** A rabbit reticulocyte factor which stimulates protein synthesis in several mammalian cell-free systems, *Biochim. Biophys. Acta,* 825, 57—69, 1985.
78. **Munoz, A., Alonso, M. A., and Carrasco, L.,** Synthesis of heat-shock proteins in HeLa cells: Inhibition by virus infection, *Virology,* 137, 150—159, 1984.
79. **Nover, L.,** *Heat Shock Response of Eukaryotic Cells,* Springer-Verlag, Berlin, 1984.
80. **Nover, L. and Scharf, K.-D.,** Synthesis, modification and structural binding of heat shock proteins in tomato cell cultures, *Eur. J. Biochem.,* 139, 303—313, 1984.

81. **Nover, L., Scharf, K.-D., and Neumann, D.,** Formation of cytoplasmic heat shock granules in tomato cell cultures and leaves, *Mol. Cell. Biol.*, 3, 1648—1655, 1983.
82. **Nover, L., Scharf, K.-D., and Neumann, D.,** Cytoplasmic heat shock granules are formed from precursor particles and contain a specific set of mRNAs, *Mol. Cell. Biol.*, 9, 1298—1308, 1989.
83. **Nover, L., Neumann, D., and Scharf, K.-D., Eds.,** *Heat Shock and Other Stress Response Systems of Plants,* Springer-Verlag, Berlin, 1990.
84. **O'Farrell, P. H.,** High resolution two-dimensional electrophoresis of proteins, *J. Biol. Chem.*, 250, 4007—4021, 1975.
85. **Oleinick, N.,** The initiation and elongation steps in protein synthesis: Relative rates in Chinese hamster ovary cells during and after hyperthermic and hypothermic shocks, *J. Cell. Physiol.*, 98, 185—192, 1979.
86. **Olsen, A. S., Feeney-Triemer, D., and Sanders, M. M.,** Dephosphorylation of S6 and expression of the heat shock response in *Drosophila melanogaster, Mol. Cell. Biol.*, 3, 2017—2027, 1983.
87. **Ornelles, D. A., Fey, E. G., and Penman, S.,** Cytochalasin release mRNA from the cytoskeletal framework and inhibits protein synthesis, *Mol. Cell. Biol.*, 6, 1650—1662, 1986.
88. **Palatnik, C. M., Wilkins, C., and Jacobson, A.,** Translational control during early *Dictyostelium* development: Possible involvement of poly(A) sequences, *Cell,* 36, 1017—1025, 1984.
89. **Palen, E. and Traugh, J. A.,** Phosphorylation of ribosomal protein S6 by cAMP-dependent protein kinase and mitogen-stimulated S6 kinase differentially alter translation of globin mRNA, *J. Biol. Chem.*, 262, 3518—3523, 1987.
90. **Panniers, R. and Henshaw, E. C.,** Mechanism of inhibition of polypeptide chain initiation in heat-shocked Ehrlich ascites tumor cells, *Eur. J. Biochem.*, 140, 209—214, 1984.
91. **Panniers, R., Stewart, E. B., Merrick, W. C., and Henshaw, E. C.,** Mechanism of inhibition of polypeptide chain initiation in heat-shocked Ehrlich cells involves reduction of eucaryotic initiation factor 4F activity, *J. Biol. Chem.*, 260, 9648—9653, 1985.
92. **Pekkala, D. and Silver, J. C.,** Heat-shock-induced changes in the phosphorylation of ribosomal and ribosome-associated proteins in the filamentous fungus *Achlya ambisexualis, Exp. Cell Res.*, 168, 325—337, 1987.
92a. **Pelletier, J. and Sonnenberg, N.,** Insertion mutagenesis to increase secondary structure within the 5′ noncoding region of a eukaryotic mRNA reduces translational efficiency, *Cell,* 40, 515—526, 1985.
93. **Petersen, N. S. and Mitchell, H. K.,** Recovery of protein synthesis after heat shock: Prior heat treatment affects the ability of cells to translate mRNA, *Proc. Natl. Acad. Sci. U.S.A.*, 78, 1708—1711, 1981.
93a. **Piper, P. W., Curran, B., Davies, M. W., Hirst, K., Lockheart, A., Ogden, J. E., Stanway, C. A., Kingsman, A. J., and Kingsman, S. M.,** A heat shock element in the phosphoglycerate kinase gene promoter of yeast, *Nucl. Acids Res.*, 16, 1333—1348, 1988.
93b. **Piper, P. W., Curran, B., Davies, M. W., Hirst, K., Lockheart, A., and Seward, K.,** Catabolite control of the elevation of pgk messenger RNA levels by heat shock in *Saccharomyces cerevisiae, Mol. Microbiol.*, 2, 353—362, 1988.
94. **Plesset, J., Foy, J. J., Chia, L.-L., and McLaughlin, C. S.,** Heat shock in *Saccharomyces cerevisiae:* Quantitation of transcriptional and translational effects, *Dev. Biochem.*, 24, 495—514, 1982.
95. **Ralhan, R. and Johnson, G. S.,** Destablilization of cytoplasmic mouse mammary tumor RNA by heat shock: Prevention by cycloheximide pretreatment, *Biochem. Biophys. Res. Commun.*, 137, 1028—1033, 1986.
95a. **Rhoads, R. E.,** Cap recognition and the entry of mRNA into the protein synthesis initiation cycle, *Trends Biochem. Sci.*, 13, 52—56, 1988.
96. **Richter, W. W., Zang, K. D., and Issinger, O. G.,** Influence of hyperthermia on the phosphorylation of ribosomal protein S6 from human skin fibroblasts and meningioma, *FEBS Lett.*, 153, 262—266, 1983.
97. **Sachs, A. B. and Davis, R. W.,** The poly(A) binding protein is required for poly(A) shortening and 60S ribosomal subunit-dependent translation initiation, *Cell,* 58, 857—867, 1989.
98. **Sachs, A. B., Bond, M. W., and Kornberg, R. D.,** A single gene from yeast for both nuclear and cytoplasmic polyadenylate-binding proteins: Domaine structure and expression, *Cell,* 45, 827—835, 1986.
98a. **Sadis, S., Hickey, E., and Weber, L. A.,** Effect of heat shock on RNA metabolism in HeLa cells, *J. Cell. Physiol.*, 135, 377—386, 1988.
99. **Safer, B.,** 2B or not 2B: Regulation of the catalytic utilization of eIF-2, *Cell,* 33, 7—8, 1983.
100. **Sanders, M. M.,** Identification of histone H2b as a heat-shock protein in *Drosophila, J. Cell Biol.*, 91, 579—583, 1981.
101. **Sanders, M. M., Triemer, D. F., and Olsen, A. S.,** Regulation of protein synthesis in heat-shocked *Drosophila* cells. Soluble factors control translation in vitro, *J. Biol. Chem.*, 261, 2189—2196, 1986.
102. **Schamhart, D. H. J., Van Walraven, H.S., Wiegant, F. A. C., Linnemans, W. A. M., Van Rijn, J., Van Den Berg, J., and Van Wijk, R.,** Thermotolerance in cultured hepatoma cells: Cell viability, cell morphology, protein synthesis, and heat shock proteins, *Radiat. Res.*, 98, 82—95, 1984.
102a. **Scharf, K.-D. and Nover, L.,** Control of ribosome biosynthesis in plant cell cultures under heat shock conditions. II. Ribosomal proteins, *Biochim. Biophys. Acta,* 909, 44—57, 1987.

103. **Scharf, K.-D. and Nover, L.,** Heat shock induced alterations of ribosomal protein phosphorylation in plant cell cultures, *Cell,* 30, 427—437, 1982.

103a. **Scharf, K.-D.,** unpublished.

104. **Schiebel, W., Chayka, T. G., De Vries, A., and Rusch, H. P.,** Decrease of protein synthesis and breakdown of polysomes by elevated temperature in *Physarum polycephalum, Biochem. Biophys. Res. Commun.,* 35, 338—345, 1969.

104a. **Schoeffl, F., Rieping, M., Baumann, G., Bevan, M., and Angermueller, S.,** The function of plant heat shock promoter elements in the regulated expression of chimaeric genes in tansgenic tobacco, *Mol. Gen. Genet.,* 217, 246—253, 1989.

105. **Schönfelder, M., Horsch, A., and Schmid, H.-P.,** Heat shock increases the synthesis of the poly(A)-binding protein in HeLa cells, *Proc. Natl. Acad. Sci. U.S.A.,* 82, 6884—6888, 1985.

106. **Scott, M. P., Fostel, J. M., and Pardue, M. L.,** A new type of virus from cultured *Drosophila* cells: Characterization and use in studies of the heat-shock response, *Cell,* 22, 929—941, 1980.

107. **Scott, M. P. and Pardue, M. L.,** Translational control in lysates of *Drosophila melanogaster* cells, *Proc. Natl. Acad. Sci. U.S.A.,* 78, 3353—3357, 1981.

108. **Spadoro, J. P., Copertino, D. W., and Strausbaugh, L. D.,** Differential expression of histone sequences in *Drosophila* following heat shock, *Dev. Genet.,* 7, 133—148, 1986.

109. **Spena, A., Hain, R., Ziervogel, U., Saedler, H., and Schell, J.,** Construction of a heat-inducible gene for plants. Demonstration of heat-inducible activity of the *Drosophila* hsp70 promoter in plants, *EMBO J.,* 4, 2739—2743, 1985.

110. **Spradling, A., Penman, S., and Pardue, M. L.,** Analysis of *Drosophila* mRNA by in situ hybridization: Sequences transcribed in normal and heat shocked cultured cells, *Cell,* 4, 395—404, 1975.

111. **Spradling, A., Pardue, M. L., and Penman, S.,** Messenger RNA in heat-shocked *Drosophila* cells, *J. Mol. Biol.,* 109, 559—587, 1977.

112. **Storti, R. V., Scott, M. P., Rich, A., and Pardue, M. L.,** Translational control of protein synthesis in response to heat shock in *D. melanogaster* cells, *Cell,* 22, 825—834, 1980.

113. **Subjeck, J. R. and Shyy, T. T.,** Stress protein systems of mammalian cells, *Am. J. Physiol.,* 250, C1—C17, 1986.

114. **Subjeck, J. R., Shyy, T., Shen, J., and Johnson, R. J.,** Association between the mammalian 110,000-dalton heat-shock protein and nucleoli, *J. Cell Biol.,* 97, 1389—1395, 1983.

115. **Sunitha, I. and Slobin, L. I.,** Inhibition of poly(A)-binding protein synthesis in Friend erythroleukemia cells subsequent to heat shock, *Biochim. Biophys. Acta,* 825, 214—226, 1985.

116. **Tabarini, D., Heinrich, J., and Rosen, O. M.,** Activation of S6 kinase activity in 3T3-L1 cells by insulin and phorbol ester, *Proc. Natl. Acad. Sci. U.S.A.,* 82, 4369—4373, 1985.

117. **Tanguay, R. M., Camato, R., Lattre, F., and Vincent, M.,** Expression of histone genes during heat shock and in arsenite-treated *Drosophila* Kc cells, *Can. J. Biochem. Cell Biol.,* 61, 414—420, 1983.

118. **Tas, P. W. L. and Martini, O. H. W.,** Regulation of ribosomal protein S6 phosphorylation in heat-shocked HeLa cells, *Eur. J. Biochem.,* 163, 553—559, 1987.

119. **Terao, K. and Ogata, K.,** Proteins of small subunits of rat liver ribosomes that interact with poly(U), *J. Biochem.,* 86, 587—617, 1979.

119a. **Theodorakis, N. G., Banerji, S. S., and Morimoto, R. I.,** HSP70 mRNA translation in chicken reticulocytes is regulated at the level of elongation, *J. Biol. Chem.,* 263, 14579—14585, 1988.

120. **Thomas, G. P., Siegmann, M., Kubler, A.-M., Gordon, J., and De Asua, L. J.,** Regulation of 40S ribosomal protein S6 phosphorylation in Swiss mouse 3T3 cells, *Cell,* 19, 1015—1023, 1980.

121. **Thomas, G. P. and Mathews, M. B.,** Alterations of transcription and translation in HeLa cells exposed to amino acid analogs, *Mol. Cell. Biol.,* 4, 1063—1072, 1984.

122. **Tissieres, A., Mitchell, H. K., and Tracy, U. M.,** Protein synthesis in salivary glands of *D. melanogaster.* Relation to chromosome puffs, *J. Mol. Biol.,* 84, 389—398, 1974.

122a. **Tomasovic, S. P.,** Functional aspects of the mammalian heat-stress protein response, *Life Chem. Rep.,* 7, 33—63, 1989.

123. **Vierling, E. and Key, J. L.,** Ribulose 1,5-bisphosphate carboxylase synthesis during heat shock, *Plant Physiol.,* 78, 155—162, 1985.

124. **Vincent, M. and Tanguay, R. M.,** Different intracellular distributions of heat shock and arsenite-induced proteins in *Drosophila* Kc cells. Possible relation with the phosphorylation and translocation of a major cytoskeletal protein, *J. Mol. Biol.,* 162, 365—378, 1982.

125. **Voellmy, R., Ahmed, A., Schiller, P., Bromley, P., and Rungger, D.,** Isolation and functional analysis of a human 70,000-dalton heat shock protein gene segment, *Proc. Natl. Acad. Sci. U.S.A.,* 82, 4949—4953, 1985.

126. **Wolffe, A.-P., Glover, J. F., and Tata, J. R.,** Culture shock. Synthesis of heat-shock-like proteins in fresh primary cell culture, *Exp. Cell Res.,* 154, 581—590, 1984.

127. **Wool, I. G.,** The structure and function of eukaryotic ribosomes, *Annu. Rev. Biochem.,* 48, 719—754, 1979.

127a. **Wu, C. H., Warren, H. L., Sitaraman, K., and Tsai, C. Y.,** Translational alterations in maize leaves responding to pathogen infection, paraquat treatment, or heat shock, *Plant Physiol.*, 86, 1323—1329, 1988.

128. **Yura, Y., Terashima, K., Iga, H., Kondo, Y., Yanagawa, T., Yoshida, H., Hayashi, Y., and Sato, M.,** Macromolecular synthesis at the early stage of herpes simplex virus type 2(HSV-2) latency in a human neuroblastoma cell line IMR-32: Repression of late viral polypeptide synthesis and accumulation of cellular heat-shock proteins, *Arch. Virol.*, 96, 17—28, 1987.

129. **Zumbe, A., Staehli, C., and Trachsel, H.,** Association of a Mr 50,000 cap binding protein with the cytoskeleton in baby hamster kidney cells, *Proc. Natl. Acad. Sci. U.S.A.*, 79, 2927—2931, 1982.

Chapter 12

PROCESSING, INTRACELLULAR TOPOGENESIS, AND DEGRADATION OF PROTEINS

L. Nover

TABLE OF CONTENTS

12.1. PROTEINOGEN PROCESSING AND MODIFICATION

Many proteins need co- and/or posttranslational processing and modifications for their proper function and intracellular localization. Although in most cases the mechanisms and biological implications are obscure, many examples of heat shock (hs) induced effects have been reported. With respect to protein modification, these are summarized in Table 12.1. Not included are modifications of heat shock proteins (HSPs) themselves, which are discussed separately in Section 2.3 (see Table 2.5). Because of the characteristic changes at the transcriptional and translational levels, three types of proteins have received special attention: nuclear proteins (histones, nonhistones), ribosomal proteins, and translation factors. Their putative roles in transcriptional and translational reprogramming are discussed at the appropriate places in this book (see Sections 9.4 and 11.5). Considering the well-known role of Tyr-phosphorylation for malignant transformation, it may be an intriguing observation that a broad range of proteins were phosphorylated at Tyr residues after a 1-h hs treatment at 45 to 46°C of different mammalian cell lines.[37b]

Few examples of an altered or modified processing and intracellular transport of precursor proteins have been reported:

1. In *Drosophila melanogaster* cell cultures infected with picornaviruses, the induced HSP synthesis may be inhibited. However, synthesis of virus-specific proteins is unaffected except for the accumulation of unusual precursor proteins,[41] which under normothermic conditions are only observed in cultures treated with amino acid analogs or iodoacetamide.[40] Evidently the normal processing of picornavirus polyproteins is, at least partially, inhibited by hs.
2. A differential effect on glycoprotein synthesis was observed in rabbit brain cells after whole body hyperthermia to 42°C evoked by lysergic acid diethylamide (LSD) application. The induced synthesis of HSPs 70 and 90 coincides with an undiminished production of synaptic glycoproteins.[24] These are synthesized in the cell body and transported by fast axonal transport to the nerve terminals. In contrast, synthesis of other glycoproteins and nonglycosylated control proteins is severely repressed under hs conditions. The mechanism of discrimination between both classes of glycoproteins was not investigated. Peculiarities of the glycosyl moieties and/or of the corresponding mRNAs may be involved.
3. In plants, stress-induced alterations in the synthesis and/or processing of secreted proteins were found. In developing soybean cotyledons, even a severe hs at 46°C can not decrease the ongoing synthesis of 7S and 11S storage proteins but the maturation of the 11S-type of protein is affected, resulting in the accumulation of the 60-kDa precursor protein.[38] On the other hand, phytohemaglutinin (PHA) synthesis in developing bean seeds (*Phaseolus vulgaris*) is blocked after transfer into the ER. The pre-PHA molecule is maintained in the high mannose form in the ER, which undergoes an enlargement of its luminal part with accumulation of electron-dense material.[14] Finally, in the rice seed scutellum synthesis of α-amylase is strongly decreased during hs at 42°C. However, a special form of the enzyme (R-form) with an altered glycosyl side chain is preferentially secreted. This R-type can also be formed in the presence of excess Ca^{2+} in the surrounding medium.[39]

12.2. THE ROLE OF STRESS PROTEINS AS MOLECULAR CHAPERONES

With our increasing knowledge about the multiplicity and general properties of stress proteins, especially the HSP70 and HSP60 families, it becomes apparent that they play an

TABLE 12.1
Heat Shock Induced Changes of Protein Modifications

Organism/cell	Modifications	Ref.
Nuclear proteins, histones (see Section 9.4)		
Achlya ambisexualis	Phosphorylation of H3 increased	48, 53
Tetrahymena pyriformis	Phosphorylation of H1	25
Drosophila melanogaster	Decreased phosphorylation of H3 and H4	27
	Deubiquitination of H2A and H2B	27, 37
	H3: decreased Lys-methylation and methylation of Arg residues; H2B: strongly increased methylation of N-terminal Pro	11, 16—18
	Methylation of H3 and H4 blocked, H2B methylation increased; all four core histones deacetylated	2
Chicken embryo fibroblasts	Deubiquitination of H2A	8
Rat hepatoma cells	Deubiquitination of H2A	46
Nuclear nonhistone proteins (see Section 9.4)		
A. ambisexualis	Phosphorylation of 43 kDa actin-like protein increased	48, 53
D. melanogaster	Hypophosphorylation of nuclear lamin (pp76)	53a
CHO cells	Phosphorylation of nucleolar pp95 and nucleoplasmic pp54 increased, pp35 dephosphorylated	10, 10a
Rat hepatoma cells	Increased phosphorylation of NHP95	56, 57
Rat fibroblasts	Dephosphorylation of 21-kDa cofilin is connected with its migration to the nucleus and formation of actin rods	44a
Ribosomal proteins (see Section 11.5.2)		
A. ambisexualis	Dephosphorylation of small subunit protein S6 (M_r 30 kDa), increased phosphorylation of L proteins with M_r 22 and 24.5 kDa	47
Tomato leaves and cell cultures	Dephosphorylation of S6	51
D. melanogaster cell cultures	Dephosphorylation of S6	26, 45
HeLa, BHK cells	Dephosphorylation of S6; L14 (27 kDa) phosphorylated	36, 55
Human fibroblasts and meningioma cells	Dephosphorylation of S6	50
Translation factors (see Section 11.5.3)		
A. ambisexualis	Decreased phosphorylation of KC1 wash proteins with M_r 50, 21, 20, and 19 kDa, increased phosphorylation of proteins with 32 and 23.5 kDa	47
Rabbit reticulocytes	Phosphorylation of eIF-2α by hemin-regulated protein kinase	9, 21, 38a
HeLa cells	Phosphorylation of eIF-2α by hemin-regulated PK	15, 21
HeLa cells	Decreased phosphorylation of p28 subunit of cap-binding factor eIF-4F and of eIF-4B, phosphorylation of eIF-2α	19, 19a, 20

important catalytic role in the frequently neglected posttranslational part of gene expression. This includes transient stabilization of immature polypeptides, their proper folding and intracellular topogenesis, and, finally, the ordered assembly of subunits into multimeric complexes. The rapidly extending experimental evidence in support of a function as ''molecular chaperones'' was summarized by Ellis and co-workers[20a,20b] and Rothman[50a] (for detailed references to original papers, see Sections 2.1.2, 2.3.2, 2.3.3, and 4.3).

The considerable degree of structure conservation between prokaryotic and eukaryotic members of the HSP70 and HSP60 families (Figures 2.10 and 2.11) also includes remarkable similarities of their functions in prokaryotic and eukaryotic cells. In the latter, major sites of action of molecular chaperones are the ER/Golgi system (GRPs 94 and 78, see Section 4.3), chloroplasts and mitochondria harboring distinct members of both HSP70 and HSP60 families (see Section 16.5), and the cytosol (HSP70, HSC70). Other non-stress-related, but also stress-related, processes shown to involve members of the HSP70 and HSP60 and probably also of the HSP90 families are usually not included in the chaperone concept. But, *de facto,* they may reflect the same or similar basic activities of these proteins. Such is the role of the vertebrate HSC70 as clathrin uncoating ATPase (see Section 2.3.2), the interaction of HSP70 and HSP90 with elements of the cytoskeleton (see Sections 14.3 and 16.6), the association of HSP90 with steroid receptors (Sections 2.3.1 and 16.8) or protein kinases (Sections 11.5.3), and the binding of HSP70 to abnormal onc proteins (Sections 9.3 and 20.6) and its transient accumulation in the nucleus/nucleolus, probably caused by stress-induced aggregation of nuclear proteins or RNP (Sections 9.3, 10.5, and 16.2). Generally speaking, HSP70 and HSP60 proteins may function as repair enzymes to remove and/or reactivate proteins denatured and aggregated as a result of stress conditions.[37a,50a]

Some remarkable properties of these stress proteins are probably closely connected with their role as molecular chaperones:

1. All members of the HSP70 and HSP60 families are ATP binding and exhibit ATPase activity upon association with other proteins or protein complexes (see Sections 2.3.2 and 2.3.3).
2. Members of the HSP70 family are characterized by high-affinity, salt-resistant binding to distinct peptide motifs that can be released in the presence of ATP, i.e., recognition of unfolded proteins may be brought about by exposition of these peptide motifs.[13a,23a]
3. Disregarding characteristic differences with repect to the sites of action within eukaryotic cells, HSP70 and HSP60 proteins were found associated with immature, unfolded, or partially denatured proteins. HSP70-type proteins are required for processing, assembly, and intracellular transport of proteins within the ER/Golgi system or from cytoplasm into the organelles (see Sections 2.3.2 and 4.3), whereas HSP60-type proteins act in chloroplasts and mitochondria (see Section 16.5). Direct evidence for the ATP-dependent folding and/or assembly capacity of HSP60 (GroEL) was provided for dihydrofolate reductase in isolated *Neurospora* mitochondria[45a] and for partially denatured ribulose-bisphosphate carboxylase subunits *in vitro.*[29b] Close cooperation of different chaperones, e.g., of the organellar HSP60 and HSP70 or of GRP78, GRP94, and protein disulfide isomerase in the ER, in a protein maturation and assembly complex ("foldosome") is very likely.[24b,45a]
4. Mutants defective in the hsp60 (GroEL)[13,36b] or in hsp70 genes[13b,15a] exhibit corresponding defects in protein sorting and/or assembly. But similar results were also obtained by decreasing the GRP78 level in the ER by antisense RNA.[18a]

12.3. PROTEIN DEGRADATION

The particular interest in changes of protein degradation under hs conditions resulted from the early idea of Kelley and Schlesinger[35] and Hightower[31] that the accumulation of abnormal proteins may be a central part of the signal transduction chain and that hs, chemical stressors, and viral inducers may be linked in their common activity of producing aberrant proteins and thus, to increase HSP synthesis (see Section 1.4). As an extension of this hypothesis, Munro and Pelham[43] proposed that, due to the overload of the proteolytic system, an otherwise short-lived heat shock transcription factor (HSF) accumulates and thus causes the differential activation of hs genes.[5] Although the characterization of HSF from different systems (see Section 7.6) disproved this part of the model, i.e. the HSF so far identified are not short-lived, the intriguing facts on the role of abnormal proteins for the induction process (see Figure 1.4) and on changes of distinct parts of the proteolytic system remain.

An essential event in the degradation of short-lived or abnormal proteins in eukaryotic cells is the tagging by covalent modification with a small, highly conserved protein of 76 amino acid residues (ubiquitin) in form of a branched-chain polyubiquitin moiety (see reviews by Rechsteiner,[49] Schlesinger and Bond,[52] and Schlesinger and Hershko[52a]). Good substrates for polyubiquitination in yeast and mammals are short-lived proteins marked by certain N-terminal amino acid residues, e.g., E, Q, D, N, I, L, F, W, Y, H, R, and K plus an internal Lys-residue in the close vicinity, which serves as acceptor for the ubiquitin.[2a] It is an intriguing observation that ubiquitin may not only be used as a tag for proteolytic substrates but possesses intrinsic proteolytic activity itself, which might be relevant for protein degradation or for processing of the polyubiquitin precursor molecules.[24a]

A particular role of polyubiquitination starting reaction of protein breakdown for the stress response was suggested by four types of results:

1. In amino acid analog-treated rabbit reticulocytes and mouse ascites tumor cells a tenfold increase of ubiquitin-conjugated proteins was found.[30]
2. A heat sensitive cell cycle mutant of mouse cells was shown to originate from a ts mutation of the gene coding for the ubiquitin conjugating enzyme. At the nonpermissive temperature (40°C), HSP synthesis was induced, despite the fact that this temperature was not sufficient by itself to induce the stress response.[22] However, the simple explanation for this phenomenon was questioned when a similar ts-mutant of the same enzyme was isolated from Chinese hamster lung fibroblasts.[36a] In this case, no difference in the threshold of hs induction of HSP synthesis was found when compared to the wild-type cell line.
3. Selective induction of hs gene transcription can be achieved by microinjection of abnormal proteins into *Xenopus* oocytes[1] (see Figure 1.4) or by transformation of *Escherichia coli*[28,34] and *Drosophila*[32,33] with genes coding for abnormal proteins (see Section 1.4).
4. In yeast, two closely related ubiquitin-conjugating enzymes (UBC4 and UBC5) are induced by hs. They are responsible for the generation of polyubiquitin conjugates of short-lived or abnormal proteins. Double mutants (ubc4/ubc5) are inviable at elevated temperatures or in the presence of amino acid analogs. But, as a consequence of the defective protein degradation, they exhibit constitutive HSP synthesis.[52b]

Disregarding the particular situation of the ts mutant cell line of mouse,[14b,22] ubiquitin transferase is not rate limiting for protein degradation under hs conditions. Microinjection of radioactively labeled ubiquitin into HeLa cells[12] or analysis of the free and bound ubiquitin in chicken embryo fibroblasts[8], rat hepatoma cells[46] and the green alga *Chlamydomonas*[52c] demonstrated that ubiquitin conjugation rather increased. At the same time, the pool of free ubiquitin and the level of ubiquitinated histones markedly decreased (see Table 12.1). However, the actual effects on protein degradation are not clear. Proteolysis in rat hepatoma cells is transiently enhanced at 43°C, but inhibited at 45°C.[46] Such a decrease of protein degradation was also observed by Carlson et al.[12] in HeLa and by Munro and Pelham[42] in monkey COS cells. In summary, the most heat sensitive step of protein breakdown is not the tagging with polyubiquitin but rather conformational changes of proteins (partial denaturation), which make them substrates of the ubiquitin pathway. As it is true for all other parts of the hs response, the actual state of the proteolytic system depends on the severity of the stress treatment.

A remarkable contribution towards an understanding of proteolysis as part of the stress response system is the identification of distinct polyubiquitin genes as hs genes, e.g., of

Trypanosoma,[54a] yeast,[23,44,56b] *Dictyostelium,*[41a] plants,[14a,36b] *Xenopus,*[45b] chicken,[6,7] mammals,[3,23b] and *Drosophila.*[36c] The increase of ubiquitin synthesis during hs fills up the depleted pools of free ubiquitin and may be responsible for the significant rise of protein degradation in the recovery period, i.e. in chicken cells.[8] Yeast deletion mutants lacking the heat-inducible polyubiquitin gene (UBI 4) grow normally under normothermic conditions but were hypersensitive to heat or other stress treatments.[23] Interestingly, these mutants were also defective in sporulation induced by nutrient starvation. In bacteria, ubiquitin does not exist. Instead, the stress-induced part of the proteolytic system is the ATP-dependent Lon protease[3,28,28a,29] (see Table 2.2), which has been implicated in the degradation of amino acid analog-substituted proteins in *E. coli.* But other HSPs (Dnak, DnaJ, GrpE, and GroEL; see Table 2.2) are required for effective energy-dependent protein degradation by the Lon protease as well.[54]

Recently a member of the HSP70 family (p73) was isolated from mammalian cytosol by its affinity for the marker sequence of short-lived proteins, i.e., KFERQ for RNase A. Under conditions with increased lysosomal protein breakdown (24 h serum deprivation) p73 levels increase dramatically. Furthermore, in the presence of ATP, isolated p73 stimulates lysosomal protein degradation in two *in vitro* test systems.[13a]

The evident potential of hs and chemical stressors to damage native proteins and to destabilize multimeric protein complexes led Minton et al.[38b] to suggest that mass accumulation of HSPs might help to stabilize proteins under stress conditions. In support of this, experiments with mouse and *Drosophila* cells[43a] and with soybean seedlings[34a] demonstrated a protective effect of hs-induced cytoplasmic factors (HSPs?) against inactivation and insolubilization of proteins by hs. It is an open question whether HSPs directly stabilize native or partially denaturated proteins[38b] or whether HSP70 or others help to resolubilize and reactivate complexes of damaged proteins, as suggested by Lewis and Pelham[37a] (see Section 12.2 and review by Pelham[48a]).

Taken together, these interesting results, although still fragmentary, demonstrate an increased demand in all types of cells on the proteolytic systems, probably to remove abnormal or partially denatured proteins accumulated during the stress period. However, regarding the starting point of this section, the missing link between the increased levels of defective proteins and the changing activity state of the HSF remains to be found (see Section 1.4).

REFERENCES

1. **Ananthan, J., Goldberg, A. L., and Voellmy, R.,** Abnormal proteins serve as eukaryotic stress signals and trigger the activation of heat shock genes, *Science,* 232, 522—524, 1986.
2. **Arrigo, A. P.,** Acetylation and methylation patterns of core histones are modified after heat or arsenite treatment of *Drosophila* tissue culture cells, *Nucl. Acids Res.,* 11, 1389—1404, 1983.

2a. **Bachmair, A. and Varshavsky, A.,** The degradation signal in a short-lived protein, *Cell,* 56, 1019—1032, 1989.

3. **Baker, Z. T. and Board, P. G.,** The human ubiquitin gene family. Structure of a gene and pseudogenes from the UbB subfamily, *Nucl. Acids Res.,* 15, 443—463, 1987.
4. **Baker, T. A., Grossman, A. D., and Gross, C. A.,** A gene regulating the heat shock response in *Escherichia coli* also affects proteolysis, *Proc. Natl. Acad. Sci. U.S.A.,* 81, 6779—6783, 1984.
5. **Bienz, M. and Pelham, H. R. B.,** Mechanisms of heat-shock gene activation in higher eucaryotes, *Adv. Genet.,* 24, 31—72, 1987.
6. **Bond, U. and Schlesinger, M. J.,** Ubiquitin is a heat shock protein in chicken embryo fibroblasts, *Mol. Cell. Biol.,* 5, 949—956, 1985.
7. **Bond, U. and Schlesinger, M. J.,** The chicken ubiquitin gene contains a heat shock promoter and express an unstable mRNA in heat-shocked cells, *Mol. Cell. Biol.,* 6, 4602—4610, 1986.

8. **Bond, U., Agell, N., Haas, A. L., Redman, K., and Schlesinger, M. J.,** Ubiquitin in stressed chicken embryo fibroblasts, *J. Biol. Chem.*, 263, 2384—2388, 1988.
9. **Bonanou-Tziedaki, S., Sohl, M. K., and Arnstein, H. R. V.,** Regulation of protein synthesis in reticulocyte lysates. Characterization of the inhibitor generated in the post-ribosomal supernatant by heating at 44°C, *Eur. J. Biochem.*, 114, 69—77, 1981.
10. **Caizergues-Ferrer, M., Bouche, G., and Amalric, F.,** Phosphorylated protein involved in the regulation of rRNA synthesis in CHO cells recovering from heat shock, *FEBS Lett.*, 116, 261—264, 1980.
10a. **Caizergues-Ferrer, M., Bouche, G., Amalric, F., and Zalta, J. P.,** Effect of heat shock on nuclear and nucleolar protein phosphorylation in Chinese hamster ovary cells, *Eur. J. Biochem.*, 108, 399—407, 1980.
11. **Camato, R. and Tanguay, R. M.,** Changes in the methylation pattern of core histones during heat-shock in *Drosophila* cells, *EMBO J.*, 1, 1529—1532, 1982.
12. **Carlson, N., Rogers, S., and Rechsteiner, M.,** Microinjection of ubiquitin: Changes in protein degradation in HeLa cells subjected to heat-shock, *J. Cell Biol.*, 104, 547—555, 1987.
13. **Cheng, M. Y., Hartl, F.-U., Martin, J., Pollock, R. A., Kalousek, F., Neupert, W., Hallberg, E. M., Hallberg, R. L., and Horwich, A. L.,** Mitochondrial heat-shock protein hsp60 is essential for assembly of proteins imported into yeast mitochondria, *Nature*, 337, 620—625, 1989.
13a. **Chiang, H. L., Terlecky, S. R., Plant, C. P., and Dice, J. F.,** A role for a 70-kilodalton heat shock protein in lysosomal degradation of intracellular proteins, *Science*, 246, 382—385, 1989.
13b. **Chirico, W. J., Waters, M. G., and Blobel, G.,** 70k heat shock related proteins stimulate protein translocation into microsomes, *Nature*, 332, 805—810, 1988.
14. **Chrispeels, M. and Greenwood, J. S.,** Heat stress enhances phytohemaglutinin synthesis but inhibits its transport out of the endoplasmic reticulum, *Plant Physiol.*, 83, 778—784, 1987.
14a. **Christensen, A. H. and Quail, P. H.,** Sequence analysis and transcriptional regulation by heat shock of polyubiquitin transcripts from maize, *Plant Mol. Biol.*, 12, 619—632, 1989.
14b. **Ciechanover, A., Finley, D., and Varshavsky, A.,** Ubiquitin dependence of selective protein degradation demonstrated in the mammalian cell cycle mutant ts85, *Cell*, 37, 57—66, 1984.
15. **de Benedetti, A. and Baglioni, C.,** Activation of hemin-regulated initiation factor-2 kinase in heat-shocked HeLa cells, *J. Biol. Chem.*, 261, 338—342, 1986.
15a. **Deshaies, R. J., Koch, B. D., Werner-Washburne, M., Craig, E. A., and Schekman, R.,** A subfamily of stress proteins facilitates translocation of secretory and mitochondrial precursor polypeptides, *Nature*, 332, 800—805, 1988.
16. **Desrosier, R. and Tanguay, R. M.,** The modification in the methylation patterns of H2B and H3 after heat shock can be correlated with the inactivation of normal gene expression, *Biochem. Biophys. Res. Commun.*, 133, 823—829, 1985.
17. **Desrosiers, R. and Tanguay, R. M.,** Further characterization of the posttranslational modifications of core histones in response to heat and arsenite stress in *Drosophila, Biochem. Cell Biol.*, 64, 750—784, 1986.
18. **Desrosiers, R. and Tanguay, R. M.,** Methylation of *Drosophila* histones at proline, lysine and arginine residues during heat shock, *J. Biol. Chem.*, 263, 4686—4692, 1988.
18a. **Dorner, A. J., Krane, M. G., and Kaufman, R. J.,** Reduction of endogenous GRP78 levels improves secretion of a heterologous protein in CHO cells, *Mol. Cell. Biol.*, 8, 4063—4070, 1988.
19. **Duncan, R. and Hershey, J. W. B.,** Heat shock-induced translational alterations in HeLa cells, *J. Biol. Chem.*, 259, 11882—11889, 1984.
19a. **Duncan, R. F. and Hershey, J. W. B.,** Protein synthesis and protein phosphorylation during heat stress, recovery, and adaptation, *J. Cell Biol.*, 109, 1467—1481, 1989.
20. **Duncan, R., Milburn, S. C., and Hershey, J. W. B.,** Regulated phosphorylation and low abundance of HeLa cell initiation factor eIF-4F suggest a role in translational control. Heat shock effects on eIF-4F, *J. Biol. Chem.*, 262, 380—389, 1987.
20a. **Ellis, R. J., Van Der Vies, S. M., and Hemmingsen, S. M.,** The molecular chaperone concept, *Biochem. Soc. Symp.*, 55, 145—153, 1989.
20b. **Ellis, R. J. and Hemmingsen, S. M.,** Molecular chaperones: proteins essential for the biogenesis of some macromolecular structures, *Trends Biochem. Sci.*, 14, 339—342, 1989.
21. **Ernst, V., Baum, E. Z., and Reddy, P.,** Heat shock protein phosphorylation, and the control of translation in rabbit reticulocytes, reticulocyte lysates, and HeLa cells, in *Heat Shock from Bacteria to Man*, Schlesinger, M. J., Ashburner, M., and Tissieres, A., Eds., Cold Spring Harbor Laboratory, Cold Spring Harbor, NY, 1982, 215—225.
22. **Finley, D., Ciechanover, A., and Varshavsky, A.,** Thermolability of ubiquitin-activating enzyme from the mammalian cell cycle mutant ts85, *Cell*, 37, 43—55, 1984.
23. **Finley, D., Özkaynak, E., and Varshavsky, A.,** The yeast polyubiquitin gene is essential for resistance to high temperatures, starvation and other stresses, *Cell*, 48, 1035—1046, 1987.
23a. **Flynn, G. C., Chappell, T. G., and Rothman, J. E.,** Peptide binding and release by proteins implicated as catalysts of protein assembly, *Science*, 245, 385—390, 1989.

23b. **Fornace, A. J., Alamo, I., Hollander, M. C., and Lamoreaux, E.,** Ubiquitin mRNA is a major stress-induced transcript in mammalian cells, *Nucl. Acids Res.*, 17, 1215—1230, 1989.

24. **Freedman, M. S., Clark, B. D., Cruz, T. F., Gurd, J. W., and Brown, I. R.,** Selective effects of LSD and hyperthermia on the synthesis of synaptic proteins and glycoproteins, *Brain Res.*, 207, 129—145, 1981.

24a. **Fried, V. A., Smith, H. T., Hildebrandt, E., and Weiner, K.,** Ubiquitin has intrinsic proteolytic activity: Implications for cellular regulation, *Proc. Natl. Acad. Sci. U.S.A.*, 84, 3685—3689, 1987.

24b. **Gething, M.-J. and Sambrook, J.,** Protein folding and intracellular transport: Studies on influenza virus haemagglutinin, *Biochem. Soc. Symp.*, 55, 155—166, 1989.

25. **Glover, C. V. C., Vavra, K. J., Guttman, S. D., and Gorovsky, M. A.,** Heat shock and deciliation induce phosphorylation of histone H1 in *T. pyriformis, Cell,* 23, 73—77, 1981.

26. **Glover, C. V. C.,** Heat shock induces rapid dephosphorylation of a ribosomal protein in *Drosophila, Proc. Natl. Acad. Sci. U.S.A.*, 79, 1781—1785, 1982.

27. **Glover, C. V. C.,** Heat-shock effects on protein phosphorylation in *Drosophila,* in *Heat Shock from Bacteria to Man,* Schlesinger, M. J., Ahsburner, M., and Tissieres, A., Eds., Cold Spring Harbor Laboratory, Cold Spring Harbor, NY, 1982, 227—234.

28. **Goff, S. A. and Goldberg, A. L.,** Production of abnormal proteins in *E. coli* stimulates transcription of lon and other heat shock genes, *Cell,* 41, 587—595, 1985.

28a. **Goff, S. A. and Goldberg, A. L.,** An increased content of protease La, the lon gene product, increases protein degradation and blocks growth in *Escherichia coli, J. Biol. Chem.*, 262, 1508—1515, 1987.

29. **Goff, S. A., Casson, L. P., and Goldberg, A. L.,** Heat shock regulatory gene htpR influences rates of protein degradation and expression of the lon gene in *Escherichia coli, Proc. Natl. Acad. Sci. U.S.A.*, 81, 6647—6651, 1984.

29a. **Goloubinoff, P., Gatenby, A. A., and Lorimer, G. H.,** GroE heat shock proteins promote assembly of foreign prokaryotic ribulose bisphosphate carboxylase oligomers in *Escherichia coli, Nature,* 337, 44—47, 1989.

29b. **Goloubinoff, P., Christeller, J. T., Gatenby, A. A., and Lorimer, G. H.,** Reconstitution of active dimeric ribulose bisphosphate carboxylase from an unfolded state depends on two chaperonin proteins and MgATP, *Nature,* 342, 884—888, 1989.

30. **Hershko, A., Eytan, E., Ciechanover, A., and Haas, A. L.,** Immunochemical analysis of the turnover of ubiquitin-protein conjugates in intact cells. Relationship to the breakdown of abnormal proteins, *J. Biol. Chem.*, 257, 13964—13970, 1982.

31. **Hightower, L. E.,** Cultured animal cells exposed to amino acid analogues or puromycin rapidly synthesize several polypeptides, *J. Cell. Physiol.*, 102, 407—427, 1980.

32. **Hiromi, Y. and Hotta, Y.,** Actin gene mutations in *Drosophila* heat shock activation in the direct flight muscles, *EMBO J.*, 4, 1681—1687, 1985.

33. **Hiromi, Y., Okamoto, H., Gehring, W. J., and Hotta, Y.,** Germline transformation with *Drosophila* mutant actin genes induces constitutive expression of heat shock genes, *Cell,* 44, 293—301, 1986.

34. **Ito, K., Akiyama, Y., Yura, T., and Shiba, K.,** Diverse effects of the MalE-LacZ hybrid protein on *Escherichia coli* cell physiology, *J. Bacteriol.*, 167, 201—204, 1986.

34a. **Jinn, T. L., Yeh, Y. C., Chen, Y. M., and Lin, C. Y.,** Stabilization of soluble proteins in vitro by heat shock proteins-enriched ammonium sulfate fraction from soybean seedlings, *Plant Cell Physiol.*, 30, 463—469, 1989.

35. **Kelley, P. M. and Schlesinger, M. J.,** The effect of amino acid analogues and heat shock on gene expression in chicken embryo fibroblasts, *Cell,* 15, 1277—1286, 1978.

36. **Kennedy, I. M., Burdon, R. H., and Leader, D. P.,** Heat shock causes diverse changes in the phosphorylation of the ribosomal proteins of mammalian cells, *FEBS Lett.*, 169, 267—273, 1984.

36a. **Kulka, R. G., Raboy, B., Schuster, R., Parag, H. A., Diamond, G., Ciechanover, A., and Marcus, M.,** The heat-shock response in a Chinese hamster cell mutant with a temperature-sensitive ubiquitin-activating enzyme, E1, in *The Ubiquitin System,* Schlesinger, M. and Hershko, A., Eds., Cold Spring Harbor Press, Cold Spring Harbor, NY, 1988, 195—200.

36b. **Kusukawa, N., Yura, T., Ueguchi, C., Akiyama, Y., and Ito, K.,** Effects of mutations in heat shock genes groES and groEL on protein export in *Escherichia coli, EMBO J.*, 8, 3517—3521, 1989.

36c. **Lee, H., Simon, J. A., and Lis, J. T.,** Structure and expression of ubiquitin genes of *Drosophila melanogaster, Mol. Cell. Biol.*, 8, 4727—4735, 1988.

37. **Levinger, L. and Varshavsky, A.,** Selective arrangement of ubiquitinated and Dl protein-containing nucleosomes within the *Drosophila* genome, *Cell,* 28, 357—385, 1982.

37a. **Lewis, M. J. and Pelham, H. R. B.,** Involvement of ATP in the nuclear and nucleolar functions of the 70 kd heat shock protein, *EMBO J.*, 4, 3137—3143, 1985.

37b. **Maher, P. A. and Pasquale, E. B.,** Heat shock induces protein tyrosine phosphorylation in cultured cells, *J. Cell Biol.*, 108, 2029—2035, 1989.

38. **Mascarenhas, J. P. and Altschuler, M.,** Responses to environmental heat stress in plant embryo, in *Changes in Eukaryotic Gene Expression in Response to Environmental Stress,* Atkinson, B. G. and Walden, D. B., Eds., Academic Press, Orlando, FL, 1985, 315—326.
38a. **Matts, R. L. and Hurst, R.,** Evidence for the association of the heme-regulated eIF-kinase with the 90-kDa heat shock protein in rabbit reticulocyte lysate in situ, *J. Biol. Chem.,* 264, 15542—15547, 1989.
38b. **Minton, K. W., Karmin, P., Hahn, G. M., and Minton, A. P.,** Nonspecific stabilization of stress-susceptible proteins by stress-resistant proteins: A model for the biological role of heat shock proteins, *Proc. Natl. Acad. Sci. U.S.A.,* 79, 7107—7111, 1982.
39. **Mitsui, T. and Akazawa, T.,** Preferential secretion of R-type alpha-amylase molecules in rice seed scutellum at high temperatures, *Plant Physiol.,* 82, 880—884, 1986.
40. **Moore, N. F. and Pullin, J. S. K.,** Heat shock used in combination with amino acid analogs and protease inhibitors to demonstrate the processing of proteins of an insect picornavirus (*Drosophila.*-C virus), in *Drosophila melanogaster* cell, *Ann. Virol.,* 134, 285—292, 1983.
41. **Moore, N. F., Pullin, J. S. K., and Reavy, B.,** Inhibition of the induction of heat shock proteins in *Drosophila melanogaster* cells infected with insect picornavirus, *FEBS Lett.,* 128, 93—96, 1981.
41a. **Mueller-Taubenberger, A., Hagmann, J., Noegel, A., and Gerisch, G.,** Ubiquitin gene expression in *Dictyostelium* is induced by heat and cold shock, cadmium, and inhibitors of protein synthesis, *J. Cell Sci.,* 90, 51—58, 1988.
42. **Munro, S. and Pelham, H. R. B.,** Use of peptide tagging to detect proteins expressed from cloned genes: Deletion mapping functional domains of *Drosophila* hsp70, *EMBO J.,* 3, 3087—3093, 1984.
43. **Munro, S. and Pelham, H.,** What turns on heat shock genes?, *Nature,* 317, 477—478, 1985.
43a. **Nguyen, V. T., Morange, M., and Bensaude, O.,** Protein denaturation during heat shock and related stress—*Escherichia coli* beta-galactosidase and *Photinus pyralis* luciferase inactivation in mouse cells, *J. Biol. Chem.,* 264, 10487—10492, 1989.
44. **Özkaynak, E., Finley, D., Solomon, M. J., and Varshavsky, A.,** The yeast ubiquitin genes: A family of natural gene fusions, *EMBO J.,* 6, 1429—1439, 1987.
44a. **Ohta, Y., Nishida, E., Sakai, H., and Miyamoto, E.,** Dephosphorylation of cofilin accompanies heat shock-induced nuclear accumulation of cofilin, *J. Biol. Chem.,* 264, 16143—16148, 1989.
45. **Olsen, A. S., Feeney-Triemer, D., and Sanders, M. M.,** Dephosphorylation of S6 and expression of the heat shock response in *Drosophila melanogaster, Mol. Cell. Biol.,* 3, 2017—2027, 1983.
45a. **Ostermann, J., Horwich, A. L., Neupert, W., and Hartl, F.-U.,** Protein folding in mitochondria requires complex formation with hsp60 and ATP hydrolysis, *Nature,* 341, 125—130, 1989.
45b. **Ovsenek, N. and Heikkila, J. J.,** Heat shock-induced accumulation of ubiquitin messenger RNA in *Xenopus laevis* embryos is developmentally regulated, *Dev. Biol.,* 129, 582—585, 1988.
46. **Parag, H. A., Raboy, B., and Kulka, R. G.,** Effect of heat shock on protein degradation in mammalian cells: Involvement of the ubiquitin system, *EMBO J.,* 6, 55—63, 1987.
47. **Pekkala, D. and Silver, J. C.,** Heat-shock-induced changes in the phosphorylation of ribosomal and ribosome-associated proteins in the filamentous fungus *Achlya ambisexualis, Exp. Cell Res.,* 168, 325—337, 1987.
48. **Pekkala, D., Heath, I. B., and Silver, J. C.,** Changes in chromatin and the phosphorylation of nuclear proteins during heat shock of *Achlya ambisexualis, Mol. Cell. Biol.,* 4, 1198—1205, 1984.
48a. **Pelham, H. R. B.,** Heat shock and the sorting of luminal ER proteins, *EMBO J.,* 8, 3171—3176, 1989.
49. **Rechsteiner, M.,** Ubiquitin-mediated pathways for intracellular proteolysis, *Annu. Rev. Cell Biol.,* 3, 1—30, 1987.
50. **Richter, W. W., Zang, K. D., and Issinger, O. G.,** Influence of hyperthermia on the phosphorylation of ribosomal protein S6 from human skin fibroblasts and meningioma, *FEBS Lett.,* 153, 262—266, 1983.
50a. **Rothman, J. E.,** Polypeptide chain binding proteins: Catalysts of protein folding and related processes in cells, *Cell,* 59, 591—601, 1989.
51. **Scharf, K.-D. and Nover, L.,** Heat shock induced alterations of ribosomal protein phosphorylation in plant cell cultures, *Cell,* 30, 427—437, 1982.
52. **Schlesinger, M. J. and Bond, U.,** Ubiquitin genes, in *Oxford Survey of Eukaryotic Genes,* 4, 77, 1987.
52a. **Schlesinger, M. and Hershko, A., Eds.,** *The Ubiquitin System, Curr. Commun. Mol. Biol.,* Cold Spring Harbor Press, Cold Spring Harbor, NY, 1988.
52b. **Seufert, W. and Jentsch, S.,** Ubiquitin-conjugating enzymes UBC4 and UBC5 mediate selective degradation of short-lived and abnormal proteins, *EMBO J.,* 9, 543—550, 1990.
52c. **Shimogawara, K. and Muto, S.,** Heat shock induced change in protein ubiquitination in *Chlamydomonas, Plant Cell Physiol.,* 30, 9—16, 1989.
53. **Silver, J. C., Andrews, D. R., and Pekkala, D.,** Effect of heat shock on synthesis and phosphorylation of nuclear and cytoplasmatic proteins in the fungus *Achlya, Can. J. Biochem. Cell Biol.,* 61, 447—455, 1983.

53a. **Smith, D. E., Gruenbaum, Y., Berrios, M., and Fisher, P. A.,** Biosynthesis and interconcersion of *Drosophila* nuclear lamin isoforms during normal growth and in response to heat shock, *J. Cell Biol.,* 105, 771—790, 1987.

54. **Straus, D. B., Walter, W. A., and Gross, C. A.,** *Escherichia coli* heat shock gene mutants are defective in proteolysis, *Genes Dev.,* 2, 1851—1858, 1988.

54a. **Swindle, J., Ajioka, J., Eisen, H., Sanwal, B., Jaqument, C., Browder, Z., and Buck, G.,** The genomic organization and transcription of the ubiquitin genes of *Trypanosoma cruzi, EMBO J.,* 7, 1121—1127, 1988.

54b. **Tanaka, K., Matsumoto, K., and Toh-E, A.,** Dual regulation of the expression of the polyubiquitin gene by cyclic AMP and heat shock in yeast, *EMBO J.,* 7, 495—502, 1988.

55. **Tas, P. W. L. and Martini, O. H. W.,** Regulation of ribosomal protein S6 phosphorylation in heat-shocked HeLa cells, *Eur. J. Biochem.,* 163, 553—559, 1987.

56. **Van Dongen, G., Geilenkirchen, W., Van Rijn, J., and Van Wijk, R.,** Increase of thermoresistance after growth stimulation of resting Reuber H35 hepatoma cells, *Exp. Cell Res.,* 166, 427—441, 1986.

57. **Van Dongen, G., Schamhart, D., and Van Wijk, R.,** Heat-induced alterations in non-histone chromosomal proteins, chromatin associated heat shock protein and the occurence of thermotolerance in Reuber H35 hepatoma cells, *Int. J. Hyperthermia,* 113, 252—267, 1988.

58. **Vierstra, R. D., Burke, T. J., Callis, J., Hatfield, P. M., Jabben, M., Shanklin, J., and Sullivan, M. L.,** Characterization of the ubiquitin-dependent proteolytic pathway in higher plants, in *The Ubiquitin System,* Schlesinger, M. and Hershko, A., Eds., Cold Spring Harbor Press, Cold Spring Harbor, NY, 1988, 119—125.

Chapter 13

RECOVERY OF GENE EXPRESSION PATTERNS

L. Nover

TABLE OF CONTENTS

Attempts to analyze mechanisms of induced thermotolerance (see Chapter 17) and to find a possible role of heat shock proteins (HSPs) in this respect led to the two main aspects of the heat shock (hs) response: (1) structural and functional conservation under hyperthermic conditions and (2) the potential to normalize cellular activities soon after or even during the stress period (recovery). Due to the spectacular and rapid changes, which occur at all levels at the onset of the stress period, the recovery phase frequently has been neglected. Thus, despite of the fragmentary information available, I decided to review the relevant data on gene expression in a separate chapter, although most of them are also presented in different contexts in other chapters of the book (see Sections 7.3, 8.7, and 11.4). A characteristic of the hs response system is that practically all changes concern the specificity rather than the efficiency of the gene expression apparatus.

13.1. TRANSCRIPTION: DECAY AND NEW SYNTHESIS OF mRNAs

In all kinds of organisms, induction of HSP synthesis depends on the formation of new mRNAs, i.e., on the redistribution of RNA polymerase II from previous sites of gene activity to the newly activated hs genes. The activity of this enzyme so well as of the other two RNA polymerases (I and III) is not primarily affected by hs (see Section 9.1). Restoration of the normal pattern of gene transcription is mandatory, especially in those systems without effective mechanisms for mRNA storage, i.e., bacteria so well as yeast and related microorganisms.

The molecular mechanisms of recovery are particularly evident in *Escherichia coli.* At 42 to 46°C, induced HSP synthesis is transient and usually declines after 10 to 20 min.[26,27] Central elements of the induction and recovery process are two transcription factors, which actually are subunits of RNA polymerase (σ-factors) required for promoter recognition (for details see Section 7.3). The constitutive σ-factor with M_r 70,000 (σ^{70}, rpoD gene) is encoded by a complex operon (MMS operon), whose genes are involved in key steps of macromolecular synthesis (DNA synthesis, transcription, and translation).[20] The other σ-factor (M_r 32,000) is encoded by the rpoH gene. At the onset of hs, a rapidly increasing level of σ^{32} causes a replacement of σ^{70} in the RNA polymerase core enzyme. The reciprocal changes of σ^{70} and σ^{32} levels are directly proportional to the extent of induced HSP synthesis.[10,14,49]

Recovery is preprogrammed by two mechanisms. (1) There is an internal promoter of the MMS operon in front of the rpoD gene which is responsive to hs, i.e., it is recognized by the σ^{32}-containing RNA polymerase[4,40] (see also Figure 7.1). As a consequence, the deficiency of σ^{70} is rapidly overcome by its hs-induced *de novo* synthesis. (2) The transient increase of σ^{32} at the onset of hs is reversed with the increase of the HSP70 level (DnaK protein). It is not observed in a defective dnaK mutant.[14] Two different mechanisms are evidently operative.[39a] After temperature shift-down from 42 to 30°C, an immediate inhibition of the function of σ^{32} is followed by a much slower decrease of its level. Clearly, the rapid and reversible reprogramming of protein synthesis in *E. coli* during hs and recovery is the result of an inbuilt program for interconversion of these two forms of RNA polymerase (see Section 7.3).

Similar to the results with *E. coli,* the hs-induced alterations of the transcription patterns in yeast are short-lived. Complete restoration of the pre-hs pattern is observed within 90 to 120 min during a 36°C stress period.[15a,21-23] Unlike other eukaryotes, the changing protein patterns of yeast and probably of related microorganisms are controlled mostly or exclusively at the transcriptional level.[18] No storage of mRNA was observed. Instead of translational discrimination, decay and new synthesis of control protein mRNAs are decisive for these effects, and recovery is dependent on protein (HSP?) synthesis during the stress period. Isolation and characterization of the yeast transcription factor[38,38a,39,47a] (HSF) (Figure 7.5) provided evidence that the changing transcription patterns depend on the phosphorylation state of this factor (see Section 7.6).

What is true for the bulk of mRNAs in yeast may also apply to the decay of individual mRNAs, mostly coding for specialized proteins, in other eukaryotes (Section 11.2). This is illustrated in Figure 11.4 for α-amylase mRNA induced by the phytohormone gibberellic acid in barley aleurone cells.[3] The complete destruction of α-amylase mRNA during a 3 h hs at 40°C coincides with maximum HSP synthesis and, vice versa, when HSP synthesis declines after 12 to 24 h at 40°C, α-amylase synthesis reappears.

An interesting example for the coexistence of transcriptional vs. translational control mechanisms for different subunits of the same plant protein was given by Vierling and Key.[43] Ribulosebisphosphate carboxylase (Rubisco) is the most abundant protein in nature, making up about 40% of cellular proteins in green parts of plants. It is localized within the chloroplast and catalyzes the initial step of CO_2 fixation. Synthesis of Rubisco proceeds from a large subunit (LSU, M_r ~55 kDa) produced in the chloroplast and a small subunit (SSU, M_r 15 kDa) encoded in the nuclear genome and imported from the cytoplasm.

Changing synthesis of Rubisco during hs and recovery in a green cell suspension culture of soybean was analyzed at the protein and mRNA levels (Figure 13.1A, B, D, and E). For comparison, corresponding data for HSP17 are included (Figure 13.1C and F). It is apparent that the immediate block of Rubisco synthesis under hs is controlled at the translation level in the chloroplast compartment (LSU). Disregarding the temperature of treatment (38° vs. 40°C), the mRNA is stable, and recovery proceeds within 6 h at 25°C (parts A, D). In contrast, inhibition of SSU synthesis in the cytoplasm is connected with a degradation of the mRNA. Recovery requires new synthesis, which is rapid after 38°C hs, but seriously delayed after 2 h at 40°C (parts B, E). The translational control of LSU synthesis in the chloroplast is remarkable because the gene expression machinery in this compartment is basically prokaryotic, i.e., similar in many properties to that of *E. coli*. Unfortunately, data on the hs response in the "prokaryotic inhabitants" of eukaryotic cells (chloroplasts, mitochondria) are practically nil.

13.2. NUCLEAR RNP PROCESSING AND RESTORATION OF NUCLEAR STRUCTURE

The most heat sensitive nuclear function is evidently not transcription but processing of pre-RNP preceding export to the cytoplasm. This is discussed in Section 10.1 for preribosomal RNP and in Section 9.5 for splicing of pre-mRNP. In both cases the almost immediate block of processing is readily reversed in the recovery period. Prerequisite for the conservation of the processing machinery and its pre-RNP substrates accumulating under hs is presumably the accumulation of HSP70 in the nuclear compartment (see Sections 10.6 and 16.2). At least, the reversibility of the splicing block in *Drosophila* and HeLa cells is considerably improved by a prior conditioning treatment.[4a,50,51]

In addition to its putative function in protection, HSP70 may act in a kind of shuttle mechanism to clear the nuclei from unwanted protein or RNP material in the recovery period.[17,32] The implications of such a mechanism are illustrated by two facts: (1) in vertebrate cells, hs leads to an unusual accumulation of proteins bound to the nuclear matrix and to chromation.[11,19,34,35,41] Recovery of the normal nuclear protein: DNA ratio in the post-hs period precedes resumption of DNA repair and cell cycle activities.[35,45] (2) As demonstrated for tomato cell cultures (Section 10.5), the accumulation of unprocessed preribosomal RNP under appropriate hs conditions causes a characteristic deformation of the nucleolus resulting in an increasing inhibition of RNA polymerase I activity. Effective processing of intact preribosomes and disposal of the preribosomal wastage is only possible in the recovery period, presumably aided by HSP70 which is found in close association with this rRNP material.

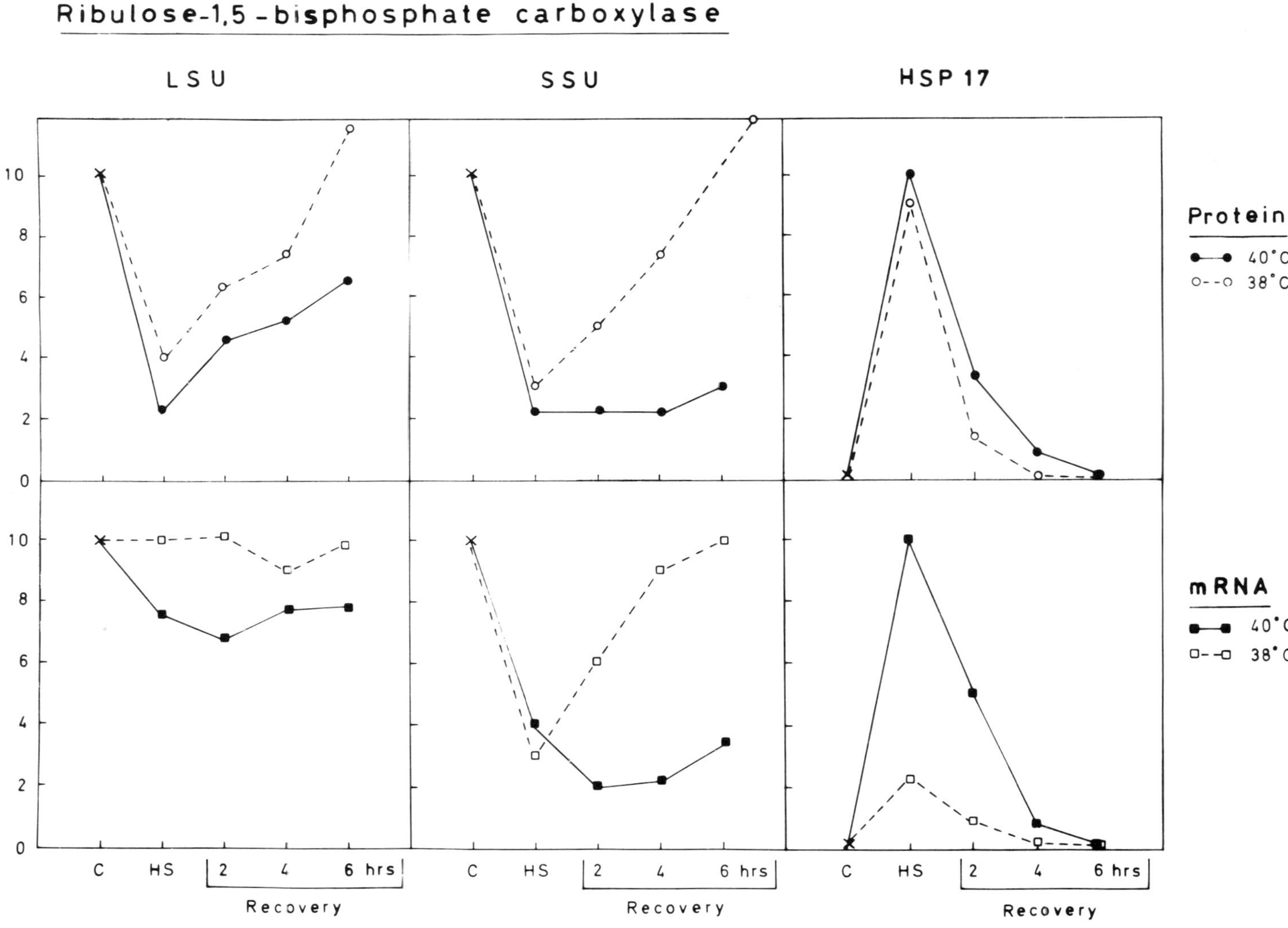
Ribulose-1,5-bisphosphate carboxylase
LSU
SSU
HSP 17
Protein
40°C
38°C
mRNA
40°C
38°C
10
8
6
4
2
0
C
HS
2
4
6 hrs
Recovery

FIGURE 13.1. Protein labeling and mRNA levels of ribulose bisphosphate carboxylase and HSP17 in green cell suspension cultures of soybean (*Glycine max*). Cells were treated by a 2 h heat shock (hs) at 38°C (open symbols) or 40°C (closed symbols) followed by a 6 h recovery period. Proteins were labeled with ^{35}S-Met always in the last hour before harvesting. Incorporation of radioactivity into the large and small subunits of ribulose bisphosphate carboxylase (LSU, SSU) and into a prominent small heat shock protein (HSP) was determined after two-dimensional electrophoretic separation. mRNA levels were determined with corresponding clones using Northern and dot blot analyses. (Modified from Vierling, E. and Key, J. L., *Plant Physiol.*, 78, 155, 1985.)

It is intriguing that the restoration of normal nucleolar morphology in transformed monkey COS cells is improved in the presence of high levels of the heterologous *Drosophila* HSP70[25,31] and that release of the tightly bound nucleolar HSP70 requires adenosinetriphosphate (ATP).[17] However, it remains to be shown that HSP70 really assists in nuclear export of unwanted proteins.

13.3. TRANSLATION

The changing patterns of protein synthesis occupy a central role in the analysis of the hs response. The initial, transient dominance of HSP synthesis is gradually overcome with restoration of control protein synthesis in later stages of hs or in the recovery period. In many eukaryotic organisms, e.g., plants, insects, and vertebrates, there is good experimental evidence for an essential role of HSPs also in this part of the recovery program. The results are the following:

1. Provided optimum conditions for HSP synthesis, e.g., during a moderate hs treatment or a severe hs preceded by a conditioning treatment, restoration of control protein synthesis is markedly enhanced[5,13,24,29,33,36] and proceeds even during the stress period.[9a,29]
2. In plants the conservation of untranslated control mRNAs is intimately connected with a hs-specific macromolecular complex involving perinuclear cytoskeletal aggregates and hs granules.[30] The latter are mainly formed of small HSPs (see Section 16.7). Reactivation of protein synthesis in close vicinity to these complexes can be directly visualized in the electron microscope (Figure 13.2). Although corresponding analyses are lacking, the detection of similar structures in *Drosophila*[1,9,15] and vertebrate cells[2,6,6a] may indicate a general mechanism of mRNA storage in different types of eukaryotic cells.[30] Certainly, this is not the only way to improve recovery of the normal pattern of protein synthesis. In their early investigation on the recovery of control protein synthesis in *Drosophila* pupae, Petersen and Mitchell[33] demonstrated a direct protection of the translation machinery by a prior conditioning treatment. The mRNA levels were not changed irrespective of whether the severe hs at 40.1°C was preceded or not by a 35°C pretreatment.
3. There is a striking correlation between the maximum amount of HSP70 induced by a certain stress treatment and the level of thermotolerance, as expressed, e.g., by restoration of the translation pattern.[8,42,47] For *Drosophila* cell cultures, this is illustrated in Figure 13.3. The recovery of actin synthesis on the basis of constant mRNA levels correlates with decreasing HSP70 synthesis. As outlined in Section 8.7., this decline results from autorepressive control at the transcriptional and translational level operative when a certain amount of HSP70 is formed and is released from its tight nuclear binding during the recovery period.[7,8,42] At present a reasonable explanation for this correlation is lacking. Redistribution of HSP70 from the nuclear to the cytoplasmic compartments of rat embryo fibroblasts led to an association with phase dense material and ribosomes.[46]
4. The phase dense material may in fact represent part of the reforming cytoskeleton. Collapse of the intermediate size filament (IF) cytoskeleton in the early phase of the hs response is part of the translational reprogramming (see Section 11.5.1). It can be hypothesized that an important property of the hs translational machinery and its preference for hs mRNAs are cap-independent initiation and its independence of structural binding to an intact IF system.[12,28] It is an intriguing observation[35a] that polysomes from control *Drosophila* cells always contain considerable amounts of the major IF protein with M_r 46 kDa, whereas in polysomes from hs cells this protein is lacking.

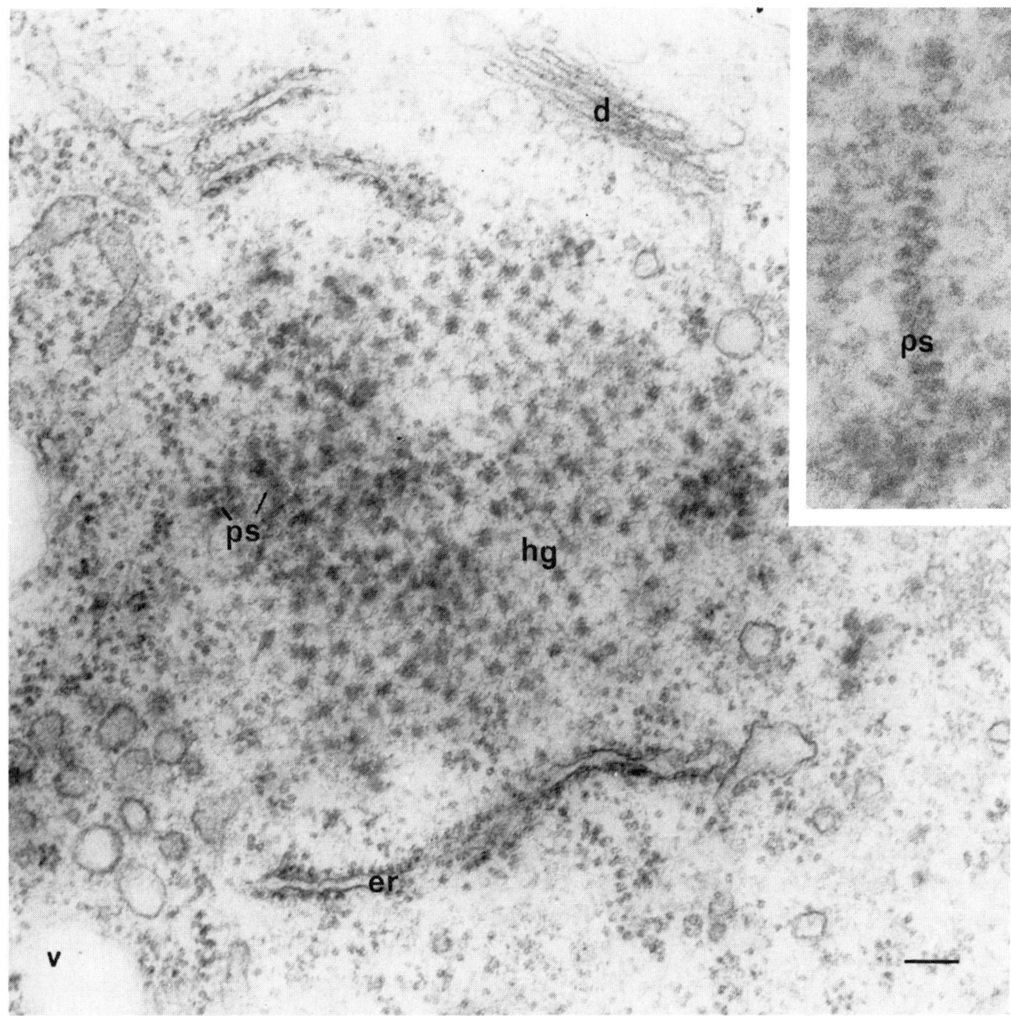

FIGURE 13.2. Active protein synthesis in close vicinity of heat shock granules in tomato cells recovering from hs. Preinduced cultures were heat-shocked for 2 h at 40°C and then allowed to recover for 1 h at 25°C. Regions with heat shock granules (HSG) aggregates are surrounded by many polysomes, active Golgi stacks, and rER. hg, heat shock granules; d, Golgi apparatus; ps, polysomes; v, vacuole; er, rough endoplasmic reticulum. Bar: 0.1 μm. Inset, polysome within a HSG complex at higher magnification.

Hence, rebuilding of the normal IF network may be crucial not only for the release of protected mRNA from the perinuclear complex but also for their effective reincorporation into polysomes. At any rate, the reversibility of cytoskeletal rearrangements is markedly improved in thermotolerant cells,[6,47,48] and HSP70 was shown to bind to this IF cytoskeletal system in vertebrate cells (see Table 16.1).[16,37,44]

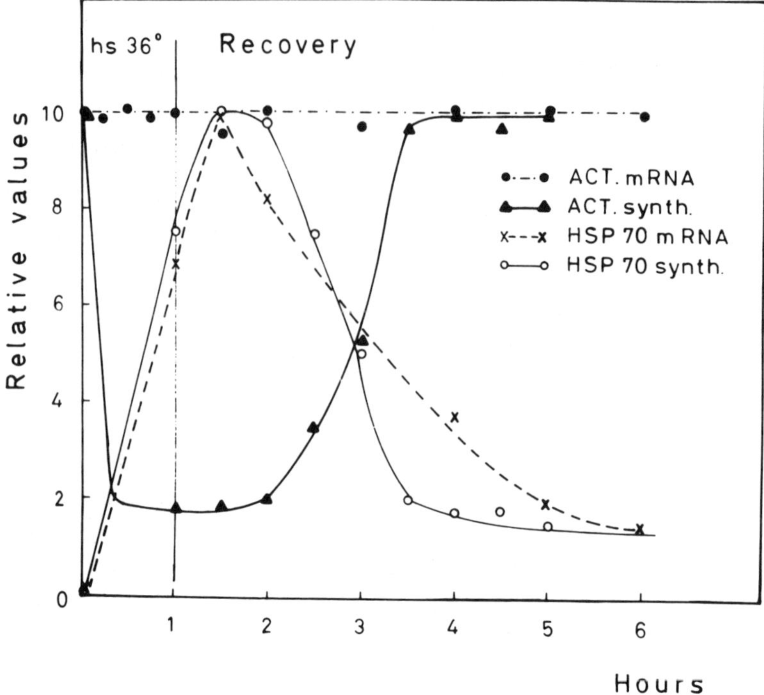

FIGURE 13.3. Changing mRNA levels and protein synthesis in *Drosophila* cell cultures during 1 h heat shock (hs) and 6 h recovery. Protein (HSP70 and actin) synthesis was analyzed after pulse labeling with ^{3}H-leucine and quantitative evaluation of fluorographs. The corresponding mRNAs were quantified with labeled probes using Northern blots. The changing levels of HSP70 synthesis directly reflect synthesis and decay of the mRNA, whereas the transient block of actin synthesis by the hs treatment results from translational control. (Compiled from Di Domenico, B. J., Bugaisky, G. E., and Lindquist, S., *Proc. Natl. Acad. Sci. U.S.A.*, 79, 6181, 1982.)

REFERENCES

1. **Arrigo, A.-P.,** Cellular localization of HSP23 during *Drosophila* development and following subsequent heat shock, *Dev. Biol.*, 122, 39—48, 1987.
2. **Arrigo, A.-P. and Welch, W. J.,** Characterization and purification of the small 28,000 dalton mammalian heat shock protein, *J. Biol. Chem.*, 262, 15359—15369, 1987.
3. **Belanger, F. C., Brodl, M. R., and Ho, T.-H. D.,** Heat shock causes destabilization of specific mRNAs and destruction of endoplasmic reticulum in barley aleurone cells, *Proc. Natl. Acad. Sci. U.S.A.*, 83, 1354—1358, 1986.
4. **Bloom, M., Skelly, S., Van Bogelen, R., Neidhardt, F., Brot, N., and Weissbach, H.,** In vitro effect of the *Escherichia coli* heat-shock regulatory protein on expression of heat shock genes, *J. Bacteriol.*, 166, 380—384, 1986.

4a. **Bond, U.,** Heat shock but not other stress inducers leads to the disruption of a sub-set of snRNPs and inhibition of in vitro splicing in HeLa cells, *EMBO J.*, 7, 3509—3518, 1988.

5. **Chomyn, A., Moller, G., and Mitchell, H. K.,** Patterns of protein synthesis following heat shock in pupae of *Drosophila melanogaster, Dev. Genet.*, 1, 77—95, 1979.
6. **Collier, N. C. and Schlesinger, M. J.,** Induction of heat-shock proteins in the embryonic chicken lens, *Exp. Eye Res.*, 43, 103—117, 1986.

6a. **Collier, N. C., Heuser, J., Levy, M. A., and Schlesinger, M. J.,** Ultrastructural and biochemical of the stress granule in chicken embryo fibroblasts, *J. Cell Biol.*, 106, 1131—1139, 1988.

7. **Di Domenico, B. J., Bugaisky, G. E., and Lindquist, S.,** The heat shock response is self-regulated at both the transcriptional and posttranscriptional levels, *Cell,* 31, 593—603, 1982.
8. **Di Domenico, B. J., Bugaisky, G. E., and Lindquist, S.,** Heat shock and recovery are mediated by different translational mechanisms, *Proc. Natl. Acad. Sci. U.S.A.,* 79, 6181—6185, 1982.
9. **Duband, J. L., Lettre, F., Arrigo, A. P., and Tanguay, R. M.,** Expression and localization of hsp-23 in upstressed and heat-shocked *Drosophila* cultured cells, *Can. J. Genet. Cytol.,* 28, 1088—1092, 1987.
9a. **Duncan, R. F. and Hershey, J. W. B.,** Protein synthesis and protein phosphorylation during heat stress, recovery, and adaptation, *J. Cell Biol.,* 109, 1467—1481, 1989.
10. **Erickson, J. W., Vaughn, V., Walter, W. A., Neidhardt, F. C., and Gross, C. A.,** Regulation of the promoters and transcripts of rpoH, the *Escherichia coli* heat shock regulatory gene, *Genes Devel.,* 1, 419—432, 1987.
11. **Evan, G. I. and Hancock, D. C.,** Studies on the interaction of the human c-myc protein with cell nuclei: $p62^{c\text{-}myc}$ as a member of a discrete subset of nuclear proteins, *Cell,* 43, 253—261, 1985.
12. **Falkner, F. G., Saumweber, H., and Biessmann, H.,** Two *Drosophila melanogaster* proteins related to intermediate filament proteins of vertebrate cells, *J. Cell Biol.,* 91, 175—183, 1981.
13. **Galego, L. and Rodrigues-Pousada, C.,** Regulation of gene expression in *Tetrahymena pyriformis* under heat-shock and during recovery, *Eur. J. Biochem.,* 149, 571—588, 1985.
14. **Grossman, A. D., Straus, D. B., Walter, W. A., and Gross, C. A.,** Sigma 32 synthesis can regulate the synthesis of heat shock proteins in *Escherichia coli, Genes Devel.,* 1, 179—184, 1987.
15. **Haass, C., Falkenburg, P. E., and Kloetzel, P.-M.,** The molecular organization of the small heat shock proteins in *Drosophila,* in *Stress-Induced Proteins,* Pardue, M. L., Feramisco, J. R., and Lindquist, S., Eds., Alan R. Liss, NY, 1989, 175—185.
15a. **Herruer, M. H., Mager, W. H., Raue, H. A., Vreken, P., Wilms, E., and Planta, R. J.,** Mild temperature shock affects transcription of yeast ribosomal protein genes as well as the stability of their messenger RNA, *Nucl. Acids Res.,* 16, 7917—7930, 1988.
16. **La Thangue, N. B.,** A major heat-shock protein defined by a monoclonal antibody, *EMBO J.,* 4, 1871—1879, 1984.
17. **Lewis, M. J. and Pelham, H. R. B.,** Involvement of ATP in the nuclear and nucleolar functions of the 70 kd heat shock protein, *EMBO J.,* 4, 3137—3143, 1985.
18. **Lindquist, S.,** Regulation of protein synthesis during heat shock, *Nature,* 293, 311—314, 1981.
19. **Littlewood, T. D., Hancock, D. C., and Evan, G. I.,** Characterization of a heat shock-induced insoluble complex in the nuclei of cells, *J. Cell Sci.,* 88, 65—72, 1987.
20. **Lupski, J. R. and Godson, G. N.,** The rspU-dnaG-rpoD macromolecular synthesis operon of *E. coli, Cell,* 39, 251—252, 1984.
21. **McAlister, L. and Finkelstein, D. B.,** Alterations in translatable ribonucleic acid after heat shock of *Saccharomyces cerevisiae, J. Bacteriol.,* 143, 603—612, 1980.
22. **Miller, M. J., Xuong, N. H., and Geiduschek, E. P.,** A response of protein synthesis to temperature shift in the yeast *Saccharomyces cerevisiae, Proc. Natl. Acad. Sci. U.S.A.,* 76, 5222—5225, 1979.
23. **Miller, M. J., Xuong, N. H., and Geiduschek, E. P.,** Quantitative analysis of the heat shock response of *Saccharomyces cerevisiae, J. Bacteriol.,* 151, 311—327, 1982.
24. **Mizzen, L. A. and Welch, W. J.,** Characterization of the thermotolerant cell. I. Effects on protein synthesis activity and the regulation of HSP70 expression, *J. Cell Biol.,* 106, 1105—1116, 1988.
25. **Munro, S. and Pelham, H. R. B.,** Use of peptide tagging to detect proteins expressed from cloned genes: Deletion mapping functional domains of *Drosophila* hsp70, *EMBO J.,* 3, 3087—3093, 1984.
26. **Neidhardt, F. C. and Van Bogelen, R. A.,** Positive regulatory gene for temperature-controlled proteins in *Escherichia coli, Biochem. Biophys. Res. Commun.,* 100, 894—900, 1981.
27. **Neidhardt, F. C., Van Bogelen, R. A., and Vaughn, V.,** The genetics and regulation of heat-shock proteins, *Annu. Rev. Genet.,* 18, 295—329, 1984.
28. **Nover, L., Ed.,** *Heat Shock Response of Eukaryotic Cells,* Springer-Verlag, Berlin, 1984.
29. **Nover, L., and Scharf, K.-D.,** Synthesis, modification and structural binding of heat shock proteins in tomato cell cultures, *Eur. J. Biochem.,* 139, 303—313, 1984.
30. **Nover, L., Scharf, K.-D., and Neumann, D.,** Cytoplasmic heat shock granules are formed from precursor particles and contain a specific set of mRNAs, *Mol. Cell. Biol.,* 9, 1298—1308, 1989.
31. **Pelham, H. R. B.,** Hsp 70 accelerates the recovery of nucleolar morphology after heat shock, *EMBO J.,* 3, 3095—3100, 1984.
32. **Pelham, H. R. B.,** Speculations on the functions of the major heat shock and glucose-regulated proteins, *Cell,* 46, 959—961, 1986.
33. **Petersen, N. S. and Mitchell, H. K.,** Recovery of protein synthesis after heat shock: Prior heat treatment affects the ability of cells to translate mRNA, *Proc. Natl. Acad. Sci. U.S.A.,* 78, 1708—1711, 1981.
33a. **Plesovski-Vig, N. and Brambl, R.,** Two developmental stages of *Neurospora crassa* utilize singular mechanism for responding to heat shock but contrasting mechanisms for recovery, *Mol. Cell. Biol.,* 7, 3041—3048, 1987.

34. **Roti-Roti, J. L. and Winward, R. T.,** Factors affecting the heat-induced increase in protein content of chromatin, *Radiat. Res.,* 81, 138—144, 1980.
35. **Roti-Roti, J. L., Uygur, N., and Higashikubo, R.,** Nuclear protein following heat shock: Protein removal kinetics and cell cycle rearrangements, *Radiat. Res.,* 107, 250—261, 1986.

35a. **Sanders, M. M.,** personal communication.

36. **Schamhart, D. H. J., Van Walraven, H. S., Wiegant, F. A. C., Linnemans, W. A. M., Van Rijn, J., Van Den Berg, J., and Van Wijk, R.,** Thermotolerance in cultured hepatoma cells: Cell viability, cell morphology, protein synthesis, and heat shock proteins, *Radiat. Res.,* 98, 82—95, 1984.
37. **Schlesinger, M. J., Aliperti, G., and Kelley, P. M.,** The response of cells to heat shock, *Trends Biochem. Sci.,* 7, 222—225, 1982.
38. **Sorger, P. K. and Pelham, H. R. B.,** Purification and characterization of a heat shock element binding protein from yeast, *EMBO J.,* 6, 3035—3042, 1987.

38a. **Sorger, P. K. and Pelham, H. R. B.,** Yeast heat shock factor is an essential DNA-binding protein that exhibits temperature-dependent phosphorylation, *Cell,* 54, 855—864, 1988.

39. **Sorger, P. K., Lewis, M. J., and Pelham, H. R. B.,** Heat shock factor is regulated differently in yeast and HeLa cells, *Nature,* 329, 81—84, 1987.

39a. **Straus, D. B., Walter, W. A., and Gross, C. A.,** The activity of sigma32 is reduced under conditions of excess heat shock protein production in *Escherichia coli, Genes Dev.,* 3, 2003—2010, 1989.

40. **Taylor, W. E., Straus, D. B., Grossman, A. D., Burton, Z. F., Gross, C. A., and Burgess, R. R.,** Trancription from heat-inducible promoter causes heat shock regulation of the sigma subunit of *E. coli* RNA polymerase, *Cell,* 38, 371—381, 1984.
41. **Tomasovic, S. P., Turner, G. N., and Dewey, W. C.,** Effect of hyperthermia on nonhistone proteins isolated with DNA, *Radiat. Res.,* 73, 535—552, 1978.
42. **Velazquez, J. M. and Lindquist, S.,** Hsp70: Nuclear concentration during environmental stress and cytoplasmic storage during recovery, *Cell,* 36, 655—662, 1984.
43. **Vierling, E. and Key, J. L.,** Ribulose 1,5-bisphosphate carboxylase synthesis during heat shock, *Plant Physiol.,* 78, 155—162, 1985.
44. **Wang, C., Asai, D. J., and Lazarides, E.,** The 68,000-dalton neurofilament-associated polypeptide is a component of nonneuronal cells and of skeletal myofibrils, *Proc. Natl. Acad. Sci. U.S.A.,* 77, 1541—1545, 1980.
45. **Warters, R. L., Brizgys, L. M., Sharma, R., and Roti-Roti, J. L.,** Heat shock (45°C) results in an increase of nuclear matrix protein mass in HeLa cells, *Int. J. Radiat. Biol.,* 50, 253—268, 1986.
46. **Welch, W. J. and Suhan, J. P.,** Cellular and biochemical events in mammalian cells during and after recovery from physiological stress, *J. Cell Biol.,* 103, 2035—2052, 1986.
47. **Welch, W. J. and Mizzen, L. A.,** Characterization of the thermotolerant cell. II. Effects on the intracellular distribution of HSP70, intermediate filaments, and snRNPs, *J. Cell Biol.,* 106, 1117—1130, 1988.

47a. **Wiederrecht, G., Seto, D., and Parker, C. S.,** Isolation of the gene encoding the *S. cerevisiae* heat shock transcription factor, *Cell,* 54, 841—853, 1988.

48. **Wiegant, F. A. C., Van Bergen En Henegouven, P. M. P., Van Dongen, G., and Linnemans, W. A. M.,** Stress-induced thermotolerance of the cytoskeleton, *Cancer Res.,* 47, 1674—1680, 1987.
49. **Yamamori, T., Osawa, T., Tobe, T., Ito, K., and Yura, T.,** *Escherichia coli* gene (hin) controls transcription of heat-shock operons and cell growth at high temperature, in *Heat Shock from Bacteria to Man,* Schlesinger, M. J., Ashburner, M., and Tissieres, A., Eds., Cold Spring Harbor Laboratory, Cold Spring Harbor, NY, 1982, 131—137.
50. **Yost, H. J. and Lindquist, S.,** RNA splicing is interrupted by heat shock and is rescued by heat shock protein synthesis, *Cell,* 45, 185—193, 1986.
51. **Yost, H. J. and Lindquist, S.,** Translation of unspliced transcripts after heat shock, *Science,* 242, 1544—1548, 1988.

Chapter 14

HEAT SHOCK-INDUCED CHANGES OF CELL ULTRASTRUCTURE

D. Neumann and L. Nover

TABLE OF CONTENTS

14.1. JULIUS SACHS (1864) AND FERRUCCIO RITOSSA (1962)

Important milestones in the history of the heat shock (hs) response are marked by two reports whose publication is separated by almost 100 years. Although not immediately comparable, both are concerned with morphological observations on changes of cell ultrastructure visualized in the light microscope.

In 1864, Julius Sachs[117] concluded from investigations on a hs-induced, reversible block of the protoplasmic streaming and changing permeability of the vacuolar membrane in plants that hyperthermic conditions overcome the molecular forces, which are essential for maintenance of the internal fine structure of the cell. Depending on the severeness of the stress treatment, collapse of the intracellular order ultimately may cause cell death. Sachs' very sensitive experimental criteria for the changing internal structure, in fact, revealed effects on the cytoskeleton, on the one hand, and the leakage of anthocyanins from vacuoles of red beet cell, on the other hand. It is evident from the compilation of Alexandrov[1,2] for leaf cells of *Tradescantia fluminensis* (Figure 14.1), that the typical protoplasmic streaming of plant cells is particularly heat sensitive. It is dependent on the interaction and normal fine structure of the cytoskeleton, mainly of the actin system. The hs arrest of protoplasmic streaming is readily reversed after a few minutes or hours of recovery. In contrast, the tonoplast membrane is rather stable. Leakage of anthocyanins is irreversible and observed only at the border to a lethal hs.

One hundred years later (1962), Ritossa[111] reported on changing patterns of gene activity after hs. Using polytene chromosomes of salivary glands of *Drosophila* larvae investigated by light microscopy, he detected the formation of new hs-induced puffs while other sites of gene activity regressed. At that time, the striking site-specific decondensation of the genetic material of polytene chromosomes (puffing) had been shown to be connected with vigorous synthesis of RNA.[104] Consequently, by incorporation of ^{3}H-cytidine Ritossa[111] also demonstrated that the newly forming hs puffs were sites of RNA synthesis. The important implications of this purely cytological data on transcription reprogramming are repeatedly stated in different chapters of this book.

Despite these long-term experiences with hs-induced changes of cell ultrastructure, our knowledge today is still rather fragmentary. On the one hand, there may be special problems with the experimental object. Although occuring in most or all cells in a similar way, changes may be readily detectable only in certain objects or even certain cells. Of course, not only the object but also the application of appropriate microscopic techniques, possibly in conjunction with specific staining or immunological techniques, are decisive for the outcome. In other cases, even now, only data on a changed or defective function lead to conclusions on alterations of fine structures of a given cellular compartment, whose morphological equivalent remains to be documented.

14.2. COMPONENTS OF THE GENE EXPRESSION APPARATUS

Reprogramming of gene expression after the onset of a hs is in the center of all preceding chapters. Ultrastructural changes observed in connection with this were already discussed in more or less detail. Hence, they are only briefly summarized once more with references to figures present in other chapters.

14.2.1. NUCLEUS

Since the early electronmicroscopic characterization of the nucleolus of mammalian cells as a particularly heat sensitive structure,[125] many papers are concerned with the dynamic changes of this compartment and the reversible halt of preribosome processing (see Chapter 10). Although there are characteristic differences in morphological details, e.g., between

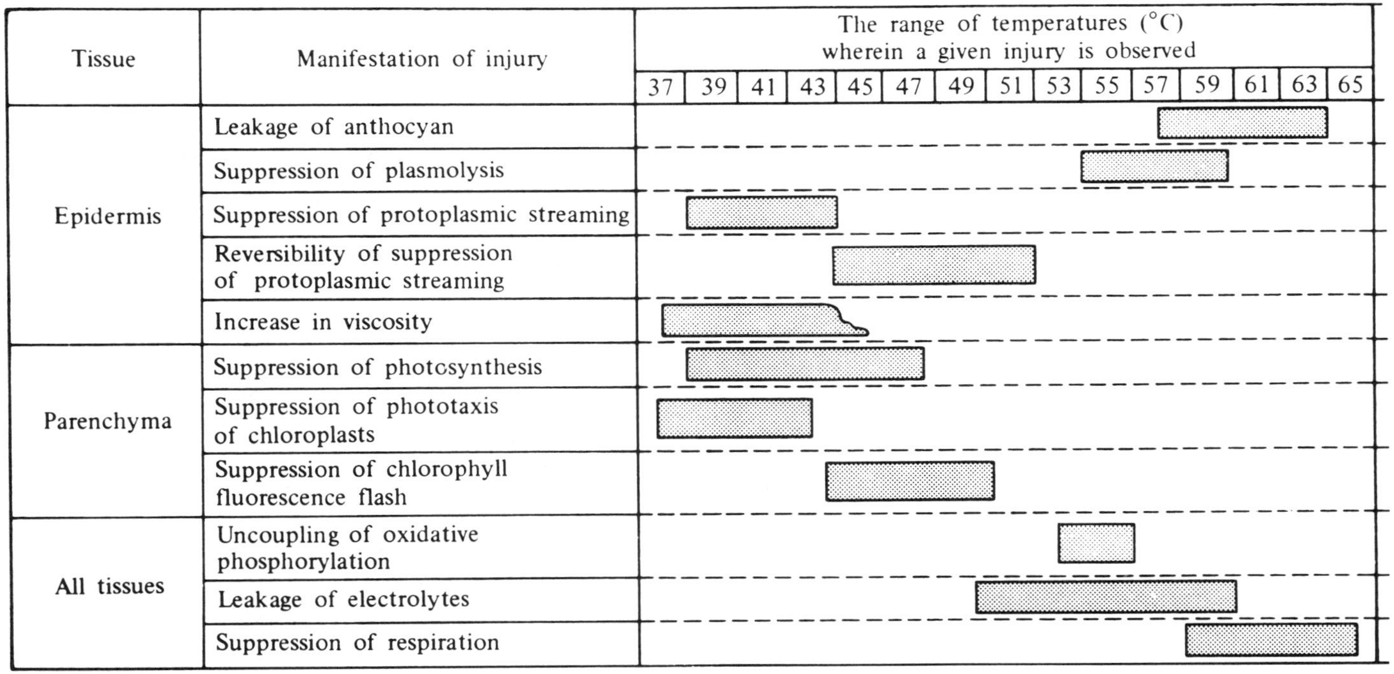

Tissue	Manifestation of injury	The range of temperatures (°C) wherein a given injury is observed														
		37	39	41	43	45	47	49	51	53	55	57	59	61	63	65
Epidermis	Leakage of anthocyan															
	Suppression of plasmolysis															
	Suppression of protoplasmic streaming															
	Reversibility of suppression of protoplasmic streaming															
	Increase in viscosity															
Parenchyma	Suppression of photosynthesis															
	Suppression of phototaxis of chloroplasts															
	Suppression of chlorophyll fluorescence flash															
All tissues	Uncoupling of oxidative phosphorylation															
	Leakage of electrolytes															
	Suppression of respiration															

FIGURE 14.1. Influence of 5 min heat shock (hs) treatments at the indicated temperatures on cytoskeletal and membrane functions of *Tradescantia fluminensis* leaves. (From Alexandrov, V. Ya., *Cells, Molecules and Temperature,* Springer-Verlag, Heidelberg, 1977. With permission.)

plant and mammalian cells, the data are summarized as follows: the usually observed fine granular material of the nucleoli disappears (degranulation), and the continued synthesis of preribosomes with reduced or lack of processing leads to an increasing amount of larger pre-rRNP granules surrounding the fibrillar centers (microsegregation of nucleolus and formation of perichromatin granules,[42,55] see Figure 10.5). Basically similar are the hs-induced alterations of the nucleoplasmic compartment, but synthesis and processing of pre-mRNP are usually affected only at significantly higher temperatures (see Section 9.5).

Other, more subtle changes of the nuclear fine structure usually escape the observation even in the electron microscope. This is true for the reprogramming of transcription, initially observed by Ritossa,[111] as well as for the formation of large, tightly bound protein aggregates at the nuclear matrix (see Section 9.3). Use of antibodies to key nuclear proteins may help to visualize part of these protein redistributions. Thus, the rapid association of RNA polymerase II and the DNA topoisomerase II with the newly formed hs puffs was visualized in *Drosophila* (see Sections 9.1 and 9.2).

In addition, by means of snRNP antibodies several groups[86,147,157a] documented subtle changes in the immune fluorescence indicative of the amount and intracellular distribution of these particles, which are evidently involved in the splicing of hnRNP (see Section 9.5) Relevant in this respect are also the considerable number of papers demonstrating the recompartmentation of HSP70 from its largely cytoplasmic sites to nucleolar sites under stress conditions (see Section 16.2).

14.2.2. POLYSOMES

The decay of polysomes, connected with the discontinued synthesis of many control proteins (see Section 11.1), was also recognized at the ultrastructural level.[58,86,125] From the biochemical analyses (see Figures 8.1 and 11.1), it is evident that the ratio of monosomes to polysomes depends on the severity of the stress treatment and the time point of observation. After an initial breakdown of all polysomes (Figures 11.1 and 11.2), reformation of hs polysomes is eventually followed by a resumption of control protein synthesis at later stages of the stress period (Figure 8.1).

These alterations are also observed at the level of ER-bound polysomes, although the disappearance of the rER may be delayed (see Figure 11.4). In the case of aleurone layers of germinating barley seeds, it depends on the slow decay of the corresponding mRNAs, e.g., for α-amylase.[9,19a] This clearly contrasts to the rapid decay of "free" polysomes, which proceeds with preservation of the corresponding mRNAs (see Section 11.4). Details of the fate of membrane and export protein synthesis are virtually lacking. It is interesting to notice that in vertebrate cells and plants the Golgi apparatus as a processing and sorting site for these proteins is also disrupted or disappears during hs.[88,120,136,148] A special demand for an increased synthesis of membrane proteins after a stress period may be the cause for the striking accumulation of rER and Golgi vesicles in close vicinity of the plant heat shock granules (HSG) (Figure 13.2) which are presumably sites of mRNA storage[101] (see Section 11.4).

14.2.3. HEAT SHOCK GRANULES

Besides the microsegregation of the nucleolus usually connected with accumulation of the HSP70, the formation of cytoplasmic HSG in the perinuclear region is another remarkable change of the cell ultrastructure (see Figure 16.4). The term was originally introduced for heat shock protein (HSP)-containing aggregates detected and purified from tomato cell cultures.[97,100] However, formation of similar "dense material" in heat shocked cells was detected much earlier in root tips of onion[110] and maize[47] and possibly also in cultures of mammalian cells.[85a] Meanwhile, HSGs were recognized as typical stress-related HSP aggregates in all types of plant, insect, and vertebrate cells so far investigated (see Section

16.7). They are reversibly formed from precursor-HSG and are probably embedded in the collapsed cytoskeletal complex around the nucleus.[32,74] Isolation and biochemical analysis of HSGs were only reported for tomato cell cultures. They are dominant sites of accumulation of mainly the small HSPs. In addition, the HSG complex contains a specific subset of cytoplasmic mRNAs not translated during hs. Thus, at least in plant cells, they may have an essential function for the protection of transiently not translated mRNAs during the hs period.[99,101]

14.3. CYTOSKELETAL REARRANGEMENTS

Many aspects of the fine structure and function of eukaryotic cells depend on at least three cytoskeletal systems interacting with each other and spanning the whole cell with a highly dynamic, elaborate network connected to the nuclear matrix and to the plasma membrane.[26,45,119,123] The three main types of cytoskeletal systems are composed of ~25 nm microtubules (tubulin), 8 to 10 nm intermediate filaments and the 5 to 6 nm microfilaments (actin).[73,105,119,121,144,157] The nature of the intermediate filament proteins was mainly studied in vertebrates. They are composed of a constant internal domain and variable domains of different length at the C- and N-termini, which are characteristic of the cell-specific protein type[45,119,123,126] (desmin, keratins, vimentin etc.). The cytoskeleton is evidently involved in many cellular functions including internal fine structure, cell-cell and cell-substrate contacts, movements, and shape of the cell, but it is also essential for the organization of complex enzymatic chains, for the redistribution of proteins and particles within the cell, for virus multiplication, and for translation of a defined subset of mRNAs at the intermediate size filament system.[28,75,76,91,92]

As already mentioned, hs induced changes of the cytoskeletal structure and function were among the earliest observations:[117] stop of protoplasmic streaming in plant cells is readily recognized under comparatively mild hs conditions. It was broadly used by plant physiologists to characterize stress effects and their reversibility.[1,2] In cultures of animal cells, convenient indicators are the changing cell structure (rounding up) and loss of cell-cell and cell-substrate contacts,[81,118] but only the generation of specific antibodies against distinct proteins of the cytoskeletal systems allowed more detailed studies, almost exclusively restricted to *Drosophila* and vertebrate cells. The results are exemplified for mouse neuroblastoma cells (Figure 14.2).

14.3.1. MICROTUBULES

The network of microtubule changes[152,155] (Figure 14.2D and E). Studies with isolated microtubules from rabbit brain showed a high degree of cross-linking after lysergic acid diethylamide (LSD)-induced hyperthermia.[30] The interference of hs with the normal microtubular structure and function was considered as the main reason for the generation of tetraploid, proliferationally dead Chinese hamster ovary cells,[34] and for the block of cell division in the ciliate *Tetrahymena*.[112] In the Northwest rough-skinned newt (*Taricha granulosa*), the nonkinetochore microtubules are immediately destroyed, whereas the kinetochore microtubules form hexagonal, closely packed structures after 15 min treatment at 33 to 36°C.[109]

14.3.2. INTERMEDIATE FILAMENTS

Most dramatic is the collapse of the intermediate filament cytoskeletal system. Shortly after the onset of the hs, it forms a cage-like aggregate around the nucleus (Figure 14.2A and B). Before visualization was possible by means of antibodies,[13,31,44,74,135,147-150] Falkner and Biessmann[43] and Vincent and Tanguay[139] had observed the rapid redistribution of two closely related proteins with M_r 39 and 46 kDa from the cytoplasm to the nuclear fraction

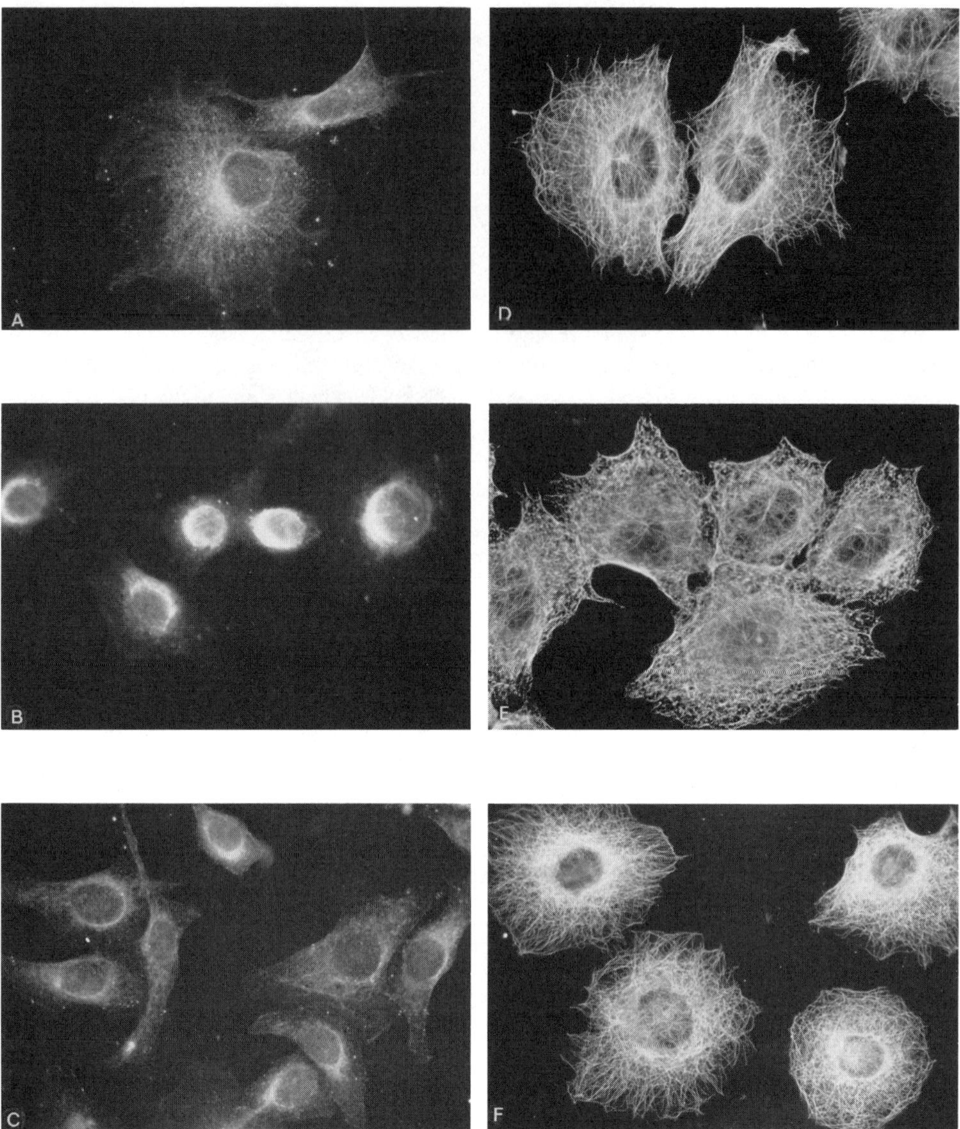

FIGURE 14.2. Heat shock (hs)-induced changes of cytoskeletal systems in mouse neuroblastoma N_2A cells. Distribution of the vimentin (A to C) and tubulin (D to F) cytoskeletal systems were analyzed by immunofluorescence using appropriate antibodies. A and D are control cells (37°C); B and E show hs without conditioning (30 min 43°C); C and F show thermotolerant cells with subsequent hs (30 min 42°C → 4 h 37°C → 30 min 43°C). The hs-induced alterations in B and E are largely absent in thermotolerant cells (C and F). (From Wiegant, F. A. C., Blok F. J., and Linnemans, W. A. M., *Cancer Res.*, 47, 1674, 1987. With permission.)

of heat-shocked *Drosophila* cells. The *Drosophila* 46-kDa protein is homologous to vertebrate vimentin, and its isolation with the nuclear fraction reflects the collapse of the IF system shortly after the onset of hs.[44] It is remarkable that, presumably due to their association with the intermediate filaments, many cell organelles are found concentrated in the juxtanuclear cell space of heat shocked vertebrate cells.[15,124a] However, this is evidently also true for mRNAs bound to the IF system.[61] Their protection from degradation during the stress period probably depends on the formation of large HSP complexes (HSG) attached to the collapsed cytoskeleton (see Sections 11.4 and 16.7).

The fine structure of the intermediate filaments in this aggregated state may be different depending on the cell type. Franke et al.[46] described protofilament-like threads and formation of large aggregate bodies after transient disintegration of the IF system in vertebrates. In agreement with this, densely packed bundles of filaments were observed after hs in chicken[32] and mammalian cells;[148] however, in *Drosophila* cells the collapsed IF system forms homogenous granular aggregates without evidence for fibrillar organization.[141,142] No details are known for plant systems. Much further work is needed to evaluate the significance of these structural differences especially with respect to the formation of HSG and the possible storage of mRNAs.

14.3.3. MICROFILAMENTS

Hs treatments reduce the number of "stress fibers" in vertebrate cells.[52,63,134,148] In CHO cells, more than 90% of all actin microfilaments disappeared after a 5 min hs at 45°C.[52] Reduction of the cytoplasmic stress fibers may contribute to the formation of intranuclear bundles of actin fibers observed in mouse and rat fibroblasts so well as in HeLa cells[63,148] but also in the water mold *Achlya*.[103]

14.3.4. REVERSIBILITY OF CYTOSKELETAL REORGANIZATION

Despite extensive investigations many details of the mechanism and functional implications of cytoskeletal reorganization after stress remain to be elaborated; however, few results deserve notice:

1. Induction of HSP synthesis so well as the characteristic translocation of HSP70 to the nucleus are not dependent on the changes of the cytoskeletal systems.[63,135,146,151]
2. In mouse neuroblastoma and rat hepatoma cells, inhibitors of the Ca^{2+}/calmodulin system potentiate hyperthermic cell killing and prevent cytoskeletal rearrangements.[71,151] In this respect, it is a remarkable finding that HSP90 and HSP70 bind Ca^{2}/calmodulin so well as to different parts of the cytoskeleton.[68a,127a] Details of the calmodulin effects remain to be established. At any rate, it is not the hs-induced increase of intracellular Ca^{2+} levels per se but rather its recompartmentation, which trigger the stress effects on cytoskeletal structure.[41,71,155]
3. As illustrated in Figures 14.2C and F, a characteristic thermotolerance effect is also observed at the level of the cytoskeleton: part of the fine structure is maintained, if the hs is applied to conditioned cultures. Consequently, rapid restoration of the cytoskeleton after hs-induced disruption requires protein (HSP?) synthesis.[52,134,152-154]
4. The tight association of members of the HSP70 and HSP90 families with parts of the cytoskeletal systems (see summary and references in Table 16.1F) suggest an intricate role in the dynamic adaptations during and after stress treatments. For mouse neuroblastoma and rat hepatoma cells, a good correlation was observed between the capability and velocity to reorganize the cytoskeleton and the rates of survival. This agrees with the effects after application of calmodulin inhibitors: together with the inhibition of the necessary cytoskeletal adaptations hyperthermic cell killing is enhanced.[151,153,154] These results support the central role of the cytoskeleton for many aspects of the hs response. Protection of the cytoskeletal fine structure by hs proteins and their constitutively expressed homologs, in fact, may be the basis for many thermotolerance effects observed, e.g., at the level of translation and mRNA conservation (Chapter 11), of membrane stability and functions (Section 14.4), intracellular redistribution of proteins and organelles (protoplasmic streaming), membrane stability,

cell shape, and contacts with substrate or neighbored cells (Section 14.4) or cell cycle activities (Chapter 15).

In addition to the large HSPs, ubiquitin, whose synthesis is also induced by hs (see Section 2.3.5), was recently characterized as part of the microtubular system of baby hamster kidney cells,[93a] but tubulin itself is not ubiquitinated. A function for the stability or rapid reorganization of microtubules can be discussed. Alternatively, the presence of ubiquitin may indicate a role of this cytoskeletal system in proteolysis.[93a] So far, stress-related effects were not analyzed.

14.4. MEMBRANE STRUCTURE AND FUNCTION

Besides the cytoskeletal systems, the second major structural element of cells are membranes. Their peculiar state as a highly ordered but permanently changing crystalline fluid makes them primary targets of hs effects.[5] Moreover, because of their large pores required for intracellular transport of macromolecules, membranes of organelles (mitochondria, chloroplasts, nuclei) may be particularly heat sensitive. Brock[18,19] considers the breakdown of the integrity of organellar membranes to be the main reason for the relatively low temperature limits for growth of eukaryotic organisms as compared to prokaryotes. More than with other elements of the intracellular fine structure, hs-induced changes of membranes can be rarely observed by direct microscopical inspection. They are rather deduced from biochemical or functional changes. As is true for other hs effects, controversial data reported in the literature frequently result from differences of the hs regimes. Research workers interested in the hyperthermic killing of cancer cells find grossly changed membranes at the border to lethality, including the block of amino acid uptake, leakage of electrolytes, or massive formation of large membrane blebs. In contrast, more subtle effects are observed in the temperature range of a more physiological, readily reversible hs.

14.4.1. MEMBRANE FLUIDITY

Extensive analyses of facultative thermophilic bacteria showed that in the transition period from growth at normal temperature to high temperature a transient permeability of the outer membrane[132a] is rapidly repaired[88c] and followed by a homeoviscous adaptation of the membrane, i.e., unsaturated fatty acids in the membrane lipids are replaced by more saturated ones.[17-19,84,106,116] Direct manipulation of the membrane viscosity of a fatty acid dependent mutant of *E. coli* by feeding on an oleic acid (18:1) vs. linoleic acid (18:3) diet proved the important role of membrane microviscosity for the hyperthermic sensitivity: bacteria on the oleic acid diet survived higher hs temperatures,[39,160] and the instability of the cell membrane could be increased in the presence of procaine.[160]

Homeoviscous adaptation is also observed in eukaryotic cells.[3 5,60,106,156] Quinn[106] stressed the particular role of an increased cholesterol level for maintenance of the ordered membrane state under hyperthermic conditions, but the experimental results in this respect are equivocal. Cress and Gerner[35] and Harms-Ringdahl et al.[56] reported on positive evidence for higher cholesterol content/synthesis, whereas essentially no changes were observed by Cress et al.[36] and Burns et al.[21] However, it should be noted that homeoviscous adaptation of membranes is a relatively slow process[3,4] and, contrasting to the model for heat death by excessive fluidization of membranes,[17,160] dramatic changes of membrane fluidity are evidently not the primary cause for hyperthermic cell death.[67,77,78] At any rate, the usually applied hs regimes are not connected with typical transition points in the state of membrane phospholipids.[78,96] However, there are marked effects on the mutal interaction of membrane components and on the state of membrane proteins. By means of transparanaic acid as an indicator, Lepock et al.[78] revealed conformational transitions reminiscent of heat-induced denaturation of soluble proteins.

Membrane fluidity decisively influences the extent of hs-induced changes. Thus, in mouse ascites tumor, cell membranes are more resistant if the animals are fed on a saturated fatty acid diet and, vice versa, hs effects are potentiated by addition of lipotropic local anesthetics, e.g., of ethanol, lidocaine, and procaine.[40,78a,93,108,161] Finally, when comparing metastasizing and nonmetastasizing lines of rat mammary tumors, Yatvin et al.[164] found an increased thermosensitivity of the former connected with a higher degree of fatty acid unsaturation in plasma membrane lipids.

14.4.2. MICROSORTING OF MEMBRANE PROTEINS

An important factor for the state, order, and function of membrane proteins may be the connection with the peripheral membrane-attached cytoskeleton, which is also affected by the hs.[153] Aggregation of the cortical cytoskeleton disrupts the fine network of contacts with membrane proteins. It may be responsible for a microsorting of membrane components with formation of intramembranous particles, as observed in Chinese hamster cells by staining with the fluorescent membrane probe *N*-ϵ-dansyl-L-lysine and by freeze-fracture electron microscopy.[5a,108] Moreover, scanning electronmicrographs revealed plasma membrane evaginations (blebs), which initially are small and reversible (microblebs), but later become larger (macroblebs) marking cells with irreversible membrane damages.[6,7,15,33,64,93,153] In support of these profound local alterations of membrane structure, Shivers et al.[124] reported on the disappearance of tight junctions and formation of intramembrane particles in brain capillary endothelium.

14.4.3. MEMBRANE LEAKINESS

Microsorting and blebbing as indicators of hyperthermia-induced instability are paralleled by disturbances in the function of membranes as a permeability barrier towards the surrounding milieu. This may be the basis for the potentiation of drug cytotoxicity by hyperthermia, as reported for bleomycin and adriamycin. A significant increase of bleomycin uptake is observed under heat stress conditions, but not if hyperthermia precedes bleomycin treatment.[17a,55a,82a,86a,88a] On the other hand, leakiness for solutes was reported for different types of organisms. This is true for the myxomycete *Physarum,*[11] the ciliate *Tetrahymena,*[25] for mammalian cells,[107] and plants.[2,10,83,158,159] Langridge[72] summarizes many examples of conditional auxotrophy for amino acids, vitamins, electrolytes, etc. in bacteria, yeast, and plants after high temperature-induced lesions in the biosynthesis and/or leakiness. The nutrient requirements increase during or shortly after hs, and addition of the deficiency factor increases the heat resistance of the culture. The rapid loss of polyamines (spermine, spermidine) from Chinese hamster ovary cells was considered as the primary cause for the stop of DNA synthesis during hs.[49,50] In mouse fibroblasts, the extent of leakage of potassium to the surrounding milieu correlates with the reduction of viability. It is increased by lipotropic drugs (procaine) and decreased by erythritol.[113,114] Finally, although at present the details are not yet clarified, it is reasonable to assume that the rapid loss of newly synthesized soluble proteins from *E. coli* cells during hs at 48°C may reflect a transient destabilization of the plasma membrane, which is repaired in the course of the hs response.[88c,162,163]

14.4.4. ION TRANSPORT

Characterization of more subtle functions of membranes under hs conditions is still very limited. An early paper demonstrating the dramatic loss of the Na^+/K^+-ATPase activity of HeLa cells[20] may, in fact, indicate effects of a very severe stress treatment. Ruifrok et al.[115] found the ATPase activity essentially unchanged in mouse lung fibroblasts and HeLa cells (40 min 44°C). Even a short-term activation of the Na^+/K^+ transport system was reported for rat hepatoma cells,[14] but it does not lead to significant changes of the ion content and balance.[14,16,127] In contrast to this, there is evidently a stimulation of Ca^{2+}-influx in mammals[24,128,129] and plants.[65] In mammalian cells, the effect may be tightly coupled to the stimulation of inositol trisphosphate release, which together with Ca^{2+} are part of membrane associated systems for signal transformation. Transient, marked increases of the cAMP levels of mammalian cells can also be considered as indicators of a disturbed signal transformation system, e.g., by uncoupling of the enzymatic and regulatory subunits of the adenylyl cyclase complex.[23,82]

14.4.5. MEMBRANE RECEPTORS

Direct investigations of the signal receptor molecules and their fate during hs are restricted to few examples only, and the results are not yet conclusive. Magun[87] and Magun and Fennie[87a] reported on a decreased affinity of the EGF receptor in rat embryo fibroblasts after a 30 min heating to 45°C. Although receptor-mediated endocytosis of the growth factor was not affected, the intracellular degradation of the internalized EGF-receptor complex was significantly slowed down. The normal state of this particular signal system with normal levels of the high affinity EGF receptor at the cell surface was restored only 12 h after the stress period. In complementation of these data, Wiegant et al.[153] found that the coupling of the EGF receptor with the cortical cytoskeletal system was unchanged. Normal velocity of internalization with reduced intracellular processing may be the reason for a depletion of membrane receptors, as found for the EGF receptor in rat cells[87,87a] and for the insulin receptor in CHO cells.[22] The response of CHO cells to mitogenic factors (insulin, transferin, FGF) was heavily impaired and, finally, may contribute to the phenomenon of hyperthermic cell killing.[79,80] Whether there is a link between the defective response to growth factors and the changed levels of phosphoinositol, cAMP, and Ca^{2+}-ions is unclear.[23,24,82,128,129] At any rate, all data together indicate profound disturbances in the membrane-associated signal reception and transformation systems.

14.4.6. NUTRIENT UPTAKE

Although the kinetic state and efficiency of nutrient transport systems may represent sensitive indicators of perturbations of the membrane state, relevant studies in connection with hs effects are scarce. A hs-induced stimulation of hexose uptake in baby hamster kidney cells and chicken embryo fibroblasts results from activation and not increased *de novo* synthesis of the transport system, i.e., it is not inhibited in the presence of cycloheximide or actinomycin D.[48,143] Lin et al.[82] reported on a biphasic response of amino acid transport (α-aminoisobutyric acid) in rat thymocytes (hs at 43°C): an initial increase of AIB uptake is followed by a marked reduction. The V_{max} of the Na^+-dependent component of the neutral amino acid uptake system is primarily affected.[70,82] Our investigations with tomato cell cultures (Figure 14.3) showed that methionine uptake was essentially unchanged except at the border to a lethal hs. Interestingly, the strong decrease after 30 min at 40°C was not observed, if the cell cultures were made thermotolerant by a prior conditioning treatment.

14.4.7. THERMOTOLERANCE

As it is true for most other parts of the hs response, structure and function of cell membranes are protected from heat damage by a prior conditioning treatment (Figure 14.3). Although based on different criteria for membrane integrity (membrane leakiness), basically similar results were reported for pig kidney cells,[107] pear cell cultures,[158,159] or soybean seedlings.[83] Thermotolerance phenomena of mammalian cells with respect to the changes of the membrane-associated cytoskeleton, bleb formation and rounding up of cells were studied by Carr and De Pomerai,[27] Schamhart et al.,[118] and Wiegant et al.[153]

Evidence for the role of HSPs in these phenomena are not conclusive. On the one hand, members of the HSP70 family may play an essential role as interfacing proteins between different cytoskeletal systems and also for anchorage to membrane proteins,[62,95] and they

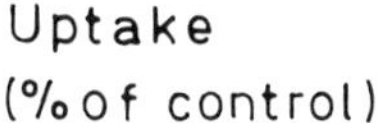

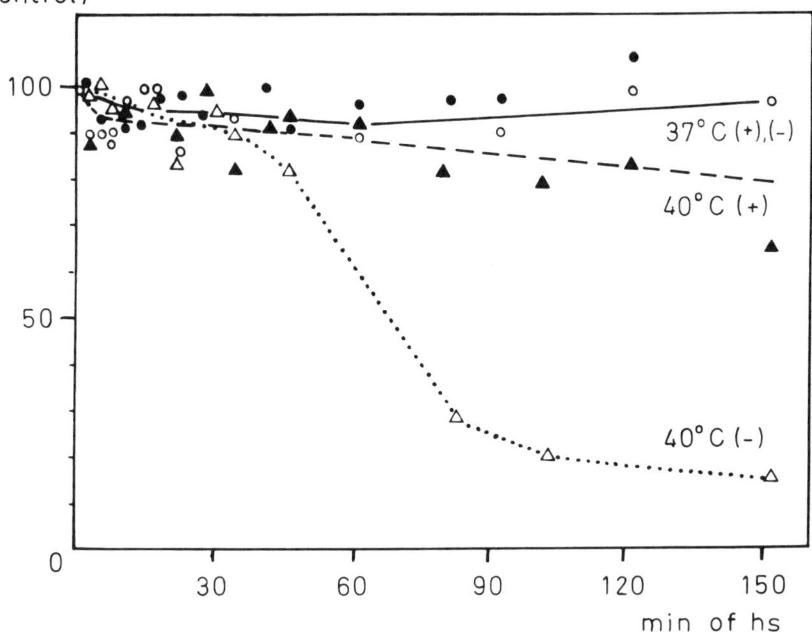

FIGURE 14.3. Changing amino acid uptake in heat-shocked tomato cells. Tomato cell cultures were preinduced (+) by pulse hs treatment (15 min, 40°C) or not preinduced (−). Methionine uptake was measured 3 h later during a mild hs at 37°C (circles) and severe hs at 40°C (triangles), respectively. Values are plotted as % of untreated control. The rapid decrease of methionine uptake at 40°C (dotted line) is not observed, if cultures were made thermotolerant by hs conditioning (dashed line). (From Nover, L., Ed., *Heat Shock Response of Eukaryotic Cells,* G. Thieme Verlag, 1984. With permission.)

represent the clathrin uncoating ATPase essential for receptor internalization and membrane traffic.[29,38,102,133] In this connection, it is certainly worth recalling that HSC71, purified from rat brain and liver, was found tightly coupled to equimolar amounts of palmitic and stearic acids.[53,54] The noncovalently bound fatty acid component may be important for interaction with cell membranes. On the other hand, two hs-induced membrane proteins were described: (1) in soybean seedlings the protection from solute leakage by hs conditioning treatment is connected with accumulation of a unique small HSP (HSP15).[83] (2) In chicken embryo fibroblasts synthesis of a phosphorylated membrane glycoprotein of M_r 47 kDa is enhanced after mile heat stress of 42°C. It is the major collagen-binding protein, i.e., it is responsible for cell-substrate contacts.[94,94a]

14.5. CHLOROPLASTS

Although not connected with visible changes of their structure, chloroplasts are among the most heat sensitive parts of the cell. As summarized by Alexandrov[1,2] (Figure 14.1) photosynthesis *in toto,* chloroplast phototaxis, and chlorophyll fluorescence are affected under relatively mild heat stress conditions. These data compiled for *Tradescantia fluminensis* were confirmed with many other plants.[8,12,37,57,89,90,130-132,140,145] The primary site of hs damage is the thylakoid membrane exerting a secondary effect on the photosynthetic dark reactions in the stroma.[145] Volger and Santarius[140] found a release of thylakoid membrane proteins in spinach leaves. Evidently long-term changes are observed in the composition of the thylakoid membrane lipids,[12,130] resulting in an increased amount of polar lipids containing saturated fatty acids. Although the functional significance is far from clear, this hs-induced slow-adaptation of chloroplast membranes is probably complemented by the increased synthesis of chloroplast-specific HSPs. Among them are proteins of the HSP70 (DnaK-like, see Section 2.3.2),[2a,88b] HSP60 (GroEL-like, see Section 2.3.3),[59,69] and HSP20 families (see Section 2.3.4). The latter are evidently nuclear encoded proteins, which are synthesized as cytoplasmic precursors imported into the chloroplasts.[51,66,130,137,138] In the green alga *Chlamydomonas*[122] and pea leaves[51,66] part of the HSP22 is tightly bound to thylakoid membranes under conditions of heat stress connected with low light. There is evidence for a reversible aggregation of membrane proteins, but a protective role of HSP22 for the conservation of membrane structure and function remains to be proved.

REFERENCES

1. **Alexandrov, V. Ya.,** *Cells, Molecules and Temperature,* Springer-Verlag, Berlin, 1977.
2. **Alexandrov, V. Ya.,** Cell reparation of non-DNA injury, *Int. Rev. Cytol.*, 60, 223—269, 1979.

2a. **Amir-Shapira, D., Leustek, T., Dalie, B., Weissbach, H., and Brot, N.,** Hsp70 proteins, similar to *Escherichia coli* DnaK, in chloroplasts and mitochondria of *Euglena gracilis, Proc. Natl. Acad. Sci. U.S.A.*, 87, 1749—1752, 1990.

3. **Anderson, R. L. and Parker, R.,** Analysis of membrane lipid composition of mammalian cells during the development of thermotolerance, *Int. J. Radiat. Biol.*, 42, 57—69, 1982.
4. **Anderson, R. L., Minton, K. W., Li, G. C., and Hahn, G. M.,** Temperature-induced homeoviscous adaptation of Chinese hamster ovary cells, *Biochim. Biophys. Acta,* 641, 334—348, 1981.
5. **Anghileri, L. J.,** Role of tumor cell membrane in hyperthermia, in *Hyperthermia in Cancer Treatment,* Anghileri, L. J. and Roberts, J., Eds., CRC Press, Boca Raton, FL, 1986, 1—36.

5a. **Arancia, G., Malorni, W., Mariutti, G., and Trovalusci, P.,** Effect of hyperthermia on the plasma membrane structure of Chinese hamster V79 fibroblasts: A quantitative freeze-fracture study, *Radiat. Res.*, 106, 47—55, 1986.

6. **Bass, H., Moore, J. L., and Coakley, W. T.,** Lethality in mammalian cells due to hyperthermia under oxic and hypoxic conditions, *Int. J. Radiat. Biol.*, 33, 57—67, 1978.
7. **Bass, H., Coakley, W. T., Moore, J. L., and Tilley, D.,** Hyperthermia-induced changes in the morphology of CHO-K1 cells and their refractile inclusions, *J. Therm. Biol.*, 7, 231—242, 1982.
8. **Bauer, A. and Senser, M.,** Photosynthesis in ivy leaves *(Hedera helix* L.) after heat stress. II. Activity of ribulose biophosphate carboxylase, Hill reaction, and chloroplast ultrastructure, *Z. Pflanzenphysiol.*, 91, 359—369, 1979.
9. **Belanger, F. C., Brodl, M.R., and Ho, T.-H. D.,** Heat shock causes destabilization of specific mRNAs and destruction of endoplasmic reticulum in barley aleurone cells, *Proc. Natl. Acad. Sci. U.S.A.*, 83, 1354—1358, 1986.
10. **Benzioni, A. and Itai, C.,** Short- and long-term effects of higher termperature (47 to 49°C) on tobacco leaves. III. Efflux and 32-P incorporation into phospholipids, *Physiol. Plant.*, 28, 493—497, 1973.
11. **Bernstam, V. A. and Arndt, S.,** Effects of supraoptimal temperatures on the myxomycete *Physarum polycephalum.* I. Protoplasmic streaming, respiration and leakage of protoplasmic substances, *Arch. Mikrobiol.*, 92, 251—261, 1973.
12. **Berry, J. A. and Björkman, O.,** Photosynthetic response and adaptation to temperature in higher plants, *Annu. Rev. Plant Physiol.*, 31, 491—543, 1980.
13. **Biessmann, H., Falkner, F. G., Saumweber, H., and Walter, M. F.,** Disruption of the vimentin cytoskeleton may play a role in heat-shock response, in *Heat Shock from Bacteria to Man,* Schlesinger, M. J., Ashburner, M., and Tissieres, A., Eds., Cold Spring Harbor Laboratory, Cold Spring Harbor, NY, 1982, 275—281.
14. **Boonstra, J., Schamhart, D. H. J., De Laat, S. W., and Van Wijk, R.,** Analysis of K+ and Na+ transport and intracellular contents during and after heat shock and their role in protein synthesis in rat hepatoma cells, *Cancer Res.*, 44, 955—960, 1984.
15. **Borelli, M. J., Wong, R. S. L., and Dewey, W. C.,** A direct correlation between hyperthermia induced membrane blebbing and survival in synchronous G1 CHO cells, *J. Cell. Physiol.*, 126, 181—190, 1986.
16. **Borelli, M. J., Carlini, W. C., Ransom, B. R., and Dewey, W. C.,** Ion-sensitive microelectrode measurements of free intracellular chloride and potassium concentrations in hyperthermia-treated neuroblastoma cells, *J. Cell. Physiol.*, 129, 175—184, 1986.

17. **Bowler, K.**, Heat death and cellular heat injury, *J. Therm. Biol.*, 6, 171—178, 1981.

17a. **Braun, J. and Hahn, G. M.**, Enhanced cell killing by bleomycin and 43°C hyperthermia and the inhibition of recovery from potentially lethal damage, *Cancer Res.*, 35, 2921—2927, 1975.

18. **Brock, T. D.**, *Thermophilic Microorganismus and Life at High Temperatures*, Springer-Verlag, Berlin, 1978.

19. **Brock, T. D.**, Life at high tempertures, *Science*, 230, 132—138, 1985.

19a. **Brodl, M. R., Belanger, F. C., and Ho, T. H. D.**, Heat shock proteins are not required for the degradation of α-amylase messenger RNA and the delamellation of endoplasmic reticulum in heat-stressed barley aleurone cells, *Plant Physiol.*, 92, 1133—1141, 1990.

20. **Burdon, R. H. and Cutmore, C. M. M.**, Human heat shock gene expression and the modulation of plasma membrane sodium-potassium ATPase activity, *FEBS Lett.*, 140, 45—48, 1982.

21. **Burns, C. P., Lambert, B. J., Haugstad, B. N., and Guffy, M. M.**, Influence of rate of heating on thermosensitivity of L1210 leukemia: Membrane lipids and Mr 70,000 heat shock protein, *Cancer Res.*, 46, 1882—1887, 1986.

22. **Calderwood, S. K. and Hahn, G. M.**, Thermal sensitivity and resistance of insulin-receptor binding, *Biochim. Biophys. Acta*, 756, 1—8, 1983.

23. **Calderwood, S. K., Stevenson, M. A., and Hahn, G. M.**, Cyclic AMP and the heat shock response in Chinese hamster ovary cells, *Biochem. Biophys. Res. Commun.*, 126, 911—916, 1985.

24. **Calderwood, S. K., Stevenson, M. A., and Hahn, G. M.**, Heat stress stimulates inositol trisphosphate release and phosphorylation of phosphoinositides in CHO and BALB/C 3T3 cells, *J. Cell. Physiol.*, 130, 369—376, 1987.

25. **Cann, J. R.**, On the leakage of ultraviolet absorbing materials and alanine from synchronized *Tetrahymena pyriformis*, *C. R. Trav. Lab.*, Carlsberg, 36, 319—325, 1968.

26. **Capco. D. G., Wan, K. M., and Penman, S.**, The nuclear matrix: Three dimensional architecture and protein composition, *Cell*, 29, 847—858, 1982.

27. **Carr, A. and De Pomerai, D. I.**, Stress protein and crystallin synthesis during heat shock and transdifferentiation of embryonic chicken neural retina cells, *Dev. Growth Differ.*, 27, 435—445, 1985.

28. **Cervera, M., Dreyfuss, G., and Penman, S.**, Messenger RNA is translated when associated with the cytoskeletal framework in normal and VSV-infected HeLa cells, *Cell*, 23, 113—120, 1981.

29. **Chappell, T. G., Welch, W. J., Schlossman, D. M., Palter, K. B., Schlesinger, M. J., and Rothman, J. E.**, Uncoating ATPase is a member of the 70 kilodalton family of stress proteins, *Cell*, 45, 3—13, 1986.

30. **Clark, B. D. and Brown, I. R.**, Altered expression of a heat shock protein in the mammalian nervous system in the presence of agents which affect microtubule stability, *Neurochem. Res.*, 12, 819—823, 1987.

31. **Collier, N. C. and Schlesinger, M. J.**, The dynamic state of heat shock proteins in chicken embryo fibroblasts, *J. Cell Biol.*, 103, 1495—1507, 1986.

32. **Collier, N. C., Heuser, J., Levy, M. A., and Schlesinger, M. J.**, Ultrastructural and biochemical analysis of the stress granule in chicken embryo fibroblasts, *J. Cell Biol.*, 106, 1131—1139, 1988.

33. **Coss, R. A., Dewey, W. C., and Bamburg, J. R.**, Effects of hyperthermia (41.5°C) on Chinese hamster ovary cells analyzed in mitosis, *Cancer Res.*, 39, 1911—1918, 1979.

34. **Coss, R. A., Dewey, W. C., and Bamburg, J. R.**, Effects of hyperthermia on dividing Chinese hamster ovary cells and on microtubules in vitro, *Cancer Res.*, 42, 1059—1071, 1982.

35. **Cress, A. E. and Gerner, E. W.**, Cholesterol levels inversely reflect the thermal sensitivity of mammalian cells in culture, *Nature*, 283, 677—679, 1980.

36. **Cress, A. E., Culver, P. S., Moon, T. E., and Gerner, E. W.**, Correlation between amounts of cellular membrane components and sensitivity to hyperthermia in a variety of mammalian cell lines in culture, *Cancer Res.*, 42, 1716—1721, 1982.

37. **Daniell, J. W., Chappell, W. E., and Couch, H. B.**, Effect of sublethal and lethal temperatures on plant cells, *Plant Physiol.*, 44, 1684—1689, 1969.

38. **Davis, J. Q., Dansereau, D., Johnston, R. M., and Bennett, V.**, Selective externalization of an ATP-binding protein structurally related to the clathrin-uncoating ATPase/heat shock protein in vesicles containing terminal transferrin receptors during reticulocyte maturation, *J. Biol. Chem.*, 261, 15368—15371, 1986.

39. **Dennis, W. H. and Yatvin, M. B.**, Correlation of hyperthermic sensitivity and membrane microviscosity in *E. coli* K 1060, *Int. J. Radiat. Biol.*, 39, 265—271, 1981.

40. **Djordjevic, B.**, Variable interaction of heat and procaine in potentiation of radiation lethality in mammalian cells of neoplastic origin, *Int. J. Radiat. Biol.*, 43, 399—409, 1983.

41. **Drummond, I. A. S., Livingstone, D., and Steinhardt, R. A.**, Heat shock protein synthesis and cytoskeletal rearrangements occur independently of intracellular free calcium increases in *Drosophila* cells and tissues, *Radiat. Res.*, 113, 402—413, 1988.

42. **Fakan, S. and Puvion, E.**, The ultrastructural visualization of nucleolar and extranucleolar RNA synthesis and distribution, *Int. Rev. Cytol.*, 65, 255—299, 1980.

43. **Falkner, F. G. and Biessmann, H.**, Nuclear proteins in *Drosophila melanogaster* cells after heat-shock and their binding to homologous DNA, *Nucl. Acids Res.*, 8, 943—955, 1980.

44. **Falkner, F. G., Saumweber, H., and Biessmann, H.,** Two *Drosophila melanogaster* proteins related to intermediate filament proteins of vertebrate cells, *J. Cell Biol.*, 91, 175—183, 1981.
45. **Franke, W. W.,** Nuclear lamins and cytoplasmic intermediate filament proteins: A growing multigene family, *Cell*, 48, 3—4, 1987.
46. **Franke, W. W., Schmid, E., Grund, C., and Geiger, B.,** Intermediate filament proteins in monofilamentous structures. Transient disintegration and inclusion of subunit proteins in granular aggregates, *Cell*, 30, 103—113, 1982.
47. **Fransolet, S., Deltour, R., Bronchart, R., and Van De Walle, C.,** Changes in ultrastructure and transcription by elevated temperatures in *Zea mays* embryonic root cells, *Planta*, 146, 7—18, 1979.
48. **Garry, R. F. and Bostick, D. A.,** Induction of the stress response alteration in membrane-associated transport systems and protein modification in heat shocked or Sindbis virus-infected cells, *Virus Res.*, 8, 245—260, 1987.
49. **Gerner, E. W. and Russel, D. H.,** The relationship between polyamine accumulation and DNA replication in synchronized Chinese hamster ovary cells after heat shock, *Cancer Res.*, 37, 482—489, 1977.
50. **Gerner, E. W., Holmes, D. K., Stickney, D. G., Noterman, J. A., and Fuller, D. J. M.,** Enhancement of hyperthermia-induced cytotoxicity by polyamines, *Cancer Res.*, 40, 432—438, 1980.
51. **Glaczinski, H., Ohad, I., and Kloppstech, K.,** Temperature-dependent binding to the thylakoid of nuclear-coded chloroplast heat-shock proteins, *Eur. J. Biochem.*, 173, 579—583, 1988.
52. **Glass, J. R., De Witt, R. G., and Cress, A. E.,** Rapid loss of stress fibers in Chinese hamster ovary cells after hyperthermia, *Cancer Res.*, 45, 258—262, 1985.
53. **Guidon, P. T. and Hightower, L. E.,** Purification and initial characterization of the 71-kilodalton rat heat-shock protein and its cognate as fatty acid binding proteins, *Biochemistry*, 25, 3231—3239, 1986.
54. **Guidon, P. T. and Hightower, L. E.,** The 73 kilodalton heat shock cognate protein purified from rat brain contains nonesterified palmitic and stearic acids, *J. Cell. Physiol.*, 128, 239—245, 1986.
55. **Hadjiolov, A. A.,** *The Nucleolus and Ribosome Biogenesis. Cell Biol. Monogr.*, Vol. 12, Springer-Verlag, Vienna, 1985.
55a. **Hahn, G. M., Braun, J., and Har-Kedar, I.,** Thermochemotherapy: synergism between hyperthermia (42-43°C) and adriamycin (or bleomycin) in mammalian cell inactivation, *Proc. Natl. Acad. Sci. U.S.A.*, 72, 937—940, 1975.
56. **Harms-Ringdahl, M., Anderstam, B., and Vaca, C.,** Heat-induced changes in the incorporation of (H3) acetate in membrane lipids. *Int. J. Radiat. Biol.*, 52, 315—324, 1987.
57. **Havaux, M., Canaani, O., and Malkin, S.,** Rapid screening for heat tolerance in *Phaseolus* species using the photoacoustic technique, *Plant Sci.*, 48, 143—149, 1987.
58. **Heine, U., Sverak, L., Kondratick, J., and Bonar, R. A.,** The behaviour of HeLa-S3 cells under the influence of supranormal temperatures, *J. Ultrastruct. Res.*, 34, 375—396, 1971.
59. **Hemmingsen, S. M., Woolford, C., Van Der Vies, S. M., Tilly, K., Dennis, D. T., Georgopoulos, C. P., Hendrix, R. W., and Ellis, R. J.,** Homologous plant and bacterial proteins chaperone oligomeric protein assembly, *Nature*, 333, 330—335, 1988.
60. **Hochachka, P. W. and Somero, G. N.,** *Biochemical Adaption*, Princeton University Press, Princeton, NY, 1984.
61. **Howe, J. G. and Hershey, J. W. B.,** Translation initiation factor and ribosome association with the cytoskeletal framework fraction from HeLa cells, *Cell*, 37, 85—93, 1984.
62. **Hughes, E. N. and August, J. T.,** Coprecipitation of heat shock proteins with a cell surface glycoprotein, *Proc. Natl. Acad. Sci. U.S.A.*, 79, 2305—2309, 1982.
63. **Iida, K., Iida, H., and Yahara, I.,** Heat shock induction of intranuclear actin rods in cultured mammalian cells, *Exp. Cell Res.*, 165, 207—215, 1986.
64. **Kapiszewska, M. and Hoopwood, L. E.,** Changes in bleb-formation following hyperthermia treatment of CHO-cells, *Radiat. Res.*, 105, 405—412, 1986.
65. **Klein, J. D. and Ferguson, I. B.,** Effect of high temperature on calcium uptake by suspension-cultured pear fruit cells, *Plant Physiol.*, 84, 153—156, 1987.
66. **Kloppstech, K., Meyer, G., Schuster, G., and Ohad, I.,** Synthesis, transport and localization of a nuclear coded 22-kd heat-shock protein in the chloroplast membranes of peas and *Chlamydomonas reinhardi*, *EMBO J.*, 4, 1901—1909, 1985.
67. **Konings, A. T. W. and Ruifrok, A. C. C.,** Role of membrane lipids and membrane fluidity in the thermosensitivity and thermotolerance of mammalian cells, *Radiat. Res.*, 102, 86—98, 1985.
68. **Koyasu, S., Nishida, E., Kadowaki, T., Matsuzaki, F., Iida, K., Harada, F., Kasuga, M., Sakai, H., and Yahara, I.,** Two mammalian heat shock proteins, HSP90 and HSP100, are actin-binding proteins, *Proc. Natl. Acad. Sci. U.S.A.*, 83, 8054—8058, 1986.
68a. **Koyasu, S., Nishida, E., Miyata, Y., Sakai, H., and Yahara, I.,** HSP100, and 100-kDa heat shock protein, is a Ca^{2+}-calmodulin-regulated actin-binding protein, *J. Biol. Chem.*, 264, 15083—15087, 1989.
69. **Krishnasamy, S., Mannar-Mannan, R., Krishnan, M., and Gnanam, A.,** Heat shock response of the chloroplast genome in *Vigna sinensis*, *J. Biol. Chem.*, 263, 5104—5109, 1988.

70. **Kwock, L., Lin, P. S., Hefter, K., and Wallach, D. F. H.,** Impairment of Na^+-dependent amino acid transport in a cultured human T-cell line by hyperthermia and irradiation, *Cancer Res.,* 38, 83—87, 1978.
71. **Landry, J., Crete, P., Lamarche, S., and Chretien, P.,** Activation of Ca^{2+}-dependent processes during heat shock: Role in cell thermoresistance, *Radiat. Res.,* 113, 426—436, 1988.
72. **Langridge, J.,** Biochemical aspects of temperature response, *Annu. Rev. Plant Physiol.,* 14, 441—462, 1963.
73. **Lazarides, E.,** Intermediate filaments as mechanical intergrators of cellular space, *Nature,* 283, 249—256, 1980.
74. **Leicht, B. G., Biessmann, H., Palter, K. B., and Bonner, J. J.,** Small heat shock proteins of *Drosophila* associate with the cytoskeleton, *Proc. Natl. Acad. Sci. U.S.A.,* 83, 90—94, 1986.
75. **Lenk, R. and Penman, S.,** The cytoskeletal framework and poliovirus metabolism, *Cell,* 16, 289—301, 1979.
76. **Lenk, R., Ransom, L., Kaufmann, Y., and Penman, S.,** A cytoskeletal structure with associated polyribosomes obtained from HeLa cells, *Cell,* 10, 67—78, 1977.
77. **Lepock, J. R., Massicotte-Nolan, P., Rule, G. S., and Kruuv, J.,** Lack of correlation between hyperthermic cell killing, thermotolerance and membrane lipid fluidity, *Radiat. Res.,* 87, 300—313, 1981.
78. **Lepock, J. R., Cheng, K. H., Al-Qusi, H., and Kruuv, J.,** Thermotropic lipid and protein transitions in Chinese hamster lung cell membranes: Relationship to hyperthermic cell killing, *Can. J. Biochem. Cell Biol.,* 61, 421—427, 1983.
78a. **Li, G. C., Shu, E. C., and Hahn, G. M.,** Similarities in cellular inactivation by hyperthermia or by ethanol, *Radiat. Res.,* 82, 257—268, 1980.
79. **Lin, P. P. and Hahn, G. M.,** Growth factors and hyperthermia. I. The relationship between hyperthermic cell killing and the mitogenic response to serum and growth factors, *Radiat. Res.,* 113, 501—512, 1988.
80. **Lin, P. P. and Hahn, G. M.,** Growth factors and hyperthermia. II. Viability of Chinese hamster ovary HA-1 cells during serum starvation and hyperthermia, *Radiat. Res.,* 113, 513—525, 1988.
81. **Lin, P. S., Wallach, D. F. H., and Tsai, S.,** Temperature-induced variations in the surface topology of cultured lymphocytes are revealed by scanning electron microscopy, *Proc. Natl. Acad. Sci. U.S.A.,* 70, 2492—2496, 1973.
82. **Lin, P. S., Kwock, L., Hefter, K., and Wallach, D. F. H.,** Modification of rat thymocyte membrane properties by hyperthermia and ionizing radiation, *Int. J. Radiat. Biol.,* 33, 371—382, 1978.
82a. **Lin, P. S., Hefter, K., and Jones, M.,** Hyperthermia and bleomycin schedules on V79 Chinese hamster cell cytotoxicity in vitro, *Cancer Res.,* 43, 4557—4561, 1983.
83. **Lin, C.-Y., Chen, Y.-M., and Key, J. L.,** Solute leakage in soybean seedlings under various heat shock regimes, *Plant Cell Physiol.,* 26, 1493—1498, 1985.
84. **Ljungdahl, L. G.,** Physiology of thermophilic bacteria, *Adv. Microb. Physiol.,* 19, 150—243, 1979.
85. **Lomagin, A. G.,** Repair of functional and ultrastructural alterations after thermal injury of *Physarum polycephalum, Planta,* 142, 123—134, 1978.
85a. **Love, R., Soriano, R. Z., and Walsh, R. J.,** Effect of hyperthermia on normal and neoplastic cells in vitro, *Cancer Res.,* 30, 1525—1533, 1970.
86. **Lutz, Y., Jacob, M., and Fuchs, J. P.,** The distribution of two hnRNP-associated proteins defined by a monoclonal antibody is altered in heat-shocked HeLa cells, *Exp. Cell Res.,* 175, 109—124, 1988.
86a. **Magin, R. L., Sikic, B. I., and Lysyk, R. L.,** Enchancement of bleomycin activity against Lewis lung tumors in mice by local hyperthermia, *Cancer Res.,* 39, 3792—3795, 1979.
87. **Magun, B. E.,** Inhibition and recovery of macromolecular synthesis, membrane transport, and lysosomal function following exposure of cultured cells to hyperthermia, *Radiat. Res.,* 87, 657—669, 1981.
87a. **Magun, B. E. and Fennie, C. W.,** Effects of hyperthermia on binding, internalization, and degradation of epidermal growth factor, *Radiat. Res.,* 86, 133—146, 1981.
88. **Mansfield, M. A., Lingle, W. L., and Key, J. L.,** The effects of lethal heat shock on nonadapted and thermotolerant root cells of *Glycine max, J. Ultrastruct. Mol. Struct. Res.,* 99, 96—105, 1988.
88a. **Marmor, J. B.,** Interactions of hyperthermia and chemotherapy in animals, *Cancer Res.,* 39, 2269—2276, 1979.
88b. **Marshall, J. S., de Rocher, A. E., Keegstra, K., and Vierling, E.,** Identification of hsp70 homologues in chloroplasts, *Proc. Natl. Acad. Sci. U.S.A.,* 87, 374—378, 1990.
88c. **Marvin, H. J. P., Ter Beest, M. B. A., and Witholt, B.,** Release of outer membrane fragments from wild-type *Escherichia coli* and several *E. coli* lipopolysaccharide mutants by EDTA and heat shock treatments, *J. Bacteriol.,* 171, 5262—5267, 1989.
89. **Mohanty, N., Murthy, S. D. S., and Mohanty, P.,** Reversal of heat induced alterations in photochemical activities in wheat primary leaves, *Photosynth. Res.,* 14, 259—268, 1987.
90. **Monfort, C., Ried, A., and Ried, J.,** Abstufungen der funktionellen Waermeresistenz bei Meeresalgen in ihrer Beziehung zu Umwelt und Erbgut, *Biol. Zbl.,* 76, 257—289, 1957.
91. **Moon, R. T., Nicosia, R. F., Olsen, C., Hille, M. B., and Jeffery, W. R.,** The cytoskeletal framework of sea urchin eggs and embryos: Developmental changes in the association of messenger RNA, *Dev. Biol.,* 95, 447—458, 1983.

92. **Moyer, S. A., Baker, S. C., and Lessard, J. L.,** Tubulin: A factor necessary for the synthesis of both Sendai virus and vesicular stomatitis virus RNAs, *Proc. Natl. Acad. Sci. U.S.A.*, 83, 5405—5409, 1986.
93. **Mulcahy, R. T., Gould, M. N., Hidvergi, E., Elson, C. E., and Yatvin, M. B.,** Hyperthermia and surface morphology of P388 ascites tumor cells: Effects of membrane modifications, *Int. J. Radiat. Biol.*, 39, 95—106, 1981.
93a. **Murti, K. G., Smith, H. T., and Fried, V. A.,** Ubiquitin is a component of the microtubule network, *Proc. Natl. Acad. Sci. U.S.A.*, 85, 3019—3023, 1988.
94. **Nagata, K., Saga, S., and Yamada, K. M.,** A major collagen-binding protein of chicken embryo fibroblasts is a novel heat shock protein, *J. Cell Biol.*, 103, 223—230, 1986.
94a. **Nakai, A., Hirayoshi, K., and Nagata, K.,** Transformation of BALB/3T3 cells by simian virus-40 causes a decreased synthesis of a collagen-binding heat shock protein (Hsp47), *J. Biol. Chem.*, 265, 992—999, 1990.
95. **Napolitano, E. W., Pachter, J. S., and Liem, R. K. H.,** Intracellular distribution of mammalian stress proteins. Effects of cytoskeletal-specific agents, *J. Biol. Chem.*, 262, 1493—1504, 1987.
96. **Nelles, A.,** Das Membranpotential von Maiskoleoptilen unter dem Einfluss von Hitzestress, *Biochem. Physiol. Pflanz.*, 180, 459—463, 1985.
97. **Neumann, D., Scharf, K.-D., and Nover, L.,** Heat shock induced changes of plant cell ultrastructure and autoradiographic localization of heat shock proteins, *Eur. J. Cell Biol.*, 34, 254—264, 1984.
98. **Nishida, E., Koyasu, S., Sakai, H., and Yahara,, I.,** Calmodulin-regulated binding of the 90 kDa heat shock protein to actin filaments, *J. Biol. Chem.*, 261, 16033—16037, 1986.
98a. **Nover, L., Ed.,** *Heat Shock Response of Eukaryotic Cells,* Springer-Verlag, Berlin, 1984.
99. **Nover, L. and Scharf, K.-D.,** Synthesis, modification and structural binding of heat shock proteins in tomato cell cultures, *Eur. J. Biochem.*, 139, 303—313, 1984.
100. **Nover, L., Scharf, K.-D., and Neumann, D.,** Formation of cytoplasmic heat shock granules in tomato cell cultures and leaves, *Mol. Cell. Biol.*, 3, 1648—1655, 1983.
101. **Nover, L., Scharf, K.-D., and Neumann, D.,** Cytoplasmic heat shock granules are formed from precursor particles and contain a specific set of mRNAs, *Mol. Cell. Biol.*, 9, 1298—1308, 1989.
102. **Pan, B. T. and Johnston, R. M.,** Selective externalization of the transferrin receptor by sheep reticulocytes *in vitro, J. Biol. Chem.*, 259, 9776—9782, 1984.
103. **Pekkala, D., Heath, I. B., and Silver, J. C.,** Changes in chromatin and the phosphorylation of nuclear proteins during heat shock of *Achlya ambisexualis, Mol. Cell. Biol.*, 4, 1198—1205, 1984.
104. **Pelling, G.,** Chromosomal synthesis of ribonucleic acid as shown by incorporation of uridine labelled with tritium, *Nature,* 184, 655—656, 1959.
105. **Powell, A. J., Peace, G. W., Slabas, A. R., and Leoyd, C. W.,** The detergent-resistant cytoskeleton of higher plant protoplasts contains nucleus-associated fibrillar bundles in addition to microtubules, *J. Cell Sci.*, 56, 319—335, 1982.
106. **Quinn, P. J.,** The fluidity of cell membranes and its regulation, *Prog. Biophys. Mol. Biol.*, 38, 1—104, 1981.
107. **Reeves, R. O.,** Mechanisms of acquired resistance to acute heat shock in cultured mammalian cells, *J. Cell. Physiol.*, 79, 157—170, 1972.
108. **Rice, G. C., Fisher, K. A., Fisher, G. A., and Hahn, G. M.,** Correlation of mammalian cell killing by heat shock to intramembranous particle aggregation and lateral phase separation using fluorescence-activated cell sorting, *Radiat. Res.*, 112, 351—364, 1987.
109. **Rieder, C. and Bajer, A. S.,** Heat induced reversible hexagonal packing of spindle microtubules, *J. Cell Biol.*, 74, 717—725, 1977.
110. **Risueno, M. C., Stockert, J. C., Gimenez-Martin, G., and Diez, J. L.,** Effect of supraoptimal temperatures on meristematic cell nucleoli, *J. Microscopie,* 16, 87—94, 1973.
111. **Ritossa, F.,** A new puffing pattern induced by heat shock and DNP in *Drosophila, Experientia,* 18, 571—573, 1962.
112. **Ron, A. and Zeuthen, E.,** Tubulin synthesis and heat-shock-induced cell synchrony in *Tetrahymena, Exp. Cell Res.*, 128, 303—309, 1980.
113. **Ruifrok, A. C. C., Kanon, B., and Konings, A. W. T.,** Correlation between cellular survival and potassium loss in mouse fibroblasts after hyperthermia alone and after a combined treatment with X-rays, *Radiat. Res.*, 101, 326—331, 1985.
114. **Ruifrok, A. C. C., Kanon, B., and Konings, A. W. T.,** Correlation of colony forming ability of mammalian cells with potassium content after hyperthermia under different experimental conditions, *Radiat. Res.*, 103, 452—454, 1985.
115. **Ruifrok, A. C. C., Kanon, B., and Konings, A. W. T.,** Na^+/K^+ ATPase activity in mouse lung fibroblasts and HeLa S3 cells during and after hyperthermia, *Int. J. Hyperthermia,* 2, 51—59, 1986.
116. **Russell, N. J.,** Mechanisms of thermal adaptation in bacteria: Blueprints for survival, *Trends Biochem. Sci.*, 9, 108—112, 1984.

117. **Sachs, J.,** Ueber die obere Temperatur-Gränze der Vegetation, *Flora,* 47, 5—12, 24—29, 33—39, 64—75, 1864.
118. **Schamhart, D. H. J., Van Walraven, H. S., Wiegant, F. A. C., Linnemans, W. A. M., Van Rijn, J., Van Den Berg, J., and Van Wijk, R.,** Thermotolerance in cultured hepatoma cells: Cell viability, cell morphology, protein synthesis, and heat shock proteins. *Radiat. Res.,* 98, 82—95, 1984.
119. **Schliwa, M.,** *The Cytoskeleton. Cell Biol. Monogr.,* Vol. 13, Springer-Verlag, New York, 1986.
120. **Schnepf, E. and Schmitt, U.,** Destruction and reconstitution of the dictyosome in the chrysophycean flagellate, *Poterioochromonas malhamensis,* after heat-shock, and other heat-shock effects, *Protoplasma,* 106, 261—271, 1981.
121. **Schröder, M., Wehland, J., and Weber, K.,** Immunofluorescence microscopy of microtubules in plant cells: Stabilization by dimethylsulfoxide, *Eur. J. Cell Biol.,* 38, 211—218, 1985.
122. **Schuster, G., Even, D., Kloppstech, K., and Ohad, I.,** Evidence for protection by heat-shock proteins against photoinhibition during heat-shock, *EMBO J.,* 7, 1—6, 1988.
123. **Shay, W.,** *Cell and Molecular Biology of the Cytoskeleton,* Plenum Press, New York, 1986.
124. **Shivers, R. R., Pollock, R., Bowman, P. D., and Atkinson, B. G.,** The effect of heat-shock on primary cultures of brain capillary endothelium inhibition of assembly of *zonulae occludentes* and the synthesis of heat shock proteins, *J. Cell Biol.,* 46, 181—195, 1988.

124a. **Shyy, T.-T., Asch, B. B., and Asch, H. L.,** Concurrent collapse of keratin filaments, aggregation of organelles, and inhibition of protein synthesis during the heat shock response in mammary epithelial cells, *J. Cell Biol.,* 108, 997—1008, 1989.

125. **Simard, R. and Bernhard, W.,** A heat-sensitive cellular function located in the nucleolus, *J. Cell Biol.,* 34, 61—76, 1967.
126. **Steinert, P. M., Steven, A. C., and Roop, D. R.,** The molecular biology of intermediate filaments, *Cell,* 42, 411—419, 1985.
127. **Stevenson, A. P., Galey, W. R., and Tobey, R. A.,** Hyperthermia-induced increase in potassium transport in Chinese hamster cells, *J. Cell. Physiol.,* 115, 75—86, 1983.

127a. **Stevenson, M. A. and Calderwood, S. K.,** Members of the 70-kilodalton heat shock protein family contain a highly conserved calmodulin-binding domain, *Mol. Cell. Biol.,* 10, 1234—1238, 1990.

128. **Stevenson, M. A., Calderwood, S. K., and Hahn, G. M.,** Rapid increases in inositol trisphosphate and intracellular calcium after heat shock, *Biochem. Biophys. Res. Commun.,* 137, 826—833, 1986.
129. **Stevenson, M. A., Calderwood, S. K., and Hahn, G. M.,** Effect of hyperthermia (45°C) on calcium flux in Chinese hamster ovary HA-1 fibroblasts and its potential role in cytotoxicity and heat resistance, *Cancer Res.,* 47, 3712—3717, 1987.
130. **Süss, K. H. and Jordanov, I. T.,** Biosynthetic cause of in vivo acquired thermotolerance of photosynthetic light reactions and metabolic responses of chloroplasts to heat stress, *Plant Physiol.,* 81, 192—199, 1986.
131. **Thebaud, R. and Santarius, K. A.,** Effects of high temperature stress on various biomembranes of leaf cells in situ and in vitro, *Plant Physiol.,* 70, 200—206, 1982.
132. **Tischner, R. and Lorenzen, H.,** Physiologische Auswirkungen von Hitzeschocks auf synchrone Chlorellen im empfindlichsten Entwicklungsstadium, *Biochem. Physiol. Pflanzen,* 168, 233—245, 1975.

132a. **Tsuchido, T., Aoki, I., and Takano, M.,** Interaction of the fluorescent dye 1-N-phenylnaphthylamine with *Escherichia coli* cells during heat stress and recovery from heat stress, *J. Gen. Microbiol.,* 135, 1941—1947, 1989.

133. **Ungewickell, E.,** The 70-kd mammalian heat shock proteins are structurally and functionally related to the uncoating protein that releases clathrin triskelia from coated vesicles, *EMBO J.,* 4, 3385—3391, 1985.
134. **Van Bergen en Henegouwen, P. M. P. and Linnemans, W. A. M.,** Heat shock gene expression and cytoskeletal alterations in mouse neuroblastoma cells, *Exp. Cell Res.,* 171, 367—375, 1987.
135. **Van Bergen en Henegouwen, P. M. P., Jordi, W. J. R. M., Van Dongen, G., Ramaekers, F. C. S., Amesz, H., and Linnemans, W. A. M.,** Studies on a possible relationship between alterations in the cytoskeleton and induction of heat shock protein synthesis in mammalian cells, *Int. J. Hyperthermia,* 1, 69—83, 1985.
136. **Van Bergen en Henegouwen, P. M. P., Berbers, G., Linnemans, W. A. M., and Van Wijk, R.,** Cytoplasmic and nuclear localization of the 84,000 dalton heat shock protein in mouse neuroblastoma cells, *Eur. J. Cell Biol.,* 43, 469—478, 1987.
137. **Vierling, E., Mishkind, M. L., Schmidt, G. W., and Key, J. L.,** Specific heat shock proteins are transported into chloroplasts, *Proc. Natl. Acad. Sci. U.S.A.,* 83, 361—365, 1986.
138. **Vierling, E., Nagao, R. T., De Rocher, A. E., and Harris, L. M.,** A heat shock protein localized to chloroplasts is a member of a eukaryotic superfamily of heat shock proteins, *EMBO J.,* 7, 575—581, 1988.
139. **Vincent, M. and Tanguay, R. M.,** Different intracellular distributions of heat shock and arsenite-induced proteins in *Drosophila* Kc cells. Possible relation with the phosphorylation and translocation of a major cytoskeletal protein, *J. Mol. Biol.,* 162, 365—378, 1982.
140. **Volger, H. and Santarius, K. A.,** Release of membrane proteins in relation to heat injury of spinach chloroplasts, *Physiol. Plant,* 51, 195—200, 1981.

141. **Walter, M. F. and Biessmann, H.,** Intermediate-sized filaments in *Drosophila* tissue culture cells, *J. Cell Biol.,* 99, 1468—1477, 1984.
142. **Walter, M. F. and Biessmann, H.,** A non-filamentous configuration of intermediate-sized filament proteins in *Drosophila* Kc tissue culture cells, *Cell. Dev. Biol.,* 23, 453—458, 1987.
143. **Warren, A. P., James, M. H., Menzies, D. E., Widnell, C. C., Whitaker-Dowling, P. A., and Pasternak, C. A.,** Stress induces an increased hexose uptake in cultured cells, *J. Cell. Physiol.,* 128, 383—388, 1986.
144. **Weber, K. and Glenney, J. R.,** Microfilament-membrane interaction: The brush border of intestinal epithelial cells as a model, *Philos. Trans. R. Soc. London,* 299, 207—213, 1982.
145. **Weis, E.,** Reversible heat-inactivation of the Calvin cycle: A possible mechanism of the temperature regulation of photosynthesis, *Planta,* 151, 33—39, 1981.
146. **Welch, W. J. and Feramisco, J. R.,** Disruption of the three cytoskeletal networks in mammalian cells does not affect transcription, translation, or protein translocation changes induced by heat shock, *Mol. Cell. Biol.,* 5, 1571—1581, 1985.
147. **Welch, W. J. and Mizzen, L. A.,** Characterization of the thermotolerant cell. II. Effects on the intracellular distribution of HSP70, intermediate filaments, and snRNP's, *J. Cell Biol.,* 106, 1117—1130, 1988.
148. **Welch, W. J. and Suhan, J. P.,** Morphological study of the mammalian stress response: Characterization of changes in cytoplasmic organelles, cytoskeleton, and nucleoli, and appearance of intranuclear actin filaments in rat fibroblasts after heat-shock treatment, *J. Cell Biol.,* 101, 1198—1211, 1985.
149. **Welch, W. J. and Suhan, J. P.,** Cellular and biochemical events in mammalian cells during and after recovery from physiological stress, *J. Cell Biol.,* 103, 2035—2052, 1986.
150. **Welch, W. J., Feramisco, J. R., and Blose, S. H.,** The mammalian stress response and the cytoskeleton: Alterations in intermediate filaments, *Ann. Acad. Sci. N.Y.,* 455, 57—67, 1985.
151. **Wiegant, F. A. C., Tuyl, M., and Linnemans, W. A. M.,** Calmodulin inhibitors potentiate hyperthermic cell killing, *Int. J. Hyperthermia,* 1, 157—170, 1985.
152. **Wiegant, F. A. C., Van Bergen En Henegouwen, P. M. P., Van Dongen, A. A. M. S., and Linnemans, W. A. M.,** Stress-induced thermotolerance of the cytoskeleton, *Cancer Res.,* 47, 1674—1680, 1987.
153. **Wiegant, F. A. C., Blok, F. J., and Linnemans, W. A. M.,** Hyperthermia and the membrane-associated cytoskeleton of mammalian cells, Proefschrift University, Utrecht, 1987, 103—124.
154. **Wiegant, F. A. C., Blok, F. J., Van Bergen en Henegouwen, P. M. P., and Linnemans, W. A. M.,** Heat shock induced cytoskeletal reorganization is related to cell survival in mammalian cells, Proefschrift University, Utrecht, 1987, 79—94.
155. **Wiegant, F. A. C., Van Bergen en Henegouwen, P. M. P., and Linnemans, W. A. M.,** Studies on the mechanism of heat shock induced cytoskeletal reorganization in mouse neuroblastoma N2A cells, Proefschrift University, Utrecht, 1987, 141—158.
156. **Wodtke, E.,** Adaptation of biological membranes to temperature: modifications and their mechanisms in the eurythermic carp, in *Temperature, Relations in Animals and Man, Biona-Report,* Vol. 4, Laudien, H., Ed., Fischer, Stuttgart, 1986, 129—138.
157. **Wolosewick, J. J. and Porter, K. R.,** Microtrabecular lattice of the cytoplasmic ground substance. Artifact or reality, *J. Cell Biol.,* 82, 114—139, 1979.
157a. **Wright-Sandor, L. G., Reichlin, M., and Tobin, S. L.,** Alteration by heat shock and immunological characterization of *Drosophila* small nuclear ribonucleoproteins, *J. Cell Biol.,* 108, 2007—2016, 1989.
158. **Wu, M.-T. and Wallner, S. J.,** Heat stress responses in cultured plant cells. Development and comparison of viability tests, *Plant Physiol.,* 72, 817—820, 1983.
159. **Wu, M.-T. and Wallner, S. J.,** Heat stress responses in cultured plant cells. Heat tolerance induced by heat shock versus elevated growing temperature, *Plant Physiol.,* 75, 778—780, 1984.
160. **Yatvin, M. B.,** The influence of membrane lipid composition and procaine on hyperthermic death of cells, *Int. J. Radiat. Biol.,* 32, 513—521, 1977.
161. **Yatvin, M. B., Clifton, K. B., and Dennis, W. H.,** Hyperthermia and local anesthetics: Potentiation of survival of tumor-bearing mice, *Science,* 205, 195—196, 1979.
162. **Yatvin, M. B., Smith, K. M., and Siegel, F. L.,** Translocation of nascent non-signal sequence protein in heated *Escherichia coli, J. Biol. Chem.,* 261, 8070—8075, 1986.
163. **Yatvin, M. B., Clark, A. W., and Siegel, F. L.,** Major *E. coli* heat-stress protein do not translocate: Implications for cell survival, *Int. J. Radiat. Biol.,* 52, 603—614, 1987.
164. **Yatvin, M. B., Vorpahl, J. W., Ghosh, S. K., Kine, U., and Elson, C. E.,** Heat sensitivity and membrane properties of metastasizing and non-metastasizing rat mammary tumors, *Radiat. Environ. Biophys.,* 26, 89—101, 1987.

Chapter 15

CELL CYCLING AND DNA SYNTHESIS

L. Nover

TABLE OF CONTENTS

Perturbations by heat shock (hs) of activities connected with the ordered progression through the cell cycle result from various facts. The immediate reprogramming of gene expression presumably stops the supply of short-lived control proteins. The rearrangement of the cytoskeleton, especially of the tubulin system (see Section 14.3) interferes with the normal distribution of the genetic material to the daughter cells. DNA synthesis is greatly reduced or disturbed, and hs, as other stress factors, contributes to damages of the genetic material (genomic stress). The induction of heat shock protein (HSP) synthesis evidently proceeds equally well in all phases of the cell cycle.[45,68] In contrast, the apparent differences of sensitivity to heat killing between cells in the G_1/G_0 phases (more resistant) vs. those in the G_2/S phase (more sensitive) are due to intrinsic factors unrelated to induced HSP synthesis[2,3,23,39,44,45,51,55,69] (see Section 17.4). Whether the formation of certain HSPs under the control of cell cycle factors (see Section 8.5) has any influence on this intrinsic heat sensitivity remains to be analyzed.

Hyperthermic treatment of cancer, especially in combination with radiotherapy and/or chemotherapy, initiated numerous studies on DNA damage as the primary cause of cell killing.[78] As is true for many other parts of the hs response, these effects, evoked by severe or lethal stress, are different from those observed under moderate stress conditions with formation of thermotolerant states. Hence, it will be recognized from the following overview that the presently rather fragmentary results, elaborated with different experimental systems under barely comparable conditions, are difficult to combine to a uniform picture. We will deal exclusively with eukaryotic organisms. The remarkable role of bacterial HSP for DNA replication is documented in Section 2.1.2.

15.1. CELL CYCLE BLOCK AND SYNCHRONIZATION

Depending on the severity of the hs treatment and on the actual proliferative state, cell cycling is blocked and/or subsequent phases are delayed. This phenomenon was characterized for a broad range of cell types and organisms.[1a,13,14,17-19,22,25,26,37,39,40,42,43,51,53,56,63,68,72a,80] The effect is reversible and can be used for synchronization. Optimum procedures with repeated, correctly spaced hs pulses were elaborated for two eukaryotic microorganisms, i.e., for *Schizosaccharomyces*[19] and *Tetrahymena pyriformis*[80] (Figure 15.1). Effective synchronization of *Tetrahymena* depends on 30 min hs pulses at 34°C given at early G_2 phases when the characteristic oral structure of the ciliate is assembled. The transient discontinuation of synthesis and proper organization of tubulin may be decisive for these effects.[49]

So far attempts to synchronize vertebrate and plant cells by similar procedures were much less successful.[25,43] Studies with CHO cells showed that even a short hs treatment of 5 to 15 min at 45.5°C led to a complete disruption of the microtubular apparatus, which did not reform readily after return to 37°C. The very low level of colony-forming cells compared to the much higher fraction of surviving cells was evidently due to the prevention of cytokinesis resulting in formation of proliferatively dead tetraploid cells.[7,18]

Another example of the interference of relatively mild hyperthermia with the proper functioning of the mitotic apparatus was reported by Mackey et al.[23b] Synchronized CHO cells were heated for 12 h to 41.5°C, beginning in the early S phase, and then returned to 37°C. Cell cycle progression continues during hyperthermia, though at a reduced rate, resulting in reproductive cell death. Characteristic abnormalities are chromosome breaks, defects in nuclear envelope formation with nuclear fragmentation, and metaphase-associated prematurely condensed chromatin.

In tomato cell cultures, the onset of the hs period is accompanied by a rapid decline of the mitotic index from 7 to 10% of exponentially growing cells to less than 1.5%.[53] Following a long-term cell cycle arrest at 37°C, there is a wave of mitoses in the recovery period with maximum values of 15 to 20% of cells in the M-phase. However, a further improvement of this partial synchronization effect was not achieved by different hs treatments.[54]

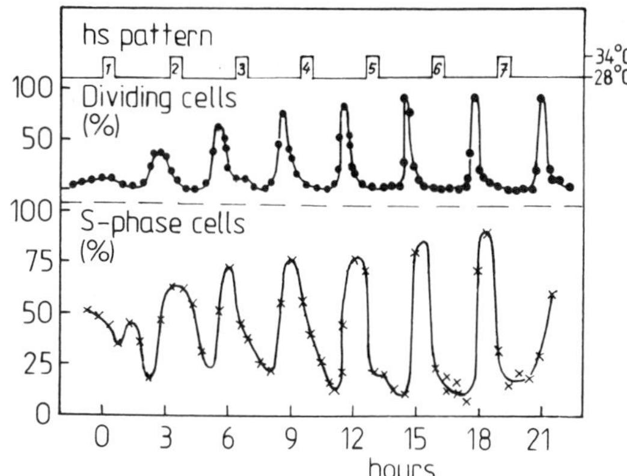

FIGURE 15.1. Synchronization of *Tetrahymena pyriformis* cultures by repetitive heat shock (hs) treatments. A repetitive hs pattern of 30 min hs at 34°C and 160 min recovery at 28°C was used to synchronize *Tetrahymena* cultures. Synchrony was analyzed by visual estimation of dividing cells (full circles) or by 15 min pulse labeling with ^{14}C-thymidine and autoradiography (crosses). The synchronizing hs periods are just spaced at the interval of a normal cell generation time at 28°C. They hit the cells at early G_2 phase. (Modified from Zeuthen, E., *Exp. Cell Res.*, 68, 49, 1971; Nover, L., Ed., *Heat Shock Response of Eukaryotic Cells,* G. Thieme Verlag, 1984. With permission.)

15.2. DNA SYNTHESIS

The rapid and profound effects on cell cycle progression but, above all, the particular heat sensitivity of S-phase cells to hyperthermic killing led to detailed studies on DNA synthesis mainly in mammalian cells. The results are summarized by Wong and Dewey.[78] The inhibition of thymidine incorporation, i.e., the overall reduction of DNA synthesis[4,10,12,63] is connected with a reduction of replicon initiation, of chain elongation and processing of DNA fragments to the high molecular weight products.[8,75,77] Depending on the severity of the stress applied, normalization of DNA synthesis takes several hours. A depletion of CHO cells of polyamines was considered as an essential factor for the block of DNA synthesis.[11] In support of this, membrane damage by procaine together with hs potentiates the inhibitory effect on DNA synthesis.[77]

The conservation of DNA polymerase activities is evidently tightly connected with the capability and velocity to recover from heat damage. Studies with fractionated hyperthermia of CHO and mouse neuroblastoma cells demonstrated that DNA synthesis becomes thermotolerant.[9,20,36,69] The reversible changes of DNA synthesis correlate with the accumulation of excess, nonhistone proteins bound to the nuclear matrix (see Section 9.3) and their slow removal in the recovery period,[50] probably by interaction with HSP70 (see Sections 2.3.2 and 9.3).

15.3. DNA DAMAGE AND REPAIR

15.3.1. ENHANCED DAMAGE AND REDUCED REPAIR

With respect to the clinical treatment of cancer, the combination of radiotherapy with hyperthermia is based on the particular heat sensitivity of the otherwise radioresistant S-phase cells and on the decreased repair of radiation-evoked DNA damages.[15,29,41,73,74] On the one hand, this is evidently due to a hs effect on the activity of repair enzymes, e.g., of DNA polymerase 3.[5b,9,36,64] On the other hand, the massive increase of nonhistone proteins in the nucleus (see Section 9.3) may render the chromatin inaccessible to repair enzymes. Evidence for this "substrate effect" was provided by Clark et al.[6] and Warters et al.[76] using mammalian cells. The only example with a nonmammalian system is the hs inhibition of excision repair of X-ray induced DNA damage in *Drosophila*.[35]

Other important aspects of cancer therapy are synergistic interactions of hyperthermic treatments with chemotherapeutics intercalating into DNA. Thus, a potentiation of mutagenic effects (chromosome damage, sister chromatid exchanges) was reported for mitomycin C[4a] and for bleomycin,[21,65,70] but the extent of chromosome damage can also be influenced by addition of inhibitors of gene expression. In Chinese hamster ovary cells, actinomycin D strongly increases and cycloheximide markedly reduces the number of chromosomal aberrations observed after hs. Altered cell cycle distribution of the cell population and/or transition through the S phase may be responsible for these effects.[57a]

Similar examples of increased mutagenicity after hs were reported for the green phytoflagellate *Euglena gracilis*[5a,23c] and for the ascomycete *Neurospora crassa*.[66a] In the former system, pretreatment for 15 min at 42°C sensitizes the *Euglena* cells to mutagenic effects of nitrofurans, nitrosomethylurea, and nitrosoguanidine. In the latter, a 5- to 100-fold enhancement of mutagenesis by different alkylating agents, but not by UV irradiation, was observed after 60 min hs at 43°C. Addition of cycloheximide does not affect the expression of the sensitive state but markedly inhibits its reversion in the recovery period. Tanaka et al.[66a] discuss increased uptake of mutagens as the major cause of enhanced mutation rates in *Neurospora*.

15.3.2. HEAT SHOCK-INDUCED RESISTANCE AND REPAIR

In contrast to the former examples, the following deal with thermotolerance as observed at the DNA level or with cross-resistance induced by hs or mutagen treatment. As is true for other parts of the hs response, the "genomic stress" triggers mechanisms which reduce DNA damage and enhance repair activities.

1. As a result of genomic instability after heat stress (4 min at 48°C) Taylor and Holliday[67] observed a more than 50-fold increase of the mitotic recombination in the smut fungus of maize (*Ustilago maydis*). Under these conditions, the surviving fraction is less than 10^{-4}. However, if the fungus was pretreated for 10 min at 42°C and allowed to recover for 2 h at 32°C, survival of the challenge treatment at 48°C was greatly improved to almost 100% and the frequency of recombination dropped to normal. A certain protection against heat killing could also be induced by preceding irradiation with UV light (200 J/m^2).
2. Indicators of genomic stress are also extensive transpositions of copia-like mobile elements observed in heat-shocked males of *Drosophila melanogaster*.[16] The effect probably reflects the general genomic instability and the existence of a hs element in the copia LTR region, which may be responsible for the increased transcription under conditions of environmental stress.[66] The results of Junakovic et al.[16] were not confirmed by Arnault and Biemont[1] analyzing genomic positions of four different mobile elements by *in situ* hybridization and Southern analyses.
3. In proliferative germ cells of the fish *Oryzias latipes*, a 30 min hs at 41°C, but not pretreatment with Cd^{2+} or arsenite, causes resistance to X-irradiation. The effect needs about 2 h to develop and is inhibited in the presence of cycloheximide.[58-62] Tests were cell survival and mitotic index after X-irradiation.
4. Studies on metaphase chromatid aberrations were used to characterize a special type of induced cross-resistance (clastogenic adaptation) in *Vicia faba* root tips. A pulse hs of 1 to 10 min at 40°C plus 1 h recovery protects against chromatid aberrations evoked by maleic hydrazide (MH), bleomycin (B), triethylene melamine (TEM), or *N*-methyl-*N*-nitrosourea (MNU). The profound differences in the mechanism of action of these mutagens, i.e., formation of oxygen radicals (MH) vs. intercalation (B) vs.

DNA alkylation (TEM, MNU), suggests that induction of a repair system is responsible for the protective effect. In keeping with this, low doses of MH or MNU can also be used to trigger clastogenic adaptation, but this is also true for chemical stressors like Cd^{2+}, Zn^{2+}, Hg^{2+}, or H_2O_2. Development of the resistant state is abolished in the presence of cycloheximide and, in the case of pretreatment with low doses of MH and MNU, also by inhibitors of "G_2-repair" (hydroxyurea, 5-fluorodeoxyuridine). Other plants show similar effects.[28,45a-48,57]

5. The molecular biology of the hs-induced UV resistance in *E. coli* was studied by genetic and biochemical experiments[37a,b] (see summary by Gottesman[11a]). A DNA damage-induced regulatory protein (Sul1A) is responsible for a transient block of cell division. Its rapid degradation after the UV-stress period depends on the Lon protease. Mutants defective in the lon gene are particularly UV-sensitive because long, non-septated *E. coli* filaments are formed. Hs induction of Lon activity evidently protects from the deleterious effects of Sul1A. Correspondingly, induced UV resistance was not observed in lon^- mutants.
6. On the basis of extensive studies on a hs-induced radiation resistance (X-ray, UV light) in yeast, Mitchel and Morrison[30-33] proposed the existence of an inducible error-free recombinational repair system. Similar to the situation in *Vicia faba* root tips, the protective effect by hs is abolished in the presence of cycloheximide. On the other hand, Mitchel and Morrison[34] also provided evidence for induction of an error-prone repair in the presence of the alkylating mutagens, e.g., of *N*-methyl-*N'*-nitro-*N*-nitrosoguanidine and methylnitrosourea. In this case, however, a hs treatment of yeast cells at 38°C during the mutagen treatment significantly reduces the inductive effect. Finally, expression of some of the DNA damage responsive genes (DDR genes) of yeast[24,27] and *Drosophila*[71] are induced by hs or UV-irradiation. Their relation to any of the poorly defined DNA repair systems is unclear at present.
7. Possibly related to the hs-induced radiation resistance in yeast are observations on an enhanced reactivation of UV-irradiated adenovirus in HeLa cells[38] or herpesvirus in African green monkey kidney epithelial cells.[76a,79] Besides hs (10 min 45.5°C), the conditioning of HeLa cells was also possible by UV-irradiation (5 J/m^2) or addition of 5% ethanol or 50 μM arsenite 18 h before the infection with the irradiated adenovirus. Interestingly, Brunet and Giacomoni[5] and Williams et al.[76a] reported on the marked induction of hsp70 expression after UV irradiation of mouse and monkey cells. In the latter, the maximum accumulation of HSP70 in the nucleus 4 to 9 h after hs induction or 19 to 24 h after UV induction correlates with optimum reactivation of irradiated herpessimplex virus.[76a] In addition, it was discussed by Piperakis and McLennan[38] and Yager et al.[79] that the halt of DNA synthesis by chemical or physical stressors combined with the disruption of replication control mechanisms may induce a DNA repair activity similar to the bacterial SOS-repair.[72]

Although the situation is far from clear, the seemingly controversial results with inhibited DNA repair during hyperthermic treatment of cancer vs. different types of induced repair in the above seven examples may simply reflect two dose-dependent aspects of genomic stress evoked by hs. Considering the restricted list of non-hs inducers and the time course of induction, the major HSPs are presumably not directly involved in DNA repair. On the other hand, it is remarkable, in addition to the induction by UV irradiation,[5,76a] that two antineoplastic drugs, causing interstrand cross-links, single strand breaks, and DNA-protein cross-links in human adenocarcinoma cells, were found to cause a strong increase of HSP70 mRNA levels and synthesis.[52] These are 1,3-bis-(2-chloroethyl)-1-nitrosourea (BCNU) and 2-(2-chloroethyl)-3-cyclohexyl-1-nitrosourea (CCNU). However, nine other structurally non-related neoplastic drugs with similar damaging effects on DNA structure were ineffective in HSP70 induction.

REFERENCES

1. **Arnault, C. and Biemont, C.,** Heat shocks do not mobilize mobile elements in genomes of *Drosophila melanogaster, J. Mol. Evol.*, 28, 388—390, 1989.
1a. **Boon-Niermeijer, E. K.,** Morphogenesis after heat shock during the cell cycle of *Lymnaea:* A new interpretation, *Wilhelm Roux's Arch. Dev. Biol.*, 180, 241—252, 1976.
2. **Boon-Niermeijer, E. K. and Van De Scheur, H.,** Thermosensitivity during embryonic development of *Lymnaea stagnalis* (Mollusca), *J. Therm. Biol.*, 9, 265—269, 1984.
3. **Boon-Niermeijer, E. K., De Waal, A. M., Souren, J. E. M., and Van Wijk, R.,** Heat-induced changes in thermosensitivity and gene expression during development, *Devel. Growth Differ.*, 30, 705—715, 1988.
4. **Boonstra, J., Sybesma, F., and Van Wijk, R.,** Effect of external K^+ on protein and DNA synthesis during and after heat shock in rat hepatoma cells, *Int. J. Hyperthermia,* 1, 255—263, 1985.
4a. **Brkic, G., Tuschl, H., Kovac, R., and Altmann, H.,** Heat shock potentiates the effect of mitomycin-C on SCE induction, *Stud. Biophys.*, 120, 227—234, 1987.
5. **Brunet, S. and Giacomoni, P. U.,** Heat shock messenger RNA in mouse epidermis after UV irradiation, *Mutation Res.*, 219, 217—224, 1989.
5a. **Chreno, O., Krajcovic, J., Ebringer, L., and Polonyi, J.,** Effect of heat shock on the mutagenicity of mutagens and carcinogens in *Euglena gracilis, Teratogen. Carcinogen. Mutagen.*, 8, 161—168, 1988.
5b. **Chu, G. L. and Dewey, W. C.,** Effect of cycloheximide on heat induced cell killing, radiosensitization, and loss of cellular DNA polymerase activities in Chinese hamster ovary cells, *Radiat. Res.*, 112, 575—580, 1987.
6. **Clark, E. P., Dewey, W. C., and Lett, J. T.,** Recovery of CHO cells from hyperthermic potentiation to X rays: Repair of DNA and chromatin, *Radiat. Res.*, 85, 302—313, 1981.
6a. **Coias, R., Galego, L., Barchona, I., and Rodrigues-Pousada, C.,** Destabilization of tubulin mRNA during heat-shock in *Tetrahymena pyriformis, Eur. J. Biochem.*, 175, 467—474, 1988.
7. **Coss, R. A., Dewey, W. C., and Bamburg, J. R.,** Effects of hyperthermia on dividing Chinese hamster ovary cells and on microtubules in vitro, *Cancer Res.*, 42, 1059—1071, 1982.
8. **Davis, J. Q., Browden, G. T., and Cress, A. E.,** The effect of heat and radiation on the initiation and elongation processes of DNA synthesis, *Int. J. Radiat. Biol.*, 43, 379, 1983.
9. **Dewey, W. C. and Esch, J. L.,** Transient thermal tolerance: Cell killing and DNA polymerase activities, *Radiat. Res.*, 92, 611—614, 1982.
10. **Evenson, D. P. and Prescott, D. M.,** Disruption of DNA synthesis in *Euplotes* by heat shock, *Exp. Cell Res.*, 63, 245—252, 1970.
11. **Gerner, E. W. and Russel, D. H.,** The relationship between polyamine accumulation and DNA replication in synchronized Chinese hamster ovary cells after heat shock, *Cancer Res.*, 37, 482—489, 1977.
11a. **Gottesman, S.,** Genetics of proteolysis in *Escherichia coli, Annu. Rev. Genet.*, 23, 163—198, 1989.
12. **Graziosi, G., Micali, F., Marzari, R., Decristini, F., and Savoini, A.,** Variability of response of early *Drosophila* embryos to heat shock, *J. Exp. Zool.*, 214, 141—145, 1980.
13. **Graziosi, G., Decristini, F., Di Marcotullio, A., Marzari, R., Micali, F., and Savoini, A.,** Morphological and molecular modifications induced by heat shock in *Drosophila melanogaster* embryos, *J. Embryol. Exp. Morphol.*, 77, 167—182, 1983.
14. **Johnston, G. C. and Singer, R. A.,** Ribosomal precursor RNA metabolism and cell division in the yeast *Saccharomyces cerevisiae, Mol. Gen. Genet.*, 178, 357—360, 1980.
15. **Jorritsma, J. B. M. and Konings, A. W. T.,** Inhibition of repair of radiation-induced strand breaks by hyperthermia, and its relationship to cell survival after hyperthermia alone, *Int. J. Radiat. Biol.*, 43, 505—516, 1983.
16. **Junakovic, N., Di Franco, C., Barsanti, P., and Palumbo, G.,** Transposition of copia-like nomadic elements can be induced by heat shock, *J. Mol. Evol.*, 24, 89—93, 1987.
17. **Kal, H. B., Hatfield, M., and Hahn, G. M.,** Cell cycle progression of murine sarcoma cells after X irradiation or heat shock, *Radiology,* 117, 215—217, 1975.
18. **Kase, K. R. and Hahn, G. M.,** Comparison of some responses to hyperthermia by normal human diploid cells and neoplastic cells from the same origin, *Eur. J. Cancer,* 12, 481—491, 1976.
19. **Kramhoft, B. and Zeuthen, E.,** Synchronization of cell division in the fission yeast, *Schizosaccharomyces pombe,* using heat shocks, *C. R. Trav. Lab. Carlsberg,* 38, 351—368, 1971.
20. **Li, G. C. and Mivechi, N. F.,** Thermotolerance in mammalian systems: A review, in *Hyperthermia in Cancer Treatment,* Anghilert, L. J. and Robert, J., Eds., CRC Press, Boca Raton, FL, 1986, 59—77.
21. **Lin, P. S., Hefter, K., and Jones, M.,** Hyperthermia and bleomycin schedules on V79 Chinese hamster cell cytotoxicity in vitro, *Cancer Res.*, 43, 4557—4561, 1983.
22. **Lomagin, A. G.,** Repair of functional and ultrastructural alterations after thermal injury of *Physarum polycephalum, Planta,* 142, 123—134, 1978.

23. **Ludwig, J. R., Foy, J. J., Elliot, S. G., and McLaughlin, C. S.,** Synthesis of specific identified, phosphorylated, heat shock and heat stroke proteins through the cell cycle of *Saccharomyces cerevisiae, Mol. Cell. Biol.*, 2, 117—126, 1982.

23a. **Lytle, C. D. and Carney, P. G.,** Heat shock and Herpes virus: enhanced reactivation without untargeted mutagenesis, *Environ. Mol. Mutagen.*, 12, 201—208, 1988.

23b. **Mackey, M. A., Morgan, W. F., and Dewey, W. C.,** Nuclear fragmentation and premature chromosome condensation induced by heat shock in S-phase Chinese hamster ovary cells, *Cancer Res.*, 48, 6478—6483, 1988.

23c. **Macor, M., Ebringer, L., and Siekel, P.,** Hyperthermia and other factors increasing sensitivity of *Euglena* to mutagens and carcinogens, *Teratogen. Carcinogen. Mutagen.*, 5, 329—337, 1985.

24. **Maga, J. A., McClanahan, T. A., and McEntee, K.,** Transcriptional regulation of DNA damage responsive DDR genes in different RAD mutant strains of *Saccharomyces cerevisiae, Mol. Gen. Genet.*, 205, 276—284, 1986.

25. **Martin, R. J. and Schloerb, P. R.,** Induction of mitotic synchrony by intermittent hyperthermia in the Walker 256 rat carcinoma, *Cancer Res.*, 24, 1997—2000, 1964.

26. **Matsumoto, S., Gugg, S., and Lafontaine, J. G.,** Ultrastructural investigation of the effects of supraoptimal temperature on late interphase nuclei of *Physarum polycephalum* plasmodia, *Biol. Cell,* 60, 87—96, 1987.

27. **McClanahan, T. and McEntee, K.,** DNA damage and heat shock dually regulate genes in *Saccharomyces cerevisiae, Mol. Cell. Biol.*, 6, 90—96, 1986.

27a. **McClintock, B.,** The significance of responses of the genome to challenge, *Science,* 226, 792—801, 1984.

28. **Michaelis, A., Takehisa, S., Rieger, R., and Aurich, O.,** Ammonium chloride and zinc sulfate pretreatments reduce the yield of chromatid aberrations induced by triethylene amine and maleic hydrazide in *Vicia faba Mutat. Res.*, 173, 187—192, 1986.

29. **Mills, M. D. and Meyn, R. E.,** Effects of hyperthermia on repair of radiation induced DNA strand breaks, *Radiat. Res.*, 87, 314—328, 1981.

30. **Mitchel, R. E. J. and Morrison, D. P.,** Heat-shock induction of ionizing radiation resistance in *Saccharomyces cerevisiae,* and correlation with stationary growth phase, *Radiat. Res.*, 90, 284—291, 1982.

31. **Mitchel, R. E. J. and Morrison, D. P.,** Heat-shock induction of ionizing radiation resistance in *Saccharomyces cerevisiae*. Transient changes in growth cycle distribution and recombinational ability, *Radiat. Res.*, 92, 182—187, 1982.

32. **Mitchel, R. E. J. and Morrison, D. P.,** Assessment of the role of oxygen and mitochondria in heat shock induction of radiation and thermal resistance in *Saccharomyces cerevisiae, Radiat. Res.*, 96, 113—117, 1983.

33. **Mitchel, R. E. J. and Morrison, D. P.,** Heat-shock induction of ultraviolet-light resistance in *Saccharomyces cerevisiae, Radiat. Res.*, 96, 95—99, 1983.

34. **Mitchel, R. E. J. and Morrison, D. P.,** Inducible error-prone repair in yeast. Suppression by heat shock, *Mutat. Res.*, 159, 31—39, 1986.

35. **Mittler, S.,** Effect of hyperthermia upon gamma-ray induced crossing-over in an excision repair deficient male *Drosophila melanogaster, Experientia,* 43, 931—932, 1987.

36. **Mivechi, N. F. and Dewey, W. C.,** DNA polymerase alpha and beta activities during the cell cycle and their role in heat radiosensitization in Chinese hamster ovary cells, *Radiat. Res.*, 103, 337—350, 1985.

37. **Mosser, D. D., Heikkila, J. J., and Bols, N. C.,** Temperature ranges over which rainbow trout fibroblasts survive and synthesize heat-shock proteins, *J. Cell. Physiol.*, 128, 432—440, 1986.

37a. **Pardasani, D. and Fitt, P. S.,** Strain-dependent induction by heat shock of resistance to ultraviolet light in *Escherichia coli, Curr. Trends Microbiol.*, 18, 99—103, 1989.

37b. **Pardasani, D., Sharma, N., and Fitt, P. S.,** Dependence on the lon gene of the thermal induction of resistance to UV light in *Escherichia coli, Curr. Trends Microbiol.*, 19, 129—134, 1989.

38. **Piperakis, S. M. and McLennan, A. G.,** Enhanced reactivation of UV-irradiated adenovirus 2 in HeLa cells treated with nonmutagenic chemical agents, *Mutat. Res.*, 142, 83—86, 1985.

39. **Plesset, J., Ludwig, J. R., Cox, B. S., and McLaughlin, C. S.,** Effect of cell cycle position on the thermotolerance in *Saccharomyces cerevisiae, J. Bacteriol.*, 169, 779—784, 1987.

40. **Polanshek, M. M.,** Effects of heat shock and cycloheximide on growth and division of the fission yeast *Schizosaccharomyces pombe, J. Cell Sci.*, 23, 1—23, 1977.

41. **Radford, I. B.,** Effects of hyperthermia on the repair of X-ray induced DNA double strand breaks in mouse L cells, *Int. J. Radiat. Biol.*, 5, 551—557, 1983.

42. **Rao, P. N. and Engelberg, J.,** HeLa cells: Effects of temperature on the life cycle, *Science,* 148, 1092—1094, 1965.

43. **Rao, P. N. and Engelberg, J.,** Effects of temperature on the mitotic cycle of normal and synchronized mammalian cells, in *Cell Synchrony,* Cameron, I. L. and Padilla, G. M., Eds., Academic Press, Orlando, FL, 1966, 332—352.

44. **Read, R. A., Fox, M. H., and Bedford, J. S.,** The cell cycle dependence of thermotolerance. I. CHO cells heated at 42°C. *Radiat. Res.*, 93, 93—106, 1983.

45. **Rice, G., Laszlo, A., Li, G., Gray, J., and Dewey, W.,** Heat shock proteins within the mammalian cell cycle: Relationship to thermal sensitivity, thermal tolerance and cell cycle progression, *J. Cell. Physiol.,* 126, 291—297, 1986.

45a. **Rieger, R.,** Stress-induced protective effects against DNA damage, in *Heat Shock and Other Stress Response Systems of Plants,* Nover, L., Neumann, D., and Scharf, K.-D., Eds., Springer-Verlag, Berlin, 1990, 105—112.

45b. **Rieger, R. and Michaelis, A.,** Heat shock protection against induction of chromatid aberrations is dependent on the time span between heat shock and clastogen treatment of *Vicia faba* root tip meristem cells, *Mutat. Res.,* 209, 141—144, 1988.

46. **Rieger, R., Michaelis, A., and Schubert, I.,** Heat-shock prior to treatment of *Vicia faba* root tip meristems with maleic hydrazide or TEM reduce the yield of chromatid aberrations, *Mutat. Res.,* 143, 79—82, 1985.

47. **Rieger, R., Michaelis, A., and Nicoloff, H.,** Effects of stress factors on the clastogen response of *Vicia faba* root tip meristems — Clastogenic adaptation, *Biol. Zbl.,* 105, 19—28, 28, 1986.

48. **Rieger, R., Michaelis, A., and Schubert, I.,** Reduction by heat shock of maleic hydrazide-induced aberration yield is dependent on temperature and duration of heat pretreatment, *Mutat. Res.,* 174, 199—204, 1986.

49. **Ron, A. and Zeuthen, E.,** Tubulin synthesis and heat-shock-induced cell synchrony in *Tetrahymena, Exp. Cell Res.,* 128, 303—309, 1980.

50. **Roti-Roti, J. L., Uygur, N., and Higashikubo, R.,** Nuclear protein following heat shock: Protein removal kinetics and cell cycle rearrangements, *Radiat. Res.,* 107, 250—261, 1986.

51. **Sapareto, S. A., Hopwood, L. E., Dewey, W. C., Raju, M. R., and Gray, J. W.,** Effects of hyperthermia on survival and progression of Chinese hamster ovary cells, *Cancer Res.,* 38, 393—400, 1978.

52. **Schaefer, E. L., Morimoto, R. I., Theodorakis, N. G., and Seidenfeld, J.,** Chemical specificity for induction of stress-response genes by DNA-damaging drugs in human adenocarcinoma cells, *Carcinogenesis,* 9, 1733—1738, 1988.

53. **Scharf, K.-D. and Nover, L.,** Heat shock induced alterations of ribosomal protein phosphorylation in plant cell cultures, *Cell,* 30, 427—437, 1982.

54. **Scharf, K.-D. and Nover, L.,** unpublished.

55. **Schenberg-Frascino, A. and Monstacchi, E.,** Lethal and mutagenic effects of elevated temperatures on haploid yeast. I. Variations in sensitivity during cell cycle, *Mol. Gen. Genet.,* 115, 243—257, 1972.

56. **Scherbaum, O. H. and Zeuthen, E.,** Induction of synchronous cell division in cultures of *Tetrahymena pyriformis, Exp. Cell Res.,* 6, 221—227, 1954.

57. **Schubert, I., Rieger, R., and Michaelis, A.,** Effects of G2-repair inhibitors on "clastogenic adaptation" in *Vicia faba, Mol. Gen. Genet.,* 204, 174—179, 1986.

57a. **Sherwood, S. W., Dagett, A. S., and Schimke, R. T.,** Interaction of hyperthermia and metabolic inhibitors on the induction of chromosome damage in Chinese hamster ovary cells, *Cancer Res.,* 47, 3584—3588, 1987.

58. **Shimada, Y.,** Heat-shock induction of radiation resistance in primordial germ cells of the fish, *Oryzias latipes, Int. J. Radiat. Biol.,* 48, 189—196, 1985.

59. **Shimada, Y.,** Influence of the thermal conditioning on the heat-induced radioresistance in primordial germ cells of the fish *Oryzia latipes, Int. J. Radiat. Biol.,* 48, 423—430, 1985.

60. **Shimada, Y.,** Induction of thermotolerance in fish embryos *Oryzias latipes, Comp. Biochem. Physiol.,* 80A, 177—181, 1985.

61. **Shimada, Y., Shima, A., and Egami, N.,** Effects of heat, release from hypoxia, cadmium and arsenite on radiation sensitivity of primordial germ cells in the fish *Oryzias latipes, J. Radiat. Res.,* 26, 411—417, 1985.

62. **Shimada, Y., Shima, A., and Egami, N.,** Effects of dose fractionation and cycloheximide on the heat-shock induction of radiation resistance in primordial germ cells of the fish *Oryzias latipes, Radiat. Res.,* 104, 78—82, 1985.

63. **Sisken, J. E., Morasca, L., and Kirby, S.,** Effects of temperature on the kinetics of the mitotic cycle of mammalian cells in culture, *Exp. Cell Res.,* 39, 103—116, 1965.

64. **Spiro, E. J., Denman, D. L., and Dewey, W. C.,** Effect of hyperthermia on CHO DNA polymerases alpha and beta, *Radiat. Res.,* 89, 134—149, 1982.

65. **Stephanou, G. and Demopoulos, N. A.,** Heat shock phenomena in *Aspergillus nidulans.* II. Combined effects of heat and bleomycin on heat shock protein synthesis, survival rate and induction of mutations, *Curr. Genet.,* 12, 443—448, 1987.

66. **Strand, D. J. and McDonald, J. F.,** Copia is transcriptionally responsive to environmental stress, *Nucl. Acids Res.,* 13, 4401—4410, 1985.

66a. **Tanaka, S., Ishii, C., and Inoue, H.,** Effects of heat shock on the induction of mutations by chemical mutagens in *Neurospora crassa, Mutation Res.,* 223, 233—242, 1989.

67. **Taylor, S. Y. and Holliday, R.,** Induction of thermotolerance and mitotic recombination by heat-shock in *Ustilago maydis, Curr. Genet.,* 9, 59—64, 1984.

68. **Van Dongen, G. and Van Wijk, R.,** Evidence for a role of heat shock proteins in proliferation after heat shock treatment of synchronized mouse neuroblastoma cells, *Radiat. Res.*, 113, 252—267, 1988.
69. **Van Dongen, G., Van De Zande, L., Schamhart, D., and Van Wijk, R.,** Comparatative studies on the heat-induced thermotolerance of protein synthesis and cell division in synchronized mouse neuroblastoma cells, *Int. J. Radiat. Biol.*, 46, 759—769, 1984.
70. **Vig, B. K.,** Hyperthermic enhancement of chromosome damage and lack of effect on sister-chromatid exchange induced by bleomycin in Chinese hamster cells in vitro, *Mutat. Res.*, 61, 309—317, 1979.
71. **Vivino, A. A., Smith, M. D., and Minton, K. W.,** A DNA damage-responsive *Drosophila melanogaster* gene is also induced by heat shock, *Mol. Cell. Biol.*, 6, 4767—4769, 1986.
72. **Walker, G. C.,** Inducible DNA repair system, *Annu. Rev. Biochem.*, 54, 425—457, 1985.
72a. **Walsh, D. A. and Morris, V. B.,** Heat shock affects cell cycling in the neural plate of cultured rat embryos: A flow cytometric study, *Teratology,* 40, 583—592, 1989.
73. **Warters, R. L. and Roti Roti, J. L.,** Production and excision of 5′6′-dihydroxy-dihydrothymine type products in the DNA of preheated cells, *Int. J. Radiat. Res.*, 34, 381—384, 1978.
74. **Warters, R. L. and Roti Roti, J. L.,** Excision of X-ray-induced thymine damage in chromatin from heated cells, *Radiat. Res.*, 79, 113—121, 1979.
75. **Warters, R. L. and Stone, O. L.,** Macromolecule synthesis in HeLa cells after thermal shock, *Radiat. Res.*, 96, 646—652, 1983.
76. **Warters, R. L., Brizgys, L. M., and Lyons, B. W.,** Alteration in the nuclear matrix protein mass correlates with heat-induced inhibition of DNA single-strand-break repair, *Int. J. Radiat. Biol.*, 52, 299—314, 1987.
76a. **Williams, K. J., Landgraf, B. E., Whiting, N. L., and Zurlo, J.,** Correlation between the induction of heat shock protein-70 and enhanced viral reactivation in mammalian cells treated with ultraviolet light and heat shock, *Cancer Res.*, 49, 2735—2742, 1989.
77. **Wong, R. S. L. and Dewey, W. C.,** Molecular studies on the hyperthermic inhibition of DNA synthesis in Chinese hamster ovary cells, *Radiat. Res.*, 92, 370—395, 1982.
78. **Wong, R. S. L. and Dewey, W. C.,** Effect of hyperthermia on DNA synthesis, in *Hyperthermia in Cancer Treatment,* Anghileri, J. L. and Robert, J., Eds., CRC Press, Boca Raton, FL, 1986, 79—91.
79. **Yager, J. D., Zurlo, J., and Penn, A. L.,** Heat-shock-induced enhanced reactivation of UV-irradiated herpesvirus, *Mutat. Res.*, 146, 121—128, 1985.
80. **Zeuthen, E.,** Synchrony in *Tetrahymena* by heat shocks spaced a normal cell generation appart, *Exp. Cell Res.*, 68, 49—60, 1971.

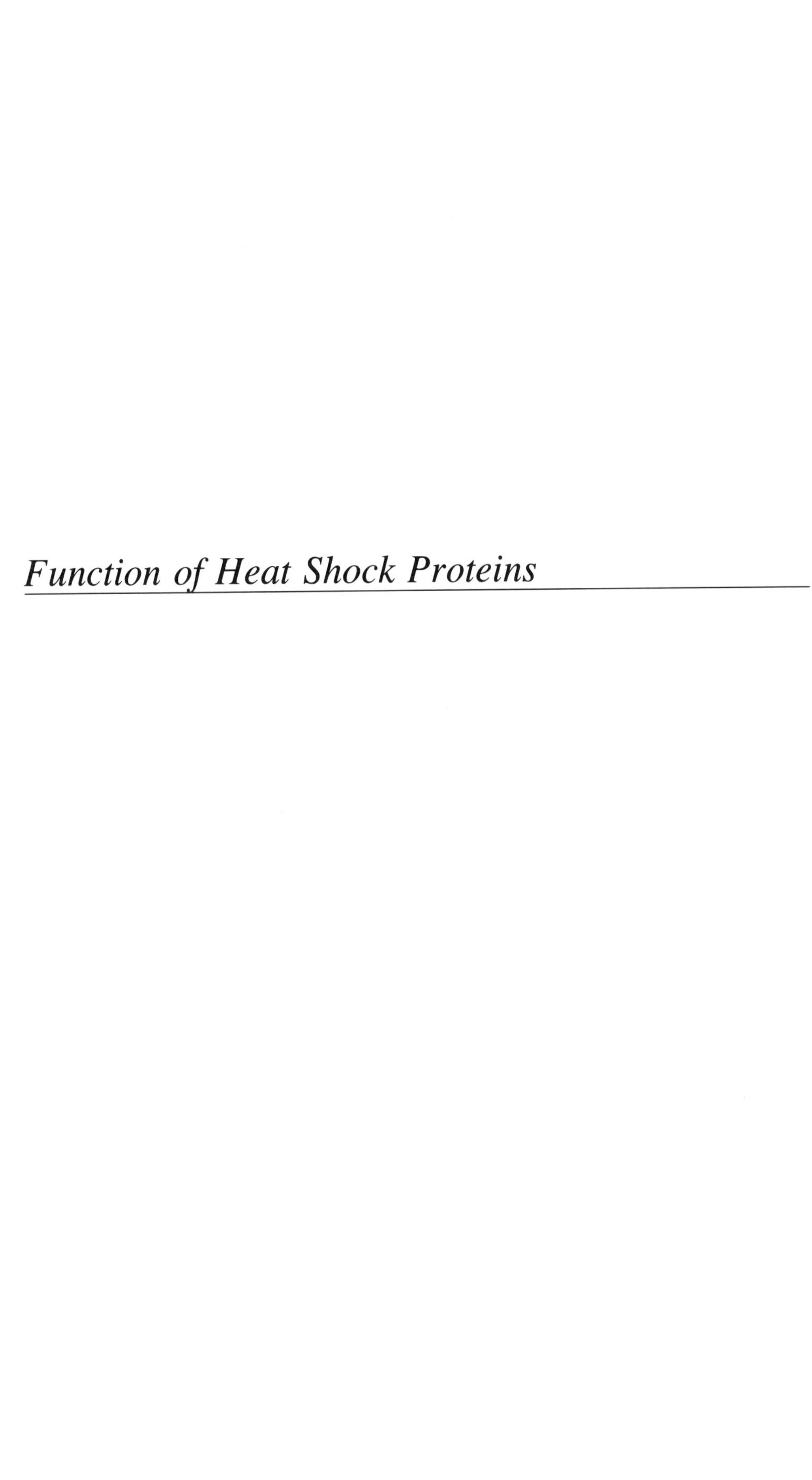

Function of Heat Shock Proteins

Chapter 16

INTRACELLULAR LOCALIZATION AND RELATED FUNCTIONS OF HEAT SHOCK PROTEINS

L. Nover, D. Neumann, and K.-D. Scharf

TABLE OF CONTENTS

16.1. METHODS AND PROBLEMS

Studies on the intracellular localization of heat shock proteins (HSPs) are an important part of the search for their possible functions within the stressed cell. Three general aspects emerge (for a summary, see Neumann et al.[83]). On the one hand, protein aggregation during stress is observed in different parts of the cell. HSPs are frequently involved and may be essential for the reversibility of this process. On the other hand, members of the HSP families act as catalysts or helper proteins in proteinogen processing, intracellular topogenesis, and assembly of heteromeric protein complexes (see Section 12.2). Finally, they modulate the function of regulatory or enzymatic proteins. Many papers have contributed to our knowledge in this field (Table 16.1); however, the experimental approach to this end encounters three basic problems:

1. Only part of the results concern stress-specific, mostly reversible associations of HSPs, e.g., HSP70 in the nucleus/nucleolus (Section 16.2) or the small HSPs in heat shock granules (hsg) (Section 16.7). Frequently, however, binding to a given compartment or structure is not *a priori* stress-related. It was rather defined for constitutive members of the HSP families under control temperature conditions, e.g., HSP110 in the nucleoli of vertebrates, the HSP90 as part of steroid hormone receptor complexes (Section 16.8), GRPs 94 and 78 in the ER/Golgi system (Sections 4.3 and 16.4) or members of the HSP70 family associated with cytoskeletal elements or onc gene products (Sections 16.6 and 16.9). Unfortunately, only in some of the latter examples, which include well-defined heterologous protein complexes, we have a fairly detailed knowledge of the function of the HSP involved. In contrast, although the hs-dependent translocation of HSPs to nuclei and hs granules correlates with cellular structures or functions protected from heat damage, there is no proof for a causal connection between both events. In the following parts, we briefly comment on selected data from Table 16.1 and try to correlate them with the corresponding analyses on induced thermotolerance (see Chapter 17).
2. The second type of problem results from limitations of the methods used to study intracellular localization of proteins in general. Abbreviations for the method applied in a given paper are indicated in Table 16.1. Cell fractionation procedures (CF) disrupt the native state of cellular organization and may be connected with the loss of proteins from sensitive structures and unspecific gain respectively. Yet, provided careful control of the conditions and integrity of purified subcellular fractions, it can give valuable insights into intracellular localization of proteins. An illustrative example of difficulties encountered with cell fractionation is the equivocal localization of small HSPs in the nucleus and other organelles of *Drosophila* and plant cells (see Section 16.7). In fact, most of them are associated with a very stabile perinuclear complex of the collapsed cytoskeleton with hsg. To improve reliability of cell fractionation, the procedure was coupled with *in situ* cross-linking, e.g., of RNP material,[50] prior to the disruption of the cell structure or with immunocoprecipitation (IP) of HSPs in a complex with other proteins (see Sections 16.8 and 16.9). In addition, to discriminate between proteins residing within chloroplasts, mitochondria, or ER vesicles and those adhering to them from outside, treatment with trypsin is a valuable tool. To this aim, immobilized trypsin was even used for nuclear fractions.[28] Under appropriate experimental conditions, even trace amounts of a given protein can be detected by cell fractionation procedures.

   In contrast to this are methods for *in situ* localization in the intact cell. On the one hand, a peculiarity of the heat shock (hs) response, i.e., the almost exclusive labeling of HSPs under conditions of severe hs (Figure 2.1), allows the direct local-

TABLE 16.1
Intracellular Localization of Heat Shock Proteins

Organism/cell	Method	Remarks	Ref.
A: Nucleus, nucleolus			
Saccharomyces cerevisiae (yeast)	IF	HSP26 accumulates in nucleus after hs, depends on culture conditions	108b
Achlya ambisexualis	CF	HSPs 43 and 23—28 enriched in nuclear fraction	122
Tetrahymena pyriformis	CF	HSPs 29c, 73, 75a, and 75b are bound to nucleus under hs	43
Dictyostelium discoideum	CF	8 small HSPs of 26—32 kDa firmly bound to chromatin	67
Drosophila melanogaster			
Larval salivary glands	IF	HSP70 in nucleus under stress conditions, function for hnRNP and chromatin stability?	137
Primary embryonic cells	IF	HSP70 and HSC70 migrate to nucleus under stress conditions	94
Cell cultures (Kc, S2 cells)	IF	HSP70 in nucleolus under hs	131a
	AR	Major HSP (HSP70?) in nucleus/nucleolus of stressed cells	7, 138
	CF[a]	HSPs 70, 68, 27, 26, 23, and 22 firmly bound to nuclear fractions (constituents of nuclear matrix) of hs cells	7, 60, 74, 123, 133, 142
	CF, UV-cross-linking[a]	HSPs 70, 68 plus small HSPs are part of hnRNP	50
	IF, CF	HSP27 in nucleus, stably bound during hs, loosely bound at 25°C	10a, 10c
Chironomus tentans, salivary glands	CF	HSPs 68 and 34 in nuclei, isolated by microdissection	132, 141
Strongylocentrotus lividus, embryos	CF	HSP70 in nucleus of hs embryos	108
Chicken embryo fibroblasts	CF, IF	HSP70 in nucleus of stressed cells, removed by Nonidet treatment from nucleoplasm but not from nucleolus	28
Mouse neuroblastoma cells	IF, I-Au	HSP90 in nucleus during long-term hs	136
Mouse L cells	IF	*Drosophila* HSP70, expressed in mouse cells, is localized in nucleolus	75

TABLE 16.1 (continued)
Intracellular Localization of Heat Shock Proteins

Organism/cell	Method	Remarks	Ref.
Rat embryo fibroblasts	IF, Microinjection, I-Au	HSP72 and HSC73 migrate to nucleus during stress	58, 148, 149, 150, 151
BHK cells	CF	HSP70 bound to nuclear ''matrix'' (nucleolus?) of hs cells	101, 148, 149
CHO cells	IF	HSP70 in nucleolus, HSC72 in nucleoplasma of hs cells	90
Gerbil fibroma cells, rat kidney cells	IF	HSP72 in nucleolus of hs cells, but not after arsenite or AzC stress	148, 149
Rabbit retina	CF	HSP74 associated with nuclear fraction in recovery period after hs	25
Monkey COS cells	IF	*Drosophila* HSP70 expressed in COS cells is localized in nucleolus, improved recovery of nucleolar function	75, 96
	IF, CF	HSPs72/73 form complexes with mutant p53 found in nucleus	126
Human foreskin fibroblasts	IF, I-PO	Part of HSP27 in large aggregates in nuclei of hs cells	9a
HeLa cells	IF, CF	HSP72 in nucleolus of hs cells	17, 91, 150
	IF	HSP70 synthesized in S-phase of cell cycle transiently accumulates in nucleus	73
CHO cells, mouse liver and brain cells	IF	HSP110 constitutive major protein of nucleoli	16, 121, 128
Tomato cell cultures	IF, I-Au, AR	HSP70 bound to preribosomal material in nucleoli of hs cells	81—83, 87, 88
Soybean seedling	CF[a]	HSP70 and small HSPs in nuclear fraction	64
B: Cell periphery, plasma membrane (PM)			
Drosophila Kc cells	IF	HSP83 concentrated in region of PM in hs cells	19
Rat brain, mouse 3T3 cells	CF, IP	Mirotubule-associated HSP68 bound to plasma membrane coprecipitates with PM glycoprotein of 90 kDa	47, 63
Mouse, human cells	IF	GRP95 associated with PM and pericellular matrix fibers	69
Adenovirus transformed hamster cells	IP	HSP74 coprecipitated with antibodies against tumor T and S antigens bound to cell membrane	105

	Maturing mammalian reticulocytes	CF	HSPs 71/72 associated with transferrin receptor containing vesicles to be secreted	31, 32, 95
	Mammalian cells	CF, IP	HSC71—73 is the clathrin-uncoating ATPase	23, 45e, 109, 135
	Soybean seedlings	CF	HSP15 but no other HSPs associated with PM fraction	65
	Maize root	CF	HSPs 70 and 18 cosediment with PM fraction	30
C:	ER/Golgi system (see Chapter 4)			
	Mammalian cells	IF	GRP78 major sol. protein of ER, synthesized as precursor protein with N-terminal leader sequence	22, 76
	Lymphoid cells	IP	GRP78 transiently associates with newly formed Ig H-chains in ER	13, 44
		IP	"GRP78" associates with aberrant proteins in ER	34, 45b, 120
	Mammalian cells	CF, IF	GRP94 is a major glycoprotein of ER/Golgi	54, 62a, 125, 152
	Maize seedling roots	CF	HSPs 72 and 25 associated with ER fraction	30
	Yeast	CF, IF	GRP78 (=KAR2 protein) in perinuclear ring of ER/Golgi	108a
D:	Mitochondria			
	Tetrahymena thermophila	IF, CF	HSP58 forms 20—25S particles in mitochondria, homologous to Ec-GroEL	70, 71
	Yeast, *Xenopus*, mammalian and plant cells	CF, WB	Member of the HSP60 family detected by cross-reaction with antibodies to Tt-HSP58	23b, 70, 71, 98, 101a, 101d
	Yeast, human cells, trypanosomes, nematode	CF, WB	Constitutive DnaK-like members of the HSP70 family identified and genes sequenced	30a, 37a, 45d, 59a
	Plant leaves, roots, and suspension cultures	I-Au, WB	HSP68 constitutive and hs-induced mitochondrial protein (DnaK-like)	83, 83a
	Soybean seedling	CF	HSPs 15—18, 22, and 24 detected in purified mitochondria, related to stability of mitochondria at 42.5°C	23c, 64
	Zea mays seedling	CF	Isolated mitochondria synthesize HSP60 after 40 min hs at 37°C; result may be due to bacterial contamination[84]	80, 124
		CF	HSP29 in mitochondria	30
	Neurospora crassa	CF	HSPs 34 and 30 in mitochondria	98
		CF, IP	HSP60 (GroEL) identified in 12-nm particles	47a

TABLE 16.1 (continued)
Intracellular Localization of Heat Shock Proteins

	Organism/cell	Method	Remarks	Ref.
E:	Chloroplasts			
	Phaseolus vulgaris	CF	HSP22 is part of stroma	129
	Pisum sativum (pea), seedling	CF	Two HSPs 21 are part of chloroplast membranes under hs, but in stroma under control condition, both proteins synthesized as precursor (30 and 26 kDa) with N-terminal transit peptide	23a, 41, 52, 139
		CF, WB	Three constitutive members of HSP70 family in outer membrane and stroma	68b
	Lycopersicon esculentum, leaves	I-Au	HSPs 21 and 22 in stroma, independent of stress conditions	83
	Glycine max (soybean), seedling	CF	HSP22 chloroplast protein, formed as 28 kDa precursor with N-terminal transit peptide	139, 140, 140a
	Vigna sinensis, *Sorghum vulgare*	CF	Isolated chloroplasts synthesize HSPs 85, 70, 60 and 23; HSP60 is the major constituent	57
	Chlamydomonas reinhardi	CF	HSP22 is tightly bound to grana lamellae, no precursor protein detected; may be involved in protection of photosystem II	42, 52, 118
	Acetabularia mediterranea	CF	Major HSP70 is coded and localized in chloroplasts	53
	Euglena gracilis, pea, spinach, maize	CF, WB	DnaK-like HSP70 indentified by immunological methods	2a, 68b
F:	Cytoskeleton			
	Drosophila primary embryonic cells	IF	HSC70 bound to perinuclear cytoskeleton under control conditions, but migrates to nucleus under hs	94
	Avian and mammalian cells	CF, IF	HSP70/HSC73 associated with microtubules and intermediate filament system	24, 58, 63, 114, 144-146, 153
	Rabbit brain, retina	CF	HSC74 associated with purified microtubules and intermediate size filaments	25, 26
	Mammalian cells	CF	HSP70 associated with cytoskeleton in control and hs cells (interfacing proteins of cytoskeletal systems)	41a, 78, 79

	Mouse 3T3 cells	IP	HSP70/HSC72 coprecipitated with membrane 90 kDa glycoprotein	47
	Murine mastocytoma cells	CF	HSP70 copurifies with microtubules	92
	Rat neurons, hamster fibroblasts	CF	Identification of MAP68 with HSC73 (peptide mapping)	152a, 153
	Mammalian cells	CF	HSP90 and GRP94 bind to actin under polymerizing conditions, calmodulin dependent	55, 55a, 85
		IP, IF	HSP90 colocalization with microtubules	101e, 113b
G:	Cytoplasmic RNP aggregates			
	Tetrahymena pyriformis	CF	Small HSPs form aggregates sedimenting at 16,000 × g	43
	Drosophila melanogaster			
	Larvae	CF	Developmentally induced HSPs 23, 26, and 27 contained in 16S RNP	45, 117
	Cell cultures	CF	Small HSPs contained in 19—20S RNP, which form larger aggregates during hs	3, 4, 5, 7
		CF, IF	Under hs small HSPs form large perinuclear aggregates together with vimentin-like protein p46	36a, 59, 131a, 133
	Mammalian cells	CF, IF	HSP27 forms 16S RNP, which under hs aggregate in perinuclear region and nucleus	6, 9a
	Chicken cells	CF, IF, I-Au	HSP24 contained in 16S RNP, forming perinuclear aggregates during hs	28, 29
	Plants			
	Tomato cell cultures	CF, IF, I-Au	HSP70 and small HSPs contained in 16S complexes, which under hs form large perinuclear aggregates (hs granules, see Figure 16.4) with cytoskeletal elements, role for reversible mRNA storage?	81—83, 88, 89
	Different plant species and organs thereof	Cell ultrastructure	Hs granules are formed under hs in all parts of plants	83
H:	Inactive hormone receptor complexes			
	Murine cells	CF, IP	HSP90 is part of nonactivated glucocorticoid receptor complex	72, 72a, 93a, 111—113a, 156
	Chicken oviduct	CF, IP	HSP90 is part of inactive 8S progesterone and estrogen receptor complexes	12, 20, 35, 40, 48, 54a, 103, 110, 116
	Rat liver	CF, IP	HSP90 is part of steroid hormone receptor complex, substrate of type II casein kinase	36

TABLE 16.1 (continued)
Intracellular Localization of Heat Shock Proteins

	Organism/cell	Method	Remarks	Ref.
	Calf uterus	CF	Inactive estrogen receptor complex contains divalent cation and HSP90	101c, 110
	Achlya ambisexualis	CF	HSP85 induced by hs and antheridiol, is part of the antheridiol receptor complex in cytoplasm and nuclear fraction	15, 15a
	Rat liver, human placenta	CF, IP	Dioxin receptor is related to hormone receptor, also complexed with HSP90	33, 153b
I:	Complexes with onc gene products			
	Avian cells	CF, IP	HSP90, together with pp50, participates in a transient maturation complex with the Rous sarcoma virus pp60src (see Figure 16.8) and other retroviral onc-proteins	1, 14, 57b, 66, 92a, 119, 154
		CF, IP	HSP90 and pp50 in long-lived complexes with members of the Tyr kinase oncogene family (src, fes, fgr)	156
	Cat embryo fibroblasts	CF, IP	HSP90 and pp50 form stable complexes with Tyr kinase fusion proteins of feline sarcoma virus (gag-fes, gag-fgr)	155
	Rat cells, monkey COS cells	IP	HSP70/HSC74 form complexes with transformation-associated p53, higher affinity for mutant p53	27, 38, 46, 100, 126
	Mouse 3T3 cells, human 293 cells	CF, IP	HSC73 associated with mutant polyoma medium T-antigen	93c, 143
	Mouse 3T3 cells	IP	HSC73 forms high salt-resistant complexes with SV40 large T-antigen, dissociated by ATP	113c
	Escherichia coli	CF, IP	p53 forms complexes with DnaK in cells transformed with p53 gene	27

Note: The following abbreviations for methods are used: CF, cell fractionation; IF, immunofluorescence; AR, autoradiography; I-Au, I-PO, immunelectron microscopy with protein A-gold or peroxidase, respectively; IP, immune coprecipitation; WB, western blot.

[a] These entries notify papers reporting on the erroneous localization of small HSPs in nuclei; they are part of the perinuclear hsg complex (see text).

ization of the newly formed HSPs by microautoradiography (see Figure 16.4F). On the other hand, individual HSPs are detected by corresponding antibodies either in the light microscope (Figures 16.1 to 16.3) or in the electron microscope (Figure 16.5). These immunological studies can be complemented by microinjection of an isolated and appropriately labeled HSP.[148,150] Usually the immunological methods for *in situ* localization need sufficiently large amounts of the protein to be detected. In addition, the necessary structural conservation for electron microscopy, brought about by fixation with formaldehyde and/or glutaraldehyde, may seriously reduce the antigenicity of HSPs.[81] It is evident that only the time consuming combination of all available methods can help to provide a complete picture on the intracellular locale of HSPs and on the dynamics of their recompartmentation. Only very few examples of such a concise analysis were reported.

3. Finally, the hs response is connected with profound changes of the intracellular protein distribution in general. This is observed in "microscale", e.g., the migration of RNAP II from control genes to hs-activated genes (Section 9.1) or the binding of the activated cytoplasmic heat shock transcription factor (HSF) to nuclear hs promoter sites (Sections 7.4 and 7.5), but also in macroscale. Examples are the reorganization of the cytoskeleton (Section 14.3), the massive accumulation of nonhistone proteins in the chromatin (Section 9.3) or the formation of hsg (Section 16.7). Restoration of the normal pattern of protein distribution is an essential part of the recovery period. It is improved in thermotolerant cells and precedes the recovery of normal cellular functions. As already mentioned, fractionation of hs cells can certainly help to identify trace HSPs, observed only after separation from the bulk of unrelated proteins, but due to the hs-induced redistribution of proteins, new proteins detected in a given compartment need not be hs proteins. Part of the "prompt" HSPs of HeLa cells, described by Reiter and Penman,[102] may in fact represent non-hs proteins enriched by binding to the complex of nuclear matrix with the perinuclear aggregate of the intermediate size filament system.[93] The same may be true for some of the "new proteins" induced in the nuclear matrix of CHO cells.[18]

16.2. NUCLEAR HEAT SHOCK PROTEINS

The reversible binding of HSP70 to the nucleus/nucleolus of hs cells is generally observed (see Table 16.1A). This is illustrated by immunfluorescence pictures for *Drosophila* larvae (Figure 16.1), Chinese hamster HA-1 cells (Figure 16.2) and tomato cell cultures (Figure 16.3, right part). In vertebrate and plant cells, the HSP70 is predominantly in the nucleolus, whereas in *Drosophila* it is in the nucleoplasm. Contrasting to the former two, the antibody used for the experiments with *Drosophila* cells is specific for the HSP70, i.e., the mass of HSC70, contained in control cells, is not detected. The functional difference between both proteins so well as the special demand for HSP70 is not clear. There is preliminary evidence[94] that HSC70 also migrates to the nucleus of stressed *Drosophila* cells. The same is true for the constitutively expressed HSC73 of vertebrates[150] or the human HSP70 expressed in the S-phase of the cell cycle.[73]

It is evident that not the presence of HSP70 per se but rather the particular situation of a stressed cell is responsible for the tight binding of this protein to the nuclear compartment. Induction of HSP synthesis by pulse hs (see Figure 16.3) or by mild chemical stressors, e.g., arsenite[91,142,151] is not connected with nuclear binding. The extent of HSP70 translocation to the nucleus corresponds to the severity of stress[137,151] and, in vertebrates and plants, it is correlated with the changes of nucleolar morphology and function (see Section 10.5).

It is remarkable that the *Drosophila* HSP70, if expressed in mouse cells or monkey COS cells after transformation with the Dm-hsp70 gene, behaves "vertebrate-like", i.e., it is

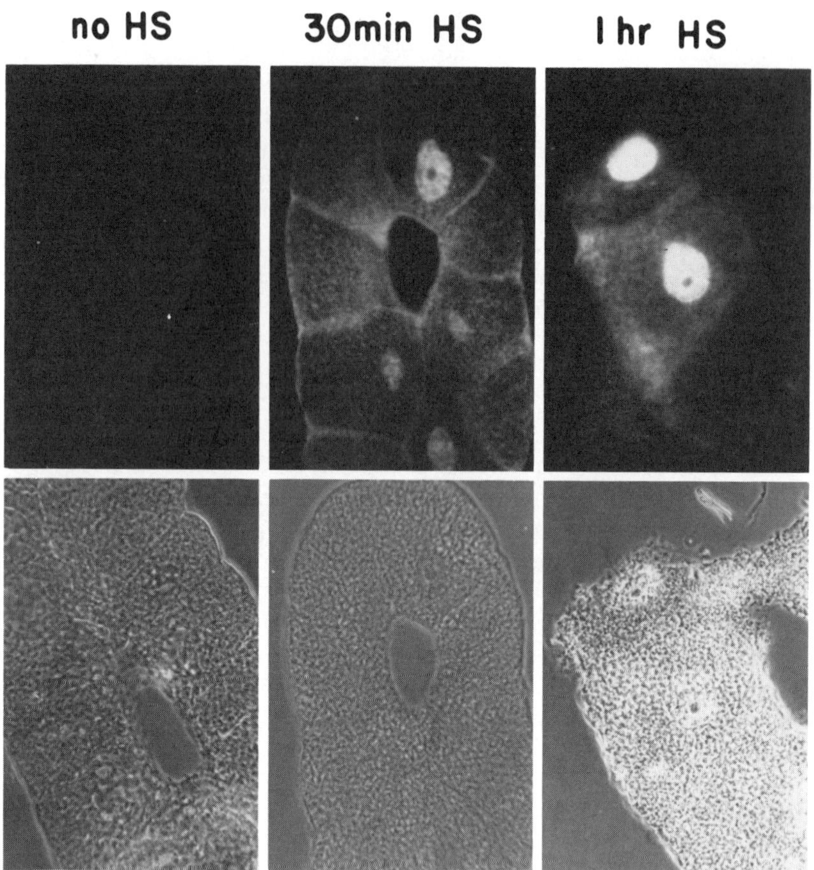

FIGURE 16.1. Immunofluorescence localization of HSP70 in *Drosophila* larvae. Cryostat-sectioned salivary glands were stained by indirect immunofluorescence. Top series shows the distribution of HSP70 in control and heat-shocked cells (30 min and 1 h 36.5°C), whereas the three pictures below represent phase-contrast micrographs of the same sections. Note that HSP70 is virtually absent in control cells and highly concentrates in the nucleus under heat shock (hs) conditions. The antibodies used do not cross-react with the constitutive form of the HSP70 (see Section 2.3.2). (From Velazquez, J. M. and Lindquist, S., *Cell,* 36, 655, 1984. With permission.)

transported to the nucleolus after hs treatment.[75,96] Truncation of the hsp70 gene by removing coding parts for the C terminus and N-terminus, respectively, showed that the nucleolar binding depends on the conserved N-terminal domain and is independent of the stress induced migration to the nucleus.[61,75] Evidently two binding states can be discriminated in COS cells, a loose binding to the nucleoplasm and a very firm binding to the nucleolus. Using an *in vitro* assay with isolated nuclei from hs cells, Lewis and Pelham[61] demonstrated that the release of HSP70 from its nuclear binding sites needs adenosine triphosphate (ATP). There is no release in the presence of adenosine diphosphate (ADP) or nonhydrolyzable analogs of ATP. They discuss a kind of shuttle function for ATP to remove unwanted protein aggregates from the nucleus/nucleolus. This correlates with the property of all members of the HSP70 family to acquire ATPase activity by binding to heterologous proteins[11,97,97a] (see Section 2.3.2). Possible substrates for such a transport are the nucleolar rRNP material (Section 10.5), the excess nonhistone proteins bound to chromatin (Section 9.3), and possibly also nuclear onc-protein complexes (Sections 9.3 and 16.9).

Sequence comparison allows the tentative assignment of two regions in the N-terminal half to the ATPase function and to the nuclear recognition signal, respectively (see Section

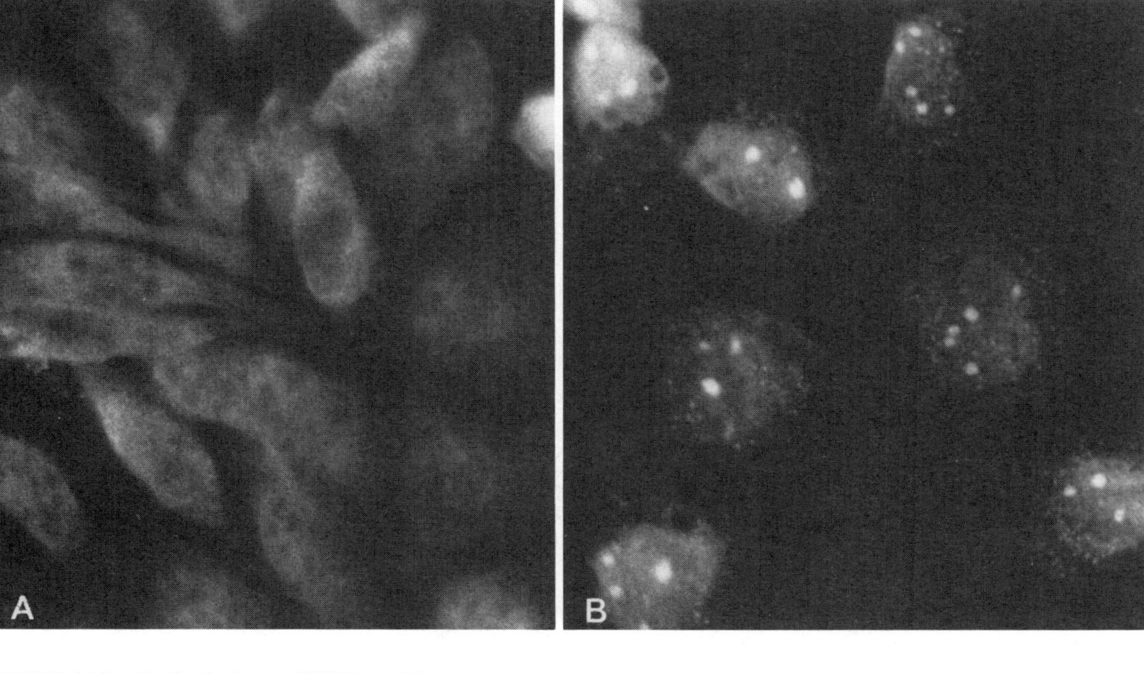

FIGURE 16.2. Redistribution of HSP70 in Chinese hamster HA-1 cells shown by immunofluorescence. HSP70 was detected by appropriate antibodies in control cells (A) and heat-shocked cells (B, 15 min 45°C → 12 h 37°C → 60 min 43°C). B shows the bright nucleolar staining, which contrasts to the general cytoplasmic distribution under control conditions (A). (Courtesy of K. Ohtsuka and A. Laszlo.)

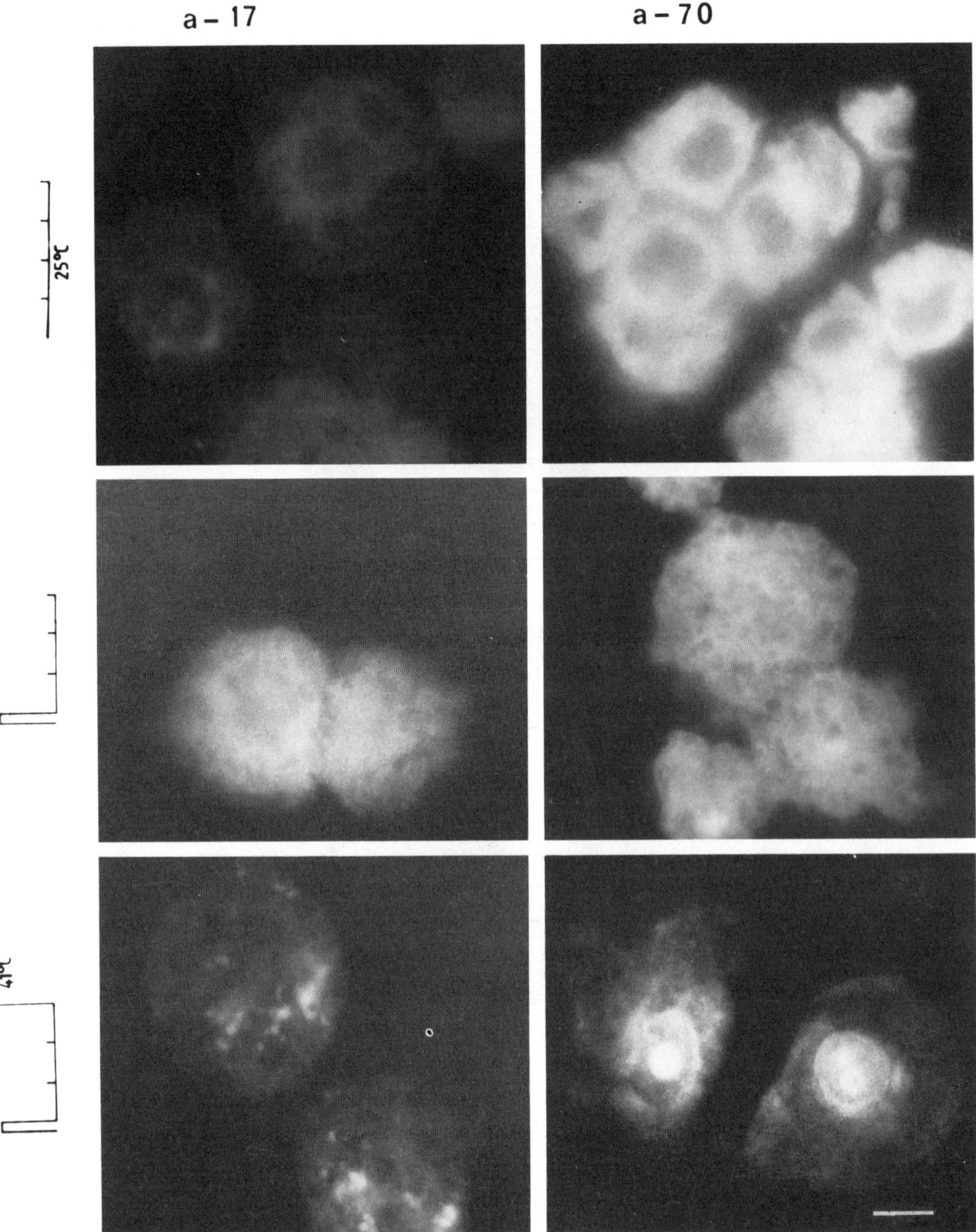

FIGURE 16.3. Immunofluorescence localization of HSP17 and HSP70 in tomato cells. Protoplasts prepared from tomato cell cultures were stained with anti-HSP17 (left) and anti-HSP70 antibodies (right) and counterstained with fluorescein-labeled anti-antibodies. Pretreatment of cultures are illustrated by pictographs on the left margin, i.e., control cells (above), preinduced cells (middle), and hs cells (below). In the latter HSP17 is mainly contained in cytoplasmic granular material (hsg) in the perinuclear region, whereas most of the HSP70 is found in the nucleolus. (From Neumann, D., Zur Nieden, U., Manteuffel, R., Walter, G., Scharf, K.-D., and Nover, L., *Eur. J. Cell Biol.*, 43, 71, 1987. With permission.)

2.3.2 and summary by Neumann et al.[83]). Interestingly, the latter is identified by an uninterrupted sequence of six basic amino acid residues and, together with the following 20 amino acid residues, exhibits homology to certain homeobox-containing nuclear transcription factors (see Section 2.3.2). Transfer of the corresponding peptide to the N-terminus of the chicken cytoplasmic pyruvate kinase directs this hydrid protein into the nucleus of transformed COS cells.[30b] But in contrast to the natural situation with HSP70, this nuclear transfer is hs-independent.

In addition to HSP70, HSP110 of mammalian cells is a constitutive and moderately hs-induced constituent of nucleoli. No hs-dependent translocation is observed.[16,121,127,128] Corresponding analyses for other organisms are lacking.

Entries in Table 16.1 labeled by a star describe the association of small HSPs with nuclear fractions in *Drosophila* and plant cells. These findings very likely represent the formation of heat shock granules in the perinuclear region (see Section 16.7). However, the results are not unequivocal. Beaulieu et al.[10c] reported on a hs-dependent tight binding of the *Drosophila* HSP27 to the nucleus. Though their data, obtained by immunofluorescence and cell fractionation, support their conclusions, these data are in contrast to earlier results (see Section 16.7), and the necessary electronmicroscopic studies were not done. In HeLa cells, the HSP27 complexes are found associated with a network of Golgi membranes surrounding the nucleus. However, after hs, part of the HSP27 forms large aggregates in the nucleus.[9a] In this case, immuno-electronmicroscopic investigations document the hs-induced translocation.

The situation in eukaryotic microorganisms, studied for localization of HSPs is unclear. This is true for *Achlya ambisexualis,*[122] *Tetrahymena pyriformis,*[43] and *Dictyostelium discoideum.*[67] Recent studies on the HSP26 localization in yeast showed a nuclear accumulation after hs of log-phase cultures grown on glucose, but not of cultures under a variety of other conditions.[108b]

16.3. PLASMA MEMBRANE

Several papers report on changing structure and functions of plasma membranes after treatment of cells with hs or chemical stressors (see Section 14.4), and there is evidence of thermotolerance effects also at this level (see Figure 14.3). Hs-induced proteins specifically associated with plasma membrane were detected in plants.[30,65] In soybean seedlings, the presence of HSP15 was considered necessary for protection against solute leakage after hs.[65]

Although the relationship of any aspect of the hs response is not evident at present, the identification of the mammalian HSC71-73 as the clathrin uncoating ATPase is remarkable.[23,109,135] In complex with clathrin cages, the HSC71-73 acquires ATPase activity and thus helps to uncoat coated vesicles[45e] (see Section 2.3.2). This process is an essential part of the receptor-mediated endocytosis which serves the signal and substrate supply of mammalian cells. Probably related to this is the identification of a member of the HSP70 family in small membrane vesicles containing transferrin receptors and secreted by mammalian reticulocytes in final stages of their maturation.[31,32,95]

16.4. THE ER/GOLGI SYSTEM

Members of two HSP families are prominent constitutive components of the ER/Golgi system (see Table 16.1C). These are the two glucose-regulated proteins, GRP94 and GRP78 (see Chapter 4). Similar to the clathrin uncoating ATPase, their function is not *a priori* hs-related. However, as a member of the HSP70 family, GRP78 acquires ATPase activity when associated with heterologous proteins. A function in protein assembly and possibly in the stabilization of defective or immature subunits of protein complexes is very likely.[76,97-97b]

Both GRPS are formed as preproteins with an N-terminal signal peptide (see Figures 2.8 and 2.9). In addition, the C-terminal tetrapeptide –Lys–Asp–Glu–Leu (–KDEL) is responsible for their retention in the lumen of the ER.[76,125] Transfer of the last 6 codons of the grp78 gene to the chicken lysozyme gene and expression of the hybrid gene in monkey COS cells leads to a lysozyme variant which is retained in the ER instead of being secreted.[76] On the other hand, removal of the N-terminal signal sequence generates a truncated GRP78 derivative which behaves HSP70-like, i.e. it migrates to the nucleus, if the COS cells are subjected to hs.[76] These authors also reported on a fusion of the signal peptide sequence to the *Drosophila* hsp70 gene. The resulting hybrid HSP70 was detected in the ER in a heavily glycosylated form.

Meanwhile, a GRP78 was also characterized from yeast.[85a,108a] The protein coded by the KAR2 gene ends with –HDEL and is localized in the perinuclear ER complex, and KAR2 mutants can be substituted by the mammalian grp78 gene. Induction of GRP78 is evidently connected with stress by the accumulation of abnormal proteins in the ER/Golgi system (Section 4.1). Its biological role in such a stress situation is underlined by the characterization of distinct peptide tags recognized by GRP78 or other members of the HSP70 family,[38a] by the stable association of GRP78 with abnormal proteins in the ER,[45b,48a] and by the observation that incompletely processed proteins escape the ER if GRP78 levels are reduced by introduction of anti-sense mRNA.[34]

16.5. CHLOROPLASTS AND MITOCHONDRIA

Proteins from three HSP families (HSP70, HSP60, and HSP20 families) are characteristic of chloroplasts and mitochondria.

By immunological criteria and sequence homology, the constitutively expressed members of the HSP70 family in chloroplasts and mitochondria belong to the prokaryotic subfamily characterized by the DnaK protein. Three genes coding for mitochondrial proteins were sequenced (see Section 2.3.2), i.e., the yeast SSC1 gene[30a] and HSC70 genes of *Caenorhabditis elegans,*[45d] *Trypanosoma cruzi,*[37a] and *Leishmania major.*[118a] Using corresponding antibodies, DnaK-like proteins were described for HeLa and calf liver cell mitochondria[59a] (P71) for *Euglena* and plant chloroplasts,[2a,68b] and for plant mitochondria.[83,83a] In pea chloroplasts, two additional members of the HSP70 family with M_r 75 kDa are localized at the inner surface of the outer membrane and in the stroma.[68b] Interestingly, the *Trypanosoma* HSC71 is tightly connected with the kinetoplast DNA, i.e., a DnaK-like function for DNA replication is very likely.

McMullin and Hallberg[70,71] described a constitutive protein of *Tetrahymena* mitochondria, which is selectively synthesized and accumulated after hs. It is a representative of the HSP60 family. Similar proteins were found in mitochondria of yeast, human, *Xenopus*, and plant cells. Moreover, by immunological and structural criteria, HSP60 is homologous to the bacterial GroEL heat shock protein (see Figure 2.11 and Table 2.2). In all cases investigated, proteins of the HSP60 family were found to form homooligomeric complexes with hollow-core morphology, which can be isolated as 12.5 nm particles.[23b,45c,47a,70,71,98,101a]

From the sequence data in Figure 2.11 and functional complementation, it is evident that the assembly protein for large and small subunits of the chloroplast ribulose-bisphosphate carboxylase is a plant GroEL protein.[41a,45a,77a,108a,108c] Sequences of GroEL-type organellar proteins were reported for mitochondria of yeast,[101d] human,[47c] and Chinese hamster cells,[98] as well as for plant chloroplasts[45a] (see Figure 2.11). Plants contain an additional HSP60 in their mitochondria.[101a] All GroEL-type proteins of bacteria, chloroplasts and mitochondria have similar functions. In an ATP-dependent process they act as morphopoietic proteins in the ordered assembly of complex multiprotein structures from the preformed subunits (see Section 12.2).[23b,45a,45c,93b,108c]

The third group of organellar proteins are tentatively collected together in one HSP20 subfamily (see sequences compiled in Figure 2.13). Two mitochondrial HSPs 34 and 30 of *Neurospora crassa* were described by Plesofsky-Vig and Brambl.[99,99a] By treating isolated mitochondria with protease, both proteins were shown to reside within the organell, although the sequence of the hsp30 gene shows no N-terminal transit peptide (Figure 2.13C). The same is true for the HSP22 attached to the membrane fraction of *Chlamydomonas reinhardi* chloroplasts[42,52,118] (Figure 2.13A). In contrast, N-terminal transit peptides of 5 to 8 kDa and cleavage of the cytoplasmic precursor protein during uptake into chloroplasts were found for the HSP21/22 of *Arabidopsis,* soybean and pea (Figure 2.13B).[23a,41,52,139,140,140a]

Remarkable are observations on a stress-dependent binding of HSPs 21/22 to chloroplast membranes in pea[52] and on a protective function of HSPs 22/29 for photosystem II of *Chlamydomonas reinhardi,* if the alga is subjected simultaneously to light and heat stress.[118] However, in view of the technical difficulties encountered with the analysis of these fractions, especially in *Chlamydomonas,* much further work is needed to elaborate the details of this interaction. At least, reexamination of the fine localization of HSP21 in pea chloroplasts by Chen et al.[23a] demonstrated that also under hs conditions (5 h 38 to 40°C) 80% was not membrane-bound but was in the stroma fraction. This agrees with electron-microscopic data from our laboratories.[83]

Whenever investigated in detail, the organellar HSPs are coded by nuclear genes and usually synthesized as precursors with corresponding transit peptides at their N-termini. It is also evident from the published sequences of mitochondrial and chloroplast genomes that, at least in these cases, there are no open reading frames left for HSP-coding genes. Though basic differences between different types of organisms cannot be excluded at present, reports on HSPs synthesized in chloroplasts or mitochondria of the green alga *Acetabularia* and some plants (for references, see Table 16.1, parts D, E) need careful reinvestigation with cloning of the corresponding genes. Moreover, bacterial contaminations must be excluded rigorously.[84]

16.6. THE CYTOSKELETON

Constitutively expressed and induced members of the HSP70 family are tightly associated with microtubules and the intermediate size filament cytoskeletal system of vertebrate cells (see Table 16.1F). HSP70-type proteins were described as microtubule-associated proteins (MAP),[63,147,152a,153] β-internexin,[41b,78,79] as intermediate filament-associated protein (NAPA-73),[24] or as Tau proteins.[26] The interaction is specific and resists purification of the cytoskeletal systems, e.g., by repeated cycles of dissociation and reassociation of microtubules.[26,63,92,147] A role as interfacing protein connecting different types of cytoskeletal systems with each other and possibly also with the plasma membrane is discussed.[26,47,79,145] In view of the rapid reorganization of the cytoskeleton induced by hs and its multifold influence on different stress-related phenomena (see Section 14.3), the binding of members of the HSP70 family is very suggestive. In fact, HSP70 antibodies were shown to decorate specifically stress fibers of chicken cells.[114] However, so far there is no evidence that the requirement of HSP70 for the integrity and function of the cytoskeleton increases or changes during hs.

Members of the HSP90 family are also connected with cytoskeletal systems of mammalian cells. The protein was coprecipitated with tubulin antibodies and, using immunofluorescence methods, both tubulin and HSP90 were colocalized in different mouse, rat, hamster, and primate cells.[101e,113b] Moreover, HSP90 and GRP94 bind to actin filaments, and this association is disrupted by *Ca^{2+}-calmodulin*. A role for the hs-dependent reorganization of the actin network can be discussed.[55,55a]

16.7. CYTOPLASMIC HEAT SHOCK GRANULES

16.7.1. STRUCTURAL BINDING OF SMALL HSPs

It is a common observation in all types of eukaryotic cells that small HSPs largely exist in rapidly sedimenting structure-bound form under hs and in ''soluble'' form under normothermic conditions. Reports on related effects stem from experiments with *Drosophila,*[7,60,123,133] *Tetrahymena,*[43] vertebrates,[6,9a,28] and plants.[30,64,68a,88,104] The initial characterization of this stress-specific state of small HSPs in *Drosophila* and plant cells was performed by cell fractionation procedures leading to the erroneous conclusion that nuclei, mitochondria, chloroplasts, and polysomes are cellular binding sites of small HSPs.[7,60,64,68a,104,123,133] In fact, the small HSPs of *Drosophila* may behave differentially. Recent immunofluorescence and cell fractionation studies demonstrate that, in contrast to the HSP23, HSP27 associates with the nucleus of heat-shocked or ecdysone-stimulated cells (see Section 16.2).[10a]

A major step towards an understanding of the changing compartmentation of HSPs during hs and recovery was the detection and characterization of large cytoplasmic granular aggregates (hsg) in tomato cell cultures[88] (Figure 16.4B and C). Their formation is not a consequence of HSP synthesis per se but is strictly dependent on the hyperthermic state, i.e., in preinduced cultures a reversible shift of preformed HSPs from ''soluble'' form to a structure-bound form (hsg) can be induced by subsequent hs, even if concommitant synthesis of new HSPs is blocked by cycloheximide.[81,82] The mass accumulation of hsg in tomato cell cultures facilitated isolation and characterization. They are resistant to 30 m*M* EDTA, 500 m*M* KCl or NaCl and 1% Nonidet P 40 or 0.2% Sarcosyl. They are mainly composed of HSPs (Figure 2.2C and D) containing 50 to 80% of the total amount of small HSPs and about 5% of the HSP70. Other prominent HSPs of the tomato cell cultures, e.g., HSPs80 and 95, are not bound.[87,88] In confirmation of the biochemical data the hs-dependent accumulation of HSP70 and the major small HSP (HSP17) was confirmed by autoradiography[88] (Figure 16.4F) and by immune staining of tomato protoplasts using immunofluorescence in combination with light microscopy (Figure 16.3) or the protein A-gold technique in combination with electron microscopy (Figure 16.5). Details are described by Neumann et al.[82]

Under appropriate hs conditions, hsg are formed in all types of higher plants (monocots and dicots) and in many different cell types including leaves[83] (Figure 16.4A), but the amount is usually far below that encountered in tomato cell cultures. This makes isolation from other plant parts much more difficult. Hsg are very prominent in few, actively dividing cell layers in the vicinity of root tips. In fact, early reports on further unspecified ''dense corpuscules'' or ''dense bodies'' in the cytoplasm of heat-shocked plant cells dealt with hsg in root tips of corn[39] or onion.[107]

Ultrastructure and staining behavior (Figure 16.4C and D) suggest that hsg are RNP complexes. First, they can be labeled with uridine.[81] Second, isolated, highly purified hsg were banded by CsCl centrifugation after formaldehyde fixation at a density of 1.3 g/cm^3. Third, purified hsg contains mRNA. Comparison of the polysomal and hsg mRNA patterns by *in vitro* translation in a wheat germ system make it very likely that hs in plant cells is connected with a reversible shift of control mRNA species from polysomes to the hsg fraction.[89]

Formation and function of hsg is presumably closely connected with cytoskeletal systems. This is evident from negative staining pictures of isolated hsg,[89] but also by *in situ* localization using whole mount pictures obtained from detergent-extracted protoplasts (Figure 16.4E). At early stages of hsg assembly, precursor aggregates can be found in the close vicinity of cytoskeletal bundles. This may also explain the high concentration of hsg in the perinuclear region (see Figure 16.4F and the immunofluorescence picture in Figure 16.3).

Our experience on the heterogeneity, stability, sedimentation behavior, and putative

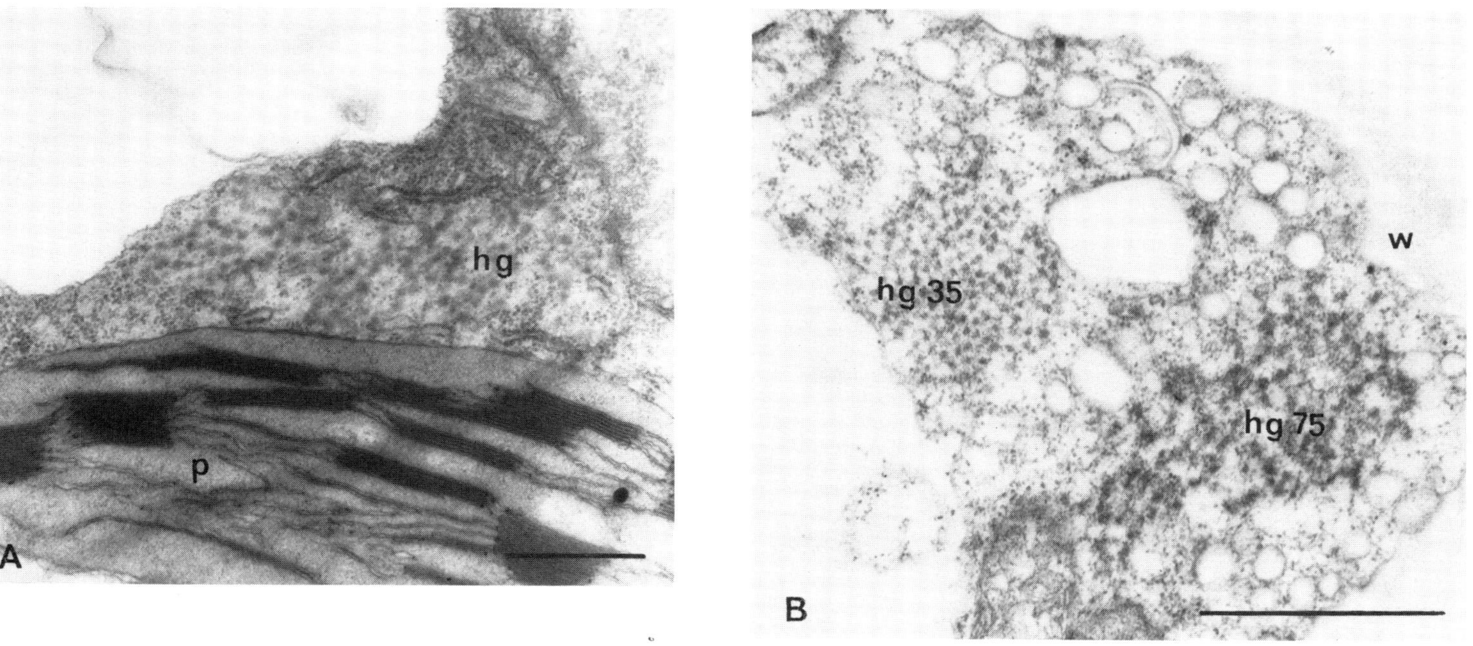

FIGURE 16.4. Cytoplasmic heat shock granules (hsg) of plants. Heat-shocked tomato cell cultures (B to F) or tobacco leaf discs (A) were fixed with glutaraldehyde and stained with osmium tetroxide (A, B, D to F) or with tannic acid (C). hg, hs granules, c, cytoskeleton, m, mitochondria, n, nucleus, no, nucleolus, p, plastids. Bars are 0.5 μm, and in D, 0.1 μm. (A) hsg in leaves of tobacco *(Nicotiana rustica);* (B) two types of hsg of 30 to 40 nm and 70 to 80 nm in tomato cell cultures *(Lycopersicon peruvianum);* (C) hsg cluster in tomato cell cultures after glutaraldehyde tannic acid fixation; (D) high resolution picture of B showing irregular fine structure of hsg;(E) a whole mount picture of tomato protoplasts after extraction with 2.5% Triton, shows hsg in close connection with cytoskeleton; (F) microautoradiographing of tomato cell cultures after ^{3}H-leucin labeling under stringent hs conditions. Note the exclusive labeling of hsg and nucleolus as the major sites of HSP accumulation.

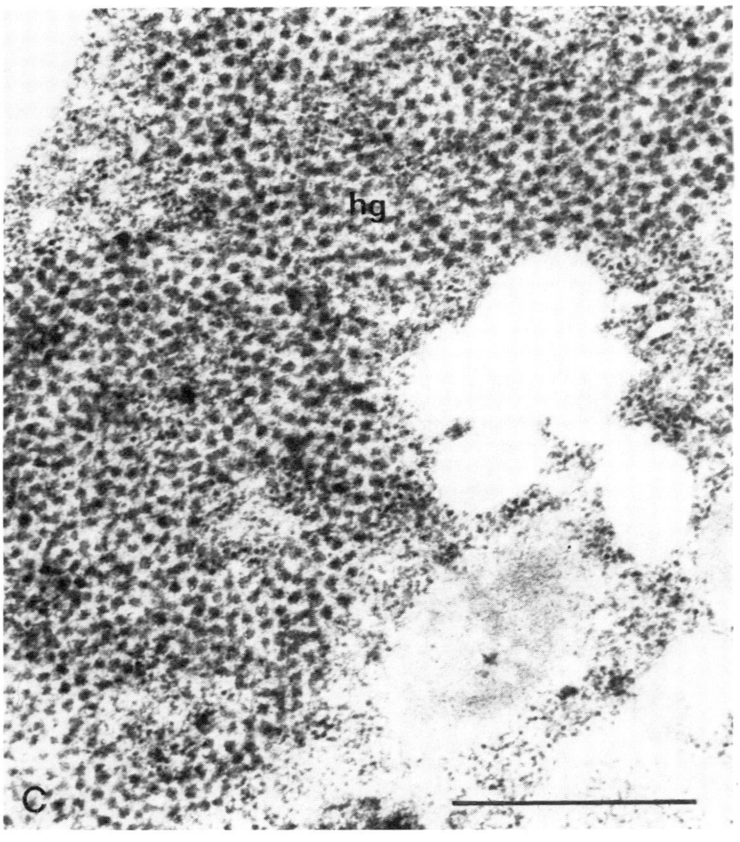

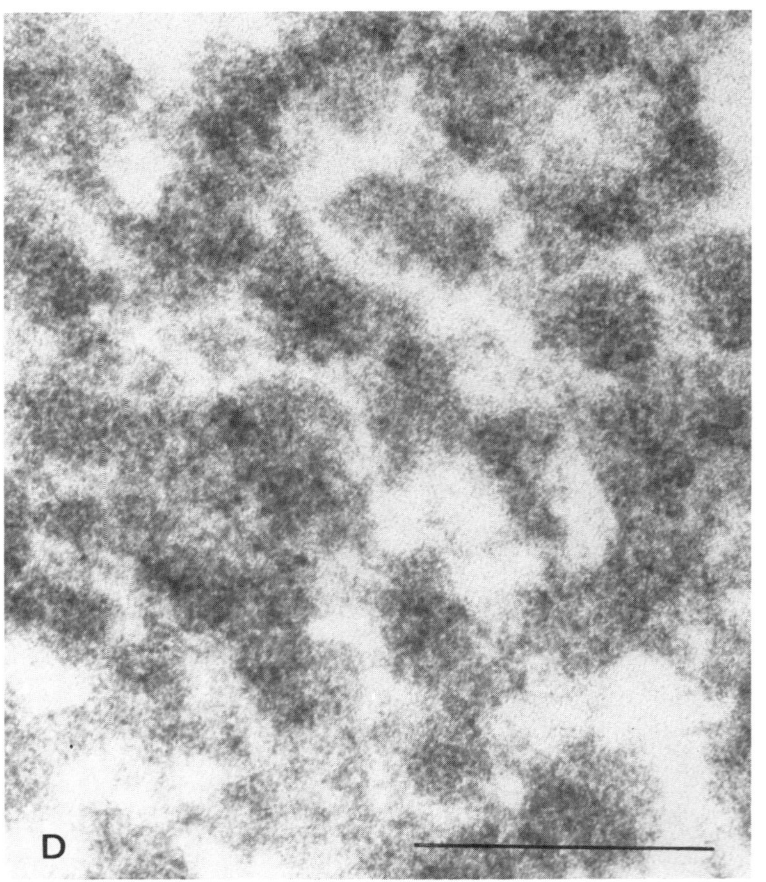

FIGURE 16.4. (continued).

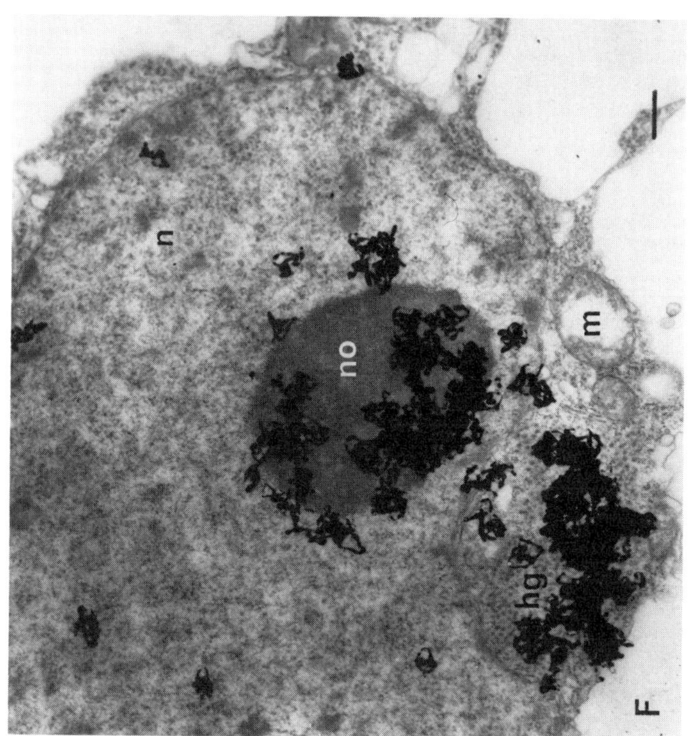

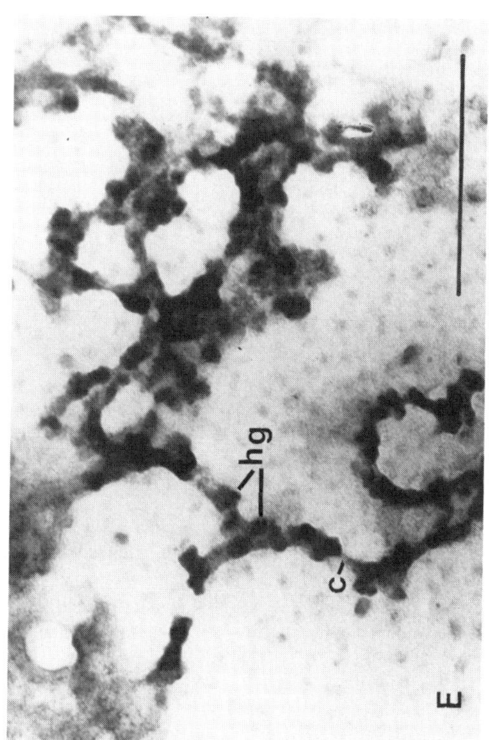

FIGURE 16.4. (continued)

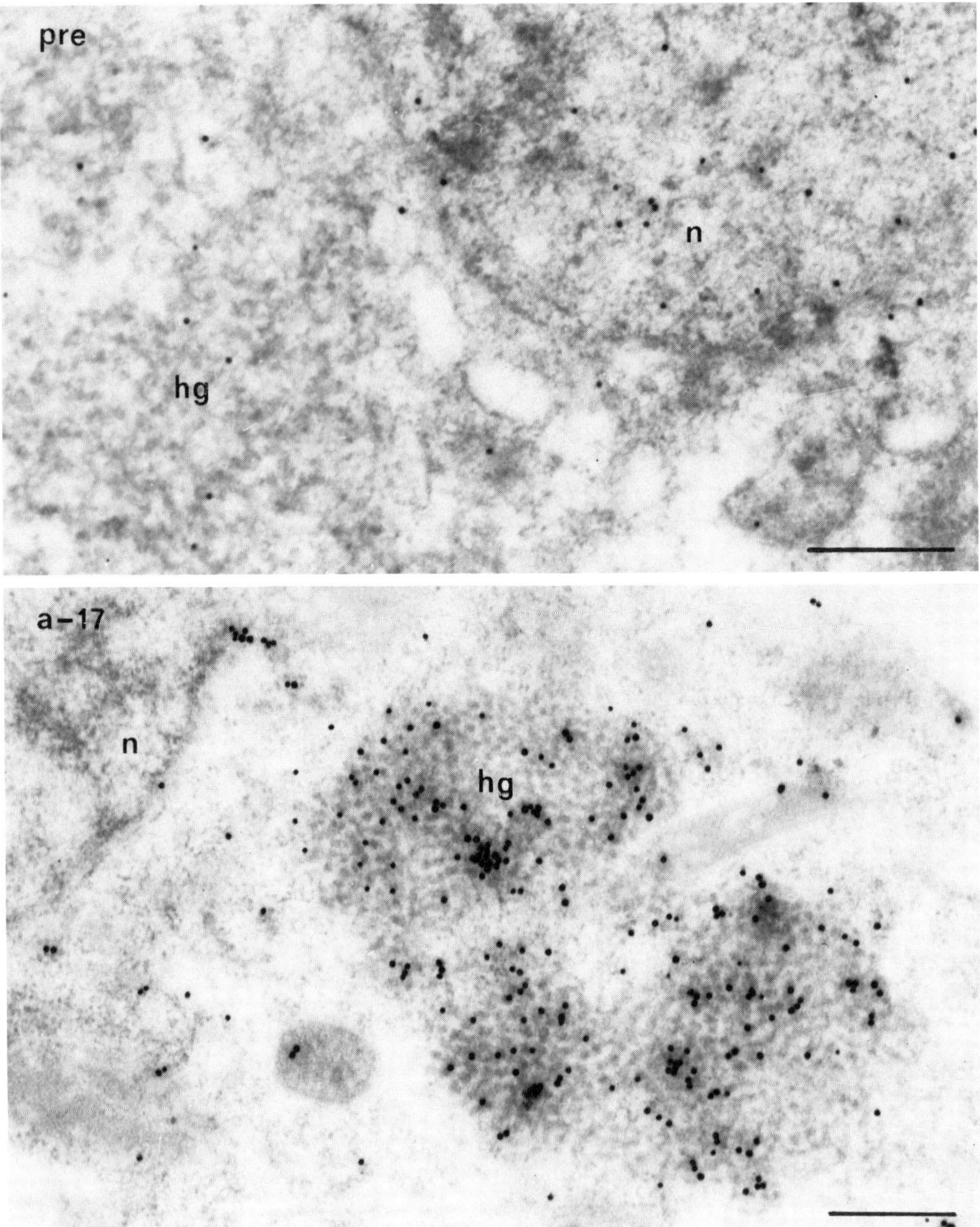

FIGURE 16.5. Immune staining of heat shock granules (hsg) in tomato cell cultures. Electron microscopic techniques were described by Neumann et al.[82] Ultrathin sections from Lovicryl-embedded, heat-shocked cell cultures were stained with preimmune serum (pre), HSP17 antiserum (a-17), or HSP70 antiserum (a-70). Counterstaining was with protein A coupled to gold particles of 16 nm (black dots). Regions with hsg (hg) are heavily labeled after staining with HSP17 and HSP70 antibodies but not with preimmune serum. The staining of the whole cytoplasm after application of HSP 70 antibodies reflects the more general distribution of this heat shock protein (see also Figure 16.3). The bar is 0.5 μm.

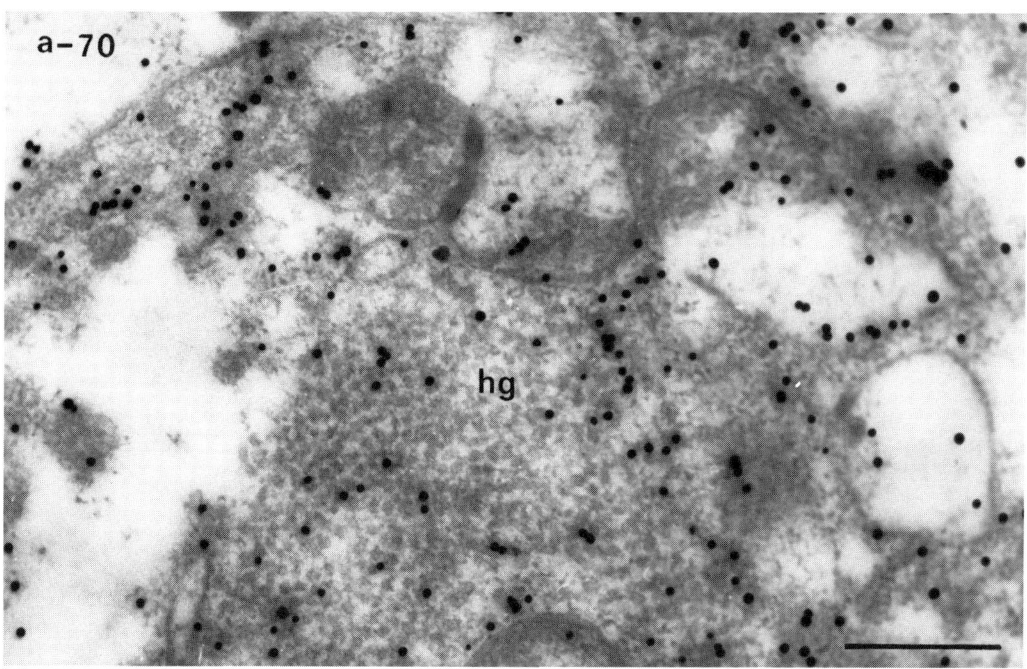

FIGURE 16.5. (continued).

cytoskeletal association in the perinuclear region may explain why small HSPs were found in close association with nuclear, mitochondrial, chloroplast and polysomal fractions. Insufficient characterization and purification of subcellular fractions are major obstacles to the reliable localization of HSPs by cell fractionation. Especially for objects with low levels of HSPs and consequently few hsgs or with massive hsg formation restricted to few cell layers (e.g., in roots), autoradiographic methods combined with immunological studies should be applied.

Formation of hs granules is evidently not restricted to plant cells. Early investigations revealed similar heterogeneously sized granular material in the cytoplasm of heat-shocked mammalian cells.[21,68] The reversible shift of small HSPs between "soluble" and structure-bound form is a long-known phenomenon of hs treatments in *Drosophila,*[4,7,142] but recently similar results were also reported for the only small HSP (HSP27) of vertebrate cells.[6,28,29] In fact, reinvestigation of the results with *Drosophila* demonstrated that small HSPs of heat-shocked cells are not in the nucleus but rather in perinuclear aggregates closely associated, but not identical, with the collapsed cytoskeleton.[36a,59] The immunofluorescence pictures obtained with *Drosophila* (Figure 16.6) and also with mammalian cells (Fugure 16.7) are strikingly similar to those for tomato cell cultures. Moreover, the freeze fracture image in Figure 16.7B shows a chicken hsg aggregate embedded in the cytoskeletal network surrounding the nucleus.[29] Further electronmicroscopic and biochemical characterization of hsg from *Drosophila* and vertebrate cells must show how far this homology between plant and animal systems can be extended.

16.7.2. PRECURSOR HSG

What is the "soluble" form of small HSPs, found in preinduced cells or after hs recovery? From immunofluorescence data (Figure 16.3) and also from biochemical analyses,[3,89] an almost quantitative and reversible shuttle between "soluble" and aggregate forms is evident. A preliminary characterization of small HSP complexes in *Drosophila* cells pointed to the

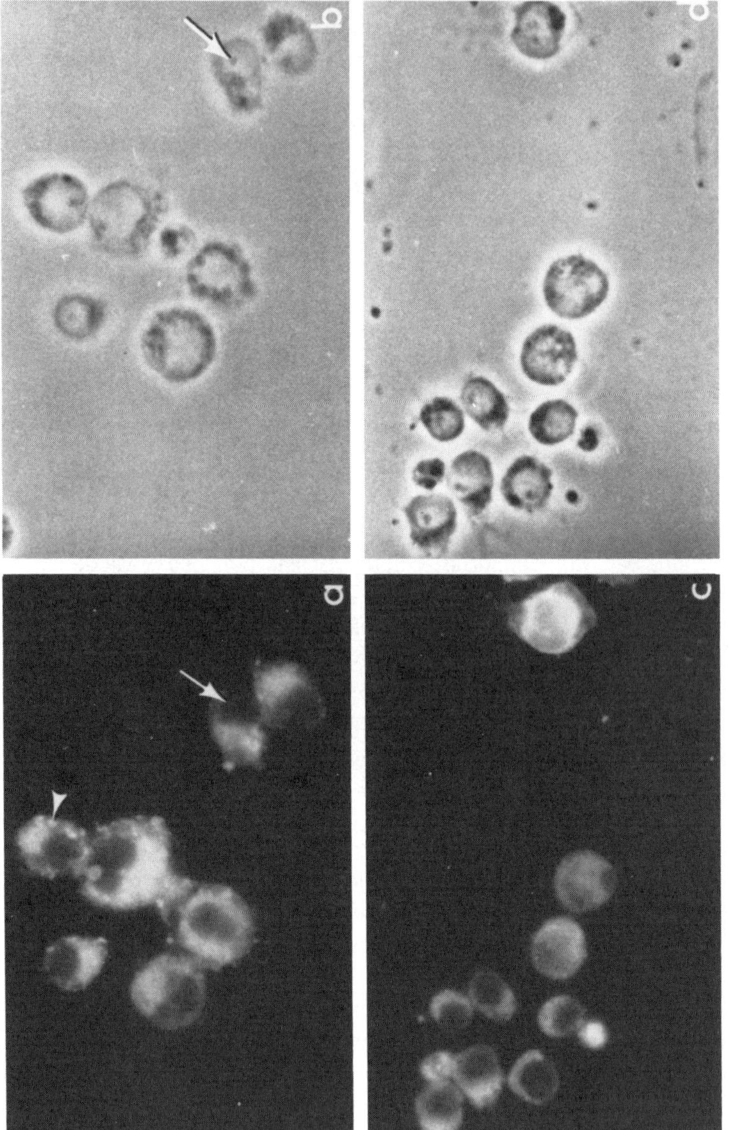

FIGURE 16.6. Localization of HSP23 in *Drosophila* Kc cells by immunofluorescence. Cells were heat-shocked for 1 h at 37°C and allowed to recover for 3 h (a and b) and 5 h (c and d), respectively. The immunofluorescence pictures on the left show the dense fluorescent granular material in the cytoplasm (arrow head in a) which slowly disappears with ongoing recovery. Nuclei (arrows) are essentially free of HSP23. (From Duband, J. L., Lettre, F., Arrigo, A. P., and Tanguay, R. M., *Can. J. Genet. Cytol.*, 28, 1088, 1987. With permission.)

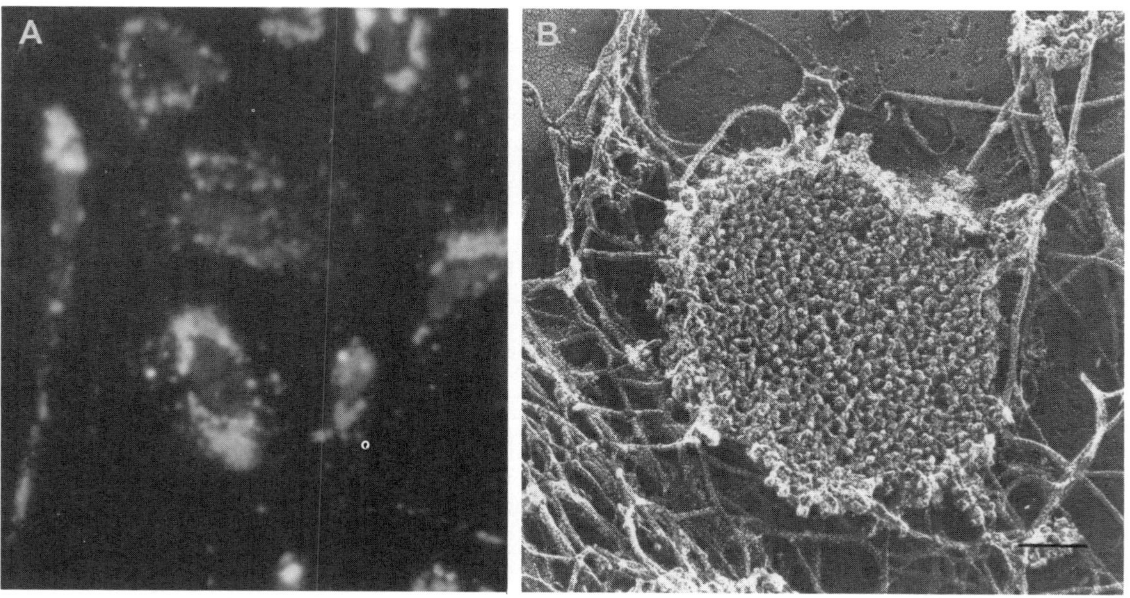

FIGURE 16.7. Aggregation of HSP24 in the perinuclear region of chicken embryo fibroblasts. Chicken embryo fibroblasts were preinduced (3 h 45°C) recovered over night and then stressed a second time for 3 h at 45°C. Heat shock granules (hsg) in chicken are mainly formed of HSP24. A is the immunofluorescence picture after staining with anti-HSP24 antibodies. The dark region in the center of the cells contain the nucleus, which are surrounded by a halo of brightly fluorescing granular material. B is the freeze fracture image using a deep etch replica formation technique showing a hsg in close association with the cytoskeletal complex. Bar is 0.1 μm. (From Collier, N. C. and Schlesinger, M. J., *J. Cell Biol.*, 103, 1495, 1986, and Collier, N. C., Heuser, J., Levy, M. A., and Schlessinger, M. J., *J. Cell Biol.*, 106, 1131, 1988. Reproduced from *The Journal of Cell Biology*, by copyright permission of the Rockefeller University Press.)

existence of 15- to 20 S RNP particles.[4] Later on this led to speculations that the fraction of "soluble" small HSPs exist in prosome-like particles.[8,117] Prosomes[115] were identified as 12 nm detergent and high salt-resistant, ubiquitous 19S RNP particles with a highly conserved protein composition and small cytoplasmic RNAs. This is true for duck, mouse, and HeLa cells,[115] for *Drosophila,*[51] sea urchin,[2] and plants.[56,89] Prosomes, at least in vertebrate cells, are evidently parts of a high-molecular-weight, ATP-requiring protease complex of 26S originally isolated from rat liver[131] but meanwhile also characterized from other mammalian cells, from *Drosophila,* and from yeast.[9,37b,68c,68d].

The "soluble" cytoplasmic RNPs containing the small HSPs are clearly different from prosomes with respect to sedimentation behavior in sucrose gradients (15 to 16S), protein composition and detergent stability (no resistance to 1% sarcosyl). To avoid confusion with prosomes, we call them precursor-heat shock granules (pre-hsg).[89] Isolation of pre-hsg from *Drosophila* cells[45] gave 12 nm hollow core particles of 16S (sucrose gradients) and $\rho = 1.34$ in Cs_2SO_4/DMSO gradients. They contained scRNAs of 80 to 120 nucleotides. The HSP27 of vertebrate cells is also found in 15S particles with a Stokes radius of 67 Å and an estimated molecular weight of 500 to 700 kDA.[6,9a,154a] Finally, by gel filtration experiments the yeast HSP26 was shown to exist in large oligomeric complexes of molecular mass > 500 kDa.[108b]

Hs-induced formation of aggregate hsg from "soluble" pre-hsg is very rapid,[3,6,28,45,82,88] but recovery of the original state takes several hours.[3,89] The *Drosophila* small HSPs exist in the 15S complexes irrespective of whether their synthesis is induced by hs or by developmental signals.[5,45]

Speculations about the role of the hsg formation in the hs response are based on the reversibility of their aggregation, on their association with the cytoskeleton and on their similarity to and scRNPs: (1) they may function in the protection of mRNAs, other constituents of the machinery for protein synthesis;[28,86,89] (2) they are HSP storage particles;[86] (3) they form a "cellular matrix in which other cell components/organelles are trapped and protected from heat damage."[49] The demonstration of a defined subset of transiently non-translated mRNAs associated with highly purified hsg from plants[89] is in agreement with the first function, but only the direct comparison of hsg from different organisms and characterization of their composition and state in the hs cells will help to clarify these questions. At any rate, the protein composition of pre-hsgs is very different depending on the organism studied. In plants, they contain a complex set of small HSPs (see Figure 2.2C and D), in *Drosophila* their composition may vary with the inducer applied, e.g., ecdysterone vs. hs, whereas in vertebrates the only small HSP (HSP27) is incorporated (see legend to Figure 16.7).

16.8. HORMONE RECEPTOR COMPLEXES

By immune fluorescence studies, members of the HSP90 family were always found evenly distributed in the cytoplasm of insect and vertebrate cells without apparent changes during hs.[19,57a,136,147a] However, there are several types of "soluble" protein complexes, whose functions are dependent on the presence or absence of HSP90 (Table 16.1, parts H and I), and a temperature-dependent oligomerization of HSP90 was reported for mouse cells.[57c]

Table 16.1, part H summarizes data about the role of HSP90 as an integral part of the inactive steroid hormone receptor complex in animals and also in the water mould *Achlya ambisexualis,*[15,15a] whose sexual development is controlled by steroid hormones. Receptors to all types of steroid hormones in animals exist in the cytoplasm as inactive 8- to 9S complexes formed of two molecules of a p90 and one molecule of the 100-kDa hormone receptor. Both are phosphoproteins. Extensive purification, preparation of corresponding antibodies and sequence analyses revealed that the receptor-associated protein is identical with HSP90.[35,46a,80a,101c,111-113a,130]

Upon binding of the hormone ligand, HSP90 is released and the active 4S hormone × receptor complex is formed. This transformation requires SH groups.[134] The subsequent interaction of the 4S hormone × receptor complex with DNA at the promoter binding sites depends on Zn^{2+}-containing protein domains[10a,10b,53a,110] (zinc-fingers). In the 8S complex the HSP90 shields the integrated Zn^{2+} ions from accessibility to *o*-phenanthrolin, i.e., it masks the DNA-binding site.[46a,110] Two properties of HSP90 may be important for its regulatory function as an inhibitor subunit of hormone receptors. On the one hand, purification by hydrophobic affinity chromatography indicates a highly hydrophobic surface that becomes inaccessible upon interaction with a hormone receptor.[47b] On the other hand, a "DNA-like" α-helical domain, probably formed by the large hydrophilic region (K/E in Figure 2.9), may be another determinant for the formation of the 8S complex with the hormone receptor.[10a,12a] At least the interaction with HSP90 is restricted to a short region of the receptor molecule at the border of the DNA binding domain. The whole C-terminal steroid-binding domain can be removed by trypsin treatment without release of the HSP90.[10b,33a] No net change of phosphorylation is connected with binding or release of HSP90 from the hormone receptor.[93a]

Steroid antagonists, frequently used as pharmacologically active agents, can be classified into those stabilizing the 8S complex and those interfering with DNA binding of the 4S receptor complex.[10] So far, there is no evidence for any stress-related function of HSP90 in its interaction with the steroid-hormone receptor proteins, but an increased HSP90 level may lead to a transient "hormone resistance" during the stress period.[9b]

Interestingly, the dioxin-induced expression of cytochrome P450 is based on a similar mechanism. The dioxin receptor is complexed in a cytoplasmic 9S complex with HSP90. Upon interaction with the inducer, a 6S form without HSP90 is formed, which is capable of binding to the xenobiotic response element in the cytochrome P450 promoter region.[33,80b,153b]

16.9. COMPLEXES WITH ONC-GENE PRODUCTS

Similar to the situation with the inactive hormone receptor, examples compiled in part I of Table 16.1 on the interaction of proteins of the HSP90 and HSP70 families with onc-proteins presumably describe general functions of HSPs. By cell fractionation and immune coprecipitation, HSP89 was detected as part of the cytoplasmic complexes with onc-gene products of the tyrosine kinase family. In chicken cells, this complex of HSP89 with the immature $pp60^{src}$ and a cytosolic pp50 is thought to represent a necessary but transient intermediate in the maturation and membrane binding of the onc-gene product (Figure 16.8). The onc-protein exhibits its typical tyrosine kinase activity only after autophosphorylation of the Tyr residues and subsequent binding to the plasma membrane.[14,154] Normally, the half-life of the maturation complex is in the range of 15 min. This contrasts to the more stabile complexes of different retroviral Tyr-kinases characterized by Ziemicki[155] and Ziemicki et al.[156] in avian and mammalian cells. It can be speculated that, in addition to the maturation, HSP90 may function in the activity control of retroviral onc-gene products.

The particular interest in the HSP70 complexes with onc-gene products concentrates on the phosphorylated nuclear onc-protein p53. Usually, this is a short-lived protein expressed in a cell-cycle dependent manner. Elevated levels of p53, were observed in cells transformed by chemicals or viruses. Moreover, transformation with p53 cDNA leads to immortalization of cells. Hence increase of the cellular level of p53 is evidently an important factor in the multistage process of carcinogenesis. Stabilization of p53 results from binding to viral tumor proteins, e.g., to the SV40 large T-antigen, and/or to proteins of the HSP70 family.[38,46,100,125a] Interestingly, the interaction of HSP70 with p53 was mainly reserved to activating mutant forms of the protein[38,126] or to partially denatured p53 under stress conditions.[125a] It is an

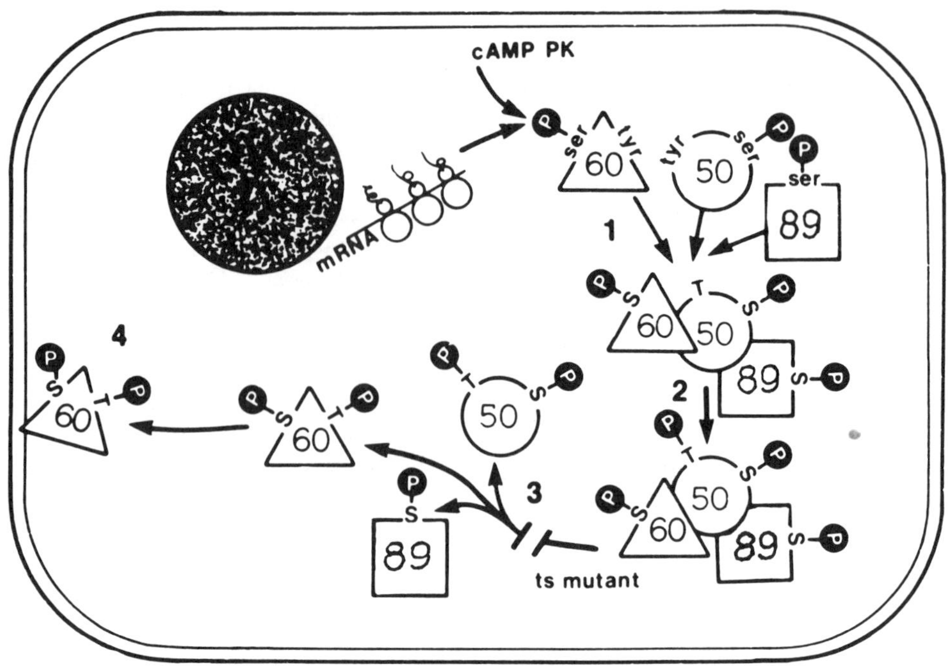

FIGURE 16.8. Maturation of the pp60^{src} protein kinase involves a transient complex with HSP 89. The inactive precursor form of the Rous sarcoma virus transforming protein pp60^{src} is translated on cytosolic polysomes. Phosphorylation at Ser-residues precedes association with the cytoplasmic proteins pp50 and HSP 89. The complex serves the Tyr-phosphorylation and may function as a transport shuttle to bring the mature pp60^{src} to the inner surface of the plasma membrane. The three participating proteins are identified by their M_r in kD. A temperature-sensitive mutant with altered src^{60} gene is blocked in the dissociation of the maturation complex. (From Yonemoto, W., Lipsich, L. A., Darrow, D., Brugge, J. S., *Heat Shock from Bacteria to Man,* Cold Spring Harbor Laboratories, 1982, 289; Nover, L., Ed., *Heat Shock Response of Eukaryotic Cells,* G. Thieme Verlag, 1984. With permission.)

intriguing possibility that p53 normally acts as a negative control element of cell division and that mutation or stress-dependent alteration of its structure interferes with this control, leading to malignant transformation.[37] Similar complexes with HSP70-type proteins were also reported for nuclear proteins of DNA-tumor viruses, i.e., for the SV40 large T-antigen,[113c] the polyomavirus middle T-antigen,[93c,143] and the adenovirus E1A protein.[153a]

A HSC72/p53 complex of 660,000 mol wt was purified from rat cells by rapid immunaffinity chromatography.[27] It can be dissociated by ATP but not by nonhydrolyzable analogs of ATP. Remarkably, if the murine p53 is expressed in *Escherichia coli,* it associates with the DnaK protein and, similar to the native situation in mammalian cells, this complex dissociates in the presence of ATP.[27] Although limited to a few cases only, these findings certainly include the intriguing possibility of a stress-promoted cancerogenesis by stabilization of p53 and other immortalizing proteins (see Section 20.6.2).

REFERENCES

1. **Adkins, B., Hunter, T., and Sefton, B. M.** The transforming proteins of PRC II virus and Rous sarcoma virus form a complex with the same two cellular phosphoproteins, *J. Virol.*, 43, 448—455, 1982.
2. **Akhayat, O., Grossi De Sa, F., and Infante, A. A.**, Sea urchin prosome: Characterization and changes during development, *Proc. Natl. Acad. Sci. U.S.A.* 84, 1595-1599, 1987.

2a. **Amir-Shapira, D., Leustek, T., Dalie, B., Weissbach, H., and Brot, N.**, Hsp70 proteins, similar to *Escherichia coli* DnaK, in chloroplasts and mitochondria of *Euglena gracilis, Proc. Natl. Acad. Sci. U.S.A.*, 87, 1749—1752, 1990.

3. **Arrigo, A.-P.**, Cellular localization of HSP23 during (*Drosophila*) development and following subsequent heat shock, *Dev. Biol.*, 122, 39—48, 1987.
4. **Arrigo, A.-P. and Ahmad-Zadeh, C.**, Immunofluorescence localization of a small heat shock protein (hsp23) in salivary gland cells of *Drosophila melanogaster, Mol Gen. Genet.*, 184, 73—79, 1981.
5. **Arrigo, A.-P. and Pauli, D.**, Characterization of HSP27 and three immunologically related polypeptides during (*Drosophila*) development, *Exp. Cell Res.*, 175, 169—183, 1988.
6. **Arrigo, A.-P. and Welch, W. J.**, Characterization and purification of the small 28,000 dalton mammalian heat shock protein, *J. Biol. Chem.*, 262, 15359—15369, 1987.
7. **Arrigo, A.-P., Fakan, S., and Tissieres, A.**, Localization of the heat shock-induced proteins in *Drosophila melanogaster* tissue culture cells, *Dev. Biol.*, 78, 86—103, 1980.
8. **Arrigo, A.-P., Darlix, J. L., Khandjian, E. W., Simon, M., and Spahr, P. F.**, Characterization of the prosome from *Drosophila* and its similarity to the cytoplasmic structures formed by the low molecular weight heat-shock proteins, *EMBO J.*, 4, 399—406, 1985.
9. **Arrigo, A.-P., Tanaka, K., Goldberg, A. L., and Welch, W. I.**, Identity of the 19S 'prosome' particle with the large multifunctional protease of mammalian cells (the proteasome), *Nature,* 331, 192—194, 1988.

9a. **Arrigo, A.-P., Suhan, J. P., and Welch, W. J.**, Dynamic changes in the structure and intracellular locale of the mammalian low-molecular-weight heat shock protein, *Mol. Cell. Biol.*, 8, 5059—5071, 1988.

9b. **Baulieu, E.-E.**, personal communication.

10. **Baulieu, E.-E.**, Steroid hormone antagonists at the receptor level: A role for the heat-shock protein MW 90,000 (hsp90), *J. Cell Biochem.*, 35, 161—174, 1987.

10a. **Baulieu, E.-E. and Catelli, M.-G.**, Steroid hormone receptors and heat shock protein Mr 90,000 (HSP90): A functional interaction?, in *Stress-Induced Proteins,* Pardue, M. L., Feramisco, J. R., and Lindquist, S., Eds., Alan R. Liss, New York, 1989, 203—219.

10b. **Beato, M.**, Gene regulation by steroid hormones, *Cell,* 56, 335—344, 1989.

10c. **Beaulieu, Y. F., Arrigo, A.-P., and Tanguay, R. M.**, Interaction of *Drosophila* 27,000-dalton heat-shock protein with the nucleus of heat shocked and ecdysone stimulated culture cells, *J. Cell Sci.*, 92, 29—36, 1989.

11. **Bienz, M. and Pelham, H. R. B.**, Mechanisms of heat-shock gene activation in higher eukaryotes, *Adv. Genet.*, 24, 31—72, 1987.
12. **Binart, N.**, Interactions du recepteur de la progesterone de l'oviducte de poulet avec la proteine de choc thermique hsp90, *Biochimie,* 68, 223—227, 1986.

12a. **Binart, N., Chambraud, B., Dumas, B., Rowlands, D. A., Bigogne, C., Levin, J. M., Garnier, J., Baulieu, E.-E., and Catelli, M. G.**, The cDNA-derived amino acid sequence of chick heat shock protein Mr 90,000 (HSP90) reveals a "DNA like" structure: Potential site of interaction with steroid receptors, *Biochem. Biophys. Res. Commun.*, 159, 140—147, 1989.

13. **Bole, D. G., Hendershot, L. M., and Kearny, J. F.**, Posttranslational association of immunoglobulin heavy chain binding protein with nascent heavy chains in non-secreting and secreting hybridomas, *J. Cell Biol.*, 102, 1558—1566, 1986.
14. **Brugge, J., Yonemoto, W., and Darrow, D.**, Interaction between the Rous sarcoma virus transforming protein and two cellular phosphoproteins: Analysis of the turnover and distribution of this complex, *Mol. Cell. Biol.*, 3, 9—19, 1983.
15. **Brunt, S. A. and Silver, J. C.**, Cellular localization of steroid hormone-regulated proteins during sexual development in *Achlya, Exp. Cell Res.*, 165, 306—319, 1986.

15a. **Brunt, S. A., Riehl, R., and Silver, J. C.**, Steroid hormone regulation of the *Achlya ambisexualis* 85-kilodalton heat shock protein, a component of the *Achlya* steroid hormone receptor, *Mol. Cell. Biol.*, 10, 273—281, 1990.

16. **Bugler, B., Caizergues-Ferrer, M., Bouche, G., Bourbon, H., and Amalric, F.**, Detection and localization of a class of proteins immunologically related to a 110kDa nucleolar protein, *Eur. J. Biochem.*, 128, 475—480, 1982.
17. **Burdon, R. H., Slater, A., McMahon, M., and Cato, A. C. B.**, Hyperthermia and the heat-shock proteins of HeLa cells, *Br. J. Cancer,* 45, 953—963, 1982.

18. **Caizergues-Ferrer, M., Dousseau, F., Gas, N., Bouche, G., Stevens, B., and Amalric, F.,** Induction of new proteins in the nuclear matrix of CHO cells by heat shock: Detection of a specific set in the nucleolar matrix, *Biochem. Biophys. Res. Commun.*, 118, 444—450, 1984.
19. **Carbajal, M. E., Duband, J. L., Lettre, F., Valet, J. P., and Tanguay, R. M.,** Cellular localization of *Drosophila* 83-kilodalton heat shock protein in normal, heat-shocked, and recovering cultured cells with a specific antibody, *Biochem. Cell Biol.*, 64, 816—825, 1986.
20. **Catelli, M. G., Binart, N., Jung-Testas, I., Renoir, J. M., Baulieu, E. E., Feramisco, J. R., and Welch, W. J.,** The common 90-kd protein component of non-transformed '8S' steroid receptors is a heat-shock protein, *EMBO J.*, 4, 3131—3135, 1985.
21. **Cervera, J.,** Effects of thermic shock on HEp-2 cells. An ultrastructural and high-resolution autoradiographic study, *J. Ultrastruct. Res.*, 63, 51—63, 1978.
22. **Chang, S. C., Wooden, S. K., Nakaki, T., Kim, Y. K., Lin, A. Y., Kung, L., Attenello, I. W., and Lee, A. S.,** Rat gene encoding the 78-kDa glucose-regulated protein GRP78: Its regulatory sequences and the effect of protein glycosylation on its expression, *Proc. Natl. Acad. Sci. U.S.A.*, 84, 680—684, 1987.
23. **Chappell, T. G., Welch, W. J., Schlossman, D. M., Palter, K. B., Schlesinger, M. J., and Rothman, J. E.,** Uncoating ATPase is a member of the 70 kilodalton family of stress proteins, *Cell,* 45, 3—13, 1986.

23a. **Chen, R., Lanzon, L. M., De Rocher, A. E., and Vierling, E.,** Accumulation, stability and localization of a major chloroplast heat shock protein, *J. Cell Biol.*, 110, 1873—1883, 1990.

23b. **Cheng, M. Y., Hartl, F.-U., Martin, J., Pollock, R. A., Kalonsek, F., Neupert, W., Hallberg, E. M., Hallberg, R. L., and Horwich, A. L.,** Mitochondrial heat-shock protein hsp60 is essential for assembly of proteins imported into yeast mitochondria, *Nature,* 337, 620—625, 1989.

23c. **Chou, M., Chen, Y.-M., and Lin, C.-Y.,** Thermotolerance of isolated mitochondria associated with heat shock proteins, *Plant Physiol.*, 89, 617—621, 1989.

24. **Ciment, G., Ressler, A., Letourneau, P. C., and Weston, J. A.,** A novel intermediate filament-associated protein, NAPA-73, that binds to different filament types at different stages of nervous system development, *J. Cell Biol.*, 102, 246—251, 1986.
25. **Clark, B. D. and Brown, I. R.,** A retinal heat shock protein is associated with elements of the cytoskeleton and binds to calmodulin, *Biochem. Biophys. Res. Commun.*, 139, 974—981, 1986.
26. **Clark, B. D. and Brown, I. R.,** Altered expression of a heat shock protein in the mammalian nervous system in the presence of agents which effect microtubule stability, *Neurochem. Res.*, 12, 819—823, 1987.,
27. **Clark, C. F., Cheng, K., Frey, A. B., Stein, R., Hinds, P. W., and Levine, A. J.,** Purification of complexes of nuclear oncogene p53 with rat and Escherichia coli heat shock proteins. In vitro dissociation of Hsc 70 and DnaK from murine p53 by ATP, *Mol. Cell. Biol.*, 8, 1206—1215, 1988.
28. **Collier, N. C. and Schlesinger, M. J.,** The dynamic state of heat shock proteins in chicken embryo fibroblasts, *J. Cell Biol.*, 103, 1495—1507, 1986.
29. **Collier, N. C., Heuser, J., Levy, M. A., and Schlesinger, M. J.,** Ultrastructural and biochemical analysis of the stress granule in chicken embryo fibroblasts, *J. Cell Biol,* 106, 1131—1139, 1988.
30. **Cooper, P. and Ho, T.-H.-D.,** Intracellular localization of heat shock proteins in maize, *Plant Physiol.*, 84, 1197—1203, 1987.

30a. **Craig, E. A., Kramer, J., Shilling, J., Werner-Washburne, M., Holmes, S., Kosic-Smithers, J., and Nicolet, C. M.,** SSC1, an essential member of the yeast HSP70 multigene family, encodes a mitochondrial protein, *Mol. Cell. Biol.*, 9, 3000—3008, 1989.

30b. **Dang, Ch. V. and Lee, W. F.,** Nuclear and nucleolar targeting sequences of c-erb-A, c-myb, N-myc, p53, HSP70, and HIV tat proteins, *J. Biol. Chem.*, 264, 18019—18023, 1989.

31. **Davis, J. Q. and Bennett, V.,** Human erythrocyte clathrin and clathrin-uncoating protein, *J. Biol. Chem.*, 260, 14850—14856, 1985.
32. **Davis, J. Q., Dansereau, D., Johnstone, R. M., and Bennett, V.,** Selective externalization of an ATP-binding protein structurally related to the clathrin-uncoating ATPase/heat shock protein in vesicles containing terminal transferrin receptors during reticulocyte maturation, *J. Biol. Chem.*, 261, 15368—15371, 1986.
33. **Denis, M., Cuthill, S., Wikstrom, A.-C., Poellinger, L., and Gustafsson, J. A.,** Association of the dioxin receptor with the Mr 90,000 heat shock protein. A structural kinship with the glucocorticoid receptor, *Biochem. Biophys. Res. Commun.*, 155, 801—807, 1988.

33a. **Denis, M., Gustafsson, J. A., and Wikstrom, A.-C.,** Interaction of the Mr 90,000 heat shock protein with the steroid-binding domain of the glucocorticoid receptor, *J. Biol. Chem.*, 263, 18520—18523, 1988.

34. **Dorner, A. J., Krane, M. G., and Kaufman, R. J.,** Reduction of endogenous GRP78 levels improves secretion of a heterologous protein in CHO cells, *Mol. Cell. Biol.*, 8, 4063—4070, 1988.
35. **Dougherty, J. J., Puri, R. K., and Toff, D. O.,** Polypeptide components of two 8S forms of chicken oviduct progesterone receptor, *J. Biol. Chem.* 259, 8004—8009, 1984.
36. **Dougherty, J. J., Rabideau, D. A., Iannotti, A. M., Dullivan, W. P., and Toft, D. O.,** Identification of the 90 kDa substrate of rat liver type II casein kinase with the heat shock protein which binds steroid receptors, *Biochim. Biophys. Acta,* 927, 74—80, 1987.

36a. **Duband, J. L., Lettre, F., Arrigo, A.-P., and Tanguay, R. M.,** Expression and localization of hsp-23 in unstressed and heat-shocked *Drosophila* cultured cells, *Can. J. Genet. Cytol.*, 28, 1088—1092, 1987.

37. **Eliyahn, D., Michalovitz, D., Eliyahn, S., Pinhasi-Kimhi, O., and Oren, M.,** Wild-type p53 can inhibit oncogene-mediated focus formation, *Proc. Natl. Acad. Sci. U.S.A.*, 86, 8763—8767, 1989.

37a. **Engman, D. M., Kirchhoff, L. V., and Donelson, J. E.,** Molecular cloning of mtp70, a mitochondrial member of the hsp70 family, *Mol. Cell. Biol.*, 9, 5163—5168, 1989.

37b. **Eytan, E., Ganoth, D., Armon, T., and Hershko, A.,** ATP-dependent incorporation of 20S proteasome into the 26S complex that degrades proteins conjugated to ubiquitin, *Proc. Natl. Acad. Sci. U.S.A.*, 86, 7751—7755, 1989.

38. **Finlay, C. A., Hinds, P. W., Tan, T.-H., Eliyahn, D., Oren, M., and Levine, A. J.,** Activating mutations for transformation by 53p produce a gene product that forms an hsc 70-p53 complex with an altered half-life, *Mol. Cell. Biol.*, 8, 531—539, 1988.

38a. **Flynn, G. C., Chappell, T. G., and Rothman, J. E.,** Peptide binding and release by proteins implicated as catalysts of protein assembly, *Science*, 245, 385—390, 1989.

39. **Fransolet, S., Deltour, R., Bronchart, R., and Van De Walle, C.,** Changes in ultrastructure and transcription by elevated temperatures in *Zea mays* embryonic root cells, *Planta*, 146, 7—18, 1979.

40. **Gasc, J. M., Renoir, J. M., Randanyi, C., Joab, I., Tuohimaa, P., and Baulieu, E.-E.,** Progesterone receptor in the chick oviduct: An immunohistochemical study with antibodies to distinct receptor components, *J. Cell Biol.*, 99, 1193—1201, 1984.

40a. **Georgopoulos, C., Tilly, V., Ang, D., Chandrasekhar, G. N., Fayet, O., Spence, J., Ziegelhoffer, T., Liberek, K., and Zylicz, M.,** The role of the *Escherichia coli* heat shock proteins in bacteriophage lambda growth, in *Stress-Induced Proteins*, Pardue, M. L., Feramisco, J. R., and Lindquist, S., Eds., Alan R. Liss, New York, 1989, 37—47.

41. **Glaczinski, H. and Kloppstech, K.,** Temperature-dependnet binding to the thylakoid membranes of nuclear-coded chloroplast heat shock proteins, *Eur. J. Biochem.*, 173, 579—583, 1988.

41a. **Goloubinoff, P., Gatenby, A. A., Lorimer, G. H.,** GroE heat shock proteins promote assembly of foreign prokaryotic ribulose bisphosphate carboxylase oligomers in *Escherichia coli*, *Nature*, 337, 44—47, 1989.

41b. **Green, L. A. D. and Liem, R. K. H.,** Beta-internexin is a microtubule-associated protein identical to the 70-kDa heat shock cognate protein and the clathrin uncoating ATPase, *J. Biol. Chem.*, 264, 15210—15215, 1989.

42. **Grimm, B., Ish-Shalom, D., Even, D., Glasczinski, H., Ottersbach, P., Ohad, I., and Kloppstech, K.,** The nuclear-coded chloroplast 22kDa heat shock protein of *Chlamydomonas*. Evidence for translocation into the organelle without a processing step, *Eur. J. Biochem.*, 182, 539—546, 1989.

43. **Guttman, S. D., Glover, C. V. C., Allis, C. D., and Gorovsky, M. A.,** Heat shock, deciliation and release from anoxia induce the synthesis of the same set of polypeptides in starved *T. pyriformis*, *Cell*, 22, 299—307, 1980.

44. **Haas, I. G. and Wabl, M.,** Immunoglobulin heavy chain binding protein, *Nature*, 306, 387—389, 1983.

45. **Haass, C., Falkenburg, P. E., and Kloetzel, P. M.,** The molecular organization of the small heat shock proteins in *Drosophila*, in *Stress-Induced Proteins*, Pardue, M. L., Feramisco, J. R., and Lindquist, S., Eds., Alan R. Liss, New York, 1989, 175—185.

45a. **Hemmingsen, S. M., Woolford, C., Van Der Vies, S. M., Tilly, K., Dennis, D. T., Georgopoulos, C. P., Hendrix, R. W., and Ellis, R. J.,** Homologous plant and bacterial proteins chaperone oligomeric protein assembly, *Nature*, 333, 330—335, 1988.

45b. **Hendershot, L. M., Ting, J., and Lee, A. S.,** Identity of the immunoglobulin heavy-chain-binding protein with the 78,000-dalton glucose-regulated protein and the role of posttranslational modifications in its binding function, *Mol. Cell. Biol.*, 8, 4250—4256, 1988.

45c. **Hendrix, R. W. and Tsui, L.,** Role of the host in virus assembly: Cloning of the *Escherichia coli* groE gene and identification of its protein product, *Proc. Natl. Acad. Sci. U.S.A.*, 75, 136—139, 1978.

45d. **Heschl, M. F. P. and Baillie, D. L.,** Characterization of the hsp70 multigene family of *Caenorhabditis elegans*, *DNA*, 8, 233—243, 1989.

45e. **Heuser, J. and Steer, C. J.,** Trimeric binding of the 70-kDa uncoating ATPase to the vertices of clathrin triskelia: A candidate intermediate in the vesicle uncoating reaction, *J. Cell Biol.*, 109, 1457—1466, 1989.

46. **Hinds, P. W., Finlay, C. A., Frey, A. B., and Levine, A. J.,** Immunological evidence for the association of p53 with a heat shock protein, Hsc 70, in p53-plus-ras-transformed cell lines, *Mol. Cell. Biol.*, 7, 2863—2869, 1987.

46a. **Howard, K. J. and Distelhorst, C. W.,** Evidence for intracellular association of the glucocorticoid receptor with the 90-kDa heat shock protein, *J. Biol. Chem.*, 263, 3474—3481, 1988.

46b. **Howard, K. J. and Distelhorst, C. W.,** Effect of the 90-kDa heat shock protein on glucocorticoid receptor binding to DNA-cellulose, *Biochem. Biophys. Res. Commun.*, 151, 1226—1232, 1988.

47. **Hughes, E. N. and August, J. T.,** Coprecipitation of heat shock proteins with a cell surface glycoprotein, *Proc. Natl. Acad. Sci. U.S.A.*, 79, 2305—2309, 1982.

47a. **Hutchinson, E. G., Tichelaar, W., Hofhaus, G., Weiss, H., and Leonard, K. R.,** Identification and electron microscopic analysis of a chaperonin oligomer from *Neurospora crassa* mitochondria, *EMBO J.*, 8, 1485—1490, 1989.

47b. **Iwasaki, M., Saito, H., Yamamoto, M., Korach, K. S., Hirogome, T., and Sugano, H.,** Purification of heat shock protein90 from calf uterus and rat liver and characterization of the highly hydrophobic region, *Biochim. Biophys. Acta,* 992, 1—8, 1989.

47c. **Jindal, S., Dudani, A. K., Singh, B., Harley, C. B., and Gupta, R. S.,** Primary structure of a human mitochondrial protein homologous to the bacterial and plant chaperonins and to the 65-kd mycobacterial antigen, *Mol. Cell. Biol.*, 9, 2279—2283, 1989.

48. **Joab, I., Radanyi, C., Renoir, M., Buchou, T., Catelli, M. G., Binart, N., Mester, J., and Baulieu, E. E.,** Common non-hormone binding component in non-transformed chick oviduct receptors of four steroid hormones, *Nature,* 308, 850—853, 1984.

48a. **Kassenbrock, C. K., Garcia, P. D., Walter, P., and Kelly, R. B.,** Heavy-chain binding protein recognizes aberrant polypeptides translocated in vitro, *Nature,* 333, 90—93, 1988.

49. **Kimpel, J. A. and Key, J. L.,** Heat shock in plants, *Trends Biochem. Sci.*, 10, 353—357, 1985.

49a. **Kishore, R. and Upadhyaya, K. C.,** Heat shock proteins of pigeon pea *(Cajanus cajan), Plant Cell Physiol.*, 29, 517—521, 1988.

50. **Kloetzel, P. and Bautz, E. K. F.,** Heat-shock proteins are associated with hnRNP in *Drosophila melanogaster* tissue culture cells, *EMBO J.*, 2, 705—710, 1983.

51. **Kloetzel, P.-M., Falkenburg, P.-E., Hoessl, P., and Glaetzer, K. H.,** The 19S ring-type particles of *Drosophila.* Cytological and biochemical analysis of their intracellular association and distribution, *Exp. Cell Res.*, 170, 204—213, 1987.

52. **Kloppstech, K., Meyer, G., Schuster, G., and Ohad, I.,** Synthesis, transport and localization of a nuclear coded 22-kd heat-shock protein in the chloroplast membranes of peas and *Chlamydomonas reinhardi, EMBO J.*, 4, 1901—1909, 1985.

53. **Kloppstech, K., Ohad, I., and Schweiger, A.-G.,** Evidence for an extranuclear coding site for a heat-shock protein in Acetabularia, *Eur. J. Cell Biol.*, 42, 239—245, 1986.

53a. **Klug, A. and Rhodes, D.,** Zinc fingers: A novel protein motif for nucleic acid recognition, *Trends Biochem. Sci.*, 12, 464—469, 1987.

54. **Koch, G., Smith, M., Macer, D., Webster, P., and Mortara, R.,** Endoplasmic reticulum contains a common, abundant calcium-binding glycoprotein, endoplasmin, *J. Cell Sci.*, 86, 217—232, 1986.

54a. **Kost, S. L., Smith, D. F., Sullivan, W. P., Welch, W. J., and Toft, D. O.,** Binding of heat shock proteins to the avian progesterone receptor, *Mol. Cell. Biol.*, 9, 3829—3838, 1989.

55. **Koyasu, S., Nishida, E., Kadowaki, T., Matsuzaki, F., Iida, K., Harada, F., Kasuga, M., Sakai, H., and Yahara, I.,** Two mammalian heat shock proteins, HSP90 and HSP100, are actin-binding proteins, *Proc. Natl. Acad. Sci. U.S.A.*, 83, 8054—8058, 1986.

55a. **Koyasu, S., Nishida, E., Miyata, Y., Sakai, H., and Yahara, I.,** Hsp-100, a 100-kDa heat shock protein, is a Ca^{2+}-calmodulin-regulated actin-binding protein, *J. Biol. Chem.*, 264, 15083—15087, 1989.

56. **Kremp, A., Schliephacke, M., Kull, U., and Schmid, H.-P.,** Prosome exist in plant cells too, *Exp. Cell Res.*, 166, 553—557, 1986.

57. **Krishnasamy, S., Manar Mannan, R., Krishnan, M., and Gnanam, A.,** Heat shock response of the chloroplast genome in *Vigna sinensis, J. Biol. Chem.*, 263, 5104—5109, 1988.

57a. **Lai, B.-T., Chin, N. W., Stanek, A. E., Keh, W., and Lanks, K. W.,** Quantitation and intracellular localization of the 85K heat shock protein by using monoclonal and polyclonal antibodies, *Mol. Cell. Biol.*, 4, 2802—2810, 1984.

57b. **Lanks, K. W., Kasambalides, E. J., Chinkers, M., and Brugge, J. S.,** A major cytoplasmic glucose-regulated protein is associated with the Rous sarcoma virus $pp60^{src}$ protein, *J. Biol. Chem.*, 257, 8604—8607, 1982.

57c. **Lanks, K. W.,** Temperature-dependent oligomerization of HSP85 in vitro, *J. Cell. Physiol.*, 140, 601—607, 1989.

58. **La Thangue, N. B. A.,** A major heat-shock protein defined by a monoclonal antibody, *EMBO J.*, 4, 1871—1879, 1984.

59. **Leicht, B. G., Biessmann, H., Palter, K. B., and Bonner, J. J.,** Small heat shock proteins of *Drosophila* associates with the cytoskeleton, *Proc. Natl. Acad. Sci. U.S.A.*, 83, 90—94, 1986.

59a. **Leustek, T., Dalie, B., Amir-Shapira, D., Brot, N., and Weissbach, H.,** A member of the hsp70 family is localized in mitochondria and resembles *Escherichia coli* DnaK, *Proc. Natl. Acad. Sci. U.S.A.*, 86, 7805—7808, 1989.

60. **Levinger, L. and Varshavsky, A.,** Selective arrangement of ubiquitinated and D1 protein-containing nucleosomes within the *Drosophila* genome, *Cell,* 28, 375—385, 1982.

61. **Lewis, M. J. and Pelham, H. R. B.,** Involvement of ATP in the nuclear and nucleolar functions of the 70 kd heat shock protein, *EMBO J.*, 4, 3137—3143, 1985.

62. **Lewis, M., Helmsing, P. J., and Ashburner, M.,** Parallel changes in puffing activity and patterns of protein synthesis in salivary glands of *Drosophila, Proc. Natl. Acad. Sci. U.S.A.*, 72, 3604—3608, 1975.

62a. **Lewis, M. J., Mazzarella, R. A., and Green, M.,** Structure and assembly of the endoplasmic reticulum. The synthesis of three major endoplasmic reticulum proteins during lipopolysaccharide induced differentiation of murine lymphocytes, *J. Biol. Chem.*, 260, 3050—3057, 1985.

63. **Lim, L., Hall, C., Leung, T., and Whatley, S.,** The relationship of the rat brain 68kDa microtubule-associated protein with synaptosomal plasma membrane and with the *Drosophila* 70 kDa heat-shock protein, *Biochem. J.*, 224, 677—680, 1984.

64. **Lin, C.-Y., Roberts, J. K., and Key, J. L.,** Acquisition of thermotolerance in soybean seedlings, *Plant Physiol.*, 74, 152—160, 1984.

65. **Lin, C.-Y., Chen, Y.-M., and Key, J. L.,** Solute leakage in soybean seedlings under various heat shock regimes, *Plant Cell Physiol.*, 26, 1493—1498, 1985.

66. **Lipsich, L. A., Cutt, R. J., and Brugge, J. S.,** Association of the transforming proteins of Rous, Fujinami and Y73 avian sarcoma viruses with the same two cellular proteins, *Mol. Cell. Biol.*, 2, 875—880, 1982.

67. **Loomis, W. F. and Wheeler, S. A.,** Chromatin-associated heat shock proteins in *Dictyostelium, Dev. Biol.*, 90, 412—418, 1982.

68. **Love, R., Soriano, R. Z., and Walsh, R. J.,** Effect of hyperthermia on normal and neoplastic cells in vitro, *Cancer Res.*, 30, 1525—1533, 1970.

68a. **Mansfield, M. A., and Key, J. L.,** Cytoplasmic distribution of heat shock proteins in soybean, *Plant Physiol.*, 86, 1240—1246, 1988.

68b. **Marshall, J. S., Derocher, A. E., Keegstra, K., and Vierling, E.,** Identification of hsp70 homologues in chloroplasts, *Proc. Natl. Acad. Sci. U.S.A.*, 87, 374—378, 1990.

68c. **Martins de Sa, C., Rollet, E., Grossi de Sa, M.-F., Tanguay, R. M., Best-Belpomme, M., and Scherrer, K.,** Prosomes and heat shock complexes in *Drosophila melanogaster* cells, *Mol. Cell. Biol.*, 9, 2672—2681, 1989.

68d. **Matthews, W., Tanaka, K., Driscoll, J., Ichihara, A., and Goldberg, A. L.,** Involvement of the proteasome in various degradative processes in mammalian cells, *Proc. Natl. Acad. Sci. U.S.A.*, 86, 2597—2601, 1989.

69. **McCormick, P. J., Millis, A. J. T., and Babiarz, B.,** Distribution of a 100 kdalton glucose-regulated cell surface protein in mammalian cell cultures and sectioned tissues, *Exp. Cell Res.*, 138, 63—72, 1982.

70. **McMullin, T. W. and Hallberg, R. L.,** A normal mitochondrial protein is selectively synthesized and accumulated during heat shock on *Tetrahymena thermophila, Mol. Cell. Biol.*, 7, 4414—4423, 1987.

71. **McMullin, T. W. and Hallberg, R. L.,** A highly evolutionarily conserved mitochondrial protein is structurally related to the protein encoded by the *E. coli* gene groEL, *Mol. Cell. Biol.*, 8, 371—380, 1988.

72. **Mendel, D. B., Bodwell, J. E., Gametchu, B., Harrison, R. W., and Munck, A.,.** Molybdate-stabilized nonactivated glucocorticoidreceptor complexes contain a 90-kDa non-steroid-binding phosphoprotein that is lost on activation, *J. Biol. Chem.*, 261, 3758—3763, 1986.

72a. **Mendel, D. B. and Orti, E.,** Isoform composition and stoichiometry of the 90kDa heat shock protein associated with glucocorticoid receptors, *J. Biol. Chem.*, 263, 6695—6702, 1988.

73. **Milarski, K. L. and Morimoto, R. I.,** Expression of human HSP70 during the synthetic phase of the cell cycle, *Proc. Natl. Acad. Sci. U.S.A.*, 83, 9517—9522, 1986.

74. **Mitchell, H. and Lipps, L.,** Rapidly labeled proteins on the salivary gland chromosomes of *Drosophila melanogaster, Biochem. Genet.*, 13, 585—627, 1975.

75. **Munro., S. and Pelham, H. R. B.,** Use of peptide tagging to detect proteins expressed from cloned genes: Deletion mapping functional domains of *Drosophila* hsp 70, *EMBO J.*, 3, 3087—3093, 1984.

76. **Munro, S. and Pelham, H. R. B.,** An hsp70-like protein in the ER: Identity with the 78 kd glucose-regulated protein and immunoglobulin heavy chain binding protein, *Cell*, 46, 291—300, 1986.

77. **Munro, S. and Pelham, H. R. B.,** A C-terminal signal prevents secretion of luminal ER proteins, *Cell*, 48, 899—907, 1987.

77a. **Musgrove, J. E. and Ellis, R. J.,** The Rubisco large subunit binding protein, *Philos. Trans. R. Soc. Lond. [Biol]*, 312, 419—428, 1986.

78. **Napolitano, E. W., Pachter, J. S., Chin, S. S. M., and Liem, R. K. H.,** β-Internexin, a ubiquitous intermediate filament-associated protein, *J. Cell Biol.*, 101, 1323—1331, 1985.

79. **Napolitano, E. W., Pachter, J. S., and Liem, R. K. H.,** Intracellular distribution of mammalian stress proteins. Effects of cytoskeletal-specific agents, *J. Biol. Chem.*, 262, 1493—1504, 1987.

80. **Nebiolo, C. M. and White, E. M.,** Corn mitochondrial protein synthesis in response to heat shock, *Plant Physiol.*, 79, 1129—1132, 1985.

80a. **Nemoto, T., Ohara-Nemoto, Y., and Ota, M.,** Purification and characterization of a nonhormone-binding component of the nontransformed glucocorticoid receptor from rat liver, *J. Biochem.*, 102, 513—524, 1987.

80b. **Nemoto, T., Mason, B. G. F., Wilhelmsson, A., Cuthill, S., Hapgood, J., Gustafsson, J. A., and Poellinger, L.,** Activation of the dioxin and glucocorticoid receptors to a DNA binding state under cell-free conditions, *J. Biol. Chem.*, 265, 2269—2277, 1990.

81. **Neumann, D., Scharf, K.-D., and Nover, L.**, Heat shock induced changes of plant cell ultrastructure and autoradiographic localization of heat shock proteins, *Eur. J. Cell Biol.*, 34, 254—264, 1984.
82. **Neumann, D., Zur Nieden, U., Manteuffel, R., Walter, G., Scharf, K.-D., and Nover, L.**, Intracellular localization of heat shock proteins in tomato cells cultures, *Eur. J. Cell Biol.*, 43, 71—81, 1987.
83. **Neumann, D., Zur Nieden, U., Scharf, K.-D., Wollgiehn, R., and Nover, L.**, Heat shock and other stress response systems of plants, *Biol. Zbl.*, 108 (6), 1—156, 1989.
83a. **Neumann, D.**, et al., in preparation.
84. **Nieto-Sotelo, J. and Ho, T.-H. D.**, Absence of heat shock protein synthesis in isolated mitochondria and plastids from maize, *J. Biol. Chem.*, 262, 12288—12292, 1987.
85. **Nishida, E., Koyasu, S., Sakai, H., and Yahara, I.**, Calmodulin-regulated binding of the 90 kDa heat shock protein to actin filaments, *J. Biol. Chem.*, 261, 16033—16037, 1986.
85a. **Normington, K., Kohno, K., Kozutsumi, Y., Gething, M.-J., and Sambrook, J.**, *S. cerevisiae* encodes an essential protein homologous to mammalian BiP, *Cell,* 57, 1223—1236, 1989.
86. **Nover, L., Ed.**, *Heat Shock Response of Eukaryotic Cells,* Springer-Verlag, Berlin, 1984.
87. **Nover, L. and Scharf, K.-D.**, Synthesis, modification and structural binding of heat shock proteins in tomato cell cultures, *Eur. J. Biochem.*, 139, 303—313, 1984.
88. **Nover, L., Scharf, K.-D., and Neumann, D.**, Formation of cytoplasmic heat shock granules in tomato cell cultures and leaves, *Mol. Cell. Biol.*, 3, 1648—1655, 1983.
89. **Nover, L., Scharf, K.-D., and Neumann, D.**, Cytoplasmic heat shock granules are formed from precursor particles and contain a specific set of mRNAs, *Mol. Cell. Biol.*, 9, 1298—1308, 1989.
90. **Ohtsuka, K. and Laszlo, A.**, personal communication.
91. **Ohtsuka, K., Nakamura, H., and Sato, C.**, Intracellular distribution of 73000 and 72000 dalton heat shock proteins in HeLa cells, *Int. J. Hyperthermia,* 2, 267—276, 1986.
92. **Ohtsuka, K., Tanabe, K., Nakamura, H., and Sato, C.**, Possible cytoskeletal association of 69,000- and 68,000-Dalton heat shock proteins and structural relations among heat shock proteins in murine mastocytoma cells, *Radiat. Res.*, 108, 34—42, 1986.
92a. **Oppermann, H., Levinson, W., and Bishop, J. M.**, A cellular protein that associates with the transforming protein of Rous sarcoma virus is also a heat-shock protein, *Proc. Natl. Acad. Sci. U.S.A.*, 78, 1067—1071, 1981.
93. **Ornelles, D. A. and Penman, S.**, Prompt heat-shock and heat-shifted proteins associated with the nuclear matrix intermediate filament scaffold in *Drosophila melanogaster* cells, *J. Cell Sci.*, 95, 393—404, 1990.
93a. **Orti, E., Mendel, D., and Munck, A.**, Phosphorylation of glucocorticoid receptor-associated and free forms of the 90-kDa heat shock protein before and after receptor activation, *J. Biol. Chem.*, 264, 231—237, 1989.
93b. **Ostermann, J., Horwich, A. L., Neupert, W., and Hertl, F.-U.**, Protein folding in mitochondria requires complex formation with hsp60 and ATP hydrolysis, *Nature,* 341, 125—130, 1989.
93c. **Pallas, D. C., Morgan, W., and Roberts, T. M.**, The cellular proteins which can associate specifically with polyomavirus middle T-antigen in human 293 cells include the major human 70-kDa heat shock proteins, *J. Virol.*, 63, 4533—4539, 1989.
94. **Palter, K. B., Watanabe, M., Stinson, L., Mahowald, A. P., and Craig, E. A.**, Expression and localization of *Drosophila melanogaster* hsp70 cognate proteins, *Mol. Cell. Biol.*, 6, 1187—1203, 1986.
95. **Pan, B. T. and Johnston, R. M.**, Selective externalization of the transferrin receptor by sheep reticulocytes in vitro, *J. Biol. Chem.*, 259, 9776—9782, 1984.
96. **Pelham, H. R. B.**, Hsp 70 accelerates the recovery of nucleolar morphology after heat shock, *EMBO J.*, 3, 3095—3100, 1984.
97. **Pelham, H. R. B.**, Speculations on the functions of the major heat shock and glucose-regulated proteins, *Cell,* 46, 959—961, 1986.
97a. **Pelham, H.**, Heat shock proteins. Coming in from the cold, *Nature,* 332, 776—777, 1988.
97b. **Pelham, H. R. B.**, Heat shock and the sorting of luminal ER protcins, *EMBO J.*, 8, 3171—3176, 1989.
98. **Picketts, D. J., Mayanil, C. S. K., and Gupta, R. S.**, Molecular cloning of a Chinese hamster mitochondrial protein related to the 'chaperonin' family of bacterial and plant proteins, *J. Biol. Chem.*, 264, 12001—12008, 1989.
99. **Plesovski-Vig, N. and Brambl, R.**, Heat shock response of *Neurospora crassa:* Protein synthesis and induced thermotolerance. *J. Bacteriol.*, 161, 1083—1091, 1985.
99a. **Plesofski-Vig, N. and Brambl, R.**, submitted.
100. **Pinhasi-Kimhi, O., Michalovitz, D., Ben-Zeev, A., and Oren, M.**, Specific interaction between the p53 cellular tumor antigen and major heat shock proteins, *Nature,* 320, 182—185, 1986.
101. **Pouchelet, M., Pierre, E. S., Bibor-Hardy, V., and Simard, R.**, Localization of the 70,000 dalton heat-induced protein in the nuclear matrix of BHK cells, *Exp. Cell Res.*, 149, 451—459, 1983.
101a. **Prasad, T. K. and Hallberg, R. L.**, Identification and metabolic characterization of the *Zea mays* mitochondrial homolog of the *Escherichia coli* groEL protein, *Plant Mol. Biol.*, 12, 609—618, 1989.
101b. **Ramachandran, C., Catelli, M. G., Schneider, W., and Shyamala, G.**, Estrogenic regulation of uterine 90-kilodalton heat shock protein, *Endocrinology,* 123, 956—961, 1988.

101c. **Ratajczak, T., Brockway, M. J., Hahnel, R., Moritz, R. L., and Simpson, R. J.,** Sequence analysis of the nonsteroid binding component of the calf uterine estrogen receptor, *Biochem. Biophys. Res. Commun.*, 151, 1156—1163, 1988.

101d. **Reading, D. S., Hallberg, R., and Myers, A. M.,** Characterization of the yeast HSP60 gene coding for a mitochondrial assembly factor, *Nature,* 337, 655—659, 1989.

101e. **Redmond, T., Sanchez, E. R., Bresnick, E. H., Schlesinger, M. J., Toft, D. O., Pratt, W. B., and Welsh, M. J.,** Immunofluorescence colocalization of the 90-kDa heat-shock protein and microtubules in interphase and mitotic mammalian cells, *Eur. J. Cell Biol.*, 50, 66—75, 1989.

102. **Reiter, T. and Penman, S.,** ''Prompt'' heat shock proteins: Translationally regulated synthesis of new proteins associated with nuclear matrix-intermediate filaments as an early response to heat shock, *Proc. Natl. Acad. Sci. U.S.A.*, 80, 4737—4741, 1983.

103. **Renoir, J. M., Mester, J., Buchou, T., Catelli, M. G., Tuohimaa, P., Binart, N., Joab, I., Radanyi, C., and Baulieu, E. E.,** Purification by affinity chromatography and immunological characterization of a 110 kDa component of the chick oviduct progesterone receptor, *Biochem. J.*, 217, 685—692, 1984.

104. **Restivo, F. M., Tassi, F., Maestri, E., Lorenzoni, C., Puglisi, P. P., and Marmiroli, N.,** Identification of chloroplast associated heat-shock proteins in *Nicotiana plumbaginifolia* protoplasts, *Curr. Genet.*, 11, 145—151, 1986.

105. **Ribeiro, G. and Vasconcelos-Costa, J.,** A heat-shock protein is associated with adenovirus type 12S antigen at the membrane of tumor cells, *Cienc. Biol. Mol. Cell. Biol.*, 9, 227—231, 1984.

106. **Riehl, R. M., Sullivan, W. P., Vroman, B. T., Bauer, V. J., Pearson, G. R. and Toft, D. O.,** Immunological evidence that the non hormone binding component of avian steroid receptors exists in a wide range of tissues and species, *Biochemistry,* 24, 6586—6591, 1985.

107. **Risueno, M. C., Stockert, J. C., Gimenez-Martin, G., and Diez, J. L.,** Effect of supraoptimal temperatures on meristematic cell nucleoli, *J. Microscopie,* 16, 87—94, 1973.

108. **Roccheri, M. C., Sconzo, G., Dibernhardo, M. G., Albanese, I., Dicarlo, M., and Giudice, G.,** Heat shock proteins in sea urchin embryos. Territorial and intracellular location, *Acta Embryol. Morphol. Exp.*, 2, 91—96, 1981.

108a. **Rose, M. D., Misra, L. M., and Vogel, J. P.,** KAR2, a karyogamy gene, is the yeast homolog of the mammalian BiP/GRP78 genes, *Cell,* 57, 1211—1221, 1989.

108b. **Rossi, J. M. and Lindquist, S.,** The intracellular location of yeast HSP26 varies with metabolism, *J. Cell. Biol.*, 108, 425—439, 1989.

108c. **Roy, H.,** Rubisco assembly: A model system for studying the mechanism of chaperonin action, *Plant Cell,* 1, 1035—1042, 1989.

108d. **Roy, H., and Cannon, S.,** Ribulose bisphosphate carboxylase assembly: what is the role of the large subunit binding protein, *Trends Biochem. Sci.*, 13, 163—165, 1988.

109. **Rothman, J. E. and Schmid, S.L.,** Enzymatic recycling of clathrin from coated vesicles, *Cell,* 46, 5—9, 1986.

110. **Sabbah, M., Redeulich, G., Secco, C., and Baulieu, E.-E.,** The binding activity of estrogen receptor to DNA and heat shock protein (Mr 90,000) is dependent on receptor-bound metal, *J. Biol. Chem.*, 262, 8631—8635, 1987.

111. **Sanchez, E. R., Toft, D. O., Schlesinger, M. J., and Pratt, W. B.,** Evidence that the 90-kDa phosphoprotein associated with the untransformed L-cell glucocorticoid receptor is a murine heat shock protein, *J. Biol. Chem.*, 260, 12398—12401, 1985.

112. **Sanchez, E. R., Housley, P. R., and Pratt, W. B.,** The molybdate-stabilized glucocorticoid binding complex of L-cells contains a 98-100 kdalton steroid-binding phosphoprotein and a 90 kdalton nonsteroid-binding phosphoprotein that is part of the murine heat-shock complex, *J. Steroid Biochem.*, 24, 9—18, 1986.

113. **Sanchez, E. R., Meshinchi, S., Tienrungroj, W., Schlesinger, M. J., Toft, T. O., and Pratt, W. B.,** Relationship of the 90-kDa murine heat-shock protein to the untransformed and transformed states of the L cell glucocorticoid receptor, *J. Biol. Chem.*, 262, 6986—6991, 1987.

113a. **Sanchez, E. R., Meshinchi, S., Schlesinger, M. J., and Pratt, W. B.,** Demonstration that the 90-Kilodalton heat shock protein is bound to the glucocorticoid receptor in its 9S nondeoxynucleic acid binding form, *Mol. Endocrinol.*, 1, 908—912, 1987.

113b. **Sawai, E. T., and Butel, J. S.,** Association of a cellular heat shock protein with Simian virus 40 large T-antigen in transformed cells, *J. Virol.*, 63, 3961—3973, 1989.

113c. **Sanchez, E. R., Redmond, T., Scherrer, L. C., Bresnick, E. H., Welsh, M. J., and Pratt, W. B.,** Evidence that the 90-kDa heat shock protein is associated with tubulin-containing complexes in L cell cytosol and in intact PtK cells, *Mol. Endocrinol.*, 2, 756—760, 1988.

114. **Schlesinger, M. J., Aliperti, G., and Kelley, P. M.,** The response of cells to heat shock, *Trends Biochem. Sci.*, 7, 222—225, 1982.

115. **Schmid, H. P., Akhayat, O., Martins De Sa, C., Puvion, F., Koehler, K., and Scherrer, K.,** The prosome: An ubiquitous morphologically distinct RNP particle associated with repressed mRNPs and containing specific scRNA and a characteristic set of proteins, *EMBO J.*, 3, 29—34, 1984.

116. **Schuh, S., Yonemoto, W., Brugge, J., Bauer, V. J., Riehl, R. M., Sullivan, W. F., and Toft, D. O.,** A 90,000-dalton binding protein common to both steroid receptors and the Rous sarcoma virus transforming protein, pp60v-sarc, *J. Biol. Chem.* 260, 14292—14296, 1985.
117. **Schuldt, C. and Kloetzel, P. M.,** Analysis of cytoplasmic 19S ringtype particles in *Drosophila* which contain hsp23 at normal growth temperature, *Dev. Biol.,* 110, 65—74, 1985.
118. **Schuster, G., Even, D., Kloppstech, K., and Ohad, I.,** Evidence for protection by heat-shock proteins against photoinhibition during heat shock, *EMBO J.,* 7, 1—6, 1988.
118a. **Searle, S., Campos, A. J. R., Coulson, R. M. R., Spithill, T. W., and Smith, D. F.,** A family of heat shock protein 70-related genes are expressed in the promastigotes of *Leishmania major, Nucl. Acids Res.,* 17, 5081—5095, 1989.
119. **Sefton, B. M., Beemon, K., and Hunter, T.,** Comparison of the expression of the src gene of Rous sarcoma virus in vitro and in vivo, *J. Virol.,* 28, 957—971, 1978.
120. **Sharma, S., Rodgers, L., Brandsma, J., Gething, M.-J., and Sambrook, J.,** SV40 T antigen and the exocytotic pathway, *EMBO J.,* 4, 1479—1489, 1985.
121. **Shyy, T. T., Subjeck, J. R., Heinaman, R., and Anderson, G.,** Effect of growth state and heat shock on nucleolar localization of the 110,000-Da heat shock protein in mouse embryo fibroblasts, *Cancer Res.,* 46, 4738—4745, 1986.
122. **Silver, J. C., Andrews, D. R., and Pekkala, D.,** Effect of heat shock on synthesis and phosphorylation of nuclear and cytoplasmic proteins in the fungus Achlya, *Can. J. Biochem. Cell Biol.,* 61, 447—455, 1983.
123. **Sinibaldi, R. M. and Morris, P. W.,** Putative function of *Drosophila melanogaster* heat shock proteins in the nucleoskeleton, *J. Biol. Chem.,* 256, 10735—10738, 1981.
124. **Sinibaldi, R. M. and Turpen, T.,** A heat shock protein is encoded within mitochondria of higher plants, *J. Biol. Chem.,* 260, 15382—15385, 1985.
125. **Sorger, P. K. and Pelham, H. R. B.,** The glucose-regulated protein grp94 is related to heat shock protein hsp90, *J. Mol. Biol.,* 194, 341—344, 1987.
125a. **Soussi, T., Defromentel, C. C., Stuerzbecher, H. W., Ullrich, S., Jenkins, J., and May, P.,** Evolutionary conservation of the biochemical properties of p53. Specific interaction of *Xenopus laevis* p53 with Simian virus 40 large T-antigen and mammalian heat shock proteins-70, *J. Virol.,* 63, 3894—3901, 1989.
126. **Stürzbecher, H.-W., Chumakov, P., Welch, W. J., and Jenkins, J. R.,** Mutant p53 proteins bind hsp72/73 cellular heat shock-related proteins in SV40-transformed monkey cells, *Oncogene,* 1, 201—211, 1987.
127. **Subjeck, J. R. and Shyy, T. T.,** Stress protein systems of mammalian cells, *Am. J. Physiol.,* 250, C1—C17, 1986.
128. **Subjeck, J. R., Shyy, T., Shen, J., and Johnson, R. J.,** Association between the mammalian 110,000-dalton heat-shock protein and nucleoli, *J. Cell Biol.,* 97, 1389—1395, 1983.
129. **Süss, K. H. and Jordanov, I. T.,** Biosynthetic cause of in vivo acquired thermotolerance of photosynthetic light reactions and metabolic responses of chloroplasts to heat stress, *Plant Physiol.,* 81, 192—199, 1986.
130. **Sullivan, W. P., Vroman, B. T., Bauer, V. J., Puri, R. K., Riehl, R. M., Pearson, G. R., and Toft, D. O.,** Isolation of steroid receptor binding protein from chicken oviduct and production of monoclonal antibodies, *Biochemistry,* 24, 4214—4222, 1985.
131. **Tanaka, K., Kunio, L., and Akira, I.,** A high molecular weight protease in the cytosol of rat liver, *J. Biol. Chem.,* 261, 15197—15203, 1986.
131a. **Tanguay, R. M.,** Intracellular localization and possible functions of heat shock proteins, in *Changes in Eukaryotic Gene Expression in Response to Environmental Stress,* Atkinson, B. G. and Walden, D. B., Academic Press, Orlando, FL, 1985, 91—113.
132. **Tanguay, R. M. and Vincent, M.,** Biosynthesis and characterization of heat shock proteins in Chironomus tentans salivary glands, *Can. J. Biochem.,* 59, 67—73, 1981.
133. **Tanguay, R. M. and Vincent, M.,** Intracellular translocation of cellular and heat shock induced proteins upon heat shock in *Drosophila* Kc cells, *Can. J. Biochem.,* 60, 306—315, 1982.
134. **Tienrungroj, W., Meshinchi, S., Sanchez, E. R., Pratt, S. E., Grippo., J. F., Holmgren, A., and Pratt, B.,** The role of sulfhydryl groups in permitting transformation and DNA binding of the glucocorticoid receptor, *J. Biol. Chem.,* 262, 6992—7000, 1987.
135. **Ungewickell, E.,** The 70-kd mammalian heat shock proteins are structurally and functionally related to the uncoating protein that releases clathrin triskelia from coated vesicles, *EMBO J.,* 4, 3385—3391, 1985.
136. **Van Bergen En Henegouwen, P. M. P., Berbers, G., Linnemans, W. A. M., and Van Wijk, R.,** Subcellular localization of the 84,000 dalton heat-shock protein in mouse neuroblastoma cells: Evidence for a cytoplasmic and nuclear location, *Eur. J. Cell Biol.,* 43, 469—478, 1987.
137. **Velazquez, J. M. and Lindquist, S.,** Hsp70: Nuclear concentration during environmental stress and cytoplasmic storage during recovery, *Cell,* 36, 655—662, 1984.
138. **Velazquez, J. M., Didomenico, B. J., and Lindquist, S.,** Intracellular localization of heat shock proteins in *Drosophila, Cell,* 20, 679—689, 1980.

139. **Vierling,E., Mishkind, M. L., Schmidt, G. W., and Key, J. L.,** Specific heat shock proteins are transported into chloroplasts, *Proc. Natl. Acad. Sci. U.S.A.*, 83, 361—365, 1986.

140. **Vierling, E., Nagao, R. T., De Rocher, A. E., and Harris, L. M.,** A heat shock protein localized to chloroplasts is a member of a eukaryotic superfamily of heat shock proteins, *EMBO J.*, 7, 578—582, 1988.

140a. **Vierling, E., Harris, L. M., and Chen, Q.,** The major low molecular weight heat shock protein in chloroplasts shows antigenic conservation among diverse higher plant species, *Mol. Cell. Biol.*, 9, 461—468, 1989.

141. **Vincent, M. and Tanguay, R. M.,** Heat-shock induced proteins present in the cell nucleus of *Chironomus tentans* salivary gland, *Nature,* 281, 501—503, 1979.

142. **Vincent, M. and Tanguay, R. M.,** Different intracellular distributions of heat shock and arsenite-induced proteins in *Drosophila* Kc cells. Possible relation with the phosphorylation and translocation of a major cytoskeletal protein, *J. Mol. Biol.*, 162, 365—378, 1982.

143. **Walter G., Carbone, A., and Welch, W. J.,** Medium tumor antigen of polyomavirus transformation-defective mutant NG59 is associated with 73-kilodalton heat shock protein, *J. Virol.*, 61, 405—410, 1987.

144. **Wang, C. and Lazarides, E.,** Arsenite-induced changes in methylation of the 70,000 dalton heat shock proteins in chicken embryo fibroblasts, *Biochem. Biophys. Res. Commun.*, 119, 735—743, 1984.

145. **Wang C., Asai, D. J., and Lazarides, E.,** The 68,000-dalton neurofilament-associated polypeptide is a component of nonneuronal cells and of skeletal myofibrils, *Proc. Natl. Acad. Sci. U.S.A.*, 77, 1541—1545, 1980.

146. **Wang, C., Gomer, R. H., and Lazarides, E.,** Heat shock proteins are methylated in avian and mammalian cells, *Proc. Natl. Acad. Sci. U.S.A.*, 78, 3531—3535, 1981.

147. **Weatherbee, J. A., Luftig, R. B., and Weiling, R. R.,** Purification and reconstitution of HeLa cell microtubules, *Biochemistry,* 19, 4116—4123, 1980.

147a. **Welch, W. J. and Feramisco, J. R.,** Purification of the major mammalian heat shock proteins, *J. Biol. Chem.*, 257, 14949—14959, 1982.

148. **Welch, W. J. and Feramisco, J. R.,** Nuclear and nucleolar localization of the 72,000 dalton heat shock protein in heat-shocked mammalian cells, *J. Biol. Chem.*, 259, 4501—4513, 1984.

149. **Welch, W. J. and Feramisco, J. R.,** Rapid purification of mammalian 70,000-dalton stress proteins: Affinity of the proteins for nucleotides, *Mol. Cell. Biol.*, 5, 1229—1237, 1985.

150. **Welch, W. J. and Mizzen, L. A.,** Characterization of the thermotolerant cell. II. Effects on the intracellular distribution of HSP70, intermediate filaments, and snRNP's, *J. Cell Biol.*, 106, 1117—1130, 1988.

151. **Welch, W. J. and Suhan, J. P.,** Cellular and biochemical events in mammalian cells during and after recovery from physiological stress, *J. Cell Biol.*, 103, 2035—2052, 1986.

152. **Welch, W. J., Garrels, J. I., Thomas, G. P., Lin, J. J. C., and Feramisco, J. R.,** Biochemical characterization of the mammalian stress proteins and identification of two stress proteins as glucose- and calcium ionophore-regulated proteins, *J. Biol. Chem.*, 258, 7102—7111, 1983.

152a. **Weller, N. K.,** A 70 kDa microtubule-associated protein in Nil8 cells comigrates with the 70 kDa heat shock protein, *Biol. Cell,* 63, 307—318, 1988.

153. **Whatley, S. A., Leung, T., Hall, C., and Lim, L.,** The brain 68-kilo-dalton microtubule-associated protein is a cognate form of the 70-kilodalton mammalian heat-shock protein and is present as a specific isoform in synaptosomal membranes, *J. Neurochem.*, 47, 1576—1583, 1986.

153a. **White, E., Spector, D., and Welch, W.,** Differential distribution of the adenovirus E1A proteins and colocalization of E1A with the 70-kDa cellular heat shock protein in infected cells, *J. Virol.*, 62, 4153—4166, 1988.

153b. **Wilhelmsson, A., Cuthill, S., Dehis, M., Wikstrom, A. C., Gustafsson, J. A., and Poellinger, L.,** The specific DNA binding activity of the dioxin receptor is modulated by the 90kD heat shock protein, *EMBO J.*, 9, 69—76, 1990.

154. **Yonemoto, W., Lipsich, L. A., Darrow, D., and Brugge, J. S.,** An analysis of the interaction of the Rous sarcoma virus transforming protein, pp60sarc, with a major heat-shock protein, In *Heat Shock from Bacteria to Man,* Schlesinger, M. J., Ashburner, M., and Tissieres, A., Eds., Cold Spring Harbor Laboratory, Cold Spring Harbor, NY, 1982, 289—298.

154a. **Zantema, A., De Jong, E., Lardenoije, R., and Van der Eb, A. J.,** The expression of heat shock protein HSP27 and a complexed 22-kilodalton protein is inversely correlated with oncogenicity of adenovirus-transformed cells., *J. Virol.*, 63, 3368—3375, 1989.

155. **Ziemiecki, A.,** Characterization of the monomeric and complex-associated forms of the gag-onc fusion proteins of three isolates of feline sarcoma virus: Phosphorylation, kinase activity, acylations and kinetics of complex formation, Virology, 151, 265—273, 1986.

156. **Ziemiecki, A., Catelli, M.-G., Joab, I., and Moncharmont, B.,** Association of the heat shock protein HSP90 with steroid hormone receptors and tyrosine kinase oncogene products, *Biochem. Biophys. Res. Commun.*, 3, 1298—1307, 1986.

Chapter 17

INDUCED THERMOTOLERANCE

L. Nover

TABLE OF CONTENTS

17.1. SURVEY OF SYSTEMS AND METHODS

The capability of adapting to hyperthermic stress is common to all living cells (Table 17.1). The term thermotolerance was coined by Henle and Dethlefsen[84b] in connection with a review on hyperthermic treatment of cancer, but, in addition to its important medical implications, the phenomenon of acquired tolerance to heat,[254,255] of heat hardening[4] or of adapted heat resistance[52,115c,196] was well known to plant physiologists long before induced synthesis of heat shock (hs) proteins was detected. Many aspects of heat hardening in about 60 species of higher plants were studied by Alexandrov's group at the Botanical Institute in Leningrad.[1a-4] The particular interest in plants results from the trivial fact that, unlike animals, they are fixed in their natural habitat and thus are much less capable of minimizing their exposure to stressful conditions.

Even older are the roots of studies on thermoadaptation in bacteria[28,43,186] which were mainly concerned with alterations of the cell membrane by substitution with saturated fatty acids and with the unusual stability and temperature optima of enzymes in thermophilic bacteria. They are not considered in this chapter.

The rapidly increasing interest in the molecular mechanisms of induced thermotolerance (TT) and the intriguing role of HSPs brought about a great number of papers summarized by Burdon,[29a] Hahn and Li,[70] Carper et al.,[33] Li,[129] Li and Mivechi,[133] Landry,[111] Nover,[165] Subjeck and Shyy,[217a] and Tomasovic.[223b] The relevant data are compiled in Table 17.1. Basically four types of procedures (conditioning treatments) are applied to make a system tolerant to the challenge with an otherwise highly damaging or even lethal hs.

1. During a long-term, moderate hs, cells start to restore structures and functions initially affected by the stress treatment. This "reparatory adaptation"[2] only functions below a certain threshold temperature and mainly concerns the most heat sensitive parts of the cell. If survival curves of animal cells are monitored under these conditions, a triphasic behavior is observed.[187] An initial phase of cell killing is followed by a much longer phase with cells surviving, but arrested in G_1 or G_2 phase (thermotolerance zone), and finally by a phase with slow resumption of cell cycling and killing due to aberrant DNA synthesis (see Section 15.2).
2. Closer to natural conditions is the method of a gradual heat adaption from very moderate to severe stress conditions (exemplified in Figure 2.1).
3. For experimental reasons, preinduction is the most convenient and frequently used method. Cells are treated by a pulse of hs and afterwards allowed to recover for a given time, i.e., usually several hours.
4. Related to preinduction is the pretreatment with chemical stressors. The effect is based on the phenomenon of induced cross-tolerance (see Figure 17.2).

The profound influence of hs on all parts of cellular structure and function is reflected in the multiplicity of protective effects observed in thermotolerant cells (Table 17.1). The extension of the analysis of thermotolerance phenomena from survival to distinct subcellular structures or functions reveals a considerable variability of the optimum temperature ranges required to observe a given effect. Finding this range is a prerequisite for analysis. This situation reflects the continuum of hs dependent changes as illustrated in Section 1.1 for *Drosophila* (Figure 1.1), but basically similar data were also compiled for plants[2] (see Figure 14.1). Methods used for analysis of thermotolerance effects are listed here once more to provide the basis for the following section on the role of HSPs in this respect.

17.1.1. CELL SURVIVAL AND GROWTH

It is the most general test for TT and is usually estimated from a clonogenic assay (plating efficiency). It is based on long-term recovery of cells from heat damage and is

TABLE 17.1
Induced Thermotolerance and Related Phenomena

Organism/cell	Conditioning treatment	Challenge treatment	Test	Remarks	Ref.
Bacteria					
Escherichia coli	htpR mutant	1 h 42°C	HSP synthesis (Table 2.2), survival	Not observed in htpR mutant (see Section 17.3)	160, 223, 253
	dnaK mutant, 5 min 42°C	3 h 42°C or 5 min 52°C	Inviable, prolonged HSP synthesis, but induction of TT normal	Dnak required for survival and autorepression (see Section 8.7)	175, 186a
	56 μ*M* H_2O_2 for 35 min	52°C	Improved survival	Resistance also to H_2O_2 and NEM	68
	30 min 42°C	50—55°C or ethanol	Improved survival	CT	161, 253
	Growth at 41°C	4 h 46—48°C	Improved survival	Increased level of saturated fatty acids in membranes	44, 256
	42°C	60 min 50°C	Improved survival	TT also induced by Eth, Cd, H_2O_2;	227
Salmonella thompson	30 min 48°C	1—3 h 54—60°C	Improved survival	HSPs not involved (see Section 17.3)	143a
Salmonella typhimurium	60 μ*M* H_2O_2	50°C or 10 m*M* H_2O_2, 1 m*M* *N*-ethylmaleinimide, 50 m*M* menadione	Induced resistance correlates with synthesis of enzymes under oxy^R control	—	37c
Myxococcus xanthus	1 h 36°C	1—2 h 40°C	Improved survival	—	162
Sulfolobus spec. strain B12	1 h 70° → 88°C	4 h at 92°C	Survival increased by six orders of magnitude	—	224b
Eukaryotic microorganisms					
Saccharomyces cerevisiae (yeast)	120 min 36°C	—	Recovery of control protein synthesis improved	Inhibited by CH or in ts mutant	147
	90 min 36°C	5 min 52°C	Protect from cell killing	Not observed in rna 1 mutant	148
	1.5 h 36°C or nutrient starvation	5 min 52°C	100-fold increased survival	HSP48 most prominent	94

TABLE 17.1 (continued)
Induced Thermotolerance and Related Phenomena

Organism/cell	Conditioning treatment	Challenge treatment	Test	Remarks	Ref.
	30 min 37°C	5 min 55°C or 10—12% ethanol	Increased survival	No protein synthesis required, HSP level not correlated with TT level	35, 238
	1—2 h at 36°C	4 min 52°C	Increased survival	Low cAMP levels required for TT induction	211
	60 min 37°C, in presence of CH, Can, or pF-Phe	8 min 52°C	Increased survival	No HSP synthesis required	73
	hs at 37°C, also with HSP 70- defective strains (see Table 2.4)	5 min 52°C	Increased survival	None of the HSP 70 isoforms required	240
	1 h 36°C or 1.55 *M* ethanol	10 min 50—52°C	Improved survival, more cells in G_0 state	CT	183, 184
	30—120 min 36°C	4 min 52°C or X-irradiation	Survival, induced resistance to hs or irradiation	CT	153, 154
		UV irradiation (254 nm)	Induced resistance	—	156
	30—120 min 38°C or 100 μg/ml CH	10 μg/ml MNNG or 2 mg/ml MNU	Suppresssion of mutation effects, → no induction of error-prone repair	—	157
	60 min 40°C	8 min 50.4°C	Improved survival	Correlates with markedly increased trehalose level	91
	40 min 25 → 38°C	10 min 50—52°C	Improved survival	Haploid more TT than diploid and tetraploid cells	180
	ubc4/ubc5 mutants	5 min 52°C	Tenfold increased rate of survival compared to wild-type cells	Constitutive HSP synthesis as a result of defective protein degradation	206a
	1 h 37°C or 5 h 1 mg/ml paronomycin	10 min 50°C	Improved survival	Both treatments induce HSP synthesis	66a
	1 h 37°C or 0.7*M* NaCl	60 min 48°C	Improved survival	CT	224c
Bremia lactucae (spores)	2 h 15° → 28°C	2 h 37°C	Improved germination at 15°C	—	99a
Candida albicans	30 min 44—46°C	20 min 55°C	Improved survival	Inhibited by trichodermin	258a

Dictyostelium discoideum	3 h 30°C	2—6 h at 34°C	Improved survival	Inhibited by CH	142
	3 h 30°C with mutant HL 122	2—6 h at 34°C	No protection	Mutant unable to synthesis small HSPs	143
Blastocladiella emersonii	30 min 34°C plus 30 min 38°C	30 min 42°C	Improved protein synthesis and zoospore development at 27°C	Inhibited by CH	212
Ustilago maydis (smut fungus)	10 min 42°C plus 2 h 32°C	2—8 min 48°C	Decreased DNA damage (frequency of recombination reduced)	—	221
Neurospora crassa (germinating conidiospores)	60 min 45°C	50°C	Improved survival	Inhibited by Ch	182
	60 min 48°C	30 min 2 m*M* H_2O_2	CT to H_2O_2, induced peroxidase	—	100—100b
	15 min 48° or H_2O_2, Ars, Cd^{2+}	40 min 51.5°C	Improved survival of mycelium	Correlates with induced synthesis of peroxidase	100a,b
	60 min 43°C	2 h 49°C	Survival of conidiospores	Inhibited by CH	220b
Tetrahymena thermophila	60 min 41°C in peptone medium plus 4 h recovery	90 min 41°C in starvation medium	Improved survival, no ribosome degradation	Role of rprotein p22, see Section 10.7	75, 150
	1 hr 40°C or 0.5 μg/ ml CH	20—90 min 43°C	Moderate stress resistance, improved survival	No HSP synthesis required, not inhibited by CH	74
	30 min 40°C	4—10 min 46°C	Resistance to severe stress, improved survival, stabilization of ribosomes	HSP synthesis required, inhibited by CH or emetine	74, 108
	30 min at 34°C, but not 200 μ*M* Ars	Long-term hs at 34°C or 90 min 39°C	Improved survival	Inhibited by CH or Act D	6
Poterioochromonas malhamensis (*Chrysophyceae*)	Growth at supraoptimal temperature	—	Reduced hs sensitivity	—	201
Acetabularia mediterranea (green alga)	1 h 30 or 36°C plus 2 h 26°C	1 h 37°C	Protects against heat damage of protoplasma	Effect inhibited by CH, also observed in enucleated cells	107
Chlamydomonas eugametos (green alga)	5 min 41°C plus 24 h recovery	5 min 37—45°C	Protection of cell movement	—	2
Chlamydomonas reinhardtii	2 h 40°C in the dark	40—42°C in the light	HSPs in chloroplast protect against photoinhibition of PSII	Inhibited by CH	202a

TABLE 17.1 (continued)
Induced Thermotolerance and Related Phenomena

Organism/cell	Conditioning treatment	Challenge treatment	Test	Remarks	Ref.
Animals					
Coelenteratae					
Hydra atenuata	2 h 18° → 30°C plus 4 h 18°C	12 h 34°C	Increased viability of polyps	TT also induced by Cd or azide	26a
Crustaceae					
Artemia salina	60 min 40°C	30 min 42°C	Delay of hatching	Cysts are more resistant than larvae	151a, 151b
Molluscs					
Lymnaea stagnalis (pond snail, larvae)	5 min 33—35°C vs. 30—60 min >37°C, or 3.5 vs. 9.1% ethanol	1 h 40°C or 3.5 min 43.6°C	Improved survival and development	Two states of TT defined (see Table 17.3)	23—25
Insects					
Drosophila melanogaster					
(Larvae)	30 min 35°C	20 min 40.1°C	Recovery of protein synthesis improved	—	178
	25 min 36°C plus 30 min 28°C	25 min 40.5°C or 90 min anoxia	Survival improved, rapid translocation of HSP70 to nuclei	CT	231
	30 min 36°C plus recovery	5 m*M* chloramphenicol	Protection from membrane depolarization	—	14
(Pupae)	10—30 sec 40.5°C plus recovery	35—43 min 40.5°C	Reduced phenocopy induction	—	151
	45 min 34—37°C	1 h 41°C	Rapid redistribution of small HSPs, reduced development defects	—	11
	40 min 40.2°C plus 2—5 h recovery or 60 min 34—35°C	40 min 40.2°C	Improved HSP synthesis, protection from phenocopy induction	—	38, 158, 178a
(S cells)	10—60 min 36°C	1 h 40°C	Protection of hnRNP splicing (Figure 9.2)	—	257
	Ecdysterone-induced synthesis of small HSPs		Recovery of protein synthesis improved	Only small HSPs required	15, 16

	Gradual heat adaptation	38°C	HSP synthesis and mRNA stability at high temperature improved, rapid recovery of control protein synthesis		47, 140, 141
	30 min 33°C plus 1—6 h recovery	Long-term cultivation at 33°C and 37°C	Improved survival	—	224
(Kc-cells)	1 h 37°C	—	Recovery of rRNA synthesis dependent on preceding HSP synthesis	Inhibited by emetine	9
	1 h 37°C	—	Recovery of normal pattern of histone modification needs preceding HSP synthesis	Inhibited by CH or emetine	10
(Primary embryonic cell cultures)	30 min 35°C	2 h 40.2°C, 0.1 m*M* diphenylhydantoin or 1 m*M* coumarol	Protection from teratogenic effects	CT	27, 31
Ceratitis capitata (larvae)	30—60 min 35°C	45°C	Improved survival	—	214, 215
Calpodes ethlius (larvae)	1—2 h 37°C	1 h 45°C	Improved survival and development	—	42
Echinoderms					
Paracentrotus lividus, Sphaerechinus granularis, Arbacia punctulata (sea urchins)	1 h 31°C	1 h 35°C	Survival and normal development dependent on HSP 70/72 synthesis	Early developmental stages unable to induce HSP70/72 synthesis	83, 192, 193, 205
Fishes					
Rainbow trout (fibroblasts)	Culture at 28°C	4 h 32°C	Improved survival	Inhibited by Act D	159, 159a
Pimephales promelas (FHM cells)	10 min 35°C plus 1 h 18°C	5 min 43°C	Improved survival	—	200a
Tilapia mossambica, T. nilotica (ovary cells)	15 min 40°C plus recovery	20 min 43°C	Improved survival	Correlates with HSP synthesis	36a
Oryzias latipes					
(Primordial germ cells)	30 min 37—42°C	γ-irradiation at 2.5 Gy/min	Improved survival	No resistance induced by Ars, Cd, or anoxia	207, 207a, 209, 210
(Embryos)	10 min 43°C plus 4 h recovery	1—2 h 43°C	Improved hatchability	Inhibited by CH	208

TABLE 17.1 (continued)
Induced Thermotolerance and Related Phenomena

Organism/cell	Conditioning treatment	Challenge treatment	Test	Remarks	Ref.
Amphibia					
Xenopus laevis (embryos)	20 min 35°C	20 min 35°C	Developmental defects in early embryonic stages without HSP synthesis, mRNA degradation	No induced TT before midblastula stage (see Figure 17.2)	82, 163
Rana temporaria (embryos)	2 min 37°C plus 24 h 15°C	8 min 37°C	Protection from developmental defects	—	51
Reptilia					
Eurycea bislineata (salamander)	1 h 31.2°C plus 90 min recovery	34.2°C	Extension of critical thermal maximum of tetanic paralysis	No correlation with HSP synthesis	49a, 194a
Birds					
Chicken					
(Embryo fibroblasts)	30 min 47°C plus 5 h 37°C	1—4 h 47°C	Improved survival, HSP synthesis	—	114
	2 h 45°C plus recovery	—	Recovery of IF morphology requires HSP synthesis	Inhibited in presence of Act D	39
(Neural retina explants)	Growth at 43°C	2—3 h 46°C	Increased HSP synthesis, maintenance of substrate contact	—	34
Mammals					
Rat					
(Embryos, *in utero*)	30 min 42°C plus 1 h 37°C	30 min 43°C	Protection from growth retardation and teratogenic effects	—	152
	10 min 42°C plus 15 min 38.5°C	7.5 min 43°C	Protection from teratogenic effects	—	235
(Embryo fibroblasts)	1.5 h 43°C plus 8 h recovery	30 min 45°C	Improved recovery of protein synthesis, conservation of cytoskele-	—	159, 239

			ton and nuclear snRNP distribution		
(L 132 cells)	1 h 42°C plus 10 h recovery	1 h 44.5°C	Improved survival	Not induced by 10 μg/ml Cd	36
(Hepatoma cells)	30 min 42°C	60—180 min 42°C	Improved survival and protein synthesis, reduced loss of intercellular contacts	—	197
	30 min 42°C	43—44°C	Protection from rounding up and collapse of IF and tubulin cytoskeleton	Requires HSP synthesis	244, 245
	15—30 min 43°C plus 5 h 37°C	2.5 h 43°C	Improved survival	Also induced by 6% Eth but not by 50 μ*M* Ars, not inhibited by CH	112, 113
	Ca^{2+} - deprivation	Long-term hs at 43°C	Improved survival	Inhibited HSP synthesis	110
	Serum stimulation of G_0 cells	1 hr 42.5°C	Elevated HSP levels, improved survival	—	228
(Mammary carcinoma cells)	20 min 45°C plus recovery or long-term 42°C	20—80 min 45°C	Improved survival	Poor correlation with level of HSP synthesis	224a
(Rat-1 cells)	20 min 45°C plus recovery	Permanent 45°C	Protection from cell killing	Not inhibited by CH	242
(Primary culture of hepatocytes)	30 min 42°C plus 16 h recovery	30 min 44°C	In TT cells no hs effects on substrate contact, protein level and activity of membrane K^+/Na^+ pump	—	197a
Mouse					
(Ear, *in situ*)	20—40 min 43.5°C plus 1—4 d recovery	Hyperthermia ± irradiation	Reduction of TER	—	122, 123
(Mammary carcinoma cells)	12 min 45°C plus 8 h 37°C	27 min 45°C	Improved survival	—	219
	30 min 43°C plus 5 h 37°C	30 min 45—47°C ± irradiation	Improved survival	—	80, 81
	30 min 43°C plus 2h 37°C	15 min 45°C	Improved maintenance of IF and resumption of protein synthesis	—	211a
(Mammary carcinoma, implanted)	30 min 43°C plus 1—7 d 37°C	1—4 h local hyperthermia at 44°C	Tumor growth, reduction of "tumor bed effect"	—	247

TABLE 17.1 (continued)
Induced Thermotolerance and Related Phenomena

Organism/cell	Conditioning treatment	Challenge treatment	Test	Remarks	Ref.
(RIF-1 tumor cells)	10 min 45°C plus recovery	20—60 min 45°C	Improved survival	Selection of resistant cell line	71
(L cells)	Slow heat adaption	60—240 min 43°C	Improved survival	No change of membrane lipids	30
	15 min 45°C plus 20 h 37°C	80 min 45°C	Improved stability of reporter proteins (β-galactosidase)	—	162a
(Plasmacytoma cells)	10 min 44°C plus 2 h 37°C	4 h 43°C	Improved survival	Induced HSP70 not required	11b
(3T3 cells, fibroblasts)	Mild stress vs. severe stress conditioning	30 min 46°C	Protection from rounding up and contraction	Two states of TT defined (see Table 17.3)	229
	20 min 43.7°C plus 3 h 37°C	2 h 46°C	Improved survival and increased cell cycle activity in recovery	—	230
(Embryo cells)	15 min 45°C plus 6 h 37°C	1—4 h 45°C	Improved survival	Correlates with increased levels of SOD	172
(Embryo fibroblasts)	4 h 38.5°C	45 min 42°C	No formation of intranuclear actin rods	—	95a
	10 min 44°C plus 5 h 37°C	2 h 44°C	Improved survival	Effect abolished by 3 μ*M* amiloride	194a
(Implanted squamous cell carcinoma and fibrosarcoma)	15 min 43°C *in situ*	45 min 45°C	Improved colony assay of isolated tumor cells	TT correlates with HSP70/68 levels	132
(Neuroblastoma cells)	20 min 42°C	43—44°C	Protection from rounding up and collapse of cytoskeletal systems	Requires HSP synthesis	244, 245
(Melanoma cells)	Selection of resistant cell lines	20—120 min 45°C	Survival of resistant cell lines	Not correlated to HSP levels	7
(Implanted tumors in feet)	30 min 44°C *(in situ)* plus 24 h recovery	90 min 44°C	Resistance to cell killing	—	41

(Ehrlich ascites tumor cells)	10 min 44°C plus 24 h 37°C	40 min 44°C	Improved survival	No increase of cholesterol level in TT cells	6b
(Foot skin)	30 min 43°C plus 6—24 h recovery	60 min 44°C ± irradiation	Resistance to cell damage	CT	246
(Small intestinal mucosa)	40—90 min 42°C plus recovery	γ-irradiation at 42.3°C	Reduced radiosensitization	—	92, 144a
(Exteriorized loop of small intestine)	60 min 42°C plus 10 h recovery	35 min 43°C	No destruction of intestinal crypts	—	91a
(Testis)	30—60 min 40°C plus 4 h recovery	30 min 41.5°C	Developmental retardation and weight loss reduced	—	144b
(Foot)	22.5 min 45.5°C plus 24 h 37°C	Long-term hs at 45.5°C	Reduced necrotization of toes	—	188b
	30 min 43°C plus 168 h recovery	Hyperthermia plus irradiation	Reduced TER	—	246a
Chinese hamster					
(Exteriorized loop of small intestine)	8 min 44.5°C plus 24 h recovery	Continuous hs at 44.5°C	LD_{50} of challenge hs increases from 6 to 21 min	—	151c
(Ovary fibroblasts, CHO cells)	Continuous hs at 42.4°C	Irradiation	With ongoing hs TER decreases from 2.5 to 1.0	—	89a
	10 min 45.5°C plus 12 h 37°C	20—120 min 45.5°C	Improved survival	Effect inhibition by CH and puromycin	124—124c
	1.5 h 43°C plus recovery	20—40 min 45°C	Improved recovery of protein synthesis	—	158b
	Permanent hs at 42.2°C	—	Transient TT between 4—15 h (survival, stability of DNA polymerase activity)	—	44a
	10—30 min 44°C plus 5—24 h 37°C	1—8 h at 44°C	Improved survival	Effect also observed in respiration deficient cells	37b, 62a, 115
	12 min 45.5°C plus 5—10 h recovery	Incubation at 45.5°C	Improved survival	TT expressed equally well in G_1, G_2, or S phase cells	189
	2 h 40°C	20—30 min 45°C	Improved survival	Inhibited by CH	185

TABLE 17.1 (continued)
Induced Thermotolerance and Related Phenomena

Organism/cell	Conditioning treatment	Challenge treatment	Test	Remarks	Ref.
	10—15 min 45°C plus 7 h 37°C	Long-term hs at 45°C	Improved survival	Inhibited by CH, if added after pulse hs	85—87, 195
	Selection of resistant cell line HR-01	20 min 46°C	Improved survival	Increased level of HSP90, but not of HSP70	252
	10 min 45°C plus recovery	45 min 45°C	Improved survival	Induced TT inhibited in presence of 85% D_2O	55, 135-137
	80h 42°C	—	Stop of cell killing in G_1 cells after 4 h, G_2 after 12 h, S phase cells no stop	—	187
	5—12 min 45°C plus 10 h recovery	27 min 45°C	Improved survival and protein synthesis	No good correlation with HSP levels	204, 218
	6 min 46°C or 20 min 45°C plus recovery	45 min 45°C	Improved survival	Only HSP 70 level correlates with state of TT, also induced by Eth, Ars (see Figure 17.3)	129, 134
	10^{-6} *M* glucocorticoids for 20 h	45 min 45°C	Improved survival	No increase of HSP levels, other steroids not active	56
	Addition of amino acid analogs (Can, AzC)	45 min 45°C	No induced TT, despite increased HSP synthesis	Reversible heat sensitization	120, 121b, 131
	10 min 45°C plus recovery	45 min 45°C	Protection from cell killing	Not related to cAMP level	32
	Alcohols (C_2 to C_8), local anesthetics, DMSO, DMF	45 min 45°C	Protection from cell killing	Correlates with induced HSP synthesis, CT	72
	30 min 43°C plus 6 h 37°C		Protection from cell killing	Depletion of polyamines not inhibitory to induction of TT	58

	10—20 min 45°C plus 9 h 37°C	45 min 45°C	Improved survival	Correlates with increased Ca^{2+} levels	216, 217
	15 min 45°C plus 4 h 37°C vs. 6 h 41°C or gradual heating to 43°C	45 min 45°C	Improved survival and protein synthesis under hs conditions	Two states of TT defined, only indution at 45°C needs protein (HSP70) synthesis, also induced by Ars	7a, 121b, c
	1 h 0.4 m*M* diamide or 15 min 43°C	2 hrs 43°C or 0.8 m*M* diamide	Improved survivial	Correlates with increased HSP levels, inhibited by CH, CT	56a—c, 124d
(Lung cells)	20 min 44°C plus 10 h recovery	4 h 44°C	Improved survival	Thermoresistance correlates with HSP27 level	115a
	20 min 44.5°C plus 16 h 37°C	50 min 44.5°C	Improved survival	No influence of membrane fluidity	125
	20 min 44°C plus 24 h 37°C	50—70 min 45°C	Improved survival	Induced TT correlates with increased cholesterol synthesis	78
	2 h 42°C	10—60 min 44°C	Improved survival	Correlates with level of preformed HSP68	170
	1 h 43°C plus 5 h 37°C	3 h 43°C	Improved survival	Correlates with increased glutathione levels	158a
Pig, kidney cells	Selection of resistant cell line	20—90 min 46°C	Improved survival and membrane stability	—	188
African green monkey (kidney epithelial cells, Vero cells)	30 min 45°C	—	Enhanced reactivation of UV-damaged Herpes virus	Also induced by UV irradiation	251
Monkey, COS cells	Transformation with *Drosophila* hsp70 gene	45 min 43.5°C plus recovery	Enhanced recovery of nucleolar ultrastructure	—	177a
Man					
(HeLa cells)	1 h 44°C plus 5 h 37°C	1 h 44°C	Improved survival	—	61
	1.5 h 43°C plus recovery	20—40 min 45°C	Recovery of protein synthesis improved	Also induced by 80 μM Ars or 15 m*M* AzC	159
	1.5 h 43°C plus 8 h 37°C	30 min 45°C	Conservation of IF cytoskeleton and of nuclear snRNP distribution	—	239
	10 min 45.5°C	—	Improved reactivation of UV-damaged adenovirus 2	Also induced by 5% Eth, 50 μM Ars or UV irradiation	181

TABLE 17.1 (continued)
Induced Thermotolerance and Related Phenomena

Organism/cell	Conditioning treatment	Challenge treatment	Test	Remarks	Ref.
	1m*M* Na-butyrate or dibutyryl-cAMP	6 h 42°C	Improved thermoresistance	No influence on HSP synthesis, but increased level of HSC 90	79, 89b
	Cultivation at 39°C	20 h at 42°C or 3 h at 43°C	Improved survival and recovery	Selection of resistant cell lines	206
	15 min 45°C plus 5 h 37°C	Continuous hs at 45°C	Improved survival and recovery of normal nuclear protein content (see Section 9.3)	—	99b
	1 h 42.5°C plus 3 h 37°C	1 h 45°C	*In vitro* splicing activity of nuclear extracts preserved	—	18a
(Surgical specimens of breast carcinoma, malignant melanoma, and squamous cell carcinoma)	1 h 43.5°C plus 24 h 37°C	2—6 h at 43.5°C	Improved survival	Heat sensitivity and TT levels greatly different between individual tumors	193a
(Skin fibroblasts)	20 min 45—47°C	CO_2 laser irradiation	Protection from laser damage	Correlates with induced HSP synthesis	184a
Plants					
Conifers					
Abies alba (plantlets)	30 min 44°C plus 8 d at 20°C	39—42°C	CO_2 fixation in TT plantlets continued	—	13a
Monocotyledons					
Tradescantia palludosa (germinating pollen)	Slow temperature adaptation	30 min 41°C	No growth arrest	No HSP synthesis	5, 250
Tradescantia fluminensis (leaf cells)	3 h 33—35°C	5 min 43.7°C	Protection of protoplasma streaming	—	4
		5 min ≤ 44.2°C	Protection of chloroplast phototaxis	—	

		5 min ≤ 59.4°C	Protection of plasmolysis capacity	—	
		5 min ≤ 60.9°C (see Figure 14.1)	Protection of tonoplast membrane (anthocyanin leakage)	—	
Hordeum vulgare (barley, seedling)	30—60 min 37°C	4 h 45°C	Growth at 25°C improved	—	145
Triticum aestivum (wheat, seedling) and suspension cultures	2 h 41°C or 12 h 37°C	1—3 h 48—51°C	Growth at 25°C improved, TTC reduction	High intrinsic heat resistance in early phase of germination, genotype-specific differences	1, 109a, 235a, 259
Sorghum bicolor (seedling)	15—30 min 40—45°C plus 2 h at 35°C	2 h 45°C	Growth at 25°C improved	—	173
Panicum spec. (millet, seedling)	2 h 40°C	40—50°C	Improved ^{3}H-Leu incorporation	—	105
Lilium longiflorum (growing pollen)	800 μM proline in medium	10 min 45°C	Protection from heat damage, conservation of control protein synthesis	HSP synthesis inhibited	90
Dactylis glomerata (leaf epidermis)	18 h 37.5°C	5 min 46.9°C	Maintenance of protoplasmic streaming	—	4
Zea mays (corn, seedling)	6 h pretreatment with PEG or 4 d with 50—200 μM Cu^{2+}, Cd^{2+}, or Zn^{2+}	3 h 40 or 45°C	Diminished growth retardation, analyzed after 3 d	Not correlated with increased HSP levels	21a
Saccharum officinarum (sugarcane, cells)	2 h 36°C plus 8 h 25°C	15 min 54°C	Improved survival	TT induction connected with HSP synthesis	158c
Dicotyledons					
Glycine max (soybean) (Seedlings)	2 h 40°C	10 min 47.5°C or 2—5 h 45°C	Sugar and amino acid leakage decreased	HSP15 in plasma membrane	139
	Slow heat adaption		Protein synthesis extended to higher temperature, survival of cells	—	5, 106
	2 h 40°C	40—50°C	Improved ^{3}H-Leu incorporation	—	105

TABLE 17.1 (continued)
Induced Thermotolerance and Related Phenomena

Organism/cell	Conditioning treatment	Challenge treatment	Test	Remarks	Ref.
	10—15 min 40—45°C plus 2 h 28°C or 3 h 50 μ*M* Ars	2 h 45°C	Protein synthesis at 45°C and development improved	See Table 17.2	138
	2 h 38°C plus 0.5 h 42.5°C	—	Respiratory coupling between O_2 uptake and ATP generation in isolated mitochondria at 42.5°C	Protection correlates with HSPs 15—18, 22, and 24 levels in isolated mitochondria	37a
(Roots)	3 h 40°C plus 3 h 28°C	1 h 45°C	Improved survival and preservation of cell morphology	—	144
Vicia faba (broad bean, root tip)	10—30 min 40°C plus 10—60 min 25°C	Treatment with maleic hydrazide, triethylene-immino triazine, bleomycin, X-rays, trenimon	Reduced number of chromatide aberrations	Hs-induced SOS repair (?), inhibited by CH	190—191b
Pisum sativum (pea, seedling)	2 h 37.5°C	40—50°C	Improved ^{3}H-Leu incorporation	—	105
Phaseolus vulgaris (bean)					
(Seedling)	3 times 3 h/day at 37, 45 and 47°C, respectively	5 h 50°C	Recovery of chloroplast electron transport and protein synthesis	—	219a
(Leaves)	25 s 50°C plus 36 h recovery	12 s 55°C	Survival improved	Similar results with leaves of cowpea, cucumber, fig, and tobacco	254, 255
Vigna unguiculata (cowpea, cell culture)	Cultivation at 36°C	Growth at 38—40°C	Fresh weight gain improved	Only HSPs70 and 80 markedly increased	89
Vigna radiata (cowpea, seedling)	1 h 40°C	45°C	Survival and growth at 25°C improved	Act D and CH inhibition, addn. of gibberellic acid improves	36b

Pear (suspension culture)	6 d at 30°C or 20 min 38°C plus 6 h recovery	20 min 43°C	Survival improved, electrolyte leakage decreased	—	248
Agave deserti, Carnegiea gigantea, Ferocactus acanthoides (desert succulents)	10 d adaptation at 50°C day/ 40°C night	60—70°C	Survival of chlorenchyma cells (vital staining)	—	104
Nicotiana rustica (tobacco, leaves)	2 min 46°C plus 3—24 h 20°C	2 min 53—54°C	Leaf damage decreased	Cytokinins improve protection	52
Lycopersicon peruvianum (tomato, cell cultures)	Slow heat adaptation or 15 min 40°C + recovery	≥ 40°C	Protein synthesis extended up to 43°C, protection of control mRNAs	—	166, 168, 198
		≥ 40°C	Recovery of protein S6 phosphorylation at 25°C	—	198
		150 min 40°C	Amino acid uptake stabilized (Figure 14.3)	—	165
		1—2 h 40°C	Synthesis of preribosomes and processing in recovery period (Figure 10.2)	HSP70 in nucleolus (Section 10.6)	167, 199
Leucanthemum vulgare (epidermal cells of leaves)	10 s 49.5°C plus 3 min recovery	5 min 50—60°C	Protection of cell membranes, capable for plasmolysis	—	4
Campanula persicifolia (leaf epidermis cells)	5 min 45°C plus 24 h recovery	5 min 50—54°C	Protection of protoplasma streaming	—	2
Zebrina pendula (leaf epidermis cells)	5 min 46°C plus 24 h recovery	5 min 50—54°C	Protection of protoplasma streaming	—	2
Atriplex lentiformis	Growth at 43/30°C (day/night) instead of 23/18°C	46°C	No reduction of photosystem II activity	—	177

Note: Abbreviations are: Act D, actinomycin D; AzC, azetidine carboxylic acid; Ars, arsenite; Can, canavanine; Cd, Cd^{2+} salts; CH, cycloheximide; CT, cross tolerance; DMF, dimethylformamide; DMSO, dimethylsulfoxide; Eth, ethanol; IF, intermediate size filaments; MNU, methyl nitroso urea; MNNG, *N*-methyl-*N'*-nitro-*N*-nitrosoguanidine; NEM, N-ethylmaleimide; PEG, polyethylene glycol; pF-Phe, *p*-fluoro-phenylalamine; SOD, superoxide dismutase; ts, temperature-sensitive (mutant); TT, thermotolerance; and TER, thermal enhancement ratio.

TABLE 17.2
Thermotolerance of Soybean Seedlings Induced by hs Preconditioning

		Size range of seedlings (%) (cm)			Average length (cm)
		<5	5—10	>10	
1	Control (28°C)	0	0	100	21.2
2	2 h at 40°C	0	0	100	19.4
3	2 h at 45°C	100	0	0	3.0
4	15 min 40°C + 2 h 45°C	43	40	17	6.8
5	1 h 40°C + 2 h 45°C	0	3	97	12.9
6	15 min 40°C + 4 h 28°C + 2 h 45°C	11	48	41	8.8
7	10 min 45°C + 0.5 h 28°C + 2 h 45°C	70	30	0	4.3
8	10 min 45°C + 1 h 28°C + 2 h 45°C	47	40	13	6.8
9	10 min 45°C + 2 h 28°C + 2 h 45°C	0	17	83	11.2
10	3 h 50 μ*M* arsenite	—	—	—	15.9
11	3 h 50 μ*M* arsenite + 2 h 45°C	—	—	—	9.8

Note: Germinating seedlings of *Glycine max*. var. Wayne at 1-cm length of the embryonic axis were subjected to the indicated temperature treatments and afterwards kept for 3 d in the dark at 28°C. Protective effect against strongly deleterious effects of 2 h 45°C treatment (1 vs. 3) are optimal by 1 h 40°C conditioning (5) or 10 min 45°C followed by 2 h 28°C (9). Necessary is an extensive period of heat shock protein (HSP) synthesis. Arsenite (50 μ*M*) has a comparable protective effect. From Lin, C.-Y., Roberts, J. K., and Key, J. L., *Plant Physiol.*, 74, 152, 1984.

almost certainly the least suitable test to gain insights into the molecular levels of putative HSP functions. Somewhat related to this are experiments with developing or growing organisms. This is exemplified in Table 17.2 for soybean seedlings. The severe retardation of seedling growth by 2 h 45°C is readily overcome by a conditioning treatment of 1 h at 40°C or 15 min at 45°C followed by 2-h recovery.[138] Very similar are the protective effects in developing rat embryos (Figure 17.1). The thermotolerant state is connected with a protection from teratogenic effects by the challenge treatment at 43°C.[235]

17.1.2. DEVELOPMENT

In rare cases of hs-induced developmental deviations, e.g., phenocopies of *Drosophila*, the defect can be traced back to a single gene expressed at a given time. Correspondingly, protection from phenocopy induction requires stabilization of the expression and/or biological action of this gene product (see Section 18.3). Use of developing organisms provides a particularly sensitive assay because minute changes in the determined program may give rise to relatively large effects by multiplication during the development of cell lineages.[158,178a]

17.1.3. MEMBRANE FUNCTIONS

The prerequisite for survival is the integrity of the plasma membrane, which is considered as a primary site of hs effects (see Section 14.4). Structure and function of the plasma membrane were checked by vital staining, leakiness to solutes (amino acids, electrolytes, nucleosides), amino acid uptake, hormone-dependent signal transformation, membrane potential, and microviscosity measurements using fluorescence polarization, as well as aspects of membrane biosynthesis. In plants, a special type of intracellular membrane (tonoplast) separates the usually cytotoxic vacuolar content from the cytoplasma. Integrity of the tonoplast was followed by plasmolysis and by the retention of vacuolar constituents, e.g., anthocyanins.[2]

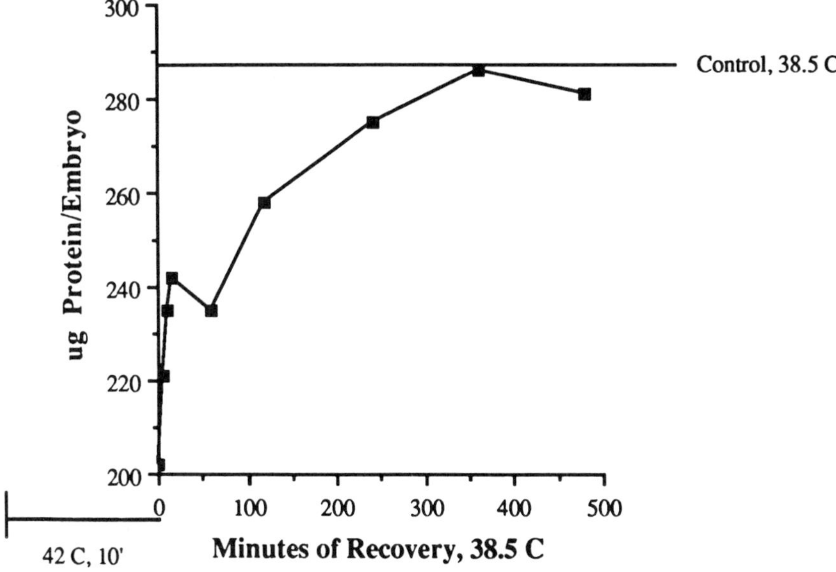

FIGURE 17.1. Acquisition of thermotolerance in developing rat embryos. Isolated, 9.5-gestational-day rat embryos were heated at 42°C for 10 min followed by recovery at 38.5°C. After the given period, they were challenged by 7.5 min heating to 43°C, normally a teratogenic dose in an unprotected embryo. After 48 additional hours of *in vitro* cultivation at 38.5°C, they were investigated for defects,[235] and the development was monitored by the protein content. Optimum protection was achieved after 5-h recovery between the two hyperthermic treatments. (From Walsh, D. A. and Hightower, L. E., unpublished.)

17.1.4. CYTOSKELETAL SYSTEMS

The rapid hs-induced rearrangements of cytoskeletal systems (see Figure 14.2) thoroughly influence many aspects of cellular activities, from membrane functions to intracellular transport systems, diverse parts of gene expression, and to cell-cell or cell-substrate contacts. Although all three types of networks (see Section 14.3) are affected, the intermediate-size filament system (7- to 10-nm fibers) is evidently most important. Maintenance and/or rapid recovery of its normal distribution can be monitored by means of immunofluorescence methods using appropriate antibodies (Figure 14.2). Presumably intimately related to the changing network of microfilaments is the interruption of protoplasma streaming, i.e., the well-known and easily observed movement of cell organelles along the thin film of protoplasma surrounding the extensive vacuolar system of plant cells.[2,4]

17.1.5. CHLOROPLASTS

Regarding survival and growth of plants under stressful conditions, the preservation of chloroplasts is of particular importance. Protection was documented by Pearcy et al.[177] and Süss and Jordanov,[220] investigating photosynthetic electron transport, changes of membrane composition or chlorophyll fluorescence.

17.1.6. DNA DAMAGE AND REPAIR

Deleterious effects of stress treatments are minimized in thermotolerant cells. Simultaneously, an increased resistance to certain mutagenic treatments is observed (see Section 15.3). These effects were extensively studied in connection with hyperthermia in cancer therapy (see Chapter 20).

17.1.7. GENE EXPRESSION

Many data rely on the analysis of protected structure and function of the gene expression machinery. This is detailed in different parts of the book. Examples are: processing of intron-containing pre-mRNP (see Section 9.5), ribosome biosynthesis and the reversible changes of nucleolar structure (see Chapter 10), stability of ribosomes in *Tetrahymena* (Section 10.7), maintenance of translation even at high temperatures (see Figure 2.1), the recovery of control protein synthesis (Section 13.3), and protection of untranslated mRNAs (Section 11.4).

17.1.8. PROTEIN DENATURATION

Effects of hs or chemical stressors on secondary and tertiary structure of proteins were frequently investigated. The intricate role of abnormal proteins for induction of hs genes (see Section 1.4) and the characterization of proteases and ubiquitin as HSP (see Sections 2.3.5 and 12.2) underline the importance of this part of the hs response. In support of an early suggestion by Minton et al.,[151d] plant HSPs were reported to stabilize soluble proteins against heat denaturation *in vitro* at 55°C.[97a] On the other hand, inactivation of reporter enzymes (β-galactosidase, luciferase) in transformed mouse cells was reduced by a prior conditioning treatment of 15 min at 45°C followed by 20 h recovery at 37°C.[162a] The recognition of defined peptide motifs exposed by partial unfolding of proteins and the ATP-dependent renaturation are postulated functions of HSP70 and HSP60 proteins in their role as 'molecular chaperones' (see Section 12.2).[49b,194]

17.2. EVIDENCE FOR THE ROLE OF HEAT SHOCK PROTEINS

Attempts to prove or disprove the function of heat shock proteins (HSPs) for thermotolerance are multifold, and the outcome is frequently equivocal. This is not only due to the complexity of the matter, i.e., the multiplicity of cellular functions to be considered, but also to the highly restricted experimental procedures in many papers which contrasts to the far reaching conclusions drawn from the results. Experimental manipulation of the HSP content is the main method applied. However, it should be kept in mind that not only the hs-induced, but also the constitutively expressed isoforms of stress proteins are essential for the overall effect, which, in fact, may require cooperation between both.

17.2.1. DEVELOPMENTAL ACQUISITION OF HSP SYNTHESIS AND THERMOTOLERANCE

A strong argument in favor of a general role of HSPs in thermotolerance comes from hs studies on developing embryos (Figure 17.2). In the earliest stages of embryo development, hs is particularly toxic, whereas late blastula to gastrula stage embryos can survive the same dosis of stress. The decisive difference between both stages may be the capability to synthesize HSPs, which appear in midblastula embryos. In the early stages, induced HSP synthesis is lacking because of the general inactivity of the embryonic genome.[6a,49] The results, compiled in Figure 17.2 for *Drosophila*[17,67] and *Xenopus,*[29,82] are equally valid for sea urchin[64,192,193,205] and snail embryos.[23,26]

A peculiarity was noticed for *Drosophila* larval and pupal stages.[15,16] As detailed in Section 8.4, synthesis of small HSPs is induced during certain developmental stages (Figure 8.4). This selective, hs-independent induction leads also to a thermotolerant state analyzed by improved survival and recovery of control protein synthesis. However, the authors conclusion that, in *Drosophila,* only the small HSPs are required for thermotolerance is certainly premature since the existence of the constitutive isoform of HSP70 was neglected. However, it can be speculated that the protection of mRNA in the perinuclear heat shock granules (hsg) may be particularly effective in these developmental stages (see Sections 11.4 and 16.7).

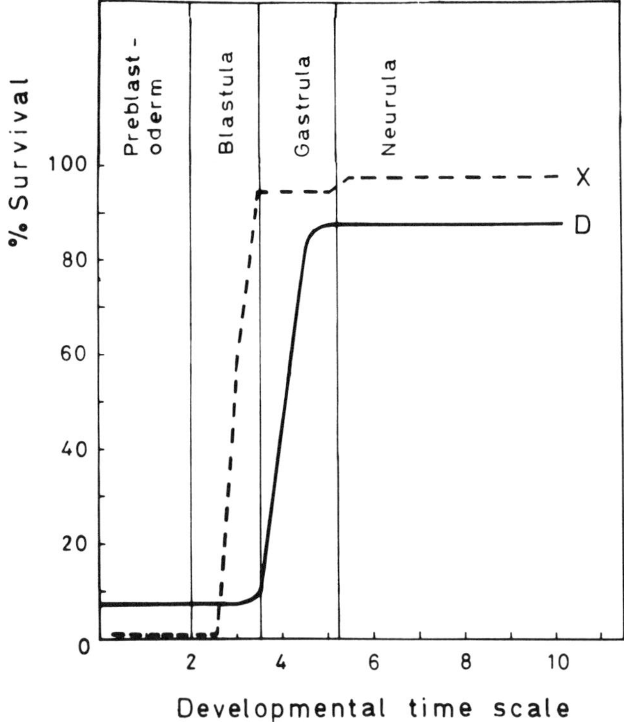

FIGURE 17.2. Developmental induction of thermotolerance in *Drosophila* (D) and *Xenopus* (X). Data were compiled from Bergh and Arking[17] *(Drosophila)* and Heikkila et al.[82] *(Xenopus)*, respectively. Embryonic development proceeded at 22°C. Survival was monitored after a challenge hs of 40 min 37°C for *Drosophila* and 20 min 35°C for *Xenopus*. The arbitrary time scale at the bottom is defined by the developmental stages given above. The appearing resistance in the blastula *Xenopus* and gastrula stages *Drosophila* correlates with the acquired capability for induced HSP synthesis.

The role of developmentally regulated HSPs and other components increasing the intrinsic heat resistance was also stressed by Abernethy et al.[1] and Helm et al.[84a] In the very early phase of germination of wheat seeds (0 to 12 h), a remarkably high intrinsic heat resistance could not be increased by an hs conditioning treatment (2 h 42°C). But thermotolerance was inducible later in the germination process. No difference between the two states was found with respect to hs-induced HSP synthesis but rather in the constitutive synthesis of HSPs 60, 58, 46, 40, and 14 between hours 0 and 6 of germination.

17.2.2. EXPERIMENTAL INDUCTION AND INHIBITION OF THERMOTOLERANCE

The frequently used preinduction protocol to make cells thermotolerant corresponds to the three phases of thermotolerance:[113,130] (1) a few minutes *induction* by a pulse hs is followed by, (2) several hours of *expression* under control conditions, and then by (3) a slow *decay* (several hours to days). This three-phasic behavior correlates perfectly with all that is known about HSP synthesis and stability. A pulse hs is sufficient to trigger their expression. Following this, they accumulate to appreciable levels within a few hours, and their stability is usually sufficiently high to maintain the elevated levels over several days. In keeping with this, optimum conditions for protein synthesis are required in the expression phase.[70,113,128,130,148,158,178,218] Addition of cycloheximide or other inhibitors of gene expression prevent the acquisition of thermotolerance (see References 9, 56b, 56c, 85, 112, 124a, 124c, 142, 185, 212, and 244). No inhibition by cycloheximide was reported by other authors. This is discussed in Section 17.5.

Direct evidence for the role of HSP70 for hs survival came from microinjection experiments with affinity-purified antibodies and rat fibroblasts. The antibodies inhibited translocation of the HSP70 to the nucleus and reduced survival of even a short incubation at 45°C.[188a] Other approaches to selective manipulation of HSP synthesis arose with advancing gene technology. They may provide the means for a more direct experimental access to the role of HSPs for distinct aspects of thermotolerance. Pelham[177a] used the *Drosophila* hsp70 gene fused to the adenovirus major late promoter to transform monkey COS and mouse cells. The constitutively expressed, heterologous HSP70 was correctly bound to the nucleolus during hs and enhanced the recovery of nucleolar structure after the stress period. Very similar is the experimental strategy followed by Schöffl et al.[202] in tobacco. They fused a small HSP-coding gene of soybean to the very active 35S-promoter of cauliflower mosaic virus. Transgenic tobacco plants obtained by transformation had high levels of hsp17.5 mRNA at 25°C. Analyses of the protein level and eventual effects on thermotolerance are lacking.

Following a similar line, Landry et al.[115b] transformed CHO and mouse cells with the human hsp27 gene resulting in a constitutive accumulation of this stress protein and the generation of a permanent thermoresistant phenotype. The particular role of HSP27 and its phoshorylated isoforms for heat stress survival is also supported by an hs-resistant cell line of CHO cells selected after mutagenesis with ethyl methane sulfonate.[37b] However, the role of the HSP27 may be somehow coupled to the Ca^{2+}-calmodulin level. At least, by increasing the calmodulin content of mouse mammary tumor cells by gene transfer, Evans et al.[53a] found a reduction of the HSP27 synthesis without effects on the intrinsic heat sensitivity nor on the induced thermotolerance.

On the other hand, generation of antisense RNA *in vivo*[149] and *in vitro*[164] was shown to block specifically the translation of the corresponding HSP mRNA. Finally, transcription of hsp coding genes was decreased after transforming cells with an excess of plasmids containing polymers of the hs promoter element[97b,249] (see Figure 7.4). The highly selective effect depends on the competition for the limiting amount of hs transcription factor (see Section 7.5). In Chinese hamster ovary cells, the reduction of hs-induced HSP70 synthesis coincides with a dramatic decline of survival of a 30-min treatment at 45°C.[97b]

17.2.3. INDUCTION OF CROSS-TOLERANCE BY CHEMICAL STRESSORS

Induction of HSP synthesis by numerous chemical stressors is a widely studied phenomenon (see Table 1.1). In vertebrate cells, a considerable number of the hs-like inducers was shown to induce also thermotolerance.[70,72,98,112,121b,124d,134,135,158b] This phenomenon of cross-tolerance is illustrated in Figure 17.3. Hamster fibroblasts survive an otherwise killing hs (45 min 45°C), irrespective of the mode of conditioning treatment, i.e., by pulse hs, arsenite, cadmium chloride, or ethanol. No systematic experiments in this direction were done with other cell systems. Lin et al.[138] reported on arsenite-induced thermotolerance in soybean (see Table 17.2) and resistance to a 52°C killing hs was found in *Escherichia coli* after pretreatment with hydrogen peroxide.[68] Finally, yeast cells became thermotolerant, if HSP synthesis was induced by the antibiotic paronomycin[66a] or if cells were made osmotolerant by treatment with 0.7*M* NaCl,[224b] but it can be anticipated that the molecular basis of both states of TT will be found to be different.

Usually induced cross-tolerance is one-sided, i.e., chemical stressors confer thermotolerance, but, vice versa, hs conditioning does not make resistant to chemical stressors.[131a] A few notable exceptions have been reported:

1. Hs-treated vertebrate, plant, and yeast cells are more resistant to ethanol.[2,135,237a,238]
2. Hs protects *Drosophila* cells from damaging effects of long-term anoxia.[231]
3. Improved resistance to hydrogen peroxide stress after hs conditioning of vertebrate

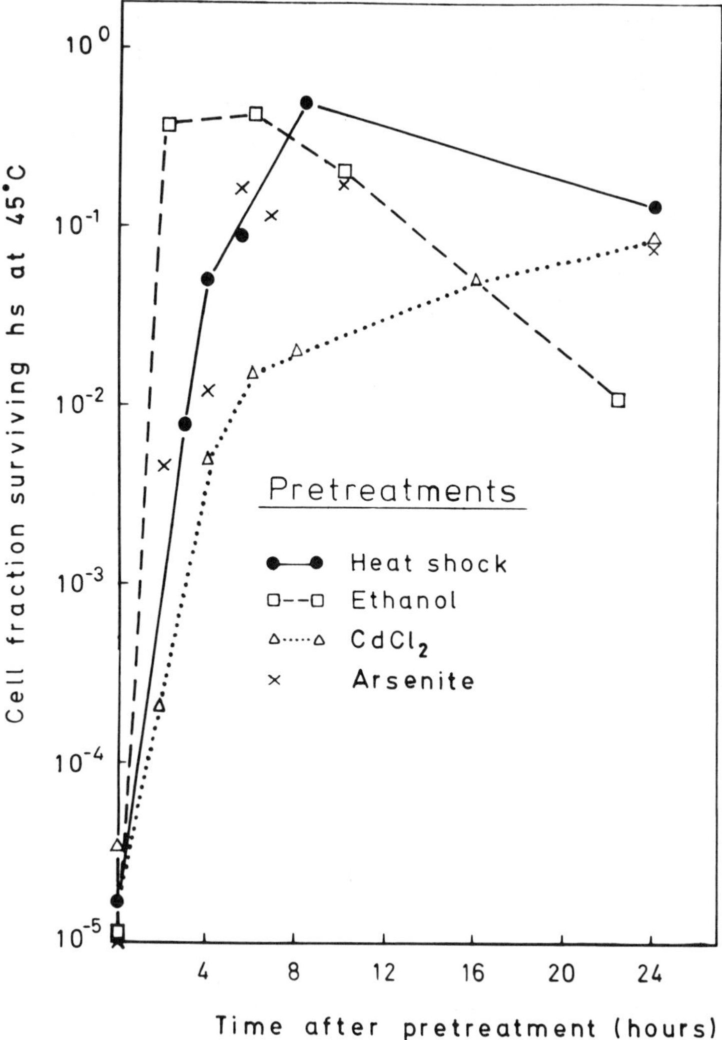

FIGURE 17.3. Induction of thermotolerance by different stress factors. Chinese hamster fibroblasts were pretreated by 15 min hs at 45°C (closed circles) or for 1 h at 37°C in the presence of 100 μ*M* sodium arsenite (crosses), 100 μ*M* cadminum chloride (open triangles) or 6% ethanol (open squares). After this, cells were allowed to express thermotolerance at 37°C. At the indicated time points, hs survival was challenged by 45 min exposure to 45°C. All types of pretreatments result in a dramatic increase of the resistance towards the killing hs. (Modified from Li, G. C., Shrieve, D. C., and Werb, Z., *Heat Shock from Bacteria to Man*, Cold Spring Harbor Laboratories, 1982, 395; Nover, L., *Heat Shock Response of Eukaryotic Cells,* G. Thieme Verlag, 1984. With permission.)

cells[77,172,213] and *Neurospora*[100-100b] is evidently due to increased levels or activities of peroxidase or superoxide dismutase.

4. A special type of cross-tolerance concerns resistance to lethal DNA damage probably by increased repair activities after hs (see Section 15.3). This is true for plants,[84,191-191b] vertebrates,[181,246,251] fishes,[207,207a,209,210] and yeast.[153-157]
5. Heat shock pretreatment protects germinating conidiospores of *Neurospora*[69a] or *Drosophila* larvae[30a] against damages by low temperature stress.
6. Increased diamide resistance of CHO cells after hs pretreatment is evidently not connected with the protection from diamide oxidation of SH-proteins but rather with

improved survival despite the accumulation of damaged proteins.[56c]

7. Of clinical interest is the observation that induced thermotolerance is combined with temporary or permanent resistance to anticancer drugs, e.g. to adriamycin, bleomycin, or cis-platinum. Reduced uptake of the drugs as a consequence of membrane modifications[72a] and/or increased excretion of cytotoxic compounds by hs induction of the multi drug-resistance pump may contribute to this effect.[37,240b]

Concerning the central question about the correlation of increasing HSP levels with the acquisition of thermotolerance, Li and Laszlo[131a] summarized the relavant data. All hs-like chemical stressors reported to evoke thermotolerance are also known as inducers of HSP synthesis. With respect to the general criterion of survival (see Figure 17.3), a careful analysis for vertebrate cells shows that only the amount of HSP70, but not of HSPs 110, 90, or 27, evidently correlates with the level of thermotolerance.[129,131a,132] Though the decisive role of HSP70 for many aspects of induced thermotolerance is unabated, the exclusion of HSP27 is certainly not rectified.[37b,53a,115b] (see Section 17.2.2).

An interesting confirmation for the intricate role of HSPs comes from the induction with amino acid analogs. In contrast to other chemical stressors, they need incorporation into proteins for induction. Consequently, they are also incorporated into the newly formed HSPs, which are made nonfunctional.[46,99,131,158b] The result is a highly increased HSP level with a drastically decreased intrinsic thermoresistance because of the overload with aberrant proteins.[131] However, as to be expected, cells become thermotolerant in the recovery period, when the removal of the inducing analog allows the formation of functional HSPs[158b] or if HSP synthesis is induced prior to the analog treatment.[120]

17.2.4. FUNCTION OF HSPs IN SUBCELLULAR COMPARTMENTS

Search for functions of HSPs is an important aspect when analyzing the relation between thermotolerance phenomena and the increased synthesis of these proteins. However, it should be kept in mind that functions defined for constitutive representatives of HSPs under normothermic conditions need not be identical with their role under hyperthermic conditions (see summary by Nover et al.[168a]). In this respect, studies on newly formed HSPs and their intracellular recompartmentation during hs provided more direct insights into their possible sites of action. In this section, we only summarize some relevant facts from Table 16.1, illustrating the close connection of thermotolerance phenomena observed in a given subcellular compartment with the accumulation of HSP(s) in it. This compilation is not intended to indicate a causal relationship between both, which is yet to be proved in most cases. Details and references are given in the indicated sections of this book.

1. In all cells studied so far, HSP70 is strikingly concentrated in the nuclear compartment of stressed cells (Table 16.1A). A role for protection of pre-mRNP processing (see Section 9.5) and/or of the nucleolar machinery involved in ribosome biosynthesis (see Section 10.6) is very plausible. In addition, a function as an energy-dependent shuttle for removal of unwanted nuclear protein or RNP aggregates in the recovery period was discussed.[127]
2. Small HSPs of plant, *Drosophila,* and vertebrate cells form large perinuclear complexes with cytoskeletal elements (Table 16.1G). At least in plants, they are presumably involved in the preservation of untranslated mRNAs[168] (see Section 11.4). This mechanism may explain the frequently observed effect of an improved recovery of control protein synthesis in thermotolerant cells (see Section 13.3).
3. Members of the HSP70 and HSP90 families are closely associated with cytoskeletal systems, whose rapid and reversible changes profoundly influence many parts of cellular activities (see Table 16.1F and Section 14.3).

4. Protection of the outer membranes was analyzed by amino acid uptake (Figure 14.3) or by leakage of solutes. Lin et al.[139] found a single HSP15 associated with plasma membrane fractions from thermotolerant soybean seedlings.
5. Constitutively expressed members of the HSP70 and HSP60 families are detected in mitochondria and chloroplasts (Table 16.1, parts D, E). They are presumably involved in protein import, folding and assembly (see Section 12.2). Stress-dependent synthesis and import of new HSPs were described for plants, algae, and *Neurospora*. In chloroplasts, HSPs 21 to 25, accumulating in the stroma fraction, are discussed as necessary components to stabilize the thylakoid membranes under certain stress conditions.[202a] Chou et al.[37a] prepared mitochondria from soybean seedlings and tested respiratory coupling of O_2 uptake with ATP generation *in vitro* at 42.5°C. Normal ATP generation was maintained only if seedlings were conditioned by a pretreatment for 2 h at 38°C plus 0.5 h at 42.5°C. These mitochondrial fractions are enriched in HSPs 15 to 18, 22, and 24.
6. Particularly clear-cut is the situation for ubiquitin, characterized as an hs-induced protein in human,[12,200] chicken,[19,20] *Trypanosoma cruzi*,[220a] yeast,[169] and other organisms (see Section 2.3.5). The pronounced changes of protein-ubiquitin conjugation in the initial phase of hs are presumably the result of altered transcription patterns, i.e., deubiquitination of histone H2A (see Section 12.1) and of the accumulation of aberrant proteins. Hs-induced ubiquitin synthesis may be essential for restoration of normal activities.[21,176] In yeast, deletion of the hs-inducible polyubiquitin gene makes the cells defective in sporulation and hypersensitive to hs and other stress regimes.[54a] Ubiquitin does not exist in bacteria. But synthesis of a special ATP-dependent protease, coded by the lon gene, is induced by hs in *E. coli*.[12a,65,161] Thus, support of the proteolytic system by distinct HSPs seems to be a general trait of all types of organisms.

17.3. MUTANTS

17.3.1. MAMMALIAN CELLS

Early reports on a greater heat sensitivity of cancer cells as compared to normal cells[48,63,103,226,233] led to several attempts to investigate HSP levels in cell lines selected for their heat resistance. The outcome was equivocal. Laszlo and Li[121] described Chinese hamster fibroblast lines, in which the increased resistance correlated with a higher constitutive level of HSC70 and enhanced inducibility of HSP110, 90, and 70. Interestingly, these cell lines were also resistant to the heat-sensitizing effect of the amino acid analogs canavanine and azetidine carboxylic acid.[121a,131] Using a similar starting material and ethyl methane sulfonate mutagenesis, Yahara et al.[252] selected a Chinese hamster ovary cell line with tenfold increased heat resistance. But in this case, the only HSP with a significantly higher level was HSP90. To complete the list of controversial results, Chretien and Landry[37b] reported on a heat-resistant CHO cell line with markedly increased constitutive synthesis of HSP27. The level was comparable to that in wild-type cells made thermotolerant by preinduction.

The outcome was totally negative with myeloma cell lines from mouse[7] and man[54] as well as with SV40 transformed mouse embryo cells.[171] In the latter case, even an inverse relation between HSP levels and heat resistance was observed. The transformed cell lines combined an increased heat sensitivity with increased constitutive levels of HSP70 and HSP90. Instead, Anderson et al.[7b] described a new member of the HSP70 family from heat-resistant mouse tumor cell lines, and Omar et al.[172] discussed a possible role of antioxidant enzymes, such as the copper/zinc-containing superoxide dismutase or peroxidase, for intrinsic heat sensitivity and induced thermoprotection. When considering the variability of intrinsic heat resistance of cells and the multiplicity of factors influencing this state (see Section 17.4), it is not surprising that a simple correlation between survival and constitutive or inducible HSP levels could not be established by these experiments.

17.3.2. *DICTYOSTELIUM DISCOIDEUM*

More promising are attempts to generate or select mutants with defined changes in the HSP coding genes or in their expression. Such a mutant of *Dictyostelium discoideum* (H122) was found unable to synthesize the small HSPs. Despite a normal HSP70 synthesis, it fails to develop resistance to a killing hs at 34°C after a prior conditioning treatment at 30°C.[143,193b] The particular role of the small HSPs was not further investigated. Intriguing, but not analyzed with respect to their physiological defects, are the deletion mutants of *Drosophila melanogaster,* which lost both loci (87A7 and 87C1) with hs-inducible hsp70 genes.[59,96,97]

17.3.3. *ESCHERICHIA COLI*

Due to the favorable genetic background, the situation is much clearer in *Escherichia coli*. The regulatory gene controlling the hs regulon (rpoH) codes for a special σ-subunit of the RNA polymerase (see Sections 7.1 to 7.3). rpoH-defective mutants are unable to synthesize HSPs and to grow at temperatures above 20°C.[223,225,253,258,260] Reversion of the mutant phenotype to grow at 42°C was achieved by constructing and/or selecting cells with hs-independent overproduction of GroE and DnaK. Evidently these proteins play a dominant role for this aspect of thermotolerance.[109b]

In addition, mutation of individual HSP-coding genes demonstrate their particular role in different aspects of the hs response (for a survey, see Table 2.2):

1. The special functions of the Lon protease have already been mentioned (see Section 2.1.2).[12a,65,66a]
2. Deletion of the HSP83 gene (htpG), though not lethal, causes slight growth disadvantages, which increase at higher temperatures.[13]
3. Tests with an *E. coli* strain harboring a mutant dnaK gene (dnaK 756) revealed multiple effects,[186a] i.e., no growth at 43.5°C, reduced survival after heat inactivation at 52°C, and sharply reduced ATPase activity of the purified DnaK 756 protein. But there was no influence on the induction of thermotolerance as compared to the wild-type strain.
4. Mutants with defective grpE, groEL/S, and/or dnaK genes are restricted in DNA and RNA synthesis at 42°C.[8,174,234]

In fact, for optimum growth, all of these genes are required under all temperature conditions (see Section 2.1.2).[53a,60a,103a,109c] This is even true for the hs-specific sigma factor.[222a,260] Moreover, functional DnaK and Lon proteins are involved in the rapid reversibility of the hs response, i.e., in autorepression exerted presumably via the level of the RpoH protein (see Section 7.3).[66a,69] On the other hand, there are findings of heat resistance or induced thermotolerance in *E. coli,* which is independent of HSP synthesis. Delaney[42a] described a CyA deletion mutant with increased thermotolerance but without higher levels of HSP synthesis, and van Bogelen et al.[227] reported on the induction of HSP synthesis at normal termperatures by overproduction of σ^{32}. No thermotolerance was observed. Chemical stressors known to induce thermotolerance in *E. coli* (ethanol, $CdCl_2$, or H_2O_2) were also tested. Again, there was no reasonable correlation to the induced HSP pattern, which was very different for each stressor.[227] It can be concluded that other factors, e.g., adaptation of membrane microviscosity[28,43,44] (see Section 14.4), are much more important for survival of *E. coli* after a challenge hs at >50°C.

17.3.4. YEAST

In yeast, nine HSP70-type genes were identified and characterized with respect to their regulatory behavior under hs and control conditions (see data summarized in Table 2.4). Site-directed mutagenesis allowed the elimination of individual members of the hsp70 gene family or groups of them[40,50,240] and the testing of the viability and biochemical peculiarities of the mutant strains. Among the four members of group A, the two genes with high constitutive expression (SSA1 and SSA2) are required for growth at higher temperatures, whereas the strongly hs-induced SSA3 and SSA4 genes seem to be dispensable. Double mutants were normal in growth, sporulation, and induced thermotolerance at 23, 30, or 37°C. Evidently HSP70 of types A1 and A2 are sufficient to meet all normal and stress-dependent needs of the cell. The constitutively expressed and hs-repressed genes of group B (SSB1 and SSB2) are required for growth under low temperature conditions. Finally, gene SSC1 codes for a mitochondrial protein and the KAR2 gene codes for a GRP78 localized in the perinuclear ER system, whereas gene SSD1 could not be characterized so far (see details in Table 2.4).

It should be noted that this elegant method of gene elimination provides important insights into the significance of individual gene products for growth or survival at a given temperature. But this is only the first step on the way to analyze the molecular details, to define the cellular site and the process that is protected by HSP70. Basically similar methods were applied to generate yeast mutants lacking the only gene coding for HSP26.[179,220] In this case, neither induction of thermotolerance nor ethanol resistance, sporulation, spore germination, or survival after long-term storage at 23, 30, or 37°C were altered.

Interesting but not yet fully defined, hs response mutants of yeast were described by Shin et al.[211] (Figure 17.4). Induction of HSP synthesis and thermotolerance by transfer of cells from 23 to 36°C is connected with a rapid decrease in the intracellular cAMP levels and a transient arrest in the G_1 phase of the cell cycle. Two mutants with altered thermotolerance were characterized (Figure 17.4). Mutant cyr 1-2 has a very low constitutive level of cAMP and requires an exogenous supply of the nucleotide for growth. Without cAMP it is arrested in the G_1 phase and exhibits a high level of thermotolerance connected with increased constitutive synthesis of HSPs 72A, 72B, and 41. In contrast, mutant bcy-l has lost its cAMP control of protein phosphorylation, i.e., it has a high level of cAMP-independent protein kinase activity. This trait coincides with a loss of hs inducibility. Moreover, the same phenotype can be generated by adding cAMP to wild-type cells before the hs. In all examples with high thermotolerance, increased levels of HSPs 72A, 72B, and 41 and G_1 arrest were observed. Two members of the yeast hsp70 gene family were reported to be induced under conditions of cAMP depletion (cyr 1 mutants or wild-type cells in the stationary phase.) These are SSA1[220c] and SSA3.[240a] In both cases cAMP regulation is independent of the heat shock promoter elements.

However, the emerging picture on the role of the cAMP level in yeast (Figure 17.4) may be too simple. Search for revertants of the bcy-1 phenotype (defective regulatory subunit) revealed mutants with attenuated catalytic subunit of the cAMP-dependent protein kinase (tpk^w).[32a] The lacking increase of the thermoresistance under conditions of nutrient starvation in bcy-1 strains is restored in the bcy-1/tpk^w revertants even in strains defective in cAMP synthesis (e.g., in cyr-1 mutants), i.e., the normal cell cycle variation of the intrinsic heat sensitivity is independent of cAMP.[32a,223a]

Another type of hs response mutant of yeast (hsrl) was generated by ethylmethane sulfonate mutagenesis.[94] The mutant is recessive to the wild-type allele. Homozygous mutant strains had an extended G_1 phase and constitutively synthesized HSPs 48A and 48B so well as two G_0 phase-specific proteins of yeast. Growing mutant cells were resistant to cell killing by a 5 min hs at 52°C to the same extent as conditioned wild-type cells. Thus, a decisive influence of HSPs on this aspect of thermotolerance is not very likely, i.e., the situation is very similar to that found in *E. coli.* Resistance to cell killing is limited by factors other than HSP levels, but this does not preclude that other aspects of thermotolerance are critically dependent on HSP synthesis (see Section 17.5). In this respect, a thorough analysis of the gene expression chain in heat-shocked yeast is required.

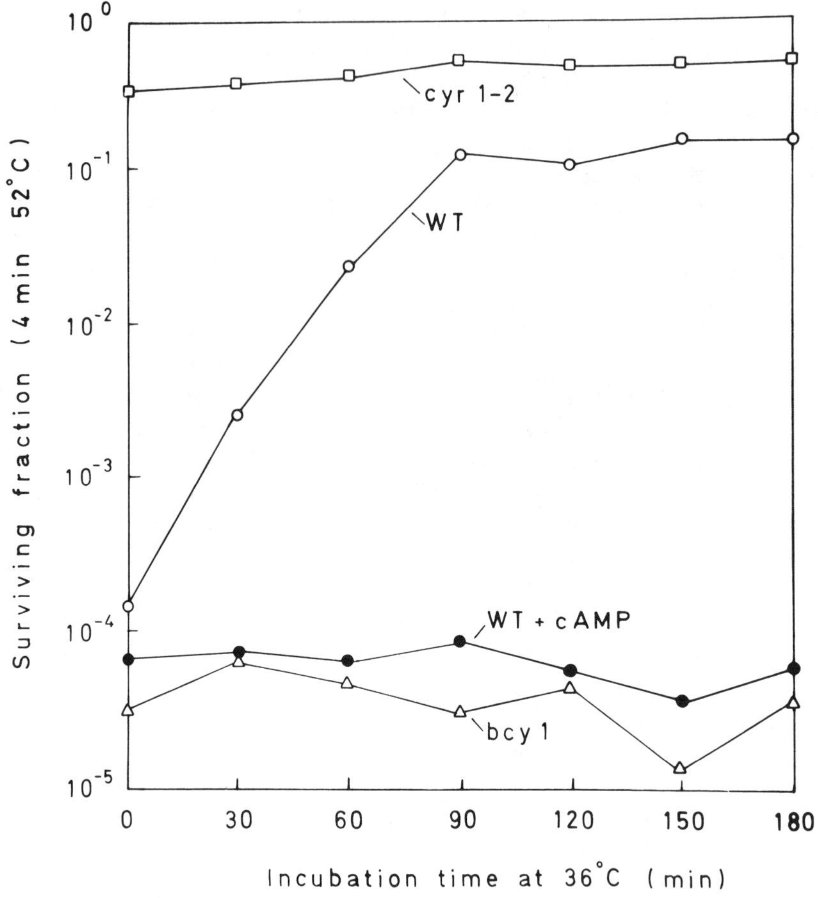

FIGURE 17.4. Cyclic AMP-dependent protein phosphorylation and induced thermotolerance in yeast. Thermotolerance (survival at 52°C) was induced by conditioning treatment at 36°C using three yeast strains: (1) wild type (WT, normal levels of cAMP, induced synthesis of heat shock proteins); (2) cyr 1-2 (low constitutive level of cAMP, constitutive expression of heat shock proteins); (3) bcy 1 (cAMP-independent protein kinase, no heat shock [hs]-induced formation of HSPs). Thermotolerance induction is connected with a decrease of cAMP-dependent protein phosphorylation during hs and the increased synthesis of HSPs 72 A/B and 41. Presence of 2 m*M* cAMP during the conditioning treatment blocks development of thermotolerance. (From Shin, D.-Y., Matsumoto, K., Iida, H., Uno, I., and Ishikawa, T., *Mol. Cell. Biol.*, 7, 244, 1987. With permission.)

17.3.5. *TETRAHYMENA*

A very special aspect of induced thermotolerance, which is independent of HSP synthesis, was described by Hallberg and co-workers for *Tetrahymena thermophila.* Stabilization of the ribosomal apparatus against degradation during the hs period (see Section 10.7) is achieved by prior conditioning treatment, but it proceeds in the complete absence of HSP synthesis.[74] Tight binding of a preformed basic protein to the small ribosomal subunit[150] and hs-induced synthesis of a ribosome-associated scRNA[109] are essential. The mutant strain MC-3[108] is defective in this stabilization, despite a normal set of HSPs being synthesized after hs conditioning.

17.4. VARIATION OF INTRINSIC HEAT SENSITIVITY

The term "intrinsic heat sensitivity" is introduced to stress the fact that the cytotoxicity of a given hyperthermic treatment is dependent on the metabolic situation of the cell. Many factors influencing this cellular state may contribute positively or negatively to the survival rate after heat killing. Alexandrov et al.[4] described this phenomenon as "primary heat resistance" and opposed it to "adaptive heat hardening". Intrinsic heat sensitivity is not directly related to the HSP level; however, induction of thermotolerance may very well include metabolic changes increasing the intrinsic heat sensitivity. Thus, the difficulties encountered with the description of the putative role of HSPs for thermotolerance are not due only to the apparent experimental limitations when cell survival is taken as a criterion. We want to emphasize once more the necessity of using more defined subcellular test systems for the analysis of protective effects exerted by HSPs (see Sections 10.6 and 11.4). In addition, the brief summary of factors with marked influence on the intrinsic heat sensitivity of cells illustrate a field which is of utmost importance for hyperthermic treatment of cancer.

1. Cell cycle-dependent variation of heat sensitivity is well known. The preferential killing of the radioresistant S-phase cells is particularly noteworthy and presumably results from aberrant DNA replication.[18,45,187,195,241] G_1 phase cells are the least sensitive.[189] The optimum for cell killing must not be identical with that for other hs-effects. Thus, during early developmental cleavages of embryos of the pond snail *Lymnaea stagnalis*[22,60] hs induction of aberrant development is most efficient during mitosis, more precisely in the metaphase (see Figure 18.1). Most other parts of the cell cycle are highly resistant, and this variation of heat sensitivity is independent of HSP synthesis. In yeast a considerable increase of thermoresistance is connected with the transition of cells from growing to a G_0-like resting state, e.g., after nutrient starvation or in temperature-sensitive mutants under nonpermissive conditions.[32a,199a,223a] This particular type of cell cycle arrest is not correlated with marked HSP synthesis,[93-95,184,211] except expression of SSA1 and SSA3 genes.[220c,240a]
2. In mammalian cells, survival of a given stress dosage is thoroughly influenced by the metabolic situation. Deprivation of amino acids, glucose, or oxygen increases heat sensitivity.[32,62,66,117,118,232] The effect of amino acid deprivation can be overcome by addition of 1 m*M* of the nonmetabolizable α-amino-isobutyric acid,[232] whereas heat sensitization by glucose deficiency in mouse L929 cells was reversed by 1 m*M* uridine,[119] in hamster fibroblasts by α-keto acids,[66] and in CHO cells in the presence of 10^{-6} *M* hydrocortisone.[56] In the latter case, development of thermotolerance by hs conditioning was not altered by the hormone and, vice versa, the protecting effect after 20 h treatment with glucocorticoids was not connected with a significant increase of HSP levels. Intrinsic thermosensitivity of mouse L929 cells was found to be considerably increased by addition of insulin or 2-cyanocinnamic acid.[118,236,237] Although the effect could be reversed by addition of pyruvate, it is very likely that depletion of reduced pyridine nucleotides is the primary cause for the high cytotoxicity of hs in the presence of insulin or cyanocinnamic acid.
3. Cellular calcium homeostasis is essential for survival. Uncontrolled Ca^{2+} influx during hs may contribute to cell death.[217] Consequently, calmodulin-inhibiting drugs or increase of the extracellular Ca^{2+}-level potentiate hyperthermic cell killing.[115a,143b,243] In this connection, it is essential to notice that Ca^{2+}-deprivation of rat hepatoma cells, while inducing GRP synthesis (see Chapter 4), makes the cells highly resistant to severe heat stress. At the same time inducibility of HSPs is repressed, i.e., the protective effect is exclusively due to the prevention of a Ca^{2+} imbalance.[110]
4. Cytotoxicity of a hyperthermic treatment can be increased by two orders of magnitude,

TABLE 17.3
Two States of Thermotolerance Exemplified for *Lymnea stagnalis* Larvae

	α-State	β-State
Response	Immediate	Delayed
Decay	Rapid	Slow
Inhibitors of protein synthesis	No effect	Inhibit
HSP synthesis	Not required	Required
Inducing conditions	Mild	Severe
	33°C,	39°C,
	3.5% Ethanol	9.1% Ethanol
	2.5 m*M* DNP	—

Compiled from References 24 to 26 and 229.

if CHO cells were pretreated with 1 m*M* α-difluoromethylornithine, a potent inhibitor of polyamine biosynthesis.[57,58] Thermosensitization is particularly pronounced 24 to 48 h after removal of the drug with maximum depletion of intracellular polyamine pools. At the same time hs-induced thermotolerance develops normally, although the upper level of resistance achieved is much lower under these conditions, i.e., both effects are clearly independent.

5. In plants, threshold temperatures of stress effects on a given cellular function and the speed of recovery are influenced by the water supply[101] or by seasonal variations of the cellular state[4] (maximal daily air temperature, water supply, metabolic state). In addition, a diurnal variation of the intrinsic heat tolerance was observed in Crassulacean species with an optimum in the light and a minimum in the dark period.[102,203] The minimum correlates with the transient accumulation of organic acids in the night. But this diurnal rhythm is general, and it even applies to the hs inducibility of hs mRNAs.[172a]

17.5. TWO STATES OF THERMOTOLERANCE

A decisive contribution towards a better understanding of thermotolerance and the seemingly controversial results on the role of HSPs came from van Wijk and Boon-Niermeijer.[229] They reported on a comparative analysis of hs effects on survival of mouse 3T3 cells and larvae of the pond snail, *Lymnaea stagnalis*. Two different states of thermotolerance were defined (Table 17.3). A transient α-state does not require HSP synthesis, whereas the slowly decaying β-state is expressed after severe stress and depends on HSP synthesis. Similar observations on differential effects evoked by severe vs. mild hs were published for plants,[1,2,4,84a,248] for *Tetrahymena*,[76,99,109] for fish FHM cells,[150a] for HeLa cells,[48a] and for Chinese hamster ovary cells.[7a,13b,86,121c,124,124a-c,189] Although the details are probably not directly comparable, the response to hyperthermic conditions is also very different in *E. coli*, i.e., hs at 42°C vs hs at 50°C[52a,53,175] and in yeast[180] (48°C vs 50 to 52°C).

The molecular basis has not been established. In particular, a participation of constitutively expressed HSPs in the α-state cannot be excluded. Thus, specific low temperature stress proteins characterized for mouse FM3A and HeLa cells[79a,89a] may, in fact, represent increased synthesis of the constitutively expressed HSC72/73 and HSC90. In this respect, the report on a mild hs-induced 47-kDa membrane glycoprotein of chicken embryo fibroblasts deserves special notice. It is phosphorylated and represents the major collagen-binding protein (see Section 2.3.6).[159a] The results demonstrate an additional complexity in the stress-induced adaptation phenomena, which were repeatedly found to proceed equally well or, in some cases, even better in the presence of cycloheximide or similar inhibitors of protein synthesis.[33,73,74,112,113,124a-c,170,238,242] It is likely that accumulation of nonproteinaceous stress metabolites may exert protective functions. Examples are prolin or glycine betaine in plants,[90,126] glycerol in vertebrates[88,146] or the nonreducing disaccharide trehalose in yeast.[11a,91]

In a recent study, Lee et al.[124e] described the differences of the thermotolerant state of CHO cells induced either by 10 min 45°C plus recovery or 1 h 100 μ*M* arsenite plus recovery. Effects on survival and protein synthesis during a 5 h challenge treatment at 43°C were measured. Both inducers caused a comparable degree of thermotolerance with respect to survival. But hs pretreatment led also to a complete protection of protein synthesis against the challenge treatment, even if cycloheximide was added during the recovery (expression) period. In contrast, arsenite-induced thermotolerance did not protect protein synthesis and was inhibited by cycloheximide. Resistance mechanisms to oxidation stress may be characteristic of the α-state. In a CHO cell line selected for resistance to H_2O_2 survival of a long-term challenge at 43°C was markedly improved, whereas survival after a 45°C hs was not affected.[213a].

REFERENCES

1. **Abernethy, R. H., Thiel, D. S., Petersen, N. S., and Helm, K.,** Thermotolerance is developmentally dependent in germinating wheat seed, *Plant Physiol.,* 89, 569—576, 1989.

1a. **Alexandrov, V. YA.,** *Cells, Molecules and Temperature,* Springer-Verlag, Berlin, 1977.

2. **Alexandrov, V. YA.,** Cell reparation of non-DNA injury, *Int. Rev. Cytol.,* 60,223—269, 1979.

3. **Alexandrov, V. YA.,** *Cell Reactivity and Proteins* (in Russian), Nauka Publisher, Leningrad, 1985.

4. **Alexandrov, V. YA., Lomagin, A. G., and Feldman, N. L.,** The responsive increase in thermostability of plant cells, *Protoplasma,* 69, 417—458, 1970.

5. **Altschuler, M. and Mascarenhas, J. P.,** The synthesis of heat-shock and normal proteins at high temperatures in plants and their possible roles in survival under heat stress, in *Heat Shock from Bacteria to Man,* Schlesinger, M. J., Ashburner, M., and Tissieres, A., Eds., Cold Spring Harbor Laboratory, Cold Spring Harbor, N.Y., 1982, 321—327.

6. **Amaral, M. D., Galego, L., and Rodrigues-Pousada, C.,** Stress response of *Tetrahymena pyriformis* to arsenite and heat shock: Differences and similarities, *Eur. J. Biochem.,* 171, 463—470, 1988.

6a. **Anderson, K-V. and Lengyel, J. A.,** Rates of synthesis of major class of RNA in *Drosophila* embryos, *Dev. Biol.,* 70, 217—231, 1979.

6b. **Anderson, R. L. and Parker, R.,** Analysis of membrane lipid composition of mammalian cells during the development of thermotolerance, *Int. J. Radiat. Biol.,* 42, 57—69, 1982.

7. **Anderson, R. L., Tao, T. W., Betten, D. A., and Hahn, G. M.,** Heat shock protein levels are not elevated in heat-resistant B16 melanoma cells, *Radiat. Res.,* 105, 240—246, 1986.

7a. **Anderson, R. L., Herman, T. S., van Kersen, I., and Hahn, G. M.,** Thermotolerance and heat shock protein induction by slow rates of heating, *Int. J. Radiat. Oncol. Biol. Phys.,* 15, 717—725, 1988.

7b. **Anderson, R. L., van Kersen, I., Kraft, P. E., and Hahn, G. M.,** Biochemical analysis of heat-resistant mouse tumor cell strains. A new member of the hsp70 family, *Mol. Cell. Biol.,* 9, 3509—3516, 1989.

8. **Ang, D., Chandrasekhar, G. N., Zylicz, M., and Georgopoulos, C.,** *Escherichia coli* grpE gene codes for heat shock protein B25.3, essential for both lambda DNA replication at all temperatures and host growth at high temperature, *J. Bacteriol.,* 167, 25—29, 1986.

9. **Arrigo, A. P.,** Investigation of the function of the heat shock proteins in *Drosophila melanogaster* tissue culture cells, *Mol. Gen. Genet.,* 178, 517—524, 1980.

10. **Arrigo, A. P.,** Acetylation and methylation patterns of core histones are modified after heat or arsenite treatment of *Drosophila* tissue culture cells, *Nucl. Acids Res.,* 11, 1389—1404, 1983.

11. **Arrigo, A.-P.,** Cellular localization of HSP23 during *Drosophila* development and following subsequent heat shock, *Dev. Biol.,* 122, 39—48, 1987.

11a. **Attfield, P. V.,** Trehalose accumulates in *Saccharomyces cerevisiae* exposure to agents that induce heat shock response, *FEBS Lett.,* 225, 259—263, 1987.

11b. **Aujame, L. and Firko, H.,** The major inducible heat shock protein hsp68 is not required for acquisition of thermal resistance in mouse plasmacytoma cell lines, *Mol. Cell. Biol.,* 8, 5486—5494, 1988.

12. **Baker, Z. T. and Board, P. G.,** The human ubiquitin gene family structure of a gene and pseudogenes from the UbB subfamily, *Nucl. Acids Res.,* 15, 443—463, 1987.

12a. **Baker, T. A., Grossman, A. D., and Gross, C. A.,** A gene regulating the heat shock response in *Escherichia coli* also affects proteolysis, *Proc. Natl. Acad. Sci., U.S.A.*, 81, 6779—6783, 1984.

13. **Bardwell, J. C. A. and Craig, E. A.,** Ancient heat shock gene is dispensable, *J. Bacteriol.*, 170, 2977—2983, 1988.

13a. **Bauer, H.,** CO_2-Gaswechsel nach Hitzestress bei *Abies alba* Mill. und *Acer pseudoplatanus* L, *Photosynthetica*, 6, 424—434, 1972.

13b. **Bauer, K. D. and Henle, K. J.,** Arrhenius analysis of heat survival curves from normal and thermotolerant CHO cells, *Radiat. Res.*, 78, 251—263, 1979.

14. **Behnel, H. J. and Wekbart, G.,** Induced stabilization of the transmembrane potential of *Drosophila* cells by heat shock and periodic applications of chloramphenicol, *J. Cell Sci.*, 87, 197—201, 1987.

15. **Berger, E. M.,** The regulation and function of small heat-shock protein synthesis, *Dev. Genet.*, 4, 255—265, 1984.

16. **Berger, E. M. and Woodward, M. P.,** Small heat shock proteins in *Drosophila* may confer thermal tolerance, *Exp. Cell Res.*, 147, 437—442, 1983.

17. **Bergh, S. and Arking, R.,** Developmental profile of the heat shock response in early embryos of *Drosophila*, *J. Exp. Zool.*, 231, 379—391, 1984.

18. **Bhuyan, B. K.,** Kinetics of cell killing by hyperthermia, *Cancer Res.*, 39, 2277—2284, 1979.

18a. **Bond, U.,** Heat shock but not other stress inducers leads to the disruption of a sub-set of snRNPs and inhibition of in vitro splicing in HeLa cells, *EMBO J.*, 7, 3509—3518, 1988.

19. **Bond, U. and Schlesinger, M. J.,** Ubiquitin is a heat shock protein in chicken embryo fibroblasts, *Mol. Cell. Biol.*, 5, 949—956, 1985.

20. **Bond, U. and Schlesinger, M. J.,** The chicken ubiquitin gene contains a heat shock promoter and expresses an unstable mRNA in heat-shocked cells, *Mol. Cell. Biol.*, 6, 4602—4610, 1986.

21. **Bond, U., Agell, N., Haas, A. L., Redman, K., and Schlesinger, M. J.,** Ubiquitin in stressed chicken embryo fibroblasts, *J. Biol. Chem.*, 263, 2384—2388, 1988.

21a. **Bonham-Smith, P. C., Kapoor, M., and Bewley, J. D.,** Establishment of thermotolerance in maize by exposure to stresses other than a heat shock does not require heat shock protein synthesis, *Plant Physiol.*, 85, 575—580, 1987.

22. **Boon-Niermeijer, E. K.,** Morphogenesis after heat shock during the cell cycle of *Lymnaea*: A new interpretation, *Wilhelm Roux's Arch. Dev. Biol.*, 180, 241—252, 1976.

23. **Boon-Niermeijer, E. K. and Van De Scheur, H.,** Thermosensitivity during embryonic development of *Lymnaea stagnalis* (Mollusca), *J. Therm. Biol.*, 9, 265—269, 1984.

24. **Boon-Niermeijer, E. K., Tuyl, M., and Van De Scheur, H.,** Evidence for two states of thermotolerance, *Int. J. Hyperthermia*, 2, 93—105, 1986.

25. **Boon-Niermeijer, E. K., Souren, J. E. M., and van Wijk, R.,** Thermotolerance induced by 2,4-dinitrophenol, *Int. J. Hyperthermia*, 3, 133—141, 1987.

26. **Boon-Niermeijer, E. K., De Waal, A. M., Souren, J. E. M., and van Wijk, R.,** Heat-induced changes in thermosensitivity and gene expression during development, *Dev. Growth Differ.*, 30, 705—715, 1988.

26a. **Bosch, T. C. G., Krylow, S. M., Bode, H. R., and Steele, R. E.,** Thermotolerance and synthesis of heat shock proteins: these responses are present in *Hydra attenuata* but absent in *Hydra oligactis*, *Proc. Natl. Acad. Sci. U.S.A.*, 85, 7927—7931, 1988.

27. **Bournias-Vardiabasis, N. and Buzin, C. H.,** Developmental effects of chemicals and the heat shock response in *Drosophila* cells, *Teratog. Carcinog. Mutag.*, 6, 523—537, 1986.

28. **Brock, T. D.,** *Thermophilic Microorganismus and Life at High Temperatures*, Springer Verlag, New York, 1978.

29. **Browder, L. W., Pollock, M., Nickells, R. W., Heikkila, J. J., and Winning, R. S.,** Developmental regulation of the heat shock response, in *Developmental Biology: A Comprehensive Synthesis*, Vol. 6, Di Berardina, M. A. and Etkin, L., Eds., Plenum Press, New York, 97—147, 1989.

29a. **Burdon, R. H.,** Thermotolerance and the heat shock proteins, *Symp. Soc. Exp. Biol.*, 41, 269—283, 1987.

30. **Burns, C. P., Lambert, B. J., Haugstad, B. N., and Guffy, M. M.,** Influence of rate of heating on thermosensitivity of L1210 leukemia: Membrane lipids and Mr 70,000 heat shock protein, *Cancer Res.*, 46, 1882—1887, 1986.

30a. **Burton, V., Mitchell, H. V., Young, P., and Petersen, N. S.,** Heat shock protection against cold stress of *Drosophila melanogaster*, *Mol. Cell. Biol.* 8, 3550—3552, 1988.

31. **Buzin, C. H. and Bournias-Vardiabasis, N.,** The induction of a subset of heat-shock proteins by drugs that inhibit differentiation in *Drosophila* embryonic cell cultures, in *Heat Shock from Bacteria to Man*, Schlesinger, M. J., Ashburner, M., and Tissieres, A., Eds., Cold Spring Harbor Laboratory, Cold Spring Harbor, NY., 1982, 387—394.

32. **Calderwood, S. K., Stevenson, M. A., and Hahn, G. M.,** Cyclic AMP and the heat shock response in Chinese hamster ovary cells, *Biochem. Biophys. Res. Commun.*, 126, 911—916, 1985.

32a. **Cameron, S., Levin, L., Zoller, M., and Wigler, M.,** cAMP-independent control of sporulation, glycogen metabolism, and heat shock resistance in *S. cerevisiae*, *Cell*, 53, 555—566, 1988.

33. **Carper, S. W., Duffy, J. J., and Gerner, E. W.,** Heat shock proteins in thermotolerance and other cellular processes, *Cancer Res.*, 47, 5249—5255, 1987.
34. **Carr, A. and De Pomerai, D. I.,** Stress protein and crystallin synthesis during heat shock and transdifferentiation of embryonic chick neural retina cells, *Dev. Growth Differ.*, 27, 435—445, 1985.
35. **Cavicchioli, R. and Watson, K.,** Loss of heat-shock acquisition of thermotolerance in yeast is not correlated with loss of heat shock proteins, *FEBS Lett.*, 207, 149—152, 1986.
36. **Cervera, J.,** Induction of self-tolerance and enhanced stress protein synthesis in L-132 cells by cadmium chloride and by hyperthermia, *Cell Biol. Int. Rep.*, 9, 131—142, 1985.
36a. **Chen, J. D., Yew, F. H., and Li, G. C.,** Thermal adaptation and heat shock response of *Tilapia* ovary cells, *J. Cell. Physiol.*, 134, 189—199, 1988.
36b. **Chen, Y.-M., Kamisaka, S., and Masuda, Y.,** Enhancing effects of heat shock and gibberellic acid on the thermotolerance in etiolated *Vigna radiata*. I. Physiological aspects on thermotolerance, *Physiol. Plant.*, 66, 595—601, 1986.
37. **Chin, K.-V., Tanaka, S., Darlington, G., Pastan, I., and Gottesman, M. M.,** Heat shock and arsenite increase expression of the multidrug resistance MDR1 gene in human renal carcinoma cells, *J. Biol. Chem.*, 265, 221—226, 1990.
37a. **Chou, M., Chen, Y.-M., Lin, C.-Y.,** Thermotolerance of isolated mitochondria associated with heat shock proteins, *Plant Physiol.*, 89, 617—621, 1989.
37b. **Chretien, P. and Landry, J.,** Enhanced constitutive expression of the 27-kDa heat shock proteins in heat-resistant variants from Chinese hamster cells, *J. Cell. Physiol.*, 137, 157—166, 1988.
37c. **Christman, M. F., Morgan, R. W., Jacobson, F. S., and Ames, B. N.,** Positive control of a regulon for defenses against oxidative stress and some heat-shock proteins in *Salmonella typhimurium, Cell,* 41, 753—762, 1985.
38. **Chomyn, A., Moller, G., and Mitchell, H. K.,** Patterns of protein synthesis following heat shock in pupae of *Drosophila melanogaster, Dev. Genet.*, 1, 77—95, 1979.
39. **Collier, N. C. and Schlesinger, M. J.,** The dynamic state of heat shock proteins in chicken embryo fibroblasts, *J. Cell Biol.*, 103, 1495—1507, 1986.
40. **Craig, E. A. and Jacobsen, K.,** Mutations in cognate genes of *Saccharomyces cerevisiae* hsp70 result in reduced growth rates at low temperatures, *Mol. Cell. Biol.*, 5, 3517—3524, 1985.
41. **Crile, G.,** The effects of heat and radiation on cancers implanted on the feet of mice, *Cancer Res.*, 23, 372—380, 1963.
42. **Dean, R. L. and Atkinson, B. G.,** The acquisition of thermal tolerance in larvae of *Calpodes ethlius* (Lepidoptera) and the *in situ* and *in vitro* synthesis of heat-shock proteins, *Can. J. Biochem., Cell Biol.*, 61, 472—479, 1983.
42a. **Delaney, J. M.,** A cyA deletion mutant of *Escherichia coli* develops thermotolerance but does not exhibit a heat-shock response, *Genet. Res.*, 55, 1—6, 1990.
43. **De Mendoza, D. and Cronoan, J. E.,** Thermal regulation of membrane lipid fluidity in bacteria, *Trends Biochem. Sci.*, 8, 49—52, 1983.
44. **Dennis, W. H. and Yatvin, M. B.,** Correlation of hyperthermic sensitivity and membrane microviscosity in *E. coli* K 1060, *Int. J. Radiat. Biol.*, 39, 265—271, 1981.
44a. **Dewey, W. C. and Esch, J. L.,** Transient thermal tolerance: Cell killing and polymerase activities, *Radiat. Res.*, 92, 611—614, 1982.
45. **Dewey, W. C., Sapareto, S. A., and Betten, D. A.,** Hyperthermic radiosensitization of synchronous Chinese hamster cells: Relationship between lethality and chromosome aberrations, *Radiat. Res.*, 76, 48—59, 1978.
46. **Di Domenico, B. J., Bugaisky, G. E., and Lindquist, S.,** The heat shock response is self-regulated at both the transcriptional and posttranscriptional levels, *Cell,* 31, 593—603, 1982.
47. **Di Domenico, B. J., Bugaisky, G. E., and Lindquist, S.,** Heat shock and recovery are mediated by different translational mechanisms, *Proc. Natl. Acad. Sci., U.S.A.*, 9, 6181—6185, 1982.
48. **Dietzel, F.,** *Tumor und Temperatur,* Urban und Schwarzenberg, Muenchen, 1975.
48a. **Duncan, R. F. and Hershey, J. W. B.,** Protein synthesis and protein phosphorylation during heat stress, recovery, and adaptation, *J. Cell Biol.*, 109, 1467—1481, 1989.
49. **Dura, J. M.,** Stage dependent synthesis of heat shock induced proteins in early embryos of *Drosophila melanogaster, Mol. Gen. Genet.*, 184, 381—385, 1981.
49a. **Easton, D. P., Rutledge, P. S., and Spotila, J. R.,** Heat shock protein induction and induced thermal tolerance are independent in adult salamanders, *J. Exp. Zool.*, 241, 263—267, 1987.
49b. **Ellis, R. J., van der Vies, S. M., and Hemmingsen, S. M.,** The molecular chaperone concept, *Biochem. Soc. Symp.*, 55, 145—153, 1989.
50. **Ellwood, M. S. and Craig, E. A.,** Differential regulation of the 70K heat shock gene and related genes in *Saccharomyces cerevisiae, Mol. Cell. Biol.*, 4, 1454—1459, 1984.
51. **Elsdale, T. and Davidson, D.,** Timekeeping by frog embryos, in normal development and after heat shock, *Development,* 99, 41—49, 1987.

52. **Engelbrecht, L. and Mothes, K.,** Weitere Untersuchungen zur experimentellen Beeinflussung der Hitzewirkung bei Blättern von *Nicotiana rustica, Flora,* 154, 279—298, 1964.

52a. **Erickson, J. W. and Gross, C. A.,** Identification of the sigmaE subunit of *Escherichia coli* RNA polymerase: a second alternate sigma factor involved in high-temperature gene expression, *Genes Dev.,* 3, 1462—1471, 1989.

53. **Erickson, J. W., Vaughn, V., Walter, W. A., Neidhardt, F. C., and Gross, C. A.,** Regulation of the promoters and transcripts of rpoH, the *Escherichia coli* heat shock regulatory gene, *Genes Devel.,* 1, 419—432, 1987.

53a. **Evans, D. P., Simonette, R. A., Rasmussen, C. D., Means, A. R., and Tomasovic, S. P.,** Altered synthesis of the 26-kDa heat stress protein family and thermotolerance in cell lines with elevated levels of calcium-binding proteins, *J. Cell. Physiol.,* 142, 615—627, 1990.

53b. **Fayet, O., Ziegelhoffer, T., and Georgopoulos, C.,** The groES and groEL heat shock gene products of *Escherichia coli* are essential for bacterial growth at all temperatures, *J. Bacteriol.,* 171, 1379—1385, 1989.

54. **Ferrini, U., Falcioni, R., Delpino, A., Cavaliere, R., Zupi, G., and Natali, P. G.,** Heat-shock proteins produced by two human melanoma cell lines: Absence of correlation with thermosensitivity, *Int. J. Cancer,* 34, 651—655, 1984.

54a. **Finley, D., Ciechanover, A., and Varshavsky, A.,** Thermolability of ubiquitin-activating enzyme from the mammalian cell cycle mutant ts85, *Cell,* 37, 43—55, 1984.

55. **Fisher, G. A., Li, G. C., and Hahn, G. M.,** Modification of the thermal response by deuterium oxide I. Cell survival and the temperature shift, *Radiat. Res.,* 92, 530—540, 1982.

56. **Fisher, G. A., Anderson, R. L., and Hahn, G. M.,** Glucocorticoid-induced heat resistance in mammalian cells, *J. Cell. Physiol.,* 128, 127—132, 1986.

56a. **Freeman, M. L., Scidmore, N. C., Malcolm, A. W., and Meredith, M. J.,** Diamide exposure, thermal resistance, and synthesis of stress (heat shock) proteins, *Biochem. Pharmacol.,* 36, 21—29, 1987.

56b. **Freeman, M. L., Scidmore, N. C., and Meredith, M. J.,** Inhibition of heat shock protein synthesis and thermotolerance by cycloheximide, *Radiat. Res.,* 112, 564—574, 1987.

56c. **Freeman, M. L. and Meredith, M. J.,** Modulation of diamide toxicity in thermotolerant cells by inhibition of protein synthesis, *Cancer Res.,* 49, 4493—4498, 1989.

57. **Fuller, D. J. M. and Gerner, E. W.,** Delayed sensitization to heat by inhibitors of polyamine-biosynthetic enzymes, *Cancer Res.,* 42, 418—437, 1982.

58. **Fuller, D. J. M. and Gerner, E. W.,** Sensitization of Chinese hamster ovary cells to heat shock by alpha-difluoromethylornithine, *Cancer Res.,* 47, 816—820, 1987.

59. **Gausz, J., Bencze, G., Gyurkovics, H., Ashburner, M., Ish-Horowicz, D., and Holden, J. J.,** Genetic characterization of the 87C region of the third chromosome of *Drosophila melanogaster, Genetics,* 93, 917—934, 1979.

60. **Geilenkirchen, W. L. M.,** Cell division and morphogenesis of *Limnaea* eggs after treatment with heat pulses at successive stages in early division cycles, *J. Embryol. Exp. Morphol.,* 16, 321—337, 1966.

60a. **Georgopoulos, C., Tilly, V., Ang, D., Chandrasekhar, G. N., Fayet, O., Spence, J., Ziegelhoffer, T., Liberek, K., and Zylicz, M.,** The role of the *Escherichia coli* heat shock proteins in bacteriophage lambda growth, in *Stress-Induced Proteins,* Pardue, M. L., Feramisco, J. R., and Lindquist, S., Eds., Alan R. Liss, New York, 1989, 37—47.

61. **Gerner, E. W. and Schneider, M. J.,** Induced thermal resistance in HeLa cells, *Nature,* 256, 500—502, 1975.

62. **Gerweck, L. E. and Delaney, T. F.,** Persistence of thermotolerance in slowly proliferating plateau phase cells, *Radiat. Res.,* 97, 365—372, 1984.

62a. **Gerweck, L. E. and Majima, H.,** Variability in the kinetics of thermotolerance decay in three cell lines, *Radiat. Res.,* 112, 365—373, 1984.

63. **Giovanella, B. C., Stehlin, J. S., and Morgan, A. C.,** Selective lethal effect of supranormal temperatures on human neoplastic cells, *Cancer Res.,* 36, 3944—3950, 1976.

64. **Giudice, G.,** Heat shock proteins in sea urchin development, in *Changes in Eukaryotic Gene Expression in Response to Environmental Stress,* Atkinson, B. G. and Walden, D. B., Eds., Academic Press, Orlando, FL, 1985, 115—133.

65. **Goff, S. A., Casson, L. P., and Goldberg, A. L.,** Heat shock regulatory gene htpR influences rates of protein degradation and expression of the lon gene in *Escherichia coli, Proc., Natl. Acad., Sci., U.S.A.,* 81, 6647—6651, 1984.

66. **Gomes, M. I., Kim, W. J., Lively, M. K., and Amos, H.,** Heat-shock treatment lethal for mammalian cells deprived of glucose and glutamine: Protection by alpha-keto acids, *Biochem. Biophys. Res. Commun.,* 131, 1013—1019, 1985.

66a. **Gottesman, S.,** Genetics of proteolysis in *Escherichia coli, Annu. Rev. Genet.,* 23, 163—198, 1989.

66b. **Grant, C. M., Firoozan, M., and Tuite, M. F.,** Mistranslation induces the heat-shock response in the yeast *Saccharomyces cerevisiae, Mol. Microbiol,* 3, 215—220, 1989.

67. **Graziosi, G., Micali, F., Marzari, R., De Cristini, F., and Savoini, A.,** Variability of response of early *Drosophila* embryos to heat shock, *J. Exp. Zool.*, 214, 141—145, 1980.
68. **Greenberg, J. T. and Demple, B.,** Glutathione in *Escherichia coli* is dispensable for resistance to H_2O_2 and gamma radiation, *J. Bacteriol.*, 168, 1026—1029, 1986.
69. **Grossman, A. D., Straus, D. B., Walter, W. A., and Gross, C. A.,** Sigma 32 synthesis ion regulates the synthesis of heat shock proteins in *Escherichia coli, Genes Devel.*, 1, 179—184, 1987.
69a. **Guy, C. L., Plesofsky-Vig, N., and Brambl, R.,** Heat shock protects germinating conidiospores of *Neurospora crassa* against freezing injury, *J. Bacteriol.*, 167, 124—129, 1986.
70. **Hahn, G. M. and Li, G. C.,** Thermotolerance and heat shock proteins in mammalian cells, *Radiat. Res.*, 92, 452—457, 1982.
71. **Hahn, G. M. and Van Kersen, I.,** Thermoresistant RIF tumor cell strains, I. Isolation and initial characterization, *Cancer Res.*, 48, 1803—1807, 1988.
72. **Hahn, G. M., Shiu, E. C., West,, B., Goldstein, L., and Li, G. C.,** Mechanistic implications of the induction of thermotolerance in Chinese hamster cells by organic solvents, *Cancer Res.*, 45, 4138—4143, 1985.
72a. **Hahn, G. M., Adwankar, M. K., Basrur, V. S., and Anderson, R. L.,** Survival of cells exposed to anticancer drugs after stress, in *Stress-Induced Proteins,* Pardue, M. L., Feramisco, J. R., and Lindquist, S., Eds., Alan R. Liss, New York, 1989, 223—233.
73. **Hall, B. G.,** Yeast thermotolerance does not require protein synthesis, *J. Bacteriol,* 156, 1363—1365, 1983.
74. **Hallberg, R. L.,** No heat shock protein synthesis is required for induced thermostabilization of translational machinery, *Mol. Cell. Biol.*, 6, 2267—2270, 1986.
75. **Hallberg, R. L., Kraus, K. W., and Findly, R. C.,** Starved *Tetrahymena thermophila* cells that are unable to mount an effective heat shock response selectively degrade their rRNA, *Mol. Cell. Biol.*, 4, 2170—2179, 1984.
76. **Hallberg, R. L., Kraus, K. W., and Hallberg, E. M.,** Induction of acquired thermotolerance in *Tetrahymena thermophila:* Effects of protein synthesis inhibitors, *Mol. Cell. Biol.*, 5, 2061—2069, 1985.
77. deleted
78. **Harms-Ringdahl, M., Anderstam, B., and Vaca, C.,** Heat-induced changes in the incorporation of (^{3}H) acetate in membrane lipids, *Int. J. Radiat. Biol.*, 52, 315—324, 1987.
79. **Hatayama, T., Honda, K., and Yukioka, M.,** Effects of sodium butyrate and dibutyryl cyclic AMP on thermosensitivity of the HeLa cells and their production of heat shock proteins, *Biochem. Int.*, 13, 793—798, 1986.
79a. **Hatayama, T., Honda, K., and Yukioka, M.,** HeLa cells synthesize a specific heat shock protein upon exposure to heat shock at 42°C but not at 45°C, *Biochem. Biophys. Res. Commun.*, 137, 957—963, 1986.
80. **Haveman, J.,** Influence of a prior heat treatment on the enhancement by hyperthermia of X-ray-induced inactivation of cultured mammalian cells, *Int. J. Radiat. Biol.*, 43, 267—280, 1983.
81. **Haveman, J., Hart, A. A. M., and Wondergem, J.,** Thermal radiosensitization and thermotolerance in cultured cells from a murine mammary carcinoma, *Int. J. Radiat. Biol.*, 51, 71—80, 1987.
82. **Heikkila, J. J., Kloc, M., Bury, J., Schultz, G. A., and Browder, L. W.,** Acquisition of the heat shock response and thermotolerance during early development of *Xenopus laevis, Dev. Biol.*, 107, 483—489, 1985.
83. **Heikkila, J. J., Browder, L. W., Gedamu, L., Nickells, R. W., and Schultz, G. A.,** Heat-shock gene expression in animal embryonic systems, *Can. J. Genet. Cytol.*, 28, 1093—1105, 1986.
84. **Heindorff, K., Rieger, R., Schubert, I., Michaelis, A., and Aurich, O.,** Clastogenic adaptation of plant cells. Reduction of the yield of clastogen-induced chromatid aberrations by various pretreatment procedures, *Mutat. Res.*, 181, 157—171, 1987.
84a. **Helm, K. W., Petersen, N. S., and Abernethy, R. H.,** Heat shock response of germinating embryos of wheat — Effects of imbibition time and seed vigor, *Plant Physiol.*, 90, 598—605, 1989.
84b. **Henle, K. J. and Dethlefsen, L. A.,** Heat fractionation and thermotolerance. A review, *Cancer Res.*, 38, 1843—1851, 1978.
85. **Henle, K. J. and Leeper, D. B.,** Modification of the heat response and thermotolerance by cycloheximide, hydroxyurea, and lucanthone in CHO cells, *Radiat. Res.*, 90, 339—347, 1982.
86. **Henle, K. J., Karamuz, J. E., and Leeper, D. B.,** Induction of thermotolerance in Chinese hamster ovary cells by high (45°C) or low (40°C) hyperthermia, *Cancer Res.*, 38, 570—574, 1978.
87. **Henle, K. J., Tomosovic, S. P., and Dethlefsen, L. A.,** Fractionation of combined heat and radiation in asynchronous CHO cells. I. Effects on radiation sensitivity, *Radiat. Res.*, 80, 369—377, 1979.
88. **Henle, K. J., Nagle, W. A., Moss, A. J., and Herman, L. S.,** Polyhydroxy compounds and thermotolerance: A proposed concatenation, *Radiat. Res.*, 92, 445—451, 1982.
89. **Heuss-Larossa, K., Mayer, R. R., and Cherry, J. H.,** Synthesis of only two heat shock proteins is required for thermoadaptation in cultured cow pea cells, *Plant Physiol.*, 85, 4—7, 1987.

89a. **Holahan, E. V., Highfield, D. P., Holahan, P. K., and Dewey, W. C.,** Hyperthermic killing and hyperthermic radiosensitization in Chinese hamster ovary cells: Effects on pH and thermal tolerance, *Radiat. Res.*, 97, 108—131, 1984.

89b. **Honda, K., Hatayama, T., and Yukioka, M.,** Characterization of a 42°C-specific heat shock protein of mammalian cells, *J. Biochem.*, 103, 81—85, 1988.

90. **Hong-Qi, Z., Croes, A. F., and Linskens, H. F.,** Qualitative changes in protein synthesis in germinating pollen of *Lilium longiflorum* after heat shock, *Plant Cell Environ.*, 7, 689—691, 1984.

91. **Hottiger, T., Boller, T., and Wiemken, A.,** Rapid changes of heat and desiccation tolerance correlate with changes of trehalose content in *Saccharomyces cerevisiae* cells subjected to temperature shifts, *FEBS Lett.*, 220, 113—115, 1987.

91a. **Hume, S. P. and Marigold, J. C. L.,** Transient, heat-induced thermal resistance in small intestine of mouse, *Radiat. Res.*, 82, 526—535, 1980.

92. **Hume, S. P. and Marigold, J. C. L.,** Time-temperature relationships for hyperthermal radiosensitization in mouse intestine: Influence of thermotolerance, *Radiother. Oncol.*, 3, 165—171, 1985.

93. **Iida, H. and Yahara, I.,** Durable synthesis of high molecular weight heat shock proteins in GO cells of the yeast and other eukaryotes, *J. Cell Biol.*, 99, 199—207, 1984.

94. **Iida, H. and Yahara, I.,** A heat shock-resistant mutant of *Saccharomyces cerevisiae* shows constitutive synthesis of two heat shock proteins and altererd growth, *J. Cell Biol.*, 99, 1441—1450, 1984.

95. **Iida, H. and Yahara, I.,** Specific early-G1 blocks accompanied with stringent response in *Saccharomyces cerevisiae* lead to growth arrest in resting state similar to the GO of higher eukaryotes, *J. Cell Biol.*, 98, 1185—1193, 1984.

95a. **Iida, K., Iida, H., and Yahara, I.,** Heat shock induction of intranuclear actin rods in cultured mammalian cells, *Exp. Cell Res.*, 165, 207—215, 1986.

96. **Ish-Horowicz, D., Holden, J. J., and Gehring, W.,** Deletions of two heat-activated loci in *Drosophila melanogaster* and their effects on heat-induced protein synthesis, *Cell,* 12, 643—652, 1977.

97. **Ish-Horowicz, D., Schedl, P., Artavanis-Tsakonas, S., and Mirault, M. E.,** Genetic and molecular analysis of the 87A7 and 87C1 heat-inducible loci of *D. melanogaster, Cell,* 18, 1351—1358, 1979.

97a. **Jinn, T. L., Yeh, Y. C., Chen, Y. M., and Lin, C. Y.,** Stabilization of soluble proteins in vitro by heat shock proteins-enriched ammonium sulfate fraction from soybean seedlings, *Plant Cell Physiol.*, 30, 463—469, 1989.

97b. **Johnston, R. N. and Kucey, B. L.,** Competitive inhibition of hsp70 gene expression causes thermosensitivity, *Science,* 242, 1551—1554, 1988.

98. **Johnston, D., Oppermann, H., Jackson, J., and Levinson, W.,** Induction of four proteins in chicken embryo cells by sodium arsenite, *J. Biol. Chem.*, 255, 6975—6980, 1980.

99. **Jones, K. A. and Findly, R. C.,** Induction of heat shock proteins by canavanine in *Tetrahymena.* No change in ATP levels measured in vivo by NMR, *J. Biol. Chem.*, 261, 8703—8707, 1986.

99a. **Judelson, H. S. and Michelmore, R. W.,** Structure and expression of a gene encoding heat-shock protein hsp70 from the oomycete fungus *Bremia lactucae, Gene,* 79, 207—217, 1989.

99b. **Kampinga, H. H., Luppes, J. G., and Konings, A. N. T.,** Heat-induced nuclear protein binding and its relation to thermal cytotoxicity, *Int. J. Hypertherm.*, 3, 459—466, 1987.

100. **Kapoor, M. and Lewis, J.,** Heat shock induces peroxidase activity in *Neurospora crassa* and confers tolerance toward oxidative stress, *Biochem. Biophys. Res. Commun.*, 147, 904—910, 1987.

100a. **Kapoor, M. and Sreenivasan, G. M.,** The heat shock response of *Neurospora crassa:* Stress-induced thermotolerance in relation to peroxidase and superoxide dismutase levels, *Biochem. Biophys. Res. Commun.*, 156, 1097—1102, 1988.

100b. **Kapoor, M., Sreenivasan, G. M., Goel, N., and Lewis, J.,** Development of thermotolerance in *Neurospora crassa* by heat shock and other stresses eliciting peroxidase induction, *J. Bacteriol.*, 172, 2798—2801, 1990.

101. **Kappen, L. and Lange, O. L.,** Die Hitzeresistenz ausgetrockneter Blätter von *Commelina africana.* Ein Vergleich zwischen zwei Untersuchungsmethoden, *Protoplasma,* 65, 119—132, 1968.

102. **Kappen, L. and Lösch, R.,** Diurnal patterns of heat tolerance in relation to CAM, *Z. Pflanzenphysiol.*, 114, 87—96, 1984.

103. **Kase, K. and Hahn, G. M.,** Differential heat response of normal and transformed human cell in tissue culture, *Nature,* 255, 228—230, 1975.

103a. **Kashlev, M. V., Gragerov, A. I., and Nikiforov, V. G.,** Heat shock response in *Escherichia coli* promotes assembly of plasmid encoded RNA polymerase beta-subunit into RNA polymerase, *Mol. Gen. Genet.*, 216, 469—474, 1989.

104. **Kee, S. C. and Nobel, P. S.,** Concomitant changes in high temperature tolerance and heat-shock proteins in desert succulents, *Plant Physiol.*, 80, 596—598, 1986.

105. **Key, J. L., Czarnecka, E., Lin, C.-Y., Kimpel, J., Mothershed, C., and Schoeffl, F.,** A comparative analysis of the heat shock response in crop plants, in *Current Topics in Plant Biochemistry and Physiology,* Randall, D. D., Blevins, D. G., and Larson, R. L., Eds., 1983, 107—118.

106. **Key, J. L., Kimpel, J., Vierling, E., Lin, C.-Y., Nagao, R. T., Czarnecka, E., and Schoeffl, F.,** Physiological and molecular analyses of the heat shock response in plants, in *Changes in Eukaryotic Gene Expression in Response to Environmental Stress,* Atkinson, B. G. and Walden, D. B., Eds., Academic Press, Orlando, FL, 327—348, 1985.
107. **Kloppstech, K., Ohad, I., and Schweiger, A.-G.,** Evidence for an extranuclear coding site for a heat-shock protein in *Acetabularia, Eur. J. Cell Biol.,* 42, 239—245, 1986.
108. **Kraus, K. W., Hallberg, E. M., and Hallberg, R.,** Characterization of a *Tetrahymena thermophila* mutant strain unable to develop normal thermotolerance, *Mol. Cell. Biol.* 6, 3854—3861, 1986.
109. **Kraus, K. W., Good, P. J., and Hallberg, R. L.,** A heat shock-induced, polymerase III-transcribed RNA selectively associates with polysomal ribosomes in *Tetrahymena thermophila, Proc. Natl. Acad. Sci. U.S.A.,* 84, 383—387, 1987.
109a. **Krishnan, M., Nguyen, H. T., and Burke, J. J.,** Heat shock protein synthesis and thermal tolerance in wheat, *Plant Physiol.,* 90, 140—145, 1989.
109b. **Kusukawa, N. and Yura, T.,** Heat shock protein GroE of *Escherichia coli:* Key protective roles against thermal stress, *Genes Devel.,* 2, 874—882, 1988.
109c. **Kusukawa, N., Yura, T., Ueguchi, C., Akiyama, Y., and Ito, K.,** Effects of mutations in heat shock genes groES and groEL on protein export in *Escherichia coli, EMBO J.,* 8, 3517—3521, 1989.
110. **Lamarche, S., Chretien, P., and Landry, J.,** Inhibition of the heat shock response and synthesis of glucose-regulated proteins in calcium-deprived rat hepatoma cells, *Biochem. Biophys. Res. Commun.,* 131, 868—876, 1985.
111. **Landry, J.,** Heat shock proteins and cell thermotolerance, in *Hyperthermia in Cancer Treatment,* Anghileri, L. J. and Robert, J., Eds., CRC Press, Boca Raton, FL, 1986, 37—58.
112. **Landry, J. and Chretien, P.,** Relationship between hyperthermia-induced heat shock proteins and thermotolerance in Morris hepatoma cells, *Can. J. Biochem. Cell Biol.,* 61, 428—437, 1983.
113. **Landry, J., Bernier, D., Chretien, P., Nicole, L. M., Tanguay, R. M., and Marceau, N.,** Synthesis and degradation of heat shock proteins during development and decay of thermotolerance, *Cancer Res.,* 42, 2457—2461, 1982.
114. **Landry, J., Chretien, P., De Muys, J. M., and Morais, R.,** Induction of thermotolerance and heat shock protein synthesis in normal and respiration-deficient chicken embryo fibroblasts, *Cancer Res.,* 45, 2240—2247, 1985.
115. **Landry, J., Samson, S., and Chretien, P.,** Hyperthermia-induced cell death, thermotolerance, and heat shock proteins in normal, respiration-deficient, and glycolysis-deficient Chinese hamster cells, *Cancer Res.,* 46, 324—327, 1986.
115a. **Landry, J., Crete, P., Lamarche, S., and Chretien, P.,** Activation of calcium-dependent processes during heat shock role in cell thermoresistance, *Radiat. Res.,* 113, 426—436, 1988.
115b. **Landry, J., Chretien, P., Lambert, H., Hickey, E., and Weber, L. A.,** Heat shock resistance conferred by expression of the human hsp27 gene in rodent cells, *J. Cell Biol.,* 109, 7—15, 1989.
115c. **Lange, O. L.,** Die Hitzeresistenz einheimischer immer- und wintergrüner Pflanzen im Jahresverlauf, *Planta,* 56, 666—683, 1961.
116. **Lange, O. L. and Schwemmle, B.,** Untersuchungen zur Hitzeresistenz vegetativer und blühender Pflanzen von *Kalanchoe blossfeldiana, Planta,* 55, 208—225, 1960.
117. **Lanks, K. W., Hitti, I. F., and Chin, N. W.,** Substrate utilization for lactate and energy production by heat-shocked L929 cells, *J. Cell. Physiol.,* 127, 451—456, 1986.
118. **Lanks, K. W., Shah, V., and Chin, N. W.,** Enhancing hyperthermic cytotoxicity in L929 cells by energy source restriction and insulin exposure, *Cancer Res.,* 46, 1382—1387, 1986.
119. **Lanks, K. W., Gao, J.-P., and Kasambalides, E. J.,** Nucleosides restore heat resistance and suppress glucose-regulated protein synthesis by glucose-deprived L929 cells, *Cancer Res.,* 48, 1442—1446, 1987.
120. **Laszlo, A. and Li, G. C.,** Thermotolerant HA-1 cells are resistant to the thermosensitization action of amino acid analogues, in *Proc. 7th Int. Congr. Radiat. Res. Tumor Biol. Ther.,* Broerse, J. J., Barendsen, G. W., Kats, H. B., and van der Kogl, A. J., Eds., Martinus Nijhoff, Amsterdam, 1983, D6—D27.
121. **Laszlo, A. and Li, H. C.,** Heat-resistant variants of Chinese hamster fibroblasts altered in expression of heat shock protein, *Proc. Natl. Acad. Sci. U.S.A.,* 82, 8029—8033, 1985.
121a. **Laszlo, A.,** Regulation of the synthesis of heat-shock proteins in heat-resistant variants of Chinese hamster fibroblasts, *Radiat. Res.,* 116, 427—441, 1988.
121b. **Laszlo, A.,** The relationship of heat-shock proteins, thermotolerance, and protein synthesis, *Exp. Cell Res.,* 178, 401—414, 1988.
121c. **Laszlo, A.,** Evidence for two states of thermotolerance in mammalian cells, *Int. J. Hyperthermia,* 4, 513—526, 1988.
122. **Law, M. P., Coultas, P. G., and Field, S. B.,** Induced thermal resistance in the mouse ear, *Br. J. Radiol.,* 52, 308—314, 1979.
123. **Law, M. P., Ahier, R. G., and Field, S. B.,** The effect of prior heat treatment on the thermal enhancement of radiation damage in the mouse ear, *Br. J. Radiol.,* 52, 315—321, 1979.

124. **Lee, Y. J. and Dewey, W. C.,** Induction of heat shock proteins in Chinese hamster ovary cells and development of thermotolerance by intermediate concentrations of puromycin, *J. Cell. Physiol.*, 132, 1—11, 1987.

124a. **Lee, Y. J. and Dewey, W. C.,** Effect of cycloheximide or puromycin on induction of thermotolerance by sodium arsenite in Chinese hamster ovary cells: Involvement of heat shock proteins, *J. Cell. Physiol.*, 132, 41—48, 1987.

124b. **Lee, Y. J. and Dewey, W. C.,** Effect of cycloheximide or puromycin on induction of thermotolerance by heat in Chinese hamster ovary cells: Dose fractionation at 45.5°C, *Cancer Res.*, 47, 5960—5966, 1987.

124c. **Lee, Y. J., Dewey, W. C., and Li, G. C.,** Protection of Chinese hamster ovary cells from heat killing by treatment with cycloheximide or puromycin: Involvement of HSPs? *Radiat. Res.*, 111, 237—253, 1987.

124d. **Lee, K.-J. and Hahn, G. M.,** Abnormal protein as the trigger for the induction of stress responses: Heat, diamide and sodium arsenite, *J. Cell. Physiol.*, 136, 411—420, 1988.

125. **Lepock, J. R., Massicotte-Nolan, P., Rule, G. S., and Kruuv, J.,** Lack of correlation between hyperthermic cell killing, thermotolerance and membrane lipid fluidity, *Radiat. Res.*, 87, 300—313, 1981.

126. **Le Rudulier, D., and Valentine, R. C.,** Genetic engineering in agriculture: Osmoregulation, *Trends Biochem. Sci.*, 7, 431—433, 1982.

127. **Lewis, M. J., and Pelham, H. R. B.,** Involvement of ATP in the nuclear and nucleolar functions of the 70 kd heat shock protein, *EMBO J.*, 4, 3137—3143, 1985.

128. **Li, G. C.,** Induction of thermotolerance and enhanced heat shock protein synthesis in Chinese hamster fibroblasts by sodium arsenite and by ethanol, *J. Cell. Physiol.*, 115, 116—122, 1983.

129. **Li, G. C.,** Elevated levels of 70,000 dalton heat shock protein in transiently thermotolerant Chinese hamster fibroblasts and their stable heat resistant variants, *Int. J. Radiat. Oncol. Biol. Phys.*, 11, 165—177, 1985.

130. **Li, G. C. and Hahn, G. M.,** A proposed operational model of thermotolerance based on effects of nutrients and the initial treatment temperature, *Cancer Res.*, 40, 4501—4508, 1980.

131. **Li, G. C. and Laszlo, A.,** Amino acid analogs while inducing heat shock proteins sensitize CHO cells to thermal damage, *J. Cell. Physiol.*, 122, 91—97, 1985.

131a. **Li, G. C. and Laszlo, A.,** Thermotolerance in mammalian cells: A possible role for heat shock proteins, in *Changes in Eukaryotic Gene Expression in Response to Environmental Stress,* Atkinson, B. G. and Walden, D. B., Eds., Academic Press, Orlando, FL, 1985, 227—254.

132. **Li, G. C. and Mak, Y.,** Induction of heat shock protein synthesis in murine tumors during the development of thermotolerance, *Cancer Res.*, 45, 3816—3824, 1985.

133. **Li, G. C. and Mivechi, N. F.,** Thermotolerance in mammalian systems: A review, in *Hyperthermia in Cancer Treatment,* Anghlieri, L. J. and Roberts, J., Eds., CRC Press, Boca Raton, FL, 1986, 59—77.

134. **Li, G. C. and Werb, Z.,** Correlation between synthesis of heat shock proteins and development of thermotolerance in Chinese hamster fibroblasts, *Proc. Natl. Acad. Sci. U.S.A.*, 79, 3218—3222, 1982.

135. **Li, G. C., Shrieve, D. C., and Werb, Z.,** Correlations between synthesis of heat shock proteins and development of tolerance to heat and to adriamycin in Chinese hamster fibroblasts: Heat shock and other inducers, in *Heat Shock from Bacteria to Man,* Schlesinger, M. J., Ashburner, M., and Tissieres, A., Eds., Cold Spring Harbor Laboratory, Cold Spring Harbor, N.Y., 1982, 395—404.

136. **Li, G. C., Petersen, N. S., and Mitchell, H. K.,** Induced thermal tolerance and heat shock protein synthesis in Chinese hamster ovary cells, *Br. J. Cancer Suppl.*, 5, 132—136, 1982.

137. **Li, G. C., Fisher, G. A., and Hahn, G. M.,** Modification of the thermal response by deuterium oxide. II. Thermotolerance and the specific inhibition of development, *Radiat. Res.*, 92, 541—551, 1982.

138. **Lin, C.-Y., Roberts, J. K., and Key, J. L.,** Acquisition of thermotolerance in soybean seedlings, *Plant Physiol.*, 74, 152—160, 1984.

139. **Lin, C.-Y., Chen, Y.-M., and Key, J. L.,** Solute leakage in soybean seedlings under various heat shock regimes, *Plant Cell Physiol.*, 26, 1493—1498, 1985.

140. **Lindquist, S.,** Translational efficiency of heat-induced messages in *Drosophila melanogaster* cells, *J. Mol. Biol.*, 137, 151—158, 1980.

141. **Lindquist, S.,** Varying patterns of protein synthesis in *Drosophila* during heat shock: Implications for regulation, *Dev. Biol.*, 77, 463—479, 1980.

142. **Loomis, W. F. and Wheeler, S.,** Heat shock response of *Dictyostelium, Dev. Biol.*, 79, 399—408, 1980.

143. **Loomis, W. F. and Wheeler, S. A.,** Chromatin-associated heat shock proteins in *Dictyostelium, Dev. Biol.*, 90, 412—418, 1982.

143a. **Mackey, B. M. and Derrick, C. M.,** The effect of prior heat shock on the thermoresistance of *Salmonella thompson* in foods, *Lett. Appl. Microbiol.*, 5, 115—118, 1987.

143b. **Malhotra, A., Kruuv, J., and Lepock, J. R.,** Sensitization of rat hepatocytes to hyperthermia by calcium, *J. Cell. Physiol.*, 128, 279—284, 1986.

144. **Mansfield, M. A., Lingle, W. L., Key, J. L.,** The effects of lethal heat shock on nonadapted and thermotolerant root cells of *Glycine max, J. Ultrastruct. Mol. Struct. Res.*, 99, 96—105, 1988.

144a. **Marigold, J. C. L. and Hume, S. P.,** Effect of prior hyperthermia on subsequent thermal enhancement of radiation damage in mouse intestine, *Int. J. Radiat. Biol.*, 42, 509—516, 1982.

144b. **Marigold, J. C. L., Hume, S. P., and Hand, J. W.,** Investigation of thermotolerance in mouse testis, *Int. J. Radiat. Biol.*, 48, 589—595, 1985.

145. **Marmiroli, N., Restivo, F. M., Odoardi Stanca, M., Terzi, V., Giovanelli, B., Tassi, F., and Lorenzoni, C.,** Induction of heat shock proteins and acquisition of thermotolerance in barley seedlings *Hordeum vulgare, Genet. Argar.*, 40, 9—25, 1986.

146. **Massicotte-Nolan, P., Glofcheski, D. J., Kruuv, J., and Lepock, J. R.,** Relationship between hyperthermic cell killing and protein denaturation by alcohols, *Radiat. Res.*, 87, 284—299, 1981.

147. **McAlister, L. and Finkelstein, D. B.,** Alterations in translatable ribonucleic acid after heat shock of *Saccharomyces cerevisiae, J. Bacteriol.*, 143, 603—612, 1980.

148. **McAlister, L. and Finkelstein, D. B.,** Heat shock proteins and thermal resistance in yeast, *Biochem. Biophys. Res. Commun.*, 93, 819—824, 1980.

149. **McGarry, T. J. and Lindquist, S.,** Inhibition of heat shock protein synthesis by heat-inducible antisense RNA, *Proc. Natl. Acad. Sci. U.S.A.*, 83, 399—403, 1986.

150. **McMullin, T. W. and Hallberg, R. L.,** Effect of heat shock on ribosome structure: Appearance of a new ribosome-associated protein, *Mol. Cell. Biol.*, 6, 2527—2535, 1986.

150a. **Merz, R. and Laudien, H.,** Two types of heat tolerance in FAM-cells. Induction by heat-shock versus elevated culturing temperature, *J. Therm. Biol.*, 12, 281—288, 1987.

151. **Milkman, R.,** On the mechanism of some temperature effects on *Drosophila, J. Gen. Physiol.*, 46, 1151—1170, 1963.

151a. **Miller, D. and McLennan, A. G.,** The heat shock response of the cryptobiotic brine shrimp *Artemia*. I. A comparison of the thermotolerance of cysts and larvae, *J. Therm. Biol.*, 13, 119—123, 1988.

151b. **Miller, D. and McLennan, A. G.,** The heat shock response of the cyrptobiotic brine shrimp *Artemia*. II. Heat shock proteins, *J. Therm. Biol.*, 13, 125—134, 1988.

151c. **Milligan, A. J., Metz, J. A., and Leeper, D. B.,** Effect of intestinal hyperthermia in the Chinese hamster, *Int. J. Radiat. Onc. Biol., Phys.*, 10, 259—263, 1984.

151d. **Minton, K. W., Karmin, P., Hahn, G. M., and Minton, A. P.,** Nonspecific stabilization of stress-susceptible proteins by stress-resistant proteins: A model for the biological role of heat shock proteins, *Proc. Natl. Acad. Sci. U.S.A.*, 79, 7107—7111, 1982.

152. **Mirkes, P. E.,** Hyperthermia-induced heat shock response and thermotolerance in postimplantation rat embryos, *Dev. Biol.*, 119, 115—123, 1987.

153. **Mitchel, R. E. J. and Morrison, D. P.,** Heat-shock induction of ionizing radiation resistance in *Saccharomyces cerevisiae,* and correlation with stationary growth phase, *Radiat. Res.*, 90, 284—291, 1982.

154. **Mitchel, R. E. J. and Morrison, D. P.,** Heat-shock induction of ionizing radiation resistance in *Saccharomyces cerevisiae*. Transient changes in growth cycle distribution and recombinational ability, *Radiat. Res.*, 92, 182—187, 1982.

155. **Mitchel, R. E. J. and Morrison, D. P.,** Assessment of the role of oxygen and mitochondria in heat shock induction of radiation and thermal resistance in *Saccharomyces cerevisiae, Radiat Res.*, 96, 113—117, 1983.

156. **Mitchel, R. E. J. and Morrison, D. P.,** Heat-shock induction of ultraviolet-light resistance in *Saccharomyces cerevisiae, Radiat. Res.*, 96, 95—99, 1983.

157. **Mitchel, R. E. J. and Morrison, D. P.,** Inducible error-prone repair in yeast. Suppression by heat shock, *Mutat. Res.*, 159, 31—39, 1986.

158. **Mitchell, H. K., Moller, G., Petersen, N. S., and Lipps-Sarmiento, L.,** Specific protection from phenocopy induction by heat shock, *Dev. Genet.*, 1, 181—192, 1979.

158a. **Mitchell, J. B., Russo, A., Kinsella, T. J., and Glatstein, E.,** Glutathione elevation during thermotolerance induction and thermosensitization by glutathione depletion, *Cancer Res.*, 43, 987—991, 1983.

158b. **Mizzen, L. A. and Welch, W. J.,** Characterization of the thermotolerant cell. I. Effects on protein synthesis activity and the regulation of HSP70 expression, *J. Cell Biol.*, 106, 1105—1116, 1988.

158c. **Moisyadi, S. and Harrington, H. M.,** Characterization of the heat shock response in cultured sugarcane cells. 1. Physiology of the heat shock response and heat shock protein synthesis, *Plant Physiol.*, 90, 1156—1162, 1989.

159. **Mosser, D. D. and Bols, N. C.,** Relationship between heat-shock protein synthesis and thermotolerance in rainbow trout fibroblasts, *J. Comp. Physiol. [B]*, 158, 457—467, 1988.

159a. **Mosser, D. D., Van Oostrom, J., and Bols, N. C.,** Induction and decay of thermotolerance in rainbow trout fibroblasts, *J. Cell. Physiol.*, 132, 155—160, 1987.

159b. **Nagata, K., Saga, S., and Yamada, K. M.,** A major collagen-binding protein of chicken embryo fibroblasts is a novel heat shock protein, *J. Cell Biol.*, 103, 223—230, 1986.

160. **Neidhardt, F. C. and Van Bogelen, R. A.,** Positive regulatory gene for temperature-controlled proteins in *Escherichia coli, Biochem. Biophys. Res. Commun.*, 100, 894—900, 1981.

161. **Neidhardt, F. C., Van Bogelen, R. A., and Vaughn, V.,** The genetics and regulation of heat-shock proteins, *Annu. Rev. Genet.*, 18, 295—329, 1984.

162. **Nelson, D. R. and Killeen, K. B.,** Heat shock proteins of vegetative and fruiting *Myxococcus xanthus* cells, *J. Bacteriol.*, 168, 1100—1106, 1986.

162a. **Nguyen, V. T., Morange, M., and Bensaude, O.,** Protein denaturation during heat shock and related stress — *Escherichia coli* β-galactosidase and *Photinus pyralis* luciferase inactivation in mouse cells, *J. Biol. Chem.*, 264, 10487—10492, 1989.

163. **Nickells, R. W. and Browder, L. W.,** Region-specific heat-shock protein synthesis correlates with a biphasic acquisition of thermotolerance in *Xenopus laevis* embryos, *Dev. Biol.*, 112, 391—395, 1985.

164. **Nicole, L. M. and Tanguay, R. M.,** On the specificity of antisense RNA to arrest in vitro translation of mRNA coding for *Drosophila* hsp23, *Biosci. Rep.*, 7, 239—245, 1987.

165. **Nover, L., Ed.,** *Heat Shock Response of Eukaryotic Cells,* Springer-Verlag, Berlin, 1984.

166. **Nover, L. and Scharf, K.-D.,** Synthesis, modification and structural binding of heat shock proteins in tomato cell cultures, *Eur. J. Biochem.*, 139, 303—313, 1984.

167. **Nover, L., Munsche, D., Ohme, K., and Scharf, K.-D.,** Ribosome biosynthesis in heat shock tomato cell cultures. I. Ribosomal RNA, *Eur. J. Biochem.*, 160, 297—304, 1986.

168. **Nover, L., Scharf, K.-D., and Neumann, D.,** Cytoplasmic heat shock granules are formed from precursor particles and contain a specific set of mRNAs, *Mol. Cell. Biol.*, 9, 1298—1308, 1989.

168a. **Nover, L., Neumann, D., and Scharf, K. D., Eds.,** *Heat Shock and Other Stress Response Systems of Plants,* Springer-Verlag, Berlin, 1990.

169. **Özkaynak, E., Finley, D., Solomon, M. J., and Varshavsky, A.,** The yeast ubiquitin genes: A family of natural gene fusions, *EMBO J.*, 6, 1429—1439, 1987.

170. **Ohtsuka, K., Furuya, M., Nitta, K., and Kano, E.,** Effect of cycloheximide on the development of thermotolerance and the synthesis of 68-kilodalton heat shock protein in Chinese hamster V79 and mouse L cells in vitro, *J. Radiat. Res.*, 27, 291—299, 1986.

171. **Omar, R. A. and Lanks, K. W.,** Heat shock protein synthesis and cell survival in clones of normal and simian virus 40-transformed mouse embryo cells, *Cancer Res.*, 44, 3976—3982, 1984.

172. **Omar, R. A., Yano, S., and Kikkawa, Y.,** Antioxidant enzymes and survival of normal and Simian virus 40-transformed mouse embryo cells after hyperthermia, *Cancer Res.*, 47, 3473—3476, 1987.

172a. **Otto, B., Grimm, B., Ottersbach, P., and Kloppstech, K.,** Circadian control of the accumulation of messenger RNAs for light-inducible and heat-inducible chloroplast proteins in pea (*Pisum sativum* L.), *Plant Physiol.*, 88, 21—25, 1988.

173. **Ougham, H. J. and Stoddart, J. L.,** Synthesis of heat-shock protein and acquisition of thermotolerance in high-temperature tolerant and high-temperature susceptible lines of sorghum, *Plant Sci.*, 44, 163—168, 1986.

174. **Paek, K.-H. and Walker, G. C.,** Defect in expression of heat-shock proteins at high temperature in xthA mutants, *J. Bacteriol.*, 165, 763—770, 1986.

175. **Paek, K.-H. and Walker, G. C.,** *Escherichia coli* dnaK null mutants are inviable at high temperature, *J. Bacteriol.*, 169, 283—290, 1987.

176. **Parag, H. A., Raboy, B., and Kulka, R. G.,** Effect of heat shock on protein degradation in mammalian cells: Involvement of the ubiquitin system, *EMBO J.*, 6, 55—63, 1987.

177. **Pearcy, R. W., Berry, J. A., and Fork, D. C.,** Effects of growth temperature on the thermal stability of the photosynthetic apparatus of *Atriplex lentiformis* (Torr.) Wats, *Plant Physiol.*, 59, 873—878, 1977.

177a. **Pelham, H. R. B.,** Hsp 70 accelerates the recovery of nucleolar morphology after heat shock, *EMBO J.*, 3, 3095—3100, 1984.

178. **Petersen, N. S. and Mitchell, H. K.,** Recovery of protein synthesis after heat shock: Prior heat treatment affects the ability of cells to translate mRNA, *Proc. Natl. Acad. Sci. U.S.A.*, 78, 1708—1711, 1981.

178a. **Petersen, N. S. and Mitchell, H. K.,** The forked phenocopy is prevented in thermotolerant pupae, in *Stress-Induced Proteins,* Pardue, M. L., Feramisco, J. R., and Lindquist, S., Eds., Alan R. Liss, New York, 1989, 235—244.

179. **Petko, L. and Lindquist, S.,** Hsp 26 is not required for growth at high temperatures, nor for thermotolerance, spore development, or germination, *Cell,* 45, 885—894, 1986.

180. **Piper, P. W., Davies, M. W., Curran, B., Lockheart, A., Spalding, A., and Tuite, M. F.,** The influence of cell ploidy on the thermotolerance of *Saccharomyces cerevisiae, Curr. Genet.*, 11, 595—598, 1987.

181. **Piperakis, S. M. and McLennan, A. G.,** Enhanced reactivation of UV-irradiated adenovirus 2 in HeLa cells treated with nonmutagenic chemical agents, *Mutat. Res.*, 142, 83—86, 1985.

182. **Plesovski-Vig, N. and Brambl, R.,** Heat shock response of *Neurospora crassa:* Protein synthesis and induced thermotolerance, *J. Bacteriol.*, 161, 1083—1091, 1985.

183. **Plesset, J., Palm, C., and McLaughlin, C. S.,** Induction of heat shock proteins and thermotolerance by ethanol in *Saccharomyces cerevisiae, Biochem. Biophys. Res. Commun.*, 108, 1340—1345, 1982.

184. **Plesset, J., Ludwig, J. R., Cox, B. S., and McLaughlin, C. S.,** Effect of cell cycle position on thermotolerance in *Saccharomyces cerevisiae, J. Bacteriol.*, 169, 779—784, 1987.

184a. **Polla, B. S. and Anderson, R. R.,** Thermal injury by laser pulses: Protection by heat shock despite failure to induce heat shock response, *Lasers Surg. Med.*, 7, 398—404, 1987.

185. **Przybytkowski, E., Bates, J. H. T., Bates, D. A., and Mackillop, W. J.,** Thermal adaptation in CHO cells at 40°C: The influence of growth conditions and the role of heat shock proteins, *Radiat. Res.*, 107, 317—331, 1986.

186. **Quinn, P. J.,** The fluidity of cell membranes and its regulation, *Prog. Biophys. Mol. Biol.*, 38, 1—104, 1981.

186a. **Ramsey, N.,** A mutant in a major heat shock protein of *Escherichia coli* continues to show inducible thermotolerance, *Mol. Gen. Genet.*, 211, 332—334, 1988.

187. **Read, R. A., Fox, M. H., and Bedford, J. S.,** The cell cycle dependence of thermotolerance I. CHO cells heated at 42°C, *Radiat. Res.*, 93, 93—106, 1983.

188. **Reeves, R. O.,** Mechanisms of acquired resistance to accute heat shock in cultured mammalian cells, *J. Cell. Physiol.*, 79, 157—170, 1972.

188a. **Riabowol, K. T., Mizzen, L. A., and Welch, W. J.,** Heat-shock is lethal to fibroblasts microinjected with antibodies against hsp70, *Science*, 242, 433—436, 1988.

188b. **Rice, L. C., Urano, M., and Maher, J.,** The kinetics of thermotolerance in the mouse foot, *Radiat. Res.*, 89, 291—297, 1982.

189. **Rice, G., Laszlo, A., Li, G., Gray, J., and Dewey, W.,** Heat shock proteins within the mammalian cell cycle: Relationship to thermal sensitivity, thermal tolerance and cell cycle progression, *J. Cell. Physiol.*, 126, 291—297, 1986.

190. **Rieger, R. and Michaelis, A.,** Heat shock protection against induction of chromatid aberrations is dependent on the time span between heat shock and clastogen treatment of *Vicia faba* root tip meristem cells, *Mutat. Res.*, 209, 141—144, 1988.

191. **Rieger, R., Michaelis, A., and Nicoloff, H.,** Effects of stress factors on the clastogen response of *Vicia faba* root tip meristems — 'Clastogenic adaptation', *Biol. Zbl.*, 105, 19—28, 1986.

191a. **Rieger, R., Michaelis, A., and Schubert, I.,** Reduction by heat shock of maleic hydrazide-induced aberration yield is dependent on temperature and duration of heat pretreatment, *Mutat. Res.*, 174, 199—204, 1986.

191b. **Rieger, R.,** Stress-induced protective effects against DNA damage, in *Heat Shock and Other Stress Response Systems of Plants*, Nover, L., Neumann, D., and Scharf, K.-D., Eds., Springer-Verlag, Berlin, 1990, 105—112.

192. **Roccheri, M. C., Di Bernardo, M. G., and Giudice, G.,** Synthesis of heat shock proteins in developing sea urchins, *Dev. Biol.*, 83, 173—177, 1981.

193. **Roccheri, M. C., Sconzo, G., Larosa, M., Oliva, D., Abrignani, A., and Giudice, G.,** Response to heat shock of different sea urchin species, *Cell Differ.*, 18, 131—135, 1986.

193a. **Rofstad, E. K.,** Heat sensitivity and thermotolerance in vitro of human breast carcinoma, malignant melanoma and squamous cell carcinoma of the head and neck. *Br. J. Cancer*, 61, 22—28, 1990.

193b. **Rosen, E., Sivertsen, A., Firtel, R. A., Wheeler, S., and Loomis, W. F.,** Heat shock genes of Dictyostelium, in *Changes in Eukaryotic Gene Expression in Response to Environmental Stress*, Atkinson, B. G. and Walden, D. B., Eds., Academic Press, Orlando, FL, 1985, 257—278.

194. **Rothman, J. E.,** Polypetide chain binding proteins: Catalysts of protein folding and related processes in cells, *Cell*, 59, 591—601, 1989.

194a. **Ruifrok, A. C. C. and Konings, A. W. T.,** Effects of amiloride on hyperthermic cell killing of normal and thermotolerant mouse fibroblast LM cells, *Int. J. Radiat. Biol.*, 52, 385—392, 1987.

194b. **Rutledge, P. S., Spotila, J. R., and Easton, D. P.,** Heat hardening in response to two types of heat shock in the lungless salamanders *Eurycea bislineata* and *Desmognathus ochrophaeus*, *J. Thermal. Biol.*, 12, 235—241, 1987.

195. **Sapareto, S. A., Hopwood, L. E., Dewey, W. C., Raju, M. R., and Gray, J. W.,** Effects of hyperthermia on survival and progression on Chinese hamster ovary cells, *Cancer Res.*, 38, 393—400, 1978.

196. **Sapper, I.,** Versuche zur Hitzeresistenz der Pflanzen, *Planta*, 23, 518—556, 1935.

197. **Schamhart, D. H. J., Van Walraven, H. S., Wiegant, F. A. C., Linnemans, W. A. M., Van Rijn, J., Van Den Berg, J., and Van Wijk, R.,** Thermotolerance in cultured hepatoma cells: Cell viability, cell morphology, protein synthesis, and heat shock proteins, *Radiat. Res.*, 98, 82—95, 1984.

197a. **Schamhart, D. H. J., Boonstra, J., Van Graft, M., Van Rijn, J., and Van Wijk, R.,** The occurence of thermotolerance in non-proliferating rat hepatocytes in primary culture, *Int. J. Radiat. Biol.*, 47, 213—218, 1985.

198. **Scharf, K.-D. and Nover, L.,** Heat shock induced alterations of ribosomal protein phosphorylation in plant cell cultures, *Cell*, 30, 427—437, 1982.

199. **Scharf, K.-D. and Nover, L.,** Control of ribosome biosynthesis in plant cell cultures under heat shock conditions. II. Ribosomal proteins, *Biochim. Biophys. Acta*, 909, 44—57, 1987.

199a. **Schenberg-Frascino, A. and Moustacchi, E.,** Lethal and mutagenic effect of elevated temperature on haploid yeast. I. Variations in sensitivity during cell cycle, *Mol. Gen. Genet.*, 115, 243—257, 1972.

200. **Schlesinger, M. J. and Bond, U.,** Ubiquitin genes, in *Oxford Survey of Eukaryotic Genes,* Vol. 4, 1987, 77.

200a. **Schmidt, J., Laudien, H., and Bowler, K.,** Acute adjustments to high temperature in FHM-cells from *Pimephales promelas (Pisces, Cyprinidae), Comp. Biochem. Physiol.,* 78A, 823—828, 1984.

201. **Schmitt, U.,** Supraoptimal growth-temperature reduces the heat-shock sensitivity of the chrysophycean flagellate *Poterioochromonas malhamensis, Protoplasma,* 123, 48—56, 1984.

202. **Schöffl, F., Rieping, M., and Baumann, G.,** Constitutive transcription of a soybean heat-shock gene by a cauliflower mosaic virus promoter in transgenic tobacco plants, *Dev. Genet.,* 8, 365—374, 1987.

202a. **Schuster, G., Even, D., Kloppstech, K., and Ohad, I.,** Evidence for protection by heat-shock proteins against photoinhibition during heat-shock, *EMBO J.,* 7, 1—6, 1988.

203. **Schwemmle, B. and Lange, O. L.,** Endogen-tagesperiodische Schwankungen der Hitzeresistenz bei *Kalanchoe blossfeldiana, Planta,* 53, 134—144, 1959.

204. **Sciandra, J. J. and Subjeck, J. R.,** Heat shock proteins and protection of proliferation and translation in mammalian cells, *Cancer Res.,* 44, 5188—5194, 1984.

205. **Sconzo, G., Roccheri, M. C., La Rosa, M., Oliva, D., Abrignani, A., and Giudice, G.,** Acquisition of thermotolerance in sea urchin embryos correlates with the synthesis and age of the heat shock proteins, *Cell Differ.,* 19, 173—179, 1986.

206. **Selawry, O. S., Goldstein, M. N., and McCormick, T.,** Hyperthermia in tissue-cultured cells of malignant origin, *Cancer Res.,* 17, 785—791, 1957.

206a. **Seufert, W. and Jentsch, S.,** Ubiquitin-conjugating enzymes UBC4 and UBC5 mediate selective degradation of short-lived and abnormal proteins, *EMBO J.,* 9, 543—550, 1990.

207. **Shimada, Y.,** Influence of the thermal conditioning on the heat-induced radioresistance in primordial germ cells of the fish *Oryzia latipes, Int. J. Radiat. Biol.,* 48, 423—430, 1985.

207a. **Shimada, Y.,** Heat-shock induction of radiation resistance in primordial germ cells of the fish *Oryzias latipes, Int. J. Radiat. Biol.,* 48, 189—196, 1985.

208. **Shimada, Y.,** Induction of thermotolerance in fish embryos *Oryzias latipes, Comp. Biochem. Physiol.,* 80a, 177—181, 1985.

209. **Shimada, Y., Shima, A., and Egami, N.,** Effects of dose fractionation and cycloheximide on the heat-shock induction of radiation resistance in primordial germ cells of the fish *Oryzias latipes, Radiat. Res.,* 104, 78—82, 1985.

210. **Shimada, Y., Shima, A., and Egami, N.,** Effects of heat, release from hypoxia, cadmium and arsenite on radiation sensitivity of primordial germ cells in the fish *Oryzias latipes, J. Radiat. Res.,* 26, 411—417, 1985.

211. **Shin, D.-Y., Matsumoto, K., Iida, H., Uno, I., and Ishikawa, T.,** Heat shock response of *Saccharomyces cerevisiae* mutants altered in cyclic AMP-dependent protein phosphorylation, *Mol. Cell. Biol.,* 7, 244—250, 1987.

211a. **Shyy, T.-T., Asch, B. B., and Asch, H. L.,** Concurrent collapse of keratin filaments, aggregation of organelles and inhibition of protein synthesis during the heat shock response in mammary epithelial cells, *J. Cell Biol.,* 108, 997—1008, 1989.

212. **Silva, A. M., Juliani, M. H., Da Costa, J. J., and Bonato, M. C. M.,** Acquisition of thermotolerance during development of *Blastocladiella emersonii, Biochem. Biophys. Res. Commun.,* 144, 491—498, 1987.

213. **Spitz, D. R., Dewey, W. C., and Li, G. C.,** Hydrogen peroxide or heat shock induces resistance to hydrogen peroxide in Chinese hamster fibroblasts, *J. Cell. Physiol.,* 131, 364—373, 1987.

213a. **Spitz, D. R. and Li, G. C.,** Heat-induced cytotoxicity in H_2O_2-resistant Chinese hamster fibroblasts, *J. Cell. Physiol.,* 142, 255—260, 1990.

214. **Stephanou, G., Alahiotis, S. N., Christodoulou, C., and Marmaras, V. J.,** Adaptation of *Drosophila* to temperature: Heat-shock proteins and survival in *Drosophila melanogaster* (genetic analysis), *Dev. Genet.,* 3, 299—308, 1983.

215. **Stephanou, G., Alahiotis, S. N., Marmaras, V. J., and Christodoulou, C.,** Heat shock response in *Ceratitis capitata, Comp. Biochem, Physiol.,* 74B, 425—432, 1983.

216. **Stevenson, M. A., Calderwood, S. K., and Hahn, G. M.,** Rapid increases in inositol trisphosphate and intracellular calcium after heat shock, *Biochem. Biophys. Res. Commun.,* 137, 826—833, 1986.

217. **Stevenson, M. A., Calderwood, S. K., and Hahn, G. M.,** Effect of hyperthermia (45°C) on calcium flux in Chinese hamster ovary HA-1 fibroblasts and its potential role in cytotoxicity and heat resistance, *Cancer Res.,* 47, 3712—3717, 1987.

217a. **Subjeck, J. R. and Shyy, T. T.,** Stress protein systems of mammalian cells, *Am. J. Physiol.,* 250, C1—C17, 1986.

218. **Subjeck, J. R., Siandra, J. J., and Johnson, R. J.,** Heat shock proteins and thermotolerance: A comparison of induction kinetics, *Br. J. Radiol.,* 55, 579—584, 1982.

219. **Subjeck, J. R., Sciandra, J. J., and Shyy, T. T.,** Analysis of the expression of the two major proteins of the 70 kilodalton mammalian heat shock family, *Int. J. Radiat. Biol.,* 47, 275—284, 1985.

219a. **Süss, K. H. and Jordanov, I. T.,** Biosynthetic cause of in vivo acquired thermotolerance of photosynthetic light reactions and metabolic responses of chloroplasts to heat stress, *Plant Physiol,* 81, 192—199, 1986.
220. **Susek, R. E. and Lindquist, S. L.,** Hsp26 of *Saccharomyces cerevisiae* is homologous to the superfamily of small heat shock proteins, but is without a demonstrable function, *Mol. Cell. Biol.,* 9, 5265—5271, 1989.
220a. **Swindle, J., Ajioka, J., Eisen, H., Sanwal, B., Jacquemot, C., Browder, Z., and Buck, G.,** The genomic organization and transcription of the ubiquitin genes of *Trypanosoma cruzi, EMBO J.,* 7, 1121—1127, 1988.
220b. **Tanaka, S., Ishii, C., and Inoue, H.,** Effects of heat shock on the induction of mutations by chemical mutagens in *Neurospora crassa, Mutation Res.,* 223, 233—242, 1989.
220c. **Tanaka, K., Yatomi, T., Matsumoto, K., and Toh-E, A.,** Transcriptional regulation of the heat shock genes by cyclic AMP and heat shock in yeast, in *Stress-Induced Proteins,* Pardue, M. L., Feramisco, J. R., and Lindquist, S. L., Eds., Alan R. Liss, New York, 1989, 63—72.
221. **Taylor, S. Y. and Holliday, R.,** Induction of thermotolerance and mitotic recombination by heat-shock in *Ustilago maydis, Curr. Genet.,* 9, 59—64, 1984.
222. **Tilly, K., McKittrick, N., Zylicz, M., and Georgopoulos, C.,** The dnaK protein modulates the heat-shock response of *Escherichia coli, Cell,* 34, 641—646, 1983.
222a. **Tilly, K., Spence, J., and Georgopoulos, C.,** Modulation of stability of the *Escherichia coli* heat shock regulatory factor sigma32, *J. Bacteriol.,* 171, 1585—1589, 1989.
223. **Tobe, T., Ito, K., and Yura, T.,** Isolation and physical mapping of temperature-sensitive mutants defective in heat-shock induction of proteins in *Escherichia coli, Mol. Gen. Genet.,* 195, 10—16, 1984.
223a. **Toda, T., Cameron, S., Sass, P., Zoller, M., Scott, J. D., McMullen, B., Hurwitz, M., Krebs, E. G., and Wigler, M.,** Cloning and characterization of BCY 1, a locus encoding a regulatory subunit of the c-AMP dependent protein kinase in *Saccharomyces cerevisiae, Mol. Cell. Biol.,* 7, 1371—1377, 1987.
223b. **Tomasovic, S. P.,** Functional aspects of the mammalian heat-stress protein response, *Life Chem. Rep.,* 7, 33—63, 1989.
224. **Tomasovic, S. P. and Koval, T. M.,** Relationship between cell survival and heat-stress protein synthesis in a *Drosophila* cell line, *Int. J. Radiat. Biol.,* 48, 635—650, 1985.
224a. **Tomasovic, S. P., Steck, P. A., and Heitzman, D.,** Heat-stress proteins and thermal resistance in rat mammary tumor cells, *Radiat. Res.,* 95, 399—413, 1983.
224b. **Trent, J. D., Osipink, J., and Pinkan, T.,** Acquired thermotolerance and heat shock in the extremely thermophilic archaebacterium *Sulfolobus* sp. strain B12, *J. Bacteriol.,* 172, 1478—1484, 1990.
224c. **Trollmo, C., Andre, L., Blomberg, A., and Adler, L.,** Physiological overlap between osmotolerance and thermotolerance in *Saccharomyces cerevisiae, FEMS Microbiol. Lett.,* 56, 321—326, 1988.
225. **Tsuchido, T., Van Bogelen, R. A., and Neidhardt, F. C.,** Heat shock response in *Escherichia coli* influences cell division, *Proc. Natl. Acad. Sci. U.S.A.,* 83, 6959—6963, 1986.
226. **Tsukeda, H., Maekawa, H., Izumi, S., and Nitta, K.,** Effect of heat shock on protein synthesis by normal and malignant human lung cells in tissue culture, *Cancer Res.,* 41, 5188—5192, 1981.
227. **Van Bogelen, R. A., Acton, M. A., and Neidhardt, F. C.,** Induction of the heat shock regulon does not produce thermotolerance in *Escherichia coli, Genes Devel.,* 1, 525—531, 1987.
228. **Van Dongen, G., Geilenkirchen, W., Van Rijn, J., and Van Wijk, R.,** Increase of thermoresistance after growth stimulation of resting Reuber H35 hepatoma cells, *Exp. Cell Res.,* 166, 427—441, 1986.
229. **van Wijk, R. and Boon-Niermeijer, E. K.,** Two types of thermotolerance in mouse 3T3 cells and 3-day old larvae of *Lymnaea, Biona-Report,* 4, 85—92, 1986.
230. **van Wijk, R., Otto, A. M., and Jimenez De Asua, L.,** Increase of epidermal growth factor stimulated cell-cycle progression and induction of thermotolerance by heat shock: Temperature and time relationship, *Int. J. Hyperthermia,* 1, 147—156, 1985.
231. **Velazquez, J. M. and Lindquist, S.,** Hsp70: Nuclear concentration during environmental stress and cytoplasmic storage during recovery, *Cell,* 36, 655—662, 1984.
232. **Vidair, C. A. and Dewey, W. C.,** Modulation of cellular heat sensitivity by specific amino acids, *J. Cell. Physiol.,* 131, 267—275, 1987.
233. **Vollmar, H.,** Ueber den Einfluss der Temperatur auf normales Gewebe und auf Tumorgewebe, *Z. Krebsforsch.,* 51, 71—99, 1941.
234. **Wada, M. and Itikawa, H.,** Participation of *Escherichia coli* K-12 grpE gene products in the synthesis of cellular DNA and RNA, *J. Bacteriol.,* 157, 694—696, 1984.
234a. **Walsh, D. A. and Hightower, L. E.,** unpublished.
235. **Walsh, D. A., Klein, N. W., Hightower, L. E., and Edwards, M. J.,** Heat shock and thermotolerance during early rat embryo development, *Teratology,* 36, 181—191, 1987.
235a. **Wang, W. C. and Nguyen, H. T.,** Thermal stress evaluation of suspension cell cultures in winter wheat, *Plant Cell Rep.,* 8, 108—111, 1989.
236. **Wang, H. and Lanks, K. W.,** 2-Cyanocinnamic acid sensitization of L929 cells to killing by hyperthermia, *Cancer Res.,* 46, 5349—5352, 1986.

237. **Wang, H., Shah, V., and Lanks, K. W.,** Use of oxidizing dyes in combination with 2-cyanocinnamic acid to enhance hyperthermic cytotoxicity in L929 cells, *Cancer Res.*, 47, 3341—3343, 1987.

237a. **Watson, K. and Cavicchiolo, R.,** Acquisition of ethanol tolerance in yeast cells by heat shock, *Biotechnol. Lett.*, 5, 683—688, 1983.

238. **Watson, K., Dunlop, G., and Cavicchioli, R.,** Mitochondrial and cytoplasmic protein synthesis are not required for heat shock acquisition of ethanol and thermotolerance in yeast, *FEBS Lett.*, 172, 299—302, 1984.

239. **Welch, W. J. and Mizzen, L. A.,** Characterization of the thermotolerant cell: II. Effects on the intracellular distribution of HSP70, intermediate filaments, and snRNP's, *J. Biol. Chem.*, 106, 1117—1130, 1988.

240. **Werner-Washburne, M., Stone, D. E., and Craig, E. A.,** Complex interactions among members of an essential subfamily of hsp70 genes in *Saccharomyces cerevisiae, Mol. Cell. Biol.*, 7, 2568—2577, 1987.

240a. **Werner-Washburne, M., Becker, J., Kosic-Smithers, J., and Craig, E. A.,** Yeast Hsp70 RNA levels vary in response to the physiological status of the cell, *J. Bacteriol.*, 171, 2680—2688, 1989.

240b. **West, I. C.,** What determines the substrate specificity of the multi-drug-resistance pump?, *Trends Biochem. Sci.*, 15, 42—46, 1990.

241. **Westra, A. and Dewey, W. C.,** Variation in sensitivity to heat shock during the cell-cycle of Chinese hamster cells in vitro, *Int. J. Radiat. Biol.*, 19, 467—477, 1971.

242. **Widelitz, R. B., Magun, B. E., and Gerner, E. W.,** Effects of cycloheximide on thermotolerance expression, heat shock protein synthesis, and heat shock protein mRNA accumulation in rat fibroblasts, *Mol. Cell. Biol.*, 6, 1088—1094, 1986.

243. **Wiegant, F. A. C., Tuyl, M., and Linnemans, W. A. M.,** Calmodulin inhibitors potentiate hyperthermic cell killing, *Int. J. Hyperthermia,* 1, 157—170, 1985.

244. **Wiegant, F. A. C., Van Bergen En Henegouwen, P. M. P., Van Dongen, A. A. M. S., and Linnemans, W. A. M.,** Stress-induced thermotolerance of the cytoskeleton, *Cancer Res.*, 47, 1674—1680, 1987.

245. **Wiegant, F. A. C., Van Bergen En Henegouven, P. M. P., and Linnemans, W. A. M.,** Studies on the mechanism of heat shock induced cytoskeletal reorganization in mouse neuroblastoma N2A cells, Proefschrift Universiteit, Utrecht, 1987, 141—158.

246. **Wondergem, J. and Haveman, J.,** A study of the effects of prior heat treatment on the skin reaction of mouse feet after heat alone or combined with X-rays: Influence of misonidazole, *Radiother. Oncol.*, 2, 159—170, 1984.

246a. **Wondergem, J. and Haveman, J.,** Thermal enhancement of the radiation damage in the mouse foot at different heat and radiation dose: Influence of thermotolerance, *Int. J. Radiat. Biol.*, 48, 337—348, 1985.

247. **Wondergem, J., Haveman, J., Schol, E., and Reinds, E.,** Influence of prior heat treatment on the effects of heat alone or combined with X-rays on mouse stomal tissue, *Int. J. Radiat. Biol.*, 51, 81—90, 1987.

248. **Wu, M.-T. and Wallner, S. J.,** Heat stress responses in cultured plant cells. Heat tolerance induced by heat shock versus elevated growing termperature, *Plant Physiol.*, 75, 778—780, 1984.

249. **Xiao, H. and Lis, J. T.,** A consensus sequence polymer inhibits in vivo expression of heat shock genes, *Mol. Cell. Biol.*, 6, 3200—3206, 1986.

250. **Xiao, C. M. and Mascarenhas, J. P.,** High temperature-induced thermotolerance in pollen tubes of *Tradescantia* and heat-shock proteins, *Plant Physiol.*, 78, 887—890, 1985.

251. **Yager, J. D., Zurlo, J., and Penn, A. L.** Heat-shock-induced enhanced reactivation of UV-irradiated Herpes virus, *Mutat. Res.*, 146, 121—128, 1985.

252. **Yahara, I., Iida, H., and Koyasu, S.,** A heat shock-resistant variant of Chinese hamster cell line constitutively expressing heat shock protein of Mr 90,000 at high level, *Cell Struct. Funct.*, 11, 65—73, 1986.

253. **Yamamori, T. and Yura, T.,** Genetic control of heat-shock protein synthesis and its bearing on growth and thermal resistance in *Escherichia coli* K-12, *Proc. Natl. Acad. Sci. U.S.A.*, 79, 860—864, 1982.

254. **Yarwood, C. E.,** Acquired tolerance of leaves to heat, *Science,* 134, 941—942, 1961.

255. **Yarwood, C. E.,** Adaptation and sensitization of bean leaves to heat, *Phytopathology,* 54, 936—940, 1964.

256. **Yatvin, M. B.,** The influence of membrane lipid composition and procaine on hyperthermic death of cells, *Int. J. Radiat. Biol.*, 32, 513—521, 1977.

257. **Yost, H. J. and Lindquist, S.,** RNA splicing is interrupted by heat shock and is rescued by heat shock protein synthesis, *Cell,* 45, 185—193, 1986.

258. **Yura, T., Tobe, T., Ito, K., and Osawa, T.,** Heat shock regulatory gene (htpR) of *Escherichia coli* is required for growth at high temperature but is dispensable at low temperature, *Proc. Natl. Acad. Sci. U.S.A.*, 81, 6803—6807, 1984.

258a. **Zeuthen, M. L. and Howard, D. H.,** Thermotolerance and heat-shock response in *Candida albicans, J. Gen. Microbiol.*, 135, 2509—2518, 1989.

259. **Zivy, M.,** Genetic variability for heat shock protein in common wheat, *Theor. Appl. Genet.*, 74, 209—213, 1987.

260. **Zhou, Y. N., Kusukawa, N., Erickson, J. W., Gross, C. A., and Yura, T.,** Isolation and characterization of *Escherichia coli* mutants that lack the heat shock sigma factor 32, *J. Bacteriol.*, 6, 3640—3649, 1988.

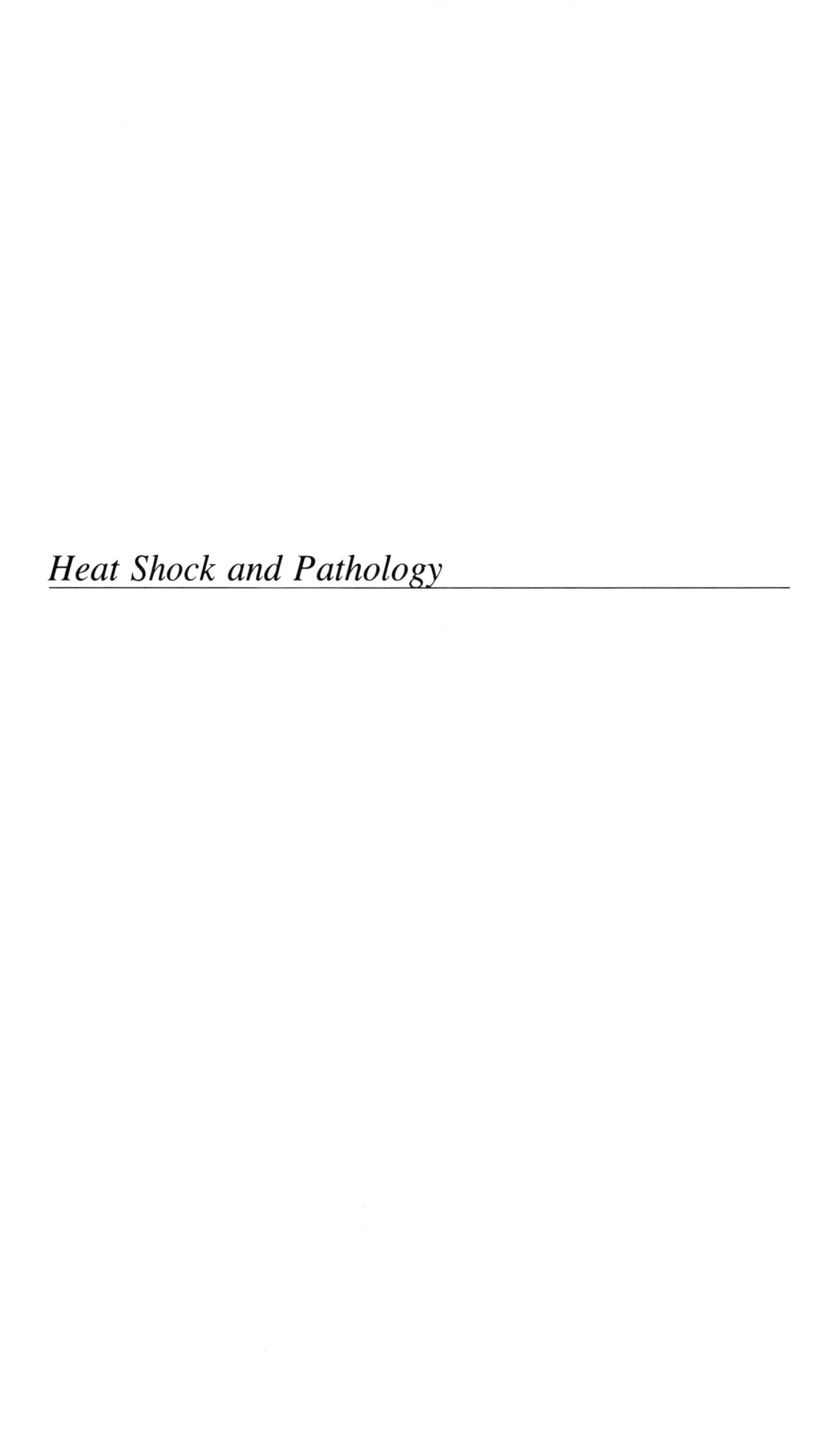

Heat Shock and Pathology

Chapter 18

HEAT SHOCK-INDUCED DEVELOPMENTAL EFFECTS

L. Nover

TABLE OF CONTENTS

The heat shock (hs) response, as it is detailed in the framework of this book, mostly refers to the cellular level and to reversible changes aimed at a rapid and complete recovery of normal metabolic activities after the stress period. However, at the border to lethal damage of the system, in particular during highly sensitive developmental stages, even minute functional defects at the cellular level become manifest at the organismic level due to amplification during embryogenesis. In a few selected cases, the cause of a developmental defect can be traced back to the cell type and to the period of gene expression required for a given phenotypic trait (see e.g., Section 18.3). In most other cases, hyperthermic stress can simply be described as teratogenic and embryotoxic. We briefly summarize this important hs related topic which is much older than our knowledge of induced heat shock proteins (HSPs) and their genes.

18.1. STRESS-INDUCED TERATOGENESIS IN VERTEBRATE EMBRYOS

For evident reasons, experimental teratology with vertebrate embryos using brief hs treatments applies to pregnant females or to isolated embryos is a broadly investigated field. Interest particularly increased with the application of hyperthermia as an adjunct to cancer therapy and with the recognition of more and more chemical stressors as inducers of HSP synthesis (see Table 1.1). Two excellent reviews by Lary[52] and German[34] are the basis of this section and are used as sources for additional references. Lary[52] discusses data from 58 papers on experimental teratology in chicken, mice, rats, hamsters, guinea pigs, rabbits, pigs, sheep, and monkeys published between 1919 and 1983. He ends with a critical evaluation of the situation in humans. German[34] develops an ''embryonic stress hypothesis'', which is essentially based on the same set of data but, in addition, he also considers the action of non-hs stressors. The relevant facts can be summarized as follows.

1. Experimental studies with different kinds of vertebrates (see above) prove that even moderate and not very prolonged periods of elevated body temperature may exert embryotoxic or teratogenic effects.
2. Most sensitive are early developmental phases preceding organogenesis. Following this period, hyperthermia may kill embryos, but does not cause gross malformations of those surviving the stress period. However, the functional maturation of systems can be affected.
3. Although all kinds of organ abnormalities were observed, aberrant development of the head and central nervous system is most frequent. Defects include ''anencephaly, exencephaly, encephalocele, anophthalmia, microphthalmia, microencephaly, facial aplasia, maxillary or mandibular agnathia and micrognathia, anotia and malformations of cranial and facial bones and teeth''.[52] Webster et al.[87] found craniofacial abnormalities in all offspring of rats exposed to only 5 min at 43°C between days nine and ten of pregnancy. Essentially the same results were obtained with rat embryos cultivated *in vitro*.[61,85]
4. Stressing of embryos *in utero* or *in vitro* is connected with increased HSP synthesis,[14,15,35,61,85,90] but contrasting to the suggestion by German,[34] increased HSP synthesis is indicative of stress but not the cause of teratogenesis. In agreement with results from experiments with *Drosophila*,[12,16] many teratogens may be defined by their activity as inducers of HSP synthesis in embryonic tissues[35] (see also Table 1.1).

   In contrast to the large group of hs-like inducers (see Section 1.2.2), they act more specifically, inducing few or even a single HSP. This was illustrated for the anticonvulsant drug sodium valproate, known as a stage-specific teratogen in vertebrates. At therapeutic concentrations of 1 m*M*, it inhibits neural tube closure in rats.

Connected with this is an inhibition of mitosis and an increase of cell-substratum adhesion of glial cells, caused by induction of a membrane-bound 43-kDa sialoglycoprotein receptor of type IV collagen, fibronectin, or laminin.[57a] Interestingly, this 43-kDa protein is also a major HSP. But valproate induction is restricted to cells of glial origin, whereas hs induction is observed in all cells investigated.

5. Searches for primary, cellular changes which might be related to the induction of developmental deviations is still in its infancy. Magun[57] described effects on epidermal growth factor binding and intracellular degradation in rat embryo fibroblasts. However, stress sensitivity of the crucial intracellular information transfer systems was also reported by others: (1) After stress treatment of *Xenopus* primary hepatocytes, the estrogen response system becomes refractory for some time.[88,88a] (2) Ashburner[1c] reported that, following a 4-h hs at 37.5°C, *Drosophila* larvae were unable to respond to ecdysterone even after 4 h recovery. Most of them degenerated or showed abnormal development. (3) Chinese hamster ovary cells respond to hs with a depletion of available insulin receptors[17] and a transient five to sixfold rise of the cAMP level.[18] (4) Sexual differentiation of the water mold *Achlya ambisexualis* is induced by the steroid hormone antheridiol. Short heat shock or arsenite treatments block hormone responsiveness for at least six hours by decreasing the level of hormone receptor in the hyphae.[77c] (5) There is increasing evidence with reference to *Drosophila* and *Xenopus* oocytes that the correct positioning of maternal determinants of the body pattern depends on the binding to elements of the cytoskeleton.[87a,88c,89] Though not yet analyzed in this respect, it is quite possible that the hs-induced reorganization of the cytoskeleton (see Sections 14.3 and 16.6) may thoroughly alter the early steps of embryonic development, with irreversible changes of the body plan.

   In a number of model systems, hs treatments were shown to influence cell differentiation directly. Thus, the dimethyl-sulfoxide-induced maturation of murine erythroleukemia cells was blocked,[75] whereas in human promyelocytic leukemia cells, hs and a number of well-known chemical stressors (ethanol, arsenite, cadmium ions, and lidocaine; see Table 1.1) induced thermotolerance and granulocyte-like differentiation.[77b] Finally, in rat embryos the normal maturation of fetal liver cells was accelerated by hs resulting in precocious synthesis of albumin.[82]

6. As is true for practically all cellular aspects of the hs response (see Chapter 17), the sensitivity of embryos to teratogenic effects of hyperthermia can be decreased by a prior conditioning treatment.[61,85] Both papers describe the protection of rat embryos in culture by a mild, nonteratogenic hs at 42°C (10 to 30 min) followed by a more severe hs at 43°C. Severe cranofacial abnormalities are induced by 7.5 min at 43°C unless embryos were made thermotolerant by the 42°C pretreatment, which is connected with an increase of HSP synthesis. However, a kinetic study by Walsh et al.[85] showed that the thermotolerance effect was observed already after 15 min. Thus, other cellular changes, rather than new synthesis of HSPs, may be important for the protective effect as well.

7. On the basis of extensive experimental studies with guinea pigs,[27,28] Edwards[29] put forward a hypothesis on the influence of hyperthermia on congenital malformations in man. As a consequence, a number of authors published retrospective and prospective epidemiological studies which partly support and partly disprove Edwards' hypothesis. The relevant data are summarized by Lary[52] and German.[34] The controversial outcome results largely from the impossibility of obtaining direct experimental data for the human. However, the largest part of the indirect data makes it very likely that early developmental phases of the human embryo are as stress sensitive as those of other mammalian species.

18.2. DEVELOPMENTAL EFFECTS IN NONVERTEBRATE ORGANISMS

Investigations of hs-induced effects on development are not restricted to vertebrate embryos (Table 18.1). Main objects are the classical favorite organisms of developmental biologists: *Drosophila, Xenopus*, sea urchin, and in addition, the pond snail *Lymnaea stagnalis*. It is a general experience with all four types of organism that cleavage stage embryos are most heat sensitive, whereas at the transition from blastula to gastrula stage, survival after the heat treatment and developmental recovery are much improved. This developmental phase coincides with the capability to become thermotolerant and to induce HSP synthesis (see Figure 17.2). Corresponding results were reported for *Lymnea*,[9,10] *Drosophila*,[5,24,26] *Xenopus*,[13,44,69] and sea urchin.[36,78,80a]

In addition to this developmental control of hs-inducibility, very early embryonic stages are not uniformly sensitive.[7,8,58] An illustrative example was given by Boon-Niermeijer[7] (Figure 18.1). The early cleavage stages of *Lymnea* allow analysis of hs effects in relation to cell cycle phases. Two effects are noteworthy. First, if cell cycle delay is used as a measure, three peaks of maximum sensitivity are observed. They are positioned at the S/G_1 transition, the end of G_2 and in the prometaphase. Second, the delay of the cell cycle is not related to the maxima of hs-induced morphogenetic effects, which are found in the metaphase exclusively. Depending on the developmental stage, the malformations are of three types. Disturbed gastrulation was frequent after hs just before the third cleavage. Beginning with the fourth cleavage, gastrulation proceeds normally, but individuals with head and shell defects are observed.

An interesting aspect of the hs response was reported by Rensing and co-workers. When investigating the circadian rhythm of conidiospore formation in *Neurospora crassa*, they observed dose-dependent shifts of the conidiation rhythm after 3 h hs at 30 to 40°C. Depending on the time point of hs treatment, the phase of conidia formation was delayed or advanced.[77a] In *Neurospora*, inducibility of HSP synthesis itself is subject to circadian variation,[20a] and the dose-dependent shift of conidiation correlates with the extent of HSP70 synthesis.[77a] On search for the hypothetical "clock protein", Rensing et al.[77a] discuss the possible role of HSPs since the daily temperature variation encountered in nature may be part of the "Zeitgeber" mechanism.

18.3. PHENOCOPY INDUCTION IN *DROSOPHILA*

An important contribution towards the understanding of developmental genetics was based on experiments of the German geneticist Richard Goldschmidt[39] published initially in 1929. He described the induction of mutant-like phenotypes of *Drosophila* by hs treatments. Six years later, this material was published in a large extended version using data from experiments with more than 500,000 individuals.[40] Part of his results are summarized in Table 18.2. The regular appearance of the developmental abnormalities in a stage- and temperature-dependent manner led Goldschmidt to the fundamentals of physiological genetics and to the definition of a new term: "Es ist im Interesse der Darstellung vielleicht wünschenswert, einen kurzen Terminus für die die Mutanten kopierenden, nicht erblichen Phänokopien zu haben. Wir wollen sie fortan *Phänokopien* nennen." (It may be desirable to create a short term for these non-inheritable phenocopies of mutants. From now on, we call them *phenocopies*.)

More than 40 years later with the rising interest in the hs response, these results were used as a basis for a reinvestigation of the phenomenon by Mitchell, Petersen, and co-workers. They described 34 phenocopies and the developmental periods for their induction.[64] In contrast to the long-term hs used by Goldschmidt (24 h 35 to 37°C), they found the

TABLE 18.1
Survey of hs-Induced Developmental Effects in Nonvertebrate Organisms

Organism, developmental stage	Heat shock treatment	Developmental effects	Ref.
Molluscs			
Lymnea stagnalis (pond snail)			
2—4-cell stage	90—180 min 37°C	Exogastrula formation and numerous head malformations	76
2—8-cell stage	10 min 38°C	Cell cycle delay, morphogenetic abnormalities due to loss of distinct cell lines, different sensitivity phases (see Figure 18.1)	33
8—16-cell stage	10 min 37—38°C	Cell cycle stage-dependent variation of developmental abnormalities (see Figure 18.1), highest sensitivity in M phase	7
Embryos, larvae (0—7 d of development)	1 h 40°C	At 1 d good survival but abnormal development, at 3—4 d few survivor but normal development, at 5—6 d good survival but abnormal development, post 6 d again high mortality, protection by 30 min 37°C conditional treatment	8
4-cell stage to 3 d trochophora larvae	2—9 min 43.6°C	Highest survival of 4-cell stage embryos, lowest of 3 d larvae, protection by 10 min 39°C conditional treatment observed in gastrula and later developmental stages, two states of TT observed (see Table 17.3)	9, 10
Insects			
Schistocerca gregaria (locust)			
Different developmental stages (1—80 h postovoposition)	15 min 48°C	Up to 30 h high mortality (except 3 h embryos), but normal development later stages develop segmentation abnormalities	58, 58a
Drosophila melanogaster			
Cellular blastoderm stage	15 min 37°C	Developmental delay	24
Blastoderm stage	15 min 37°C	Phenocopies of mutants of the bithorax complex, hs interferes with control of BX-C	25 7g
	4 h mild hs or ether	Individuals with four wings, phenocopy of bithorax mutant	38
Different developmental stages (0—20 h post fertilization)	40 min 37°C	Preblastula stage high mortality, early to late blastula survival but abnormal development, from late gastrula onward induced TT and normal development (see Figure 17.1)	5, 41
Gastrula tissues *in vitro*	2 h 40.2°C or teratogens (see Table 1.1)	Normal differentiation of myotubes and neuroblasts impaired, connected with increased synthesis of HSPs22 and 23, protection by 30 min conditioning at 35°C	12, 16
Different embryonic stages	2—3 min 42—43°C	Disrupted segmentation, germ band shortening, head defects	26

TABLE 18.1 (continued)
Survey of hs-Induced Developmental Effects in Nonvertebrate Organisms

Organism, developmental stage	Heat shock treatment	Developmental effects	Ref.
Larvae	4 h 37.5°C	Even after 4 h of recovery, larvae do not respond to ecdysterone, >75% died, survivors developed phenocopies	1c
Different larval and pupal stages ($4^1/_2$—7 d after ovoposition)	6—24 h 35—37°C	Stage-dependent induction of different phenocopies (see Table 18.2) definition of term, uncovery of recessive phenotypes by hs, investigations with ~ 500,000 individuals	39, 40
Different pupal stages	1 h 40.2°C	Stage-dependent induction of angle bristle, multiple hairs, and spairbristle, induction of double phenocopies by two hs treatments, protection by prior conditioning at 37.5°C correlates with HSP synthesis	19, 62, 63, 67
Pupae	30 min 40.7°C	Uncovery of phenocopies of recessive mutant "forked" (heterocopy)	66
Different pupal stages (20—92 h postovoposition)	35 min 40.8°C	Stage-dependent induction of 34 types of phenocopies, definition of sensitive periods	64
Different pupal stages	35 min 40.7°C	Stage-dependent induction of multihair phenocopy in different parts of the body correlates with changes in pattern of protein synthesis (Figure 18.2)	65, 67a
Different larval and pupal stages (24—168 h after ovoposition)	1 h 39°C or 4 h 38°C	Stage-dependent influence on cell size and number in wings	74
Pupae	40 min 40.5°C	Malformations of posterior crossvein, protection by prior conditioning by 10—30 s at 40.5°C	60
Amphibia			
Xenopus laevis (clawtoed frog)			
Oocytes	60 min 36°C	Hs-dependent delay of progesterone induced meiotic maturation, but continued normally in recovery period	2
Preneurula and postneurula stages	5 min 37.5°C	Somite muscle abnormalities, spatio-temporal sequence of somite block formation disturbed, neurula stage least sensitive	20
Blastula to postneurula stages	15 min 37°C	Abnormal arrangement of myofibrillar elements in somites, sensitive period traverses the neurula from head to tail, some hours preceding time of somite formation	30

Midcell blastula, fine-cell blastula, gastrula	20 min 35°C	Embryos only partially survive because of region-specific acquisition of TT, animal hemisphere develops normally, but vegetal halves undergo extragastrulation, later developmental stages are fully resistent	69
Different embryonic stages	20 min 35°C	Acquisition of TT in late blastula stage, preblastula stage embryos degenerate, postblastula stage embryos develop normally (see Figure 17.1)	44
Larvae (30—48 h postfertilization)	20 min 37°C	Damage of epidermis with cell swelling and shedding of ciliated cells, more resistance 72—96 h postfertilization	69a
Ambystoma mexicanum (axolotl, embryos after midneurulation)	Brief hs at 37—38.5°C	Abnormal development of somite primordia which just start to develop	1b
Rana temporaria (frog)			
Different embryonic stages	10—40 min 37°C	Severe developmental abnormalities with microcephaly, defects in somitogenesis, transient edema; but time-keeping within a synchronous embryo population not impaired; protection from developmental defects by conditioning (2 min 37°C)	31
Echinoderms			
Arbacia punctulata (sea urchin, unfertilized eggs)	30 min 32—33°C	Considerable number of eggs divide without fertilization, hs-induced parthenogenesis, same with eggs of annelid *(Nereis)* and clam *(Cumingia)*	44a
Paracentrotus lividus (sea urchin)			
Different development stages	1 h 31°C	High mortality and degenerative development up to early blastula stage, almost 100% survival and normal development after hatched blastula stage	36, 78
Gastrula	1 h 35°C	Immediate developmental arrest and disintegration, full protection by prior 1 h conditioning at 31°C	80a
Fishes			
Salmo salar (Atlantic salmon)			
Fertilized eggs	5 min 32°C	100% triploid with 70—90% survival, normal development	4
	6—10 min 30°C	Triploid embryos, erythrocyte fusion observed in abnormally developing embryos	32
Salmo gairdneri (rainbow trout)			
Eggs fertilized with irradiated sperm	10—20 min 24—26°C	Low level of diploid gynogenetic embryos, which may develop into transgenic fishes	21c, 76a
Microorganisms			
Naegleria gruberi (amebo-flagellate)	32—38.5°C	Induction of multiple flagella at the transition from ameba to swimming flagellate	84

TABLE 18.1 (continued)
Survey of hs-Induced Developmental Effects in Nonvertebrate Organisms

Organism, developmental stage	Heat shock treatment	Developmental effects	Ref.
Achlya ambisexualis (water mold)	1 h 37°C	Stop of sporulation in starved cells, only immature sporangia formed, resumed in recovery period	54
Neurospora crassa	3 h 30—35°C or 40°C	Depending on the time point of application and temperature, hs delays or advances circadian rhythm of conidiation	77
Dictyostelium discoideum (cellular slime mold)	⩾30°C	Arrest of fruiting body formation, resumed in recovery period	56
Acetabularia mediterranea	Development at 29°C	Decreased spacing of hair initials, 12 hairs instead of 9 developed at 18°C	43
Plants			
Sorghum bicolor (seedling)	2 h 45°C	Strong growth retardation, protection by 30 min condition at 40°C	72
Glycine max (soybean, seedling)	2 h 45°C	Strong growth retardation, protection by condition treatment either 30 min 40°C or 50 μ*M* arsenite (see Table 17.2)	55

Note: TT is thermotolerance.

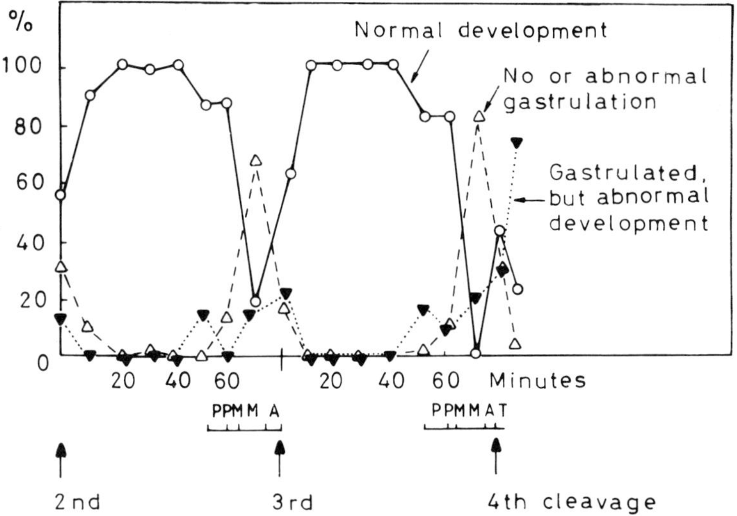

FIGURE 18.1. Changing heat shock sensitivity in early cleavage stage embryos of *Lymnea stagnalis*. Throughout the time period indicated, embryos between the second and fourth cleavage (arrows at the bottom) were treated for 10 min at 37.5°C and then returned to 25°C. Embryos with normal gastrulation were opposed to those without or with abnormal gastrulation. Most heat sensitive periods of mitosis are prometaphases (PM) and metaphases (M). P, prophase, A, anaphase, T, telophase. (Modified from Boon-Niermeijer, E. K., *Wilhelm Roux's Arch. Dev. Biol.*, 180, 241, 1976.)

TABLE 18.2
Phenocopies Induced in *Drosophila melanogaster*

	Developmental stages[b]					
	Larvae		Pupae			
Phenocopy[a]	**4.5**	**5**	**5.5**	**6**	**6.5**	**7**
Flight						
Notch-cut (35°C)	+	+	+	–	–	–
Dumpy (37°C)	–	–	+	+	+	–
Ski-curly (35—37°C)	–	–	+	+	+	+
Schmal (36—37°C)	–	–	–	–	+	+
Short	–	–	+	+	+	+
Rolled (35—37°C)	–	–	+	+	+	+
Thorax						
Trident (35—37°C)	–	–	–	–	–	+
Eyes						
Small, large (37°C)	–	–	+	+	–	–
Scutellum						
Hörnchen (35°C)	–	–	–	+	+	+
Abdomen						
Abnormal	+	+	+	+	–	–

[a] Optimum temperature of induction by a 6 to 24 h heat shock.
[b] Oregon strain, days after ovoposition

Compiled from Goldschmidt, R., *Z. Indukt. Abstamm. Vererbungsl.*, 69, 38, 1935.

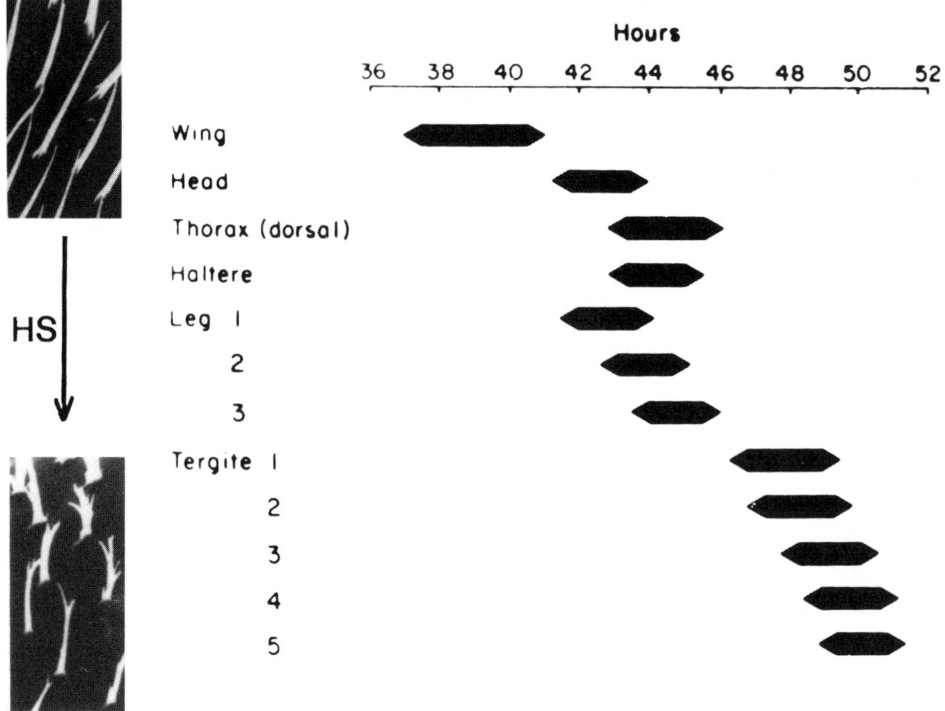

FIGURE 18.2. Multihair phenocopy induction in *Drosophila* pupae after 35 min 40.8°C. Differentiation of hairs in most epithelial cells of *Drosophila* is connected with marked changes in the pattern of protein synthesis.[67] If the process is disturbed by heat shock (hs) the normal hair morphology changes to a branched phenotype (multihair phenocopy, see left part of the figure with the two scanning electron micrographs). The hs sensitive periods vary depending on the time of hair differentiation in different parts of the pupa (right part of the figure). Thus, multihair phenocopies of wing epithelial cells are induced between 37 and 41 h of pupal development, whereas in the third leg the sensitive period is between 43.5 and 46 h., etc. (From Mitchell, H. K. and Petersen, N. S., *Dev. Biol.*, 95, 459, 1983. With permission.)

optimum for phenocopy production at shorter, more severe hs conditions (30 to 60 min 40.2 to 40.8°C). This procedure results in a long-lasting block in protein synthesis. The important implications of the method to unravel regulatory details of developmental processes were documented by Mitchell and Petersen[65] (see Figure 18.2) for the induction of the same type of phenocopy ("multihair") in different parts of the *Drosophila* body.

The unique system of the *Drosophila* wing, where about 90% of wing cells (~ 28,000 cells) differentiate synchronously to form wing hairs, allowed the definition of distinct morphological stages in correlation to changing patterns of protein synthesis and to the phases of hs-induced formation of the multihair phenocopy.[65,67a] The extrusion of wing hairs at about 34 h after puparium formation is followed by the deposition of a double layer of cuticulin to form the outer envelope of the hair. In the following phase (convolution), the surface of the hair cells increases and the hairs are pushed upright.

Optimum time for multihair phenocopies is a hs treatment at the transition between cuticulin synthesis and convolution. In particular, disruption or delay of cuticulin deposition may lead to branched hairs as cytoplasm is pushing out in places with insufficient cuticulin layers. The ultimate phenotype becomes apparent only 15 to 20 h later.[73] What occurs in detail between 38 and 40 h postpuparium formation in wing cells, evokes multihair phenocopies at other parts of the pupa at later stages (Figure 18.2), when the corresponding epithelial cells undergo the program "hair differentiation".

The close correlation of phenocopy induction with key events in the gene expression program during hair formation describes, at the molecular level, the concept of a ''physiologische Genetik'' put forward by Goldschmidt 55 years earlier.[40] A protection from phenocopy induction by severe hs causing a several hours interruption of transcription, processing and translation can be achieved by an appropriate conditioning treatment.[19,60,67] This thermotolerance effect is presumably tightly coupled to the immediate recovery of all parts of gene expression which minimizes or abolishes effects on developmental processes (for details see Chapter 17).

18.4. HEAT SHOCK-INDUCED DEVELOPMENT

18.4.1. PROTOZOA

In some pathogenic organisms affecting mammalian species but existing also in a free state or in other nonvertebrate species, the hs, experienced during invasion of the mammalian host, can act as a developmental signal. A number of parasitic protozoa of tropic and subtropic countries are the causal agents of devastating diseases to millions of humans and animals, e.g., Chagas' disease, blackfever, sleeping sickness, leishmaniasis and malaria. They belong to the classes of Flagellatae (*Trypanosoma* sp., *Leishmania* sp.) and Sporozoa, respectively, (*Eimeria* spec., *Plasmodium* sp.). Their complex life cycles between poikilothermic blood-sucking vectors, mostly insects, and their homeothermic vertebrate hosts includes not only marked changes of cell form, surface proteins, and metabolic functions,[11,42,48,71] but also a ''natural'' hs when the parasites are introduced into the host organism. In fact, the transition from the insect cell type to the cell forms found in the homeothermic host coincides with induction of HSPs or related proteins.[6,6c,83]

With *in vitro* cultures of promastigotes (insect form) of *Leishmania* sp. the differentiation could be mimicked outside of the host organism by shifting the cultivation temperature from 25° to 34 to 37°C.[1,46,53,81a,81f,83] The apparent synthesis of HSPs was connected with morphological changes of the cells, similar to the transition to the amastigote cell type observed in mammals and with increased infectivity towards vertebrate cells.[46,81f,83] It is remarkable that expression of the *Trypanosoma brucei* hsp70 gene was further stimulated tenfold, when trypomastigotes from the rat were heat-shocked *in vitro* (37 → 42°C).

The complexity of HSP-coding gene families and peculiarities was discussed in Section 3.7. Proteins of the HSP90 and HSP70 families are dominant surface antigens of protozoa. This is true for a constitutively expressed hsp90-type gene of *T. brucei.*[22] The protein product is an M_r 85 kDa surface glycoprotein with considerable homology to other members of the HSP90 family (see Figure 2.8). Basically similar is the situation for *Plasmodium falciparum* and *Eimeria bovis* containing highly expressed proteins of the HSP70 family (M_r 75 kDa) at the surface and in the cytoplasm of merozoites and other mammalian stages but not in sporozoites found in mosquitos.[1a,6,77d] So far the analysis of 5′ flanking sequences of the corresponding genes revealed no consensus with any of the known eukaryotic promoter elements (TATA, CAT, or hs boxes, see Section 3.7). As reviewed by Borst,[11] this situation reflects basic deviations in gene expression in this specialized type of protozoa.

It is worth mentioning that the characteristic variation of surface antigens in African *Trypanosoma* species, which helps the parasite to avoid destruction by the host's immune system, is also observed in free living protozoa, e.g., *Paramecium* and *Tetrahymena.* In *Tetrahymena thermophila*, three temperature-dependent serotypes were distinguished.[3] Below 20°C the surface antigen is Ser L, at 30°C it is Ser H, and at 40°C it switches to Ser T. Within 2 h of the hs treatment the 44 to 52 kDa surface proteins of the Ser H type are replaced by the 30-kDa Ser T-antigen. The change can be reversed by shifting the cells back to 28°C. In both cases, mRNA synthesis is required to replace one surface antigen by the other. Gene structures and the biological meaning of this antigen shift are unclear.

18.4.2. VOLVOX

Hs treatment (1 h 32 → 42.5°C) of the coenocytic green alga *Volvox carteri* induces synthesis of HSPs.[49] At the same time, a considerable amount of the 30 kDa sexual inducer is produced in the stressed somatic cells.[50] This glycoprotein is usually produced only in the sperm packets of males. It is released with the sperms and transforms all individuals in the surrounding milieu from asexual to sexual reproduction. The required concentration for sexual transformation is extremely low (6×10^{-17} *M*), i.e., one male produces enough inducer to transform a 1000-l culture of *Volvox carteri*. Contrasting to the normothermic situation, both asexual males and females produce the sexual inducer protein under hs conditions. This "autoinduction" of sexuality can be viewed as a mechanism to provide optimum conditions for the production of dormant zygotes, e.g., before a pond dries out in summer.[6c,50]

18.4.3. PATHOGENIC FUNGI

The human pathogenic fungi *Candida albicans*,[21] *Fonsecaea pedrosi*,[47] or *Histoplasma capsulatum*[51] exist as both saprophytic and parasitic forms. The transition is connected with a morphological change from filamentous growth to a unicellular yeast-type cell under parasitic conditions. Similar to the preceding examples with parasitic protozoans, formation of the yeast cell type of these fungi can be considered as a type of hs response,[47,51] leading to profound changes of cell morphology and function adapted to the life in or on the host organism. In *Histoplasma* strains, the mycelia-to-yeast transition at 34 to 37°C is connected with a marked increase of HSP synthesis.[18a,81c]

Infection of appropriate host plants with a phytopathogenic rust fungus starts with a clearly defined sequence of morphological structures formed by the germinating uredospore to invade the host tissue. These are the appressorium, the substomatal vesicle, and the infection hypha. In *Puccinia* spp., e.g. *Puccinia gramini tritici* (wheat stem rust) or *Puccinia coronata* (oat rust), mild hs treatments (4 h 30°C) induce formation of the characteristic infection structures together with marked changes of gene expression patterns including formation of HSPs.[23,59,82a,86] Later in the 18 to 24 h developmental cycle, HSP synthesis disappears and differentiation-specific proteins are observed.[86] The hs induction of the complete infection structure may be limited to representatives of the *Puccinia* spp. Using flax rust uredosporelings (*Melampsora lini*), Shaw et al.[81b] triggered HSP synthesis by a 2 h treatment at 31°C, but only colony formation in axenic cultures was improved.[6b] No infection structures were observed in this case.

18.5. STRESS PROTEINS AS DOMINANT ANTIGENS OF INFECTIOUS AND AUTOIMMUNE DISEASES

The abundance and strong autogenicity of stress proteins makes them excellent immune targets. The immune response to a steadily increasing number of infectious diseases is shown to be dominated by antibodies against members of different stress protein families (Table 18.3A). The well-known "common antigen" of bacterial infections was recently identified as GroEL (HSP60). Other pathogens summarized in Table 18.3 are parasitic protozoa and nematodes. They cause devastating diseases that are widespread in tropical and subtropical countries, with hundreds of millions of people affected.

Central to the immune response connected with stress antigens is a special class of T-lymphocytes (γδ-T-cells), whose surface receptors recognize a limited number of conserved antigens only. They play an important role for the immune surveillance and the controlled multiplication of immune cells.[10b,47d,50b,74b,89] Though the conservation of the stress antigens between bacteria and vertebrates (see Figures 2.8 to 2.11) makes this system very effective in primary defense and immune surveillance, it brings about the danger of autoimmune

TABLE 18.3
Stress Proteins as Dominant Antigens of Infectious and Autoimmune Disease[a]

Pathogens	Diseases	Antibodies	Ref.
A. Infectious diseases			
Bacteria			
Mycobacterium tuberculosis, M. bovis Calmette-Guerin	Tuberculosis	a-SP60 (a-SP70, a-GroES)	2a, 12a, 12b, 31a, 46a, 81d, 81e, 89a
Mycobacterium leprae, M. habana	Leprosy	a-SP60 (a-SP70, a-SP20)	10a, 15a, 50c, 68c, 81d, 89a
Coxiella burnetii	Q-fever	a-SP60, a-GroES	83b
Legionella micdadei, L. pneumophila	Legionärs' disease	a-SP60	2b, 45, 74a
Treponema palidum, T. phagendenis	Syphilis	a-SP60	45, 45c, 78a
Borrelia burgdorferi	Lyme disease	a-SP60	42a
Chlamydia trachomatis, C. psittaci	Trachomes or lymphogranulomes of the eye or urogenital system	a-SP70	21a
Protozoa			
Trypanosoma cruzi	Chagas disease	a-SP90	22
Plasmodium falciparum	Malaria	a-SP70	1a, 6, 50d, 57b, 73b
Leishmania major, L. mexicana, L. braziliensis	Leishmaniasis		81f
Nematodes			
Schistosoma mansoni	Schistosomiasis	a-SP90, a-SP70, a-SP20	43a, 47a, 68b
Brugia, Wucheria	Lymphatic filiariasis lymphadenitis, elephantiasis	a-SP70	81
B. Autoimmune disease[b]			
Salmonella, Campylobacter, Mycobacterium, Streptococcus, Yersinia	Postinfection and adjuvant arthritis	a-SP60	6a, 32a, 47b, 82c, 83a
	Rheumatoid arthritis	a-SP60	45b, 47c, 77a
	Scleroderma pigmentosum		21b
	Lupus erythematosus	a-SP90, a-SP70	60a, 60b, 70
	Spondylitis ankylosa	a-SP60	5a
	Diabetes type I	a-SP60	29a

[a] Extended version of a table compiled by Young et al.[89b]
[b] See reviews by Polla,[74b] Wraith et al.[88b] and Young and Elliott.[89]

disease (Table 18.3B).[5a,10b,70c,89] In fact a number of interesting findings indicate that γδ-T-cell clones are primarily directed to self-antigens: (1) Rajasekar et al.[75a] reported on a selective multiplication of γδ-T-cells after treating a lymphocyte population *in vitro* by a mild heat stress of 30 min at 42°C. The receptor specificity was similar to that found after mycobacterial stimulation. (2) Isolated T-lymphocytes from healthy human probands exhibited specificity for distinct determinants of the self-HSP60 from mitochondria but also for corresponding domains of the mycobacterial GroEL.[68,68a,83a] (3) Already 5% of T-lymphocytes isolated from newborn mice have GroEL-specific surface receptors.[32b,42b,70a]

Experimentally far advanced are studies on induced adjuvant arthritis in rats, which can be inhibited by prior injection of recombinant HSP60[6a,47b,82c,83a] or by T-cell "vaccination".[45a,54a] The progressive destruction of the joints in all types of arthritis is enhanced by the increased amounts of stress proteins as a result of mechanical stress and local inflammations[50c,74a,82b] and by the hs-induced formation of extracellular metalproteinases (collagenase, stromelysin).[82b] Basically similar are results with diabetes type I, characterized by a dramatic decrease of β-cells in the Langerhans islets of the pancreas, connected with an increasing number of HSP60-specific γδ-T-cells.[29a] Diabetes can be induced by transfer of a corresponding T-cell population to irradiated mice, and its progress can be retarded by intraperitoneal injection of HSP60.

REFERENCES

1. **Alcina, A. and Fresno, M.,** Early and late heat-induced proteins during *Leishmania mexicana* transformation, *Biochem. Biophys. Res. Commun.*, 156, 1360—1367, 1988.

1a. **Ardeshir, F., Flint, J. E., Richman, S. J., and Reese, R. T.,** A 75 kd merozoite surface protein of *Plasmodium falciparum* which is related to the 70 kd heat-shock proteins, *EMBO J.*, 6, 493—500, 1987.

1b. **Armstrong, J. B. and Graveson, A. C.,** Progressive patterning precedes somite segmentation in the mexican axolotl *Ambystoma mexicanum*, *Dev. Biol.*, 126, 1—6, 1988.

1c. **Ashburner, M.,** Patterns of puffing activity gland chromosomes of *Drosophila*. V. Responses to environmental treatments, *Chromosoma*, 31, 356—376, 1970.

2. **Baltus, E. and Hanocq-Quertier, J.,** Heat-shock response in *Xenopus oocytes* during meiotic maturation and activation, *Cell Differ.*, 16, 161—168, 1985.

2a. **Baird, P. N., Hall, L. M., and Coates, A. R. M.,** A major antigen from *Mycobacterium tuberculosis* which is homologous to the heat shock proteins groES from *E. coli* and the htpA gene product of *Coxiella burnetii*, *Nucl. Acids Res.*, 16, 9047, 1988.

2b. **Bangsborg, J. M., Collins, M. T., Hoiby, N., and Hindersson, P.,** Cloning and expression of the *Legionella micdadei* 'common antigen' in *Escherichi coli*, *Acta Pathol. Microbiol. Immunol. Scand.*, 97, 14—22, 1989.

3. **Bannon, G. A., Perkins-Dameron, R., and Allen-Nash, A.,** Structure and expression of two temperature-specific surface proteins in the ciliated protozoan *Tetrahymena thermophila*, *Mol. Cell. Biol.*, 6, 3240—3245, 1986.

4. **Benfey, T. J. and Sutterlin, A. M.,** Triploidy induced by heat shock and hydrostatic pressure in landlocked Atlantic salmon (*Salmo salar* L.), *Aquaculture*, 36, 359—367, 1984.

5. **Bergh, S. and Arking, R.,** Developmental profile of the heat shock response in early embryos of *Drosophila*, *J. Exp. Zool.*, 231, 379—391, 1984.

5a. **Bernstein, R. M.,** Heat shock proteins and arthritis, *Br. J. Rheumatol.*, 28, 369—371, 1989.

6. **Bianco, A. E., Favaloro, J. M., Burkot, T. R., Culvenor, J. G., Crewther, P. E., Brown, G. V., Anders, R. F., Coppel, R. L., and Kemp, D. J.,** A repetitive antigen of *Plasmodium falciparum* that is homologous to heat shock protein 70 of *Drosophila melanogaster*, *Proc. Natl. Acad. Sci. U.S.A.*, 83, 8713—8717, 1986.

6a. **Billingham, M. E. J., Carney, S., Butler, R., and Colston, M. J.,** A mycobacterial 65-Kd heat shock protein induces antigen-specific suppression of adjuvant arthritis, but is not itself arthritogenic, *J. Exp. Med.*, 171, 339—344, 1990.

6b. **Boasson, R. and Shaw, M.,** The effects of heat shock and inoculum density on growth and sporulation in axenic cultures of the flax rust fungus, *Can. J. Bot.*, 66, 189—193, 1988.

6c. **Bond, U. and Schlesinger, M.,** Heat-shock proteins and development, *Adv. Genet.*, 24, 1—29, 1987.

7. **Boon-Niermeijer, E. K.,** Morphogenesis after heat shock during the cell cycle of *Lymnaea:* A new interpretation, *Wilhelm Roux's Arch. Dev. Biol.* 180, 241—252, 1976.

8. **Boon-Niermeijer, E. K. and Van De Scheur, H.,** Thermosensitivity during embryonic development of *Lymnaea stagnalis (Mollusca)*, *J. Therm. Biol.*, 9, 265—269, 1984.

9. **Boon-Niermeijer, E. K., Tuyl, M., and Van De Scheur, H.,** Evidence for two states of thermotolerance, *Int. J. Hyperthermia*, 2, 93—105, 1986.

10. **Boon-Niermeijer, E. K., De Waal, A. M., Souren, J. E. M., and Van Wijk, R.,** Heat-induced changes in thermosensitivity and gene expression during development, *Dev. Growth Differ.*, 30, 705—715, 1988.

10a. **Booth, R. J., Harris, D. P., Love, J. M., and Watson, J. D.**, Antigenic proteins of *M. leprae*: complete sequence of the gene for the 18kD protein, *J. Immunol.*, 140, 597, 1988.

10b. **Born, W., Happ, M. P., Dallas, A., Reardon, C., Kubo, R., Shinnick, T., Brennan, P., and O'Brien, R.**, Recognition of heat shock proteins and gamma-delta-cell function, *Immunol. Today,* 11, 40—43, 1990.

11. **Borst, P.**, Discontinuous transcription and antigenic variation in *Trypanosomes, Annu. Rev. Biochem.*, 55, 701—732, 1986.

12. **Bournias-Vardiabasis, N. and Buzin, C. H.**, Developmental effects of chemicals and the heat shock response in *Drosophila* cells, *Teratog. Carcinog. Mutag.*, 6, 523—537, 1986.

12a. **Britton, W. J., Hellquist, L., Basten, A., and Raison, R. L.**, *Mycobacterium leprae* antigens involved in human immune responses. I. Identification of four antigens by monoclonal antibodies, *J. Immunol.*, 135, 4171—4177, 1985.

12b. **Britton, W. J., Hellquist, L., Basten, A., and Inglis, A. S.**, Immunoreactivity of a 70 kD protein purified from *Mycobacterium bovis* bacillus Calmette-Guerin by monoclonal antibody affinity chromatography, *J. Exp. Med.*, 164, 695—708, 1986.

13. **Browder, L. W., Pollock, M., Nickells, R. W., Heikkila, J. J., and Winning, R. S.**, Developmental regulation of the heat shock response in *Developmental Biology: A Comprehensive Synthesis,* Di Berardina, M. A. and Etkin, L. D., Eds., Vol. 6, Plenum Press, New York, 1989, 97—147.

14. **Brown, I. R.**, Hyperthermia induces the synthesis of a heat shock protein by polysomes isolated from the fetal and neonatal mammalian brain, *J. Neurochem.*, 40, 1490—1493, 1983.

15. **Brown, I. R.**, Modification of gene expression in the mammalian brain after hyperthermia, in *Gene Expression in Brain,* Zomzely-Neurath, C. and Walker, W. A., Eds., John Wiley & Sons, New York, 1985, 157—171.

15a. **Buchanan, T. M., Nomaguchi, H., Anderson, D. C., Young, R. A., Gillis, T. P., Britton, W. J., Ivanyi, J., Kolk, A. H. J., Closs, O., Bloom B. R., and Mehra, V. J.**, Characterization of antibody-reactive epitopes on the 65-kilodalton protein of *Mycobacterium leprae.* Mycobacterial species specificity, *Infect. Immunol.*, 55, 1000—1003, 1987.

16. **Buzin, C. H. and Bournias-Vardiabasis, N.**, Teratogens induce a subset of small heat shock proteins in *Drosophila* primary embryonic cell cultures, *Proc. Natl. Acad. Sci. U.S.A.*, 81, 4075—4079, 1984.

17. **Calderwood, S. K. and Hahn, G. M.**, Thermal sensitivity and resistance of insulin-receptor binding, *Biochim. Biophys. Acta,* 756, 1—8, 1983.

18. **Calderwood, S. K., Stevenson, M. A., and Hahn, G. M.**, Cyclic AMP and the heat shock response in Chinese hamster ovary cells, *Biochem. Biophys. Res. Commun.*, 126, 911—916, 1985.

18a. **Caruso, M., Sacco, M., Mendoff, G., and Maresca, B.**, Heat shock 70 gene is differentially expressed in *Histoplasma capsulatum* strains with different levels of thermotolerance and pathogenicity, *Mol. Microbiol.*, 1, 151—158, 1987.

19. **Chomyn, A., Moller, G., and Mitchell, H. K.**, Patterns of protein synthesis following heat shock in pupae of *Drosophila melanogaster, Dev. Genet.*, 1, 77—95, 1979.

20. **Cooke, J.**, Somite abnormalities caused by short heat shocks to pre-neurula stages of *Xenopus laevis, J. Embryol. Exp. Morphol.*, 45, 283—294, 1978.

20a. **Cornelius, G. and Rensing, L.**, Circadian rhythm of heat shock protein synthesis of *Neurospora crassa, Eur. J. Cell Biol.*, 40, 130—132, 1986.

21. **Dabrowa, N. and Howard, D. H.**, Heat shock and heat stroke proteins observed during germination of the blastoconidia of *Candida albicans, Infect. Immun.*, 44, 537—539, 1984.

21a. **Danilition, S. L., MacLean, I. W., Peeling, R., Winston, S., and Brunham, R. C.**, The 75-kilodalton protein of *Chlamydia trachomatis* a member of the heat shock protein 70 family, *Infect. Immunol.*, 58, 189—196, 1990.

21b. **Deguchi, Y., and Kishimoto, S.**, Elevated expression of heat shock protein gene in the fibroblasts of patients with scleroderma, *Clin. Sci.*, 78, 419—422, 1990.

21c. **Disney, J. E., Johnson, K. R., and Thorgaard, G. H.**, Intergeneric gene transfer of six isozyme loci in rainbow trout by sperm chromosome fragmentation and gynogenesis, *J. Exp. Zool.*, 244, 151—158, 1987.

22. **Dragon, E. A., Sias, S. R., Kato, E. A., and Gabe, J. D.** The genome of *Trypanosoma cruzi* contains a constitutively expressed, tandemly arranged multicopy gene homologous to a major heat shock protein, *Mol. Cell. Biol.*, 7, 1271—1275, 1987.

23. **Dunkle, L. D., Maheshwari, R., and Allen, P. J.**, Infection structurers from rust urediospores: Effect of RNA and protein synthesis inhibitors, *Science,* 163, 481—482, 1969.

24. **Dura, J. M.**, Stage dependent synthesis of heat shock induced proteins in early embryos of *Drosophila melanogaster, Mol. Gen. Genet.*, 184, 381—385, 1981.

25. **Dura, J. M. and Santamaria, P.**, Heat shock-induced phenocopies: cis regulation of the bithorax complex in *Drosophila melanogaster, Mol. Gen. Genet.*, 189, 235—239, 1983.

26. **Eberlein, S.**, Stage specific embryonic defects following heat shock in *Drosophila, Dev. Genet.*, 6, 179—197, 1986.

27. **Edwards, M. J.,** Congenital defects in guinea pigs: Fetal resorptions, abortions, and malformations following induced hyperthermia during early gestation, *Teratology,* 2, 313—328, 1969.
28. **Edwards, M. J.,** The experimental production of arthrogryposis multiplex congenita in guinea pigs by maternal hyperthermia during gestation, *J. Pathol.,* 104, 221—229, 1971.
29. **Edwards, M. J.,** Influenza, hyperthermia, and congenital malformations, *Lancet,* 1, 320—321, 1972.

29a. **Elias, D., Markovits, D., Reshef, T., van der Zee, R., and Cohen, I. R.,** Induction and therapy of autoimmune diabetes in the non-obese diabetic (Nod/LT) mouse by a 65-kDa heat shock protein, *Proc. Natl. Acad. Sci. U.S.A.,* 87, 1576—1580, 1990.

30. **Elsdale, T., Pearson, M. and Whitehead, M.,** Abnormalities in somite segmentation following heat shock to *Xenopus* embryos, *J. Embryol. Exp. Morphol.,* 35, 625—635, 1976.
31. **Elsdale, T. and Davidson, D.,** Timekeeping by frog embryos, in normal development and after heat shock, *Development,* 99, 41—49, 1987.

31a. **Emmrich, F., Thole, J., van Embden, J., and Kaufmann, S. H. E.,** A recombinant 65 kilodalton protein of *Mycobacterium bovis* bacillus Calmette-Guerin specifically stimulates human T4 clones reactive to mycobacterial antigens, *J. Exp. Med.,* 163, 1024—1029, 1986.

32. **Fox, D. P., Johnstone, R., and Durward, E.,** Erythrocyte fusion in heat-shocked atlantic salmon, *J. Fish Biol.,* 28, 491—500, 1986.

32a. **Gaston, J. S. H., Life, P. F., Jenner, P. J., Colston, M. J., and Bacon, P. A.,** Recognition of a mycobacteria-specific epitope in the 65-kD heat shock protein by synovial fluid-derived T-cell clones, *J. Exp. Med.,* 171, 831—841, 1990.

33. **Geilenkirchen, W. L. M.,** Cell division and morphogenesis of *Limnaea* eggs after treatment with heat pulses at successive stages in early division cycles, *J. Embryol. Exp. Morphol.,* 16, 321—337, 1966.
34. **German, J.,** Embryogenic stress hypothesis of teratogenesis, *Am. J. Med.,* 76, 293—301, 1984.
35. **German, J., Louie, E., and Banerjee, D.,** The heat-shock response in vivo: Experimental induction during mammalian organogenesis, *Teratog. Cancerrog. Mutag.,* 6, 555—562, 1986.
36. **Giudice, G.,** Heat shock proteins in sea urchin development, in *Changes in Eukaryotic Gene Expression in Response to Environmental Stress,* Atkinson, B. G. and Walden, D. B., Eds., Academic Press, Orlando, FL, 115—133, 1985.
37. **Glass, D. J., Polvere, R. J., and Van Der Ploeg, L. H. T.,** Conserved sequences and transcription of the hsp 70 gene family in *Trypanosoma brucei, Mol. Cell. Biol.,* 6, 4657—4666, 1986.
38. **Gloor, H.,** Phänokopie-Versuche mit Äther an *Drosophila, Rev. Suisse Zool.,* 54, 637—712, 1947.
39. **Goldschmidt, R.,** Experimentelle Mutation und das Problem der sogenannten Parallelinduktion. Versuche an *Drosophila, Biol. Zentralbl.,* 49, 437—448, 1929.
40. **Goldschmidt, R.,** Gen und Ausseneigenschaft (Untersuchungen an *Drosophila*) I. und II. *Z. Indukt. Abstamm. Vererbungsl.,* 69, 38—131, 1935.
41. **Graziosi, G., Micali, F., Marzari, R., Decristini, F., and Savoini, A.,** Variability of response of early *Drosophila* embryos to heat shock, *J. Exp. Zool.,* 214, 141—145, 1980.
42. **Hadley, T. J., Klotz, F. W., and Miller, L. H.,** Invasion of erythrocytes by malaria parasites: A cellular and molecular overview, *Annu. Rev. Microbiol.,* 40, 451—477, 1986.

42a. **Hansen, K., Bangsborg, J. M., Fjordvang, H., Pedersen, N. S., and Hindersson, P.,** Immunochemical characterization of and isolation of the gene for a *Borrelia burgdorfi* immunodominant 60-kilodalton antigen common to a wide range of bacteria, *Infect. Immunol.,* 56, 2047—2053, 1988.

42b. **Haregewoin, A., Soman, G., Hom, R. C., and Finberg, R. W.,** Human $\gamma\delta$-T-cells respond to mycobacterial heat-shock protein, *Nature,* 340, 309—312, 1989.

43. **Harrison, L. G., Snell, J., and Verdi, R.,** Turings model and pattern adjustment after temperature shock with application to *Acetabularia mediteranea* whorls, *J. Theor. Biol.,* 106, 59—78, 1984.

43a. **Hedstrom, R., Culpepper, J., Harrison, R. A., Agabian, N., and Newport, G.,** A major immunogen in *Schistosoma mansoni* infection is homologous to the heta-shock protein Hsp70, *J. Exp. Med.,* 130, 1430—1435, 1987.

44. **Heikkila, J. J., Kloc, M., Bury, J., Schultz, G. A., and Browder, L. W.,** Acquisition of the heat shock response and thermotolerance during early development of *Xenopus laevis, Dev. Biol.,* 107, 483—489, 1985.

44a. **Heilbrunn, L. V.,** Studies in artificial parthenogenesis. IV. Heat parthenogenesis, *J. Exp. Zool.,* 41, 243—260, 1925.

44b. **Hindersson, P., Knudson, J. D., and Axelsen, N. H.,** Cloning and expression of *Treponema pallidum* common antigen (Tp-4) in *E. coli* K-12, *J. Gen. Microbiol.,* 133, 587—596, 1987.

45. **Hoffman, P. S., Butler, C. A., and Quinn, F. D.,** Cloning and temperature-dependent expression in *Escherichia coli* of a *Legionella pneumophila* gene coding for a genus-common 60-kilodalton antigen, *Infect. Immunol.,* 57, 1731—1739, 1989.

45a. **Holoshitz, J., Naparstek, Y., Ben-Nun, A., and Cohen, I. R.,** Lines of T lymphocytes induce or vaccinate against autoimmune arthritis, *Science,* 219, 56—58, 1983.

45b. **Holoshitz, J., Koning, F., Coligan, J. E., Debruyn, J., and Strober, S.,** Isolation of CD4- CD8- mycobacteria-reactive T lymphocyte clones from rheumatoid arthritis synovial fluid, *Nature,* 339, 226—229, 1989.

45c. **Houston, L. S., Cook, R. G., and Norris, J.,** Isolation and characterization of a *Treponema pallidum* major 60-kilodalton protein resembling the GroEL protein of *Escherichia coli, J. Bacteriol.,* 172, 2862—2870, 1990.

46. **Hunter, K. W., Cook, C. L., and Hayunga, E. G.,** Leishmanial differentiation in vitro: Induction of heat shock proteins, *Biochem. Biophys. Res. Commun.,* 125, 755—760, 1984.

46a. **Husson, R. and Young, R. A.,** Genes for the major protein antigens of *Mycobacterium tuberculosis*: the etiologic agents of tuberculosis and leprosy share an immunodominant antigen, *Proc. Natl. Acad. Sci. U.S.A.,* 84, 1679—1683, 1987.

47. **Ibrahim-Granet, O. and De Bievre, C.,** Etude des proteines heat shock et heat-stroke chez *Fonsecaea pedrosoi*: Agent de chromomycose, *Bull. Soc. Fr. Mycol. Med.,* 15, 213—220, 1986.

47a. **Johnson, K. S., Wells, K., Bock, J. V., Nene, V., Taylor, D. W., and Cordingley, J. S.,** The 86-kilodalton antigen from *Schistosoma mansoni* is a heat shock protein homologous to yeast HSP90, *Mol. Biochem. Parasitol.,* 36, 19—28, 1989.

47b. **Kale, B., Kiesling, R., van Embden, J. D. A., Thole, J. E. R., Kumararatne, D. S., Pisa, P., Wondimu, A., and Ottenhoff, T. H. M.,** Induction of antigen-specific Cd^{4+} HLA-Dr-restricted cytotoxic lymphocytes-T as well as nonspecific nonrestricted killer cells by the recombinant mycobacterial 65-kDa heat shock protein, *Eur. J. Immunol.,* 20, 369—377, 1990.

47c. **Karlssonparra, A., Soderstrom, K., Ferm, M., Ivanyi, J., Kiessling, R., and Klareskog, L.,** Presence of human 65kD heat shock protein (HSP) in inflamed joints and subcutaneous nodules of RA patients, *Scand. J. Immunol.,* 31, 283—288, 1990.

47d. **Kaufmann, S. H. E.,** Heat shock proteins and the immune response, *Immunol. Today,* 11, 129—136, 1990.

48. **Kemp, D. J., Coppel, R. L., and Anders, R. F.,** Repetitive proteins and genes of malaria, *Annu. Rev. Microbiol.,* 41, 181—208, 1987.

49. **Kirk, M. M. and Kirk, D. L.,** Translational regulation of protein synthesis, in response to light, at a critical stage of *Volvox* development, *Cell,* 41, 419—428, 1985.

50. **Kirk, D. L. and Kirk, M. M.,** Heat shock elicits production of sexual inducer in *Volvox, Science,* 231, 51—54, 1986.

50a. **Kobayashi, H., Watanabe, T., Terashita, N., Handa, S., and Furuno, N.,** Vertebral abnormalities following heat shock in *Xenopus* embryos, *Dev. Growth Diff.,* 31, 65—70, 1989.

50b. **Koga, T., Wand-Wuerttenberger, A., Debruyn, Y., Munk M. E., Schoel, B., and Kaufmann, S. H. E.,** T cells against a bacterial heat shock protein recognize stressed macrophages, *Science,* 245, 1112—1115, 1989.

50c. **Kubo, T., Towle, C. A., Mankin, H. J., and Treadwell, B. V.,** Stress-induced proteins in chondrocytes from patients with osteoarthritis, *Arthritis Rheum.,* 28, 1140—1145, 1985.

50d. **Kumar, N., Zhao, Y., Graves, P., Folgar, J. P., Maloy, L., and Zheng, H.,** Human immune response directed against *Plasmodium falciparum* heat shock-related proteins, *Infect. Immun.,* 58, 1408—1414, 1990.

50e. **Lamb, F. I., Singh, N. B., and Colston, M. J.,** The specific 18-kilodalton antigen of *Mycobacterium leprae* is present in *Mycobacterium habana* and functions as a heat shock protein, *J. Immunol.,* 144, 1922—1925, 1990.

51. **Lambowitz, A. M., Kobayashi, G. S., Painter, A., and Medoff, G.,** Possible relationship of morphogenesis in pathogenic fungus, *Histoplasma capsulatum,* to heat shock response, *Nature,* 303, 806—808, 1983.

52. **Lary, J. M.,** Hyperthermia and teratogenicity, in *Hyperthermia in Cancer Treatment,* Anghileri, L. J. and Robert, J., Eds., CRC Press, Boca Raton, FL, 1986, 107—126.

53. **Lawrence, F. and Robert-Gero, M.,** Induction of heat shock and stress proteins in promastigotes of three *Leishmania* species, *Proc. Natl. Acad. Sci. U.S.A.,* 82, 4414—4417, 1985.

54. **LeJohn, H. B. and Braithwaite, E. E.,** Heat and nutritional shock-induced proteins of the fungus *Achyla* are different and under independent transcriptional control, *Can. J. Biochem. Cell Biol.,* 62, 837—846, 1984.

54a. **Liden, O., Karin, N., Shinitzky, M., and Cohen, I. R.,** Therapeutic vaccination against adjuvant arthritis using autoimmune T-cells treated with hydrostatic pressure, *Proc. Natl. Acad. Sci. U.S.A.,* 84, 4577—4580, 1987.

55. **Lin, C.-Y., Roberts, J. K., and Key, J. L.,** Acquisition of thermotolerance in soybean seedlings, *Plant Physiol.,* 74, 152—160, 1984.

56. **Loomis, W. F. and Wheeler, S.,** Heat shock response of *Dictyostelium, Dev. Biol.,* 79, 399—408, 1980.

57. **Magun, B. E.,** Inhibition and recovery of macromolecular synthesis, membrane transport, and lysosomal function following exposure of cultured cells to hyperthermia, *Radiat. Res.,* 87, 657—669, 1981.

57a. **Martin, M. L. and Regan, C. M.,** The anticonvulsant sodium valproate specifically induces the expression of a rat glial heat shock protein which is identified as the collagen type IV receptor, *Brain Res.*, 459, 131—137, 1988.

57b. **Mattei, D., Scherf, A., Bensaude, O., and Pereira da Silva, L.,** A heat shock like protein from human malaria parasite *Plasmodium falciparum* induces autoantibodies, *Eur. J. Immunol.*, 19, 1823—1828, 1989.

58. **Mee, J. E. and French, V.,** Disruption of segmentation in a short germ insect embryo. I. The location of abnormalities induced by heat shock, *J. Embryol. Exp. Morphol.*, 96, 245—266, 1986.

58a. **Mee, J. E. and French, V.,** Disruption of segmentation in a short germ insect embryo. II. The structure of segmental abnormalities induced by heat shock, *J. Embryol. Exp. Morphol.*, 96, 267—294, 1986.

59. **Mendgen, K. and Dressler, E.,** Culturing *Puccinia coronata* on a cell monolayer of the *Avena* coleoptile, *Phytopathol. Z.*, 108, 226—234, 1983.

60. **Milkman, R.,** On the mechanism of some temperature effects on *Drosophila, J. Gen. Physiol.*, 46, 1151—1170, 1963.

60a. **Minota, S., Cameron, B., Welch, W. J., and Winfield, J. B.,** Autoantibodies to the constitutive 73-kD member of the hsp70 family of heat shock proteins in systemic *lupus erythematosus, J. Exp. Med.*, 168, 1475—1480, 1988.

60b. **Minota, S., Koyasu, S., Yahara, I., and Winfield, J.,** Autoantibodies to the heat-shock protein hsp90 in systemic *lupus erythematosus, J. Chem. Invest.*, 81, 106—109, 1988.

61. **Mirkes, P. E.,** Hyperthermia induced heat shock response and thermotolerance in postimplantation rat embryos, *Dev. Biol.*, 119, 115—123, 1987.

62. **Mitchell, H. K. and Lipps, L. S.,** Heat shock and phenocopy induction in *Drosophila, Cell,* 15, 907—918, 1978.

63. **Mitchell, H. K. and Petersen, N. S.,** Rapid changes in gene expression in differentiating tissues of *Drosophila, Dev. Biol.*, 85, 233—242, 1981.

64. **Mitchell, H. K. and Petersen, N. S.,** Developmental abnormalities in *Drosophila* induced by heat shock, *Dev. Genet.*, 3, 91—102, 1982.

65. **Mitchell, H. K. and Petersen, N. S.,** Gradients of differentiation in wildtype and bithorax mutants of *Drosophila melanogaster, Dev. Biol.*, 95, 459—467, 1983.

66. **Mitchell, H. K. and Petersen, N. S.,** The recessive phenotype of forked can be uncovered by heat shock in *Drosophila, Dev. Genet.*, 6, 93—100, 1985.

67. **Mitchell, H. K., Moller, G., Petersen, N. S., and Lipps-Sarmiento, L.,** Specific protection from phenocopy induction by heat shock, *Dev. Genet.*, 1, 181—192, 1979.

67a. **Mitchell, H. K., Roach, J., and Petersen, N. S.,** Morphogenesis of cell hairs on *Drosophila* wings, *Dev. Biol.*, 95, 387—398, 1983.

68. **Munk, M. E., Schoel, B., Modrow, S., Karr, R. W., Young, R. A., and Kaufmann, S. H. E.,** T lymphocytes from healthy individuals with specificity to self-epitopes shared by the mycobacterial and human 65-kilodalton heat shock protein, *J. Immunol.*, 143, 2844—2849, 1989.

68a. **Munk, M. E., Shinnick, T. M., and Kaufmann, S. H. E.,** Epitopes of the mycobacterial heat shock protein-65 for human T-cells comprise different structures, *Immunobiology,* 180, 272—277, 1990.

68b. **Nene, V., Dunne, D. W., Johnson, K. S., Taylor, D. W., and Cordingley, J. S.,** Sequence and expression of a major egg antigen from *Schistosoma mansoni*. Homologies to heat shock proteins and alpha-crystallins, *Mol. Biochem. Parasitol.*, 21, 179—188, 1986.

68c. **Nerland, A. H., Mustafa, A. S., Sweetser, D., Godal, T., and Young, R. A.,** A protein antigen of *Mycobacterium leprae* is related to a family of small heat shock proteins, *J. Bacteriol.*, 170, 5919—5921, 1988.

69. **Nickells, R. W. and Browder, L. W.,** Region-specific heat-shock protein synthesis correlates with a biphasic acquisition of thermotolerance in *Xenopus laevis* embryos, *Dev. Biol.*, 112, 391—395, 1985.

69a. **Nickells, R. W., Cavey, M. J., and Browder, L. W.,** The effects of heat shock on the morphology and protein synthesis of the epidermis of *Xenopus laevis* larvae, *J. Cell Biol.*, 106, 905—914, 1988.

70. **Norton, P., Isenberg, D. A., and Latchman, D. S.,** Elevated levels of the 90kD heat shock protein in a proportion of SLE patients with active disease, *J. Autoimmun.*, 2, 187—195, 1989.

70a. **Nover, L., Neumann, D., and Scharf, K.-D., Eds.,** *Heat Shock and Other Stress Response Systems of Plants,* Springer-Verlag, Berlin, 1990.

70b. **O'Brien, R., Happ, M. P., Dallas, A., Palmer, E., Kubo, R., and Born, W. K.,** Stimulation of a major subset of lymphocytes expressing T cell receptor gamma delta by an antigen derived from *Mycobacterium tuberculosis, Cell,* 57, 667—674, 1989.

70c. **Oldstone, M. B. A.,** Molecular mimicry and autoimmune disease, *Cell,* 50, 819—820, 1987.

71. **Opperdoes, F. R.,** Compartmentation of carbohydrate metabolism in trypanosomes, *Annu. Rev. Microbiol.*, 41, 127—151, 1987.

72. **Ougham, H. J. and Stoddart, J. L.,** Synthesis of heat-shock protein and acquisition of thermotolerance in high-temperature tolerant and high-temperature susceptible lines of *Sorghum, Plant Sci.*, 44, 163—168, 1986.

73. **Petersen, N. S. and Mitchell, H. K.,** Effects of heat shock on gene expression during development: Induction and prevention of the multihair phenocopy in *Drosophila,* in *Heat Shock from Bacteria to Man,* Cold Spring Harbor Laboratory, Cold Spring Harbor, NY, 1982, 345—352.

73a. **Petersen, N. S. and Mitchell, H. K.,** The forked phenocopy is prevented in thermotolerant pupae, in *Stress-Induced Proteins,* Pardue, M. L., Feramisco, J. R., and Lindquist, S., Eds., Alan R. Liss, New York, 1989, 235—244.

73b. **Peterson, M. G., Crewther, P. E., Thompson, J. K., Corcoran, L. M., Coppel, R. L., Brown, G. V., Anders, R. F., and Kemp, D. J.,** A second antigenic heat shock protein of *Plasmodium falciparum, DNA,* 7, 71—78, 1988.

74. **Pezzolo, C., Guerra, D., Bellucci, R., and Giangrande, E.,** Cell size and cell number in *Drosophila* wing after heat shock, *Biol. Zool.,* 53, 25—28, 1986.

74a. **Plikaytis, B. B., Carlone, G. M., Pau, C.-P., and Wilkinson, H. W.,** Purified 60kD Legionella protein antigen with *Legionella*-specific and nonspecific epitopes, *J. Clin. Microbiol.,* 25, 2080, 1987.

74b. **Polla, B. S.,** A role for heat shock proteins in inflammation, *Immunol. Today,* 9, 134—137, 1988.

75. **Raaphorst, G. P., Azzam, E. I., Borsa, J., Einspenner, M., and Vadasz, J. A.,** Inhibition of DMSO-induced differentiation by hyperthermia in a murine erythroleukemia cell system, *Can. J. Biochem. Cell Biol.,* 62, 1091—1096, 1984.

75a. **Rajasekar, R., Sim, G.-K., and Augustin, A.,** Self heat shock and T-cell reactivity, *Proc. Natl. Acad. Sci. U.S.A.,* 87, 1767—1771, 1990.

76. **Raven, C. P., De Roon, A. C., and Stadhouders, A. M.,** Morphogenetic effects of heat shock on the eggs of *Limnaea stagnalis, J. Embryol. Exp. Morphol.,* 3, 142—159, 1955.

76a. **Refstie, T.,** Induction of diploid gynogenesis in Atlantic salmon and rainbow trout using irradiated sperm and heat shock, *Can. J. Zool.,* 61, 2411—2416, 1983.

77. **Rensing, L., Bos, A., Kroeger, J., and Cornelius, G.,** Possible link between circadian rhythm and heat shock response in *Neurospora crassa, Chronobiol. Int.,* 4, 543—549, 1987.

77a. **Res, P. C. M., Schaar, C. G., Breedveld, F. C., van Eden, W., van Embden, J. D. A., Cohen, I. R., and de Vries, R. R. P.,** Synovial fluid T cell reactivity against 65 kD heat shock protein of mycobacteria in early chronic arthritis, *Lancet,* 2, 478—480, 1988.

77b. **Richards, F. M., Watson, A., and Hickman, J. A.,** Investigation of the effects of heat shock and agents which induce a heat shock response on the induction of differentiation of HL-60 cells, *Cancer Res.,* 48, 6715—6720, 1988.

77c. **Riehl, R. M.,** The effects of environmental stress on steroid receptor levels and steroid-induced morphogenesis in *Achlya ambisexualis, Exp. Cell Res.,* 179, 462—476, 1988.

77d. **Robertson, N. P., Reese, R. T., Henson, J. M., and Speer, C. A.,** Heat shock-like polypeptides of the sporozoites and merozoites of *Eimeria bovis, J. Parasitol.,* 74, 1004—1008, 1988.

78. **Roccheri, M. C., Sconzo, G., Di Bernhardo, M. G., Albanese, I., Di Carlo, M., and Giudice, G.,** Heat shock proteins in sea urchin embryos. Territorial and intracellular location, *Acta Embryol. Morphol. Exp.,* 2, 91—96, 1981.

78a. **Sand-Petersen, C., Strandberg-Pedersen, N., and Axelsen, N. H.,** Isolation and characterisation of Tr-c, an antigen of the Reiter treponeme precipitating with antibodies in syphilis, *Scand. J. Immunol.,* 15, 459—465, 1982.

79. **Santamaria, P.,** Heat shock induced phenocopies of dominant mutants of bithorax complex in *Drosophila melanogaster, Mol. Gen. Genet.,* 172, 161—163, 1979.

80. **Schlesinger, M. J., Ashburner, M., and Tissieres, A., Eds.,** *Heat Shock from Bacteria to Man,* Cold Spring Harbor Laboratory, Cold Spring Harbor, NY, 1982.

80a. **Sconzo, G., Roccheri, M. C., La Rosa, M., Oliva, D., Abrignani, A., and Gindice, G.,** Acquisition of thermotolerance in sea urchin embryos correlates with the synthesis and age of the heat shock proteins, *Cell Differ.,* 19, 173—179, 1986.

81. **Selkirk, M. E., Denham, D. A., Partono, F., and Maizels, R. M.,** Heat shock cognate-70 is a prominent immunogen in brugian filariasis, *J. Immunol.,* 143, 299—308, 1989.

81a. **Shapira, M., McEwen, J. G., and Jaffe, C. L.,** Temperature effects on molecular processes which lead to stage differentiation in *Leishmania, EMBO J.,* 7, 2895—2901, 1988.

81b. **Shaw, M., Boasson, R., and Scrubb, L.,** Effect of heat shock on protein synthesis in flax rust uredosporelings, *Can. J. Bot.,* 63, 2069—2076, 1985.

81c. **Shearer, G., Birge, C. H., Yuckenberg, P. D., Kobayashi, G. S., and Mendoff, G.,** Heat shock proteins induced during the mycelial to-yeast transitions of strains of *Histoplasma capsulatum, J. Gen. Microbiol.,* 133, 3375—3382, 1987.

81d. **Shinnick, T. M., Sweetser, D., Thole, J. E. R., van Embden, J. D. A., and Young, R. A.,** The etiologic agents of leprosy and tuberculosis share immunoreactive protein antigen with the vaccine strain *Mycobacterium bovis* BCG, *Infect. Immun.,* 55, 1932—1935, 1987.

81e. **Shinnick T. M., Plikaytis, B. B., Hyche, A. D., van Landingham, R. M., and Walker, L. L.,** The *Mycobacterium tuberculosis* BCG-a protein has homology with the *Escherichia coli* GroES protein, *Nucl. Acids Res.*, 17, 1254, 1989.

81f. **Smejkal, R. M., Wolff, R., and Olenick, J. G.,** *Leishmania braziliensis* panamensis: Increased infectivity resulting from heat shock, *Exp. Parasitol*, 65, 1—9, 1988.

82. **Srinivas, U. K., Revathi, C. J., and Das, M. R.,** Heat-induced expression of albumin during early stages of rat embryo development, *Mol. Cell. Biol.*, 7, 4599—4602, 1987.

82a. **Staples, R. C., Hoch, H. C., Freve, P., and Bourett, T. M.,** Heat shock-induced development of infection structures by bean rust uredospore germlings, *Exp. Mycol.*, 13, 149—157, 1989.

82b. **Vance, B. A., Kowalski, C. G., and Brinckerhoff, C. E.,** Heat shock of rabbit synovial fibroblasts increases expression of mRNAs for two metalloproteinases, collagenase and stromelysin, *J. Cell Biol.*, 108, 2037—2043, 1989.

82c. **van den Broek, M. F., Hogervorst, E. J. M., van Bruggen, M. C. J., van Eden, W., van der Zee, R., and van den Berg, W. B.,** Protection against streptococcal cell wall-induced arthritis by pretreatment with the 65-kD mycobacterial heat shock protein, *J. Exp. Med.*, 170, 449—466, 1989.

83. **Van Der Ploeg, L. H. T., Giannini, S. H., and Cantor, C. R.,** Heat shock genes: Regulatory role for differentiation in parasitic protozoa, *Science*, 228, 1443—1446, 1985.

83a. **van Eden, W., Thole, J. E. R., van der Zee, R., Noordzij, A., van Embden, J. D. A., Hensen, E. J., and Cohen, I. R.,** Cloning of the mycobacterial epitope recognized by T lymphocytes in adjuvant arthritis, *Nature*, 331, 171—173, 1988.

83b. **Vodkin, M. H. and Williams, Y. C.,** A heat shock operon in *Coxiella burnetii* produces a major antigen homologous to a protein in both mycobacteria and *Escherichia coli*, *J. Bacteriol.*, 170, 1227—1234, 1988.

84. **Walsh, C.,** Appearance of heat shock proteins during the induction of multiple flagella in *Naegleria gruberi*, *J. Biol. Chem.*, 255, 2629—2632, 1980.

85. **Walsh, D. A., Klein, N. W., Hightower, L. E., and Edwards, M. J.,** Heat shock and thermotolerance during early rat embryo development, *Teratology*, 36, 181—191, 1987.

86. **Wanner, R., Foerster, H., Mendgen, K., and Staples, R. C.,** Synthesis of differentiation-specific proteins in germlings of the wheat stem rust fungus after heat shock, *Exp. Mycol.*, 9, 279—283, 1985.

87. **Webster, W. S., Germain, M. A., and Edwards, M. J.,** The induction of microphthalmia, encephalocele and other heat defects following hyperthermia during gastrulation process in the rat, *Teratology*, 31, 73—82, 1985.

87a. **Wharton, R. P. and Struhl, G.,** Structure of the *Drosophila* BicaudalD protein and its role in localizing the posterior determinant nanos, *Cell*, 59, 881—892, 1989.

88. **Wolffe, A.-P., Glover, J. F., and Tata, J. R.,** Culture shock. Synthesis of heat-shock-like proteins in fresh primary cell cultures, *Exp. Cell Res.*, 154, 581—590, 1984.

88a. **Wolffe, A. P., Perlman, A. J., and Tata, J. R.,** Transient paralysis by heat shock of hormonal regulation of gene expression, *EMBO J.*, 3, 2763—2770, 1984.

88b. **Wraith, D. C., McDevitt, H. O., Steinman, L., and Acha-Orbea, H.,** T cell recognition as the target for immune intervention in autoimmune disease, *Cell*, 57, 709—715, 1989.

88c. **Yisraeli, J. K. and Melton, D. A.,** The maternal mRNA Vg1 is correctly localized following injection into *Xenopus* oocytes, *Nature*, 336, 592—595, 1988.

89. **Young, R. A. and Elliott, T. J.,** Stress proteins, infection, and immune surveillance, *Cell*, 59, 5—8, 1989.

89a. **Young, D., Lathigra, R., Hendrix, R., Sweetser, D., and Young, R. A.,** Stress proteins are immune targets in leprosy and tuberculosis, *Proc. Natl. Acad. Sci. U.S.A.*, 85, 4267—4270, 1988.

89b. **Young, D. B., Lathigra, R., and Mehlert, A.,** Stress-induced proteins as antigens in infectious diseases, in *Stress-Induced Proteins*, Pardue, M. L., Feramisco, J. R., and Lindquist, S., Eds., Alan R. Liss, New York, 1989, 275—285.

90. **Zagris, N. and Matthopoulos, D.,** Differential heat shock gene expression in chick blastula, *Roux's Arch. Dev. Biol.*, 195, 403—407, 1986.

Chapter 19

HYPERTHERMIA AND VIRUS MULTIPLICATION

L. Nover

TABLE OF CONTENTS

Interest in the interaction between virus- and stress-induced systems for reprogramming of gene expression result from two facts: (1) studies of the mechanisms, used to specialize the cellular machinery for gene expression to the needs of virus replication, on the one hand, and of the stress response, on the other hand, help to elucidate the regulatory network operative in that system in general. (2) Hyperthermic conditions can potentiate or diminish virus multiplication, a situation which is of considerable importance for plant and animal pathology.

19.1. VIRUS-INDUCED STRESS PROTEINS

In several cases, the profound interference of viral infection with cell metabolism leads to increased synthesis of stress proteins. Three different examples are noteworthy:

1. As summarized in Table 4.1, RNA viruses, e.g., Newcastle disease v., Rous sarcoma v., Sendai v., Sindbis v., and vesicular stomatitis virus, stimulate synthesis of glucose-regulated proteins (GRPs94 and 78) in vertebrate cells.[9,17,23,47-49] Glucose deprivation by virus multiplication was considered as the main reason for this effect (see Chapter 4), but recent experiments with temperature sensitive mutants of the Rous sarcoma virus in chicken embryo fibroblasts showed that induction proceeds equally well in the presence or absence of glucose in the culture medium.[62b] Thus, in addition to glucose deprivation there are other perturbations of cell metabolism during virus multiplication which contribute to GRP induction. At any rate, generation of abnormal proteins in the ER seems to be the key event.[30a]
2. The abuse of a host cell regulatory system for early gene expression by certain DNA viruses is the reason for the induction of two genes of the hsp70 and one of the hsp90 family in mammalian cells,[8a,24,27,31-33,42,44,54,66] (see also Table 1.1). The mechanism was intensively studied for adenovirus. The viral E1A phosphoprotein, derived from the 13S mRNA, is responsible for the inducing effect, which is characteristic of the middle phase of virus replication (Figure 8.5). The E1A protein does not act directly by binding to hs promoters, but its effect is mediated by a cellular system of regulatory proteins engaged in cell cycle or basal level control of gene expression.[14b,22,25,40a,43,61,72] In this respect, it is worth noticing that transformation of yeast cells with the E1A gene hooked up to a gal promoter resulted in production of two E1A proteins of 60 and 62 kDa, increased synthesis of the yeast HSP71, and an elongated generation time from 2.4 to 3.9 h.[20]

   In addition to the E1A protein, other nuclear onc proteins were also found to increase hsp70 transcription, i.e., c-myc,[23a, 27a] v-myb[27b] or SV40 large T-antigen.[63a] In all cases, stimulation is independent of the heat shock transcription factor and evidently proceeds by interaction with preformed DNP complexes at the hsp70 promoter (see Section 6.5). Attempts to identify the essential promoter elements led to equivocal results.[55] Using deletion constructs of the human hsp70 promoter with an appropriate reporter gene, undiminished activity of SV40 large T-antigen and v-myb were even found with a promoter deleted to bp -84, and linker-scanning mutations in the TATA-box region had no effect.[27b,63a] In contrast to this, optimum stimulation by c-myc requires sequences upstream of bp -120.[23a]

   Even results with the same viral protein (E1A protein) are contradictory.[14b] On the one hand, Morimoto et al.[40a] and Williams et al.[71a] documented the important role of the CCAAT-box at bp -68 with its binding proteins for transcriptional regulation by the E1A protein. On the other hand, Simon et al.[62] replaced the Hs-hsp70-specific TATA-box (–TATAA–) by the SV40-derived –TATTTAT– and found that the E1A induction was abolished. If correct, this result would indicate an unexpected specificity of the TATA-binding proteins(s). By repeating and extending these experiments, Taylor and Kingston[63b,c] confirmed the specific role of the TATA-box in general. But this is evidently without effect for the E1A stimulation, which was found independent of sequence elements either in the TATA or in the CCAAT-box regions[14b] (see Section 6.5).

The independence of the E1A induction of the hs signal system was already deduced from its specificity. Only two out of three hsp70 and one out of two hsp90 genes are affected.[61] It is remarkable that the hsp70 mRNA escapes the characteristic block of nucleocytoplasmic transport of mRNAs observed after adenovirus infection,[3,40] but despite the ongoing transcription and transport of hs mRNAs, there is a precipitous decline of the mRNA levels late in the infection cycle (see Figure 8.5).[64]

3. Induction of HSPs in *Escherichia coli* after λ-phage infection[14,28] is a result of a stabilizing effect of the phage cIII protein on the hs-specific σ^{32}-factor of RNA polymerase.[4] The authors fused the λ-cIII gene to the tac promoter and transformed *E. coli* cells with the corresponding plasmid. Synthesis of all HSPs increased markedly after induction of cIII overproduction with isopropylthiogalactoside activating transcription from the tac promoter.[4] The cIII-protein evidently functions as a protein-specific proteinase inhibitor, probably stabilizing DNA-binding proteins (cII-protein, σ^{32}). At least, other short-lived proteins of *E. coli* are not affected.

19.2. HEAT SHOCK INHIBITS VIRUS MULTIPLICATION

Virus eradication from plant tissues by long-term, moderate hs followed by *in vitro* propagation of meristems with mass regeneration of virus-free plant material is a general method for maintenance of the productivity of cultural plants. Although applied in the broadest range all over the world,[26,67,69] details of the underlying mechanisms are practically unknown. Dawson[12] and Dawson and Boyd[13] reported for tobacco mosaic virus that there is an immediate block of single strand TMV RNA accumulation, whereas synthesis of viral proteins continues.

Similar to the situation in plants, mild hyperthermic conditions were also shown to interfere with virus multiplication in animal cells.[19,21,36,37] Generally, DNA viruses are more sensitive than RNA viruses.[19] Thus, heat shock (hs) is a convenient means to maintain herpes simplex virus infection in a cryptic state. This was found for human lung and rabbit kidney cells,[71] for chicken embryo fibroblasts,[44] and for human neuroblastoma cells.[76] Yura et al.[76] demonstrated that under these mild hs conditions virus attenuation results from a block of late viral polypeptide synthesis, whereas formation of immediate-early and early polypeptides is less affected. They speculate that the accumulation of HSPs may be involved in the arrest of viral multiplication and survival of the infected cell.

In 1969, Lwoff[37] summarized experimental data and hypotheses on mechanisms involved in the inhibition of virus multiplication under moderate fever conditions using poliovirus-infected HeLa cells as a model system. Two factors may be of primary importance for a successful defense: (1) the liberation of lysosomal enzymes including RNases may help to destroy viral RNAs and (2) block of the viral replication cycle, which is especially heat-sensitive. In continuation of these investigations, Yerushalmi and Lwoff[73] and Yerushalmi et al.[74] reported on a device for local hyperthermia of the nose (Rhinotherm device) for suppression of the common cold caused by rhinovirus infection.

Although the details are mostly unclear, it is evident that the levels of viral mRNAs and/or their translation may also be affected by hs. Thus, a 12 h hs at 43°C applied to SV40 infected monkey CV1 cells markedly reduced the level of viral mRNA coding for the major capsid protein (VP1).[2] Under the same conditions SV40-dependent DNA synthesis continues, and there is no detectable effect on cellular mRNAs (actin) and their translation. It was not further investigated, whether the lack of VP1 mRNA is due to a block of its synthesis and/or to a drastically enhanced decay. The latter is evidently true for the glucocorticoid-stimulated mouse mammary tumor RNA, which decays within 1—2 h after a shift of the culture temperature to 42°C.[50] Again, the effect is selective, and actin mRNA levels are maintained under these mild hs conditions. There was no significant decrease of total cellular protein synthesis. Since the MMTV RNA degradation is inhibited by pretreatment of the cells with cycloheximide, Rahlhan and Johnson[50] discuss the role of an unidentified short-lived RNase, whose activity or intracellular localization changes during hs.

Well-known candidates for activation of a specific RNase in connection with viral infections are 2′5′oligoadenylates, whose synthesis is markedly enhanced after interferon stimulation. Interestingly, Chousterman et al.[7,8] reported a more than tenfold increase of 2′5′ oligoadenylate synthetase in bovine kidney cells, but the effect was observed only late in the recovery period following a severe hs of 1 h at 45°C. It starts with a lag phase of 6 h, is inhibited in the presence of actinomycin D, and is optimum after 18 h, whereas the "normal" HSPs are synthesized in the early recovery period. Hence, a relation between the rapid degradation of the MMTV RNA[50] and increased 2′5′ oligoadenylate levels is not very likely. The stimulation of the *de novo* synthesis of oligoadenylate synthetase is evidently not mediated by a hs-induced increase of the interferon level as reported for human β-lymphocytes.[63] At least in these bovine kidney cells, antibodies to interferon did not inhibit the increase of oligoadenylate synthetase.[7,8]

In the complex network of positive and negative interactions between adenovirus infection and hs (see Section 19.1), Moore et al.[40] observed a significant reduction of late viral mRNA accumulation about 20 h after infection, whereas synthesis of early viral mRNAs was not affected. Transcription and nuclear processing of late viral mRNAs proceed normally under hs conditions. It is very likely that the adenovirus-specific device for nuclear-cytoplasmic transport of mRNA is the hs-sensitive site. A complex of the two early viral proteins E1B/E4 is involved. It is responsible for inhibition of host cell mRNA transport. Only the late mRNAs are heavily dependent on this helper protein complex. The inhibitory effect of hs on mRNA accumulation is complemented by a reduced translation of late viral mRNAs. It is important to notice that, vice versa, the complete shutdown of host cell mRNA translation in the late phase of the adenovirus multiplication cycle likewise affects translation of hsp70 mRNA.[40]

Translation discrimination was also observed in other cases. In the early phase of investigations on the hs-induced selectivity of the translation apparatus (see Sections 11.3 to 11.5), Scott and Pardue[58] used vesicular stomatitis virus mRNAs as templates for *in vitro* translation in a *Drosophila* cell lysate. It became apparent that the lysate from hs cells was able to discriminate between mRNAs coding for the N and NS proteins and the mRNA coding for the membrane (M) protein. The latter was not translated. This result, obtained with a heterologous *in vitro* system, was confirmed by investigations on measles virus mRNA translation in rat, monkey, and human cells. In all cases, translation of the M protein mRNA was selectively blocked after a temperature shift of the cultures from 35 → 39°C, whereas translation of other viral proteins (N, P, H proteins) was not diminished.[45]

The extremely mild hs, applied to selectively block measles virus M protein synthesis, contrasts to the more severe conditions required for the effective hs-discrimination of host-cell nonviral mRNAs (see Sections 11.3 and 11.4). Thus, it is possible that a temperature-dependent cascade of changes of the translation apparatus with differential effects on distinct mRNA classes is characteristic of the hs response.

At any rate, nonstructured leader sequence of the mRNA is evidently decisive for efficient translation under stress conditions, as demonstrated by artificial modification of the pre-proinsulin gene in its leader region and expression in monkey COS cells.[30] Evidence for a particular role of the leader sequence of hs mRNAs for their preferential translation is presented in Section 11.5.5. Corresponding details of viral mRNAs are unknown, but it should be recognized that essential elements of the translation control mechanism act cell-specific (Section 11.3), i.e., in addition to the role of secondary structure of the leader region[30] more specific information encoded in the leader sequence must be involved as well (see Section 11.5.5).

In many cases, an essential final step of viral protein synthesis is the processing of precursor proteins probably by a programmed endoproteolytic cleavage of polyproteins.[29] Details of this process under hs conditions were not studied so far. However, it is well known that changes of proteinogen processing are observed (see Section 12.1). The only paper directly relevant to this point[38] indeed shows that cleavage of *Drosophila* C-virus polyproteins is strongly reduced, but the accumulated precursor can be chased into mature viral proteins in the recovery period. In summary, different mechanisms may contribute to virus attenuation in heat shocked cells: impaired replication, mRNA degradation, translational discrimination of distinct viral mRNAs, and lacking or aberrant processing of precursor proteins. Details and extent of the interference with the viral life cycle are evidently dependent on the particular virus and host cell type.

19.3. VIRUS INFECTION INHIBITS HSP SYNTHESIS

The highly sophisticated strategies of certain viruses to reprogram the host cell gene expression apparatus to their own needs may also block induced stress protein synthesis. Characteristically, in the early phase of infection, host cell gene expression is needed for effective initiation of the viral replication program. Hence, the hs response is largely unaffected or even enhanced. In contrast, in the middle to late phase of the infection cycle the virus-specific modification of the gene expression apparatus may severely interfere also with stress protein synthesis. The following examples illustrate these points.

1. Significant induction of GRP synthesis by RNA virus infection (see Chapter 4) can only be observed, if avirulent strains are used, which are unable to complete their infection cycle with the usual shut-down of host cell gene expression.[9]
2. Induction of HSP synthesis after adenovirus infection (see Section 19.1) is characteristic of the early phase. In the late phase of the viral multiplication in HeLa cells, there is a rapid decline of the hsp70 mRNA level (see Figure 8.5), and translation of hs-induced mRNAs is blocked.[40]
3. Soon after picornavirus infection of animal cells, translation of capped host cell mRNAs ceases due to the degradation of the 220 kDa protein involved in cap recognition and proper positioning of the components of the initiation complex.[52,56] Infection of *Drosophila* cells with a picornavirus (Cricket paralysis virus) followed by hs treatment leads to accumulation of hs-mRNAs but there is no translation of the corresponding HSPs.[39,51] This result is in evident contradiction to the suggestion of a cap-independent internal initiation under stress conditions (see Section 11.5.5). Interestingly, the closely related but very inefficient *Drosophila* C-virus does not seriously affect hs induction. On the contrary, hs interferes with the normal processing of the viral polyproteins[39] (see Section 19.2).
4. Another class of viral infections does not fit into this scheme of a phase-dependent inhibition of host cell gene expression. Munoz et al.[41] investigated the role of poliovirus infection on HSP synthesis in HeLa cells. They found that the drastic inhibition of host cell protein synthesis also extends, though less efficiently, to HSP synthesis. The effect is observed even if viral replication is blocked by actinomycin D or prior treatment with human α-interferon. Details of the underlying mechanism were not analyzed. mRNA degradation or a translational block, as discussed by Munoz et al.,[41]

are equally possible. Infection with other RNA viruses (vesicular stomatitis v., Semliki Forest v., reovirus) gave similar results.

19.4. HEAT SHOCK TOLERATES OR PROMOTES VIRAL GENE EXPRESSION

Evidence for two types of complexes of hs proteins with proteins of animal tumor viruses was summarized in Table 16.1, part I. Their particular role for viral gene expression or protein function is unclear. On the one hand, these are transient complexes with retroviral onc-gene products of the tyrosine-kinase type.[1,6,34,57,60,75,79,80] As illustrated in Figure 16.8, proteins of the HSP90 family may be required for the maturation and intracellular topogenesis of these onc-gene products. On the other hand, binding of members of the HSP70 family to proteins of tumor viruses evidently reflects the capability of the former to recognize abnormal proteins. Such complexes were reported for the polyomavirus medium T-antigen[45a,48] for the SV40 large T-antigen,[55a,62b] for the adenovirus E1A protein,[53,69a] and for a number of nuclear onc-gene products forming large insoluble protein aggregates in heat-shocked mammalian cells.[35] These aggregates may include $p58^{v\text{-}myc}$, $p45^{v\text{-}myb}$, SV40 large T-antigen and the adenovirus E1A protein. The putative role of HSP70 for stabilization and/or resolubilization of these protein complexes is discussed in Section 2.3.2.

It is important to notice that these complexes of HSPs 70 and 90 with viral proteins give no clue for any direct effect on virus multiplication, but there are examples of an enhancement, at least on certain parts, of the viral multiplication cycle under hs conditions. Most remarkable are findings of an increased viral protein synthesis because, together with the newly synthesized mRNAs, the viral mRNAs evidently meet the special requirements of the altered translational machinery (see Section 11.5). This was reported for *Drosophila* viruses,[10,59] for tobacco mosaic virus[13] and for cauliflower mosaic virus.[56a] On the other hand, hs and a number of chemical stressors induce immediate early transcription of human cytomegalovirus, immunodeficiency virus (HIV), and Epstein-Barr virus.[18,18a,77,78] In the former two viruses, a –GGACTTTC– motif in the long terminal repeat may function as the hs box. At the least, coupling of the LTR of HIV to the bacterial chloramphenicol acetyltransferase as reporter gene results in a hs-inducible construct if tested in rat cells.[18a] These findings point to the intriguing possibility of a stress activation of latent viruses in mammals.

Clearest is the situation in *Escherichia coli*. The multiplication cycles of *E. coli* phages and the hs regulon are mutually connected. Several factors contribute to this interdependence:

1. Transition of the λ-phage from the lysogenic to the lytic state is induced by high temperature stress.
2. A special λ-protein (gpcIII), which plays a key role in the decision between both states of the λ-phage, also stabilizes the hs σ factor against rapid degradation and thus leads to increased HSP synthesis.[4]
3. Most important is the requirement of DnaK and GroEL/S proteins for phage DNA synthesis and morphogenesis in general.

All three proteins are part of the HSP set of *E. coli* (see Table 2.2).[18b,41a] This is the evident explanation for the 8- to 260-fold increased yield of T2, T4, and T6 phages in *E. coli* cells preadapted by 8 min hs at 42.8 to 44°C before phage infection.[70] However, the same may be true for the lytic development of phages λ[65] and Mu[46]. Interestingly, phage T7 multiplication was also stimulated by high temperatures, but in this case a functional rpoH gene is not required.[70] In addition to its role as control factor for increased synthesis of DnaK and GroE proteins the product of the rpoH gene (σ^{32}) exerts an ill-defined, direct effect for the maintenance of the lysogenic state of phage λ[16] and for early and late phases of phage T4 gene expression.[15]

REFERENCES

1. **Adkins, B., Hunter, T., and Sefton, B. M.,** The transforming proteins of PRC II virus and Rous sarcoma virus form a complex with the same two cellular phosphoproteins, *J. Virol.*, 43, 448—455, 1982.

ibition of simian virus 40 protein
4, 1988.

nberger, C., Effect of adenovirus
translational discrimination, *Mol.*

eorgopoulos, C. P., Induction of
phage lambda cIII protein, *Genes*

Adv. Genet., 24, 1—29, 1987.
Rous sarcoma virus transforming
distribution of this complex, *Mol.*

shock induced regulation of 2′5′

adenylate synthetase expression is
987.

f chromosomal gene expression by

llates the cellular accumulation of
2.

etle virus-infected *Drosophila* cells
1981.

es, *Virology*, 73, 319—326, 1976.
vith double-stranded RNA synthesis
3.

nally regulated by heat shock, *Plant*

n the synthesis of groE protein and

an, A. G., Reduced activity of the
-dependent protein kinase during a heat shock stress, *J. Biol. Chem.*,
264, 12165—12171, 1989.

14b. **Flint, J. and Shenk, T.,** Adenovirus E1A protein: paradigm viral transactivator, *Annu. Rev. Genet.*, 23, 141—161, 1989.

15. **Frazier, M. W. and Mosig, G.,** Roles of the *Escherichia coli* heat shock sigma factor 32 in early and late gene expression of bacteriophage T4, *J. Bacteriol.*, 170, 1384—1388, 1988.

16. **Fuerst, C. R.,** Indications of an involvement of heat-shock proteins in restoration of repression of temperature-inducible lambda prophage, *Virology*, 159, 183—186, 1987.

17. **Garry, R. F., Ulug, E. T., and Bose, H. R.,** Induction of stress proteins in Sindbis virus- and vesicular stomatitis virus-infected cells, *Virology*, 129, 319—332, 1983.

18. **Geelen, J. M. C., Boom, R., Klaver, G. P. M., Minnaar, R. P., Feltkamp, M. C. W., Van Milligen, F. J., Sol, C. J. A., and Van Der Nordaa, J.,** Transcriptional activation of the major immediate transcription unit of human cytomegalovirus by heat-shock, arsenite and protein synthesis inhibitors, *J. Gen. Virol.*, 68, 2925—2932, 1987.

18a. **Geelen, J. M. C., Minnaar, R.P., Boom, R., van der Nordaa, J., and Goudsmit, J.,** Heat-shock induction of the human immunodeficiency virus long terminal repeat, *J. Gen. Virol.*, 69, 2913—2917, 1988.

18b. **Georgopoulos, C., Tilly, V., Ang, D., Chandrasekhar, G. N., Fayet, O., Spence, J., Ziegelhoffer, T., Liberek, K., and Zylicz, M.,** The role of the *Escherichia coli* heat shock proteins in bacteriophage lambda growth, in *Stress-Induced Proteins*, Pardue, M. L., Feramisco, J. R., and Lindquist, S., Eds., Alan R. Liss, New York, 1989, 37—47.

19. **Gharpure, M.,** A heat-sensitive cellular function required for the replication of DNA but not RNA viruses, *Virology*, 27, 308—319, 1965.

20. **Handa, H., Toda, T., Tajima, M., Wada, T., and Iida, H.,** Expression of the human adenovirus E1A product in yeast, *Gene*, 58, 127—136, 1987.

21. **Hoggan, M. D. and Roizman, B.,** The effect of the temperature of inhibition on the formation and release of Herpes simplex virus in infected FL cells, *Virology*, 8, 508—524, 1959.

22. **Imperiale, M. J., Kao, H.-T., Feldman, L. T., Nevins, J. R., and Strickland, S.,** Common control of the heat shock gene and early adenovirus genes: Evidence for a cellular E1A-like activity, *Mol. Cell. Biol.*, 4, 867—874, 1984.
23. **Isaka, T., Yoshida, M., Owada, M., and Toyoshima, K.,** Alterations in membrance polypeptides of chick embryo fibroblasts induced by transformation with avian sarcoma viruses, *Virology,* 65, 226—237, 1975.
23a. **Kaddurah-Daouk, R., Greene, J. M., Baldwin, A. S., and Kingston, R. E.,** Activation and repression of mammalian gene expression by the c-myc protein, *Genes Dev.*, 1, 347—357, 1987.
24. **Kao, H.-T. and Nevins, J. R.,** Transcriptional activation and subsequent control of the human heat shock gene during adenovirus infection, *Mol. Cell. Biol.*, 3, 2058—2065, 1983.
25. **Kao, H.-T., Capasso, O., Heintz, N., and Nevins, J. R.,** Cell cycle control of the human HSP70 gene: Implications for the role of a cellular E1A-like function, *Mol. Cell. Biol.* 5, 628—633, 1985.
26. **Kassanis, B.,** Effects of changing temperature on plant virus diseases, *Adv. Virus. Res.*, 4, 221—241, 1957.
27. **Khandjian, E. W. and Türler, H.,** Simian virus 40 and polyoma virus induce synthesis of heat shock proteins in permissive cells, *Mol. Cell. Biol.*, 3, 1—8, 1983.
27a. **Kingston, R. E., Baldwin, A. S., and Sharp, P. A.,** Regulation of heat shock protein 70 gene expression by c-myc, *Nature,* 312, 280—282, 1984.
27b. **Klempnauer, K.-H., Arnold, H., and Biedenkapp, H.,** Activation of transcription by v-myb: evidence for two different mechanisms, *Genes Dev.*, 3, 1582—1589, 1989.
28. **Kochan, J. and Murialdo, H.,** Stimulation of groE synthesis in E. coli by bacteriophage lambda infection, *J. Bacteriol.*, 149, 1166—1170, 1982.
29. **Korant, B. D.,** Viral proteases — an emerging therapeutic target, *CRC Crit. Rev. Biotechnol.*, 8, 149—157, 1988.
30. **Kozak, M.,** Leader length and secondary structure modulate mRNA function under conditions of stress, *Mol. Cell. Biol.*, 8, 2737—2744, 1988.
30a. **Kozutsumi, Y., Segal, M., Normington, K., Gething, M.-J., and Sambrook, Y.,** The presence of malfolded proteins in the endoplasmic reticulum signals the induction of glucose-regulated proteins, *Nature,* 332, 462—464, 1988.
31. **Latchman, D. S., Chan, W. L., Leaver, C. E. L., Patel, R., Oliver, P., and Lathangue, N. B.,** The human Mr 90,000 heat shock protein and the *Escherichia coli* Lon protein share an antigenic determinant *Comp. Biochem. Physiol.*, B87, 961—967, 1987.
32. **La Thangue, N. B. and Latchman, D. S.,** Nuclear accumulation of a heat-shock 70-like protein during herpes simplex virus replication, *Biosci. Rep.*, 7, 475—484, 1987.
33. **La Thangue, N. B., Shriver, K., Dawson, C., and Chan, W. L.,** Herpes simplex virus infection causes the accumulation of a heat-shock protein, *EMBO J.*, 3, 267—277, 1984.
34. **Lipsich, L. A., Cutt, R. J., and Brugge, J. S.,** Association of the transforming proteins of Rous, Fujinami and Y73 avian sarcoma viruses with the same two cellular proteins, *Mol. Cell. Biol.*, 2, 875—880, 1982.
35. **Littlewood, T. D., Hancock, D. C., and Evan, G. I.,** Characterization of a heat shock-induced insoluble complex in the nuclei of cells, *J. Cell Sci.*, 88, 65—72, 1987.
36. **Lwoff, A.,** Factors influencing the evolution of viral diseases at the cellular level and in the organism, *Bacteriol. Rev.*, 23, 109—124, 1959.
37. **Lwoff, A.,** Death and transfiguration of a problem, *Bacteriol. Rev.*, 33, 390—403, 1969.
38. **Moore, N. F. and Pullin, J. S. K.,** Heat shock used in combination with amino acid analogs and protease inhibitors to demonstrate the processing of proteins of an insect picornavirus (*Drosophila* -C virus) in *Drosophila melanogaster* cells, *Ann. Virol.*, 134, 285—292, 1983.
39. **Moore, N. F., Pullin, J. S. K., and Reavy, B.,** Inhibition of the induction of heat shock proteins in *Drosophila melanogaster* cells infected with insect picornavirus, *FEBS Lett.*, 128, 93—96, 1981.
40. **Moore, M., Schaack, J., Balm, S. B., Morimoto, R. I., and Schenk, T.,** Induced heat-shock mRNAs escape the nucleo-cytoplasmic transport block in adenovirus-infected cells, *Mol. Cell. Biol.*, 7, 4505—4512, 1987.
40a. **Morimoto, R. I., Mosser, D., McClanahan, T. K., Theodorakis, N. G., and Williams, G.,** Transcriptional regulation of the human HSP70 gene, in *Stress-Induced Proteins,* Pardue, M. L., Feramisco, J. R., and Lindquist, S., Eds., Alan R. Liss, New York, 1989, 83—94.
41. **Munoz, A., Alonso, A. M., and Carrasco, L.,** Synthesis of heat-shock proteins in HeLa cells: Inhibition by virus infection, *Virology,* 137, 150—159, 1984.
41a. **Neidhardt, F. C., Van Bogelen, R. A., and Vaughn, V.,** The genetics and regulation of heat-shock proteins, *Annu. Rev. Genet.*, 18, 295—329, 1984.
42. **Nevins, J. R.,** Induction of the synthesis of a 70,000 dalton mammalian heat shock protein by the adenovirus E1A gene product, *Cell,* 29, 913—919, 1982.
43. **Nevins, J. R., Raychandhuri, P., Yee, A. S., Rooney, R. J., Kovesdi, I., and Reichel, R.,** Transactivation by the adenovirus E1A gene., *Biochem. Cell Biol.*, 66, 578—583, 1988.

44. **Notarianni, E. L. and Preston, C. M.,** Activation of cellular stress protein genes by herpes simplex virus temperature-sensitive mutants which overproduce immediate early polypeptides, *Virology,* 123, 113—122, 1982.
45. **Ogura, H., Baczko, K., Rima, B. K., and Ter Meulen, V.,** Selective inhibition of translation of the mRNA coding for measles virus membrane protein at elevated temperatures, *J. Virol.,* 61, 472—479, 1987.
45a. **Pallas, D. C., Morgan, W., and Roberts, T. M.,** The cellular proteins which can associate specifically with polyomavirus middle T-antigen in human 293 cells include the major human 70-kDa heat shock proteins, *J. Virol.,* 63, 4533—4539, 1989.
46. **Pato, M., Banerjee, M., Desmet, L., and Toussaint, A.,** Involvement of heat shock proteins in bacteriophage Mu development, *J. Bacteriol.,* 169, 5504—5509, 1987.
47. **Peluso, R. W., Lamb, R. A., and Choppin, P. W.,** Infection with paramyxoviruses stimulates synthesis of cellular polypeptides that are also stimulated in cells transformed by Rous sarcoma virus or deprived of glucose, *Proc. Natl. Acad. Sci. U.S.A.,* 75, 6120—6124, 1978.
48. **Pouyssegur, J. and Yamada, K.-M.,** Isolation and immunological characterization of a glucose-regulated fibroblast cell surface glycoprotein and its nonglycosylated precursor, *Cell,* 13, 139—150, 1978.
49. **Pouyssegur, J., Shiu, R. P. C., and Pastan, I.,** Induction of two transformation-sensitive membrane polypeptides in normal fibroblasts by a block in glycoprotein synthesis or glucose deprivation, *Cell,* 11, 941—947, 1977.
50. **Rahlhan, R. and Johnson, G. S.,** Destabilization of cytoplasmic mouse mammary tumor RNA by heat shock: Prevention by cycloheximide pretreatment, *Biochem. Biophys. Res. Commun.,* 137, 1028—1033, 1986.
51. **Reavy, B., Pullin, J. S. K., and Moore, N. F.,** Translational inhibition of heat-shock induced gene expression in picornavirus-infected *Drosophila melanogaster* cells, *Microbios,* 38, 91—98, 1983.
52. **Rhoads, R. E.,** Cap recognition and the entry of mRNA into the protein synthesis initiation cycle, *Trends Biochem. Sci.,* 13, 52—56, 1988.
53. **Ribeiro, G. and Vasconcelos-Costa, J.,** A heat-shock protein is associated with adenovirus type 12 S antigen at the membrane of tumor cells, *Cienc. Biol. Mol. Cell. Biol.,* 9, 227—231, 1984.
54. **Rose, T. M. and Khandjian, E. W.,** A 105,000-dalton antigen of transformed mouse cells is a stress protein, *Can. J. Biochem. Cell Biol.,* 63, 1258—1264, 1985.
55. **Sassone-Corsi, P.,** Pleiotropic action of the adenovirus E1A proteins, *Trends Genet.,* 1, 98, 1985.
55a. **Sawai, E. T. and Butel, J. S.,** Association of a cellular heat shock protein with Simian virus 40 large T-antigen in transformed cells, *J. Virol.,* 63, 3961—3973, 1989.
56. **Schneider, R. J. and Shenk, T.,** Impact of virus infection on host cell protein synthesis, *Annu. Rev. Biochem.,* 56, 317—332, 1987.
56a. **Schoeffl, F., Rieping, M., Baumann, G., Bevan, M., and Angermueller, S.,** The function of plant heat shock promoter elements in the regulated expression of chimaeric genes in transgenic tobacco, *Mol. Gen. Genet.,* 217, 246—253, 1989.
57. **Schuh, S., Yonemoto, W., Brugge, J., Bauer, V. J., Riehl, R. M., Sullivan, W. F., and Toft, D. O.,** A 90,000-dalton binding protein common to both steroid receptors and the Rous sarcoma virus transforming protein, pp60$^{v\text{-}sarc}$, *J. Biol. Chem.,* 260, 14292—14296, 1985.
58. **Scott, M. P. and Pardue, M. L.,** Translational control in lysates of *Drosophila melanogaster* cells, *Proc. Natl. Acad. Sci. U.S.A.,* 78, 3353—3357, 1981.
59. **Scott, M. P., Fostel, J. M., and Pardue, M. L.,** A new type of virus from cultured *Drosophila* cells: Characterization and use in studies of the heat-shock response, *Cell,* 22, 929—941, 1980.
60. **Sefton, B. M., Beemon, K., and Hunter, T.,** Comparison of the expression of the src gene of Rous sarcoma virus in vitro and in vivo, *J. Virol.,* 28, 957—972, 1978.
61. **Simon, M. C., Kitchener, K., Kao, H.-T., Hickey, E., Weber, L., Voellmy, R., Heintz, N., and Nevins, J. R.,** Selective induction of human heat shock gene transcription by the adenovirus E1A gene products, including the 12S E1A product, *Mol. Cell. Biol.,* 7, 2884—2890, 1987.
62. **Simon, M. C., Fisch, T. M., Benecke, B. J., Nevins, J. R., and Heintz, N.,** Definition of multiple functionally distinct TATA elements, one of which is a target in the hsp70 promoter for E1A regulation, *Cell,* 52, 723—729, 1988.
62a. **Soussi, T., Defromentel, C. C., Stuerzbecher, H. W., Ullrich, S., Jenkins, J., and May, P.,** Evolutionary conservation of the biochemical properties of p53. Specific interaction of *Xenopus laevis* p53 with Simian virus 40 large T-antigen and mammalian heat shock proteins70, *J. Virol.,* 63, 3894—3901, 1989.
62b. **Stoeckle, M. Y., Sugano, S., Hampe, A., Vashistha, A., Pellman, D., and Hanafusa, H.,** 78-Kilodalton glucose-regulated protein is induced in Rous sarcoma virus-transformed cells independently of glucose deprivation, *Mol. Cell. Biol.,* 8, 2675—2680, 1988.
63. **Taylor, M. W., Long, T., Matinez-Valdez, H., Downing, J., and Zeige, G.,** Induction of gamma-interferon activity by elevated temperatures in human B-lymphoblastoid cell lines, *Proc. Natl. Acad. Sci. U.S.A.,* 81, 4033—4036, 1984.

63a. **Taylor, I. C. A., Solomon, W., Weiner, B. M., Paucha, E., Bradley, M., and Kingston, R. E.,** Stimulation of the human heat shock protein 70 promoter in vitro by Simian virus 40 large T-antigen, *J. Biol. Chem.,* 264, 16160—16164, 1989.

63b. **Taylor, I. C. A. and Kingston, R. E.,** Factor substitution in a human HSP70 gene promoter: TATA-dependent and TATA-independent interactions, *Mol. Cell. Biol.,* 10, 165—175, 1990.

63c. **Taylor, I. C. A. and Kingston, R. E.,** E1a transactivation of human HSP70 gene promotor substitution mutants is independent of the composition of upstream and TATA elements, *Mol. Cell. Biol.,* 10, 176—183, 1990.

64. **Theodorakis, N. G. and Morimoto, R. I.,** Post-transcriptional regulation of HSP 70 expression in human cells: Effects of heat shock, inhibition of protein synthesis, and mRNA stability, *Mol. Cell. Biol.,* 7, 4357—4368, 1987.

65. **Waghorne, C. and Fuerst, C. R.,** Involvement of the htp-R gene product of *Escherichia coli* in phage lambda development, *Virology,* 141, 51—64, 1985.

66. **Wakakura, M., Kennedy, P. G. E., Foulds, W. S., and Clements, G. B.,** Stress protein accumulate in cultured retinal glial cells during herpes simplex viral infection, *Exp. Eye Res.,* 45, 557—568, 1987.

67. **Walkey, D. G. A. and Cooper, V. C.,** Effect of temperature on virus eradication and growth of infected tissue cultures, *Ann. Appl. Biol.,* 80, 185—190, 1975.

68. **Walter, G., Carbone, A., and Welch, W. J.,** Medium tumor antigen of polyomavirus transformation-defective mutant NG59 is associated with 73-kilodalton heat shock protein, *J. Virol.,* 61, 405—410, 1987.

69. **Wang, P. J.,** Producing pathogen-free plants using tissue culture, *Plant Tissue Cult. Newslett.,* 46, 2—8, 1985.

69a. **White, E., Spector, D., and Welch, W.,** Differential distribution of the adenovirus E1A proteins and colocalization of E1A with the 70-kDa cellular heat shock protein in infected cells, *J. Virol.,* 62, 4153—4166, 1988.

70. **Wiberg, J. S., Mowrey-McKee, M. F., and Stevens, E. J.,** Induction of heat shock regulation of *Escherichia coli* markedly increases production of bacterial viruses at high temperatures, *J. Virol.,* 62, 234—245, 1988.

71. **Wigdahl, B. L., Isom, H. C., and Rapp, F.,** Repression and activation of the genome of herpes simplex viruses in human cells, *Proc. Natl. Acad. Sci. U.S.A.,* 78, 6522—6526, 1981.

71a. **Williams, G. T., McClanahan, T. K., and Morimoto, R. I.,** E1a transactivation of the human HSP70 promoter is mediated through the basal transcription complex, *Mol. Cell. Biol.,* 9, 2574—2587, 1989.

72. **Wu, B. J., Hurst, H. C., Jones, N. C., and Morimoto, R. I.,** The E1A 13S product of adenovirus 5 activates transcription of the cellular human HSP70 gene, *Mol. Cell. Biol.,* 6, 2994—2999, 1986.

73. **Yerushalmi, A. and Lwoff, A.,** Medecine et therapeutique. Traitement du coryza infecteux et de rhinites persistantes allergiques par la thermotherapie, *C. R. Acad. Sci. Paris,* 291, 957—959, 1980.

74. **Yerushalmi, A., Karman, S., and Lwoff, A.,** Treatment of perennial allergic rhinitis by local hyperthermia, *Proc. Natl. Acad. Sci. U.S.A.,* 79, 4766—4769, 1982.

75. **Yonemoto, W., Lipsich, L. A., Darrow, D., and Brugge, J. S.,** An analysis of the interaction of the Rous sarcoma virus transforming protein, pp60src, with a major heat-shock protein, in *Heat Shock From Bacteria to Man,* Schlesinger, M. J., Ashburner, M., and Tissieres, A., Eds., Cold Spring Harbor Laboratory, Cold Spring Harbor, NY, 1982, 289—298.

76. **Yura, Y., Terashima, K., Iga, H., Kondo, Y., Yanagawa, T., Yoshida, H., Hayashi, Y., and Sato, M.,** Macromolecular synthesis at the early stage of herpes simplex virus type 2(HSV-2) latency in a human neuroblastoma cell line IMR-32: Repression of late viral polypeptide synthesis and accumulation of cellular heat shock proteins, *Arch. Virol.,* 96, 17—28, 1987.

77. **Zerbini, M., Musiani, M., and La Placa, M.,** Stimulating effect of heat shock on the early stage of human cytomegalovirus replication cycle, *Virus Res.,* 6, 211—216, 1986.

78. **Zerbini, M., Musiani, M., and La Placa, M.,** Effect of heat shock on Epstein-barr virus and cytomegalovirus expression, *J. Gen. Virol.,* 66, 633—636, 1985.

79. **Ziemiecki, A.,** Characterization of the monomeric and complex-associated forms of the gag-onc fusion proteins of three isolates of feline sarcoma virus: Phosphorylation, kinase activity, acylations and kinetics of complex formation, *Virology,* 151, 265—273, 1986.

80. **Ziemiecki, A., Catelli, M.-G., Joab, I., and Moncharmont, B.,** Association of the heat shock protein HSP90 with steroid hormone receptors and tyrosine kinase oncogene products, *Biochem. Biophys. Res. Commun.,* 138, 1298—1307, 1986.

Chapter 20

HYPERTHERMIC TREATMENT OF CANCER

L. Nover

TABLE OF CONTENTS

20.1. INTRODUCTION: HYPERTHERMIA PAST AND PRESENT

There are historical reasons, summarized by Cavaliere et al.,[18] Crile,[23] Nauts[89,90] and Overgaard[93] (see Introduction to this book) to begin this monograph with remarks on cancer therapy, and there are practical reasons to end with this topic. On the one hand, the exploding development of our knowledge on the complexity of the cellular heat shock (hs) response in the preceding 15 years has thoroughly influenced research on and application of hyperthermia in cancer therapy. At least since 1980, there is a fruitful interaction between molecular cell biology and experimental medicine in this field. This was frequently documented in preceding chapters of this book dealing with signal transformation (Sections 1.2 to 1.5), chromatin structure (Section 9.3), DNA damage and repair (Section 15.3), alterations of membrane structure and function (Section 14.4), intrinsic heat sensitivity and induced thermotolerance (Chapter 17), or hs-induced developmental effects (Section 18.1).

On the other hand, despite many intimate connections, it is important to emphasize once more the differences between these two unequal parts of high temperature stress research. Investigations on the heat shock response are concerned with the physiological range of hyperthermia, in which cells react rapidly to protect sensitive structures against heat damage and to provide means for an efficient survival and recovery after the stress period. In contrast to this, hyperthermic treatment of cancer is aimed at a selective killing of malignant tissues. The adaptive reactions just mentioned are largely avoided by particular hs regimes, and/or by combination with other cytotoxic treatments (radiotherapy, chemotherapy). To a certain extent, cellular changes after severe hyperthermia at or beyond the border to cell death are basically different from those found during or after a "physiological" hs. Moreover, in all situations of experimental research or medical application of thermotherapy to whole tumors *in situ*, the cell biological aspects are grossly complicated by the integration of the malignant cells into the surrounding tissue and the control networks of the whole organism. This organismic aspect and the individuality of each tumor[101a] considerably complicate the situation for medical application. An optimum therapeutic procedure should be adjusted to the particular needs of each tumor patient individually.

Following the observation of Busch[13] in 1866 on tumor regression after severe local inflammation caused by infection with *Streptococcus erysipelatis*, 125 years of intensive interest into hyperthermic treatment of cancer are documented in many reviews and books. (See References 1, 6, 7, 11, 18, 21, 28, 35, 37, 39, 42, 58, 59, 91, 95, 97, 102, 117, 120, 125, 126, 134, and 135.) They illustrate the constant progress, but also the considerable difficulties, still faced with the clinical application of hyperthermia as a reliable part of cancer therapy. In the frame of this book, we try only to overview this field briefly and to provide access to the relevant literature. A concise and critical discussion of all essential aspects can be found in three volumes published 1986 in this series[6] and the excellent proceedings of the fourth International Congress on Hyperthermic Oncology.[91]

Although cancer treatment is the most prominent field, thermotherapy was and is also applied to other pathological situations of human and plants. The use of bacterial toxins for local fever induction in humans, e.g., after direct injection into tumor tissue, was first described by Coley[21] in 1893, but even at present this "natural" method to evoke hyperthermia either by bacterial infections or toxins is successfully applied in cancer therapy.[89,90] It is of historical interest that in 1927 the Austrian psychiatrist Julius Wagner-Jauregg[127] was honored with a Nobel prize for the development of a surprisingly effective therapy of late stages of syphilis characterized by progressive and fatal destruction of the brain by the s pirochaetes (*dementia paralytica*). He injected patients subcutaneously with 0.1 ml of blood containing tertian malaria parasites. The following repeated cycles of severe fever attacks of 40 to 41°C eventually combined with neoarsphenamine application led to complete cure in 30 to 50% of cases. The period of broad application of this thermochemotherapy of syphilis was only stopped after 1945, when penicillin was introduced into medical practice.

Local application of hyperthermia was later proposed to stop rhinoviral infections causing the common cold,[146,147] and it is successfully used in Japan to treat a far-spread, lymphocutaneous mycosis evoked by *Sporothrix schenckii*. Daily application, one to three times, of a pocket warmer for 20 min local heating of the lesion site to 40 to 42°C frequently leads to complete cure from sporotrichosis within 5 to 13 weeks.[51,108,133] Finally, a field of broad application and great economic value is the virus eradication from different types of cultural plants by a long-term, mild "whole-body-hyperthermia" followed by mass regeneration of plants from the virus-free meristems.[62,128,129]

20.2. HEAT SENSITIVITY OF CANCER CELLS AND ITS MODULATION

Early observations by Vollmar[125] on an increased heat sensitivity of cancer cells were later confirmed by others.[24,40,61,117,118] But direct proof of this higher sensitivity by comparison of transformed and nontransformed cell lines of similar origin was only reported by Giovanella et al.[40] for different human cancers (melanoma, colon carcinoma, ovary teratocarcinoma, brain fibrosarcoma) and by Schamhart et al.[103] for rat hepatoma cell lines compared to hepatocytes. However, after many years of practical experience, it appears that this may not be generally true and that differences, if existent, are by no means sufficient as a basis for selective killing of cancer cells. A major obstacle is the considerable variation of intrinsic heat sensitivity of all types of cells (see Section 17.4) depending on their cell cycle stage and metabolic situation. Many drugs have been found to influence heat sensitivity. Some of them act as modulators of the hs response[70] (see Section 1.3).

20.2.1. DRUGS AFFECTING THE REDOX STATE AND ENERGY METABOLISM

Since the remarkable investigations of Warburg,[132] summarized in his book on the metabolism of tumor cells, it has been recognized that many tumor cells or parts of solid tumors are in a hypoxic state, which is connected with a preferential conversion of glucose to lactate and an acidification of the cytoplasm. *In situ*, the relatively low blood supply of central parts of a solid tumor[26,39,98] is certainly an important factor contributing to this effect. The particular interest in these metabolic abnormalities results from the radioresistance of hypoxic tumor cells. Different types of combination therapy with hyperthermia and/or chemotherapy were designed to make hypoxic cells more radiosensitive, to increase their sensitivity to heat killing, or to selectively damage them with specific cytotoxic drugs. To this aim, a number of electrophilic heterocyclic compounds with nitro groups (misonidazole, metronidazole) were characterized as radiosensitizers and "anaerobic antibiotics".[88,112,140] Their cytotoxicity depends on the intracellular reduction to nitroso- and hydroxylamine-compounds, which is inhibited in well oxygenized tissues, but enhanced by mild hyperthermia. Thus, combination of these nitroheterocycles with hyperthermia act as effective radiosensitizers of hypoxic cells.

Low pH and hypoxic conditions make tumor cells more heat sensitive than well oxygenized cells. The three factors are interdependent: first, hypoxia and/or hs increase cellular acidification;[123] second, killing of Chinese hamster ovary cells by combined hyperthermia/X-irradiation is more effective at pH 6.75 than at pH 7.4 in the medium;[55] finally, an important factor is the inhibition of thermotolerance development at pH 6.75. These findings reflect an intimate connection between energy balance or the type of energy metabolism, hs response, and cell survival (see Section 1.3). In this respect, illustrative examples were given by Lanks and co-workers. Lactate formation from glucose in mouse L929 cells is greatly increased during hs and/or by addition of glutamine.[68,69,71,75] Glutamine oxidation evidently drives lactate synthesis. Interestingly, these L929 cells are particularly heat sensitive if cultivated in media containing only glutamine. Induced heat shock protein (HSP) synthesis is suppressed under these conditions, and the high hyperthermic cytotoxicity can be further increased by adding insulin.[72] The opposite effect is observed in Chinese hamster ovary cells. After addition of glucocorticoids (10^{-6} *M* hydrocortisone or dexamethasone) but not of other steroid hormones, survival of a 45 min hs treatment at 45°C is markedly improved.[31]

In all likelihood, most of these more or less dramatic variations of the intrinsic heat sensitivity are independent of the levels and inducibility of heat shock proteins or their constitutive isoforms (see Section 17.4). An extreme example in this respect is observed after amino acid deprivation of Chinese hamster ovary cells. Survival of a 38 min hs at 45°C decreases from 10^{-2} to 10^{-6}, but the normal survival rate can be restored by addition of the nonmetabolizable α-aminoisobutyric acid.[124] In contrast to this, the increased heat sensitivity of mouse cells under conditions of carbon source restriction can be reversed by addition of 1 m*M* uridine or other ribonucleosides.[73] The authors discuss the central role of the pentose phosphate shunt for ribonucleoside synthesis to explain the potentiation effect on hyperthermic cytotoxicity observed after energy restriction, addition of insulin or glutathione depletion.[83,107,119] Interference with ribonucleoside synthesis and increased lactate formation are probably also responsible for the high thermosensitivity of mouse tumor cells treated with 2-cyanocinnamic acid alone or together with oxidizing dyes, e.g., with triphenyltetrazolium chloride or methylene blue.[130,131]

An important factor determining survival of heat stress periods is evidently the glutathione (GSH) level. Thus, the 10^4-fold increased cytotoxicity of 1 h hs at 43°C in mouse cells after treatment with interferon is connected with an irreversible decline of the GSH level.[119] Hs-induced polyamine oxidation and the potentiation of hyperthermic cell killing by exogenous polyamines (spermine, spermidine) are presumably also connected with a decline of intracellular GSH levels.[38,44,45,122] Consequently, thermosensitizers are found among the drugs interfering with GSH synthesis, e.g., diethylmaleate or buthionine sulfoximine and, vice versa, increased GSH levels were found in thermotolerant cells[83] (see also Section 17.5).

The toxic effects of exogenous polyamines after hyperthermia are intriguing, because polyamine leakage or depletion of intracellular polyamine pools by α-difluoromethylornithine (DMFO), an inhibitor of polyamine biosynthesis, were likewise found to contribute to the thermosensitivity of Chinese hamster ovary and lung cancer cells.[32,33,36,45a] Interestingly, after 8 h treatment with 1 m*M* DMFO, the clonogenic survival of a 90 min exposition to 43°C was reduced by two orders of magnitude compared to control cells, but expression of hs-induced thermotolerance was unaffected demonstrating once more that both aspects of stress survival are independent.[33] The seemingly controversial aspects of polyamine action on thermosensitivity may be explained by the finding of two independent hs induced pathways of polyamine oxidation. One acts on exogenous, the other on endogenous polyamines.[33a,45] Although both pathways generate oxidative stress by formation of hydrogen peroxide and aldehydes, intracellular compartmentation may be different. On the other hand, depletion of intracellular polyamine pools may have deleterious effects on many aspects of cell metabolism, gene expression, and cytoskeletal functions, which are not related to polyamine oxidation.[5,76]

20.2.2. CALCIUM HOMEOSTASIS

Calcium homeostasis is crucial to the integrity and proper function of cells. It depends on the maintenance of the balance between extracellular and intracellular Ca^{2+}, mainly by energy-dependent extrusion of the ion and on the controlled intracellular storage and release of Ca^{2+}. Many cellular functions are directly or indirectly dependent on calcium and on transient local changes of the Ca^{2+} levels, respectively. These include the stability of the plasma membrane, signal transduction processes, and the structure and adaptive rearrangements of cytoskeletal systems[5,9,16,67] (see Sections 1.5 and 14.4).

Hyperthermia induces increased Ca^{2+} influx into vertebrate cells. This effect can be enhanced by increasing extracellular Ca^{2+} concentration or by addition of lanthanum ions (1 m*M* La^{3+}). Thus, following the early hypothesis of Schanne et al.[104] on a Ca^{2+}-dependent toxic cell death, hyperthermic cytotoxicity can be markedly increased by disturbance of the Ca^{2+} homeostasis.[5,80,110,111,137] In keeping with this, Wiegant et al.[137] reported on a potentiation of thermosensitivity of mouse neuroblastoma and rat hepatoma cells after addition of calmodulin-inhibiting drugs (Trifluoperazine, Calmidazolium). These inhibitors have no influence on the inducible synthesis of hs proteins, but concomitantly with the increased cell killing, they prevent the characteristic rearrangements of the cytoskeletal systems, which are evidently necessary for survival of the stress period.[136,138,139] Probably connected with the Ca^{2+}-dependent changes of the cortical cytoskeleton are heat-induced membrane damages. Formation of blebs after microsorting of membrane components[136] and (see Section 14.4) and the transient Ca^{2+} influx[5,110,111] are typical indicators of these hyperthermic effects.

20.2.3. LIPOTROPIC DRUGS

In addition to the influence of the Ca^{2+}/calmodulin system, membrane stability and necessary adaptation during heat stress are affected by lipotropic drugs or treatments changing the lipid composition.[5] As outlined in Section 14.4, excessive fluidization of membranes at high temperatures is balanced by a long-term adaptation of the fatty acid composition to more saturated fatty acid residues. Contrary to this, deliberate fluidization of membranes by incorporation of unsaturated fatty acids potentiates hyperthermic cell killing in bacteria and mammalian cells.[41,87,144] However, fluidization with deleterious effects on survival after heat stress can also be brought about by local anesthetics (lidocaine, procaine),[25,144,145] by monohydric alcohols (methanol, ethanol, isopropanol),[3,82] or by the polyene antibiotic amphotericin B, which binds to membrane cholesterol.[43,43a]

20.3. POTENTIATION OF CHEMOTHERAPY BY HYPERTHERMIA

Application of cytotoxic drugs or antibiotics is an important part of current cancer therapy. Different aspects of a combined thermochemotherapy were discussed by Anghileri,[5] Burkhardt and Ghosh,[12] Landberg,[64] Mizuno,[85] Stratford,[112] and Wong and Dewey[143] (see the book by Anghileri and Robert[6]). Chemotherapeutics used to enhance the thermosensitivity of tumor cells, based on the knowledge about their hypoxic state and metabolic abnormalities, were already mentioned. They also include drugs enhancing thermal destabilization of membranes (alcohols, local anesthetics, amphotericin B). The majority of practically important cytotoxic compounds are inhibitors of nucleic acid and protein synthesis or mutagenic agents. Frequently their effectiveness can be increased by combination with hyperthermia. A summary of Burkhardt and Ghosh[12] documents corresponding effects of a number of alkylating agents (cyclophosphoamide, triethylenethio-phosphoramide, nitrosoureas), antimetabolites (methotrexate, 5-fluorouracil), antibiotics (actinomycin D, adriamycin, bleomycin, mitomycin), and cis-platinum complexes. In most cases the mechanism of the potentiation is not clarified. Increased drug uptake, inhibition of repair processes, enhanced drug reactivity and/or true synergism between hs and drug cytotoxicity may contribute to the therapeutic effects.

The therapeutic benefit from a combined thermochemotherapy may be abolished by induced cross-tolerance.[43a,77a] Thus, CHO cells surviving the heat treatment become thermotolerant and at the same time much more resistant to chemotherapy with adriamycin, bleomycin or cis-platinum at pH 6.8. On the other hand, a murine tumor cell line (RIF-1), selected for improved survival under hs conditions, was also resistant to the polyene membrane antibiotic amphotericin B, if applied at 43°C. The easiest explanation for all these results is adaptive changes of the plasma membrane, protecting cells from the increased uptake of cytotoxic drugs and from disruption of their outer membrane.[43a] It is an intriguing observation that the multidrug-resistance pump (MDR1) present in the membrane of cancer cells but also of normal epithelial cells is induced by heat shock.[19a] The 170-kDa membrane glycoprotein catalyzes the export of different cytotoxic drugs (vinblastine, adriamycin, colchicine).[133b] Consequently, a hs- and arsenite-induced increase of vinblastine resistance was observed in human renal adenocarcinoma cells.[19a]

20.4. RADIOTHERMOTHERAPY

As summarized by Crile,[23] experiences on the useful combination of irradiation and hyperthermia for cancer therapy go back to the beginning of this century.[86,101] The great amount of literature accumulated since then is summarized by Abe and Hiraoka,[1] Field and Bleehen,[28] Haveman,[46] Streffer et al.[113] and Suit.[117] On the one hand, hyperthermia brings about a general radiosensitization. On the other hand, S-phase cells, which are relatively radioresistant, are particularly sensitive to hyperthermic cell killing. But hyperthermia is also especially useful in the treatment of certain radioresistant tumors, e.g., malignant melanomas.[62a]

The additivity of effects of a combined radiothermotherapy is evidently mainly due to the rapid and extensive changes in the DNA/protein ratio of chromatin and the inhibition of DNA repair activities by hs (see Section 15.3). An important term used to quantify the radiosensitization by hyperthermic treatments is the ''thermal enhancement ratio'' (TER). It was defined by Robinson et al.[99] as the ratio of X-ray doses needed to cause the same extent of tissue damage or cell killing (isoeffects) *without* and combined *with* hyperthermia, respectively. Values of 2 to 7 were reported.[46,56,81] In experimental and even more in clinical practice, it may be difficult to separate the radiosensitization effect in the strict sense from direct contributions of the hyperthermic treatment to the damaging effects on the tumor tissue.

The induction of thermotolerance, which counteracts the therapeutic gain by the combination of both methods, makes the optimum timing between hyperthermia and irradiation very important. Most frequently, both are applied together or immediately following each other.[92] If cells or tissues are allowed to recover after the hs, they become thermotolerant and the value for the TER can be markedly reduced.[46-49,55,56,84,121,141,142] The thermotolerance effect is maximum after 6 to 24 h and then slowly decays within 6 to 10 d (see Section 17.2.2). In view of the metabolic nutritional peculiarities of tumor cells and the list of potential thermosensitizers (see Section 20.2), it is reasonable to increase the therapeutic gain by supporting drug treatment. Robinson et al.[99] found the radiosensitization in mouse bone marrow cells more effective under anoxic than under well-oxygenized conditions, and Holahan et al.[55] improved radiosensitization of CHO cells by decreasing the pH value of the culture medium from 7.4 to 6.75. The inhibition of thermotolerance induction under low pH conditions is evidently an additional factor contributing to the cell killing effect. Similar are the supporting effects of misonidazole, if applied prior to a combined radiothermotherapy.[140]

20.5. TECHNICAL ASPECTS OF CLINICAL APPLICATION

20.5.1. THERMAL DOSE

Application of hyperthermia could be greatly facilitated by definition of a thermal dose. Many earlier attempts in this direction[8,29,42,59,65,94,102] and the inherent difficulties were summarized by Gerner.[35] Equations which relate the extent of response, e.g., cell killing, to the product of time of treatment and the temperature are based on the general observations that ''similar'' effects can be achieved either by a short severe hs or by a longer, but moderate hs. On the basis of an ''equivalent time (t_{ref}) at a reference temperature (T_{ref})'' characterized by a given cytotoxic effect, two basic types of equations were derived. Equation 1 represents a simple time-temperature relationship:[29,94,102]

$$\Delta t_{ref} = \Sigma \Delta t \cdot R^{(T - T_{ref})} \tag{1}$$

whereas equation 2 is derived on a thermodynamic basis including the Gibbs free energy (G^{*o}):[35]

$$\Delta T_{ref} = \Delta t \cdot \exp[-G^{*o}(T/RT) + \Delta G^{*o}(T_{ref}/RT_{ref})] \tag{2}$$

Unfortunately, these mathematical formulae derived to define ''isoeffects'' involve an empirical ''constant'' (R), whose actual values change with temperature, cell type, and velocity of heating.[35] The experience that cell killing by a long-term moderate hs is different from that evoked by a short, severe hs agrees with similar observations on many other aspects of the hs response. Important factors are the greatly varying intrinsic heat sensitivity of cancer cells and, above all, the multiplicity of adaptive processes observed during a hs (induced thermotolerance, see Chapter 17). Clearly, the constantly changing biological system is opposed to a clear-cut physical definition of a thermal dose or limits its applicability to a narrow time and temperature range.

20.5.2. LOCAL HYPERTHERMIA

Either part of nonsurgical cancer treatment is based on selective killing of malignant cells without unacceptable damage to the surrounding nonmalignant tissue. Hence, preferential cytotoxicity and/or exact local application of radiation, chemotherapeutics, and hyperthermia are important prerequisites for a successful therapy. The development of local hyperthermia is intimately connected with the technical progress in the construction of suitable heating devices and temperature control systems.[1,2,19,22,77,114] Heating techniques include invasive and noninvasive methods using ultrasound, microwaves, electromagnetic energy, and radiofrequency. Some major problems are the homogeneous distribution of heat within the tumor, the accessibility of deep-seated tumors, and the complications by the surrounding tissue and changing blood supply.[26,39] Clearly, the necessary improvement of the equipment for heating and intratumor thermometry requires constant and careful testing in the clinical praxis. To this aim, the design, control, and detailed documentation of corresponding clinical trials is essential.[1,92,93,96] At the present state, failure of therapeutic response may simply result from technical problems with the proper heat application and temperature control.

20.6. HEAT SHOCK PROTEINS AND THE MALIGNANT STATE

20.6.1. INCREASED HSP LEVELS CHARACTERIZE THE MALIGNANT STATE

It was discussed in preceding chapters (see Sections 1.2, 8.5, and 19.1) that increased levels of certain members of the HSP90 and HSP70 families are found in vertebrate cells transformed by DNA tumor viruses (SV40, polyoma virus, adenovirus, herpes simplex virus and human cytomegalovirus). The inherent type of control of HSP synthesis is independent of the hs induction, but is rather mediated by a regulatory system involved in cell cycle or basal level control of HSP synthesis.[85a,117a,139a] Formation of proteins, active in viral gene expression and/or replication, e.g., of the adenovirus E1A protein, the polyoma or SV40 virus large T-antigens, is required for the inducing effect on HSP synthesis (see Figure 8.5). The detection of a cellular system with E1A-like functions[60] led Imperiale et al.[57] to the investigation of hsp70 mRNA levels in transformed and nontransformed mammalian cell lines. In many cases the transformed cells were characterized by significantly higher levels of hsp70 mRNA, which might be used as indicator of rapidly growing tumor cells.

Strongly enhanced synthesis in mitogen-stimulated or transformed cell lines was also reported for the human mitochondrial HSP60[127a] and for HSP25-27. Similar to many other organisms (see Section 2.3.4), the only member of the HSP20 family in vertebrates is expressed not at all or at very low levels in unstressed cells. However, this is not true for tumor cells. A dominant phosphoprotein with M_r 25 kDa accumulating to 1% of the total protein content of mouse Ehrlich ascites tumor cells was identified as HSP25. It was detected in different murine tumors but practically not at all in nonmalignant tissues.[8a,33c] The same is true for human mammary tumor cells exhibiting estrogen-dependent HSP27 synthesis[33b] and for common acute lymphoblastic leukemia cells.[111a] Remarkably, the phosphorylation level of the mammalian HSP25-27 is increased by hs or chemical stressors, but also by mitogenic factors and tumor necrosis factor[7a,7b,63,133a] (see Table 2.5). Robaye et al.[98a] discussed a protective effect of phosphorylated HSP27 against cytotoxic effects of tumor necrosis factor on bovine aortic endothelial cells. In keeping with this, Jaattela et al.[57a] found a protective effect of heat shock against cytolysis of Wehi-164 cells by tumor necrosis factor.

In addition to DNA tumor viruses an increasing list of chemical stressors, characterized as carcinogens or teratogens in insect or vertebrate systems, were found to be inducers of HSP synthesis. Using *Drosophila* primary embryonic cell cultures as a test system, these are coumarin, diphenylhydantoin, pentobarbital, tolbutamide, 5-azacytidine, and thalidomide.[10,14] In rat hepatocytes, application of two potent carcinogens (diethylnitrosamine and 2-acetylaminofluorene) caused an increase of HSPs 90 and 70.[17] It is worth noticing in this respect the induction of a general stress protein (SP32) not only by hs, sodium arsenite, iodoacetamide, *p*-chloromercuribenzoate, and heavy metals, but also by a considerable number of structurally unrelated compounds known to act as tumor promoters, e.g., phorbol esters, alkylating agents, indole alkaloids, or the protease inhibitors TPCK and TLCK.[52-54] SP32 was identified as heme oxygenase of mammals (see Section 2.3.2).

20.6.2. ROLE OF HSPS FOR MALIGNANT TRANSFORMATION?

Another important aspect of the relation of stress proteins to malignant transformation are complexes formed by HSP90 and HSP70 with onc-gene products. On the one hand, HSP90 plays an interesting role in the postsynthetic processing of retroviral proteins inserted into the cell membrane, e.g., of $pp60^{v\text{-}src}$ (see Figure 16.8). On the other hand, members of the HSP70 family associate with nuclear onc-gene products. This is particularly prominent during and after hs when high salt insoluble complexes of these proteins are found in nuclei of vertebrate cells. They include p53, $p62^{c\text{-}myc}$, $p66^{N\text{-}myc}$, $p58^{N\text{-}myc}$, $p45^{v\text{-}myb}$, $p75^{c\text{-}myb}$, v-rel, E1A, the SV40 large T-antigen, the polyoma virus middle T-antigen and probably others.[27,77b,78,94a,102a,108a,109,135a] It is a matter of speculation that HSP70 and ATP are required to dissociate these complexes and hence to reactivate their constituents in the recovery period (see Section 2.3.2). Stress-dependent alterations of the protein conformation may be a prerequisite for aggregation and recognition by HSP70.[108a] In support of this, mutant, but not wild-type p53 forms HSP70 complexes also under nonstress conditions. By increasing the half-life and cellular level of this immortalizing nuclear onc-protein, HSP70 may play a role in malignant transformation.[20,30,50,115,116] Infact, overexpression of wild-type p53 from a plasmid-encoded gene can markedly inhibit neoplastic transformation of rat embryo fibroblasts by mutant p53 together with the Ha-ras gene.[26a] The p53 phosphoprotein is evidently a negative control factor of cell division,[67a] and mutation or stress-dependent alteration of p53 may interfere with its normal cellular function.

Finally, there are intriguing interrelations between a number of cellular or viral onc-proteins and the stress response: (1) Increased stability of c-myc mRNA was observed under heat shock conditions in HeLa and murine melanoma cells.[14a,101b] (2) A five- to eightfold extended half-life of c-myc and c-myb proteins in a chicken bursal lymphoma cell line was attributed to their hs-induced, transient storage in insoluble nuclear aggregates.[79] (3) The increase of c-fos mRNA and protein in different mammalian cells can even be taken as a general stress indicator, e.g., caused by hs, heavy metals, arsenite, or the drug kainic acid used to evoke limbic motor seizure and *status epilepticus*.[4,21a,25a,40a,76a,121a] and (4) Nuclear onc proteins c-myc and v-myb are known as inducers of HSP70 synthesis.[59a,63a] It is tempting to speculate that, in addition to hs-activation of HSF (see Section 7.7), HSP70 expression may be modulated by changing levels of cellular onc-proteins. Though not directly related to hs, the sequential increase of c-fos, followed by c-myc and finally HSP70 mRNAs during regression of the rat ventral prostate gland after androgen withdrawal, may be a relevant example in this respect.[13a]

In summary, these results illustrate the ambivalence of hyperthermia. On the one hand, it is valuable in the concert of nonsurgical methods used for selective killing of tumor cells. In addition, hs may enhance cell-cell or -substrate adhesion in connection with increased synthesis of the membrane collagen receptor (HSP47).[63b,81a,88a,88b] On the other hand, the profound changes in gene expression, cytoskeletal structure, and cell-substrate interaction, as well as the association of stress proteins with onc gene products, may themselves contribute to developmental abnormalities (teratogenesis)[34,74] (see Section 18.1) and tumorigenesis. This is another aspect important to consider when improving the scientific and technical conditions for the practical application of hyperthermia in cancer therapy.

REFERENCES

1. **Abe, M. and Hiraoka, M.,** Localized hyperthermia and radiation in cancer therapy, *Int. J. Radiat. Biol.*, 47, 347—359, 1985.
2. **Andersen, J. B.,** Electromagnetic heating, in *Hyperthermic Oncology 1984,* Vol. 2, Overgaard, J., Ed., Taylor & Francis, London, 1985, 113—128.
3. **Anderson, R. L., Ahier, R. G., and Littleton, J. M.,** Observations on the cellular effects of ethanol and hyperthermia in vivo, *Radiat. Res.*, 94, 318—325, 1983.
4. **Andrews, G. K., Harding, M. A., Calvet, J. P., and Adamson, E. D.,** The heat shock response in HeLa cells is accompanied by elevated expression of the c-fos protooncogene, *Mol. Cell. Biol.*, 7, 3452—3458, 1987.
5. **Anghileri, L. J.,** Role of tumor cell membrane in hyperthermia, in *Hyperthermia in Cancer Treatment,* Vol. 1, Anghileri, L. J., Ed., CRC Press, Boca Raton, 1986, 1—36.
6. **Anghileri, L. J. and Robert, J., Eds.,** *Hyperthermia in Cancer Treatment,* Vols. 1—3, CRC Press, Boca Raton, FL, 1986.
7. **Arcangeli, G., Benassi, M., Cividalli, A., Lovisolo, G. A., and Mauro, F.,** Radiotherapy and hyperthermia: Analysis of clinical results and identification of prognostic variables, *Cancer,* 60, 950—956, 1987.

7a. **Arrigo, A.-P.,** Tumor necrosis factor induces the rapid phosphorylation of the mammalian heat shock protein hsp28, *Mol. Cell. Biol.*, 10, 1276—1280, 1990.

7b. **Arrigo, A.-P. and Welch, W. J.,** Characterization and purification of the small 28,000 dalton mammalian heat shock protein, *J. Biol. Chem.*, 262, 15359—15369, 1987.

8. **Atkinson, E. R.,** Hyperthermia dose definition, *J. Bioeng.*, 1, 487—492, 1977.

8a. **Benndorf, R., Kraft, R., Otto, A., Stahl, J., Boehm, H., and Bielka, H.,** Purification of the growth-related protein p25 of the Ehrlich ascites tumor and analysis of its isoforms, *Biochem. Int.*, 17, 225—234, 1988.

9. **Berridge, M. J.,** Inositol trisphosphate and diacylglycerol two interacting second messengers, *Annu. Rev. Biochem.*, 56, 159—193, 1987.
10. **Bournias-Vardiabasis, N. and Buzin, C. H.,** Developmental effects of chemicals and the heat shock response in *Drosophila* cells, *Teratog. Carcinog. Mutag.*, 6, 523—537, 1986.
11. **Bruns, P.,** Die Heilwirkung des Erysipels auf Geschwulste, *Beitr. Klin. Chir.*, 3, 443—466, 1887.

12. **Burkhardt, D. and Ghosh, P.,** Synergistic combinations of hyperthermia and inhibitors of nucleic acids and protein synthesis, in *Hyperthermia in Cancer Treatment,* Vol. 1, Anghileri, L. J., Ed., CRC Press, Boca Raton, 1986, 127—149.
13. **Busch, W.,** Ueber den Einfluss welchen heftigere Erysipeln zuweilen auf organisierte Neubildungen ausüben, *Verh. Naturh. Preuss. Rhein. Westphal.,* 23, 28—30, 1866.
13a. **Buttyan, R., Zakeri, Z., Lockshin, R., and Wolgemuth, D.,** Cascade induction of c-fos, c-myc, and heat shock 70K transcripts during regression of the rat ventral prostate gland, *Mol. Endocrinol.,* 2, 650—657, 1988.
14. **Buzin, C. H. and Bournias-Vardiabasis, N.,** Teratogens induce a subset of small heat shock proteins in *Drosophila* primary embryonic cell cultures, *Proc. Natl. Acad. Sci. U.S.A.,* 81, 4075—4079, 1984.
14a. **Cajone, F., Salina, M., and Bernelli-Zazzera, A.,** C-myc gene expression in heat-adapted and heat-shocked cells, *Cell Biol. Int. Rep.,* 12, 549—554, 1988.
15. **Caltabiano, M. M., Koestler, T. P., Poste, G., and Greig, R. G.,** Induction of 32- and 34-kDa stress proteins by sodium arsenite, heavy metals, and thiol-reactive agents, *J. Biol. Chem.,* 261, 13381—13387, 1986.
16. **Carafoli, E.,** Intracellular calcium homeostasis, *Annu. Rev. Biochem.,* 56, 395—433, 1987.
17. **Carr, B. I., Huang, T. H., Buzin, C. H., and Itakura, K.,** Induction of heat shock gene expression without heat shock by hepatocarcinogens and during hepatic regeneration in rat liver, *Cancer Res.,* 46, 5106—5111, 1986.
18. **Cavaliere, R., Ciocatto, E. C., Giovanella, B. C., Heidelberger, C., Johnson, R. O., Margottini, M., Mondovi, B., Moricca, G., and Rossifan, A.,** Selective heat sensitivity of cancer cells, *Cancer,* 20, 1351—1381, 1967.
19. **Cetas, T. C.,** Thermometry and thermal dosimetry in *Hyperthermic Oncology 1984,* Vol. 2, Overgaard, J., Ed., Taylor & Francis, London, 1985, 91—112.
19a. **Chin, K.-V., Tanaka, S., Darlington, G., Pastan, I., and Gottesman, M. M.,** Heat shock and arsenite increase expression of the multidrug resistance MDR1 gene in human renal carcinoma cells, *J. Biol. Chem.,* 265, 221—226, 1990.
20. **Clarke, C. F., Cheng, K., Frey, A. B., Stein, R., Hinds, P. W., and Levine, A. J.,** Purification of complexes of nuclear oncogene p53 with rat and *Escherichia coli* heat shock proteins. In vitro dissociation of Hsc70 and DnaK from murine p53 by ATP, *Mol. Cell. Biol.,* 8, 1206—1215, 1988.
21. **Coley, W. B.,** The treatment of malignant tumors by repeated inoculations of erysipels. With a report of ten original cases, *Am. J. Med. Sci.,* 105, 487—511, 1893.
21a. **Colotta, F., Polentarutti, N., Staffico, M., Fincato, G., and Mantovani, A.,** Heat shock induces the transcriptional activation of c-fos protooncogene, *Biochem. Biophys. Res. Commun.,* 168, 1013—1019, 1990.
22. **Cosset, J. M.,** Interstitial techniques, in *Hyperthermic Oncology 1984,* Vol. 2, Overgaard, J., Ed., Taylor & Francis, London, 1985, 309—316.
23. **Crile, G.,** The effects of heat and radiation on cancers implanted on the feet of mice, *Cancer Res.,* 23, 372—380, 1963.
24. **Dietzel, F.,** *Tumor and Temperatur. Verl.,* Urban und Schwarzenberg, München, 1975.
25. **Djordjevic, B.,** Variable interaction of heat and procaine in potentiation of radiation lethality in mammalian cells of neoplastic origin, *Int. J. Radiat. Biol.,* 43, 399—409, 1983.
25a. **Dragunow, M., Currie, R. W., Robertson, H. A., and Faull, R. L. M.,** Heat shock induces c-fos protein-like immunoreactivity in glial cells in adult rat brain, *Exp. Neurol.,* 106, 105—109, 1989.
26. **Eddy, H. A.,** Alterations in tumor microvasculature during hyperthermia, *Radiology,* 137, 515—521, 1980.
26a. **Eliyahn, D., Michalovitz, D., Eliyahn, S., Pinhasi-Kimhi, O., and Oren, M.,** Wild-type p53 can inhibit oncogene-mediated focus formation, *Proc. Natl. Acad. Sci. U.S.A.,* 86, 8763—8787, 1989.
27. **Evan, G. I. and Hancock, D. C.,** Studies on the interaction of the human c-myc protein with cell nuclei:p62c myc as a member of a discrete subset of nuclear proteins, *Cell,* 43, 253—261, 1985.
28. **Field, S. B. and Blehen, N. M.,** Hyperthermia in the treatment of cancer, *Cancer Treat. Rev.,* 6, 63—94, 1979.
29. **Field, S. B. and Morris, C. C.,** Application of the relationship between heating time and temperature for use as a measure of thermal dose, in *Hyperthermic Oncology 1984,* Vol. 1, Overgaard, J., Ed., Taylor & Francis, London, 1985, 183—186.
30. **Finlay, C. A., Hinds, P. W., Tan, T.-H., Eliyahn, D., Oren, M., and Levine, A. J.,** Activating mutations for transformation by p53 produce a gene product that forms an Hsc70-p53 complex with an altered half-life, *Mol. Cell. Biol.,* 8, 531—539, 1988.
31. **Fisher, G. A., Anderson, R. L., and Hahn, G. M.,** Glucocorticoid-induced heat resistance in mammalian cells, *J. Cell. Physiol.,* 128, 127—132, 1986.
32. **Fuller, D. J. M. and Gerner, E. W.,** Delayed sensitization to heat by inhibitors polyamine-biosynthetic enzymes, *Cancer Res.,* 42, 5046—5049, 1982.
33. **Fuller, D. J. M. and Gerner, E. W.,** Sensitization of Chinese hamster ovary cells to heat shock by alpha-difluoromethyl-ornithine, *Cancer Res.,* 47, 816—820, 1987.

33a. **Fuller, D. J. M., Carper, S. W., Clay, L., Chen, J.-R., and Gerner, E. W.,** Polyamine regulation of heat shock induced spermidine N1-acetyl-transferase activity, *Biochem. J.*, 267, 601—605, 1990.

33b. **Fuqua, S. A. W., Blumsalingaros, M., and McGuire, W. L.,** Induction of the estrogen-regulated 24K protein by heat shock, *Cancer Res.*, 49, 4126—4129, 1989.

33c. **Gaestel, M., Gross, B., Benndorf, R., Strauss, M., Schunk, W.-H., Kraft, R., Otto, A., Boehm, H., Stahl, J., Drabsch, H., and Bielka, H.,** Molecular cloning, sequencing and expression in *Escherichia coli* of the 25-kDa growth-related proteins of Ehrlich ascites tumor and its homology to mammalian stress proteins, *Eur. J. Biochem.*, 179, 209—213, 1989.

34. **German, J.,** Embryogenic stress hypothesis of teratogenesis, *Am. J. Med.*, 76, 293—301, 1984.

35. **Gerner, E. W.,** Thermal dose and time-temperature factors for biological responses to heat-shock, *Int. J. Hyperthermia*, 3, 319—328, 1987.

36. **Gerner, E. W. and Russel, D. H.,** The relationship between polyamine accumulation and DNA replication in synchronized Chinese hamster ovary cells after heat shock, *Cancer Res.*, 37, 482—489, 1977.

37. **Gerner, E. W., Connor, W. G., Boone, M. L. M., Doss, J. D., Mayer, E. G., and Miller, R. C.,** The potential of localized heating as an adjunct to radiation therapy, *Radiology*, 116, 433—439, 1975.

38. **Gerner, E. W., Holmes, D. K., Stickney, D. G., Noterman, J. A., and Fuller, D. J. M.,** Enhancement of hyperthermia-induced cytotoxicity by polyamines, *Cancer Res.*, 40, 432—438, 1980.

39. **Gerwick, L. E.,** Hyperthermia in cancer therapy: The biological basis and unresolved questions, *Cancer Res.*, 45, 3408—3414, 1985.

40. **Giovanella, B. C., Stehlin, J. S., and Morgan, A. C.,** Selective lethal effect of supranormal temperatures on human neo-plastic cells, *Cancer Res.*, 36, 3944—3950, 1976.

40a. **Gubits, R. M. and Fairhurst, J. L.,** C-fos mRNA levels are increased by the cellular stressors, heat shock and sodium arsenite, *Oncogene*, 3, 163—168, 1988.

41. **Guffy, M. M., Rosenberger, J. A., Simon, I., and Burns, C. P.,** Effect of cellular fatty acid alterations on hyperthermic sensitivity in cultured L1210 murine leukemia cells, *Cancer Res.*, 42, 3625—3630, 1982.

42. **Hahn, G. M.,** *Hyperthermia and Cancer,* Plenum Press, New York, 1982.

43. **Hahn, G. M., Li, G. C., and Shiu, E.,** Interaction of amphotericin B and 43°C hyperthermia, *Cancer Res.*, 37, 761—764, 1977.

43a. **Hahn, G. M., Adwankar, M. K., Basrur, V. S., and Anderson, R. L.,** Survival of cells exposed to anticancer drugs after stress, in *Stress-Induced Proteins,* Pardue, M. L., Feramisco, J. R., and Lindquist, S., Eds., Alan R. Liss, New York, 1989, 223—233.

44. **Harari, P. M., Tome, M. E., and Gerner, E. W.,** Heat shock-induced polyamine oxidation in mammalian cells, *J. Cell Biol.*, 103, 175a, 1986.

45. **Harari, P. M., Fuller, D. J. M., and Gerner, E. W.,** Heat shock stimulates polyamine oxidation by two distinct mechanisms in mammalian cell cultures, *Int. J. Radiat. Oncol.*, 16, 451—457, 1989.

45a. **Harari, P. M., Fuller, D. J. M., Carper, S. W., Croghan, M. K., Meyskens, F. L., Shimm, D. S., and Gerner, E. W.,** Rationale for an initial clinical application of polyamine biosynthesis inhibitors combined with systemic hyperthermia in cancer therapy, *Cancer Res.*, in press.

46. **Haveman, J.,** Enchancement of radiation effects by hyperthermia, in *Hyperthermia in Cancer Treatment,* Vol. 1, Anghileri, L. J., Ed., CRC Press, Boca Raton, 1986, 169—181.

47. **Haveman, J., Hart, A. A. M., and Wondergem, J.,** Thermal radiosensitization and thermotolerance in cultured cells from a murine mammary carcinoma, *Int. J. Radiat. Biol.*, 51, 71—80, 1987.

48. **Henle, K. J. and Dethlefsen, L. A.,** Heat fractionation and thermotolerance, A review, *Cancer Res.*, 38, 1843—1851, 1978.

49. **Henle, K. J., Tomasovic, S. P., and Dethlefsen, L. A.,** Fractionation of combined heat and radiation in asynchronous CHO cells. I. Effects on radiation sensitivity, *Radiat. Res.*, 80, 369—377, 1979.

50. **Hinds, P. W., Finlay, C. A., Frey, A. B., and Levine, A. J.,** Immunological evidence for the association of p53 with a heat shock protein, Hsc70, in p53-plus-ras-transformed cell lines, *Mol. Cell. Biol.*, 7, 2863—2869, 1987.

51. **Hiruma, M., Katch, T., Yamamoto, I., and Kagawa, S.,** Local hyperthermia in the treatment of sporotrichosis, *Mykosen,* 30, 315—321, 1987.

52. **Hiwasa, T. and Sakiyama, S.,** Increase in the synthesis of a 32,000 Mr protein in BALB/C 3T3 cells after treatment with tumor promoters, chemical carcinogens, metal salts and heat shock, *Cancer Res.*, 46, 2474—2481, 1986.

53. **Hiwasa, T., Fujimura, S., and Sakiyama, S.,** Tumor promoters increase the synthesis of a 32,000-dalton protein in BALB/c 3T3 cells, *Proc. Natl. Acad. Sci. U.S.A.*, 79, 1800—1804, 1982.

54. **Hiwasa, T., Fujiki, H., Sugimura, T., and Sakiyama, S.,** Increase in the synthesis of a Mr 32,000 protein in BALB/c 3T3 cells treated with tumor-promoting indole alkaloids or polyacetates, *Cancer Res.*, 43, 5951—5955, 1983.

55. **Holahan, E. V., Highfield, D. P., Holahan, P. K., and Dewey, W. C.,** Hyperthermic killing and hyperthermic radiosensitization in Chinese hamster ovary cells: Effects on pH and thermal tolerance, *Radiat. Res.*, 97, 108—131, 1984.

56. **Hume, S. P. and Marigold, J. C. L.,** Time-temperature relationships for hyperthermal radiosensitization in mouse intestine: Influence of thermotolerance, *Radiother. Oncol.*, 3, 165—171, 1985.
57. **Imperiale, M. J., Kao, H.-T., Feldman, L. T., Nevins, J. R., and Strickland, S.,** Common control of the heat shock gene and early adenovirus genes: Evidence for a cellular E1A-like activity, *Mol. Cell. Biol.*, 4, 867—874, 1984.
57a. **Jaattela, M., Saksela, K., and Saksela, E.,** Heat shock protects Wehi-164 target cells from the cytolysis by tumor necrosis factors alpha and beta, *Eur. J. Immunol.*, 19, 1413—1417, 1989.
58. **Jensen, C. O.,** Experimentelle Untersuchungen über Krebs bei Mäusen, *Zbl. Bakteriol.*, 34, 28—122, 1903.
59. **Jung, H.,** A generalized concept for cell killing by heat, *Radiat. Res.*, 106, 56—72, 1986.
59a. **Kaddurah-Daouk, R., Greene, J. M., Baldwin, A. S., and Kingston, R. E.,** Activation and repression of mammalian gene expression by the c-myc protein, *Genes Dev.*, 1, 347—357, 1987.
60. **Kao, H.-T., Capasso, O., Heintz, N., and Nevins, J. R.,** Cell cycle control of the human HSP70 gene: Implications for the role of a cellular E1A-like function, *Mol. Cell. Biol.*, 5, 628—633, 1985.
61. **Kase, K. and Hahn, G. M.,** Differential heat response of normal and transformed human cell in tissue culture, *Nature,* 255, 228—230, 1975.
62. **Kassanis, B.,** Effects of changing temperature on plant virus diseases, *Adv. Virus, Res.*, 4, 221—241, 1957.
62a. **Kim, J. H., Hahn, E. W., and Tokita, N.,** Combination hyperthermia and radiation therapy for cutaneous malignant melanoma, *Cancer,* 41, 2143—2148, 1978.
63. **Kim, Y.-J., Shuman, J., Sette, M., and Przybyla, A.,** Nuclear localization and phosphorylation of three 25-kilodalton rat stress proteins, *Mol. Cell. Biol.*, 4, 468—474, 1984.
63a. **Klempnauer, K.-H., Arnold, H., and Biedenkapp, H.,** Activation of transcription by v-myb: evidence for two different mechanisms, *Genes Dev.*, 3, 1582—1589, 1989.
63b. **Kurkinen, M., Taylor, A., Garrels, J. I., and Hogan, B. L. M.,** Cell surface-associated proteins which bind native type IV collagen or gelatin, *J. Biol. Chem.*, 259, 5915—5922, 1984.
64. **Landberg, T.,** Hyperthermia and cancer chemotherapy. Clinical results: A literature review, in *Hyperthermic Oncology 1984,* Vol. 2, Overgaard, J., Ed., Taylor & Francis, London, 1985, 169—179.
65. **Landry, J. and Marceau, N.,** Rate-limiting events in hyperthermic cell killing, *Radiat. Res.*, 75, 573—585, 1978.
66. **Landry, J., Samson, S., and Chretien, P.,** Hyperthermia-induced cell death, thermotolerance, and heat shock proteins in normal, respiration-deficient, and glycolysis-deficient Chinese hamster cells, *Cancer Res.*, 46, 324—327, 1986.
67. **Landry, J., Crete, P., Lamarche, S., and Cretien, P.,** Activation of calcium-dependent processes during heat shock. Role in cell thermoresistance, *Radiat. Res.*, 113, 426—436, 1988.
67a. **Lane, D. P. and Benchimol, S.,** p53: oncogene or anti-oncogene, *Genes Dev.*, 4, 1—8, 1990.
68. **Lanks, K. W.,** Studies on the mechanism by which glutamine and heat shock increase lactate synthesis by L929 cells in the presence of insulin, *J. Cell. Physiol.*, 129, 385—389, 1986.
69. **Lanks, K. W.,** Glutamine is responsible for stimulating glycolysis by L929 cells, *J. Cell. Physiol.*, 126, 319—321, 1986.
70. **Lanks, K. W.,** Modulators of the eukaryotic heat shock responses, *Exp. Cell Res.*, 165, 1—10, 1986.
71. **Lanks, K. W., Hitti, I. F., and Chin, N. W.,** Substrate utilization for lactate and energy production by heat-shocked L929 cells, *J. Cell. Physiol.*, 127, 451—456, 1986.
72. **Lanks, K. W., Shah, V., and Chin, N. W.,** Enhancing hyperthermic cytotoxicity in L929 cells by energy source restriction and insulin exposure, *Cancer Res.*, 46, 1382—1387, 1986.
73. **Lanks, K. W., Gao, J.-P., and Kasambalides, E. J.,** Nucleosides restore heat resistance and suppress glucose-regulated protein synthesis by glucose-deprived L929 cells, *Cancer Res.*, 48, 1442—1446, 1987.
74. **Lary, J. M.,** Hyperthermia and teratogenicity, in *Hyperthermia in Cancer Treatment,* Vol. 1, Anghileri, L. J., Ed., CRC Press, Boca Raton, 1986, 107—126.
75. **Lazo, P. A.,** Amino acids and glucose utilization by different metabolic pathways in ascites-tumour cells, *Eur. J. Biochem.*, 117, 19—25, 1981.
76. **Leeper, D. B.,** Molecular and cellular mechanisms of hyperthermia alone or combined with other modalities, in *Hyperthermic Oncology 1984,* Vol. 2, Overgaard, J., Ed., Taylor & Francis, London, 1985, 9—40.
76a. **LeGal LaSalle, G.,** Long-lasting and sequential increase of c-fos oncoprotein expression in kainic acid-induced status epilepticus, *Neurosci. Lett.*, 88, 127—130, 1988.
77. **Lele, P. P.,** Ultrasound: Is it the modality of choice for controlled, localized heating of deep tumors? in *Hyperthermic Oncology 1984,* Vol. 2, Overgaard, J., Ed., Taylor & Francis, London, 1985, 129—154.
77a. **Li, G. C. and Mivechi, N. F.,** Thermotolerance in mammalian systems: A review, in *Hyperthermia in Cancer Treatment,* Anghileri, L. J. and Robert, J., Eds., CRC Press, Boca Raton, FL, 1986, 59—77.
77b. **Lim, M. Y., Davis, N., Zhang, J. Y., and Bose, H. R.,** The v-rel oncogene product is complexed with cellular proteins including its proto-oncogene product and heat shock protein-70, *Virology,* 175, 149—160, 1990.

78. **Littlewood, T. D., Hancock, D. C., and Evan, G. I.,** Characterization of a heat shock-induced insoluble complex in the nuclei of cells, *J. Cell Sci.*, 88, 65—72, 1987.

79. **Lüscher, B. and Eisenman, R. N.,** C-myc and c-myb protein degradation: Effect of metabolic inhibitors and heat-shock, *Mol. Cell. Biol.*, 8, 2504—2512, 1988.

80. **Malhotra, A., Kruuv, J., and Lepock, J. R.,** Sensitization of rat hepatocytes to hyperthermia by calcium, *J. Cell. Physiol.*, 128, 279—284, 1986.

81. **Marigold, J. C. L. and Hume, S. P.,** Effect of prior hyperthermia on subsequent thermal enhancement of radiation damage in mouse intestine, *Int. J. Radiat. Biol.*, 42, 509—516, 1982.

81a. **Martin, M. L. and Regan, C. M.,** The anticonvulsant sodium valproate specifically induces the expression of a rat glial heat shock protein which is identified as the collagen type IV receptor, *Brain Res.*, 459, 131—137, 1988.

82. **Massicotte-Nolan, P., Glofcheski, D. J., Kruuv, J., and Lepock, J. R.,** Relationship between hyperthermic cell killing and protein denaturation by alcohols, *Radiat. Res.*, 87, 284—299, 1981.

83. **Mitchell, J. B., Russo, A., Kinsella, T. J., and Glatstein, E.,** Glutathione elevation during thermotolerance induction and thermosensitization by glutathione depletion, *Cancer Res.*, 43, 987—991, 1983.

84. **Miyakoshi, J., Ikebuchi, M., Furukawa, M., Yamagata, K., Sugahara, T., and Kano, E.,** Combined effects of X-irradiation and hyperthermia (42°C and 44°C) on Chinese hamster V-79 cells in vitro, *Radiat. Res.*, 79, 77—88, 1979.

85. **Mizuno, S.,** Hyperthermia enhancement of the cytotoxicity of antitumor antibiotics, in *Hyperthermia in Cancer Treatment,* Vol. 1, Anghileri, L. J., Ed., CRC Press, Boca Raton, 1986, 183—190.

85a. **Morimoto, R. I., Mosser, D., McClanahan, T. K., Theodorakis, N. G., and Williams, G.,** Transcriptional regulation of the human HSP70 gene, in *Stress-Induced Proteins,* Pardue, M. L., Feramisco, J. R., and Lindquist, S., Eds., Alan R. Liss, New York, 1989, 83—94.

86. **Müller, C.,** Die Krebskrankheit und ihre Behandlung mit Roentgenstrahlen und hochfrequenter Elektrizität resp. Diathermie, *Strahlentherapie,* 2, 170—191, 1913.

87. **Mulcahy, R. T., Gould, M. N., Hidvergi, E., Elson, C. E., and Yatvin, M. B.,** Hyperthermia and surface morphology of P388 ascites tumour cells: Effects of membrane modifications, *Int. J. Radiat. Biol.*, 39, 95—106, 1981.

88. **Mulcahy, R. T., Gipp, J. J., and Tanner, M. A.,** Enhancement of misonidazole chemopotentiation by mild hyperthermia (41°C) in vitro and selective enhancement in vivo, *Int. J. Radiat. Biol.*, 52, 57—66, 1987.

88a. **Nagata, K., Hirayoshi, K., Obara, M., Saga, S., and Yamada, K. M.,** Biosynthesis of a novel transformation-sensitive heat-shock protein that binds to collagen. Regulation by messenger RNA levels and in vitro synthesis of a functional precursor, *J. Biol. Chem.*, 263, 8344—8349, 1988.

88b. **Nakai, A., Hirayoshi, K., Saga, S., Yamada, K. M., and Nagata, K.,** The transformation-sensitive heat shock protein (hsp47) binds specifically to fetuin, *Biochem. Biophys. Res. Commun.*, 164, 259—264, 1989.

89. **Nauts, H. C.,** The beneficial effects of bacterial infections on host resistance to cancer, *Cancer Res. Inst. Monogr.*, Vol. 8, 1980.

90. **Nauts, H. C.,** Hyperthermic oncology: Historic aspects and future trends, in *Hyperthermic Oncology 1984,* Vol. 2, Overgaard, J., Ed., Taylor & Francis, London, 1985, 199—209.

91. **Overgaard, J., Ed.,** *Hyperthermic Oncology 1984,* Vol. 1 and 2, Taylor & Francis, London, 1985.

92. **Overgaard, J.,** Rationale and problems in the design of clinical studies, in *Hyperthermic Oncology 1984,* Vol. 2, Overgaard, J., Ed., Taylor & Francis, London, 1985, 325—338.

93. **Overgaard, J.,** History and heritage — an introduction, in *Hyperthermic Oncology 1984,* Vol. 2, Overgaard, J., Ed., Taylor & Francis, London, 1985, 3—8.

94. **Overgaard, J.,** Time-temperature relationship for hyperthermic cytotoxicity and radiosensitization — Implications for a thermal dose unit, in *Hyperthermic Oncology 1984,* Vol. 1, 1985, 191—194.

94a. **Pallas, D. C., Morgan, W., and Roberts, T. M.,** The cellular proteins which can associate specifically with polyomavirus middle T-antigen in human 293 cells include the major human 70-kDa heat shock proteins, *J. Virol.*, 63, 4533—4539, 1989.

95. **Palzer, R. J. and Heidelberger, C.,** Studies on the quantitative biology of hyperthermic killing of HeLa cells, *Cancer Res.*, 33, 415—421, 1973.

96. **Perez, C. A. and Meyer, J. L.,** Clinical experience with localized hyperthermia and irradiation, in *Hyperthermic Oncology 1984,* Vol. 2, Overgaard, J., Ed., Taylor & Francis, London, 1985, 181—198.

97. **Pettigrew, R. T., Galt, J. M., Ludgate, C. M., and Smith, A. N.,** Clinical effects of whole-body hyperthermia in advanced malignancy, *Br. Med. J.*, 21, 679—682, 1974.

98. **Reinhold, H. S., Wike-Hooley, J. L., Van Den Berg, A. P., and Van Den Berg-Blok, A.,** Environmental factors, blood flow and microcirculation, in *Hyperthermic Oncology 1984,* Vol. 2, Overgaard, J., Ed., Taylor & Francis, London, 1985, 41—52.

98a. **Robaye, B., Hepburn, A., Lecocq, R., Fiers, W., Boeynaems, J. M., and Dumont, J. E.,** Tumor necrosis factor-alpha induces the phosphorylation of 28kDa stress proteins in endothelial cells: Possible role in protection against cytotoxicity, *Biochem. Biophys. Res. Commun.*, 163, 301—308, 1989.

99. **Robinson, J. E., Wizenberg, M. J., and McCready, W. A.,** Radiation and hyperthermal response of normal tissue in situ, *Radiology,* 113, 195—198, 1974.
100. **Robinson, J.E., Wizenberg, M. J., and McCready, W. A.,** Combined hyperthermia and radiation suggest an alternative to heavy particle therapy for reduced oxygen enhancement ratio, *Nature,* 251, 521—522, 1974.
101. **Rodenburg, G. L. and Prime, F.,** Effect of combined radiation and heat on neoplasms, *Arch. Surg.,* 2, 116—129, 1921.
101a. **Rofstad, E. K.,** Heat sensitivity and thermotolerance in vitro of human breast carcinoma, malignant melanoma and squamous cell carcinoma of the head and neck, *Br. J. Cancer,* 61, 22—28, 1990.
101b. **Sadis, S., Hickey, E., and Weber, L. A.,** Effect of heat shock on RNA metabolism in HeLa cells, *J. Cell. Physiol.,* 135, 377—386, 1988.
102. **Sapareto, S. A. and Dewey, W. C.,** Thermal dose determination in cancer therapy, *Int. J. Radiat. Oncol. Biol. Phys.,* 10, 787—800, 1984.
102a. **Sawai, E. T. and Butel, J. S.,** Association of a cellular heat shock protein with Simian virus 40 large T-antigen in transformed cells, *J. Virol.,* 63, 3961—3973, 1989.
103. **Schamhart, D. H. J., Berendsen, W., Van Rijn, J., and Van Wijk, R.,** Comparative studies of heat sensitivity of several rat hepatoma cell lines and hepatocytes in primary culture, *Cancer Res.,* 44, 4507—4516, 1984.
104. **Schanne, F. A. X., Kane, A. B., Young, E. E., and Farber, J. L.,** Calcium dependence of toxic cell death: A final common pathway, *Science,* 206, 700—702, 1979.
105. **Shelton, K. R., Egle, P. M., and Todd, J. M.,** Evidence that glutathione participates in the induction of a stress protein, *Biochem. Biophys. Res. Commun.,* 134, 492—498, 1986.
106. **Shelton, K. R., Todd, Y. M., and Egle, P. M.,** The induction of stress-related proteins by lead, *J. Biol. Chem.,* 261, 1935—1940, 1986.
107. **Shrieve, D. C., Li, G. C., Astromoft, A., and Harris, J. W.,** Cellular glutathione, thermal sensitivity, and thermotolerance in Chinese hamster fibroblasts and their heat-resistant variants, *Cancer Res.,* 46, 1684—1687, 1986.
108. **Soh, Y.,** Treatment of sporotrichosis: The effect of topical thermotherapy, *Jpn. J. Med. Mycol.,* 16, 106—110, 1975.
108a. **Soussi, T., Defromentel, C. C., Stuerzbecher, H. W., Ullrich, S., Jenkins, J., and May, P.,** Evolutionary conservation of the biochemical properties of p53. Specific interaction of *Xenopus laevis* p53 with Simian virus 40 large T-antigen and mammalian heat shock proteins-70, *J. Virol.,* 63, 3894—3901, 1989.
109. **Staufenbiel, M. and Deppert, W.,** Different structural systems of the nucleus are targets for SV40 large T antigen, *Cell,* 33, 173—181, 1983.
110. **Stevenson, M. A., Calderwood, S. K., and Hahn, G. M.,** Rapid increases in inositol trisphosphate and intracellular calcium after heat shock, *Biochem. Biophys. Res. Commun.,* 137, 826—833, 1986.
111. **Stevenson, M. A., Calderwood, S. K., and Hahn, G. M.,** Effect of hyperthermia (45°C) on calcium flux in Chinese hamster ovary HA-1 fibroblasts and its potential role in cytotoxicity and heat resistance, *Cancer Res.,* 47, 3712—3717, 1987.
111a. **Strahler, J. R., Kuick, R., Eckerskorn, C., Lottspeich, F., Richardson, B. C., Fox, D. A., Stoolman, L. M., Hanson, C. A., Nichols, D., Tueche, H. J., and Hanash, S. M.,** Identification of two related markers for common acute lymphoblastic leukemia as heat shock proteins, *J. Clin. Invest.,* 85, 200—207, 1990.
112. **Stratford, I. J.,** Hyperthermia and hypoxic cell radiosensitizers in combination, in *Hyperthermia in Cancer Treatment,* Vol. 1, 1986, 151—167.
113. **Streffer, C., Van Beuningen, D., Dietzel, F., Roltinger, E., Robinson, J. E., Scherer, E., Seeber, S., and Trott, K. R., Eds.,** *Cancer Therapy by Hyperthermia and Radiation,* Urban und Schwarzenberg, Baltimore, MD, 1978.
114. **Strohbehn, J. W.,** Summary of physical and technical studies, in *Hyperthermic Oncology 1984,* Vol. 2, Overgaard, J., Ed., Taylor & Francis, London, 1985, 353—369.
115. **Stürzbecher, H.-W., Chumakov, P., Welch, W. J., and Jenkins, J. R.,** Mutant p53 proteins bind Hsp72/73 cellular heat shock-related proteins in SV40-transformed monkey cells, *Oncogene,* 1, 201—211, 1987.
116. **Stürzbecher, H. W., Addison, C., and Jenkins, J. R.,** Characterization of mutant p53-Hsp72/73 protein-protein complexes by transient expression in monkey COS cells, *Mol. Cell. Biol.,* 8, 3740—3747, 1988.
117. **Suit, H. D.,** Hyperthermic effects on animal tissues, *Radiology,* 123, 483—487, 1977.
117a. **Taylor, I. C. A., Solomon, W., Weiner, B. M., Paucha, E., Bradley, M., and Kingston, R. E.,** Stimulation of the human heat shock protein 70 promoter in vitro by Simian virus 40 large T-antigen, *J. Biol. Chem.,* 264, 16160—16164, 1989.
118. **Tsukeda, H., Maekawa, H., Izumi, S., and Nitta, K.,** Effect of heat shock on protein synthesis by normal and malignant human lung cells in tissue culture, *Cancer Res.,* 41, 5188—5192, 1981.
119. **Tumarkin, L., Damewood, G. P., and Sreevalsan, T.,** Potentiation of thermal injury in mouse cells by interferon, *Biochem. Biophys. Res. Commun.,* 128, 179—184, 1985.

120. **Urano, M.,** Kinetics of thermotolerance in normal and tumor tissues. A review, *Cancer Res.*, 46, 474—483, 1986.
121. **Van Rijn, J., Van Den Berg, J., Schamhart, D. H. J., and Van Wijk, R.,** Effect of thermotolerance on thermal radiosensitization in hepatoma cells, *Radiat. Res.*, 97, 318—328, 1984.
121a. **Vass, K., Berger, M. L., Nowak, T. S., Welch, W. J., and Lassmann, H.,** Induction of stress protein HSP70 in nerve cells after *status epilepticus* in the rat, *Neurosci. Lett.*, 100, 259—264, 1989.
122. **Verma, A. K. and Zibell, J.,** Hyperthermia and polyamine biosynthesis: Decreased ornithine decarboxylase induction in skin and kidney after heat shock, *Biochem. Biophys. Res. Commun.*, 126, 156—162, 1985.
123. **Vexler, A. M. and Litinskaya, L. L.,** Changes in intracellular pH induced by hyperthermia and hypoxia, *Int. J. Hyperthermia*, 2, 75—81, 1986.
124. **Vidair, C. A. and Dewey, W. C.,** Modulation of cellular heat sensitivity by specific amino acids, *J. Cell. Physiol.*, 131, 267—275, 1987.
125. **Vollmar, H.,** Über den Einfluss der Temperatur auf normales Gewebe und auf Tumorgewebe, *Z. Krebsforschung*, 51, 71—99, 1941.
126. **Vollmar, H. and Lampert, H.,** Die Bedeutung der Überwärmung für die Tumorentwicklung, *Z. Krebsforschung*, 51, 322—336, 1941.
127. **Wagner-Jauregg, J.,** The treatment of *dementia paralytica* by malaria inoculation. *Nobel Lectures, Physiology or Medicine 1922—1941*, Elsevier, Amsterdam, 1965, 159—169.
127a. **Waldinger, D., Subramanian, A. R., and Cleve, H.,** The polymorphic human chaperonine protein HuCha60 is a mitochondrial protein sensitive to heat shock and cell transformation, *Eur. J. Cell Biol.*, 50, 435—441, 1989.
128. **Walkey, D. G. A. and Cooper, V. C.,** Effect of temperature on virus eradication and growth of infected tissue cultures, *Ann. Appl. Biol.*, 80, 185—190, 1975.
129. **Wang, P. J.,** Producing pathogen-free plants using tissue culture, *Plant Tissue Cult. Newslett.*, 46, 2—8, 1985.
130. **Wang, H. and Lanks, K. W.,** 2-Cyanocinnamic acid sensitization of L929 cells to killing by hyperthermia, *Cancer Res.*, 46, 5349—5352, 1986.
131. **Wang, H., Shah, V., and Lanks, K. W.,** Use of oxidizing dyes in combination with 2-cyanocinnamic acid to enhance hyperthermic cytotoxicity in L929 cells, *Cancer Res.*, 47, 3341—3343, 1987.
132. **Warburg, O.,** *The Metabolism of Tumors*, Constable, London, 1930.
133. **Watanabe, S., Morita, Y., Sudo, N., and Takasu, T.,** Local heat therapy of sporotrichosis, *Jpn. J. Clin. Dermatol.*, 25, 1053—1059, 1971.
133a. **Welch, W. J.,** Phorbol ester, calcium ionophore, or serum added to quiescent rat embryo fibroblast cells all result in the elevated phosphorylation of two 28,000-Dalton mammalian stress proteins, *J. Biol. Chem.*, 260, 3058—3062, 1985.
133b. **West, I. C.,** What determines the substrate specificity of the multi-drug-resistance pump? *Trends Biochem. Sci.*, 15, 42—46, 1990.
134. **Westermark, F.,** Über die Behandlung des ulcerirenden Cervix carcinoms mittels konstanter Wärmen, *Zbl. Gynäkol.*, 1335—1339, 1898.
135. **Westermark, N.,** The effect of heat upon rat-tumors, *Skand. Arch. Physiol.*, 52, 257—322, 1927.
135a. **White, E., Spector, D., and Welch, W.,** Differential distribution of the adenovirus E1A proteins and colocalization of E1A with the 70-kDa cellular heat shock protein in infected cells, *J. Virol.*, 62, 4153—4166, 1988.
136. **Wiegant, F. A. C., Blok, F. J., and Linnemans, W. A. M.,** Hyperthermia and the membrane-associated cytoskeleton of mammalian cells, Proefschrift Rijks Universiteit, Utrecht, 1987, 103—123.
137. **Wiegant, F. A. C., Tuyl, M., and Linnemans, W. A. M.,** Calmodulin inhibitors potentiate hyperthermic cell killing, *Int. J. Hyperthermia*, 1, 157—170, 1985.
138. **Wiegant, F. A. C., Blok, F. J., Van Bergen En Henegouven, P. M. P., and Linnemans, W. A. M.,** Heat shock induced cytoskeletal reorganization is related to cell survival in mammalian cells, Proefschrift Rijks Universiteit, Utrecht, 1987, 79—93.
139. **Wiegant, F. A. C., Van Bergen En Henegouwen, P. M. P., and Linnemans, W. A. M.,** Studies on the mechanism of heat shock induced cytoskeletal reorganization in mouse neuroblastoma N2A cells, Proefschrift Rijks Universiteit, Utrecht, 1987, 141—158.
139a. **Williams, K. J., Landgraf, B. E., Whiting, N. L., and Zurlo, J.,** Correlation between the induction of heat shock protein70 and enhanced viral reactivation in mammalian cells treated with ultraviolet light and heat shock, *Cancer Res.*, 49, 2735—2742, 1989.
140. **Wondergem, J. and Haveman, J.,** A study of the effects of prior heat treatment on the skin reaction of mouse feet after heat alone or combined with X-rays: Influence of misonidazole, *Radiother. Oncol.*, 2, 159—170, 1984.
141. **Wondergem, J. and Haveman, J.,** Thermal enhancement of the radiation damage in the mouse foot at different heat and radiation dose: Influence of thermotolerance, *Int. J. Radiat. Biol.*, 48, 337—348, 1985.
142. **Wondergem, J., Haveman, J., Schol, E., and Reinds, E.,** Influence of prior heat treatment on the effects of heat alone or combined with X-rays on mouse stromal tissue, *Int. J. Radiat. Biol.*, 51, 81—90, 1987.

143. **Wong, R. S. L. and Dewey, W. C.,** Effect of hyperthermia on DNA synthesis, in *Hyperthermia in Cancer Treatment,* Vol. 1, Anghileri, L. J., Ed., CRC Press, Boca Raton, 1986, 79—91.
144. **Yatvin, M. B.,** The influence of membrance lipid composition and procaine on hyperthermic death of cells, *Int. J. Radiat. Biol.,* 32, 513—521, 1977.
145. **Yatvin, M. B., Clifton, K. B., and Dennis, W. H.,** Hyperthermia and local anesthetics: Potentiation of survival of tumor-bearing mice, *Science,* 205, 195—196, 1979.
146. **Yerushalmi, A. and Lwoff, A.,** Medecine et therapeutique. Traitement du coryza infecteux et de rhinites persistantes allergiques par la thermotherapie, *C. R. Acad. Sci. Paris,* 291, 957—959, 1980.
147. **Yerushalmi, A., Karman, S., and Lwoff, A.,** Treatment of perennial allergic rhinitis by local hyperthermia, *Proc. Natl. Acad. Sci. U.S.A.,* 79 4766—4769, 1982.

INDEX

A

B

C

D

H

I

O

P

R

U

V

W

X

Y

Z